Lecture Notes in Mechanical Engineering

Lecture Notes in Mechanical Engineering (LNME) publishes the latest developments in Mechanical Engineering—quickly, informally and with high quality. Original research reported in proceedings and post-proceedings represents the core of LNME. Volumes published in LNME embrace all aspects, subfields and new challenges of mechanical engineering.

To submit a proposal or request further information, please contact the Springer Editor of your location:

Europe, USA, Africa: Leontina Di Cecco at Leontina.dicecco@springer.com

China: Ella Zhang at ella.zhang@springer.com
India: Priya Vyas at priya.vyas@springer.com
Rest of Asia, Australia, New Zealand: Swati Meherishi at swati.meherishi@springer.com
Topics in the series include:

- Engineering Design
- Machinery and Machine Elements
- Mechanical Structures and Stress Analysis
- Automotive Engineering
- Engine Technology
- Aerospace Technology and Astronautics
- Nanotechnology and Microengineering
- Control, Robotics, Mechatronics
- MEMS
- Theoretical and Applied Mechanics
- Dynamical Systems, Control
- Fluid Mechanics
- Engineering Thermodynamics, Heat and Mass Transfer
- Manufacturing Engineering and Smart Manufacturing
- Precision Engineering, Instrumentation, Measurement
- Materials Engineering
- Tribology and Surface Technology

Indexed by SCOPUS, EI Compendex, and INSPEC.

All books published in the series are evaluated by Web of Science for the Conference Proceedings Citation Index (CPCI).

To submit a proposal for a monograph, please check our Springer Tracts in Mechanical Engineering at https://link.springer.com/bookseries/11693.

Irwansyah · Mohd. Iqbal · Syifaul Huzni · Akhyar
Editors

Proceedings of the 4th International Conference on Experimental and Computational Mechanics in Engineering

ICECME 2022, Banda Aceh, September 14–15, 2022

Springer

Editors
Irwansyah
Department of Mechanical Engineering
Syiah Kuala University
Darussalam, Indonesia

Mohd. Iqbal
Department of Mechanical Engineering
Syiah Kuala University
Darussalam, Indonesia

Syifaul Huzni
Department of Mechanical Engineering
Syiah Kuala University
Darussalam, Indonesia

Akhyar
Department of Mechanical Engineering
Syiah Kuala University
Darussalam, Indonesia

ISSN 2195-4356 ISSN 2195-4364 (electronic)
Lecture Notes in Mechanical Engineering
ISBN 978-981-99-7494-8 ISBN 978-981-99-7495-5 (eBook)
https://doi.org/10.1007/978-981-99-7495-5

This Springer imprint is published by the registered company Springer Nature Singapore Pte Ltd.
The registered company address is: 152 Beach Road, #21-01/04 Gateway East, Singapore 189721, Singapore

Conference Organizers and Boards

Advisory Committees

Prof. Dr. Ir. Samsul Rizal M.Eng., Universitas Syiah Kuala, Indonesia
Prof. Dr. Ir. Marwan, Universitas Syiah Kuala, Indonesia
Prof. Dr. Taufik Fuadi Abidin, S.Si., M.Tech., Universitas Syiah Kuala, Indonesia
Prof. Dr. Ir. Alfiansyah Yulianur BC, Universitas Syiah Kuala, Indonesia

Organizing Committees

Chairman
Dr. Teuku Edisah, S.T., M.Eng.Sc, Ph.D., Universitas Syiah Kuala, Indonesia
Secretary
Dr. Syarizal Fonna, S.T., M.Sc., Universitas Syiah Kuala, Indonesia
Finance
Ratna Sary, S.T., M.T., Universitas Syiah Kuala, Indonesia

Committee Members

T. Arriessa Sukhairi, S.T., M.Sc., Universitas Syiah Kuala, Indonesia
Friesca Erwan, S.T., M.Proj.Mgt., Universitas Syiah Kuala, Indonesia
Andriansyah, S.T., M.T., Universitas Syiah Kuala, Indonesia
Akram, S.T., M.T., Universitas Syiah Kuala, Indonesia
Amir Zaki Mubarak, S.T., M.T., Universitas Syiah Kuala, Indonesia
Ir. Asbar, M.T., Universitas Syiah Kuala, Indonesia
Ir. Jalaluddin, M.T., Universitas Syiah Kuala, Indonesia
Ir. Darwin Harun, M.T., Universitas Syiah Kuala, Indonesia

Ir. Masri Ali, M.Eng., Universitas Syiah Kuala, Indonesia
Ir. Masri, M.T., Universitas Syiah Kuala, Indonesia
Muhammad Tadjuddin, S.T., M.Eng.Sc., Universitas Syiah Kuala, Indonesia
Sabri, S.T., M.T., Universitas Syiah Kuala, Indonesia
Ir. Suhaeri, M.Eng., Universitas Syiah Kuala, Indonesia
Syahriza, S.T., M.Eng., Universitas Syiah Kuala, Indonesia
Teuku Firsa, S.T., M.Eng.Sc., Universitas Syiah Kuala, Indonesia
Ir. Udink Aulia, M.Eng., Universitas Syiah Kuala, Indonesia
Zulfadhli, S.T., M.T., Universitas Syiah Kuala, Indonesia
Zulfan, S.T., M.T., Universitas Syiah Kuala, Indonesia

Reviewers

Prof. Dr. Ir. Ahmad Syuhada, M.T., Universitas Syiah Kuala—USK
Prof. Dr. Ir. Khairil, M.T., Universitas Syiah Kuala—USK
Prof. Dr. Ir. Nurdin Ali, Dipl. Ing., Universitas Syiah Kuala—USK
Prof. Dr. Ir. Razali, M.Si., M.T., Universitas Syiah Kuala—USK
Dr. Zahrul Fuadi, S.T., M.Sc., Universitas Syiah Kuala—USK
Dr. M. Nizar Machmud, S.T., M.Eng., Universitas Syiah Kuala—USK
Dr. Ir. Mohd. Iqbal, M.T., Universitas Syiah Kuala—USK
Dr. Syifaul Huzni, S.T., M.Sc., Universitas Syiah Kuala—USK
Dr. Irwansyah, S.T., M.Eng., Universitas Syiah Kuala—USK
Dr.Ir. Akhyar, S.T., M.P., M.Eng., IPM., ASEAN Eng., Universitas Syiah Kuala—USK
Dr. Nasruddin S.T., M.T., Universitas Samudra Langsa—UNSAM
Dr. Muhammad Amin S.T., M.T., Universitas Samudra Langsa—UNSAM
Prof. Dr. Eng. Gunawarman, Universitas Andalas—UNAND
Ismet Hari Mulyadi, Ph.D., Universitas Andalas—UNAND
Dr. Eng. Jon Affi, Universitas Andalas—UNAND
Dr. Eng. Meifal Rusli, Universitas Andalas—UNAND
Dr. Eng. Lovely Son, Universitas Andalas—UNAND
Dedison Gasni, Ph.D., Universitas Andalas—UNAND
Dr. Adjar Pratoto, Universitas Andalas—UNAND
Dr. Eng. Ilhamdi, Universitas Andalas—UNAND
Zulkifli Amin, Universitas Andalas—UNAND
Dr. -Ing. Jhon Malta, Universitas Andalas—UNAND
Dr. Faisal, S.T., M.T., Universitas Malikulsaleh—UNIMAL
Dr. Abubakar, S.T., M.T., Universitas Malikulsaleh—UNIMAL

Preface

We are proud to present these updated research papers of the Lecture Note in Mechanical Engineering which contain 53 scientific papers. The papers discuss the latest research and advanced development works in the field of computational mechanics, metallurgy and material sciences, biomechanics, mechantronics and robotics, manufacturing processing and system, industrial and system engineering, engineering design, as well as thermal and fluid engineering. This book provides information about the research activities in various countries since the authors came from Indonesia, Malaysia, Japan, India, Thailand, and Taiwan.

The editors would like to extend their gratitude to Prof. Dr. FTWI Y. Hagos; Dr. Hendra Hermawan; Dr. Terence Essomba; Ir. Jaka Kirwanto; and Dr. Ir. M. Dirhamsyah for sharing their knowledge and insights in the keynote address. A special thanks is awarded to the reviewers and editorial team for their effort in the review and editing of the manuscripts. The editors would also like to express their appreciation to the staff of Lecture Notes in Mechanical Engineering (LNME) for their full cooperation during the preparation of this collection of research papers.

Aceh, Indonesia

Irwansyah
Chairman, Chief Editor

Contents

The Posterior Tibial Slope Effect to Gap Difference on a Single Radius System for Total Knee Arthroplasty Based on the Computer-Aided Design Approach

Darmanto Darmanto, Robin Novriansyah, Rifky Ismail, Jamari Jamari, Athanasius Priharyoto Bayuseno, and Paulus Wisnu Anggoro

Abstract Total knee arthroplasty (TKA) is one of the most successful ways of treating pain and functional limitations due to osteoarthritis. Asians and Middle Easterners need a large range of flexi angles for religious, social, and cultural activities. Prayer activities and seiza required flexible angle from 150° to 165°. To support these activities, a TKA high flexion (high-flex) type is required. The application of high-flex TKA causes a difference between extension gap and flexion gap, called gap difference. Gap difference is strongly influenced by posterior tibial slope (PTS) and single radius system, and it causes TKA instability. This study aims to analyze the stability of AKJ with GD measurement on a single radius system with a computer-aided design (CAD) approach to prayer, social, and cultural activities for Asians, especially in Indonesia. This study uses the CAD approach with stages that are implant scanning, surface processing, curve processing, components improvement, and gap difference measurement. This study has successfully tested the stability of the model against the difference between gap extension and gap flexion. TKA model

D. Darmanto (✉)
Department of Mechanical Engineering, Faculty of Engineering, Wahid Hasyim University, Semarang, Central Java, Indonesia
e-mail: darmanto@unwahas.ac.id

R. Ismail · J. Jamari · A. P. Bayuseno
Department of Mechanical Engineering, Faculty of Engineering, Diponegoro University, Semarang, Central Java, Indonesia
e-mail: rifky_ismail@ft.undip.ac.id

R. Novriansyah
Department of Orthopaedic and Traumatology, Dr. Kariadi Hospital, Semarang, Central Java, Indonesia

R. Novriansyah · R. Ismail
Center for Biomechanics Biomaterials Biomechatronics and Biosignal Processing (CBIOM3S), Diponegoro University, Semarang, Central Java, Indonesia

P. W. Anggoro
Department of Industrial Engineering, Faculty of Industrial Technology, University of Atma Jaya Yogyakarta, Yogyakarta, Indonesia

Irwansyah et al. (eds.), *Proceedings of the 4th International Conference on Experimental and Computational Mechanics in Engineering*, Lecture Notes in Mechanical Engineering, https://doi.org/10.1007/978-981-99-7495-5_1

with a maximum of PTS 6° can be recommended for the purposes of Asian activities, especially Indonesians in religious, social, and cultural activities. In the future, stability tests will be needed for the movement of anterior–posterior translation.

Keywords Total knee arthroplasty · Posterior tibial slope · Gap difference · Single radius · Computer-aided design

1 Introduction

Currently, TKA is one of the most successful ways of dealing with pain and functional limitations due to osteoarthritis (OA) [1]. The range of postoperative TKA flexion angles is an important indicator in the success of knee joint replacement [2]. Younger patients need a wider range of movement than older patients. Asians and Middle Easterners need a greater range of flexion angles for religious, social, and cultural activities when compared to westerners [3].

Normal patient activity on the chair as western habits requires TKA standard type (flexion angle < 125°), while extreme activities such as Asian and Middle Eastern habits require TKA high-flex type (flexion angle > 125°) [4]. Prayer and seiza required flexion angle from 150° to 165° [5, 6]. The development of TKA high-flex was carried out by enlarging the posterior tibial slope (PTS) [7–10], thickening the posterior femoral condyle (PFC) [10–13], and applying the single radius system at [14–16].

The application of TKA high-flex type causes the difference between extension gap (EG) and flexion gap (FG), hereinafter referred to as gap difference (GD). Extension gap is the gap between the tibia and femur in the extension condition (flexion angle = 0°), while the flexion gap in the flexion angle condition is 90° [17–19]. This GD causes unstable knee joints. Ideally between extension gap and flexion gap, it is the same, so that in the case of anterior–posterior (AP) motion, the knee remains stable [20]. The GD threshold limit is 2 mm, below 2 mm is considered insignificant (stable), while above 2 mm is considered significant (unstable) [20]. According to previous research, GD is directly proportional to PTS [17, 21]. Likewise, GD is also directly proportional to the addition of FPC [20, 22, 23].

Currently, TKA has developed a multi-radius system (MR) and single radius (SR) on the femoral part. In the MR system, femoral condyle is formed from more than a radius, while the SR system is only formed from one radius. According to previous research, SR systems have the advantage of reducing GD [24]. In this study, GD will be analyzed based on a single radius system. MR and SR systems on femoral components as shown in Fig. 1.

CAD modeling has been widely used in the medical field, and anatomical information in the body from computed tomography (CT) and magnetic resonance imaging (MRI) acquisition data can be analyzed and developed with the concept of reverse engineering (RE) [25–27]. Model reconstruction of CT and MRI data can be done using specialized medical software or general software such as SolidWorks inventor

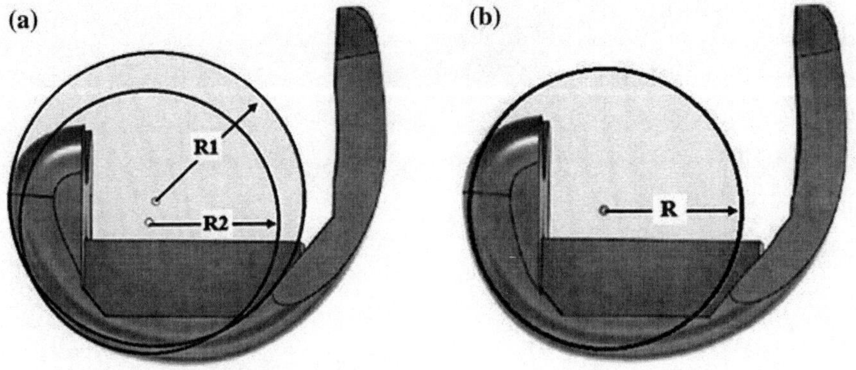

Fig. 1 Radius system on a condyle femoral component: **a** multi-radius, **b** single radius

and Autodesk fusion. The development of CAD-based AKJ models that have been widely done by researchers is to create a new design free of existing standard designs. The results of CAD modeling development can be simulated into the human body to know the success rate before manufacturing. CAD modeling has been used in bone tissue engineering, and manufacturing results using 3D printing with biomaterial material can be directly applied to the human body [26]. Other studies have applied CAD for the development of EG and FG to TKA, this application is referred to as CAD-gap, and if modification of CAD-gap is compared to actual, the results show a very small difference [28].

This study aims to analyze the stability of TKA with GD measurement on a single radius system with a CAD approach so that it is comfortable for prayer, social, and cultural activities for Asians, especially in Indonesia.

2 Materials and Methods

In this study using existing implants as the basis for development, the physical model used was a standard type of implant produced by DePuy-type CS-Revision. The femoral size on this physical model is Size 2 (S2), posterior anterior length (AP) 56 mm (AP56), and lateral medial length (ML) 60 mm (ML60). In tibial insert, components use Size 2 (S2), AP length 43 mm (AP43), and ML length 64 mm (ML64). The AKJ components are complete as seen in Fig. 2.

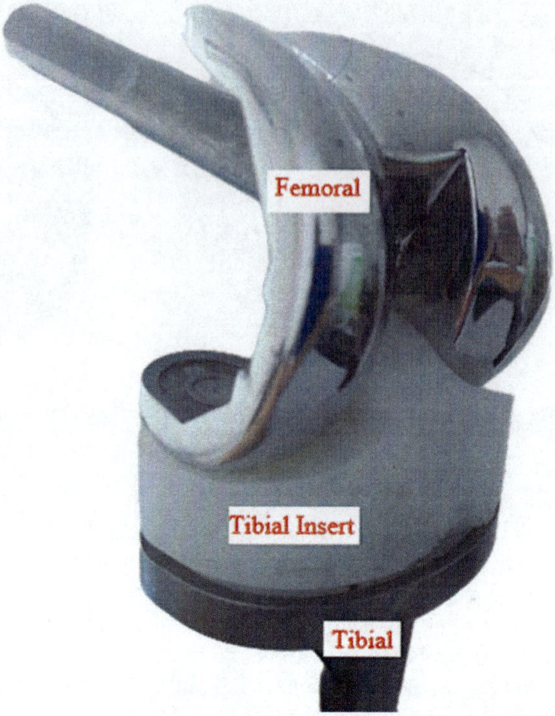

Fig. 2 Complete AKJ components

3 Image Processing

In this process, used Einscan Pro 2X Plus scanner brand: accuracy up to 0.04 mm, scan speed 1,500,000 point/s, point distance 0.2–3 mm, point depth 100 mm, work area 209 × 160 mm–312 × 240 mm, LED light source, and equipped with Solid Shining 3D Edition software. Before the scanning process, the physical model is coated with insulation whose color does not reflect light, and in certain parts, it is marked as a marker of the boundary of the pattern. The shooting process is done by rotating the physical model (fixed scan), while the scanner is fixed. Another way is by moving the desired side scanner (handheld rapid scan), while the physical model does not move. The scanning process is performed until the entire surface is detected, resulting in a complete cloud point. The next process is to close the marker and black side holes with the fill hole feature in the scanner software. Data scanning can be exported to Fusion 360 software with formats available like STL, OBJ, PLY, ASC, and P3.

4 Model Reconstruction

The scanning data has been exported, then modeled using the Autodesk Fusion 360 application, and there are steps in this process as follows:

Surface processing, this process aims to describe the same surface as the physical model. At this stage, the scanning image is made meshing to get the surface profile, too tight meshing causes the surface to be not smooth (many indentations), while it too rarely causes deviations to the original shape. Therefore, good meshing is not too tight and does not change the shape and details of the surface.

Sketch processing, to get the shape and facilitate the drawing of sketches, the data from surface processing did convert to BRep. A solid is described as a collection of surface elements connected to each other, with interior and exterior point boundaries. This method of limiting solids is a representation of the BRep system. Point sketches/ curves can be seen by losing the visibility of the model, with certain modes of visibility of the model can be displayed again. The result of the next sprocket can be reconstructed in accordance with the desired design, so it is expected to create an ideal model.

Improvement components, in this study the development of tibial inserts components with TPS settings by changing the thickness of the AP section. These settings start at − 11°, − 8°, − 5°, 0°, + 3°, + 6°, + 9°, and + 12°, with increments of about 3° [29, 30]. Reconstruction of femoral components is done by changing the thickness of FPC from 8 to 12 mm.

Measurement of GD, in this study GD measurement is carried out by calculating the difference between the EG and FG [19]. The EG is measured from the upper piece of tibia bone against the center of the condyle femoral radius system at the extension position with a flexion angle of 0° ($d1$). The FG is measured from the upper piece of the tibia bone against the center of the condyle femoral radius system at a flexion position with an angle of 90° ($d2$) [19]. GD measurement is shown in Fig. 3 and calculated by Eq. 1.

$$\text{GD} = d1 - d2 \tag{1}$$

where $d1$ is extension gap and $d2$ is flexion gap [19].

Overall, the reconstruction process of CAD-based AKJ models is seen in Fig. 4.

5 Results

The results of the complete EG and FG measurements and GD calculations are shown in Table 1. From Table 1, it can be seen that EG tends to be stable to changes in PTS, while FG is inversely proportional to PTS. Visually, the relationship between PTS

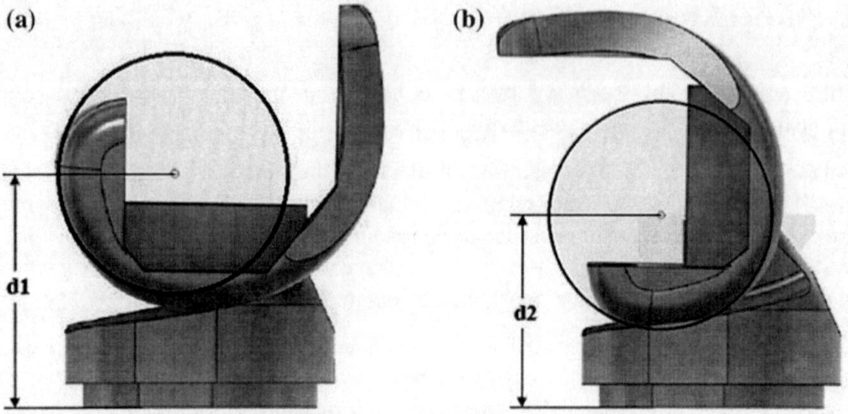

Fig. 3 Measurement of gap on TKA: **a** extension gap, **b** flexion gap

and GD is shown in Fig. 5. From Fig. 5, it can be seen that the larger the PTS, the greater the GD, and the maximum GD is 4 mm. According to the TKA standard criteria, the GD is 2 mm, so the results of the reconstruction of the model that meets the requirements are PTS − 11° to 6°.

6 Discussion

Referring to the research on CAD-gap [28], it is expected that later this study gives results that are not very different from the original knee joint. The approaches to this study have been kept as close as possible to the real condition.

The EG and FG measurements are easier to perform on SR systems because the contact radius in extension and flexion conditions is the same. While in the MR system, the radius of contact is different, so the reference center of the femoral condyle is determined by the intersection point between the extension line and the flexion, such as in Fig. 6.

The balance between EG and FG is an important indicator in the success of TKA, which causes stability in the knee joint. Instability in the TKA will reduce the range of motion (RoM), because the total knee arthroplasty does not function optimally [31, 32]. The GD threshold has been set at 2 mm [20]. GD is directly proportional to PTS (see Fig. 5), and this result is in accordance with previous studies [17, 21]. The posterior thickness of the tibial insert is affected by PTS; if the PTS is greater, the posterior tibial thickness decreases, so that the FG becomes smaller. If the FG is reduced, it will cause the GD to enlarge. In this study, the model that has stability requirements is the model with PTS from − 11° to 6°, see Table 1.

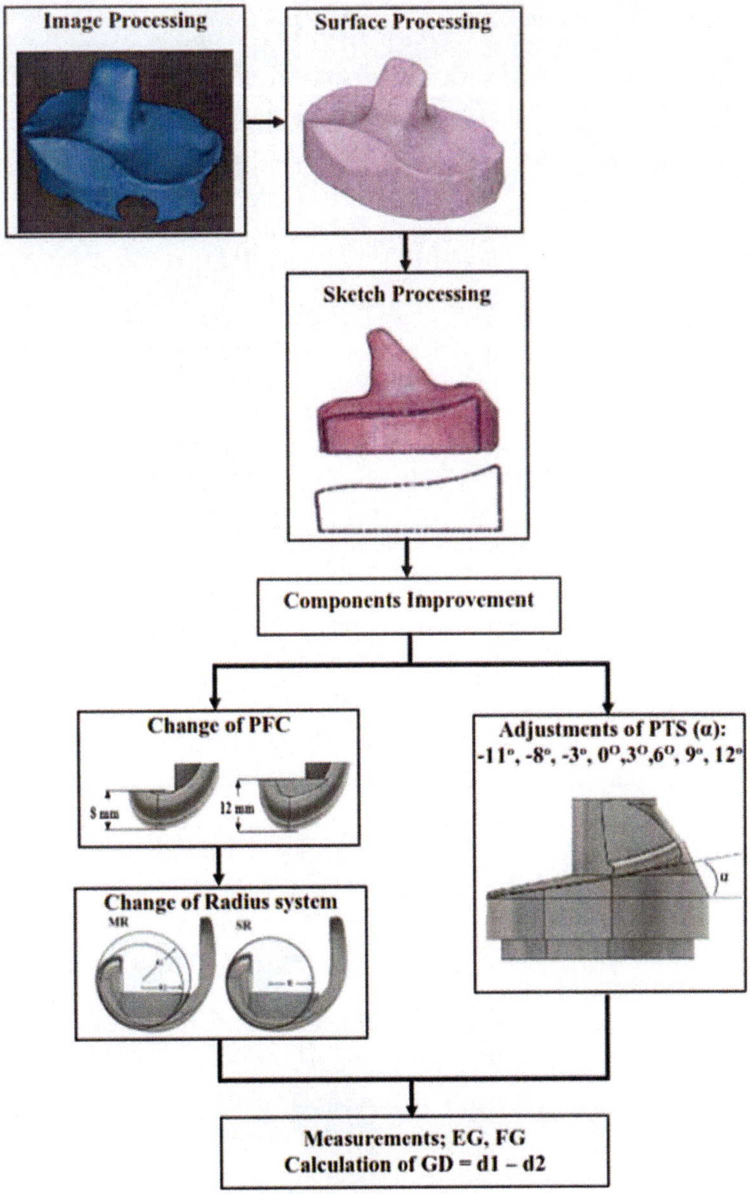

Fig. 4 TKA reconstruction process of CAD based

Table 1 Results of the complete EG and FG measurements and GD calculations

PFC (mm)	PTS (°)	EG (mm)	FG (mm)	GD (mm)
12	− 11	34.80	34.87	− 0.07
12	− 8	36.16	36.06	0.10
12	− 5	36.52	36.32	0.20
12	0	35.62	34.82	1.00
12	3	36.25	34.81	1.44
12	6	36.32	34.05	2.27
12	9	36.21	32.67	3.54
12	12	36.44	32.44	4.00

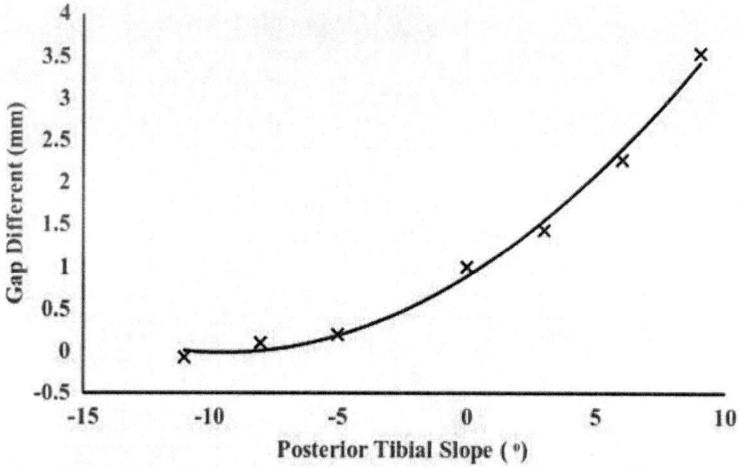

Fig. 5 Graph of the relationship between changes in PTS to GD

Fig. 6 Determining the reference point on the MR system

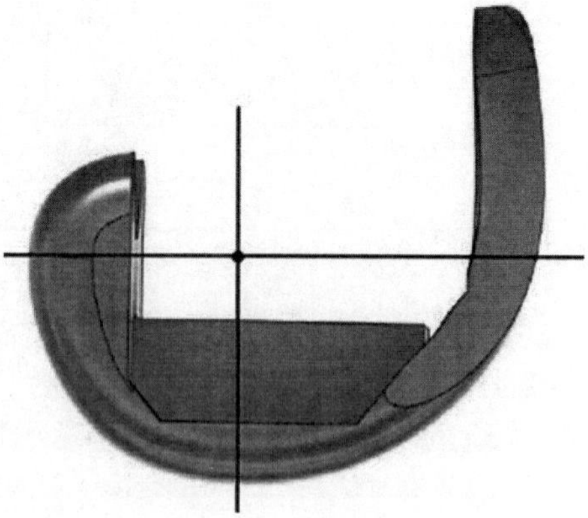

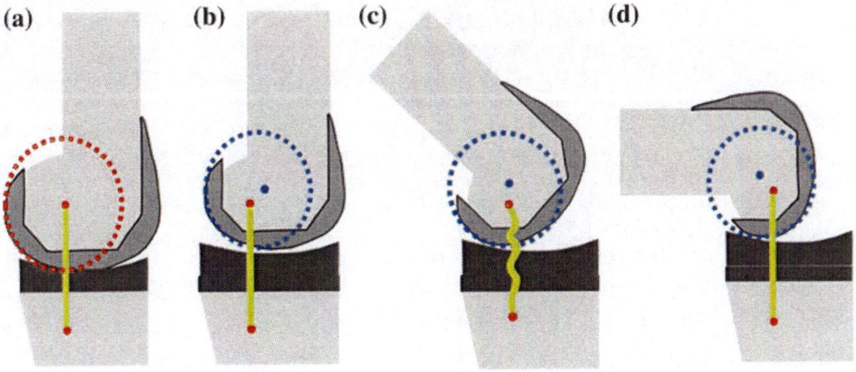

Fig. 7 Illustration of TKA instability conditions: **a** extension condition with red dot round center, **b** shift with blue dot rotation center, **c** unstable condition (saggy ligament), and **d** unstable condition (too tight ligament) [19]

The large GD (GD > 2.00 mm) causes the ligament to become saggy, while the GD < − 2.00 mm causes the ligament to become too tight, and this is what causes the instability of TKA, as an illustration can be seen in Fig. 7.

7 Conclusion

This study has successfully tested the stability of the model against the difference between gap extension and gap flexion. TKA model with a maximum of PTS 6° can be recommended for the purposes of Asian activities, especially Indonesians in religious, social, and cultural. The results of the study can be used as a reference by orthopedic doctors and biomedical engineers to perform knee joint replacement surgery in Indonesia. In the future, stability tests will be needed for the movement of anterior–posterior translation.

Acknowledgements The authors thank to Deputy for Strengthening Research and Development, the Ministry of Research and Technology and Higher Education, Republic of Indonesia, for research funding under Doctoral Research Grants of number: 017/ES/PG.02.00.PT/2022 and 18752/UN7.6.1/PP/2022.

References

1. Guo L, Yang L, Briard JL, Jun Duan X, You Wang F (2012) Long-term survival analysis of posterior cruciate-retaining total knee arthroplasty, knee surgery. Sport Traumatol Arthrosc 20:1760–1765

2. Farahini H, Moghtadaei M, Bagheri A, Akbarian E (2013) Factors influencing range of motion after total knee arthroplasty. Iran Red Crescent Med J 15:10–15
3. Park KK, Chang CB, Kang YG, Seong SC, Kim TK (2007) Correlation of maximum flexion with clinical outcome after total knee replacement in Asian patients. J Bone Jt Surg Ser B 89:604–608
4. Long WJ, Scuderi GR (2008) High-flexion total knee arthroplasty. J Arthroplasty 23:6–10
5. Acker SM, Cockburn RA, Krevolin J, Li RM, Tarabichi S, Wyss UP (2011) Knee kinematics of high-flexion activities of daily living performed by male Muslims in the middle east. J Arthroplasty 26:319–327
6. Lestari H, Izzhati DN, Rachmat N, Setyawan D, Saputra E, Ismail R (2015) Measurement of the range of motion on the knees of Indonesians as preliminary data for the design of artificial upper knee legs (original in Indonesia). Snst 64–69
7. Shi X, Shen B, Kang P, Yang J, Zhou Z, Pei F (2013) The effect of posterior tibial slope on knee flexion in posterior-stabilized total knee arthroplasty, knee surgery. Sport Traumatol Arthrosc 21:2696–2703
8. Fujito T, Tomita T, Yamazaki T, Oda K, Yoshikawa H, Sugamoto K (2018) Influence of posterior tibial slope on kinematics after cruciate-retaining total knee arthroplasty. J Arthroplasty 33:3778–3782
9. Massin P, Gournay A (2006) Optimization of the posterior condylar offset, tibial slope, and condylar roll-back in total knee arthroplasty. J Arthroplasty 21:889–896
10. Darmanto D, Novriansyah R, Ismail R, Jamari J, Anggoro PW, Bayuseno AP (2022) Reconstruction of the artificial knee joint using a reverse engineering approach based on computer-aided design. J Med Eng Technol 46:136–147
11. Yang G, Chen W, Sun X, Zhou D, Chen S, Zhang L (2016) Full-thickness cartilage-based posterior femoral condylar offset. Influence on knee flexion after posterior-stabilized total knee arthroplasty. Orthop Traumatol Surg 102:441–446
12. Kim YH, Park JW (2020) Comparison of modular conventional and high-flexion posterior-stabilized total knee arthroplasties in the same patients at a mean follow-up of 15 years. J Arthroplasty 35:1262–1267
13. Kim YH, Park JW, Kim JS (2017) Do high-flexion total knee designs increase the risk of femoral component loosening? J Arthroplasty 32:1862–1868
14. Carey BW, Harty J (2018) A comparison of clinical- and patient-reported outcomes of the cemented ATTUNE and PFC sigma fixed bearing cruciate sacrificing knee systems in patients who underwent total knee replacement with both prostheses in opposite knees. J Orthop Surg Res 13:1–5
15. Song SJ, Park CH, Liang H, Kang SG, Park JJ, Bae DK (2018) Comparison of clinical results and injury risk of posterior tibial cortex between attune and press fit condylar sigma knee systems. J Arthroplasty 33:391–397
16. Darmanto D, Novriansyah R, Ismail R, Jamari J, Bayuseno AP, Anggoro PW (2021) Computer-aided design for analyzing the influence of single radius on flexion angle of artificial knee joint, 114–118. https://doi.org/10.1109/ibitec53045.2021.9649381
17. Nowakowski AM, Kamphausen M, Pagenstert G, Valderrabano V, Müller-Gerbl M (2014) Influence of tibial slope on extension and flexion gaps in total knee arthroplasty: increasing the tibial slope affects both gaps. Int Orthop 38:2071–2077
18. Watanabe M, Tanaka Y, Kuriyama S, Nakamura S, Nishitani K, Sekiguchi K, Ito H, Matsuda S (2019) Impact of intraoperative adjustment method for increased flexion gap on knee kinematics after posterior cruciate ligament-sacrificing total knee arthroplasty. Clin Biomech 63:85–94
19. van Lieshout WAM, Koenraadt KLM, Elmans LHGJ, van Geenen RCI (2020) Flexion first balancer: description of new technique in TKA to reproduce joint line and pre-disease mechanical alignment. J Exp Orthop 7:4–9
20. Chia ZY, Pang HN, Tan MH, Yeo SJ (2018) Gap difference in navigated TKA: a measure of the imbalanced flexion-extension gap. Sicot-J. 4:30
21. Okazaki K, Tashiro Y, Mizu-uchi H, Hamai S, Doi T, Iwamoto Y (2014) Influence of the posterior tibial slope on the flexion gap in total knee arthroplasty. Knee 21:806–809

22. Matziolis G, Hube R, Perka C, Matziolis D (2012) Increased flexion position of the femoral component reduces the flexion gap in total knee arthroplasty, knee surgery. Sport Traumatol Arthrosc 20:1092–1096
23. Yoon JR, Il Jeong H, Oh KJ, Yang JH (2013) In vivo gap analysis in various knee flexion angles during navigation-assisted total knee arthroplasty. J Arthroplasty 28:1796–1800
24. Shimizu N, Tomita T, Yamazaki T, Yoshikawa H, Sugamoto K (2014) In vivo movement of femoral flexion axis of a single-radius total knee arthroplasty. J Arthroplasty 29:2407–2411
25. Piles L, Reig MJ, Seguí VJ, Pla R, Martínez F, Seguí JM (2019) Reverse engineering applied to biomodelling and pathological bone manufacturing using FDM technology. Procedia Manuf 41:739–746
26. Haleem A, Javaid M, Khan RH, Suman R (2020) 3D printing applications in bone tissue engineering. J Clin Orthop Trauma 11:S118–S124
27. Zhou F, Xue F, Zhang S (2020) The application of 3D printing patient specific instrumentation model in total knee arthroplasty, Saudi. J Biol Sci 27:1217–1221
28. Takagi H, Asai S, Kawashima F, Kato S, Sato A, Okumo T, Kanzaki K (2020) A gap balancing technique for adjusting the component gap in total knee arthroplasty using a navigation system, Asia-Pacific. J Sport Med Arthrosc Rehabil Technol 21:17–21
29. Kang KT, Kwon SK, Son J, Kwon OR, Lee JS, Koh YG (2018) Effects of posterior condylar offset and posterior tibial slope on mobile-bearing total knee arthroplasty using computational simulation. Knee 25:903–914
30. Kang KT, Koh YG, Son J, Kwon OR, Lee JS, Kwon SK (2017) Influence of increased posterior tibial slope in total knee arthroplasty on knee joint biomechanics: a computational simulation study. J Arthroplasty 33:572–579
31. Bottros J, Gad B, Krebs V, Barsoum WK (2006) Gap balancing in total knee arthroplasty. J Arthroplasty 21:11–15
32. Mullaji A, Sharma A, Marawar S, Kanna R (2009) Quantification of effect of sequential posteromedial release on flexion and extension gaps. a computer-assisted study in cadaveric knees. J Arthroplasty 24:795–805

Failure Analysis of the Minibus Crankshaft Using Experimental Methods

Husaini, Edi Saputra, and Nurdin Ali

Abstract This study aims to analyze the causes of crankshaft failure. The methods implemented to examine the value of stress acting on the crankshaft include visual observation, SEM observation, chemical composition testing, and hardness testing. Furthermore, the visual observation of the fourth crankpin fracture surface showed the presence of beachmarks and rivermarks that become characterize fatigue fractures. The SEM observations showed that the crack originated around the fillet on the fourth crankpin. Furthermore, the chemical composition test showed that the type of material used on the shaft was Steel Alloy ASTM A29 standard Grade AISI 4140. The hardness test showed that its value is not homogeneous at each test point with an average value of 99.67 HRB. This exceeds the standard hardness of the AISI 4140 material, which is 92 HRB, leading to the decrease in the toughness of the crankshaft material. The lower surface hardening in the crack initiation region triggers initial defects in the material, while there was friction with the bearing. Because the crankshaft experiences dynamic loads, over time crack propagation occurs from the initial crack to the final fracture.

Keywords Crankshaft · Failure analysis · Stress intensity factor

Husaini (✉) · E. Saputra · N. Ali
Department of Mechanical Engineering, Universitas Syiah Kuala, Darussalam, Banda Aceh 23111, Indonesia
e-mail: husainiftm@usk.ac.id

E. Saputra
e-mail: edi.s@mhs.usk.ac.id

N. Ali
e-mail: nurdin.ali@usk.ac.id

Irwansyah et al. (eds.), *Proceedings of the 4th International Conference on Experimental and Computational Mechanics in Engineering*, Lecture Notes in Mechanical Engineering, https://doi.org/10.1007/978-981-99-7495-5_2

1 Introduction

Crankshaft is the part of an engine that translates reciprocating linear piston into rotary motion [1, 2]. Furthermore, the amount of power output produced leads to a decrease in engines component life. The crankshaft continuously receives a high gas pressure in the cylinder. Therefore, the crankshafts are of materials that are strong with high durability [3]. For suitable mechanical properties, most crankshafts are made from medium carbon steel or alloy, with chrome and nickel metal using forging or casting techniques [4]. Mechanical failure can be defined as any change in size, structure, and part of an engine that makes it unable to perform its intended role satisfactorily [5, 6].

In diesel engine, the compression ratio is higher compared to gasoline engine to achieve maximum power. Therefore, its components tend to experience fatigue or decreased life span [7, 8]. Failures are usually caused by design errors, material defects, operating conditions, and faulty maintenance. The fracture originates from the initial crack which occurs on the surface of the material or in areas where stress is applied. These initial cracks gradually develop into micro-cracks, then propagate until final failure [9–11].

Farrahi et al. [12] analyzed the faults of several crankshafts of a four-cylinder diesel engine. The FEM results showed that the most critical part is the area of the last crankpin fillet closest to the flywheel. The crankshaft failure study carried out by Husaini et al. [13] showed the initial crack occurred in the fillet area of the first crankpin. Furthermore, the visual observation showed a beachmark on the fracture surface indicating the characteristic of fatigue fracture. Also, Pandey [14] investigated the crankshaft failure of a tractor diesel engine with a carbon content of 0.45%. The result showed that the high stress acting in the crankpin-web fillet area was around 175 MPa which led to the fatigue and eventual failure of the crankshaft. The crack initiation continued to propagate as long as the engine was running and finally had a final rupture. Research related to the magnitude of the stress intensity factor that occurs around the crack tip using the finite element method which has also been studied previously. That the crack will propagate if the stress intensity factor exceeds the fracture toughness of the material [15].

Therefore, this study investigated the failure of a crankshaft that occurs in diesel engine passenger minibus car. This occurred in the fourth crankpin after the car had been operating for twenty months. The failed crankshaft was obtained from a mechanic shop in the Aceh Besar District, Aceh, Indonesia. The purpose of this study is to investigate the causes of crankshaft failure.

2 Material and Methodology

Material. The material used in this study is a failure crankshaft. Figure 1 shows a failure crankshaft.

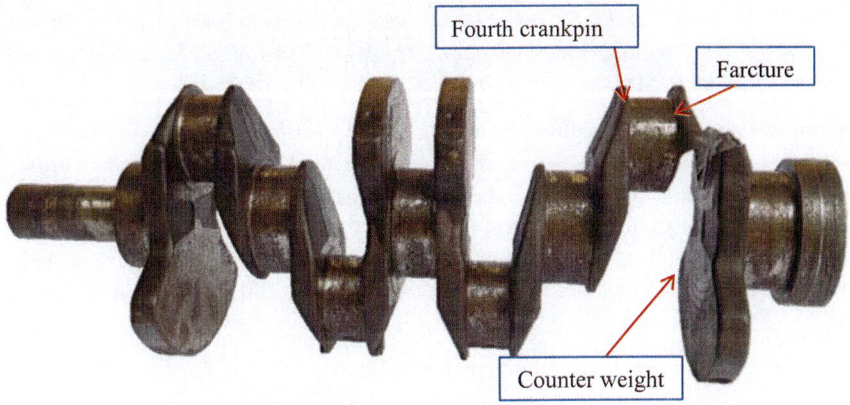

Fig. 1 Crankshaft that failed due to fracture

The failure originates with a crack as shown in Fig. 1 around the fillet area of the fourth crankpin. This propagates to the counterweight during operation and eventually failed. From some literature, it shows that the crankshaft material is made of alloy steel.

Visual observation. The macro-visual observation is the first step in identifying the causes of material failure. This aims to determine where the possible location of the crack initiation or initial defect and other indications of the cause of failure based on the topographic pattern seen on the fracture surface. Generally, the shape of the fatigue fracture is indicated by the presence of a beachmark and a rivermark on the fracture surface. Furthermore, the fatigue fracture mechanism occurs in three main stages starting with the initial crack, propagation of the crack, and final fracture [16].

The observation of fracture samples using scanning electron microscope (SEM) was carried out to examine the fracture surface, location of the crack initiation, and the direction of the crack propagation [17]. Figure 2 shows the location of initial crack for sampling SEM observations (indicated by red circle).

Fig. 2 Location of the initial crack

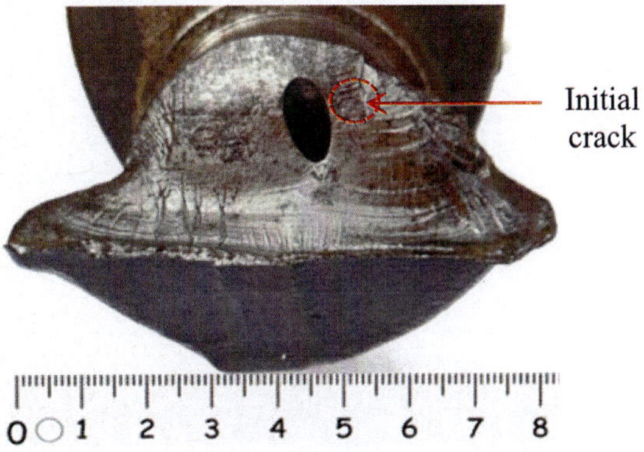

The SEM sample in Fig. 2 (indicated by red circle) is the part of the crankshaft which is the location of the crack initiation. These observations were carried out with magnification up to 500×.

Chemical composition testing. The chemical composition testing was carried out to determine the content of the chemical elements contained in the crankshaft. Furthermore, the data of this test results will be compared with the standard chemical element data for the crankshaft material to reveal the mechanical properties of the material used. The chemical composition test specimen is part of the fourth crankpin. Chemical composition testing is carried out with a PDA-700 spectrometer.

Hardness testing. The hardness testing was carried out to determine the hardness value of crankshaft failure, whether it matches the specified standards or not. Furthermore, this testing used a Rockwell-type Zwick/Roell ZHR testing equipment as referred to the ASTM 18 standard with a compressive load of 100 Kgf. The type of indentor used was 1/16 inch. Figure 3 shows hardness testing point at crankshaft cross-section.

The hardness test specimen in Fig. 3 is the fourth crankpin part where the crack initiation occurs in the crankshaft. Hardness testing was carried out at 34 points as in Fig. 3 where 17 points were in the *x*-axis and *y*-axis direction, respectively. The distance between 5 points on the edge was 1 mm, and the distance between the other points was 6.5 mm.

Fig. 3 Hardness testing point at crankshaft cross-section

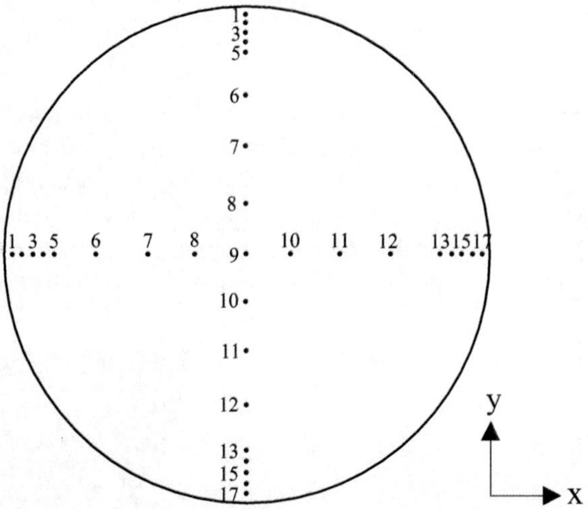

3 Result and Discussion

Visual observation. The visual observation of the crankshaft fracture surface indicated a fatigue failure process. This is characterized by crack initiation, propagation, and final rupture. Figure 4 shows the fracture surface of the crankshaft.

The results of visual observation in Fig. 4 showed that the crack originates from a defect in the fillet area that connects the fourth crankpin with the counterweight. During the combustion process, the stress concentration in the fillet area increased, while the crankshaft obtained maximum pressure in the combustion chamber. Furthermore, the initial of crack propagates during crankshaft operation. From the cracked surface, it can be observed that there is a beachmark that characterizes fatigue fractures.

SEM observations. The location of the SEM observations was carried out in the initial defect area in the fourth crankpin section with 500× magnification. Figure 5 shows the crack initiation and crack propagation observed using the SEM.

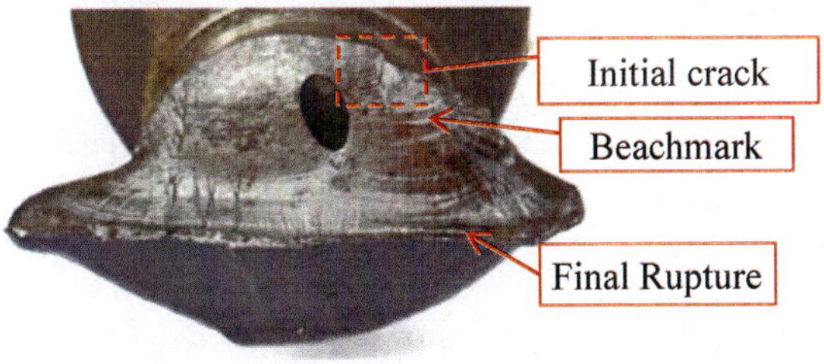

Fig. 4 Fracture surface of the crankshaft

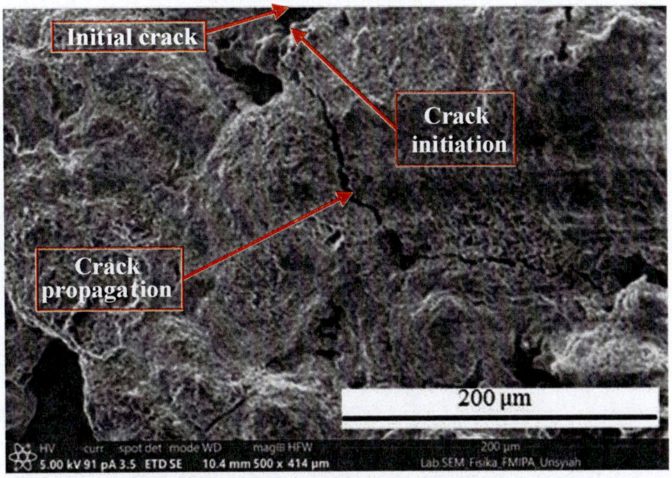

Fig. 5 Fracture surface observed by SEM near the initial crack tip

Figure 5 shows the results of the SEM observations in the crack initiation area referred to in Fig. 2, and the surface area of the crack shows the presence of crack initiation which originates from the initial defect around the fillet on the fourth crankpin and propagating to the counterweight. The crack initiation originates from fine scratches on the surface of the wearing crankpin.

Chemical composition testing results. An examination of the chemical composition of the crankshaft shows that the material used was alloy steel. Based on the ASTM A29 standard, the material used was AISI 4140. Table 1 shows the results of examining the chemical composition of the crankshaft. Table 2 shows the mechanical properties of AISI 4140 Alloy Steel.

Hardness testing results. The results of the crankshaft specimen hardness test at each point based on the hardness test specimen sketch referred to in Fig. 3 show varying values. Furthermore, the average hardness values of the x-axis outer layer at 5 mm thickness were 102.16 HRB and 94.82 HRB. Meanwhile, the average hardness values of the y-axis outer layer at 5 mm thickness were 94.54 HRB and 101.1 HRB. Figure 6 shows the graph of hardness test results.

Table 1 Chemical composition examination results

Element	Chemical composition test results (%)	Chemical composition of AISI 4140 (%)
Carbon, C	0.41	0.38–0.43
Manganese, Mn	0.86	0.75–1.0
Phosphorous, P	0.02	Max. 0.035
Sulfur, S	0.01	Max. 0.040
Silicon, Si	0.26	0.15–0.30
Cromium, Cr	0.95	0.80–1.110
Molybdenum, Mo	0.16	0.15–0.25
Nickel, Ni	0.05	Max. 0.25
Iron, Fe	97.087	96.785–97.77

Table 2 Mechanical properties of AISI 4140 alloy steel [18]

Mechanical properties	Unit
Tensile strength, (σ_B)	655 MPa
Poisons ratio, (υ)	0.27–0.30
Yield strength, (σ_{ys})	415 MPa
Young's modulus, E	190–210 GPa
Hardness Rockwell B	92 HRB
Density, ρ	7850 kg/m^3
Shear modulus, (μ)	80 GPa
Shear stress, (τ_a)	140 GPa

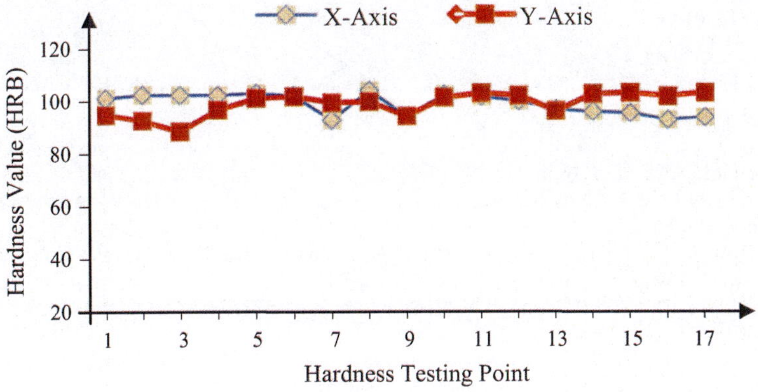

Fig. 6 Hardness values of crankpin cross-section

Based on the curve in Fig. 6, the average material hardness value of the crankshaft was 99.67 HRB. Meanwhile, the material hardness value of AISI 4140 was 92 HRB. The hardness value that exceeds the standard causes the toughness of the material to decrease, therefore leading to failure. Based on Fig. 6, the hardness depth of the crankshaft is not uniform. The non-uniformity of hardness is caused by several main factors which include heat treatment faults, irregular lubrication, which leads to overheating and wear of the crankpin surface while operating the engine. Moreover, due to high friction between the crankpin and the bearing, the wear of the crankpin material leads to an initial crack.

4 Conclusion

From the study on the causes of crankshaft failure that has been carried out, it can be concluded that the SEM observations showed the presence of crack initiation and the direction of crack propagation. Furthermore, the SEM observations in the crack initiation area show the presence of crack initiation which originates from the initial defect around the fillet on the fourth crankpin and propagating to the counterweight until the final fracture. The hardness test results showed that the average hardness value of the crankshaft was approximately 99.67 HRB which exceeds the standard of AISI 4140 material, namely 92 HRB. The high hardness led to a decrease in the toughness of the crankshaft. Additionally, the depth of hardness of the crankshaft was lower in the crack initiation area which triggered initial defects in the material while there was friction with the bearings.

Acknowledgements The authors would like to thank Universitas Syiah Kuala for providing partial financial support to this research through Research Grant No. 001/UN11.2.1/PT.01.03/ PNBP/2022.

References

1. Dindore A, Badiger G (2020) Optimization of crankshaft by modification in design and material. Int Res J Eng Technol 7:3321–3325
2. Jiao AY, Chen FL, Liu BH, Deng JW (2020) Failure analysis of a diesel engine crankshaft. Metalurgija 59:113–116
3. Myagkov LL, Mahkamov K, Chainov ND, Makhkamov I (2014) Advanced and conventional internal combustion engine materials. In: Folkson R (ed) Alternative fuels and advanced vehicle technologies for improved environmental performance: towards zero carbon transportation. Woodhead Publishing, Swaston, Cambridge, pp 370–392
4. Aliakbari K (2019) Failure analysis of four-cylinder diesel engine crankshaft. J Braz Soc Mech Sci Eng 41:1–8
5. Umar IK, Samir B (2019) Deformation and stress analysis of crankshafts for single cylinder and four cylinder IC engine using ansys. Cumhur Sci J 40:197–203
6. Husaini TE, Putra S (2020) Novriandika, study of failure analysis of a fracture crankshaft pulley used on a truck engine. IOP Conf Ser Mater Sci Eng 739:012018
7. Husaini, Ali N, Riantoni R, Putra TE, Husin H (2019) Study of leaf spring fracture behavior used in the suspension systems in the diesel truck vehicles. IOP Conf Ser Mater Sci Eng 541:012046
8. Witek L (2016) Failure and thermo-mechanical stress analysis of the exhaust valve of diesel engine. Eng Fail Anal 66:154–165
9. Wathan S, Husaini N (2021) Ali, Failure analysis on the fracture shaft of a centrifugal pump used for diesel engine cooling system. Key Eng Mater 892:107–114
10. Barizi MI, Husaini H, Ali N, Akmal R, Putra TE (2021) Failure analysis of the short drive shaft in a screw press machine. Key Eng Mater 892:74–80
11. Husaini Z (2016) Mixed mode fracture behavior of an aluminum alloy A6061 investigated by using compact tension shear specimens. Int J Technol 7:456–462
12. Farrahi GH, Hassani Gangaraj SM, Abolhassani S, Hemmati F, Sakhaei M (2011) Failure analysis of a four cylinder diesel engine crankshaft made from nodular cast iron. J Engine Res 22:21–28
13. Husaini, Putra TE, Ali N, Husin H, Zulfikar (2019) Failure analysis of the fracture surface of the crankshaft of a vehicle. IOP Conf Ser Mater Sci Eng 523:012067
14. Pandey RK (2003) Failure of diesel-engine crankshafts. Eng Fail Anal 10:165–175
15. Husaini, Kishimoto K, Notomi M (2004) Finite element analysis of rubber particles size distribution on fracture toughness of rubber-modified polymer. Key Eng Mater 261:711–716
16. Husaini, Ali N, Bakhtiar A (2017) Analysis on the surface of broken screw spring front car suspension. Proseding Seminar Nasional Rekayasa III Tahun 2016 3:111–116
17. Husaini, Kishimoto K, Hanji M, Notomi M (2016) Investigations of the mixed mode crack growth behavior of an aluminum alloy. ARPN J Eng Appl Sci 11:885–890
18. Information on http//www.azom.com/article.aspx?ArticleID=6769

Study on the Impact of Mesh Size Selection and Contact Model on Stress Distribution on Internal Fixation of Fibula Fracture Analyzed by Finite Element Method

Syifaul Huzni, Teuku Muhammad Rizki, Israr Bin M. Ibrahim, Rudi Kurniawan, Syarizal Fonna, and A. K. Arifffin

Abstract The fibula is a small bone that runs parallel to the tibia in the lower leg. The fibula receives 15% of the human body weight. Under certain conditions, fractures can occur in the fibula due to certain human activities. One of the medical solutions to treat fractures that occur in the fibula is to install internal fixation. In the case of internal fixation, the interaction between bone and bone, bone with plate, bone with bolt, bolt with bone, is complicated to study experimentally. Numerical analysis is an option to study these interactions. Several parameters that affect the analysis results need to be fully known to perform an excellent numerical analysis. The purpose of this study is to investigate the impact of mesh size and coefficient of contact model on stress distribution on the internal fixation of fibula fracture. They were analyzed by finite element method under loading conditions that represent the standing position. The stress distribution was analyzed using ANSYS software. Three mesh sizes are selected, namely coarse, medium, and fine. Friction contact models were applied to the area between the fibula and the implant. The load used in this study represents three human body weights, namely 80, 90, and 100 kg. The material chosen for the plate and screw is AISI 316. Among the three meshes used, it is seen that the fine mesh produces the highest stress and is concentrated in one particular area. Furthermore, the coefficient friction of the contact model did not produce any significant difference in the stress distribution on internal fixation.

Keywords Fibula fracture · Internal fixation · Finite element analysis · Stress distribution · Mesh size · Contact model

---

S. Huzni (✉) · T. M. Rizki · I. B. M. Ibrahim · R. Kurniawan · S. Fonna
Department of Mechanical and Industrial Engineering, Faculty of Engineering, Syiah Kuala University, 23111 Banda Aceh, Indonesia
e-mail: syifaul@unsyiah.ac.id

A. K. Arifffin
Department of Mechanical and Manufacturing Engineering, Faculty of Engineering and Built Environment, Universiti Kebangsaan Malaysia, Bangi, Malaysia

Irwansyah et al. (eds.), *Proceedings of the 4th International Conference on Experimental and Computational Mechanics in Engineering*, Lecture Notes in Mechanical Engineering, https://doi.org/10.1007/978-981-99-7495-5_3

1 Introduction

Bone fractures are a common problem, one of which is a fracture of the fibula bone. Fractures of the fibula due to stress from bodyweight are rare since the fibula only supports 15% of the body's weight [1]. Fractures of the fibula are usually caused by driving accidents or sports accidents.

A typical procedure for repairing bone fractures is the installation of internal fixation. Internal fixation is an orthopedic surgical procedure by placing temporary or permanent implants, in the form of plates and screws, on broken bones to maintain and stabilize the position of the bones to speed up the healing process [2]. In its implementation, internal fixation can fail to perform its function. Failure can occur in the fixation plate or the bolts that attach the plate to the bone. The form of failure that occurs is the fracture of the fixation plate or fixation bolt. Although the case is rare, failure of internal fixation of the fibula bone is still found [3]. The interaction between bone with bone, bone with plate, plate with screw, and screw with bone needs to be known to minimize the risk of internal fixation failure in carrying out its function. Therefore, the behavior of internal fixation attached to the fractured fibular bone needs to be studied further.

Experimental studies are complicated to investigate the failure of internal fixation for fibular fractures. The finite element method is one of the numerical methods that are widely used to overcome these difficulties. To ensure that the parameters and variables used in the finite element analysis for further investigation are appropriate, it is necessary to study the effect of the parameters that will be used on the analysis results to be achieved. Two parameters that will be tested in this study are the effect of mesh size and contact coefficient on the contact model used on the stress distribution of the finite element analysis. This research is focused on studying the effect of mesh size and contact coefficient on stress distribution on internal fixation of fibula fracture, which was analyzed by finite element method.

2 Research Methods

2.1 Models and Geometry

The shape and size of the fibula bone used in this study were adopted from the model developed by Slobodan Simic [4]. The plates and screws used were adopted from the model used by Manoharan [5]. Figures 1 and 2, respectively, show the fibula bone's shape and geometry, plate, and screw used. The mechanical properties of the bone used were referred to Ozkan [6], with an elastic modulus of 5 GPa and a Poisson ratio of 0.302. The assembly of the internal fixation of the fractured bone is shown in Fig. 3. Number 1–7 in Fig. 3 shows the serial number of the screws installed in the internal fixation to simplify the analysis. The plates and screws used in this study are 316 stainless steel with an elastic modulus of 193 GPa. The contact between bone and

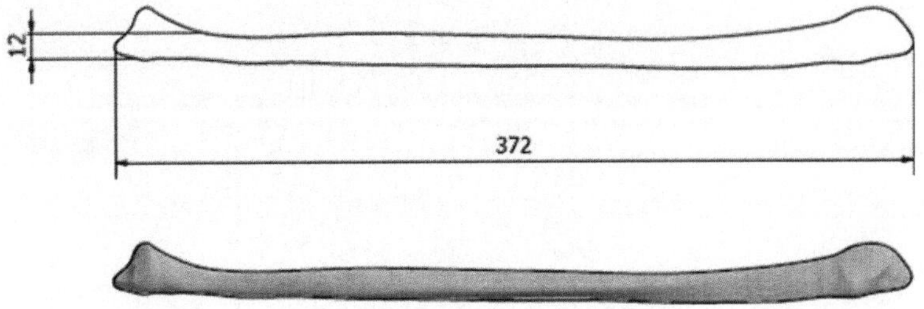

Fig. 1 Geometry of fibula

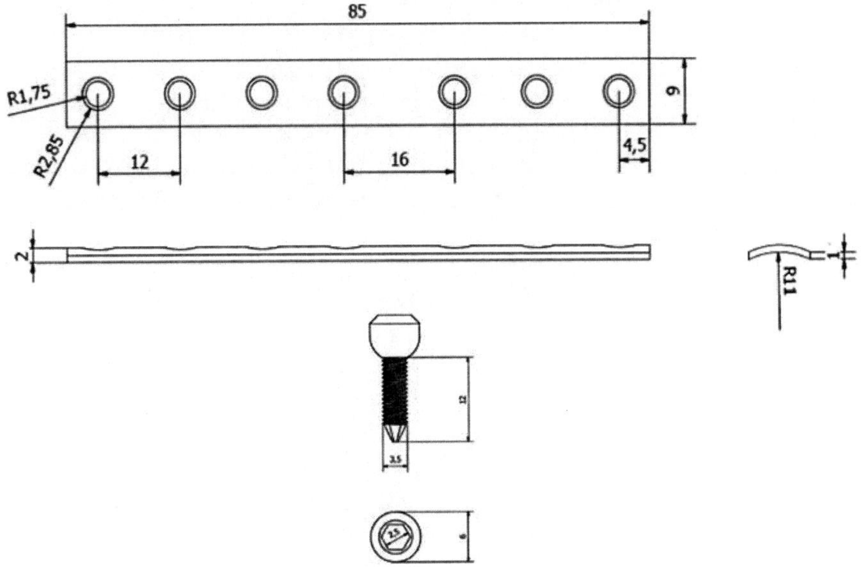

Fig. 2 Geometry of plate and screw

plate and plate and bolt is modeled by the friction contact model. Two coefficients of friction were tested for all contacts, plate-bone, plate screw, which were 0.4 [7] and 0.2 [8]. At the same time, the fracture gap between bone and bone uses a bonded contact model.

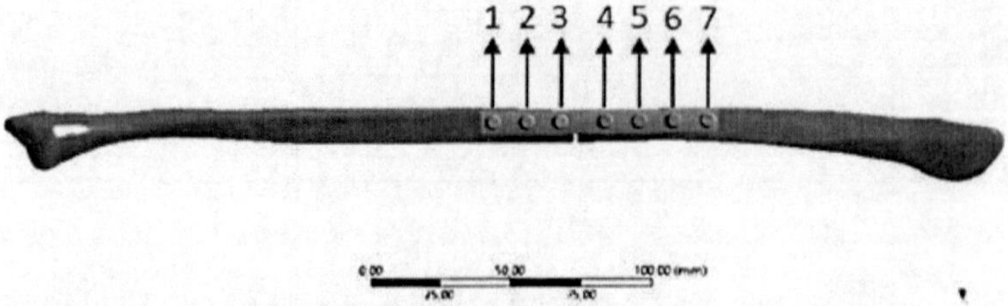

Fig. 3 Installation of plates and screws on the fibula bone

2.2 Boundary Conditions

The load used to study the effect of mesh on stress distribution is 15% of 80 kg body weight, considered to be evenly distributed on two legs, then converted and rounded to 60 N. Study on the effect of load on stress distribution: three loads were used, each representing body weight of 80, 90, and 100 kg. Assuming the load is evenly distributed on the two legs, 15% of the applied load can be written as 60 N, 68 N, and 75 N. The boundary conditions of the finite element analysis can be seen in Fig. 4.

Fig. 4 Boundary condition

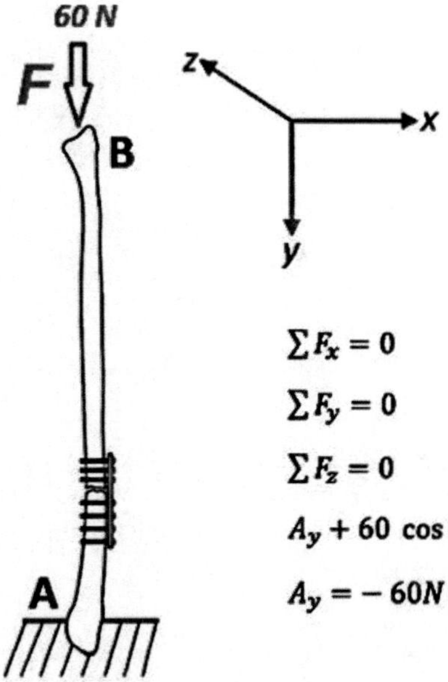

$$\sum F_x = 0$$

$$\sum F_y = 0$$

$$\sum F_z = 0$$

$$A_y + 60\cos 0 = 0$$

$$A_y = -60N\ (\uparrow)$$

2.3 Finite Element Mesh

The mesh type used in the analysis is tetrahedrons, with the software default mesh sizes as follows: coarse (± 6 mm), medium (± 4 mm), and fine (± 2 mm). Mesh is generated automatically by the software used, ANSYS.

3 Results and Discussion

3.1 Mesh Generation

Figure 5 shows the results of the mesh generation on the fracture internal fixation model used for the three selected mesh sizes. The number of elements and nodes generated for each element size can be seen in Table 1.

There is an almost 40% difference in nodes between coarse and medium meshes and almost 9% between fine and medium meshes. Likewise, with the number of elements produced, there is a difference of almost 56% between coarse mesh and medium mesh. In comparison, between medium and fine mesh, the difference is almost 8%. The number of elements and nodes will undoubtedly affect the calculation and analysis time from the computational aspect.

3.2 Von Mises Stress Distribution

The impact of mesh size selection on the stress distribution of finite element analysis results can be seen in Fig. 6. The three mesh sizes used produce different stress distribution patterns resulting from finite element analysis. Suppose the lighter color is considered to have more significant stress. In that case, it can be seen that the finer mesh size gives a broader distribution pattern than the coarser mesh. According to the theory of finite element analysis, finite modal with small element size yields high accuracy compared to the modal with large element size. However, the location of the higher stress concentrations was similar for the three mesh sizes used in the plate area around the bolt. This phenomenon can be explained that a hole area is a form of irregular of geometry that causes stress concentration around it.

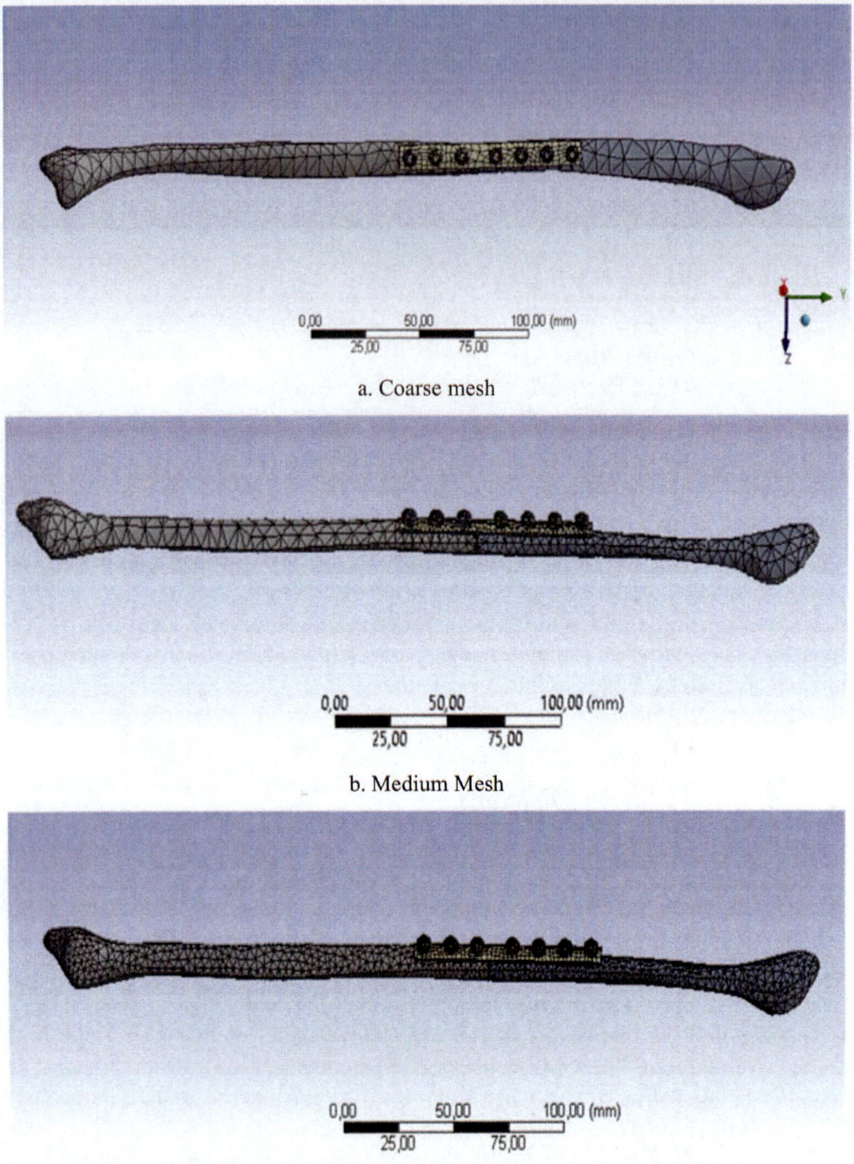

Fig. 5 Meshing generated for three different sizes of mesh

Table 1 Number of element and node

Mesh size	Nodes	Elements
Coarse	78,391	39,954
Medium	109,580	62,344
Fine	119,346	67,564

The magnitude of the von Mises stress at the location around the bolt on the internal fixation plate can be seen in Fig. 7. The average percentage difference in the magnitude of the von Mises stress resulting from the finite element analysis using coarse and medium mesh is 9%. The difference between fine and medium mesh is 12%. The difference between coarse and fine mesh is 22%. The difference in analysis results due to different mesh sizes significantly affects the results of stress analysis, so for further analysis, a fine mesh will be used.

3.3 *Coefficients of Friction*

As shown in Fig. 8, different COFs do not cause differences in the trendline of the stress calculation results. The difference in the average percentage of the von Mises stress resulting from COF 0.4 and 0.2 is 7.5%. The small effect of COF on the results is suspected because the difference in the COF magnitude used is close.

3.4 *Bodyweight*

The results of the load effect analysis, in the form of bodyweight, for standing still can be seen in Fig. 9. The trendline of the magnitude of the von Mises stress obtained is generally not affected by the applied load. The same pattern was found for all three given loads. The average percentage difference in the results of stress analysis for 60 N and 68N weight is 16%, 75 N and 68 N is 10%.

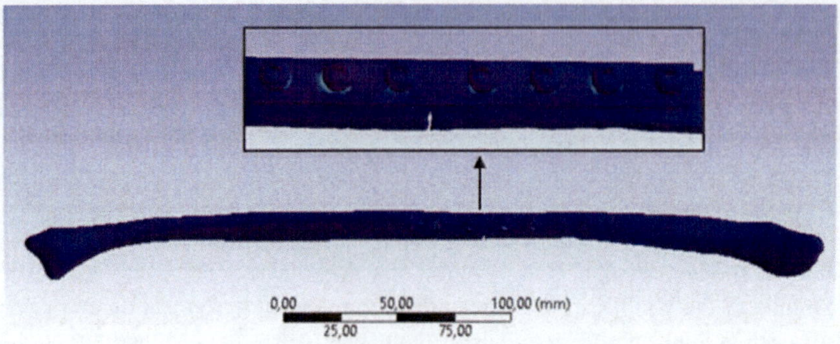

a. Coarse mesh

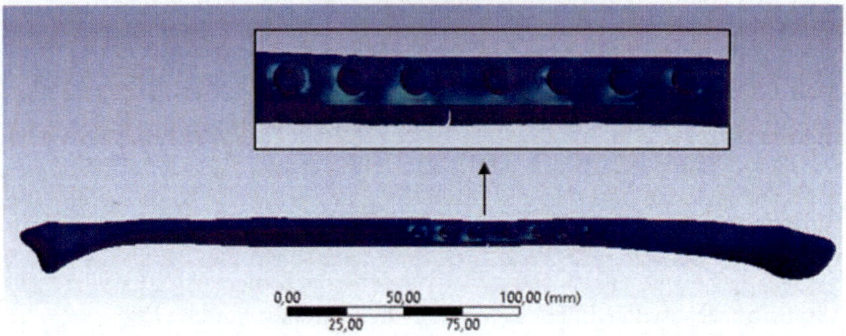

b. Medium Mesh

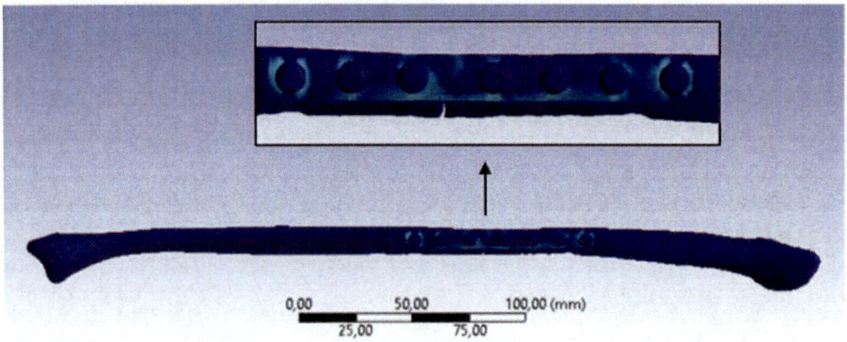

c. Fine Mesh

Fig. 6 Von Mises stress distribution results from the analysis of different mesh sizes

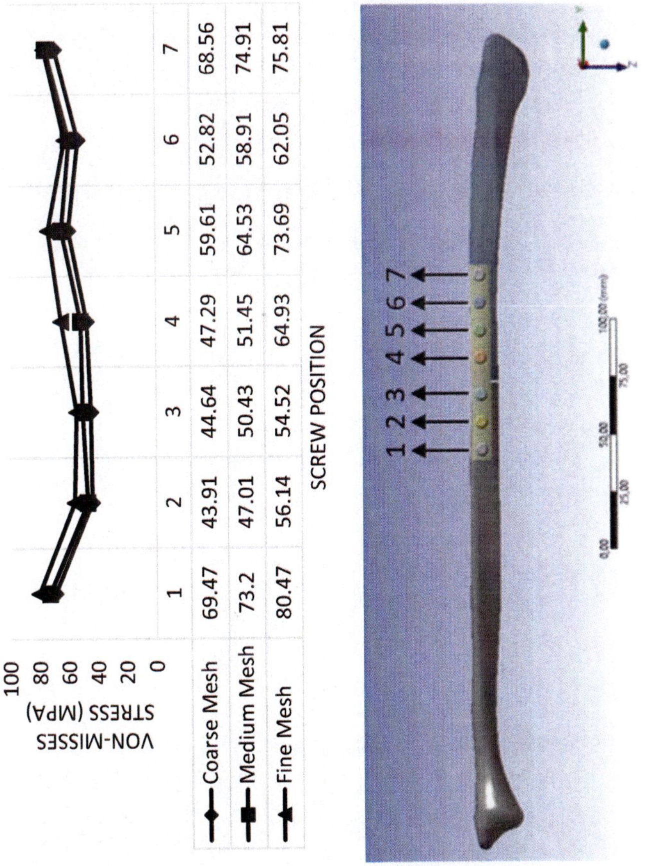

Fig. 7 Magnitude of von Mises stress measured on the screw

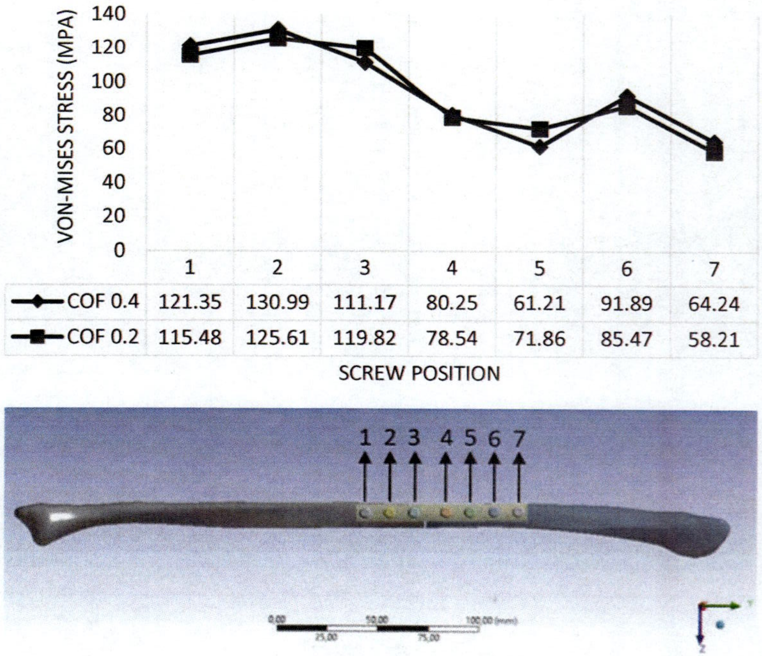

Fig. 8 Effect of friction coefficients (COF)

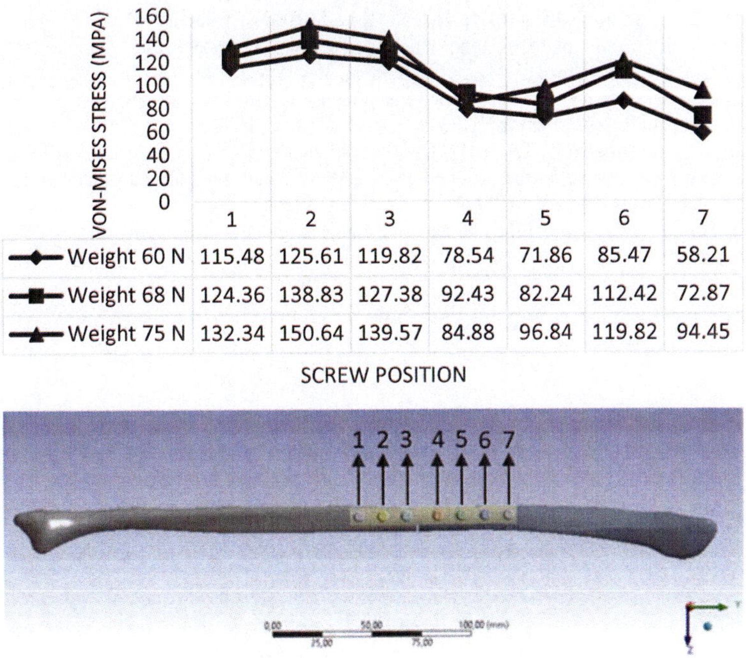

	1	2	3	4	5	6	7
Weight 60 N	115.48	125.61	119.82	78.54	71.86	85.47	58.21
Weight 68 N	124.36	138.83	127.38	92.43	82.24	112.42	72.87
Weight 75 N	132.34	150.64	139.57	84.88	96.84	119.82	94.45

Fig. 9 Magnitude of von Mises stress under different loads

4 Conclusion

The results of this study show that the mesh size used affects the distribution and magnitude of the stress resulting from the finite element analysis. Mesh with fine size gives better results than other mesh sizes. Further studies on the friction coefficient of the contact model and the applied load did not produce any significant difference in the stress distribution on internal fixation.

Acknowledgements The research was supported by Penelitian Dasar Unggulan Perguruan Tinggi Tahun Anggaran 2021, contract number Nomor: 43/UNI1.2.1/PT.O1.03/DPRM/2021.

References

1. Robertson GA, Wood AM (2017) Lower limb stress fractures in sport: optimizing their management and outcome. World J Orthop 18, 8(3):242–255
2. Thakur AJ (2012) The elements of fracture fixation, 2nd edn. Elsevier, India
3. Duan X, Kadakia AR (2017) Operative treatment of posterior malleolar fractures. Open Orthop J 11(Suppl-4, M14):732–742
4. Slobodan Simic (2019) Female Left Leg, GrabCAD Community

5. Manoharan G, Singh R, Kuiper JH, Nokes LDM (2018) Distal fibula oblique fracture fixation using one-third tubular plate with and without lag screw—a biomechanical study of stability. J Orthop 549–552
6. Ozkan A, Kisioglu Y (2013) The rotational loading capacity an effects of fibula on lower extrimity. J Eng Res Appl Sci 2(1):138–143
7. Kim SH, Chang SH, Jung HJ (2010) The finite element analysis of a fractured tibia applied by composite bone plates considering contact conditions and time-varying properties of curing tissues. Compos Struct 92(9):2109–2118
8. Das S, Sarangi SK (2014) Finite element analysis of femur fracture fixation plates. Int J Basic Appl Biol 1(1):1–5

Mechanical Properties of Quenched and Tempered Low Alloy Steel Grinding Media

Husni Usman, Syarizal Fonna, Syifaul Huzni, Sulaiman Thalib, Izzan Nur Aslam, Teuku Irvan Ramadhan, and Tommy Octaviantana

Abstract The effect of tempering temperature on mechanical properties of the quenched low alloy steel was studied. The as-received low alloy steel samples were heated to 950 °C, held for 30 min, and then oil-quenched. The as-quenched samples were finally tempered in the temperature range of 200–600 °C for 30 min. The chemical composition of as-received alloy steel was determined using spectrometer. The microstructures of the samples were observed using optical microscope. A Rockwell hardness test was done to measure the hardness of the alloy steel samples, and a Charpy impact test were conducted to measure the impact toughness of the samples. The results showed the microstructures of the as-quenched and as-tempered samples consist of martensite with different orientation and small inclusions of irregular shapes. Compared with the as-received low alloy steel, the hardness of the quenched steel increased while the impact energy decreased. In comparison with the quenched samples, increasing tempering temperature in the range of 200–600 °C would improve the impact energy of the tempered steels, and it significantly increased above 400 °C.

H. Usman (✉) · I. N. Aslam
Department of Mining Engineering, Faculty of Engineering, University of Syiah Kuala, Jl. Syech A. Rauf 7, Banda Aceh 23111, Aceh, Indonesia
e-mail: husmanphd@usk.ac.id

I. N. Aslam
e-mail: izzanaslam@unsyiah.ac.id

H. Usman · S. Fonna · S. Huzni · S. Thalib · T. I. Ramadhan
Laboratory of Materials Engineering, Department of Mechanical and Industrial Engineering, Faculty of Engineering, University of Syiah Kuala, Jl. Syech A. Rauf 7, Banda Aceh 23111, Aceh, Indonesia
e-mail: syarizal.fonna@unsyiah.ac.id

S. Huzni
e-mail: syifaul@unsyiah.ac.id

T. Octaviantana
PT. Nusa Halmahera Minerals, Gosowong Gold Mine, Halmahera Utara, Maluku Utara, Indonesia
e-mail: tommyo@nhm.co.id

Irwansyah et al. (eds.), *Proceedings of the 4th International Conference on Experimental and Computational Mechanics in Engineering*, Lecture Notes in Mechanical Engineering, https://doi.org/10.1007/978-981-99-7495-5_4

Keywords Low alloy steel · Quenching · Tempering · Microstructure · Mechanical properties

1 Introduction

Major operating costs associated with grinding mills in mining and mineral industries typically encompass energy consumption, mill liner, and grinding media wear [1, 2]. Grinding mills represent up to 40% of the direct total operating costs of mineral processing. The cost of grinding media and liner wear itself is around 15–25%. Due to incorrect liner designs, the grinding media would fall down on the liner rather than on 'the toe of the charge', resulting in less efficient energy utilization, poor grinding, and high maintenance costs [3–5]. The increased height of fall of such grinding balls in larger SAG mills results in severe impact between the grinding balls and also between the balls and the mill liners. Therefore, it is suggested to utilize tougher steels that exhibit good resistance to wear and high energy impact as well.

The hardening capacity of a steel depends mostly on the carbon content and, somewhat, on the amount of alloying elements and austenite grain size [6]. If the carbon content of martensite increases, its hardness would increase. As a quenched steel contains retained austenite, the hardness of a quenched steel is not the same as the hardness of martensite crystals. The hardness of a steel quenched from austenite temperatures reaches its maximum at a carbon concentration of 0.8–0.9 wt% C and then decreases.

The martensitic stainless steel grinding balls could be heat treated to produce microstructures at their surfaces which ranged from 100% austenite to basically all martensite [7]. Wear rate of grinding ball consisting fully of hard martensite increased under an oxygen environment. The balls that were furnace-cooled and the ice-water-quenched balls would obtain combined microstructures of soft austenite and hard martensite and thus result in less wear under an oxygen atmosphere.

The investigation of the differences in microstructure, hardness, abrasion, and impact toughness of steel media from different manufacturing sources showed that abrasive wear has a tendency to be dependent on media hardness for grinding media with a chromium content of less than 0.7% and independent of hardness for media with chromium content bigger than 0.75% [8]. Hardness and abrasion wear are slightly interrelated to the chemical composition. Charpy impact energy strongly depends on the carbon content, and alloy steels with carbon content more than about 0.2% would produce a brittle fracture at room temperature.

The previous research showed that increasing quenching temperature from 840 to 900 °C would result in an increase in hardness and tensile strength of multi-element low alloy steel [9]. The toughness increased at tempering temperature of above 300 °C, and the excellent wear resistance could be obtained by tempering at 350 °C. The optimum heat treatment conditions could be achieved by conducting quenching at 900–920 °C and tempering at 350 °C.

Furthermore, the effects of austempering and martempering processes using salt bath quenching media on wear resistance and impact toughness of Cr–Mo steels were studied, and the results were compared with compressed-air quench [10]. The samples austempered for 2 h exhibited the best impact toughness, and the samples martempered for 2 h showed the highest values of wear resistance. The austempered and martempered specimens quenched in salt bath showed a better combination of wear resistance and toughness than compressed-air quenched specimens.

In order to improve mechanical properties and wear resistance of high-carbon low alloy steel, the proper heat treatment should be established. In this research, the influence of tempering temperature on mechanical properties of quenched low alloy steel was studied.

2 Materials and Methods

The 100.75 mm diameter grinding ball was come from a mining company. The low alloy steel was melted using induction furnace and cast into sand mold. Because of insufficient information on heat treatment of the as-received grinding ball, it is believed that after casting it was followed by quenching using compressed air. A wire electro discharge machine (EDM) cutting tool was utilized to cut the as-received steel into examination and mechanical testing specimens.

The chemical composition of as-received alloy steel was determined using spectrometer. Table 1 provides the chemical composition of as-received low alloy steel that was used in this research. In terms of carbon content, the steel includes hypo-eutectoid steel.

The metallographic samples were prepared according to the standard ASTM E3-01 [11]. The mounted sample was polished by a sequence of 100, 320, 600, 1000, 1200, and 2000-grit silicon carbide papers and was finished by vibratory-polishing using an aqueous suspension of 0.01-μm alumina. The polished specimen was then etched in a solution of 3% nital at room temperature to reveal microstructure and grain boundaries. The microstructures of the samples were observed using optical microscope.

Figure 1 shows the heat treatment cycles of quenching and tempering. The samples were austenitized at 950 °C for 30 min. holding time [12]. Then, they were quenched in oil media and were finally tempered at a temperature range of 200–600 °C and held for 30 min. Both as-quenched and as-tempered specimens were also prepared for microstructural examination and mechanical tests.

Table 1 Chemical composition of as-received low alloy steel

Element	C	Mn	Si	Cr	Ni	Mo	S	P	Fe
% Weight	0.61	0.78	0.28	0.95	0.06	0.01	0.003	0.005	Bal

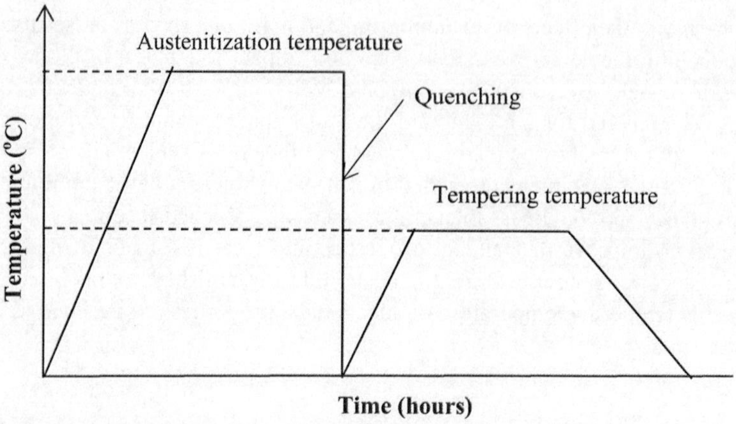

Fig. 1 Heat treatment cycles of quenching and tempering

A Rockwell hardness test was done in accordance with ASTM standard E 18-03 [13] using a Rockwell type hardness tester. A Rockwell C scale with a load of 150 kg that employed a diamond point (Brale indenter) was used. Impact Charpy test was conducted in accordance with ASTM E23-02 [14].

3 Results and Discussion

Microstructure. Figure 2 shows optical microscope images of the as-received low alloy steel and the as-quenched samples as well. The microstructure at near surface (Fig. 2a, b) consists of martensite with different orientation and small inclusions of irregular shapes. It was found in the previous research that pearlite with ferrite along prior austenite grain boundaries had been observed across most of remaining ball [12]. The measured hardness across the as-received steel ball also showed that the steel ball had been quenched but with lowered hardenability. The high carbon content of steels could also produce low hardenability during heat treatment [15, 16].

The optical microscope images of as-quenched samples are shown in Fig. 2c, d. The microstructures of the as-quenched samples are also composed of martensite with different orientation and small inclusions of irregular shapes. The brighter areas observed in Fig. 2 could be retained austenite.

Figure 3 exhibits optical images of as-tempered samples after being tempered in the range of 200–600 °C. The microstructures of the as-tempered samples also consist of martensite with different orientation and small inclusions of irregular shapes. It is very hard to differentiate the characteristics of quenched and tempered martensite with optical microscopy magnification. Scanning electron microscopy needs to be used to define martensite morphologies.

(a) As received at near surface (200x M)

(b) As received (500x M)

(c) As quenched sample (200x M)

(d) As quenched sample (500x M)

Fig. 2 Optical microscope images of as-received and as-quenched samples. Etchant: 3% nital, M: magnification

Mechanical Properties. Table 2 provides the results of Rockwell hardness and impact energy of the as-received and as-quenched low alloy steel samples. As reported in previous research, impact energy of the as-received steel ball based on the specimen location within the steel ball was independent of location [8, 12]. Compared with the as-received steel, the hardness of the as-quenched steel increases while the impact energy decreases.

Figure 4 shows impact energy of as-tempered samples after being tempered in the temperature range of 200–600 °C. Compared with the as-quenched samples (indicated at 0 °C temperature), as the tempering temperature is increased to 400 °C, the impact energy of the as-tempered steel increases, and at a higher temperature (above 400 °C), it significantly increases. It is well documented that the degree of tempering has an effect on toughness and in this study, increasing tempering temperature would produce increased impact energy. Tempered martensite embrittlement could happen for certain cases, resulting in decreased impact toughness in certain ranges of tempering temperature.

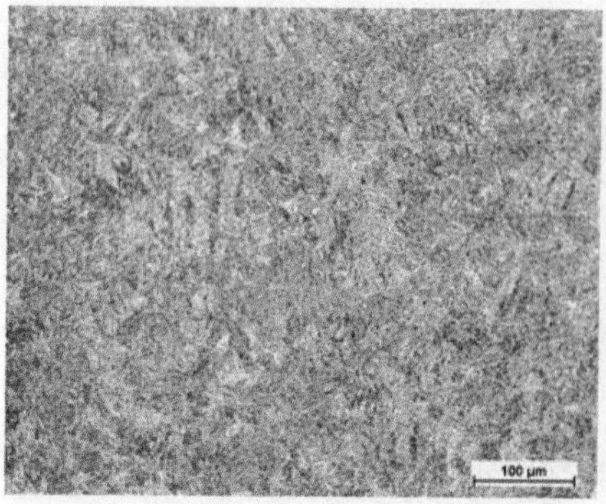

(a) As tempered at 200°C (200x M)

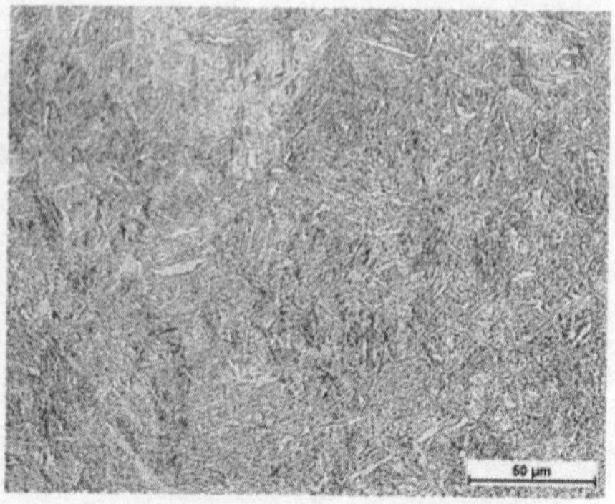

(b) As tempered at 200°C (500x)

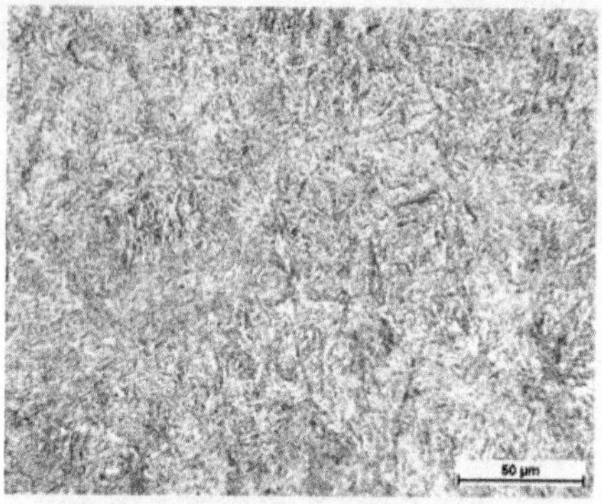

(c) As tempered at 300 °C (500x M)

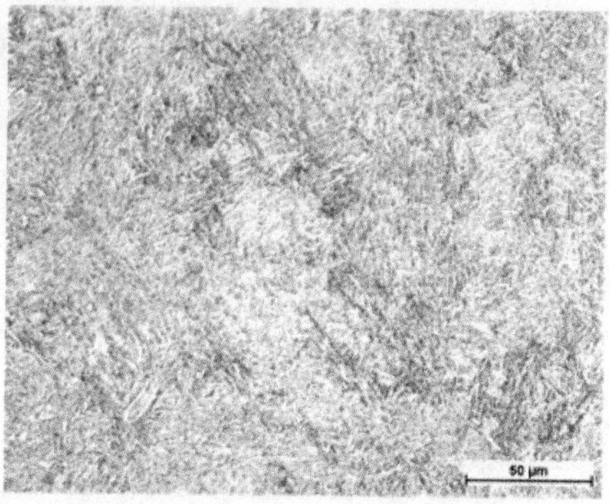

(d) As tempered at 400 °C (500x M)

(e) As tempered at 500 °C (500x M)

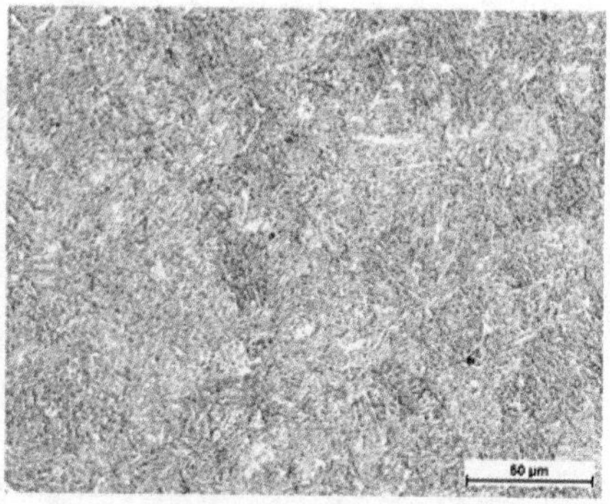

(f) As tempered at 600 °C (500x M)

Fig. 3 Optical microscope images of as-tempered samples. Etchant: 3% nital

Table 2 Mechanical properties of as-received and as-quenched samples

Samples	Testing	1st	2nd	3rd	4th	5th	Average
As-received	HRC	55.8	62.3	59.0	55.2	60.6	58.6
	Impact [J]	8	8	8			8
As-quenched	HRC	61.9	66.3	57.7	58.6	57.4	60.4
	Impact [J]	5	4	5			4.7

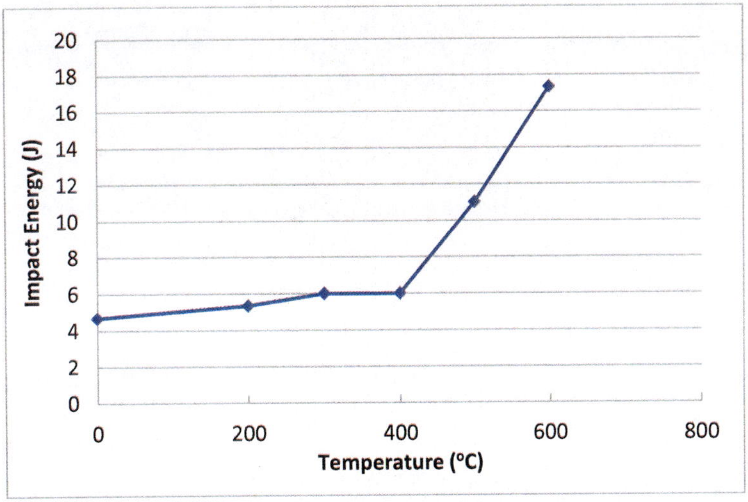

Fig. 4 Impact energy as a function of tempering temperature for as-tempered steel samples

As shown in Fig. 5, the fracture surfaces of the tested as-received, as-quenched, and also as-tempered specimens clearly exhibits brittle fracture. It seems that the mode of fracture is most likely slant brittle fracture and no ductile fracture is found on the fracture surfaces of all impact test specimens. According to the brittle fracture criteria, if the impact energy (toughness) values of all the tested specimens are found to be below the brittle fracture impact energy of 20 J, the specimens would fracture in a brittle manner [17].

4 Summary

The microstructures of the as-quenched and as-tempered samples are composed of martensite with different orientation and small inclusions of irregular shapes. Compared with the as-received low alloy steel, the hardness of the quenched steel increased while the impact energy decreased. If tempering temperature in the range of 200–600 °C increased, the impact energy of the as-tempered steels would increase,

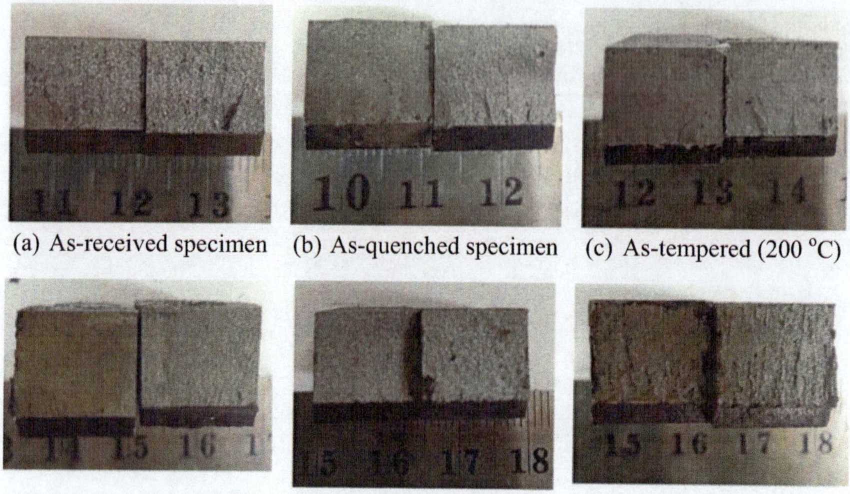

(a) As-received specimen (b) As-quenched specimen (c) As-tempered (200 °C)

(d) As-tempered (300 °C) (e) As-tempered (400 °C) (f) As-tempered (500 °C)

Fig. 5 Fracture surfaces of broken Charpy V-notch impact specimens

and it significantly increased above 400 °C. Apparently, tempering of the as-quenched low alloy steels would enhance impact toughness, and it was not found tempered martensite embrittlement resulting in decreased impact toughness in certain ranges of tempering temperature. The fracture surfaces of the broken impact as-received, as-quenched, and as-tempered specimens as well obviously shows brittle fracture.

Acknowledgements The authors would like to acknowledge the University of Syiah Kuala for providing the research fund. We are also grateful to PT Nusa Halmahera Minerals for providing low alloy steel grinding media.

References

1. Curry JA, Ismay MJL, Jameson GJ (2014) Mine operating costs and the potential impacts of energy and grinding. Miner Eng 56:70–80
2. Wills BA, Finch JA (2016) Wills' mineral processing technology: an introduction to the practical aspects of ore treatment and mineral recovery. Elsevier Ltd., Oxford
3. Powell M, Smit I, Radziszewski P, Cleary P, Rattray B, Eriksson K, Schaeffer L (2006) Selection and design of mill liners. In: SK Kawatra (ed) Advances in comminution. SME, Englewood, Colorado, pp 331–376
4. Usman H, Taylor P, Spiller DE (2017) The effects of lifter configurations and mill speeds on the mill power draw and performance. In: AIP conference proceedings, vol 1805, no 050001
5. Usman H, Fonna S, Thalib S (2020) A review on current mill liner design and performance. In: IOP conference series materials science and engineering, vol 931, no 012016
6. Sverdlin AV, Ness AR (2007) Effects of alloying elements on the heat treatment of steel. In: Totten GE (ed) Steel heat treatment handbook. Taylor & Francis Group, USA, pp 166–211

7. Jang JW, Iwasaki I (1988) Effect of martensite and austenite on grinding media wear. Wear 122:285–299
8. Sabih A, Radziszewski P, Mullany I (2017) Investigating grinding media differences in microstructure, hardness, abrasion and fracture toughness. Miner Eng 103–104:43–53
9. Fu H, Xiao Q, Fu H (2005) Heat treatment of multi-element low alloy wear-resistant steel. Mater Sci Eng A 396:206–212
10. Shaeri MH, Saghafian H, Shabestari SG (2012) Effect of heat treatment on microstructure and mechanical properties of Cr–Mo steels (FMU-226) used in mills liner. Mater Des 34:192–200
11. ASTM E3-01 (2001) Standard practice for preparation of metallographic specimens. American Standard for Testing Materials, West Conshoocken, PA, pp 1–8 (3.01)
12. Usman H, Fonna S, Huzni S, Ramadhan TI, Octaviantana T (2021) The Effect of hardening on mechanical properties of low alloy steel grinding media. In: Akhyar (ed) Proceedings of 2nd ICECME, Lecture notes in mechanical engineering. Springer, Singapore, pp 459–469
13. ASTM E18-03 (2003) Standard test methods for Rockwell hardness and Rockwell superficial hardness of metallic materials. American standard for testing materials, West Conshoocken, PA, pp 116–129 (03.01)
14. ASTM E23-02a (2003) Standard test methods for notched bar impact testing of metallic materials. American standard for testing materials, West Conshoocken, PA (3.01)
15. Elwazri AM, Wanjara P, Yuea S (2005) The effect of microstructural characteristics of pearlite on the mechanical properties of hypereutectoid steel. Mater Sci Eng A 404:91–98
16. Eshghian R, Abbasi M (2017) Wear and failure analysis of semi-autogenous grinding mill liners. J Fail Anal Preven 17:340–348
17. Hertzberg RW (1996) Deformation and fracture mechanics of engineering materials, 4th edn. Wiley, New York

Managing Defects in Steel Fabrication Process Using Failure Mode and Effect Analysis Approach

Nina Aini Mahbubah, Abdul Muid, and A. A. Atho'illah

Abstract Quality is a primary focus to both maintain and increase customers. A manufacturing production process and service business's failure impacts operational costs. Assessments are needed to instigate fabrication process failure and maintain good quality output. PT AA is a fabrication contractor firm that experienced process failures in the steel fabrication division. This study aims to evaluate the steel fabrication process to recommend improvement solutions to eliminate failures at the fabrication stage. Failure mode and effect analysis was used as the research approach. The result of the research found that two defects, namely peeling and ruuns, in the painting process gain the highest defect numbers with 117 compared to other stages. With a score of 576, the highest risk of defects ruuns the painting process. Recommendations for improvement are that the company's management regularly provides briefings and education on the importance of SOPs and installing SOP stickers on every painting equipment and tools.

Keywords FMEA · Fabrication · Defect · RPN · Ishikawa

1 Introduction

The sustainability of manufacturing and service businesses is one of the focuses of enterprises management [1]. Accurate plans and determinations are needed to sustain the quality of products and services and retain and add new customers [2, 3]. The company performs efficiently in all production and service process flows to have

N. Aini Mahbubah (✉) · A. A. Atho'illah
Industrial Engineering Department, Faculty of Engineering, Universitas Muhammadiyah Gresik, Jl. Sumatera 101 GKB, Gresik 61121, Indonesia
e-mail: n.mahbubah@umg.ac.id

A. Muid
Industrial Engineering Department, Faculty of Engineering, Universitas Negeri Malang, Jl. Semarang No. 5, Malang 65145, Indonesia
e-mail: abdul.muid.ft@um.ac.id

Irwansyah et al. (eds.), *Proceedings of the 4th International Conference on Experimental and Computational Mechanics in Engineering*, Lecture Notes in Mechanical Engineering, https://doi.org/10.1007/978-981-99-7495-5_5

an advantage in business competition. Failure of the process or production results is one of the factors that must be minimized because it results in additional costs [4–6]. The impact of this failure is an increase in production costs. Subsequently, it impacts the possibility of an increase in service or product prices and a decrease in company profits [7]. Such defect gain the probability that consumer confidence in the company will decrease and consumers will switch to another enterprise.

PT AA is a firm engaged in general contractors. This firm has been implementing a job shop as a production system. Steel is considered the primary raw material. The steel fabrication process is illustrated in Fig. 1.

It can be seen in Fig. 1. That *marking* is the initial stage: drawing the material according to specifications. Furthermore, the cutting process uses a mixture of *LPG* and *oxygen* for plate cutting using a *skater cutting machine* to precise the results. For *WF* and *CNP* materials, the cutting process uses manual tools. *Drilling using* mall tools made of zinc material is given a hole as a point or location marker on the material. This process avoids differences in the distance between one material holes with another material hole that is planned to be connected. *Sub-assembling* combines plate and *WF material* to make *rafters* and columns. The next stage is the welding process. The components assembled on the column and the *rafters* can be connected perfectly and have the strength of the building following the required specifications. After the welding process, a re-grinding process is carried out to clean the dirt caused by the welding process. *Sandblasting* is spraying *abrasives* in sand or steel grit with high pressure to remove contamination such as rust or oil. Painting is done to protect the material from the corrosion process. *Erection* is the final process, namely the assembly or installation of steel structures according to the location specified by the customer.

PT AA experienced a problem, namely a *defect,* throughout the steel fabrication process. Table 1 illustrates defects in the stages of the fabrication process that occurred for three months.

Table 1 shows that defects occurred at all stages of fabrication. There were 12 failures during the three months of the fabrication process. The highest *defect* of 117 occurred in the painting process activity. The impact of the *defect* was that rework was carried out, resulting in an increase in operational costs and causing a delay in order completion time. Quality problems that cause these losses require an approach to evaluate the quality of the steel fabrication process.

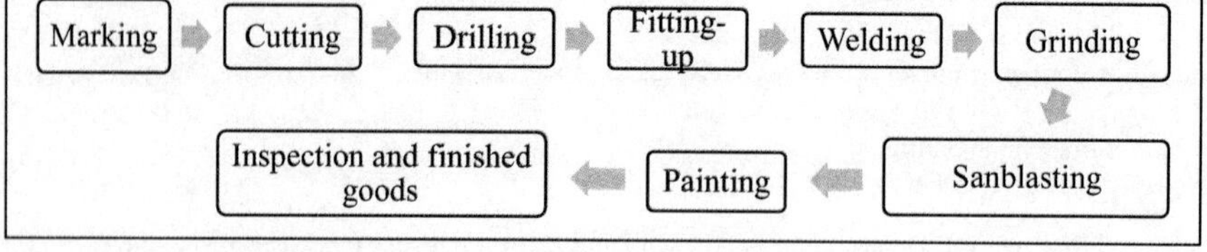

Fig. 1 Steel fabrication process

Table 1 Number of products and defects in November, December 2019 to January 2020

Process	Type of defect	Number	Total defect
Marking	The size does not match the picture	95	95 units
Cutting	Not straight	30	73 units
	Lots of crust	43	
Drilling	The hole does not fit	45	68 units
	Oval	23	
Constructing	Assembly does not match the picture	84	84 units
	Undercut	22	
Welding	Porosity	13	95 units
	Over spatter	60	
Grinding	Broken object	87	87 units
Painting	Ruuns	33	117 units
	Peeling	84	

Source Company data

Quality is a significant part of the production process. Quality management is part of the production process, which is influential in improving product quality. A strategy is needed to maintain process stability and control the process to minimize defective products [4]. One way to maintain and expand the market is by increasing supervision and quality inspection of product products [5].

Quality control is verifying and maintaining the desired level of product or process quality through careful planning, appropriate equipment, continuous inspection, and corrective action when necessary [8]. Service and manufacturing businesses always pay attention to the quality of their products and provide the best possible quality. One of the factors that must be minimized is production defects, resulting in additional production costs [9]. *Failure mode and effect analysis* (FMEA) is used to evaluate manufacturing and service business failures [3, 4, 9–11]. FMEA is a robust approach to evaluating defects in the production process of PVC pipes and pipettes [3, 12]. FMEA is an appropriate method for identifying potential defective and their effects through the product life cycle system [13]. The FMEA approach has also proven effective in recommending changes to the system to eliminate *defects* in loading and unloading machines at ports [11]. Other methods also implement FMEA, for example, *statistical process control, six sigma, fault tree analysis, six sigma,* or maintenance management methods [4, 7–9, 12–14].

This study aims to evaluate *defects* and analyse the causes of the failure of the process to get recommendations to minimize defects in the steel fabrication process at PT AA.

2 Material and Method

This research uses a *qualitative descriptive approach.* The research instrument used was the FMEA questionnaire design. Data collection technique by *walk-through survey* and *brainstorming* with four respondents from PT AA. The data was taken from November and December 2019 to January 30, 2020. The historical data from the firm was data on the number of defects and SOPs for the fabrication process.

The initial stage of the research was to conduct direct observations at the PT AA workshop on all activities and production processes, from marking to painting. The second stage was to identify, to some extent, deviations that affect production defects. At this stage, it identifies the factors that make the product defective. Determination *of severity* (*S*), *occurrence* (*O*), *and detection* (*D*) was the fourth stage. The determination of the SOD value was carried out using the brainstorming method with the head of the production and quality division and three employees. The ranking and description of SOD can be seen in Table 2.

It can be seen in Table 2 that rankings from 1 to 10 were then used to determine *the risk priority number* (*RPN*). The RPN value was the product of S and O, and D. The RPN value is then used as a reference for recommendations based on the highest to lowest values. The RPN recommendation is given to the highest score.

3 Results and Discussion

The results of data calculation result were in the form of *rework costs for the* fabrication process, a causal diagram depicted in the *Ishikawa diagram* and *failure mode* and *effect analysis.*

3.1 Rework Cost

Rework costs incurred by the company to make repairs due to process miscalculations. Rework costs include operator costs and raw material costs. The length of time for repairs for each process is as low. The marking process takes 2 min to repair, and cutting is repaired for 4 min. Drilling takes 4 min to repair. The fit-up process takes 6 min, and welding, grinding, and painting repair times are 5, 3, and 7 min, respectively. The total rework cost and the cumulative percentage of the total cost can be seen in Fig. 2.

It can be seen at Fig. 2 that cost of *rework* for the *rounds* and *peeling* process in the *painting process* was approximately 1.5 million rupiahs. The *painting process* is an activity stage that becomes the main priority for further evaluation.

Table 2 Ranking, severity, occurrence, and detection

Rank	Severity	Occurrence	Detection
10	Critical without warning Failure causes the process itself or the customer to stop entirely without warning	Constant 1 per shift or more 1 of 2 Almost or in total failure	Almost not viable No known controls are available Failure will pass
9	Critical with a warning Failure causes the process to stop completely with a few warnings	Critical Failure rate 1 in 3 The failure rate is likely to kill the process	Very remote Random checks are done at low frequency, with no prevention Failure
8	Very high Delays or additional costs incurred. Significant impact on business metrics	Very high 1 in 8 failure rate Causing significant disturbance	Isolated Random on-site inspection and prevention based on the operator noticing problems
7	Tall High level of customer dissatisfaction due to failures such as system shutdown	High 1 in 20 failure rates Frequent failure	Very low Routine on-site inspection (not 100%). Prevention was based on operator training. Failure is possible to escape
6	Very significant Failure causes significant customer dissatisfaction, disruption, or cost	Very significant 1 in 80 failures rate Current failure rate is likely to cause disruption	Low Prevention is based on operator persistence and regular training, regular checks, and significant chances of escape
5	Important Failure leads to customer dissatisfaction, for example, increasing costs	Important 1 in 400 failures rate The failure rate is quite significant	Presently Failure prevention considered, 100% checking output, medium breakout chance
4	Currently Extra effort or rework required	Moderate 1 in 2000 failure rate Medium failure rate	High enough Failure prevention built into the system, 100% recorded process audit
3	Low Failure causes minor problems that take a little time to work on	Low 1 in 15,000 failure rates Low incidence of failure with identical processes	Tall Error checking is used to try and prevent failure. 100% audit process recording
2	Very low The nature of the failure causes only minor delays or minor rework	Very low 1 in 150,000 failure rate Only isolated failures associated with identical processes	Very high Almost always control or detect failure
1	Minor Process customers will not see the effects of failure. No impact on supplier or customer processes	Very low <1 in 15,000,000 failure rate Failure is unlikely. No failure is associated with the identical process	Almost sure Controls will always prevent failures from occurring or, if they do, will always detect problems before they go away

Sources [10, 11, 13–15]

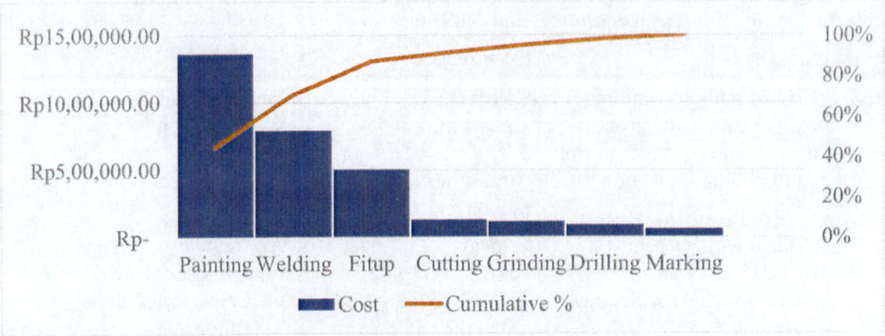

Fig. 2 Rework cost of the fabrication process (in a million rupiah)

3.2 *Cause-and-Effect Diagram of the Highest Defect*

Ishikawa diagram was a tool to outline the cause and effect of two significant defects [3]. The results of defect identification in Table 2 show that the *defects* that occur were classified into two: *defects peeling* and *defects* ruuns. Figures 3 and 4 represent the Ishikawa diagram of defect peeling and defect ruuns, respectively.

It can be seen in Fig. 3 that employee factors, raw materials, methods, and work environment were the dominant factors that caused defects peeling.

It can be seen in Fig. 4 that five factors caused defects: humans, raw materials, machines, methods, and work environment.

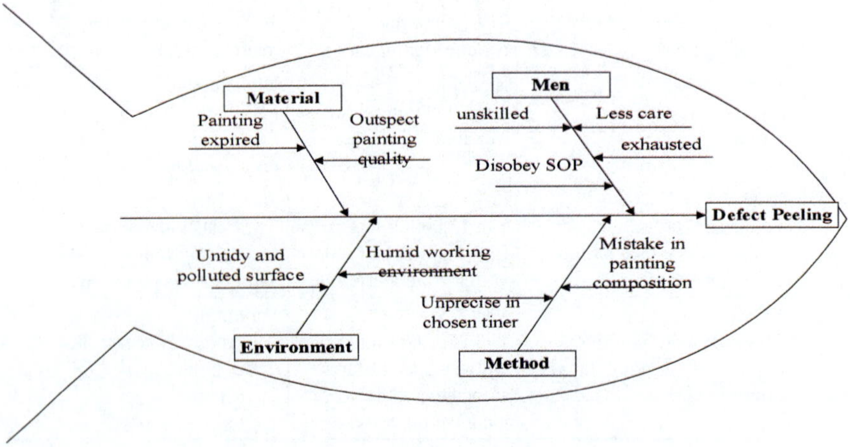

Fig. 3 Ishikawa diagram of defects peeling

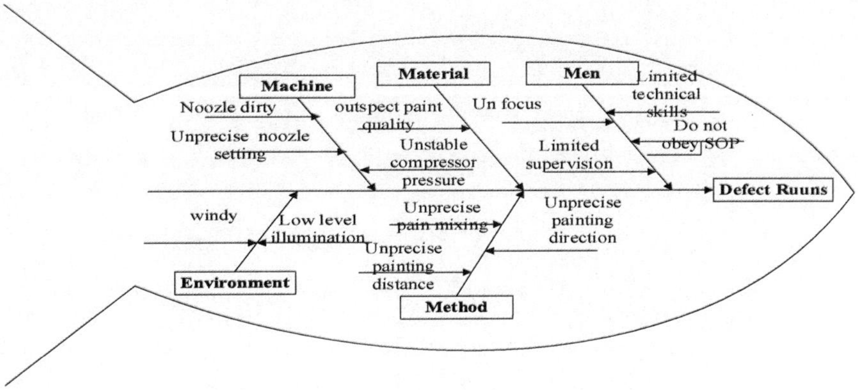

Fig. 4 Ishikawa diagram of defects ruuns

3.3 Failure Mode *Calculation and Analysis Results* and Effect Analysis

FMEA results are based on field searches, historical data, and clarification from respondents' experts. The values of severity, occurrence, detection as well as risk priority number can be seen in Table 3.

It can be seen in Table 3 that there were six effects of defect ruuns and peeling. The lowest severity score was four, and the highest was 10. While the human error was the highest potential cause with a score of 7, namely the activity of errors in the use of thinner in the painting process. In the defect identification process, it is known that the occurrence of errors in the painting process shows the most significant value compared to the other six processes. In line with the number of defects, the rework cost of the painting process is nearly fifty percent of the total rework due to defects in the fabrication process. The result of mapping the causal diagram shows that human factors, work methods, and equipment used in the fabrication process are inadequate. The highest risk priority number of defect ruuns and peels is caused by human error.

Table 3 Results of and failure mode effect analysis

Process	potential Failure Mode(s)	Potential effect (s) of failure	Severity	Potential Cause(s)	Occurrence	Current Design Controls	Detection	RPN	Recommended Action(s)
Painting	Ruuns	Rework product Schedule delay additional cost	4	Unskilled worker	4	visual	5	80	
			5	Fatigue	5	visual	6	150	
			6	Unfocused	5	visual	5	150	
			5	The composition of the paint is not right	4	visual	5	100	
			7	The paint quality is not good	6	visual	3	126	
			4	Wrong direction of painting	5	visual	4	80	
			4	Wrong painting distance	6	visual	5	120	
			6	Dirty nozzle	5	visual	3	90	
			4	Lack of supervision	4	visual	7	112	
			8	Not following the SOP	9	visual	8	576	Perform daily briefing.
			8	Lack of concern for employees	8	visual	7	448	Provide education on the importance of SOP. Installing the SOP sticker on each machine.
			5	Incorrect nozzle setting	4	visual	5	100	
			5	Compressor pressure unstable	7	visual	4	140	
			4	Lack of lighting	4	visual	5	80	
			4	Wind velocity	5	visual	4	80	

(continued)

Table 3 (continued)

Process	potential Failure Mode(s)	Potential effect (s) of failure	Severity	Potential Cause(s)	Occurrence	Current Design Controls	Detection	RPN	Recommended Action(s)
Painting	Peeling	Rework Scheduled down Costs add up	3	Less skilled	4	visual	5	60	
			6	Fatigue	5	visual	5	150	
			7	Wrong paint composition	5	visual	5	175	
			6	Poor paint quality	4	visual	5	120	
			10	Wrong use of thinner	5	visual	3	150	
			7	Paint has expired	4	visual	6	168	
			5	Dirty media surface	5	visual	5	125	
			6	Humid temperature	3	visual	3	54	
			6	Not following the SOP	7	visual	8	336	Perform daily briefing.
			7	Lack of concern for employees	6	visual	9	378	Provide education on the importance of SOP. Installing the SOP sticker on each machine.

4 Conclusion

The most critical cause of each defect based on the RPN value is employees who work not following the SOP; employees' concern for the company is lacking. The defect that occurs is ruuns and peeling. In that case, the highest defect in the manufacturing process at PT AA is the painting process, namely ruuns and peeling.

The suggestion given to the company to reduce *defects* is for the management to provide education on the importance of SOPs in every work process so that the company's operational activities run smoothly. The second recommendation is that management is transparent in explaining every policy and consideration to employees to feel involved and valued so that employee loyalty is high. The third recommendation is to provide rewards based on employee achievement.

References

1. Li H, Teixeira AP, Guedes SC (2020) A two-stage failure mode and effect analysis of offshore wind turbines. Renew Energy 162:1438–1461
2. Doshi J, Desai D (2017) Application of failure mode & effect analysis (FMEA) for continuous quality improvement—multiple case studies in automobile SMEs. Int J Qual Res 11(2):345–360
3. Hernawan A, Mahbubah NA (2021) Integrasi statistical process control dan failure mode and effect analysis guna meminimalisasi defect Pada Proses Produksi Pipa PVC. J Engine Energi Manufaktur dan Mater 5(2):65
4. Priambodo S, Mahbubah NA (2021) Implementasi metode overall equipment effectiveness berbasis six big losses guna mengevaluasi efektivitas mesin packing semen. J Serambi Eng 6(4):2363–2374
5. Prajapati DR (2012) Implementation of failure mode and effect analysis: a literature review. Int J Manage IT Eng 2(7):264–292
6. Sharma KD, Srivastava S (2018) Failure mode and effect analysis (FMEA) implementation: a literature review. J Adv Res Aeronaut Sp Sci 5(1&2):1–17
7. Hasanah TU, Wulansari T, Putra T, Fauzi M (2020) Penerapan lean manufacturing dengan Metode Takt time dan FMEA untuk Mengidentifikasi waste pada proses Produksi steril PT.XYZ. J Rekayasa Sist Ind 07(02):1–7
8. Pratama FS, Suhartini S (2019) Analisis Kecacatan Produk dengan Metode Seven Tools dan FTA dengan Mempertimbangkan Nilai Risiko berdasarkan Metode FMEA. J Senopati 1(1):41–49
9. Krisnaningsih E, Wirawati SM, Febriansyah Y (2021) Penerapan statistical process control (SPC) dan failure mode effect analysis (FMEA) pada proses Produksi Tisu Wajah. J PASTI 14(3):293
10. Khasanah SN, Mahbubah NA, Hidayat H (2021) Deteksi defect proses Produksi Sarung Menggunakan ATBM Berbasis failure mode and effect analysis. Jati Emas (Jurnal Apl Tek dan Pengabdi Masyarakat) 5(3):143–150
11. Rislamy AF, Mahbubah NA, Widyaningrum D (2020) Analisis Risiko Kerusakan Pada Alat Berat Grab Dengan Metode failure mode and effect analysis (Studi Kasus: PT Siam Maspion Terminal Gresik). Profisiensi 8(1):36–43
12. Lestari A, Mahbubah NA (2021) Analisis defect proses Produksi Songkok Berbasis Metode FMEA Dan FTA di home—Industri Songkok GSA Lamongan. J Serambi Eng 6(3):2197–2206
13. Anthony MB (2018) Analisis Penyebab Kerusakan Hot Rooler Table dengan Menggunakan Metode failure mode and effect analysis (FMEA). J INTECH Tek Ind Univ Serang Raya 4(1):1
14. Suherman A, Cahyana BJ (2019) Pengendalian Kualitas dengan metode failure mode effect and analysis (FMEA) Dan Pendekatan Kaizen untuk Mengurangi Jumlah Kecacatan dan Penyebabnya. In: Seminar Nasional Sain dan Teknologi, pp 1–9
15. Prasetya RY, Suhermanto S, Muryanto M (2021) Implementasi FMEA dalam Menganalisis Risiko Kegagalan proses Produksi Berdasarkan RPN. Performa Media Ilm Tek Ind 20(2):133–138

Design and Investigation of *Archimedes Screw* Turbine: Influence of Inclination Angle on Power Production

Suherman, Andika Pratama, M. R. Prayogi, M. F . Pane, S. Priyono, T. I. Sugatra, Faisal Azmar, and Ilmi

Abstract The Archimedes screw turbine (AST) is a type of turbine that produces micro-hydro power that can operate at low heads. In this article, the performance of AST is experimentally investigated for the angle of an inclination. This study aims to analyze the effect of variations in the turbine screw shaft angle on the shaft rotation, output power, current, and voltage generated. The AST turbine's lower inclination angle (25°) results in higher efficiency. The test results show that the optimum inclination angle is obtained at 35° for water discharge 0.011 m^3/s delivering a torque of 2 N.m.

Keywords Archimedes screw turbine · Inclination angle

1 Introduction

Increasing CO and CO_2 emissions as a consequence of fossil fuel consumption has resulted in one of many problems, such as acid rain, global warming, and ozone depletion. Climate change due to human activities requires a rapid transition in developing environmentally friendly technologies that utilize renewable energy sources [1]. Using renewable energy sources can considerably reduce these problems [2]. Only 22.8% of worldwide electricity production comes from renewable sources. Using renewable energy as a generator of electrical energy will be beneficial. Renewable energy sources such as geothermal energy, wind, solar energy, and hydropower are easy to find worldwide [3]. Compared to wind turbines and solar PV systems,

Suherman · A. Pratama · M. R. Prayogi · M. F. . Pane · S. Priyono · T. I. Sugatra · F. Azmar
Department Mechanical Engineering, Universitas Muhammadiyah Sumatera Utara, Medan, Indonesia
e-mail: suherman@umsu.ac.id

Ilmi (✉)
Department Mechanical Engineering, Universitas Sumatera Utara, Medan, Indonesia
e-mail: llmi@usu.ac.id

Irwansyah et al. (eds.), *Proceedings of the 4th International Conference on Experimental and Computational Mechanics in Engineering*, Lecture Notes in Mechanical Engineering, https://doi.org/10.1007/978-981-99-7495-5_6

the advantages of hydropower generation are low investment and operating costs [4], the capability to operate at a low flow rate, and low environmental impact [5].

In general, hydroelectric power is used to supply the world's energy [6]. A small-scale hydropower device known as an Archimedes screw turbine (AST) is often installed as a diversion (or run-of-river) system. Since the 1990s, AST has been used to generate hydroelectric energy [7]. The Archimedes screw turbine is a micro-hydro power plant (1 MW) that can be installed at a discharge of 1–10 m^3/s and with an elevation difference of 8–10 m [8]. Furthermore, the advantages of the Archimedes screw turbine (AST) are that it is more friendly to wildlife and especially fish than other hydroelectric power generation options [9] and can reduce erosion and sedimentation [10]. However, the AST will produce unstable power, especially if the river flow is much less in summer [6].

Several types of water turbines, such as the Pelton and the Kaplan turbine, have been used as power plants. However, the cost of making AST is 10% less, while the energy generated is 15% more [3]. The Archimedes screw reappeared in the latter half of the twentieth century, this occasion as a turbine in a new and inverted form. For the first time, the AST turbine device was installed in Europe in 1994 and only later in 2004 in the UK [11]. Archimedes screw turbine (AST) is a renewable power plant that converts the potential energy of fluids into suitable mechanical energy for the low head. The advantages of AST, even if the inflow fluctuations, high efficiency can still be achieved [12]. Archimedes screw turbine (AST) as a micro-hydro power plant is used to utilize renewable energy sources, which are very useful for rural development [5].

Recently, researchers have taken great interest in micro-hydro power plants as an alternative energy source. Many researchers have approached simulation and experimental to obtain the relationship between the effect of blade count and inclination angle on AST power performance. Lee and Lee [5] stated that the optimal inclination angle of the AST is 45°, producing a torque of 0.9 N.m and a rotation of 179 rpm.

Dedić-Jandrek and Nižetić [13] compared three different inclination angles on the AST efficiency, power output, and torque. The result shows maximum efficiency of (64%) with an inclination angle of 21°. Siswantara et al. [4] stated that the turbine efficiency is not affected by the inclination angle due to the increase in power due to the value of the head. The flow rate increases, and the torque and power increase. In addition, torque and power are affected by pitch length and the angle of inclination—the inclination angle of 35° results in the highest efficiency of 81.4% [14]. Almost similar results were obtained by Dellinger et al. [15], where the optimum efficiency was obtained at inclination angles of 20° and 24.5°. Shahverdi et al. [3] reported that an AST efficiency of 90.83% was obtained at an inclination angle of 20°. Next, a steep angle of 33.8 with a short screw is poor performance while a low angle of inclination (15) with a long screw is the best design [7]. Simmons et al. [16] suggest the angle of inclination for AST with a long screw in the range of 20°–25°.

Several components cause power loss in AST, such as bearing friction, outlet expansion, and outlet immersion power loss [17]. Rohmer et al. [6] reported experimental, and simulation of the AST 0.84 m diameter Archimedes turbine screw delivers torque 250-Nm. The performance of the Archimedes screw turbine was studied by developing a numerical optimization using MATLAB 2013a. The optimum inclination angle of the screw turbine is 20°, with the highest efficiency of 90.83% [18]. Edirisinghe et al. [19] conducted a simulation with a maximum efficiency of about 82% with a head of 5.2 m.

The AST performance is influenced by various factors such as inclination angle, water flow rate, the distance between screws, and turbine length. In this work, the influence of four varies the inclination angle and flow rate (0.008 to 0.01 m^3/s) to obtain optimal performance and efficiency from the developed AST (Table 1).

2 Experimental Methods

2.1 Manufacturing of AST

The turbine frame made of mild steel (30 × 30 mm) is joined by SMAW welding, as shown in Fig. 1. Plate mild steels with a thickness of 1.2 mm are cut in a circle with a diameter of 300 mm and joined with the shaft of AST by shielded metal arc welding (SMAW) (Fig. 2). The AST shaft is connected to a low-speed DC motor with a V belt type A to generate power (Fig. 4a).

The AST casing (mild steel) with a thickness of 2 mm is placed under the screw turbine so that water flows along the length of the screw. The distance between the turbine screw and the casing is 2 mm. The AST shaft tilt angle indicator is placed at an angle to facilitate variations in the shaft angle of AST. The display indicator voltage, current, and power generated by the AST are placed above the screw AST. Four light bulbs as a load are connected to a direct current (DC) generator to generate power DC to turn on the lamp. The water tank (198 L) is placed above a tower and is connected to a water pump with a pipe diameter of 3 in. The ball valve is joined at the bottom of the water tank and connected pipe (4 in.) to drain water from the tank. Main components of the AST turbine are shown in Table 2.

2.2 Experimental Procedure

The water pump with 0.0167 m^3/s used water supply from the water bath to the top of the tank during an experiment. Next, the ball valve opens for water flow into the AS turbine. Water flows with a flow rate of 0.0064–0.01 m^3/s, respectively, causing the turbine screw to rotate. The turbine shaft is connected to an electric motor using a v belt to change the mechanical force on the shaft into electrical energy. The parameters measured in this study, such as the water flow rate, used a flow meter. The rotation of the Archimedes screw turbine shaft is measured using a tachometer, and the current discharged from the Archimedes screw turbine is measured using an Avometer. The parameters to be measured are presented in Table 3.

Table 1 Parameters of AST various researchers

Parameter of AST	Performance of AST	Author
The experiment was conducted with a discharge of 0.00106 m^3, radius ratio of 0.5369, angle of 36–44°, pitch ratio of 0.1863, and length of 2.09 m	The turbine did best operation head 1.23 m at an angle of 36°	[4]
The author conducted the numerical model, and the experiment was developed AST carried out at a flow rate of 0.15 m^3/s with an inclination of 30°, pitch 0.21, outer radius 0.42 m, inner radius 0.21 m, radius r ratio 0.505, and three-blade screws	The result of the modeling and experiment shows a maximum efficiency of 72%. The flow rate is 0.25 to 6.5 m^3/s on a head from 1 to 6.5 m. The maximum power can be estimated to be between 1.7 and 300 kW	[6]
The experimental study was to find out the varying length screw (478, 617, and 952 mm) and the number of blades 4, inner diameter of 168 mm, outer diameter of 381 mm, the pitch of 381 mm, and the varying tilt angle ranges from 15.6 to 33.8°	The most extended screw (with the slightest inclination angle of 15°) performed the best	[7]
Conducted experimental study with inclination angle 20–25, outer diameter 150, inner diameter 150, inner diameter 82.5, the pitch of helical surface 300 mm, the gap with 0.0025 m, screw length 1000 mm, water flow 1–4 l/s	Inclination angles 21–30, the maximal efficiency of the AST is 64%	[13]
Experimentation carried out at the shaft angles are 30°–45°, the pitch is 70 mm, the screw length is 1000 mm, the outer diameter is 130 mm, the inlet diameter is 70 mm, the gap width is 2 mm, and the number of helices 12 m	Maximum efficiency is 81.4% at the flow rate is 1.12 l/s and the angle 35°. The best operation at the angle of 45° with maximum power obtained is 25.13 W with a flow rate is 2.065 l/s, and the efficiency was recorded at 72%	[14]
The author developed the AST by comparing experimental data with simulations on a turbine length of 1221 mm, an outer diameter of 316 mm, and an inner screw diameter of 168 mm. The inclination angle varied from 10 to 38, and the number of blades is 3, 4, and 5 pieces	Efficiency is obtained at the inclination angle of 20 and 24.5. The most power is obtained at the number of screw blades 5	[15]
CFD-based study to find out efficiency for low pitch angles. The study was carried out at 0.232 m^3/s, gap width of 0.005 m, screw outer radius of 1.2 m, screw inner radius of 0.643 m, and screw pitch of 0.846 m with a hydraulic head of 5.2 m	At the head 5.2 m and discharge 0.232 m^3/s, the result shows maximum efficiency of 82%	[19]

(continued)

Table 1 (continued)

Parameter of AST	Performance of AST	Author
The experimental study was inclination angle 20–25, water flow 20–25 l/s, head 1.26 m, load 0.3–0.9 kg, flow rate 1–4 l/s	The result shows efficiency of 74% was attained at a tilt angle of 25°. The flow rate of 4 l/s with a power of 0.00057 kW	[20]
The authors developed an AST with a length of 646 mm and an outer diameter of 142 mm. a casing diameter of 152 mm, and a gap casing to blade width of 5 mm. Varying input parameters angle of inclination (22–40°) with a water flow rate 1.2 l/s studied	An efficiency of 49% is obtained at the AST head of 0.25 m. The optimum power is obtained at an angle of 22 with a water flow rate of 1.2 l/s	[21]

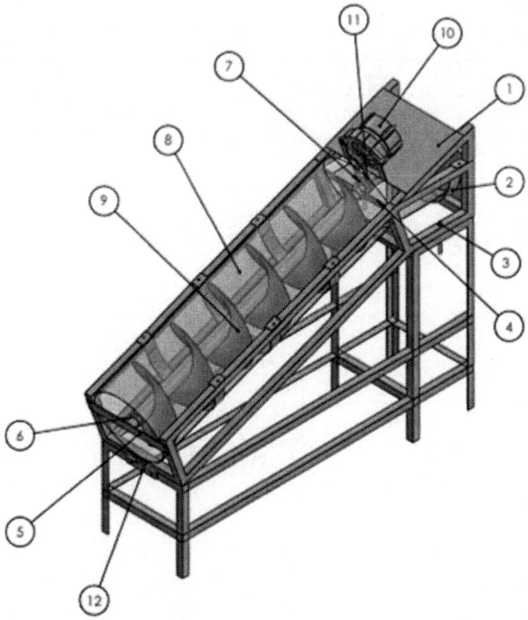

1. Chassis
2. Holder Cover Turbine
3. Bolt Adjuster
4. Holder Turbine
5. Holder Turbine
6. Bearing
7. Pulley
8. Turbine Casing
9. Turbine
10. Alternator
11. *Belt*
12. *Hinge*

Fig. 1 Design of AST test

3 Result and Discussions

The experimental setup of the Archimedes screw turbine used in this study is shown in Figs. 3 and 4. This study used four inclination angles 22°, 28°, 30°, and 35°, respectively (Table 3). The parameters measured were water flow rate, torque, mechanical power, hydraulic power, and turbine efficiency. The relationship between power production and water discharge at an inclination angle of 25–35 (Fig. 5). The angle of the shaft inclination affects the volume of water in the space between the two blades (bucket). At angles shaft above 35°, the amount of water in the bucket will decrease

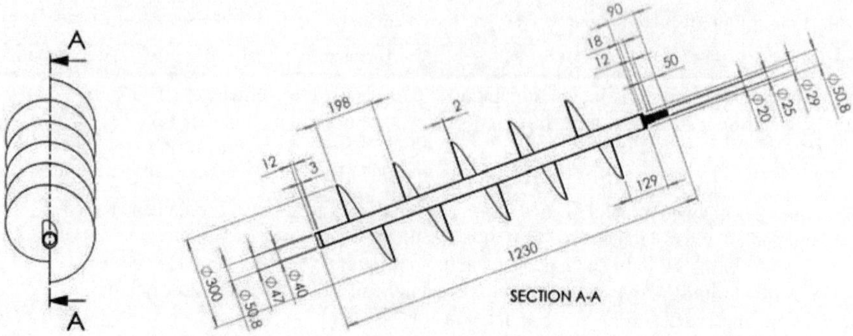

Fig. 2 Engineering drawing of prototype AST

Table 2 Main component of AST

No.	Name of part	No.	Name of part
1	Frame of AST	6	Electric fenerator
2	Casing of AST	7	V belt
3	Archimedes screw	8	Basin
4	Display volt and ampere	9	Indicator degree
5	Light		

Table 3 Geometric identities for Archimedes screw

Parameters	Variable	Units	Value
Inclination angle	B	Degree	25, 28, 30, 35
Outer diameter	D_o	M	0.260
Inner diameter	D_i	M	0.038
Pitch	P	M	0.20
Number of screws	N	Unit	1
Gap width	Gw	M	0.005
Screw length	L	M	1.230
Flow rate	Q	m^3/s	0.006, 0.0088 and 0.011
Casing diameter	Dt	M	0.290

because some of the volumes of water will flow at the top of the screw blade and not produce an optimum power generated on the shaft.

Figure 5 shows the effect of the flow of water and inclination angle on the power generated by the AST. The optimum output power is obtained at an angle of 35°, which produces a power of 6.7 W. The results obtained are similar to the statements of Kumar et al. [20].

Fig. 3 Experimental setup of AST

Fig. 4 **a** Transmission V belt for DC generator, **b** AST in final production

The effect of turbine angle and water flow on turbine shaft rotation is shown in Fig. 6. Increasing the turbine shaft angle and water flow increases the turbine shaft rotation. The optimum turbine shaft rotation is obtained at a turbine shaft angle of 35° at a water flow rate of 0.011 m^3/s (300 rpm).

Figure 7 shows the effect of inclination angle and flow rate on AST efficiency. It was found that the highest efficiency was obtained at a low inclination angle of 25°. The inclination angle highest 25° will cause a more significant water overflow than a low slope angle [11, 15]. The maximum turbine efficiency % is recorded at an angle

Fig. 5 Power production versus water discharge

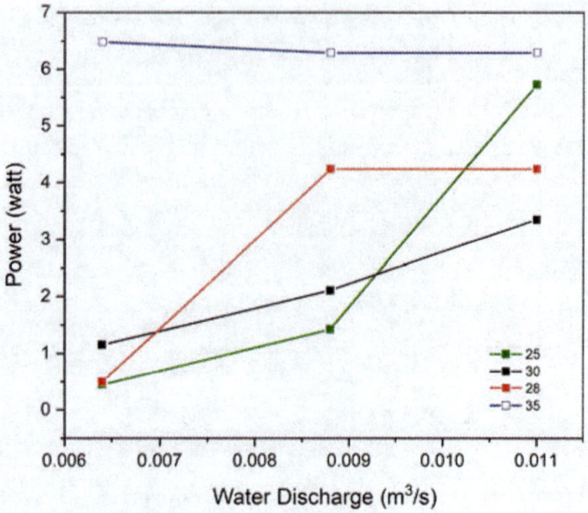

Fig. 6 Rotation shaft versus water discharge

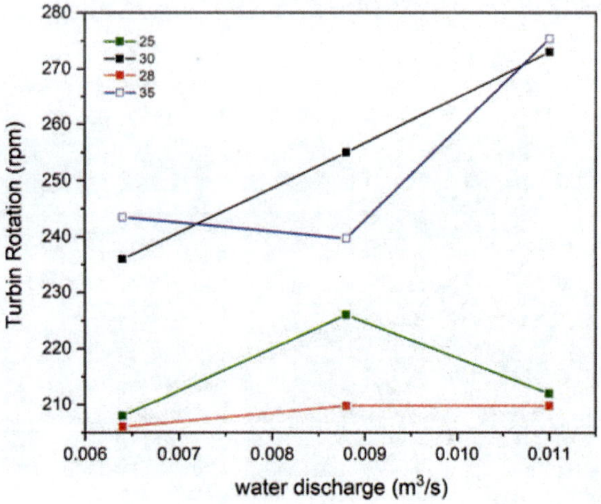

of 25° and a flow rate of 0.0064 m^3/s. The turbine efficiency obtained is very close to that of Jandrek et al. [13].

Figure 8 shows the turbine torque produced with variations in the inclination angle at water discharge from 0.0064 to 0.011 m^3/s. The optimum torque is obtained at an inclination angle of 35°, and the water flow rate is 0.011 m^3/s. The increase in the flow rate and inclination angle at the turbine will be increased the torque produced by the AST turbine [15]. Optimum torque is obtained at an inclination angle of 35° and a water flow rate of 0.01 m^3/s.

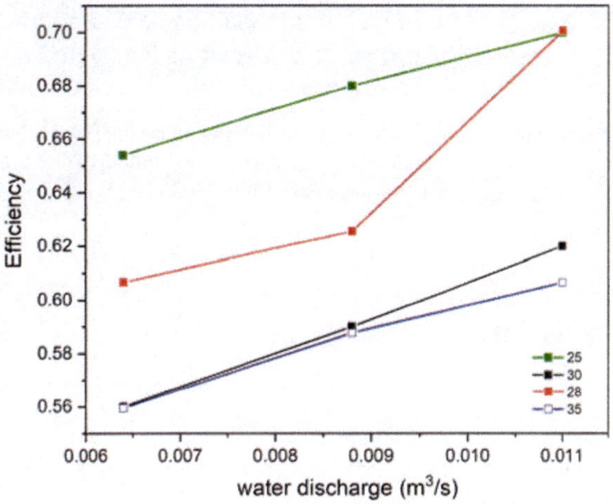

Fig. 7 Efficiency versus water discharge

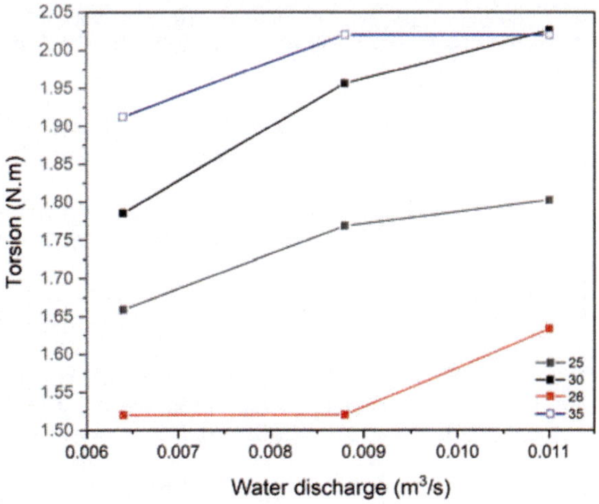

Fig. 8 Torsion versus water discharge

4 Conclusion

This work studied the effect of variations in inclination angle and flow rate on the output power, torque, rotation, and efficiency of the AST. The following conclusions can be drawn:

1. The Archimedes screw turbine (AST) with a shaft length of 1230 mm and a blade screw diameter of 300 mm with a flow rate of 0.011 m^3/s obtained a rotational speed of screw rpm of 243 rpm.
2. The optimum turbine shaft angle is obtained at a slope of 35° producing optimum torque of 2 N.m, and the power generated is 6.4 W.

3. The increase in the turbine shaft angle with a flow rate of 0.0064 to 0.011 m^3/s increases the turbine shaft's rotation but decreases AST's efficiency.

Acknowledgements Thank for Majelis Pendidikan Tinggi Penelitian dan Pengembangan Pimpinan Pusat Muhammadiyah by Basic Research funding (The Muhammadiyah Research Grant Batch V of 2021) Contract Number 0842.358/PD/I.3/C/2021.

References

1. Lavrič H, Rihar A, Fišer R (2019) Influence of equipment size and installation height on electricity production in an Archimedes screw-based ultra-low head small hydropower plant and its economic feasibility. Renew Energy 142:468–477. https://doi.org/10.1016/j.renene.2019.04.095
2. Shahverdi K et al (2019) Energy harvesting using solar ORC system and Archimedes screw turbine (AST) combination with different refrigerant working fluids. Energy Convers Manag 187(January):205–220. https://doi.org/10.1016/j.enconman.2019.01.057
3. Shahverdi K, Loni R, Ghobadian B, Gohari S, Marofi S, Bellos E (2020) Numerical optimization study of Archimedes screw turbine (AST): a case study. Renew Energy 145:2130–2143. https://doi.org/10.1016/j.renene.2019.07.124
4. Siswantara AI, Warjito, Budiarso, Harmadi R, Gumelar MH, Adanta D (2019) Investigation of the α angle's effect on the performance of an Archimedes turbine. Energy Procedia 156:458–462. https://doi.org/10.1016/j.egypro.2018.11.084
5. Lee MD, Lee PS (2021) Modelling the energy extraction from low-velocity stream water by small scale Archimedes screw turbine. J King Saud Univ—Eng Sci. https://doi.org/10.1016/j.jksues.2021.04.006
6. Rohmer J, Knittel D, Sturtzer G, Flieller D, Renaud J (2016) Modeling and experimental results of an Archimedes screw turbine. Renew Energy 94:136–146. https://doi.org/10.1016/j.renene.2016.03.044
7. Simmons S, Dellinger G, Lyons M, Terfous A, Ghenaim A, Lubitz WD (2019) Effects of varying inclination angle on Archimedes screw generator power production with constant head. J Hydraul Eng 147(3):04021001. https://doi.org/10.1061/(asce)hy.1943-7900.0001854
8. Pauwels IS et al (2020) Multi-species assessment of injury, mortality, and physical conditions during downstream passage through a large archimedes hydrodynamic screw (Albert canal, Belgium). Sustainability 12(20):1–25. https://doi.org/10.3390/su12208722
9. Piper AT, Rosewarne PJ, Wright RM, Kemp PS (2018) The impact of an Archimedes screw hydropower turbine on fish migration in a lowland river. Ecol Eng 118(March 2018):31–42. https://doi.org/10.1016/j.ecoleng.2018.04.009
10. Lubitz WD (2020) Archimedes screw turbines: a sustainable development solution for green and renewable energy generation—a review of potential and design procedures. Sustainability
11. Waters S, Aggidis GA (2015) Over 2000 years in review: revival of the Archimedes screw from pump to turbine. Renew Sustain Energy Rev 51:497–505. https://doi.org/10.1016/j.rser.2015.06.028
12. Dellinger G et al (2018) Computational fluid dynamics modeling for the design of Archimedes screw generator. Renew Energy 118:847–857. https://doi.org/10.1016/j.renene.2017.10.093
13. Dedić-Jandrek H, Nižetić S (2019) Small scale archimedes hydro power plant test station: design and experimental investigation. J Clean Prod 231:756–771. https://doi.org/10.1016/j.jclepro.2019.05.234
14. Abdullah OS, Khalil WH, Kamel AH, Shareef AJ (2020) Investigation of physical and numerical model of Archimedes screw turbine. J Power Energy Eng 8(10):26–42. https://doi.org/10.4236/jpee.2020.810003

15. Dellinger G, Simmons S, David W, Dellinger N (2019) Effect of slope and number of blades on Archimedes screw generator power output. Renew Energy 136:896–908. https://doi.org/10.1016/j.renene.2019.01.060
16. Simmons S, Dellinger G, Lyons M, Terfous A, Ghenaim A, Lubitz WD (2021) Effects of inclination angle on Archimedes screw generator power production with constant head. J Hydraul Eng 147(3). https://doi.org/10.1061/(asce)hy.1943-7900.0001854
17. Kozyn A, Lubitz WD (2017) A power loss model for Archimedes screw generators. Renew Energy 108:260–273. https://doi.org/10.1016/j.renene.2017.02.062
18. Shahverdi K, Loni R, Maestre JM, Najafi G (2021) CFD numerical simulation of Archimedes screw turbine with power output analysis. Ocean Eng 231(May):1–8. https://doi.org/10.1016/j.oceaneng.2021.108718
19. Edirisinghe DS, Yang H, Kim M, Kim B, Gunawardane SP, Lee Y (2021) Computational flow analysis on a real scale run-of-river archimedes screw turbine with a high incline angle. Energies:1–18
20. Kumar N, Thakur R, Kashyap K, Goel B (2022) Efficiency enhancement in Archimedes screw turbine by varying different input parameters—an experimental study. Mater Today Proc 52:1161–1167. https://doi.org/10.1016/j.matpr.2021.11.020
21. Erinofiardi et al (2017) Experimental study of screw turbine performance based on different angle of inclination. Energy Procedia 110(Dec 2016):8–13. https://doi.org/10.1016/j.egypro.2017.03.094

Assessment and Strengthening of Cement Plant Clinker Silo Structure Due to Corrosion of Reinforcing Bars

Azzaki Mubarak, Muttaqin Hasan, Taufiq Saidi, and Syarizal Fonna

Abstract Strengthening and repairs need to be carried out on the clinker silo of the Lhoknga Cement Plant, where there has been significant damage, making the silo unable to function at full capacity. Damage to reinforced concrete where cracks and spalling occurred as well as corrosion of reinforcing steel was assessed. The assessment was carried out using the following methods: visual inspection, rebound hammer test, ultrasonic pulse velocity test, and corrosion potential measurement using a half potential cell. From the assessment results, there have been many cracks and severe corrosion in the reinforced concrete slab, which reduces the strength of the structure. For this reason, the structural strengthening of the slab using CFRP was carried out so that the structural capacity can increase and be able to withstand the required working load. Meanwhile, in the clinker silo wall, the potential corrosion is still at a moderate level and does not pose a significant hazard to the structure in the near term.

Keywords Assessment · CFRP · Corrosion potential · Strengthening

A. Mubarak · M. Hasan (✉) · T. Saidi
Department of Civil Engineering, Universitas Syiah Kuala, Jl. Tgk. Syech Abdul Rauf, No. 7, Banda Aceh 23111, Indonesia
e-mail: muttaqin@usk.ac.id

T. Saidi
e-mail: taufiq_saidi@usk.ac.id

S. Fonna
Department of Mechanical Engineering, Universitas Syiah Kuala, Jl. Tgk. Syech Abdul Rauf, No. 7, Banda Aceh 23111, Indonesia
e-mail: syarizal.fonna@usk.ac.id

Irwansyah et al. (eds.), *Proceedings of the 4th International Conference on Experimental and Computational Mechanics in Engineering*, Lecture Notes in Mechanical Engineering, https://doi.org/10.1007/978-981-99-7495-5_7

1 Introduction

The Lhoknga Cement Plant clinker silo which is the object of this study was built using the reinforced concrete structure in 1976. The clinker silo is used for refining raw materials, which is to make Portland cement by heating the raw materials and mixing them at high temperatures [1]. The cement plant is located beside the Indian Ocean in Aceh Besar District, Indonesia, which is the most affected area during the tsunami Aceh in 2004. Many reinforced concrete structures affected by the tsunami have corrosion in reinforcing bars [2–4]. In this clinker silo, along the conveyor beltline, which is located underground, there has been damage to the reinforced concrete structure in the form of rebar corrosion, cracks, and concrete spalling as shown in Fig. 1.

As a result of the damage that has occurred, it is necessary to conduct an assessment of the current performance of the structure, so that the improvements can be planned for the structural system to be operational again. The assessment carried out consists of visual inspection; rebound hammer test, ultrasonic pulse velocity test, and corrosion potential test without damaging the concrete surface or so-called non-destructive test [5–7].

Repairs and reinforcements are carried out to restore the strength of the structure that has decreased due to the damage received, so that it has the ability to be able to sustain the load as before or the new load to be applied [8–10]. Repairs can be made using patching, grouting, and injection methods to repair concrete and using steel jacketing or FRP methods to strengthen the structural elements.

This research was conducted to determine the structural damage to the clinker silo and the potential level of corrosion that has occurred. The strengthening method

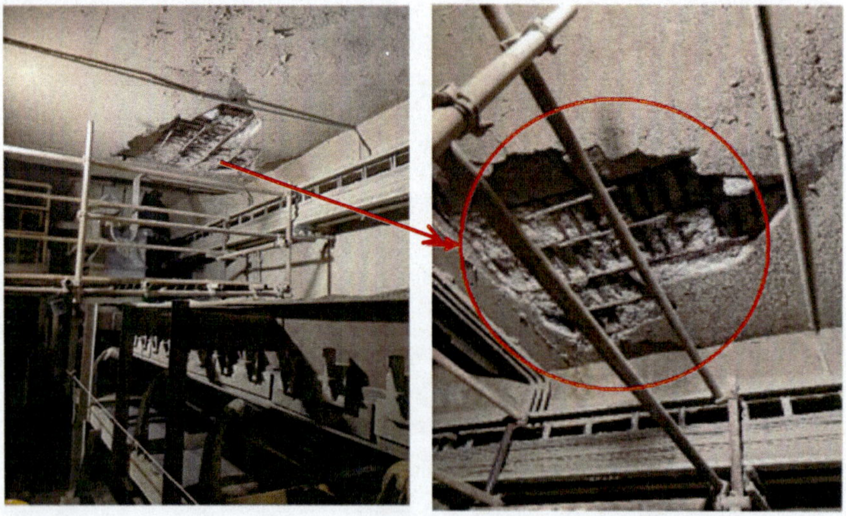

Fig. 1 The damage to the clinker silo structure

is proposed so that the silo can be functionally again with a strength capable to carry the working load.

2 Method

Description of Clinker Silo. The clinker silo area that was assessed is located on the conveyor belt with a total length of 36.40 m, a width of 3.5 m, and a height of 4.35 m from the floor as shown in Fig. 2.

Visual assessment. The crack, concrete spalling, and rebar corrosion were visually assessed in accordance with ACI 562-19 [11]. The remaining rebar diameter and other structural dimensions were measured carefully.

Rebound hammer test. The rebound hammer test was carried out by following the ASTM C805/C805M-18 procedure [12] by using an N/NR Schmid rebound hammer with a capacity of 10–70 MPa. The location of the rebound hammer test is shown in Fig. 3. Tests can be carried out in a vertical downward, vertical upward, or horizontal direction depending on the condition of the structure to be tested. At each location, 12 tests were carried out and the rebound number of the tested concrete was read and recorded, then the average value was calculated and converted into concrete compressive strength.

Ultrasonic pulse velocity test. The ultrasonic pulse velocity (UPV) test was carried out based on the ASTM C597-16 standard [13] using the UPV tool Proceq Pundit PL 200. The location of the UPV test is shown in Fig. 3. This test was carried

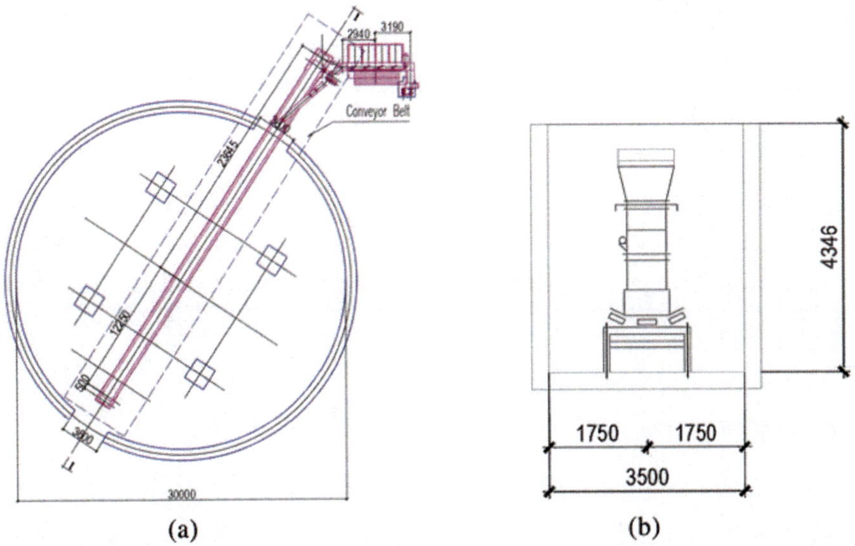

Fig. 2 Clinker silo: **a** Plan; **b** Section I-I

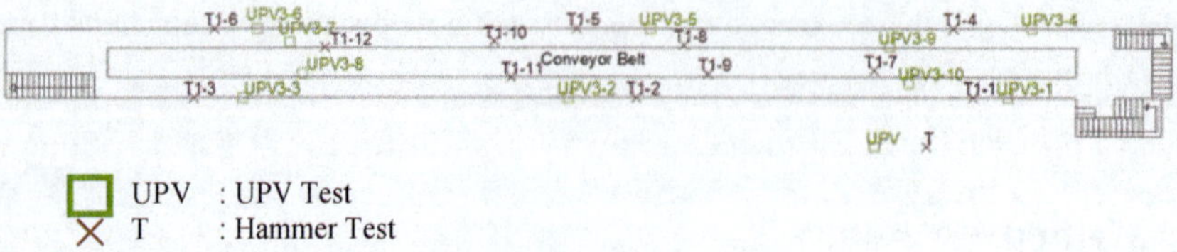

Fig. 3 Location of hammer test and UPV test

Table 1 Criteria for corrosion level of reinforcing steel based on ASTM C876-15 [15]

Cu/$CuSO_4$	Corrosion risk
> − 200 mV	Low (< 10% corrosion risk)
− 200 to − 350 mV	Medium
< − 350 mV	High (< 90% corrosion risk)
< − 500 mV	Very high

out to determine the density of concrete by using fast-wave propagation fired into the concrete, By using the relationship between longitudinal wave stresses in the concrete, the dynamic elasticity properties of concrete can be calculated using the following equation [14].

$$E_{\mathrm{d}} = V^2\rho\left(\frac{(1+\mu)(1-2\mu)}{(1-\mu)}\right) \tag{1}$$

where E_{d} = dynamic elastic modulus; V = speed of wave propagation; ρ = concrete density; and μ = Poisson's ratio.

Corrosion potential analysis. ASTM C876-15 is the standard used to measure the half-cell potential mapping method and is one way to apply half-cell potential in the field where the corrosion criteria can be seen in Table 1 [15]. The location of this corrosion potential test is shown in Fig. 4.

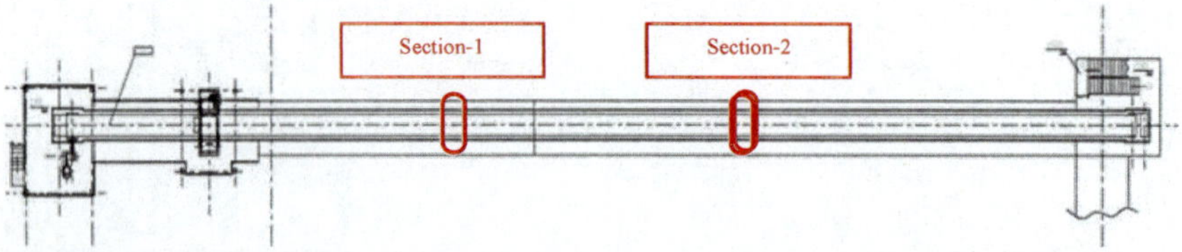

Fig. 4 Corrosion potential test location

3 Results and Discussion

Visual observation results. The results of visual observations show that the clinker silo slab reinforcing bars have severe corrosion, which had caused peeling of the concrete cover and a reduction in the diameter of the longitudinal reinforcing bars from 24 to 19.25 mm and transversal reinforcing bars from 16 to 14.5 mm. Meanwhile, the results of visual observations on the walls of the clinker silo showed that some of the walls were still in good condition, but some reinforcing bars have a low level of corrosion and the diameter of the reinforcing steel was reduced from 16 to 15.10 mm.

Compressive strength. The average concrete compressive strength of the walls and slab of the clinker silo based on the rebound hammer test is shown in Table 2. The results show that the compressive strength value is still above the minimum compressive strength value required in SNI 2847-19 [16], which is 17.5 MPa.

Dynamic elastic modulus. The results of the ultrasonic pulse velocity test of the clinker silo wall and slab structure are shown in Table 3. Based on the velocity data, the dynamic modulus of elasticity of the concrete is then calculated and shown in the same table.

The dynamic modulus of elasticity of uncracked concrete is generally greater than 25 GPa. The smaller the value of the dynamic modulus of elasticity of concrete, the concrete has experienced more severe internal cracks. The results in Table 3 show that the clinker silo structure has significant internal cracks. These cracks are more severe in the slab structure than in the wall structure. Cracks in the clinker silo structure are caused by the corrosion of reinforcing bars.

Corrosion potential. Corrosion potential testing is carried out in 2 sections, where the first section has a potential distribution indicating the magnitude of the potential

Table 2 The average compressive strength of the hammer test results

Location	Element	Average cylinder compressive strength (MPa)	Average cube compressive strength (kg/cm^2)
T1-1	Wall	27.16	333.7
T1-2	Wall	30.05	369.2
T1-3	Wall	27.55	338.4
T1-4	Wall	22.87	280.9
T1-5	Wall	26.65	327.4
T1-6	Wall	21.59	265.3
T1-7	Wall	26.02	319.7
T1-8	Wall	24.92	306.0
T1-9	Slab	25.44	312.6
T1-10	Slab	26.66	327.6
T1-11	Slab	26.12	320.9
T1-12	Slab	19.50	239.5

Table 3 Pulse velocity dan modulus elastisitas dinamis beton pada struktur clinker silo

Location	Element	Pulse velocity (m/s)	Dynamic elastic modulus (GPa)
UPV3-1	Wall	2890	18.04
UPV3-2	Wall	3289	23.37
UPV3-3	Wall	2825	17.24
UPV3-4	Wall	2252	10.95
UPV3-5	Wall	2273	11.16
UPV3-5	Wall	2016	8.78
UPV3-7	Slab	828	1.48
UPV3-8	Slab	2375	12.18
UPV3-9	Slab	154	0.05
UPV3-10	Slab	387	0.32

value between (– 347) and (– 297) mV and in Sect. 2 the distribution of the potential value between (– 517) and (– 290) mV as shown in Fig. 5.

Based on the ASTM C876-15 corrosion criteria, the clinker area measured in Section 1 has a medium corrosion level, while the measurements in Section 2, the measurement data show that this area has a medium to high corrosion level. The

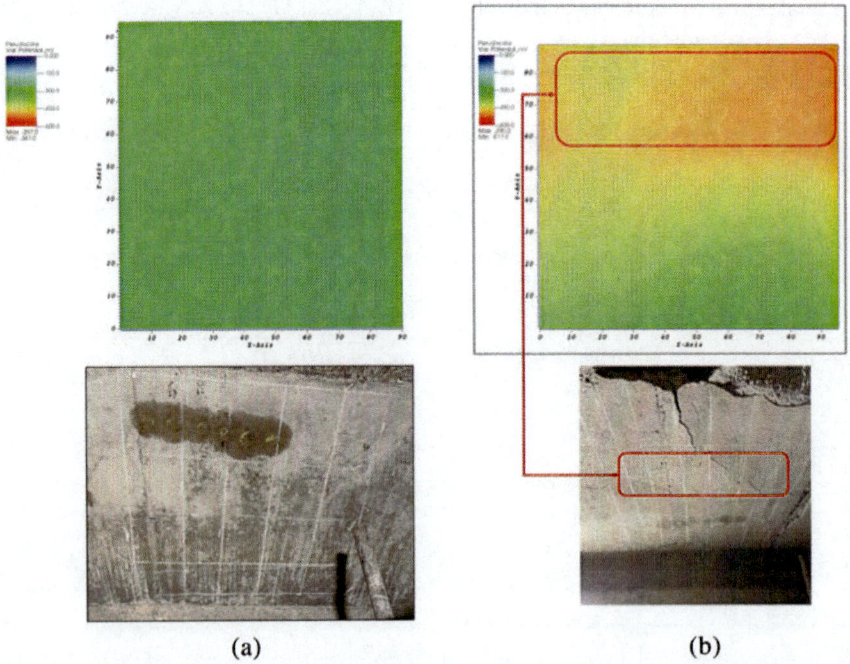

Fig. 5 Distribution of corrosion potential value of concrete surface: **a** Section 1; **b** Section 2

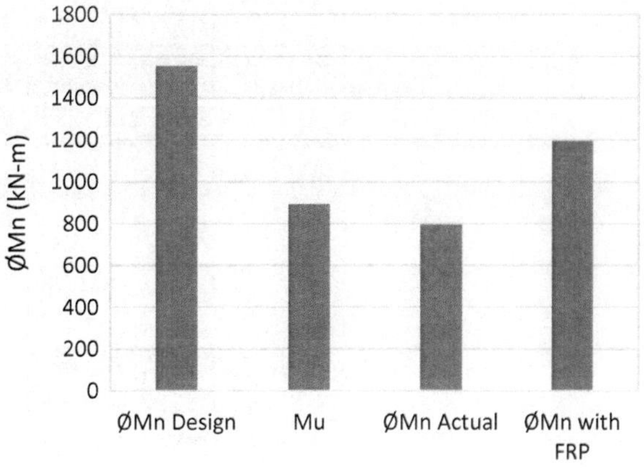

Fig. 6 Required moment and capacity of clinker silo slab before and after strengthening

medium level is shown in green in the visualization results and the high level is shown in orange.

The remaining capacity of the structure. Since only the clinker silo slab structure has cracks and rebar corrosion, then the remaining capacity of the structure was analyzed the slab only using SNI 1726 [17] and SNI 1727 [18]. The design capacity (ØMn design) was obtained using the minimum reinforcement ratio in the design, while the actual capacity (Mn actual) was obtained using the remaining reinforcement from field measurements. The required moment (Mu) was obtained by conducting a structural analysis considering all the possible loads acting on the structure. The comparison between working load and capacity is shown in Fig. 6.

Based on Fig. 6, the actual moment is much lower than the design moment which means that the corrosion of rebars caused the degradation of the clinker silo slab structure a lot. The degradation of moment capacity is due to the reduction in bar diameter, the occurrence of cracks in the concrete, and the reduction in the effective height of the slab structure. The required moment for the clinker silo slab structure is greater than the actual capacity, this will cause the possibility of the structure failing.

Strengthening of structure. Repair and strengthening of the structure need to be conducted to increase the ability of the slab to carry the applied load. Repairs are carried out to restore the strength of the concrete that has been damaged due to corrosion of reinforcement by injection of epoxy resin into the cracks and covering the loose concrete cover using mortar and wire mesh. Meanwhile, reinforcement is carried out to increase the capacity of the slab moment capacity so that it can be higher than the ultimate moment. Strengthening is proposed by using a carbon fiber-reinforced polymer (CFRP) sheet following the ACI 440-17 [19] so that the resulting capacity can increase above the required moment as shown in Fig. 6.

The reinforcing bars in the wall slab are still in good condition, there is no significant reduction in the diameter so the repairs carried out on the wall are sufficient for damaged concrete by using epoxy resin injection into the cracks and covering the loose concrete cover using mortar and wire mesh. The proposed strengthening design of the clinker silo structure is shown in Fig. 7.

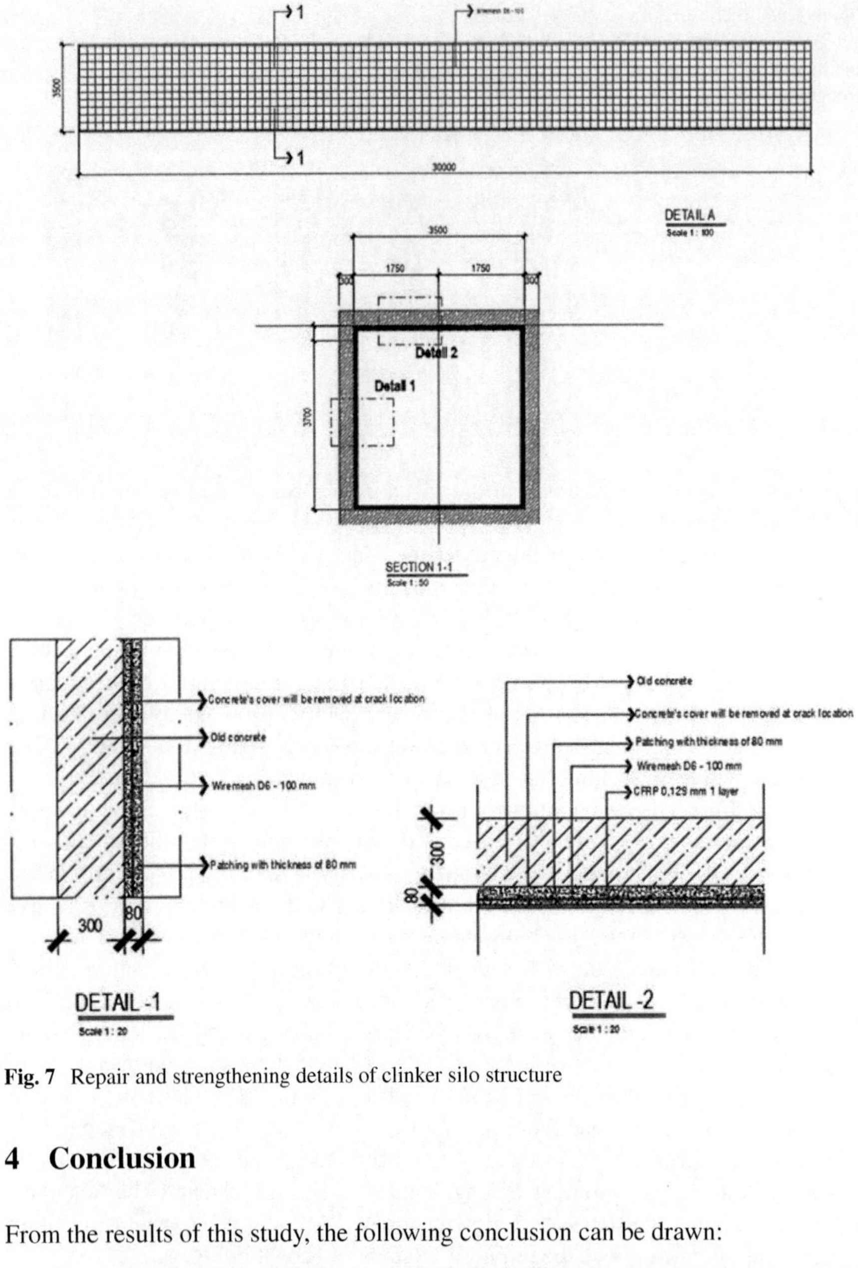

Fig. 7 Repair and strengthening details of clinker silo structure

4 Conclusion

From the results of this study, the following conclusion can be drawn:

1. The corrosion and cracks in cement plant clinker silo slab structure are more severe than in wall structure.
2. The compressive strength of the clinker silo structure meets the requirements of structural concrete based on the building code.

3. The low dynamic elastic modulus was observed indicating that the concrete structure has been cracked already.
4. The corrosion potential test results show that the corrosion level of reinforcing steel in the clinker silo slab is a high risk, while the corrosion level of reinforcing steel on the clinker silo wall is at medium risk.
5. The residual strength of the slab clinker silo structure is lower than the required strength; therefore, the strengthening of the structure is required. The repair and strengthening of the structure can be conducted by injection of epoxy resin into the cracks and by applying a carbon fiber-reinforced polymer sheet on the surface.

References

1. Ferrari L, Kaufmann J, Winnefeld F, Plank J (2012) Reaction of clinker surfaces investigated with atomic force microscopy. Constr Build Mater 35:92–96
2. Ridha M, Fonna S, Huzni S, Ariffin AK (2013) Corrosion risk assessment of public buildings affected by the 2004 tsunami in Banda Aceh. J Earthq Tsunami 7(1):1350004
3. Fonna S, Ridha M, Huzni S, Walid WA, Mulya TTD, Ariffin AK (2017) Corrosion risk of RC buildings after ten years the 2004 tsunami in Banda Aceh-Indonesia. Proc Eng 171:965–976
4. Susanto H, Huzni S, Fonna S (2018) Corrosion of reinforced concrete structures submerged by the 2004 Tsunami in West Aceh, Indonesia. Int J Corr 2018:4318434
5. Hobbs B, Kebir MT (2007) Non-destructive testing techniques for the forensic engineering investigation of reinforced concrete buildings. Forensic Sci Int 167:167–172
6. Sanchez K, Tarranza N (2014) Reliability of rebound hammer test in concrete compressive strength estimation. Int J Adv Agric Environ 1(2):198–202
7. Kurnia RDI, Suhaimi, Huzni S, Fonna S (2019) Corrosion potential of reinforced steel in reinforced concrete in Kabupaten Bireun: analysis of groundwater content used as a concrete mixture. IOP Conf Ser Mater Sci Eng 602:012076
8. Hasan M, Mahlil, Mubarak A (2020) Desain Perkuatan Struktur Bangunan Gedung Akibat Perubahan Fungsi Dan Umur Bangunan Menggunakan CFRP sheet. J Teknik Sipil 9:28–38
9. Hasan M, Saidi T, Afifuddin M, Setiawan B (2021) The assessment and strengthening proposal of building structure after the Pidie Jaya earthquake in December 2016. J King Saud Univ Eng Sci (in press, 2021). https://doi.org/10.1016/j.jksues.2021.02.007
10. Hasan M, Mubarak A, Fikri R, Mahlil, Crack and strength assessment of reinforced concrete cement plant blending silo structure, Mater Today Proc 58:1312–1318
11. ACI 562-19 (2019) Code requirements for assessment, repair, and rehabilitation of existing concrete structures and commentary, 2019
12. ASTM C805/C805M-18 (2018) Standard test method for rebound number of hardened concrete, 2018
13. ASTM C597-16 (2016) Standard test method for pulse velocity through concrete, 2016
14. Hasan M, Husin S, Nursaniah C (2016) Mechanical properties of concrete in compression exposed to sulfuric acid. Key Eng Mater 711:302–309
15. ASTM C876-15 (2015) Standard test method for corrosion potentials of uncoated reinforcing steel in concrete
16. SNI 2847 (2019) Persyaratan beton structural untuk bangunan gedung dan penjelasan, Badan Standarisasi Nasional, 2019
17. SNI 1726 (2019) Tata cara perencanaan ketahanan gempa untuk struktur bangunan gedung dan non gedung, Badan Standarisasi Nasional, 2019

18. SNI 1727 (2019) Beban desain minimum dan kriteria terkait untuk bangunan gedung dan struktur lain, Badan Standarisasi Nasional, 2019
19. ACI 440-17 (2017) Guide for the design and construction of externally bonded FRP systems for strengthening concrete structures, 2017

Performance Evaluation of *Pangium edule* Biodiesel Developed Using Palm Kernel Shell Heterogeneous Catalyst as Alternative Fuel in Multifunction Diesel Engine

Teuku Azuar Rizal, Khairil, Mahidin, and Husni Husin

Abstract The main idea of this article is to provide further knowledge related to the application of pure biodiesel in a custom power generator. Previous research has shown preliminary results regarding the potential of *Pangium edule*. Biodiesel as feedstock and has been tested to run a diesel engine with lower rotational speed. The biodiesel in this research has been prepared using a heterogeneous catalyst derived from a palm kernel shell impregnated with K_2CO_3. In this research, small alternator is driven by Yanmar TF65 R-DI multifunction diesel engine, which is easily assembled and has been utilized as a power generator to be evaluated. The engine was tested under various biodiesel blend with given load and connected to multiple instrument to record data required, i.e. engine speed (rpm), fuel flow (ml/sec), voltage (V), and current (A) readings. The data series then used to analyse the engine performances. The custom power generator has been designed to work at 5.5 kWh in prolonged operation time, suitable to supply typical household wattage in the region, ranging

T. A. Rizal (✉)
Department of Mechanical Engineering, Universitas Samudra, Meurandeh, Kota Langsa 24416, Indonesia
e-mail: azuarrizal@unsam.ac.id

Khairil (✉)
Department of Mechanical Engineering, Universitas Syiah Kuala, Darussalam, Banda Aceh 23111, Indonesia
e-mail: khairil@unsyiah.ac.id

Mahidin · H. Husin
Department of Chemical Engineering, Universitas Syiah Kuala, Darussalam, Banda Aceh 23111, Indonesia
e-mail: mahidin@unsyiah.ac.id

H. Husin
e-mail: husni_husin@unsyiah.ac.id

T. A. Rizal
Doctoral Program in Engineering, Universitas Syiah Kuala, Darussalam, Banda Aceh 23111, Indonesia

Irwansyah et al. (eds.), *Proceedings of the 4th International Conference on Experimental and Computational Mechanics in Engineering*, Lecture Notes in Mechanical Engineering, https://doi.org/10.1007/978-981-99-7495-5_8

from 0.9 up to 3 kWh. The fuel mixtures containing 20% biodiesel, or B20, is still desirable to use in general multipurposes diesel engine. While B30 show similar potential but it does not show a similar performance as in higher rotational speed diesel power generator as demonstrated earlier. However, PE biodiesel still shows a promising performance as an alternative source of diesel engine fuel.

Keywords Performance evaluation · *Pangium edule* biodiesel · Palm kernel shell · Heterogeneous catalyst · Power generator

1 Introduction

Biodiesel has become one alternative fuel to petroleum diesel which is desirable to reduce the environmental impact of emissions without modifying engines [1]. Studies have shown that biodiesel has better combustion performance than petroleum diesel due to several factors, including higher cetane numbers [2] appropriate viscosity, and density. Important parameters commonly used as benchmarking in determining the performance of diesel engines include maximum *brake thermal efficiency (BTE)*, minimal *brake-specific fuel consumption* (BSFC), and emission discharge from diesel engine exhaust [3–5]. Relatively, biodiesel is a highly oxygenated fuel that can improve combustion efficiency and can reduce unburnt hydrocarbons (HCs), carbon dioxide (CO_2), carbon monoxide (CO), sulphur dioxides (SO_2), nitric oxide (NO_x), and polycyclic aromatic HC emissions [1]. However, the study indicates that biodiesel's two primary pollutant gas emissions were generally reduced. Therefore, biodiesel was claimed to provide a more environmentally friendly, secure, and renewable approach to obtaining energy in the long run [6].

Previously, the Pangium Edule biodiesel (PEB) blend with 10% and 20% mixture was tested in Yanmar TS 50, a 5.0 HP diesel engine that works at relatively lower rotational speed [7]. The process of biodiesel production has never been specified. The research indicated that at the higher rotational speed, there is a tendency to lower the SFC and the engine's power while increasing the thermal efficiency at the same time. Much previous research related to PEB only focused on the production aspect, including the actual production of biodiesel and feedstock [8–10]. Other investigations emphasize physicochemical properties [11–13], while limited research on engine testing was implemented.

This paper intends to examine the engine performances, particularly concerning engine power, fuel consumption, and thermal efficiency of a diesel engine using PEB blend and its comparison to petroleum diesel fuel. Several types of biodiesel blends have been prepared for this experimentation, including 100% PEB (B100) and 20%, 30%, and 40% mixture of PEB to Pertamina Dex (B0). All these fuels will be discussed later in the next section.

2 Material and Methods

Biodiesel preparation and properties. Various biodiesel blends have been prepared for engine performance calculation, and they are 100% PEB or B100, B40, B30, B20, and B0. In previous research, biodiesel with a 20% mixture has been found to work at most similar to the performance of pure petroleum diesel without sacrificing the default manufacture performance itself. Pertamina Dex, later called B0, was mixed with various percentages of PEB, including 20%, 30%, 40%, and 100% PEB mixture, later labelled as B20, B30, and B40, respectively. A homogenized biodiesel blend was prepared by stirring the mixture above its cloud point temperature for at least 2–3 minutes.

Experimental setup and Instrumentation. This study used multifunction diesel engine 2200 rpm YANMARTF65di connected to a 3 kW single-phase alternator xingxing working with a power factor of 1 and at maximum rotational speed at 1500 rpm reduced by v-belt with pulley ratio at 1500: 2200 or 1: 1,4. Detailed specification of the diesel engine, YanmarTF65R in shown in Table 2. This engine was relatively small and affordable to many SMEs and households. It was also designed to run with commercial diesel in the fuel category of B0 up to B20. The engine was subjected to various biodiesel blends ranged from B0, B20, B30, B40, and B100 and later observed the engine's performance during the utilization of such biodiesel blends. Table 1 presents the detailed technical parameter of the diesel engine. This research characterized the performance of a diesel engine including power (*P*), specific fuel consumption (SCF), and thermal efficiency (*n*th). Data required for the performance calculations include voltage, current, fuel consumption, and engine rotation (Tables 2 and 3).

The experiment was conducted by pulling out all data required, i.e., voltage, current, fuel flow, and rotational speed, based on the load provided to the generator. The data were recorded along with the increase of load subjected to the engine. Each data are provided through an additional measurement instrument with 1.5 kW load at the increment of 200 rpm starting from 1500 to 2100 rpm. Figure 1 presents the schematic of the experimental setup in this research.

Table 1 Properties of diesel and biodiesel blend fuel used in this experiment

Fuel parameter	Unit	B0*	B20	B30	B40	B100
Kinematic viscosity	mm^2/s (cSt)	2.0–4.5				4.4–5.1
Density	kg/m^3	820–860	845.9	848.85	851.8	851–888
Calorific Value	MJ/Kg	56.5	53.1	51.4	49.7	39.4

*Reference: [11]

Table 2 Technical parameter of Yanmar TF65 R-DI

Model	TF 65 R-di
Engine model	4 stroke horizontal diesel engine horizontal
Combustion	Direct injection
Aspiration	Natural
No. of cylinders	1
Cyl. Bore × Stroke, mm	78 × 80
Displacement, cc	382
Cooling system	Radiator/water
Lubrication system	Forced lubrication with trochoidal pump
Starting system	Manual
Rated output, kW (PS)/min	5.5/2200
Specific fuel consumption, gr/dk jam	178
Fuel	Diesel fuel
Continuous rated output	5.5/2200
Max. output	6.5/2200
No-load Max. rotation speed, rpm (min)	2200
Compression ratio	18.1
Valve position	Over head valve
Ignition system	Pressurized air

Reference: [12]

Table 3 Specification of Xingxing alternator

Parameter	Description
Model	NES-3
Rated frequency (Hz)	50
Rated output (kVA)	3
Max. output (kVA)	5.0 5.5
Rated voltage (V)	230
Rated current (A)	13
Rated rotation speed (r/min)	1500
Phase NO	Single phase
Power factor (Cos Φ)	1.0

3 Result and Discussion

Comparison of Energy Content. The energy content of two types of biodiesel blend is generally calculated by summing energy content from a percentage of type-A biodiesel with another percentage of type-B biodiesel or expressed as in Eq. 1 [13] (Fig. 2).

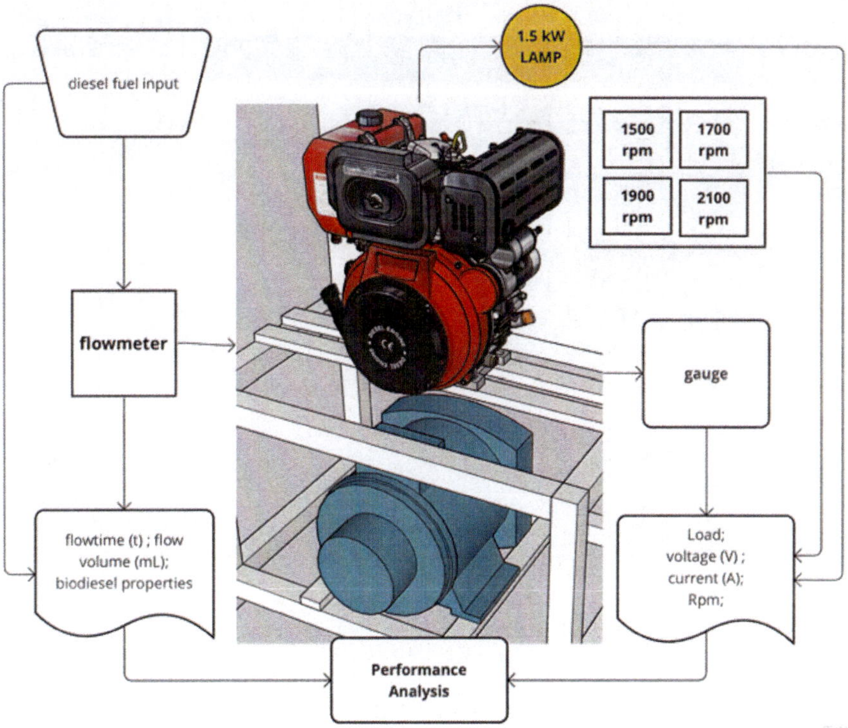

Fig. 1 Schematic illustration of the experimental flow and data recording procedure

$$H_{AB} = \#H_{\text{LHV}_A} \times \%\text{Vol}_A * + \#H_{\text{LHV}_B} \times \%\text{Vol}_B * \quad (1)$$

H_{AB} is a blended fuel energy content, while H_A and H_B are specific energy content of each fuel fraction.

Fuel Consumption. Figure 3 indicates that B20 still dominating the performance both at lower and at higher load, burned at 0.465 kg/h, which has a minuscule difference of all biodiesel blend ranged from 0.465 to 0.5 kg/h in B30. However, all biodiesel blend show a similar fuel consumption pattern which was lower to default fuel consumption model as manufactured.

Brake-Specific Fuel Consumption, BSFC. Figure 4 implies that the maximum BSFC was 0.286 kg/kWh for the BPE 20% blend at 1500 rpm. Also, the minimum BSFC, 0.222 kg/kWh, occured at a 1900 rpm load for the fuel blend of 40% biodiesel.

Brake Power. As shown in Figure 5, the brake power shows a similar phenomenon related to the amount of biodiesel flow and the load given during engine runs. The lower heating value of each fuel directly affects the total power during combustion. At a 2100 rpm, B0 performs the highest brake power at 2.7 kW out of 5.5 kW total

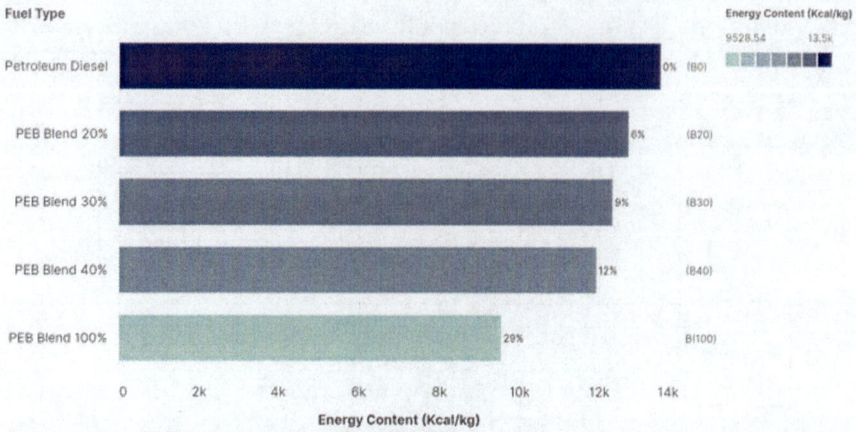

Fig. 2 Energy content of the biodiesel blend gradually decreased along with the increasing biodiesel content in the mixture

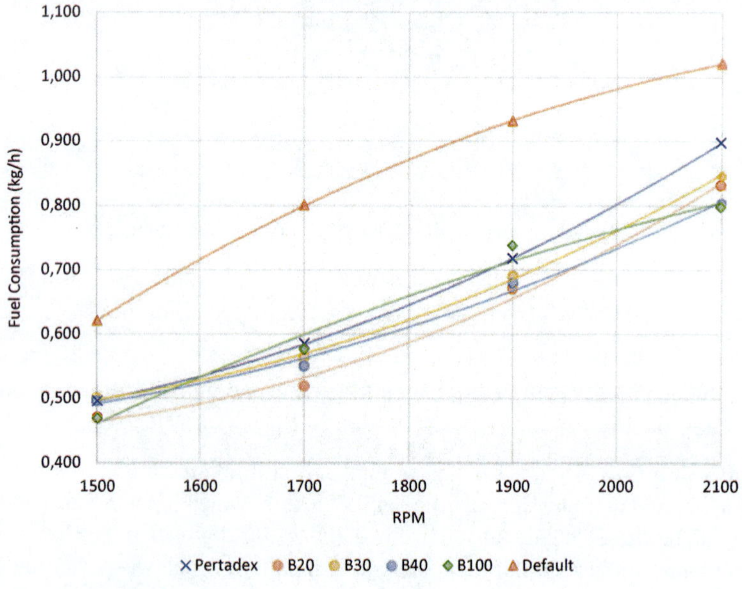

Fig. 3 Fuel consumption at different load

engine power according to the manufacturing design. BPE blend 20% enlisted in the following to deliver 2.6 kW of power at the same rpm.

Brake Thermal Efficiency, BTE. The efficiency of a diesel engine is calculated when a fuel's chemical energy is turned into valuable work, later defined as BTE [4]. Figure 6 shows biodiesel percentage and engine load effects on the BTE. In

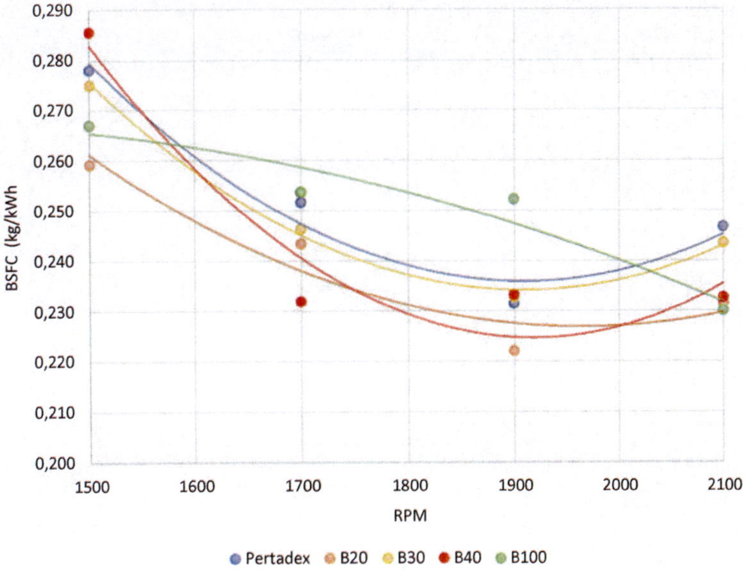

Fig. 4 BSFC at different load

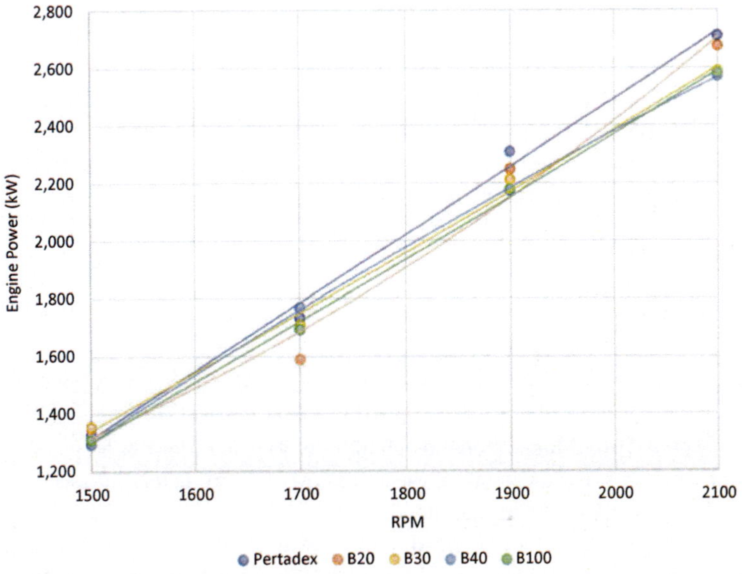

Fig. 5 Engine power at different load

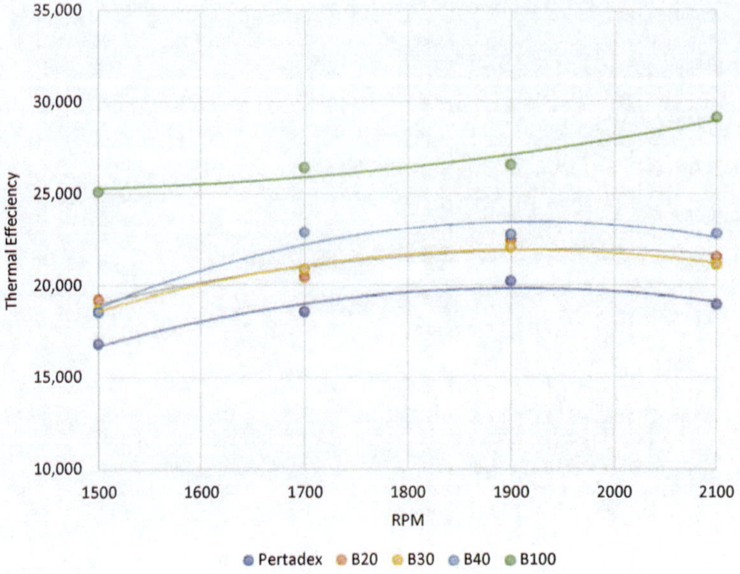

Fig. 6 BTE at different load

all fuel and blends, BTE tends to increase along with the increase of applied load. This phenomenon happens due to the reduction in heat loss and increased power developed with increased load [4]. BPE blend 100% has the highest BTE at every load given in this case. Nevertheless, the implementation of pure biodiesel in the engine application will significantly impact the cost of production and possibly stimulate significant drawbacks on the environmental issue.

4 Conclusion

The experiment on Pangium Edule biodiesel blends on YanmarTF65 R-di has been conducted successfully. The biodiesel resulting from the transesterification process with palm kernel shell heterogeneous catalyst works smoothly in a custom power generator driven by multifunctional diesel engine. Unfortunately, it does not show tremendous performance at higher blend of biodiesel content in the fuel. It is simply due to limited engine rotational speed. However, if compared to small class power generator works on higher rotational speed, i.e. > 3000 rpm, biodiesel with BPE blend 30% are show a better performance. Since there were infinitesimal changes in each variable recorded during the experiments and operational conditions, the upcoming research data and calculation should be carried out with a more state-of-the-art methodology.

Acknowledgements The authors wish to acknowledge the Laboratories of Combustion and Propulsion of Universitas Syiah Kuala and the Laboratories of chemical reaction and catalysis of Universitas Syiah Kuala to support this research.

References

1. McCarthy P, Rasul MG, Moazzem S (2011) Comparison of the performance and emissions of different biodiesel blends against petroleum diesel. Int J Low-Carbon Technol 6(4):255–260. https://doi.org/10.1093/ijlct/ctr012
2. Wood BM, Kirwan K, Maggs S, Meredith J, Coles SR (2015) Study of combustion performance of biodiesel for potential application in motorsport. J Clean Prod 93:167–173. https://doi.org/10.1016/J.JCLEPRO.2014.12.091
3. Gharehghani A, Pourrahmani H (2019) Performance evaluation of diesel engines (PEDE) for a diesel-biodiesel fueled CI engine using nano-particles additive. Energy Convers Manag 198:111921. https://doi.org/10.1016/j.enconman.2019.111921
4. Nedayali A, Shirneshan A (2016) Experimental study of the effects of biodiesel on the performance of a diesel power generator. Energy Environ 27(5):553–565. https://doi.org/10.1177/0958305X15627550
5. Altarazi YSM, Abu Talib AR, Gires E, Yu J, Lucas J, Yusaf T (2021) Performance and exhaust emissions rate of small-scale turbojet engine running on dual biodiesel blends using Gasturb. Energy 232:120971. https://doi.org/10.1016/j.energy.2021.120971
6. McCarthy P, Rasul MG, Moazzem S (2011) Analysis and comparison of performance and emissions of an internal combustion engine fuelled with petroleum diesel and different biodiesels. Fuel 90(6):2147–2157. https://doi.org/10.1016/J.FUEL.2011.02.010
7. Khairil, Thalib S, Turmizi D (2015) Fundamental study of diesel engine performance by implementing oil from tropical kepayang seeds (*Pangium edule*). Adv Mat Res 1115:480–483. https://doi.org/10.4028/www.scientific.net/AMR.1115.480
8. Norfiarli KA, Nurdin H (2013) The biodiesel characteristic of kepayang oil (*Pangium edule* Reinw). ARPN J Agric Biol Sci 8(3). Accessed: 23 May 2022 [Online]. Available: https://www.researchgate.net/publication/268982645
9. Atabani AE, Badruddin IA, Masjuki HH, Chong WT, Lee KT (2015) *Pangium edule* Reinw: a promising non-edible oil feedstock for biodiesel production. Arab J Sci Eng 40(2):583–594. https://doi.org/10.1007/s13369-014-1452-5
10. Hamzah FH, Hamzah N, Irdoni HS (2018) Potency of Picung (*Pangium edule* Reinw) endosperm utilized as a raw material in producing frying oil or biodiesel. Jurnal Agroindustri 8(1):44–52. https://doi.org/10.31186/j.agroindustri.8.1.44-52
11. SDM (2020) Spesifikasi Produk BBM, BBN & LPG [Product Specification for Fuel, Biofuel & LPG]. Spesifikasi Produk BBM, BBN & LPG, 2020. https://onesolution.pertamina.com/Product/Download?filename=20201201035120atc_spesifikasi.pdf. Accessed 20 June 2022
12. Yanmar Inc. (2022) TF series. Horizontal water-cooled diesel engines. Industrial engine. YANMAR Indonesia. https://www.yanmar.com/en_id/engine/products/diesel/h_watercooled/tfseries/#head-6. Accessed 31 March 2022
13. Khairil et al (2020) Experimental study on the performance of an SI engine fueled by waste plastic pyrolysis oil– gasoline blends. Energies (Basel) 13(16):4196. https://doi.org/10.3390/en13164196

Study the Effect of Hyperparameters on Performance of BEIA to Detect Corrosion in Reinforced Concrete

Muhibul Jamal, Syarizal Fonna, Syifaul Huzni, and Israr B. M. Ibrahim

Abstract The purpose of this study was to determine the effect of hyperparameters on the performance of BEIA in detecting corrosion in reinforced concrete. BEIA was run with fifteen data of potential measurement values on concrete surfaces. Variations in geometry size, cognitive values (a_1), and social parameters (a_2) were performed in this study. Simulation result show the hyperparameters can affect the performance of BEIA in finding the location of corrosion in reinforced concrete. Due to the effect of hyperparameter values on BEIA performance in detecting corrosion on rebar, the recommended cognitive parameter value is < 0.4; meanwhile the recommended social parameter value is $0.6 < a_2 \leq 0.9$.

Keywords Corrosion · BEM · PSO · BEIA · Detecting corrosion · Reinforced concrete

1 Introduction

Corrosion is an electrochemical metal degradation process due to redox reactions between metals and the surrounding environment [1]. Based on research in the United States, corrosion results in losses of hundreds of billions of dollars. According to the study's findings, the losses caused by corrosion in a country ranged from 1 to 5% of the Gross National Product (GNP) [2]. Corrosion that occurs in reinforced concrete can cause premature failure of infrastructure [3]. The half-cell potential mapping method is one of the techniques used to detect corrosion in concrete which was carried out extensively regarding ASTM C876 [4]. The half-cell potential mapping method performs by taking direct measurements on the reinforced concrete surface to evaluate the corrosion of the reinforcing steel. The measurement results on the concrete surface are potential values that can be used to analyze or predict corrosion

M. Jamal · S. Fonna (✉) · S. Huzni · I. B. M. Ibrahim
Department of Mechanical and Industrial Engineering, Faculty of Engineering, Universitas Syiah Kuala, Jl. Tgk. Syech Abdul Rauf No. 7, Banda Aceh 23111, Indonesia
e-mail: syarizal.fonna@usk.ac.id

Irwansyah et al. (eds.), *Proceedings of the 4th International Conference on Experimental and Computational Mechanics in Engineering*, Lecture Notes in Mechanical Engineering, https://doi.org/10.1007/978-981-99-7495-5_9

in concrete reinforcement [5]. Each potential point represents an increased chance of corrosion. Because the reinforcing steel/rebar potential is nearly the same as the value on the concrete surface in the case of uniform corrosion, this method yields better results in detecting corrosion of reinforcement in concrete [6]. However, this method has many shortcomings, namely the interpretation of data from the results of measurements in the field must be considered [7]. Sadowski [8] has developed an inverse method based on the finite element method (FEM) and the conjugate gradient method. However, corrosion only occurs on surface materials, FEM is not suitable for corrosion, and the boundary element method (BEM) is more suitable for simulating corrosion of concrete [9]. Then, corrosion investigations in reinforced concrete have also been carried out [10] using artificial neural networks (ANN). However, this technique has a lower level of accuracy in detecting the location of corrosion due to the need of large data for training.

One of the methods being developed that can overcome the weakness of half-cell potential mapping is boundary element inverse analysis (BEIA). BEIA uses a numerical simulation based on the boundary element method (BEM) and particle swarm optimization (PSO). The effect of the constant weight parameter (W) has been carried out to detect corrosion in reinforced concrete [11]. The numerical simulation shows the recommended constant value for the inertial weight (W) ($0.25 \leq W \leq 0.5$) with an error of 5%. By breaking the RC column, the location and size of the corrosion were confirmed. This method was found to be up to 95.65% accurate in detecting the location and size of corrosion. If only half-cell survey potentials are used, this information will not be obtained. This method opens the way to comprehensively determine the corrosion profile of reinforced concrete for several locations with various shapes [10].

Hyperparameter is a learning parameter whose value is selected before the learning algorithm process starts. The hyperparameter value is independent of the data, and the value is constant as long as the model is optimized. The most appropriate value for the hyperparameter is selected through the hyperparameter loop. Hyperparameters are used to customize various aspects of the learning algorithm, the effects of which on the final model and its performance can be highly variable [12]. As a result, the purpose of this research was to investigate the effect of hyperparameters on the performance of BEIA in detecting the location of corrosion in reinforced concrete in order to detect corrosion more effectively.

2 Boundary Element Inverse Analysis (BEIA)

BEIA was developed by integrating two methods: boundary element method (BEM) and particle swarm optimization (PSO). The corrosion potential value on the concrete surface is calculated using BEM. Meanwhile, PSO is used to evaluate the cost function for detecting reinforcing corrosion of steel in concrete. PSO was first introduced in 1995 by Kennedy and Eberhart [13]. The social behavior of flocks of birds and schools of fish in search of food helped inspire the PSO mechanism. In 2018, the inverse using BEM and PSO was applied in the field on reinforced concrete columns in the 2004 tsunami area, with the inverse results detecting the location and size of corrosion on reinforced concrete columns with a 5% error of the actual corrosion [14].

BEIA is executed following the flowchart depicted in Fig. 1 by determining the required constant parameters such as Z, J_{max}, ε_0, a_1, a_2, and W. Parameter Z is the number of particles, J_{max} is the maximum iteration, ε_0 is the boundary condition ($\varepsilon_0 \leq 10^{-4}$), a_1 is the value cognitive, and a_2 is social parameter W is the value of inertia weight. The word particle in this method is used for candidate solutions. Then, determine the corrosion profile or X_j at random on the concrete surface, and determine the particle velocity or V_j at random.

The potential on the concrete surface is calculated using BEM. The next step is to enter N electrical potential data obtained from field measurements on the concrete surface (N). To obtain a corrosion profile, namely the location of the corrosion on reinforcement in concrete, BEIA uses the following Eq. (1). When the maximum number of iterations is reached, the simulation is finished. If it is not reached, the iteration resumes with doing on. For equations in updating solutions for personal best (pbest) and global best (gbest) can be seen in the Eqs. (2–6). From the polarization graph, it is then converted into the boundary conditions; the boundary conditions of the corroded and uncorroded parts are $\phi_a = -10 \times i + 0.600$ V and $\phi_c = -10 \times i + 0.270$ V, where i is the current density, A/m^2. Meanwhile, the conductivity value of concrete is $k = 0.007\ \Omega^{-1}.\text{m}^{-1}$ [16].

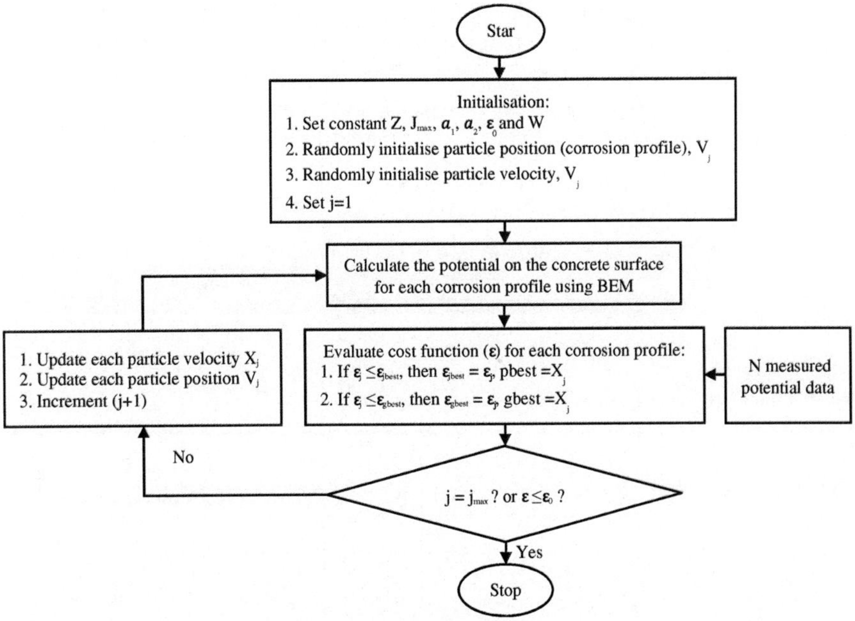

Fig. 1 Flowchart of inverse analysis using PSO [15]

$$\varepsilon(X) = \sum_{l=1}^{N} \left[\frac{\phi_l - \overline{\phi_l}}{\overline{\phi}_{\max}} \right]^2 \tag{1}$$

$$X_{j+1} = X_j + V_{j+1} \tag{2}$$

$$V_1 = a_1 r_1 (\text{pbest} - X_j) \tag{3}$$

$$V_2 = a_2 r_2 (\text{gbest} - X_j) \tag{4}$$

$$V_3 = \text{W} V_j \tag{5}$$

$$V_{j+1} = V_1 + V_2 + V_3 \tag{6}$$

3 Case Study

The geometry used is a two-dimensional shape with a rectangular geometric shape, which has a single reinforcing steel in the middle as shown in Fig. 2 and the concrete size can be seen in Table 1 with a meshing size of 1 cm. Potential data from field measurements were taken from the boundary element method (BEM) simulation in this case study, with the assumption that all corrosion sizes (CS) were 30 cm and the location of corrosion (CL) was known. This case study investigates the effect of geometry measures of values cognitive (a_1) and social parameters (a_2) on the performance of BEIA in detecting corrosion in reinforced concrete (Fig. 3).

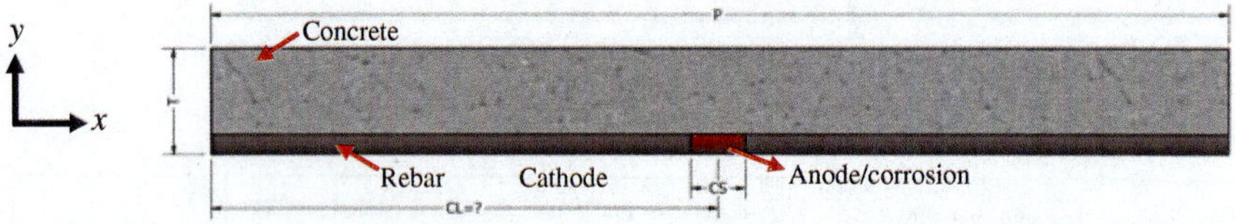

Fig. 2 Simple RC model of corrosion numerical simulation

Table 1 Dimensions of geometry

Case	P (cm)	T (cm)	Mesh size (cm)
I	560	5	1
II	1120	5	1

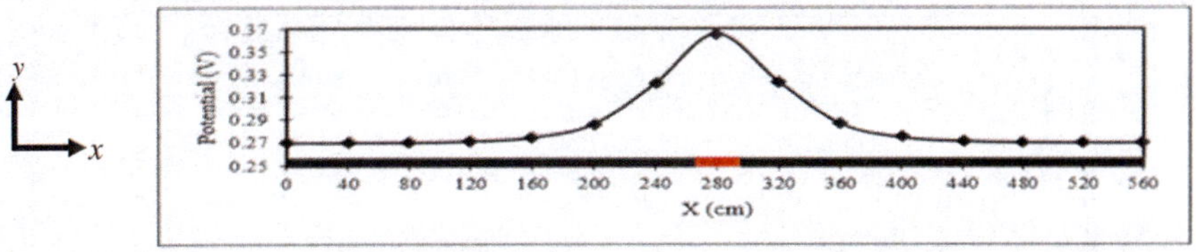

Fig. 3 Measured electrical potential data

4 Numerical Simulation Results and Discussion

The BEIA simulation results that have been obtained with variations in geometry, values cognitive (a_1), and social parameters (a_2) as shown in Tables 1, 2 and 3 can be seen in Figs. 4, 5 and 6.

Figure 4 shows the BEIA simulation for case study I with parameters N = 15, $a_1 = 0.1$, $a_2 = 0.5$, and $W = 0.25$. It can be seen that in the 12th iteration, all the particles have been collected into one location when the iteration continues until the maximum iteration of the particles is still in the same position. BEIA has been shown to detect the location of corrosion with an average error of 1% under these conditions.

Afterward, Fig. 5 provides the BEIA simulation for the case study I with variations in cognitive values (a_1), and social parameters (a_2). Figure 5a shows the results of the BEIA simulation with variations in cognitive values (a_1). Variations in cognitive value (a_1) did not affect the performance of BEIA in finding the location of corrosion in reinforced concrete. The location of corrosion was found in the 12th iteration, and

Table 2 Variations of cognitive values (a_1)

Cognitive (a_1)	Social parameter (a_2)
0.1	0.5
0.2	0.5
0.3	0.5
0.4	0.5
0.5	0.5

Table 3 Variation of social parameter values (a_2)

Cognitive (a_1)	Social parameter (a_2)
0.5	0.4
0.5	0.6
0.5	0.7
0.5	0.8
0.5	0.9

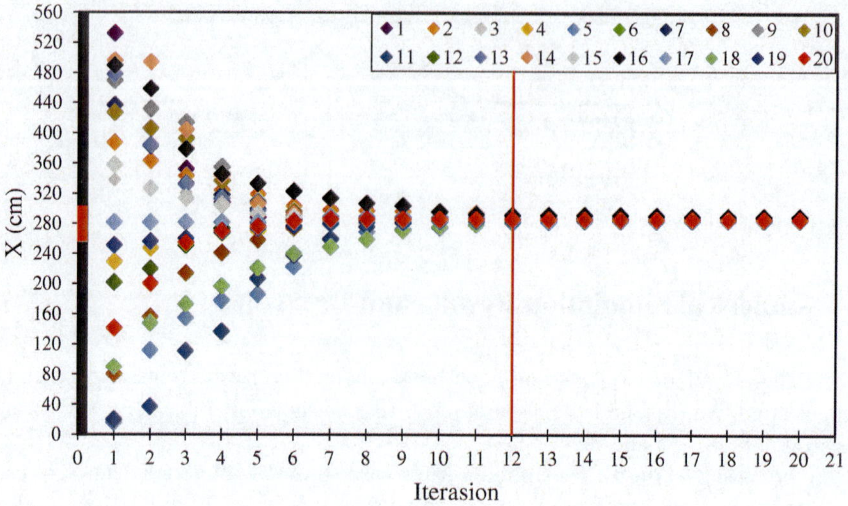

Fig. 4 Movement of particles for case I with $N = 15$, $a_1 = 0.1$, $a_2 = 0.5$, and $W = 0.25$

the variation of value cognitive (a_1) in case I showed that BEIA can detect the location of corrosion with an average error of 1%.

Furthermore, Fig. 5b gives the BEIA simulation with variations in the value of the social parameter (a_2) contained in the PSO. Variations in the value of the social parameter (a_2) can affect the performance of BEIA in finding the location of corrosion, the greater the value of the social parameter (a_2), the faster the location of corrosion is found. For social parameter values (a_2) 0.4, corrosion locations were found in the 17th iteration, for social parameter values 0.6, corrosion locations were found in the 9th iteration, while for social parameter values 0.7, 0.8, and 0.9, corrosion locations were found in the 8th iteration. For case study I, corrosion locations can be detected by BEIA with an average error of 1%.

Figure 6 shows BEIA simulation for case study II with variations in cognitive values (a_1) and social parameters (a_2) contained in PSO. Figure 6a shows the results of BEIA in case II with variations in cognitive values that can affect the performance of BEIA in finding the location of corrosion in reinforced concrete. For cognitive values (a_1) 0.1 and 0.3, corrosion locations were found in the 12th iteration, while for cognitive values (a_1) 0.2, locations of corrosion were found in the 14th iteration. For cognitive values (a_1) 0.4 and 0.5, corrosion locations were found in the 15th iteration. For case II, it shows that BEIA can detect the location of corrosion with a maximum error of 4%.

Finally, Fig. 6b depict the BEIA simulation in case study II for variations in social parameter values (a_2). Variations in the value of the social parameter (a_2) affect the performance of BEIA in finding the location of the corrosion. For social parameter values (a_2) 0.4, corrosion locations were found in the 16th iteration, while for social parameter values (a_2) 0.6, 0.8, and 0.9, corrosion locations were found in the 11th

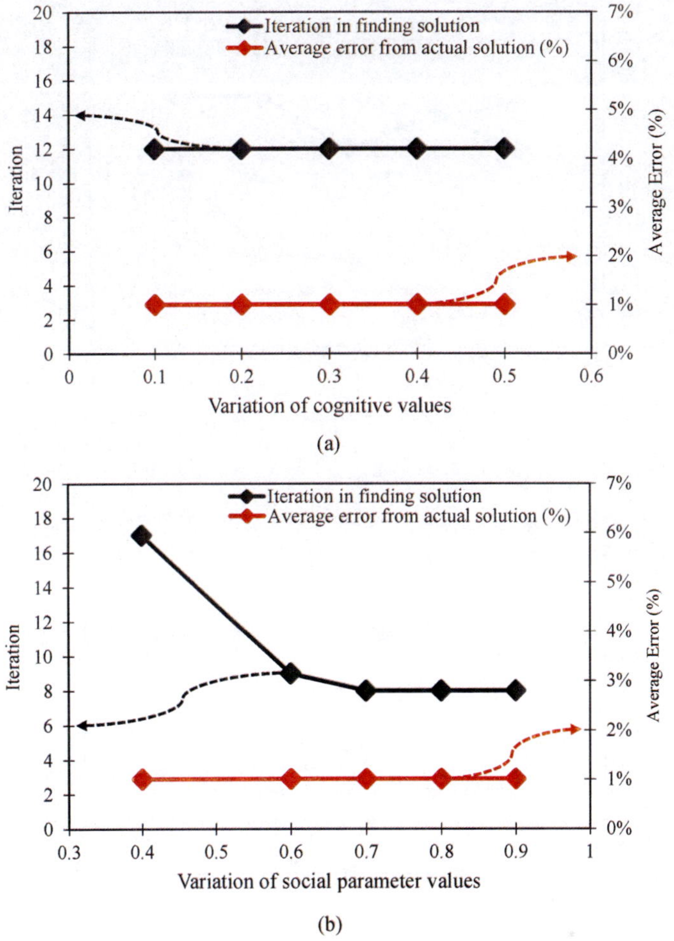

Fig. 5 BEIA simulation results for case I with a variation of the value of **a** cognitive (a_1), and **b** social parameters (a_2)

iteration. For social parameter value (a_2) 0.7, corrosion location was found in the 15th iteration. For the value of social parameters (a_2) 0.4 in case II, corrosion locations cannot be detected by BEIA, while for social parameter values of 0.6, 0.7, 0.8, and 0.9, corrosion locations can be detected by BEIA with < 5% error.

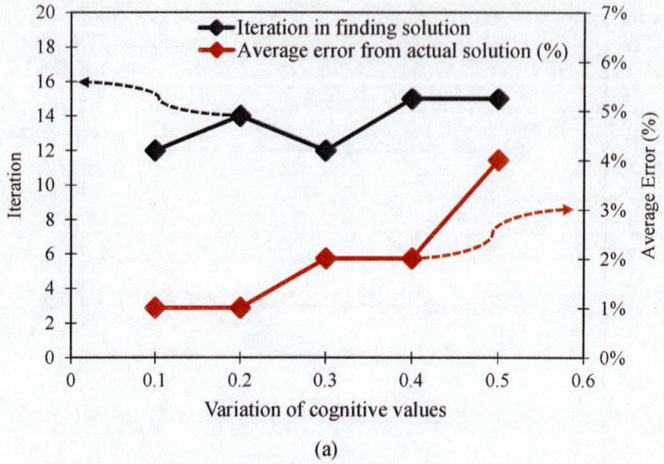

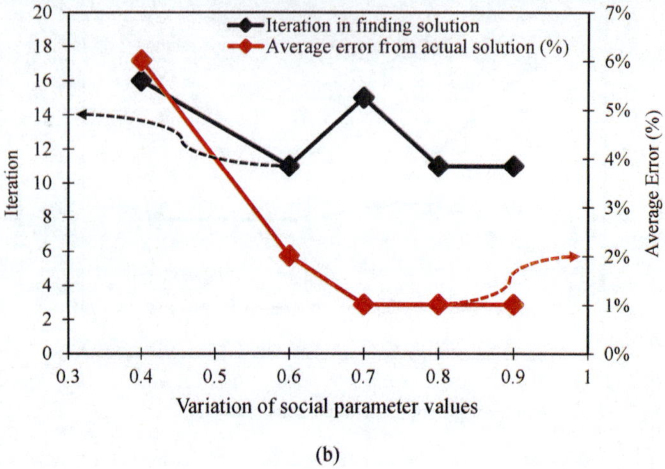

Fig. 6 BEIA simulation results for case II with a variation of the value of **a** cognitive (a_1), and **b** social parameters (a_2)

5 Conclusion

Based on the results of the BEIA simulation, it can be concluded that variations of geometry size, cognitive values (a_1) and social parameters (a_2), can influence the performance of BEIA in detecting corrosion in reinforced concrete. The recommended values for the cognitive value are < 0.4, while the values for social parameters are between $0.6 < a_2 \leq 0.9$.

Acknowledgements This work was supported by Ministry of Education, Culture, Research, and Technology, Republic of Indonesia, Penelitian Dasar (PD) No. 20/UN11.2.1/PT.01.03/DPRM/2022.

References

1. Fontana M, Greene N (1987) Mars Fontana-corrosion engineering
2. Roberge PR (2008) 59. Corrosion engineering principles and practice. McGraw-Hill, New York, US
3. Moreno E, Cobo A, Palomo G, Gonzalez MN (2014) Mathematical models to predict the mechanical behaviour of reinforcements depending on their degree of corrosion and the diameter of the rebars. Constr Build Mater 61:156–163
4. Sagüés AA, Walsh MT (2012) Kelvin Probe electrode for contactless potential measurement on concrete—properties and corrosion profiling application. Corros Sci 56:26–35
5. Hussain RR (2011) Underwater half-cell corrosion potential bench mark measurements of corroding steel in concrete influenced by a variety of material science and environmental engineering variables. Meas: J Int Meas Confederation 44(1):274–280
6. Marinier P, Isgor OB (2013) Model-assisted non-destructive monitoring of reinforcement corrosion in concrete structures. In: Büyüköztürk O et al (Ed) Nondestructive testing of materials and structure. RILEM Bookseries. Springer-Verlag, pp 719–724
7. Lan Z, Wang X, Hou B, Wang Z, Song J, Chen S (2012) Simulation of sacrificial anode protection for steel platform using boundary element method. Eng Anal Boundary Elem 36(5):903–906
8. Sadowski L (2013) Non-destructive investigation of corrosion current density in steel reinforced concrete by artificial neural networks. Arch Civ Mech Eng 13(1):104–111
9. Song H-W, Saraswathy V (2007) Corrosion monitoring of reinforced concrete structures—a review. Int J Electrochem Sci 2(June):1–27
10. Adriman R, Bin I, Ibrahim M, Huzni S, Fonna S, Kamal A (2022) Improving half-cell potential survey through computational inverse analysis for quantitative corrosion profiling. Case Stud Constr Mater 16:e00854
11. Fonna S, Huzni S, Ridha M, Ariffin AK (2013) Inverse analysis using particle swarm optimization for detecting corrosion profile of rebar in concrete structure. Eng Anal Boundary Elem 37(3):585–593
12. Claesen M, De Moor B (2015) Hyperparameter search in machine learning. February
13. Okwu MO, Tartibu LK (2021) Particle swarm optimisation. Stud Comput Intell 927:5–13
14. Fonna S, Gunawarman HS, Ariffin AK (2018) Boundary element inverse analysis for rebar corrosion detection: study on the 2004 tsunami-affected structure in Aceh. Case Stud Constr Mater 8(June 2017):292–298
15. Fonna S, Gunawarman HS, Ariffin AK (2019) The influence of some solution candidate on the performance of boundary element inverse analysis in detecting rebar corrosion. IOP Conf Ser: Mater Sci Eng 602(1)
16. Ridha M, Amaya K, Aoki S (2005) Boundary element simulation for identification of steel corrosion in concrete using magnetic field measurement. Corrosion 61(8):784–791

Manufacturing Processes of Table Tennis Racket Coconut Coir Fiber/Polyester Composite Polymer

M. Yani, Arfis Amiruddin, Balisran Islam, Indrayani, Riadini Wanty Lubis, Partaonan Harahap, and Wahyudi Pranata

Abstract Coconut oil waste is considered to be a promising prospect for use and development as natural composite materials. Before use, coconut oil fibers are treated for 24 h in an alkaline treatment of 2% NaOH. In this study, coconut coir fiber was used as a reinforcement for polymer composite materials. The purpose of this study was to make a table tennis racket from a composite polymer material reinforced with coconut fiber. Making of the table tennis racket is by using a press machine. Mixing of coconut fiber and polyester resin is based on mass fraction with a weight ratio of 10:90. Making table tennis racket is by hand layup and compressing mold. Coconut coir fiber is arranged randomly with a layer on a mold in a press machine. The results obtained are a table tennis bat with a composite material that has a thickness of 6 mm and a mass table tennis bat after rubber coating on the bat part is 203 gr.

Keywords Polymer composite · Coconut coir fiber · Resin · Table tennis racket

M. Yani (✉) · A. Amiruddin · B. Islam · W. Pranata
Mechanical Engineering Department, Faculty of Engineering, Universitas Muhammadiyah Sumatera Utara, Jl. Kapten Muchtar Basri No. 3, Medan 20238, Indonesia
e-mail: m.yani@umsu.ac.id

A. Amiruddin
e-mail: arfis@umsu.ac.id

B. Islam
e-mail: balisran@umsu.ac.id

Indrayani · R. W. Lubis
Civil Engineering Department, Faculty of Engineering, Universitas Muhammadiyah Sumatera Utara, Jl. Kapten Muchtar Basri No. 3, Medan 20238, Indonesia

P. Harahap
Electrical Engineering Department, Faculty of Engineering, Universitas Muhammadiyah Sumatera Utara, Jl. Kapten Muchtar Basri No. 3, Medan 20238, Indonesia
e-mail: partaonanharahap@umsu.ac.id

Irwansyah et al. (eds.), *Proceedings of the 4th International Conference on Experimental and Computational Mechanics in Engineering*, Lecture Notes in Mechanical Engineering, https://doi.org/10.1007/978-981-99-7495-5_10

1 Introduction

The development of materials engineering and the use of composite materials in Indonesia has not been very popular, and not many industries have developed this technology [1–5]. In modern countries, the development of composite material technology is growing rapidly. The need for materials with certain characteristics is also a driving factor. Various kinds of materials have been used, and further research continues to be carried out to obtain the right materials, one of which is composite materials.

Its ability to be easily shaped according to needs, both in terms of strength and the advantages of other properties [6–9], encourages the use of composite materials as an alternative material to replace wood in various products produced by industry, especially the manufacturing industry. Many products use wood. The more wood is produced from the forest the more often, the results we take will affect the sustainability of the forest. To maintain the ecosystem and forest sustainability, products made from wood can be changed to composite materials [7, 10]. One of them is the use of table tennis rackets, which currently still use wood. Therefore, a proto-type table tennis bat was made with composite materials with the same punch strength.

On the other hand, coconut coir, which is waste, is only piled under coconut plants and then left to rot to dry; its use is mainly for fuel and other things such as foot mats. Although coconut coir still has a fairly good economic value. Coconut coir, when decomposed, will produce coconut coir fiber [11] and cocopeat. However, the derivative of coir is coir fiber. For the development of manufacturing processes that enhance and increase the presence of composite materials in the industrial sector [12–15]. The aims of this research are to design and make moldings for a table tennis racket and to make table tennis bats by molding using a press machine.

2 Materials and Method

The process of making the design [16–18] and the prototype of the table tennis racket begins with drawing the model using SolidWorks software and then making the mold by milling machine with the aluminum alloy metal material and then determining the composition of materials, and a weight ratio of 10% fiber and 90% resin (Table 1).

This study uses the hand-laying method and is then pressed with a hydraulic compression machine. Mixing of the constituent materials is carried out in a separate mixing container, and even distribution of the composite material in the mold is done by hand layup. Then, press the material, so that it is evenly distributed throughout the tennis bat and fills the hollow part.

Table 1 Type of equipment and materials used

No.	Name	Acct	Unit	Materials
Equipment				
1	Table tennis racket mold	1	Set	Aluminum alloy
2	Glass 1000 ml	1	Ps	Glass
3	Glass100 ml	1	Pc	Glass
4	Cup	1	Pc	Glass
5	Stirrer	1	Pc	Plastic
6	Decomposer coir	1	Unit	Steel
Materials				
1	Matrix	*	gr	Unsaturated polyester
2	Coconut coir fiber	*	gr	Fiber
3	Catalyst	*	gr	MEKPO
4	Alkali	*	ml	NaOH
5	Cleaner liquid	*	ml	Acetone
6	Mold release	*	gr	Wax

*suitable composition of the product

The table tennis bat printing process is carried out with the following steps:

1. Apply a separate layer on the inside of the mold with mold release wax for easy removal of the product from the mold.
2. Preparation of the required materials, namely the fiber Yukalac BQTN 157-EX [19], which is an unsaturated polyester resin, then weighed according to the specified weight of the mixture.
3. Mix unsaturated polyester resin and catalyst first, methyl ethyl ketone peroxide (MEKPO), then stir until evenly distributed, code C1.
4. Pour some of the C1 mixtures into the mold and place the coconut coir fiber on top of the resin layer on the mold and again pour the C1 mixture into the mold as shown in Fig. 1a–c.
5. Next, press with a press machine and turn on the machine heater.
6. Allow to harden and dry. Then, remove the specimen from the mold as shown in Fig. 2a, b.
7. Cover the table tennis racket with a rubber layer in red and black colors.

(a) (b) (c)

Fig. 1 Pouring resin and coconut fiber into a mold on a press machine

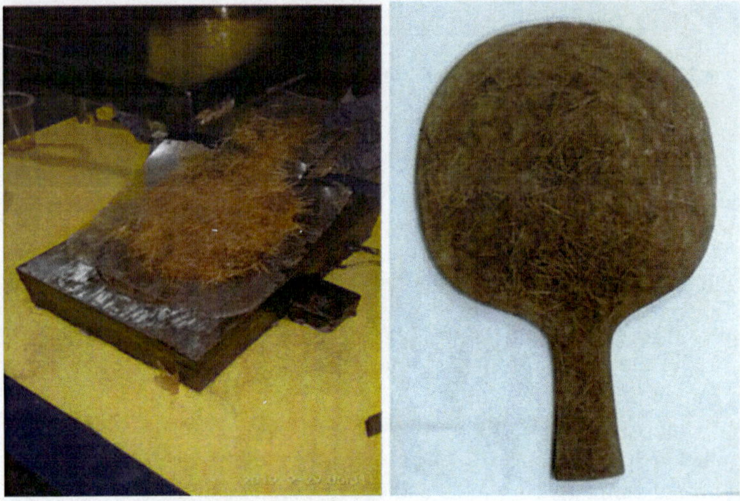

(a) (b)

Fig. 2 Removing a table tennis bat from the mold

3 Results and Discussion

The table tennis racket that will be made is made of composite mold using a mold as shown in Fig. 3.

The process of making table tennis bats is carried out as described above, and the results of the table tennis bat prototype are composite materials with a ratio of coconut fiber and polymer matrix as shown in Fig. 4. The mass of the table tennis bat prototype after the measurement is 203 g.

In the process of making conventional table tennis bats, plywood is cut and formed into table tennis bats. The plywood is made 6 mm thick, consists of 3 layers, and

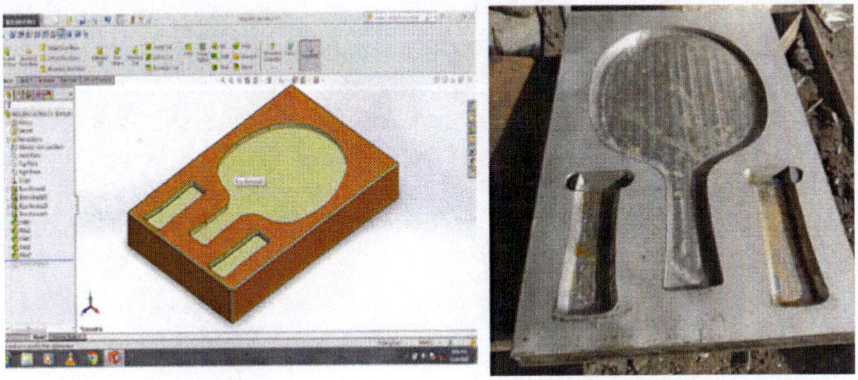

(a) (b)

Fig. 3 **a** Design and **b** molding of a table tennis racket

Fig. 4 Composite table tennis racket

then is again coated with a layer of red and black rubber on the outside. Conventional table tennis bats are shown in Fig. 5b.

Based on the SNI specifications on the mass of the table tennis racket, which is 150–210 g, the prototype table tennis bat is made of polymer composite material reinforced with coconut fiber with a fiber and matrix ratio of 10:90, which is in the SNI standard. Furthermore, if a comparison is made with the mass of table tennis

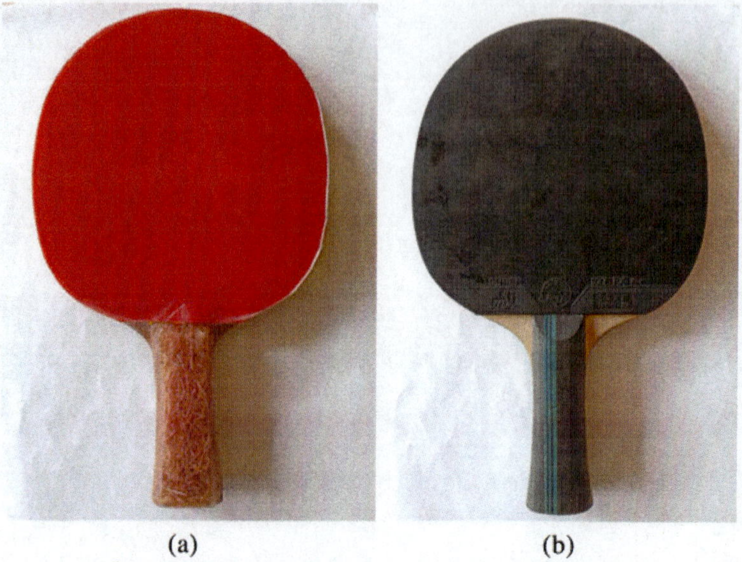

Fig. 5 Composite table tennis racket and wooden table tennis racket

Table 2 Comparison of table tennis racket of composite polymer and wooden

Table tennis racket	Mass (g)
Polymer composite	203.7
Wooden	157.9

bats on the market, the mass of this prototype table tennis bat is still heavier than the table tennis racket on the market, and the value of this comparison can be seen in Table 2.

As for the specifications of table tennis bats that follow the Indonesian National Standard (SNI), this composite table tennis racket product has reached the standard and is suitable for use.

The following table tennis bat specifications are according to SNI (Indonesian National Standard)

Total bat weight: 150–210 g

Overall length: 260–270 mm
Bat leaf width: 150–155 mm
Bat leaf thickness: 6–7 mm
Rubber layer thickness: 1.5–2 mm
Rubber bat color: Red and Black
Stem length: 105–110 mm
Stem thickness: 22–25 mm
Stem width: 28–35 mm

According to SNI 12-0799-1995
According to PTMSI (All Indonesia Table Tennis Association).

4 Conclusion

Based on the results obtained in the study, conclusions can be drawn including that the manufacture of table tennis racket molds is carried out in stages, including drawing model drawings, making molds from aluminum alloy material on the die, and punching sections with a milling machining process. The prototyping of table tennis rackets is carried out using compression molding, with a composition of a 10:90 ratio of fiber and matrix.

Acknowledgements The authors would like to thank the Universitas Muhammadiyah Sumatera Utara for providing funding for this research, through the APB UMSU for the 2020/2021 fiscal year.

References

1. Udhayasankar R, Karthikeyan B, Balaji A (2018) Coconut shell particles reinforced cardanol–formaldehyde resole resin biocomposites: effect of treatment on thermal properties. Int J Polym Anal Charact 23(3):252–259. https://doi.org/10.1080/1023666X.2018.1427187
2. Chen RS, Salleh MN, Ghani MHA, Ahmad S et al (2015) Biocomposites based on rice husk flour and recycled polymer blend: effects of interfacial modification and high fibre loading
3. Antov P, Savov V, Neykov N (2017) Utilization of agricultural waste and wood industry residues in the production of natural fiber-reinforced composite materials. Int J–Wood, Des Technol 6(1):64–71
4. Jorda JS, Barbu MC, Kral P (2019) Natural fibre reinforced veneer based products. Pro Ligno. proligno.ro
5. Debnath S (2017) Sustainable production and application of nonwoven based on natural fibers. Sustain Fibres Text
6. Yani M, Siregar MA, Suroso B, Arnita (2019) Strength of polymeric foam composite reinforced oil palm empty fruit bunch fiber subjected to impact load. IOP Conf Ser Mater Sci Eng 674(1). https://doi.org/10.1088/1757-899X/674/1/012065
7. Fangueiro R, Rana S (2016) Natural fibres: advances in science and technology towards industrial applications. Springer. https://doi.org/10.1007/978-94-017-7515-1
8. Prakash VRA, Viswanthan R (2019) Fabrication and characterization of echinoidea spike particles and kenaf natural fi bre-reinforced Azadirachta-Indica blended epoxy multi-hybrid bio composite. Compos Part A 118(Sept 2018):317–326. https://doi.org/10.1016/j.compositesa.2019.01.008
9. Adekomaya O, Majozi T (2019) Sustainability of surface treatment of natural fibre in composite formation: challenges of environment-friendly option. Int J Adv Manuf Technol. https://doi.org/10.1007/s00170-019-04581-6
10. Michelena AH, Graham-Jones J et al (2016) Eco-friendly flax fibre/epoxy resin/composite system for surfboard production. Nat Fibres. https://doi.org/10.1007/978-94-017-7515-1_20
11. Arsène MA, Bilba K, Onésippe C (2017) Treatments for viable utilization of vegetable fibers in inorganic-based composites

12. Fitri M, Mahzan S, Hidayat I, Nurato N (2021) Coir fiber powder content and hardener weight fractions on mechanical properties of an Epr-174 epoxy resin composite. In: Sinergi
13. Chen J, Lv Y, Li ZX, Chouw N (2020) Investigation of the properties of natural fibre reinforced polymer-concrete composite. In: ACMSM25. https://doi.org/10.1007/978-981-13-7603-0_43
14. Arrohman S, Mustofa ASH, Ariawan D, Diharjo K (2020) Characteristics of mechanical properties of coir-fibre/rubber composite. J Phys Conf Ser 1511(1). https://doi.org/10.1088/1742-6596/1511/1/012065
15. Lubis RW, Yani M, Amirsyah C, Siregar P. Development of cigarette butt fibre filter reinforced by opefb fiber composite material for trash can. https://doi.org/10.1088/1742-6596/2193/1/012021
16. Yani M (2017) Desain helmet sepeda half face dengan cad. 4:62–68
17. Yani M, Suroso B. Manufacturing of skate board from oil palm empty fruit bunch fiber composite. 1:2–8
18. Yani M, Lubis R, Arfis A, Putra B, Hardiansyah F. Design and manufacturing processes of half face motorcycle palm fiber reinforced composite polymer, no. Ii. https://doi.org/10.1088/1742-6596/2193/1/012011
19. Surata IW, Nindhia TGT, Yolanda WE (2020) Grain size effect on tensile and flexural strength of particulate composites reinforced with Acropora waste. Mater. Today Proc. 22(Dec):156–161. https://doi.org/10.1016/j.matpr.2019.08.031

Porous Structures Simulation Analysis: The Effect of Different Strut Geometry on the Bone Scaffold

Rochmad Winarso, Rifky Ismail, Paulus Wisnu Anggoro, Jamari Jamari, and Athanasius Priharyoto Bayuseno

Abstract Porous structures have shown promise in biomedical applications, particularly for artificial implants, because they provide a large surface area for bone ingrowth. A simulation analysis model was developed in this study to evaluate the mechanical properties of various designs based on cubic unit cell scaffolds with varying strut geometries at the same porosity. Based on the diameter and cross-section type of the struts, porous bone scaffolds with the same porosity were developed. The porous structure bone scaffold with square cross-section struts is larger based on the pore size measurements. The porous structure bone scaffold with square cross-section struts has the highest compressive strength and modulus elasticity, according to this study. All bone scaffolds with porous structures, however, may meet the trabecular bone criteria (pore size, porosity, interconnectivity, and biocompatibility), but not the cortical bone criteria. Biomaterials with higher mechanical properties, or polymers combined with ceramics, will need to be researched in the future to match cortical bone's mechanical strength.

Keywords Bone scaffold · Simulation analysis · Strut geometry

R. Winarso (✉)
Department of Mechanical Engineering, Faculty of Engineering, Universitas Muria Kudus, Kudus, Central Java, Indonesia
e-mail: rochmad.winarso@umk.ac.id

R. Winarso · R. Ismail · J. Jamari · A. P. Bayuseno
Department of Mechanical Engineering, Faculty of Engineering, Diponegoro University, Semarang, Central Java, Indonesia
e-mail: apbayuseno@lecturer.undip.ac.id

P. W. Anggoro
Department of Industrial Engineering, Faculty of Industrial Technology, University of Atma Jaya Yogyakarta, Yogyakarta 55281, Indonesia

Irwansyah et al. (eds.), *Proceedings of the 4th International Conference on Experimental and Computational Mechanics in Engineering*, Lecture Notes in Mechanical Engineering, https://doi.org/10.1007/978-981-99-7495-5_11

1 Introduction

Porous structures are used in various industries today, including aerospace, automobile manufacturing, and biomedical applications. The use of additive manufacturing methods to create porous structures has gained popularity recently [1]. Porous structures have shown promise in biomedical applications, particularly for artificial implants because they provide a large surface area for bone ingrowth. To achieve a good rate of bone ingrowth and mechanical properties, optimal morphological factors such as pore size and porosity must also be chosen [2]. The function and overall performance of a porous structure are determined by the careful selection of various morphological factors, such as average pore size and porosity, which determine the rate of bone ingrowth and interface strength [3]. The unit cell porous structure can be customized to provide appropriate mechanical properties for the biomaterial to handle physiological loadings, with regulated porosity and pore size for an optimal architectural environment for bone ingrowth [4]. Porosity should be greater than 50%, and pore size should be 100–700 mm to minimize pore occlusion and provide adequate surface area for cell attachment and load-bearing capability [5].

Polylactic acid (PLA), a biodegradable polymer that degrades to non-toxic compounds, can be used in a variety of applications such as biomedical devices, sutures, bone fixing, and food packaging [6]. Because of its unique biological and mechanical properties, PLA, a highly flexible synthetic polymer, has been used in bone scaffolds [7]. PLA is also FDA-approved for use in biomedical applications [8], which has resulted in extensive research into PLA for tissue engineering applications [9].

Recent bone scaffold research has primarily focused on three-dimensional, non-stochastic, open-cell structures. The cubic structure has received a lot of attention because of its basic geometrical design and promising results [10]. Because of the regular strut dimensions on each vertex of the cube shape, the stress distribution on each strut is constant. Previous research discovered that different scaffold geometries had varying porosity percentages. These modifications may have a significant impact on the mechanical performance of scaffold topologies, making precise comparisons difficult [11]. In other studies, finite element analysis (FEA) was used to simulate the mechanical performance of scaffolds. Using finite element analysis, Sohrabian et al. [11] investigated the effects of different lay-down patterns on the mechanical properties of bone scaffolds. Olivares et al. [12] created a computational method for relating mechanical properties to cell differentiation in relation to scaffold morphology. Shi et al. [13] used mechanical compression simulation to evaluate the mechanical properties and permeability of the scaffold. Wieding et al. [14] used numerical analysis to optimize the geometrical parameters of open-porous titanium scaffolds to match the elastic properties of human cortical bone in terms of pore size. In this work, a simulation analysis model was developed to investigate the mechanical properties of several designs based on cubic unit cell scaffolds with varying strut geometry at the same porosity.

Table 1 Properties od PLA materials [15]

Properties	Value
Solid density (g/cm^3)	1.252
Elastic modulus (MPa)	3500
Yield strength (MPa)	70
Poisson ratio	0.36

2 Materials and Methods

2.1 Materials

PLA was used as the bone scaffold material in the simulation, and its material parameters are listed in Table 1.

3 Methods

The design of the microarchitecture scaffold is the first stage of the methodology. The cube is the most fundamental porous structure unit cell shape, with all struts at 90^0 to one another, making it easier to construct. The influence of strut geometry on the mechanical performance of porous structures like bone scaffolds was investigated using three distinct strut geometries: circular, square, and hexagonal shapes. The cross-sectional type of struts in porous structures bone scaffolds is depicted in Fig. 1. To achieve an excellent clinical result, the selection criteria in this study include appropriate porosity, pore size, and interconnectivity. Increased porosity in a structure promotes bone tissue formation while decreasing structural and mechanical strength. Porosity must be greater than 50% to allow for bone ingrowth [16].

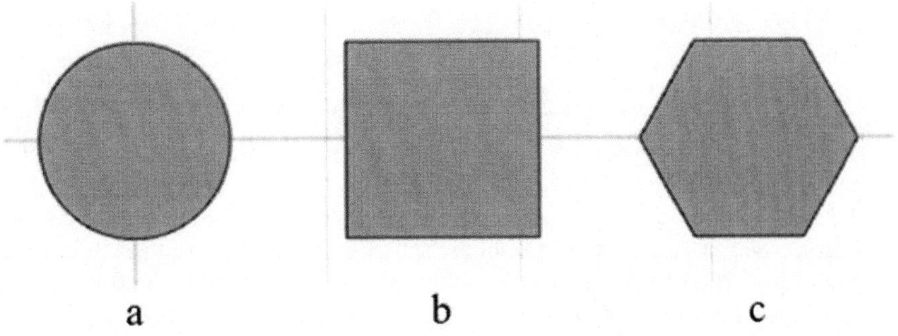

Fig. 1 **a** Circular, **b** square, and **c** hexagonal struts geometry

Fig. 2 Bone scaffolds sample based on **a** circular, **b** square, **c** hexagonal struts geometry

All samples in this study were designed with a porosity of 60%. To assess the design's suitability for implantation, the porosity of porous structures in bone scaffolds was determined. The porosity (P) of porous structures in a bone scaffold sample was calculated using Eq. 1 [13].

$$P = \frac{Ve}{V}.100\% \tag{1}$$

The pore volume of the porous structure is denoted by Ve, and the volume of the solid structure is denoted by V. The pore volume of porous and solid structures was extracted from CAD software and used to calculate porosity.

Following the selection of modeling parameters, creo-parametric software was used to design three distinct porous structures of bone scaffolds with varying strut geometry. All designs' unit cells were created using the ISO standard (ISO 13314:2011), with a sample size of 6 × 6 × 6 mm. Figure 2 shows a CAD model of three different bone scaffolds.

Pore size measurement is the second stage of the methodology. The size of the pores in a porous structure influences bone tissue regeneration significantly. A small hole can inhibit bone ingrowth by increasing pore occlusion, whereas a large hole reduces the surface area for cell adhesion and weight-bearing capability [16]. According to the literature, the optimal pore size is 100–700 mm [16]. A unit cell's pore size is defined in planar view as the diameter of the circle inscribed between the unit cell struts. Pore size was measured in this study using the ImageJ software.

The methodology concludes with simulation analysis. The simulation analysis software is used to convert all CAD model samples. Because of its mechanical and biological properties, PLA was chosen as the material for simulated sample examination. Autodesk Fusion 360 software is used in the simulated analysis technique. The mechanical properties of the materials used in the simulation are given in Table 1. In this study, the simulation method was carried out using the Autodesk Fusion 360 software, and the boundary conditions are shown in Fig. 3. The scaffold's bottom

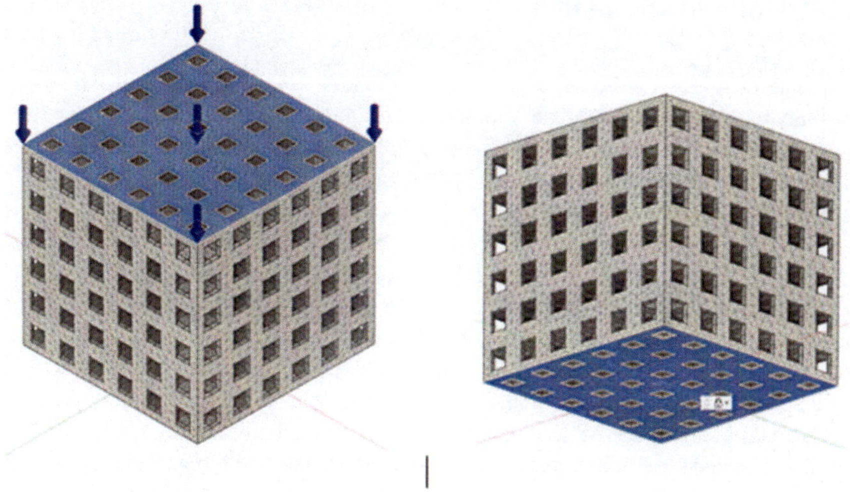

Fig. 3 Boundary conditions of porous structure bone scaffold

half was fixed for boundary conditions, while the top section was compressed twice at 25 and 250 N. Due to the geometrical complexity of the joints, the components must be precise enough to accurately describe the effects of stress concentration on the joints. However, due to the large number of struts in the sample, simulation of the entire sample took a significant amount of time. A strategy has been developed for scaling and minimizing the number of struts as well as the amount of computation. The adaptive mesh refinement method was used to determine the fewest nodes and elements needed to produce accurate results. For accurately capturing solution, adaptive mesh refinement (AMR) methods are well known and widely used for accurately capturing solution characteristics in a steady or unsteady simulation [17].

In this study, deformation was considered only in the direction of force applied. Directional deformation data were collected, and strain was calculated from deformation using Eq. 2. The stress was calculated from Eq. 3 [18]. The compressive strength of a material is defined as the stress in the material when it achieves a plastic compressive strain of 2%. The effective modulus of the scaffold can be calculated by Eq. 4 [19].

$$\varepsilon = \frac{F.l}{A.E} \tag{2}$$

$$\sigma = \frac{F}{A} \tag{3}$$

$$E = \frac{\sigma}{\varepsilon} \tag{4}$$

4 Result and Discussion

The characteristics of the sample were numerically analyzed and validated using an analytical method. Based on simulation and analytical data, Table 2 compares the mechanical properties of solid PLA samples. According to these findings, the difference between numerical and analytical values is insignificant.

The porosity of porous structure scaffolds can have a significant impact on their mechanical properties. The size of the strut cross-section has a significant impact on the porosity of the porous structure. Table 3 shows the cross-sectional size of struts that produce 60% porosity for each sample. In this study, we created a porous bone scaffold with a pore size of 500 μm because pore sizes greater than 300 μm are recommended to improve bone regeneration. The pore size of the porous structure bone scaffold with circular, square, and hexagonal cross-section struts measured using ImageJ software is 483, 556, and 551 μm, respectively.

This discovery demonstrated that the porous structure of bone scaffolds with square cross-section struts had a larger pore size. Table 4 shows the compression test simulation results, which show differences in strain caused by the same force. The results of two simulation runs can be used to derive the linear regression equation for each porous structure.

According to the linear regression equation, the maximum force for circular, square, and hexagonal shapes is 468.7, 492.62, and 454.06 N, respectively. The calculation results are based on the linear regression equation, which is then validated using a simulation, the results of which are shown in Fig. 4. The modeling results show that the stresses that occur in all porous structure samples as a result of each force are the same, which is 0.12 mm (2%). This demonstrates that the linear regression equation computation results are consistent with the simulation results.

Further, a bar chart in Fig. 5 compares the compressive strength and elastic modulus of struts with circular, square, and hexagonal cross-sections. Figure 5 clearly shows that square cross-section struts have higher compressive strength. The compressive strengths of struts with circular, square, and hexagonal cross-sections are 12.99, 13.68, and 12.61 MPa, respectively. Figure 5 shows, similarly to the compressive strength data, that the porous structure bone scaffold with square cross-section struts has the highest modulus elasticity. Struts with circular, square, and hexagonal cross-sections have elastic moduli of 649.72, 684.19, and 630.64 MPa, respectively.

These findings are consistent with those of Khodaei et al. [20], who determined the elastic coefficients of biodegradable polylactic acid polymers with 29, 49, and 69% porosity to be 502.7, 537.7, and 483.3 MPa, respectively. Furthermore, the elastic modulus of porous structure bone scaffolds with circular, square, and hexagonal cross-section struts was found to be 81, 80, and 82% lower than the solid PLA elastic modulus value of 3500 MPa. The reduction in modulus of elasticity is consistent with the findings of Cho et al. [21], who discovered that a dual-pore scaffold reduced modulus of elasticity by 84%. Sohbarian et al. show similar results, in which the elastic modulus of porous structures with lay-down patterns of 0/90 decreases by 86%. However, Tang et al. [22] found that the elastic modulus of circular and square

Table 2 Validation of analytical and simulation methods

No.	Force (N)	Analytical methods			Simulation methods			Deformation gap	Compressive stress gap
		Deformation (mm)	Strain (%)	Compressive stress (MPa)	Deformation (mm)	Strain (%)	Compressive stress (MPa)		
1	1000	0.047619	0.794	27.78	0.04746	0.791%	28.63	0.000159	0.852222
2	1750	0.083333	1.389	48.61	0.08336	1.389%	50.11	−0.000027	1.498889
3	2500	0.119048	1.984	69.44	0.1191	1.985%	71.59	−0.000052	2.145556
Average								0.000027	1.498889

Table 3 Cross-section size of struts and porosity resulted

Struts geometry	Unit cell (mm)	Cross-section size (mm)	Total volume (mm^3)	Pore volume (mm^3)	Porosity (%)
Circular	6 × 6 × 6	0.49	216	129.7131	60
Square	6 × 6 × 6	0.432	216	129.8962	60
Hexagonal	6 × 6 × 6	0.539	216	129.7348	60

Table 4 Force–deformation relationship of every geometry strut

Struts geometry	F1 (N)	Deformation (mm)	Strain (%)	F2 (N)	Deformation (mm)	Strain (%)	Linear regression equation
Circular	25	0.006413	0.107	250	0.06413	1.069	y = 23,390.x
Square	25	0.00609	0.102	250	0.0609	1.015	y = 24,631.x
Hexagonal	25	0.006607	0.110	250	0.06607	1.101	y = 22,703.x

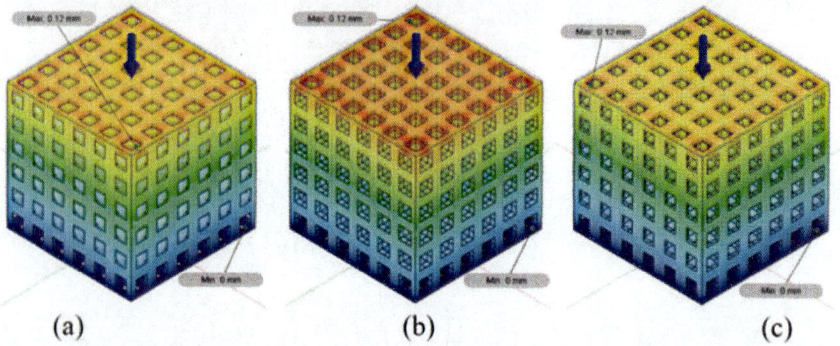

Fig. 4 Deformation caused by the applied force in the sample, **a** circular, **b** square, and **c** hexagonal

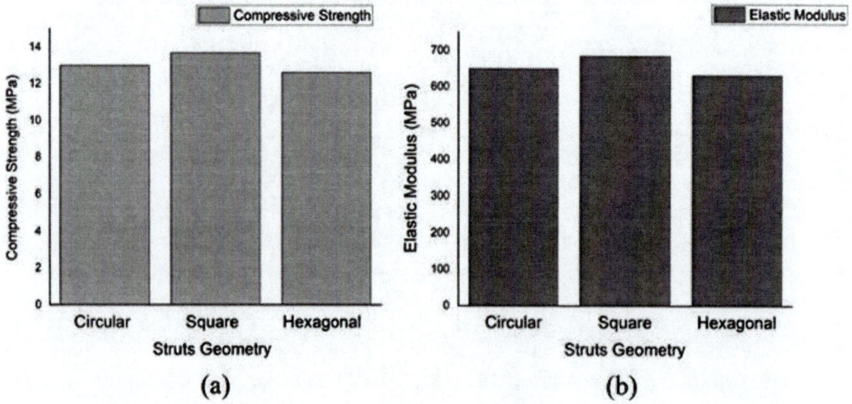

Fig. 5 Compressive strength **(a)** and elastic modulus **(b)** of porous structure bone scaffolds

porous structures decreased by 81 and 68%, respectively, when compared to solid PLA.

5 Conclusion

Implant mechanical properties must be compatible with bone mechanical properties. Porous bone scaffolds with the same porosity were created based on the diameter and cross-section type of the struts. The porous structure of bone scaffolds with square cross-section struts is larger, based on pore size measurements. According to this study, the porous structure of bone scaffolds with square cross-section struts has the highest compressive strength and modulus elasticity. However, while all of the porous structure bone scaffolds designed may meet the trabecular bone criteria (pore size, porosity, interconnectivity, and biocompatibility), they do not meet the cortical bone criteria. Cortical bone has a compressive strength of 20–190 MPa and an elastic modulus of 4–30 GPa, whereas trabecular bone has a compressive strength of 2–80 MPa [23]. In the future, biomaterials with higher mechanical properties, such as polyether ether ketone (PEEK) or compound polymers with ceramics like hydroxyapatite, will be needed to match the mechanical properties of cortical bone.

Acknowledgements The authors are grateful to the Deputy for Strengthening Research and Development, the Ministry of Research and Technology and Higher Education, Republic of Indonesia, for research funding under Doctoral Research Grants Number: 345-35/UN7.6.1/PP/2022.

References

1. Kim TB, Yue S, Zhang Z, Jones E, Jones JR, Lee PD (2014) Additive manufactured porous titanium structures: through-process quantification of pore and strut networks. J Mater Process Technol 214(11):2706–2715. https://doi.org/10.1016/j.jmatprotec.2014.05.006
2. Arabnejad S, Burnett Johnston R, Pura JA, Singh B, Tanzer M, Pasini D (2016) High-strength porous biomaterials for bone replacement: a strategy to assess the interplay between cell morphology, mechanical properties, bone ingrowth and manufacturing constraints. Acta Biomater 30:345–356. https://doi.org/10.1016/j.actbio.2015.10.048.
3. Kienapfel H, Sprey C, Wilke A, Griss P (1999) Implant fixation by bone ingrowth. J Arthroplasty 14(3):355–368. https://doi.org/10.1016/S0883-5403(99)90063-3
4. Van Bael S et al (2012) The effect of pore geometry on the in vitro biological behavior of human periosteum-derived cells seeded on selective laser-melted Ti6Al4V bone scaffolds. Acta Biomater 8(7):2824–2834. https://doi.org/10.1016/j.actbio.2012.04.001
5. Mullen L, Stamp RC, Brooks WK, Jones E, Sutcliffe CJ (2009) Selective laser melting: a regular unit cell approach for the manufacture of porous, titanium, bone in-growth constructs, suitable for orthopedic applications. J Biomed Mater Res Part B Appl Biomater 89(2):325–334. https://doi.org/10.1002/jbm.b.31219
6. Lasprilla AJR, Martinez GAR, Lunelli BH, Jardini AL, Filho RM (2012) Poly-lactic acid synthesis for application in biomedical devices—a review. Biotechnol Adv 30(1):321–328. https://doi.org/10.1016/j.biotechadv.2011.06.019

7. Tyler B, Gullotti D, Mangraviti A, Utsuki T, Brem H (2016) Polylactic acid (PLA) controlled delivery carriers for biomedical applications. Adv Drug Deliv Rev 107:163–175. https://doi.org/10.1016/j.addr.2016.06.018
8. Gupta B, Revagade N, Hilborn J (2007) Poly(lactic acid) fiber: an overview. Prog Polym Sci 32(4):455–482. https://doi.org/10.1016/j.progpolymsci.2007.01.005
9. Alizadeh-Osgouei M, Li Y, Vahid A, Ataee A, Wen C (2021) High strength porous PLA gyroid scaffolds manufactured via fused deposition modeling for tissue-engineering applications. Smart Mater Med 2:15–25. https://doi.org/10.1016/j.smaim.2020.10.003
10. Amin Yavari S et al (2015) Relationship between unit cell type and porosity and the fatigue behavior of selective laser melted meta-biomaterials. J Mech Behav Biomed Mater 43:91–100. https://doi.org/10.1016/j.jmbbm.2014.12.015
11. Sohrabian M, Vaseghi M, Khaleghi H, Dehrooyeh S, Kohan MSA (2021) Structural investigation of delicate-geometry fused deposition modeling additive manufacturing scaffolds: experiment and analytics. J Mater Eng Perform 30(9):6529–6541. https://doi.org/10.1007/s11665-021-05894-y
12. Olivares AL, Marsal È, Planell JA, Lacroix D (2009) Finite element study of scaffold architecture design and culture conditions for tissue engineering. Biomaterials 30(30):6142–6149. https://doi.org/10.1016/j.biomaterials.2009.07.041
13. Shi C, Lu N, Qin Y, Liu M, Li H, Li H (2021) Study on mechanical properties and permeability of elliptical porous scaffold based on the SLM manufactured medical Ti6Al4V. PLoS One 16(3):1–17. https://doi.org/10.1371/journal.pone.0247764
14. Wieding J, Wolf A, Bader R (2014) Numerical optimization of open-porous bone scaffold structures to match the elastic properties of human cortical bone. J Mech Behav Biomed Mater 37:56–68. https://doi.org/10.1016/j.jmbbm.2014.05.002
15. Farah S, Anderson DG, Langer R (2016) Physical and mechanical properties of PLA, and their functions in widespread applications—a comprehensive review. Adv Drug Deliv Rev 107:367–392. https://doi.org/10.1016/j.addr.2016.06.012
16. Zaharin et al (2018) Effect of unit cell type and pore size on porosity and mechanical behavior of additively manufactured Ti6Al4V scaffolds. Materials (Basel) 11(12). https://doi.org/10.3390/ma11122402
17. Srinivasan H, Tsoutsanis P (2016) Adaptive mesh refinement techniques for high-order finite-volume weno schemes. ECCOMAS Congr 2016 Proc 7th Eur Congr Comput Methods Appl Sci Eng 2:2883–2899. https://doi.org/10.7712/100016.2003.8544
18. Vakrčka P, Jíra A, Hájková P (2020) Mechanical testing and numerical modelling of porous structures improving oseintegration of implants. Acta Polytech CTU Proc 26:127–133. https://doi.org/10.14311/APP.2020.26.0126
19. Bagde AD et al (2019) Geometric modeling and finite element simulation for architecture design of 3d printed bio-ceramic scaffold used in bone tissue engineering. J Indian Inst Sci 99(3):361–374. https://doi.org/10.1007/s41745-019-00120-0
20. Khodaei M, Amini K, Valanezhad A (2020) Fabrication and characterization of poly lactic acid scaffolds by fused deposition modeling for bone tissue engineering. J Wuhan Univ Technol Mater Sci Ed 35(1):248–251. https://doi.org/10.1007/s11595-020-2250-4
21. Cho YS, Gwak SJ, Cho YS (2021) Fabrication of polycaprolactone/nano hydroxyapatite (Pcl/nha) 3d scaffold with enhanced in vitro cell response via design for additive manufacturing (dfam). Polymers (Basel) 13(9). https://doi.org/10.3390/polym13091394
22. Tang MS, Abdul Kadir AZ, Ngadiman NHA (2020) Simulation analysis of different bone scaffold porous structures for fused deposition modelling fabrication process. In: 5th International Conference on Mechanical Engineering Research 2019, ICMER 2019, vol 788. https://doi.org/10.1088/1757-899X/788/1/012023
23. Li Y, Yang C, Zhao H, Qu S, Li X, Li Y (2014) New developments of ti-based alloys for biomedical applications. Materials 7(3):1709–1800. https://doi.org/10.3390/ma7031709

Performance of Car Radiators as Alternative Electric Energy Generators

Partaonan Harahap, Benny Oktrialdi, Rimbawati, M. Yani, and Riadini Wanty Lubis

Abstract Electrical energy is one form of energy that cannot be separated in life in the modern era. Over time, energy needs continue to increase, while fossil energy reserves are decreasing and even running out. This study uses several stages, namely meeting the needs of alternative energy, one of which is by modifying a car radiator or dynamo amperage on a car that functions as a generator of electrical energy when the electrical energy is still alive to supply electrical energy to the car, especially to charge electric batteries (car batteries). This study aims to determine the output in the form of speed, voltage, current, and power generated, in order to determine efficiency and good use for homes that are far from PLN supply and to save PLN costs. The results obtained at rotational speed and load will affect the low speed output because the higher the rotational speed, the higher the voltage and frequency. The larger the load, the lower the voltage and frequency. Under no-load conditions, the tested alternator can produce an average voltage of 664 V, the average current is 30.2 A, and the average speed is 4632 rpm. While the tested alternator is able to produce an average voltage of 484 V, the average current is 100 A and the average speed is 3489 rpm.

Keywords Car radiator · Electrical energy · Genset · Alternator · Load testing

---

P. Harahap (✉) · B. Oktrialdi
Department of Electrical Engineering, Faculty of Engineering, Universitas Muhammadiyah Sumatera Utara, Jalan Mukhtar Basri No. 3, Medan, Indonesia
e-mail: partaonanharahap@umsu.ac.id

Rimbawati · M. Yani · R. W. Lubis
Department of Mechanical Engineering, Faculty of Engineering, Universitas Muhammadiyah Sumatera Utara, Jalan Mukhtar Basri No. 3, Medan, Indonesia

Irwansyah et al. (eds.), *Proceedings of the 4th International Conference on Experimental and Computational Mechanics in Engineering*, Lecture Notes in Mechanical Engineering, https://doi.org/10.1007/978-981-99-7495-5_12

1 Introduction

Generator is a tool or machine that produces electrical energy from mechanical energy obtained from the initial mover in the form of diesel as the main mover which then drives an AC generator or alternator, causing an electric voltage. The generator also functions as a power generator that works as mechanical energy that works into electrical energy which functions as a substitute for PLN electricity, namely for everything needed for electricity. Within the scope of the science of converting electrical energy, there are DC-AC converters known as inverters. Inverter is a static device that does not use moving components in the process of converting electrical voltage. To obtain high efficiency from the inverter, a sinusoidal output voltage and minimum total harmonic distortion are required. The high efficiency of an inverter will affect the inverter output voltage [1].

The ideal inverter output connection is a voltage that has a pure sine wave. There are various inverter tuning methods that are carried out to improve the performance of the inverter system. From the literature search, the number of PWM settings is one method that has proven to be good. PWM settings are implemented in the form of a pulse signal generator. Various analyzes of PWM pulse generation techniques with different algorithms have been carried out. One method that is still new and is being studied by several researchers is the harmonic elimination technique. Harmonic elimination technique or also known as harmonic elimination of PWM is a technique to reduce the harmonic level at the inverter output voltage. Basically, this technique is a combination of efficiency and inverter quality to achieve a switching pattern with very low total harmonic distortion [2].

Based on these tests, it is necessary to modify the radiator or car alternator into an alternative power plant where the electricity generated from the electrical energy conversion process commonly used is an AC generator engine. Therefore, to be able to realize, all of this requires large funds, as it is known that generators are very expensive goods [3]. From the description above, it encourages the author to use car radiators as a substitute for generators in power plants. The radiator or car amperage dynamo in the car functions as a generator of electrical energy when the engine will supply electrical energy to the car, especially to charge the battery (car battery). Therefore, the authors conducted research by trying to use used car alternators as a producer of electrical energy by relying on natural energy such as air and wind, as well as simple tools such as motors. The car alternator will be driven by this energy to be stored in the car battery. This study aims to determine the current and voltage as well as the output efficiency of a car radiator motor as an alternative electrical energy generator.

2 Literary Studies

In this study, there were several researchers before researchers who had tested electric motors, alternators, and others. According to several previous researchers' statement:

An alternative wind power plant is by utilizing a car alternator, so that it can help provide a renewable source of electricity. The test starts from the design, measuring the voltage and current on the alternator and then measuring the normal temperature on the regulator. The results obtained are known that the wind turbine as the initial mover of the alternator can produce wind power. The resulting output voltage and current is highly dependent on the wind speed generated but is very helpful in providing electricity and reducing household burdens, so that it can be used as an alternative to renewable energy.

In a vehicle (car), the source of power in the form of electric current is obtained from the battery where the battery capacity is very limited, if the car only relies on the battery only as a source of electricity, as a result in just a few hours, the battery current will run out. The amount of power generated by the alternator must be sufficient to meet the needs electricity in the vehicle. By doing an experiment calculation between the load on the vehicle's electrical needs and the power generated by the alternator with the formula $P = V \times I$, and Psis = P2-P1. At 800 rpm the output current is 12.5 amperes and voltage is 13.4 so that the power of 167.5 W plus 720 W of battery power becomes 893.75 W, while the electricity demand at full load is 367.05 W, so that electricity needs can still be met for some time according to electricity needs [4].

This study discusses the rapid development of the industrial world, especially in the automotive sector. Various kinds of findings and new breakthroughs have been widely applied, including the application of the automotive field to other fields. An example of an application that is currently widely used is the use of electronic components in car battery charging systems. These components work to control the charging voltage, so that the magnitude of the charging voltage must always be within the allowable range. Sliding mode control (SMC) is a very robust control method, so it can work well on nonlinear systems that have model or parameter uncertainties. However, in practical applications, pure SMC often appears chattering which is a high-frequency controller output oscillation. To improve system performance, pure SMC is modified with fuzzy logic control (FLC) which works qualitatively, so it is called fuzzy sliding mode control (FSMC). In this final project, research is conducted on the application of the fuzzy sliding mode control method to the battery-charging voltage control system. The results of the application of the fuzzy sliding mode control (FSMC) method on the battery-charging system conclude that the response time is faster, stronger in the face of all kinds of disturbances, and the design is not as complicated as pure SMC [5, 6].

Modification of the 12VDC car alternator into a three-phase electric generator of 220 VAC 600 rpm, so that it can be used as the main element in a pico-hydro power plant. If the potential for energy, especially pico-hydro in Indonesia is taken seriously, it will bring benefits to people in areas that do not yet have a PLN electricity network.

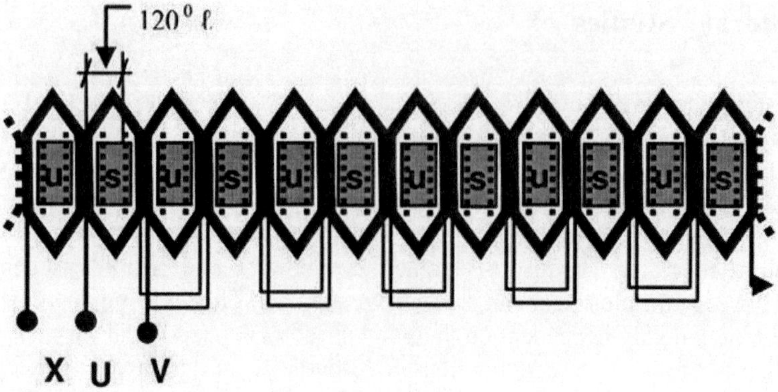

Fig. 1 Structure of the armature coil in one phase from three phases

A car alternator requires a battery, and a minimum rotation speed of 1000 rpm can only produce a voltage of 12 VDC [7]. The mains voltage E on the generator is in principle directly proportional to the amount of flux, the number of wires Z, and turns n. Based on this, it has been modified by winding the armature coil and the field coil. The goal is that with an input rotation of 600 rpm, a voltage of 220 V, 60 Hz can be generated, so that the electrical energy can be directly used for lighting lamps or other electrical devices in general. The test results of a three-phase generator with a rotation speed of around 600 rpm can produce a voltage of 230 V AC at no-load and 165 V AC at a maximum load of 34.30 W, with a frequency of 60 Hz. The generator has also been tested to be able to load three 14-W CFL lamps, each of which is equivalent to a 75-W incandescent lamp [8]. The configuration of the armature winding with field winding and the measurement method for data collection when the alternator is loaded are shown in Fig. 1 [9].

The charging system is a combination of several charging components such as generator alternator, regulator, and battery which functions to generate electricity and store electricity (Fig. 2). The battery is one of the most important components in a car unit or the engine itself because it provides enough power to the electrical parts of the car such as the starter motor, headlights, ignition, and wipers. However, the battery capacity is limited and cannot provide all the power the car needs continuously [10, 11]. Therefore, the battery must always be fully charged in order to provide the electrical power needed by the electrical parts. The battery's ability to provide electricity is limited by the battery capacity in Ampere hour (AH), to keep the battery always charged.

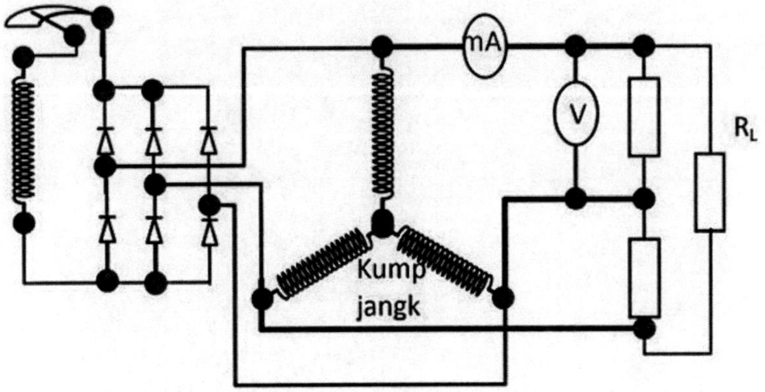

Fig. 2 Configuration of armature coils and field coils [12]

3 Research Methods

This study uses several stages that are observed, namely the need for load, the car radiator motor functions as a power generator that is used to supply electricity needs. The researcher used the Avanza 16,363-BZ010 2003–2011 12 V radiator fan motor [13] with the following specifications (Fig. 3):

Specification:
Working voltage: 12v(6v~30v DC)
No-load speed: 15000 RPM (6000~48000 RPM)
No-load current: 1A (stable!)
Step height: 4.5 mm
Step diameter: 17.5 mm
Motor diameter: 42 mm
Motor body length: 67 mm
Shaft length: 17 mm
Shaft diameter: 5 mm
Rare bolt hole: 29 mm
Bolt hole size: M4
Number of bolt holes2

In Fig. 4, the car radiator test circuit produces a voltage output of 42 V which will lower the voltage which will then be regulated to be able to support the inverter. After getting a voltage that matches the capacity of the inverter, the inverter will convert the DC voltage into AC voltage which we will use in this experiment, namely the lamp.

Fig. 3 Radiator fan motor

4 Results and Discussion

The measurement results are obtained by using current testing and analysis at the car radiator motor output as an alternative electrical energy generator. The visible results are the connection, current, and rpm at the output obtained from the results of the test and Current analysis at the output of the car radiator motor as an alternative electrical energy generator (Fig. 5).

This is in accordance with the electric engine theory which states that the rotational speed of the generator will determine the frequency of the generator. The higher the rotation speed, the higher the frequency. So theoretically when the generator is rotated at a speed of 500 rpm, it will produce a frequency of 50 Hz. The test data show that when the generator is rotated at a speed of about 615 rpm at initial start before loading (Fig. 6).

In the graphic image, the effect of the load on the alternator/generator voltage is low speed. The alternator/generator voltage is inversely proportional to the load

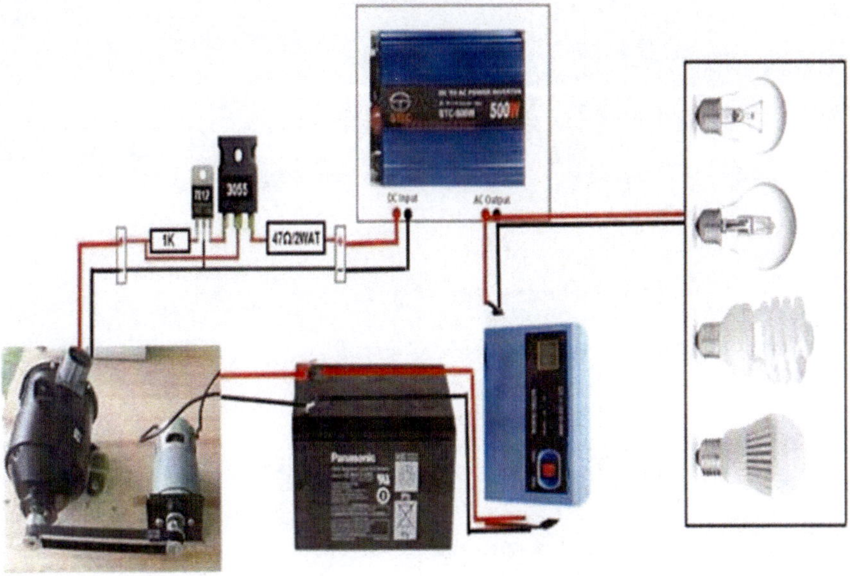

Fig. 4 Test series

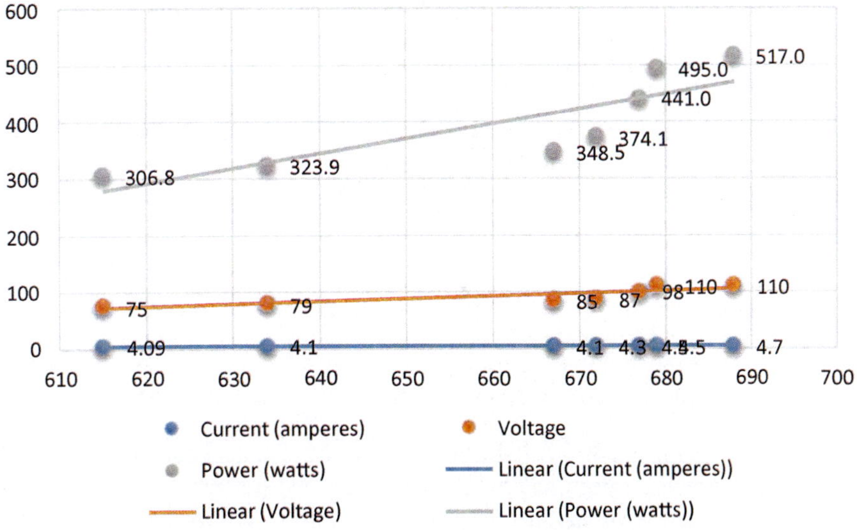

Fig. 5 Graph of measurement results before loading

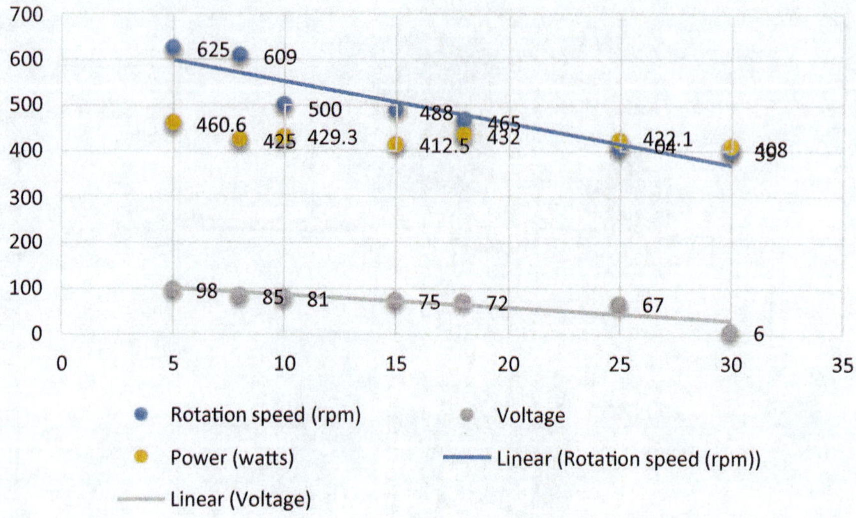

Fig. 6 Graph of measurement results when loading

power. The greater the load power, the smaller the generator voltage. Theoretically, when a load is connected to the generator terminals, current will begin to flow in the stator winding. The presence of resistance and reactance in the stator winding will result in a voltage drop. According to Ohm's law, the magnitude of this voltage drop will be directly proportional to the current flowing. The current will be determined by the load power. The greater the load power, the greater the current that needs to be supplied to the load. Thus, when the load is increased, the voltage drop will increase, so that the voltage at the generator terminals will decrease.

5 Conclusion

Based on the results of research and discussion, the following conclusions can be drawn:

1. Rotational speed and load will affect the low speed alternator output because the higher the rotation speed, the higher the voltage and frequency. The larger the load, the lower the voltage and frequency.
2. Under no-load conditions, the alternator under test can produce an average voltage of 664 V, the average current is 30.2 A, and the average speed is 4632 rpm. Meanwhile, under load conditions, the tested alternator is able to produce an average voltage of 484 V, the average current is 100 A, and the average speed is 3489 rpm.

3. That the voltage on the alternator when it is loaded will also be affected by its rotational speed. The data show that for the same load power, the higher the rotation speed, the higher the voltage.

References

1. Harahap P, Pasaribu FI, Siregar CAP, Oktrialdi B (2021) Performance of grid- connected rooftop PLTS system for households during the Covid-19 pandemic. J Electr Technol UMY 5(1):26–31. https://doi.org/10.18196/jet.v5i1.12089
2. Evalina N, Zulfikar AA (2018) Adjusting the rotational speed of a 3-phase induction motor using a programmable logic controller. J Electr Technol 3(2):73–80
3. Prasetijo H, Dharmawan B (2012) Permanent magnet generators as low rotation power generators. Din Tech 8(2):70–77
4. Profile SEE (2020) Analysis of alternator power on electricity consumption loads on 4K Kijang engines alternator power analysis on electricity consumption loads on 4K Kijang Engines alternator power analysis on electricity consumption loads on 4K Alte Kijang Engines
5. Cookson MD, Stirk PMR (2019) Simulation of alternator voltage control on a battery charging system using fuzzy sliding mode control (Fsmc) method. p 1–10
6. Balisranislam B, Sutantra IN, Sampurno B, Sufyan Hadi H (2019) Numerical study of glass cleaning robot simulation in high-rise buildings. J Electron List Telecommun Comput Inform Sis Control 1(1):25–31. https://doi.org/10.30649/je.v1i1.20
7. Rimbawati R, Harahap P, Putra KU (2019) Analysis of the effect of changes in excitation current on generator characteristics (electrical machine application laboratory, faculty of engineering-Umsu), RELE (electrical and energy engineering). J Tek Electron 2(1):37–44. https://doi.org/10.30596/rele.v2i1.3647
8. Yunus Y, Sihana S, Lukman S (2012) Modification of car alternator into 3 phase generator 220 V 600 RPM, p 1–4
9. Bandung State Polytechnic (2015) DC brushless motors 1(1):1–2
10. Theory D, Alternator P (2022) Connect to the world basic theory of alternator principles Google I/O 2022 | Play Console", p 1–6
11. Anthony Z (2018) Basic electric machines
12. Hess A (2005) CHAPTER 2 theory basic 2.1 Honeypot. p 6–26
13. https://man-es.com/applications/projectguides/4stroke/manualcontent/PG_M-III_L2330DF.pdf
14. Lubis S, Lubis F, Harahap P (2019) PLTB as an alternative to new renewable energy. Semin Not Technology [Online]. Retrieved from https://repository.unimal.ac.id/4991/1/37-Paper-SNTI-2019.pdf

Formula Optimization of Foaming Agent Using Mixture-D-Optimal Method and Application in Peat Fire Suppression

Purwo Subekti, Eliza Hambali, Prayoga Suryadarma, Bambang Hero Saharjo, Aprizal, Heri Suripto, and Isdaryanto Iskandar

Abstract Foaming agents for fire suppression on peatland were formulated from the saponification product of palm oil fatty acids, namely sodium laurate and potassium palmitate. This study aimed to obtain the best formulation of sodium laurate and potassium palmitate as raw material for foaming agent concentrate through formula selection based on physical parameters foaming ability, foam stability, and surface tension. In addition to obtaining the best formula, foam liquid performance test for the foaming agent was also conducted on the peat fire suppression applications. Formulation of sodium laurate and potassium palmitate that produce the best foaming agent concentrate was 20% sodium laurate, 5.58% potassium palmitate, and 74.42% water, which produced a foaming response of 393.58%, foam stability of 69.59%, and surface tension of 31.11 dyne/cm. Foam performance test for the mixture of water and foaming agent showed that peat fire could be extinguished in 3 h 50 min 42 s using 4.56 L/m^2 extinguishing solution. Meanwhile, fire suppression with water requires 5 h 31 min 59 s, with the application of 9.72 L/m^2 solution. Water mixed with a foaming agent concentrate can extinguish peat fires faster than using water alone, besides being faster the application of foaming agents can reduce water consumption.

Keywords Foaming agent · Palm oil · Peat fire · Suppression

P. Subekti (✉) · Aprizal · H. Suripto
Program Study of Mechanical Engineering, University of Pasir Pengaraian, Riau, Indonesia
e-mail: purwos73@gmail.com

E. Hambali · P. Suryadarma
Department of Agroindustrial Technology, IPB University, Bogor, Indonesia

E. Hambali
Surfactant and Bioenergy Research Center, IPB University, Bogor, Indonesia

B. H. Saharjo
Forest Fire Laboratory, Department of Silviculture, IPB University, Boor, Indonesia

I. Iskandar
Faculty of Engineering, Atma Jaya Catholic University of Indonesi, Jakarta, Indonesia

Irwansyah et al. (eds.), *Proceedings of the 4th International Conference on Experimental and Computational Mechanics in Engineering*, Lecture Notes in Mechanical Engineering, https://doi.org/10.1007/978-981-99-7495-5_13

1 Introduction

The foaming agent is a liquid that produces foam when added to water and stirred. One application of foaming agent concentrates is as a fire extinguisher in peatlands. The foam produced by the foaming agent works to cover, extinguish, and cool the surface of the burning peat. The foam also functions as a firebreak in the internal peat fire suppression. Firebreaks occur when foam liquid seeps in through the pores of the peat, so that the cooling effect of the foaming agent becomes a barrier to re-fire [1]. Repeated fires or flames often occur in fire suppression with water sprayed onto the surface of the peat. Fires that occur in the lower layers of the peat are very difficult to extinguish because it is difficult for the water to seep due to the change in the peat nature, which becomes hydrophobic when dry or burned. In peat fires suppression with water, a large amount of water is needed because the water must be able to flood or overburden the burned peat (about 800 L/m^3) [2, 3].

This research aimed to formulate selected raw materials from sodium lauric and potassium palmitate which were carried out to get some of the best foaming agents concentrates and selected based on physical parameters, namely foaming ability, foam stability, and surface tension. Good foam-forming ability is an important parameter in the production of foaming agent. The desired foaming agents are those that can form and produce large amounts of foam. Foam stability parameter is needed to ensure that the foam produced can withstand working conditions and can extinguish peat fires. Meanwhile, the surface tension parameter is needed to determine the performance of the foam liquid to seep into the lower layer through peat pores. The foam extinguishes the fire at the surface and bottom of the peat, and as a firebreak to prevent reigniting [4, 5].

This study aimed to obtain the best foaming agent concentrate formula from the formulation of potassium palmitate and sodium laurate, and analyze the performance of the foaming agent concentrate on peat fire suppression applications at a laboratory-scale.

2 Experimental Method

The raw material for the preparation of foaming agents for peat fire suppression applications selected from the saponification of palm oil fatty acids. The material consisted of sodium lauric and potassium palmitate. Sodium laurate has a high foam-forming ability but low foam stability, whereas potassium palmitate has a low foam-forming ability but high foam stability [6–8]. The combination of the two materials is expected to produce foaming agents that have better foam-forming properties, so that the resulting foam has a good performance in extinguishing peat fires.

The research began by determining the concentration of sodium laurate, potassium palmitate, and water as diluents, with each concentration chosen was 10–20% (w/v) for sodium laurate, 3–6% (w/v) for potassium palmitate, and 74–87% (w/v) for water. The concentration was chosen based on trial and error and referred to the study by [7, 9–12]. To determine the foaming agent formula, random concentrations of sodium lauric and potassium palmitate were randomized with the help of the Design Expert®7.1.6-Trial Mixture-D-Optimal method. Mixture-D-Optimal was chosen because the components of the raw material used had different concentration intervals [13]. Randomization resulted in several concentrations of sodium laurate and potassium palmitate, which will be formulated with water as a diluent, with test parameters included, the foam-forming ability, foam stability, and surface tension [7, 14]. After the response data have been obtained, it was continued with the determination of the best formula recommended, verification of the formula for the selected program recommendations was performed in a laboratory through the sodium laurate and potassium palmitate formulation process based on the recommended concentration. Verification was performed to check the difference between the predicted results recommended by the program with the results of the analysis on the actual conditions [15].

The next step was the performance test of foam from the mixture of water and the foaming agent of the research results (FAR) for peat fire suppression at the laboratory-scale. As a comparison, the peat fire suppression was performed using water without the addition of the foaming agent. The performance test procedure for the application of foam from water added with foaming agent and water without foaming agent, in laboratory-scale peat firefighting, is presented in Fig. 1. This step aimed to analyze the fire suppression time based on changes in temperature recorded on the computer and the amount of water added to the foaming agent concentrate and water without the addition of the foaming agent [1, 6, 8, 16]. The peat used was the fibric type with a moisture content of 12–15%, peat burning in the reactor started from the top surface (0.01 m^2) using heat from an electric current source of 80–100 W as a fire trigger [17, 18]. Combustion for each reactor was performed twice with a concentration of foaming agent solution of 3%. Concentration determination was based on trial and error and [19]. The temperature of peat in reactor 1 (water as an extinguisher) was 375–464.75 °C. The temperature is the beginning of fire suppression. In reactor 2 (FAR as an extinguisher), the temperature when fire suppression start was 389–499.75 °C. The temperature that occurs in the experiment for each peat burning was not similar, although the type of peat and the burning duration was the same. The temperature was categorized as a medium because peat from several regions can reach 700 °C [6, 17, 18].

3 Results and Discussion

In general, all formulas produced foam with good foam-forming ability, which was 234 to 418% and good foam stability, which was 52 to 86%. Furthermore, the resulting foaming agent concentrate could reduce water surface tension by 23 to 34%. Foaming ability is an important factor that must be considered in the application of foaming agent concentrates in peat fire suppression. In addition to accelerating the

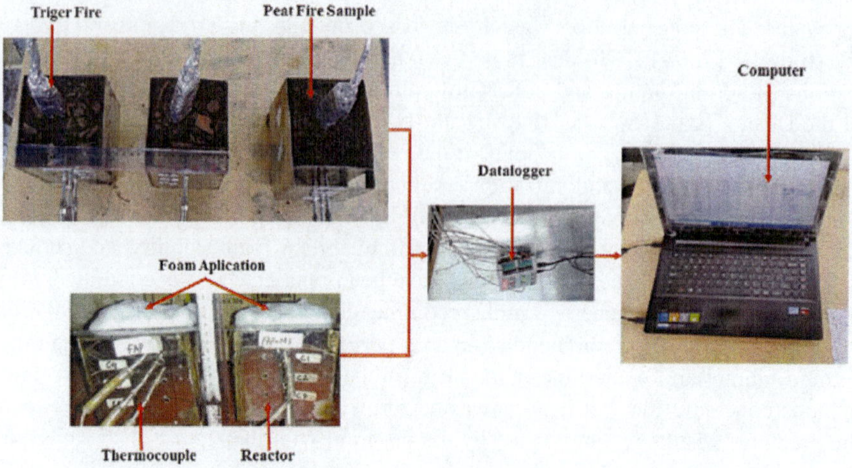

Fig. 1 Performance test procedure for the application of foam from water added with foaming agent and water without foaming agent in peat fire suppression [6, 18]

process of foaming installation and penetration to the hotspot, foam from the liquid foaming agent moves easily to cover the burned surface area and produces a lot of foam. The foam-forming ability is one of the physical properties of a foaming agent concentrate. Extinguishing time will be shorter if the concentrate is easy to produce foam when added to water [1, 17]. Then, another factor to consider is the stability of the foam. The stable foam will increase the foam's ability to inhibit oxygen supply from the surface, so that with the loss of oxygen from the firing process, the fire will be extinguished [19].

In addition to the two responses, the foaming agent concentrate could reduce the surface tension of the water by 59%. The addition of the foaming agent concentrate to the water will increase the absorption capacity of the foam liquid into the peat pores and cool and extinguish the peat litter both on the surface and in the lower layer. High surface tension will reduce water ability to seep into the burning peat litter. It is due to the change of peat's nature from hydrophilic to hydrophobic when dried or burned [2].

Analysis of Foam-Forming Ability. The foam-forming ability (Table 1) is influenced by the composition of sodium laurate and potassium palmitate and water as diluent. Sodium laurate can form a good foam, as explained in the previous chapter, while potassium palmitate has the opposite characteristics, but the resulting foam has good stability. The foaming ability of the formulated foaming agent was 234.57–418.33%. Analysis of variance using ANOVA on the foaming ability (Y_1) resulted in 0.049 of *p*-values for the three variables, where the value was smaller than α (0.05). Figure *p*-value of less than 5%, meaning that the response has a significant effect. From these values, it can be concluded that the raw materials variable of sodium laurate, potassium palmitate, and water have a significant effect on the foaming ability response [17].

Table 1 Foaming agent formulation results from palm oil fatty acids

Formula	$X_1 = A$: sodium laurate (% w/v)	$X_2 = B$: potassium palmitate (% w/v)	$X_3 = C$: water (% w/v)	Y_1: foam-forming ability (%)	Y_2: foam stability (%)	Y_3: Surface tension (dyne/cm)
1	20	6	74	319.32	71.71	28.47
2	20	4.5	75.5	386.87	60.26	25.92
3	15	3	82	325.96	54.97	30.27
4	10	6	84	234.57	86.21	26.69
5	10	3	87	317.35	56.11	28.67
6	15	6	79	246.03	78.77	30.01
7	20	3	77	418.33	51.99	25.67
8	17.5	5.25	77.25	279.12	70.66	31.47
9	20	6	74	314.61	73.21	28.77

The three components of the formulated raw materials had positive coefficient values, meaning that sodium laurate, potassium palmitate, and water had the same role in the forming ability, even with different figures. The higher the positive coefficient value, the greater the role it plays in forming foam [15]. These results confirm the previous research, sodium laurate from palm oil fatty acids had good foaming ability[5]. Furthermore, the application of lauric acid as a foaming agent for firefighting applications in vegetation could improve foaming ability [9].

Analysis of Foam Stability. The ability of the foaming agents to produce stable foam was measured by the foam stability value. The foam stability value of the formulated foaming agent was 51.99 to 86.21%. Significance analysis of the foam stability response (Y_2) produced a *p*-value for the model of $0.0093 < \alpha$ (0.05). This means that the raw materials variable of sodium laurate, potassium palmitate, and water had a significant effect on foam stability. The three components of the formulated raw material had positive coefficient values, meaning that sodium laurate, potassium palmitate, and water have the same role in the ability to produce stable foam. An increase in potassium palmitate concentration significantly influences the stability of the foam from the mix of and foaming agent concentrate, and conversely, an increase in sodium laurate concentration will reduce the foam stability. These results confirm the previous research that foam from potassium palmitate had the best stability compared to other fatty acids [11, 20].

The stability value affect the foam's ability to withstand firefighting working conditions. The higher the stability value, the harder the foam bubble to break, and conversely, the lower the stability value, the easier the foam to break [21]. Therefore, in peat firefighting applications, foam stability is important to consider. The foam that persists on the surface of the burned area will make it harder for the oxygen to interact with the fire source and, thus, ease the break of the fire triangle and speed up the extinguishing process [1, 22].

Analysis of Surface Tension. The ability of the foaming agents to reduce surface tension is a critical factor for its applications as fire extinguishers in peatland. Low surface tension facilitates the liquid to seep into the underlying layer of burning peat and cool it down. The foaming agent is similar to the surfactant that reduces the surface tension of a liquid [23, 24]. The surface tension of the foam liquid from the water added by the formulated foaming agent was 25.67–31.47 dyne/cm. Analysis of variance on the foaming agent's ability to reduce surface tension (Y_3) with p-value for the model was $0.0714 > \alpha$ (0.05), which means that the three raw material variables do not significantly influence surface tension. The ability of potassium palmitate to reduce surface tension can be seen from the high foam stability, or the foam does not break easily, because high surface tension will accelerate the bursting of foam bubbles [17].

The components of formulated raw materials had negative coefficient values, which means that sodium laurate and potassium palmitate had the same role in reducing the surface tension. The ability of potassium palmitate to reduce surface tension can be seen from the high foam stability, or the foam does not break easily, because high surface tension will accelerate the bursting of foam bubbles [17]. The contribution of raw material components in reducing surface tension is one of the advantages of potassium palmitate and sodium laurate from palm oil, so that the application continues to grow, one of which is for fire extinguisher applications [23, 25, 26]. The ability of the foaming agent concentrates to reduce the surface tension of water is one of the important characteristics in the application of foam for fire extinguisher on peatlands. It is because, in addition to the foam's function as an oxygen isolator, the dripping foam liquid will seep and enter the pores of the peat to extinguish the bottom of the burning peat [1].

Selection of the Best Formula. The best formula was determined through optimization of the physical analysis results of the foaming agent concentrate that has been discussed previously by using the Design Expert trial 7.1.6-D-Optimal. Based on the program's analysis, the recommended optimum formula was 20.00% of sodium laurate, 4.99% of potassium palmitate, and 75.01% of water. In addition to formula recommendations, the program also provided predictions of the response value of foaming ability by 357.59%, foam stability by 64.44%, and surface tension by 27.49 dyne/cm. From the program, the accuracy was 0.64. The best value is proved by the desirability value that closes to 1, while the less good is close to 0 [20]. After obtaining the best formula, then it was validated by formulating a foaming agent with sodium lauric and potassium palmitate concentrations based on the above recommendations. Validation was done to test the accuracy of the predicted results with the actual results. Verification results showed that the foaming ability was 393.58%, foam stability was 69.60%, and surface tension was 31.11 dyne/cm. If the verification results are compared with the predicted value (Table 2), then it is at a low and high predictive value.

The verification results showed that all responses could be received because they are located between the lower and upper limits and are following the results of recommendations from the design expert program.

Table 2 Comparison of the results of the prediction formula of the program with verification

Response	Formula					
	Prediction	Verification	95% CI Low	95% CI High	95% PI Low	95% PI High
Foam-forming ability (%)	357.65	393.58	306.32	409.99	273.39	441.92
Foam stability (%)	64.43	69.60	60.09	68.76	57.31	71.54
Surface tension (dyne/cm)	27.49	31.11	22.93	32.06	19.99	34.99

Evaluation of Best Formula. The best formula was 20.00% of sodium laurate, 4.99% of potassium palmitate, and 75.01% of water. Then, after verifying the best foaming agent formula, the response was obtained foaming ability was 393.58%, foam stability was 69.60%, and surface tension was 31.11 dyne/cm. Statistical analysis is needed to test the difference between the response value of foam-forming ability, foam stability, and surface tension from the optimal prediction of the Design Expert-D-Optimal program and the laboratory test results. The analysis used was an t-one sample test. The test was used to test the significance of the average difference between the two responses. The analysis showed that probability values were 0.56 > t (0.05) for foaming ability, 0.37 > t (0.05) for foam stability, and 0.52 > t (0.05) for surface tension. Based on response value the response of foaming ability, stability, and surface tension does not differ significantly between the predictions from the design Expert-D-Optimal program and the results of laboratory experiments.

Analysis of the Use of Peat Fire Extinguishers. The amount of extinguishers solution used for each experiment was different. Figure 1 shows a graph of the application of foam from FAR solution and water without the addition of foaming agents with the use of 4.56 and 9.72 L/m^2, respectively. The use of FAR could conserve water by 54.52%, while [1] was 48.57% for firefighting peat demonstration plots using foam from water added by foaming agents. Comparison of the means test on the effect of the two types of FAR extinguishers and water on the application of extinguishing solutions did not differ significantly, with a significance value of $0.23 > \alpha = 0.05$. The foam-forming ability of the foaming agent concentrate on peat fire suppression applications is a characteristic that needs to be considered because it can reduce water consumption. To suppression the fire, a large volume of water is required to flood the burned peat area [4, 27].

Analysis of Peat Fire Extinguishing Time. Extinguishing time in peat fire suppression activities is influenced by several factors, including the ability of the extinguishing media to cut off oxygen circulation, the stability of the extinguishing agent, and the ability to cool the burning material. The application of water as an extinguisher needs a longer time compared to the application of foam from the mixture of water and FAR foaming agent. This longer time is believed to be due to the slow rate of water absorption into the lower layers of the burning peat. Figure 2 shows the effect of different extinguishing materials on peat fire suppression time.

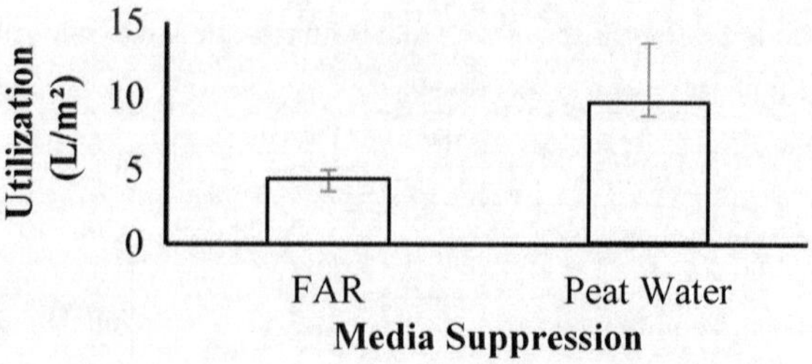

Fig. 2 Effect of the application of foam extinguisher type from water solution added with FAR, and water without the addition of foaming agents to the suppression time of peat fire at laboratory-scale

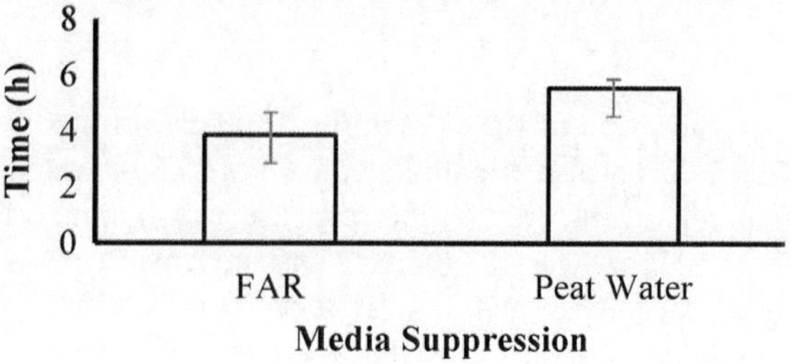

Fig. 3 Effect of foam extinguisher types from water solution added with FAR, and water without the addition of foaming agent to the amount of suppression solution used in laboratory-scale peat fire suppression applications

The FAR application required 3 h 50 min 42 s, while water required 5 h 31 min 59 s. The extinguishing time was calculated from the beginning of the fire suppression to the time when the temperature below 50 °C [27] (Fig. 3).

Theses difference needs to be analyzed with the comparison of means test to determine the effect of FAR and water on the time needed to extinguish peat fires. The analysis showed that the two ingredients had no significant effect, with a significance value of $0.09 < \alpha = 0.05$. Although the two ingredients did not significantly differ, FAR had higher fire suppression ability than water without the addition of foaming agents. This can be seen from the speed of fire suppression. FAR had a faster suppression time than water without the addition of foaming agent. FAR produces foam that can extinguish peat fires in the shortest time. This might happen because the raw material components of the product can extinguish and cool the fuel quickly [28].

4 Conclusions

The best foaming agent formula was selected based on physical parameters consisting of foam-forming ability, foam stability, and surface tension. Good foam-forming ability is an important parameter in producing foaming agents. Foam application in peatland fire suppression requires a foaming agent that can easily produce a lot of foam. Foam stability parameters are needed to ensure that the foam produced can withstand working conditions and can extinguish the fire. Meanwhile, the surface tension parameter is needed to determine the performance of the foam liquid through the ability to seep into the lower layer through peat pores. The best foaming agent formulations obtained were 20.00% of sodium laurate, 4.99% of potassium palmitate, and 75.01% of water, with the foam-forming ability of 393.58%, foam stability of 69.60%, and surface tension of 31.11 dyne/cm. Sodium laurate is the most influential factor in increasing foaming ability, while potassium palmitate affects the increase in foam stability and decreases in surface tension. Foam performance test from water added with foaming agent concentrate in peat fire suppression applications results in faster fire suppression time than using water without the addition of foaming agent, with a reduction in time by 40%. Furthermore, the application of foam from the solution of water mixed with the foaming agent can reduce water consumption by 54.52%.

References

1. JICA (2016) Japan international cooperation agency, standard operating procedure concerning fire-fighting techniques against peat-land and forest fires. Balikpapan
2. Ambak K, Melling L (2000) Management practices for sustainable cultivation of crop plants on tropical peatland; 1999 Nov. 22-24; Bogor, Indonesia. Bogor (ID): Graduate School of Environmental Earth Science Hokkaido University, Sapporo and Research and Development Center for Biology, Indonesian Institute of Sciences, pp 119–134
3. Akbar A (2016) Pemahaman dan Solusi Masalah Kebakaran Hutan di Indonesia. Forda Press, Bogor, pp 27–70
4. Nugroho TC, Oksana C, Aryanti E (2013) Analisis sifat kimia tanah gambut yang dikonversi menjadi perkebunan kelapa sawit di Kabupaten Kampar. J Agroteknol 4(1):5–30
5. Noo M, Masganti N, Agus F (2014) Pembentukan Dan Karakteristik Gambut Tropika Indonesia. Lahan Gambut Indonesia, Pembentukan, Karakteristik, Dan Potensi Mendukung Ketahanan Pangan. Badan Penelit dan Pengemb Pertan. Jakarta (ID): IAARD Pr. 2014, pp 7–32
6. Subekti P, Hambali E, Suryani A, Suryadarma P, Saharjo BH, Rivai M (2019) The effects of magnesium sulphate addition into foaming agents resulted from palm oil fatty acid saponification on the performance of peat fire suppression. Earth Environ Sci 19-20; Bogor, Indonesia. IOP. 460:012028. https://doi.org/10.1088/1755-1315/460/1/012028
7. Rivai M, Hambali E, Suryani A, Fitria R, Firmansyah S, Pradesi J (2017) Synthesis of palm oil fatty acid as foaming agent for firefighting application. Conf Ser Earth Environ Sci Okt. 10–11; Bogor, Indonesia. IOP 65: 012047. https://doi.org/10.1088/1755-1315/65/1/012047
8. Kadono T, Uezu K, Kawano T (2006) Confirming the altered toxicities of fatty acid salts in paramecium caudatum cultured in different waters. ITE Lestt Bett New Technol Med 7(6):67–70
9. Kawahara T, Hate S, Kanyama T, Ishizaki Y, Uezu K (2016) Development of eco-friendly soap-based firefighting foam for forest fire. Environ Control Biol 54(1):75–78
10. Mizuki H, Ueza K, Kawano T, Kadono T, Kobayashi M, Hatae S, Oba Y, Iwamoto S, Mitumune S, Nagatomo Y, Owari M, Umeki H, Yamaga K (2007) Novel environmental friendly soap-based fire-fighting agent. J Environ Eng Manage 17(6):403–408
11. Oguike RS (2013) Study of fire-fighting foam agent from palm oil for extinguishing of petrol fires Material Science Research Laboratory Department of Chemistry Abubakar Tafawa Balewa University Bauchi Nigeria 1(1):1–7
12. Margaretty E, Karakteristik aliran fluida newtonian dan non-newtonian dalam tangki berpengaduk (Tipe jangkar dan turbin). Bogor (ID), Institut Pertanian Bogor
13. Nugroho A (2012) Pemanfaatan software dalam penelitian. Universitas Gajah Mada, Yogyakarta (ID)

14. Wu M, Ding H, Wang S, Xu S (2006) Optimizing condition for the purification of linoleic acid from sunflower oil by urea complex fractionation. J Am Oil Chem 85:677–684
15. Subekti P, Hambali E, Suryani A, Suryadarma P, Saharjo BH, Rivai M (2019) The formulation of foaming agents from palm oil fatty acid and performance test on peat fires. J Japan Inst Energy 98:95–100
16. Design Expert 7.1.6. Minneapolis (US): Stat Ease (2005)
17. ISO (2014) International Organization for Standardization, ISO 7076–5: 2014, Fire Protection-Foam Fire Extinguishing System
18. Ramadhan ML, Palamba P, Imran FA, Kosasih EA, Nugroho YS (2017) Experimental study of the effect of water spray on the spread of smoldering in Indonesian peat fires. Fire Safety J 91:671–679
19. Limin SH (2006) Pemanfaatan Lahan Gambut dan Permasalahannya. Workshop Gambut. Kerjasama antara Badan Pengkajian dan Penerapan Teknologi dan Kementrian Koordinator Kesejahteraan Rakyat. Jakarta (ID), Univ Palangkaraya, 1–21
20. Tadros TF, Schramm LL, Platikanov D, Exerowa D (2005) Topics in Colloid and Interface Science. Weinheim (DE), WILEY-VCH
21. Maestro A, Rio E, Drenckhan W, Langevin D, Salonen A (2014) Foams stabilised by mixtures of nanoparticles and oppositely charged surfactants: relationship between bubble shrinkage and foam coarsening. Soft Matter 10:6975–6983
22. Adinugroho WC, Suryadiputra INN, Saharjo BH, Siboro L (2015) Manual for the control of fire in peatlands and peatland forest. climate change, forests and peatlands in Indonesia project. Bogor (ID), Wetlands International-Indonesia Programme
23. Rivai M, Hambali E, Suryani A, Pramuhadi G, Fitria R, Firmansyah S (2017) Physico-chemical properties tests of palm oil-based foaming agent using water solvent media from peatland physico-chemical properties tests of palm oil-based foaming agent using water solvent media from peatland. Conf Ser Earth Environ Sci Juli 24-25; Bogor, Indonesia: IOP. 209:012036.https://doi.org/10.1088/1755-1315/209/1/012036
24. Subekti P, Hambali E, Suryani A, Suryadarma P, Saharjo BH, Rivai M (2018) Selection of peat firefighting foam from palm oil fatty acid saponification with simple additive weighting (SAW) method. SNTM-BKSTM. 2018 Okt.03-05; Kupang, Indonesia. Kupang (ID). Dep. Teknik Mesin, Undana, pp 068–071
25. Subekti P, Hambali E, Suryani A, Suryadarma P (2016) Potential production of palm oil-based foaming agent as fire extinguisher of peatlands in Indonesia: Literature review. Earth Environ Sci 2016 Okt. 10–11; Bogor, Indonesia: IOP.65: 012038. https://doi.org/10.1088/1755-1315/65/1/012038
26. Hambali E, Suryani A, Rivai M (2013) Proses Pengembangan Teknologi Surfaktan Mes dari Metil Ester Minyak Sawit untuk Aplikasi Eor/Ior : Dari Skala Lab. ke Skala Pilot. Konferensi Nasional "Inovasi dan Technopreneurship"; 2013 Feb.18–19; Bogor, Indonesia. Bogor (ID), IPB. pp 18–19
27. ISO (2011) International organization for standardization, ISO-7076–3:2011. Fire extinguishing media foam concentrates specification for low-expansion foam concentrates for top application to water miscible liquids
28. Kurama H (2016) Fire retardant efficency of waste magnesia powder in cellulose insulation. J Aus Ceramic Soc 52(2):143–149

Stress Analysis on the Base of Surgeon's Chair: Frame and Truss Perspective

Gatot Santoso, S. Sugiharto, Toto Supriyono, Muhammad Imam Ammarullah, Muki Setya Permana, Randy Media Rachayu, Aris Nur Ihsan, Athanasius Priharyoto Bayuseno, Abdul Mughni, and J. Jamari

Abstract Musculoskeletal disorders frequently occur in various types of work, including laparoscopic surgeons. The surgeon profession is classified as a high-risk occupation and may have musculoskeletal disorders. Surgeons who suffer from these disorders experience illnesses ranging from mild to severe because the muscles often receive long-term static loads. These causes can be prevented when working in

G. Santoso (✉) · S. Sugiharto · T. Supriyono · M. I. Ammarullah · M. S. Permana · R. M. Rachayu · A. N. Ihsan
Department of Mechanical Engineering, Faculty of Engineering, Pasundan University, Bandung 40153, West Java, Indonesia
e-mail: gatot.santoso@unpas.ac.id

S. Sugiharto
e-mail: sugih.sugiharto@unpas.ac.id

T. Supriyono
e-mail: supriyono.toto@unpas.ac.id

M. I. Ammarullah
e-mail: imamammarullah@unpas.ac.id

M. S. Permana
e-mail: muki.satya@unpas.ac.id

R. M. Rachayu
e-mail: rachayu.183030006@mail.unpas.ac.id

A. N. Ihsan
e-mail: ihsan.183030020@mail.unpas.ac.id

G. Santoso · A. P. Bayuseno · J. Jamari
Department of Mechanical Engineering, Faculty of Engineering, Diponegoro University, Semarang 50275, Central Java, Indonesia

G. Santoso · S. Sugiharto · T. Supriyono · M. I. Ammarullah · M. S. Permana
Biomechanics and Biomedics Engineering Research Centre, Pasundan University, Bandung 40153, West Java, Indonesia

M. I. Ammarullah · J. Jamari
Undip Biomechanics Engineering & Research Centre (UBM-ERC), Diponegoro University, Semarang 50275, Central Java, Indonesia

Irwansyah et al. (eds.), *Proceedings of the 4th International Conference on Experimental and Computational Mechanics in Engineering*, Lecture Notes in Mechanical Engineering, https://doi.org/10.1007/978-981-99-7495-5_14

a seated position. In this work, the investigation being conducted is to design a chair as a surgeon's aid that consists of a base platform, seat height adjustment mechanism, seat cushion, backrest, and handrest. In this paper, a rectangular platform is analyzed using a finite element model. The analysis results by considering the structure as a truss produce the largest stress of 34 MPa compared the structure as a frame of 30 MPa. These two approaches lead to the conclusion that the structure computed as a truss is more secure than the analysis of the structure as a frame.

Keywords Chair · Platform · Frame · Truss · Stress analysis

1 Introduction

In performing operations in the operating room to treat the patient concerned, the surgeon spends many hours in a standing position [1]. This makes them experience fatigue which reduces the intensity of their focus and musculoskeletal problems. This ergonomic problem is an important thing to overcome because it can threaten patient safety in the actions taken by the surgeon [2]. Ideally, the surgeon should be in excellent health without significant fatigue to be able to perform the best course of action in each patient's situation.

One of the things that can be done to reduce the fatigue level of the surgeon is to use an assistive device in the form of a chair [3–6]. This device is used to support surgeons during surgery to avoid problems due to various ergonomic factors that arise during the operation process. In its manufacture, there are several components of the chair for the surgeon, starting from the base platform, seat height adjustment mechanism, seat cushion, backrest, and handrest [7]. In some of these components, the base platform component has a crucial role in making the surgeon's chair stand firmly.

The base platform component of the surgeon's chair needs to be investigated for its ability to have good mechanical performance when used, especially in holding heavy loads. Stress analysis has been widely adopted for the investigation of various mechanical components and is suitable for surgeon's chairs. Adopted finite element solution for stress study is very useful for the initial investigation of the current surgeon's chair, so that it does not take time and costs to manufacture tools and provide materials.

The current study attempted to examine the base component of the surgeon's chair. A computational model based on the finite element method has been developed to investigate the stress analysis of the base component.

A. Mughni
Department of Surgery, Faculty of Medicine, Diponegoro University, Semarang 50275, Central Java, Indonesia

2 Materials and Methods

Finite Element Model. A rectangular platform with a size of 900 × 600 mm for the base of surgeon's chair refers to the size of the trolley that is generally used [8]. The platform was constructed by hollow steel with cross-sectional dimensions 30 × 30 × 2 mm. The base design is static loads applied with a magnitude 1200 N which was applied based on the surgeon's weight. Fixed end for all corners of base design was considered as the boundary condition. Meshing was generated automatically using SolidWorks Premium 2016. Loading and boundary condition is explained in Fig. 1.

Materials Properties. ASTM A36 was used for materials as commonly mild steel used in the market. Also, these metallic materials are assumed to be homogeneous, isotropic, and linear elastic. The base structure of the surgeon's chair with ASTM A36 is investigated based on the frame and truss analysis method. The investigation in the frame uses von Mises stress while the truss uses axial stress. The material properties of ASTM A36 are explained in Table 1.

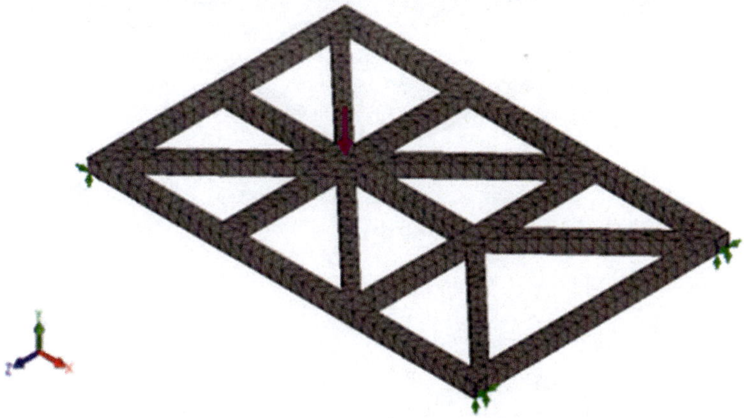

Fig. 1 Loading and boundary condition on the base of surgeon's chair

Table 1 Materials properties of ASTM A36 [9]

Material property	Value (MPa)
Yield strength (S_y)	250
Tensile strength (S_u)	400

3 Results and Discussion

The stress contours on the frame and truss can be seen in Figs. 2 and 3, respectively. From these contours, it can be seen that the stress concentration is in the area near the applied force. Meanwhile, for areas farther from the location of the applied force, the stress is relatively lower. This shows that failure is more likely to occur in the area near the applied force because the maximum stress is in that area, both in the analysis using frame and truss.

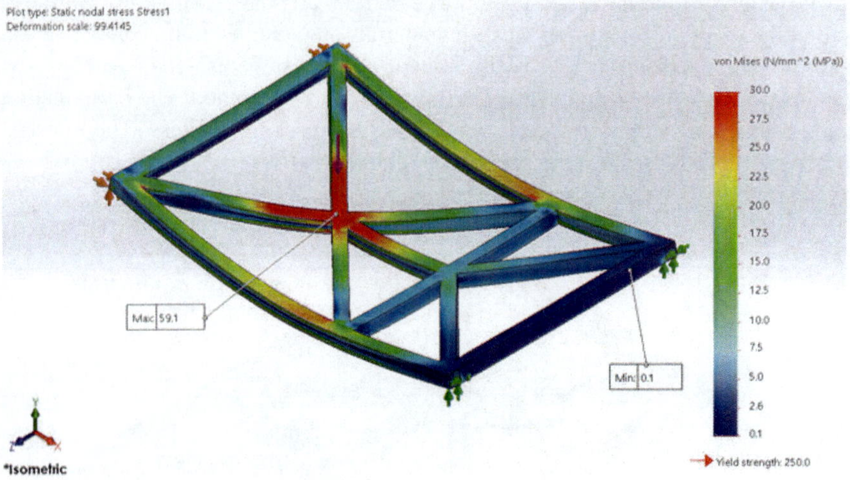

Fig. 2 Contour of von Mises stress on the frame

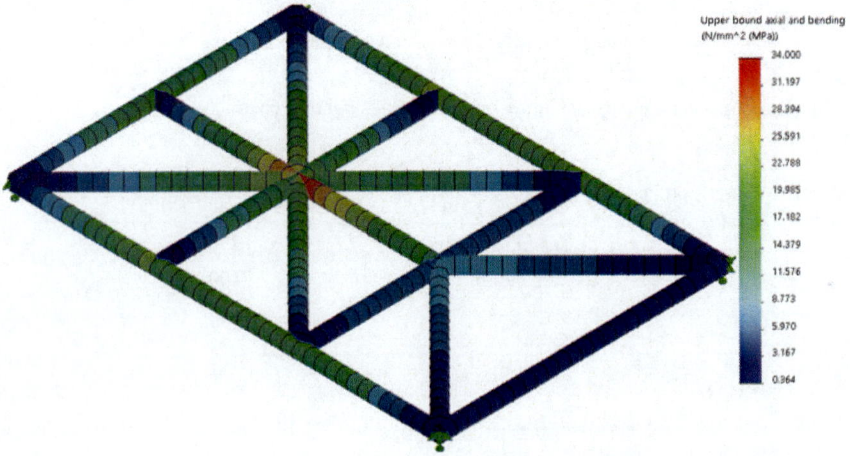

Fig. 3 Contour of axial stress on the truss

Table 2 Comparison between frame and truss analysis

Analysis method	Stress (MPa)
Frame	30
Truss	34

Furthermore, the results of the maximum stress on the frame and truss are presented in Table 2. The results of the analysis with the frame are lower than the results of the analysis with the truss. The difference in stress values between the two analysis methods is 4 MPa, where the results of the analysis using the truss 13.3% larger than the frame. This gives an understanding that failure analysis would be better to adopt the truss method. Because the truss method provides a higher stress value, we can be more certain in making sure the base component of the surgical chair is safe or not. There is a possibility that adoption of frame analysis has been safe, but when studying it using the truss perspective, it would be a failure.

4 Conclusions

The computational simulation of the base component of the surgeon's chair by comparing the analyses using the frame and truss has been successfully carried out. It was found that the axial stress in the analysis using the truss method was greater than the von Mises stress in the analysis using the frame method. It can be concluded that the study by adopting the truss will be safer than the frame, especially in the investigation of the surgeon's chair.

References

1. Takeuchi T, Yamagishi K, Konishi K, Sano H, Takahashi M, Ichimura S, Kono H, Hasegawa M, Hosogane N (2022) Radiological evaluation of combined anteroposterior fusion with vertebral body replacement using a minimally invasive lateral approach for osteoporotic vertebral fractures: verification of optimal surgical procedure. J Clin Med 11:629
2. Santoso G, Sugiharto S, Mughni A, Ammarullah MI, Bayuseno AP, Jamari J (2022) Chairless chairs for orthopedic surgery purpose—a literature review, open access maced. J Med Sci 10:146–152
3. Kim N, Kim KH, Choi YS, Song SH, Choi SH (2022) Effect of Dexmedetomidine on early postoperative cognitive function in patients undergoing arthroscopic shoulder surgery in beach chair position: a randomized double-blind study. J Clin Med 11:2970
4. Morisawa T, Saitoh M, Otsuka S, Takamura G, Tahara M, Ochi Y, Takahashi Y, Iwata K, Oura K, Sakurada K, Takahashi T (2022) Hospital-acquired functional decline and clinical outcomes in older cardiac surgical patients: a multicenter prospective cohort Study. J Clin Med 11:640
5. Sandali O, Tahiri Joutei Hassani R, Armia Balamoun A, El Sanharawi M, Borderie V (2022) Facilitating role of the 3D viewing system in tilted microscope positions for cataract surgery in patients unable to lie flat. J Clin Med 11:1865

6. Gazal G, Al-Samadani KH, Alsaidalani HM, Karbouji GA, Alharbi AM (2022) A comparison of pre-emptive co-amoxiclav, postoperative amoxicillin, and metronidazole for prevention of postoperative complications in dentoalveolar surgery: a randomized controlled trial. Int J Environ Res Public Health 19:4178
7. Nag PK (2019) Office buildings: health, safety and environment
8. Houston WWK (1983) Motorised mechanical soil sieve. Aust J Entomol 22:243–245
9. Kang S, Kim H, Kim J, Kim H, Jang J, Kwak B, Choi K, Jang H (2021) Transient structural analysis of a skid mounted on a hydrogen tube trailer under shock and vibration induced by road irregularities. Appl Sci 11:3779

Determination of the Optimal Composition of the Kenaf/Ramie Composite Material Based on the Taguchi and the Surface Response Methodology

Haniel, Baju Bawono, and Paulus Wisnu Anggoro

Abstract The use of composite materials with natural fiber reinforcement has experienced rapid development in the automotive industry to apply synthetic materials that are expensive and not environmentally friendly. Biocomposite is a type of composite consisting of a polymer matrix and natural fiber reinforcement. Natural fiber materials are used instead of conventional non-renewable materials. In this study, the vacuum-assisted resin infusion (VARI) process was used for composite fabrication using woven kenaf and ramie fibers. The Taguchi experimental design method with the $L_9 3^2$ orthogonal array matrix in this study was used to obtain optimum composite fabrication parameters with natural fiber reinforcement. Material test such as impact test and flexural test is used to obtain experimental responses. The results of the study using the RSM method obtained the optimal combination of manufacturing parameters, namely by choosing a low concentration of NaOH (4%) and a high post-curing temperature (120 °C). With these optimal parameters, the optimal flexural and impact strengths that can be achieved are 0.261 J/mm^2 and 79.658 MPa.

Keywords Composite materials · Automotive industry · Natural fiber materials

Haniel
Department of Industrial Engineering (Madiun City Campus), Faculty of Engineering, Widya Mandala Surabaya Catholic University, Manggis Street No.15-17, Madiun 63131, East Java, Indonesia

B. Bawono · P. W. Anggoro (✉)
Department of Industrial Engineering, Faculty of Industrial Technology, Universitas Atma Jaya Yogyakarta, Babarsari Street No. 44, Yogyakarta 55281, Indonesia
e-mail: pauluswisnuanggoro@ymail.com

Irwansyah et al. (eds.), *Proceedings of the 4th International Conference on Experimental and Computational Mechanics in Engineering*, Lecture Notes in Mechanical Engineering, https://doi.org/10.1007/978-981-99-7495-5_15

1 Introduction

The Indonesian government is currently aggressively pushing the Indonesian state toward Green Technology in the automotive industry to reduce carbon emissions. The government wants the automotive industry in the country to immediately become a leading sector in the development of environmentally friendly technologies such as electric vehicles [1]. One of the efforts to support the government's program in Indonesia's efforts toward Green Technology is to replace conventional non-renewable materials with alternative materials such as natural fiber reinforced composites. The development of the automotive industry is currently starting rapidly, so many researchers develop their research in the automotive industry related to the development of composite materials. The use of synthetic materials in the production process on a large scale has a negative impact on the environment and requires high production costs [2]. However, there are drawbacks to natural fibers, and namely they cannot be applied to high material strength requirements, high water absorption, and are not resistant to high temperatures [3].

Composite components with natural fiber reinforcement produced in this study will go through the impact and bending testing stages to produce experimental responses. To determine the optimum composition of the experimental response from the parameters of the composite manufacturing process with natural fiber reinforcement, this study used two experimental methods, namely the Taguchi method and surface response.

Based on this description, this paper discusses the Taguchi method approach and surface response to obtain the optimum composition of kenaf/ramie composite materials. This optimum composition is used as a reference for manufacturing composite-based vehicle components with natural fiber reinforcement.

2 Material and Methodology

This research is divided into three stages, including identify the problem, experiment stage, optimization, and evaluation stage. In the first stage (identify the problem), it is used to obtain data from VARI manufacturing engineering, natural fiber reinforcement composite material. The second stage (experimental stage) is used to obtain data from experimental design using the Taguchi method and response of material strength. The last stage (optimization and evaluation stage) is used to obtain data from response optimization, optimal parameter settings, result, and conclusion. Briefly, these stages can be visualized through Fig. 1.

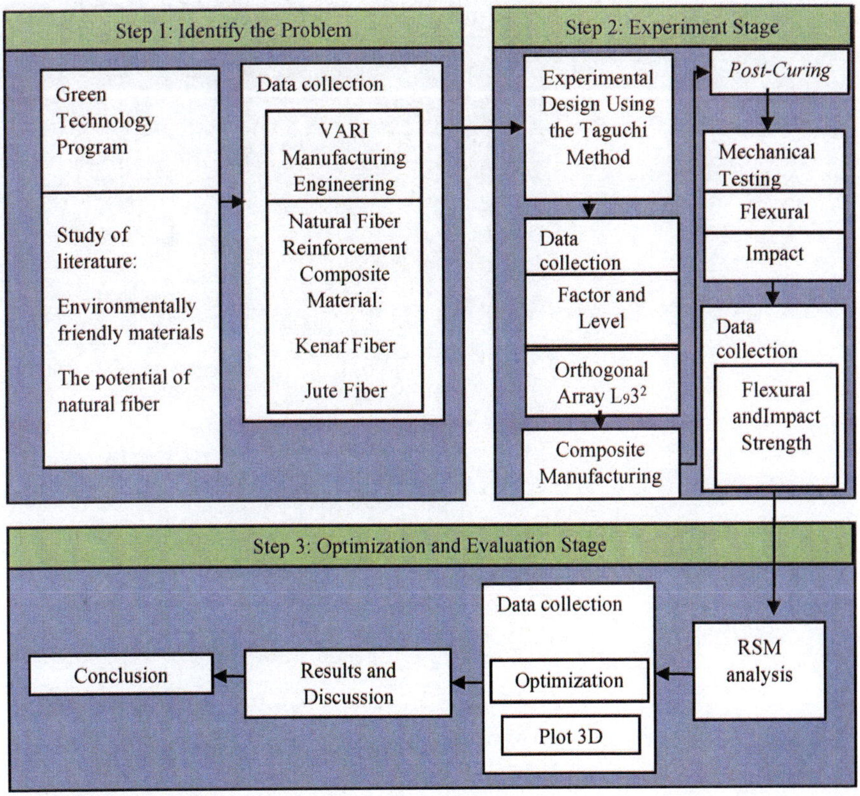

Fig. 1 Research flowchart

3 Composite Material

Fiber. The fibers used in this study were kenaf and ramie fibers in the form of woven. The fiber is cut according to the needs and the number of specimens to be formed. The natural fibers used are shown in Figs. 2 and 3.

Epoxy Resin. The matrix used in this research is bisphenol-A epoxy resin. Epoxy resin and hardener are the main ingredients in the manufacture of composite materials. The ratio of these resin types is 1:1 (by weight) when mixing the ingredients (Fig. 4).

Fig. 2 Kenaf fiber

Fig. 3 Rami fiber

4 Manufacturing Method Vacuum-Assisted Resin Infusion (VARI)

Vacuum-assisted resin infusion (VARI) can be used as a fabrication method in this study. The fiber layer is prepared on the surface of the mold in a predetermined arrangement. The prepared mold is covered with bagging film using sealant tape.

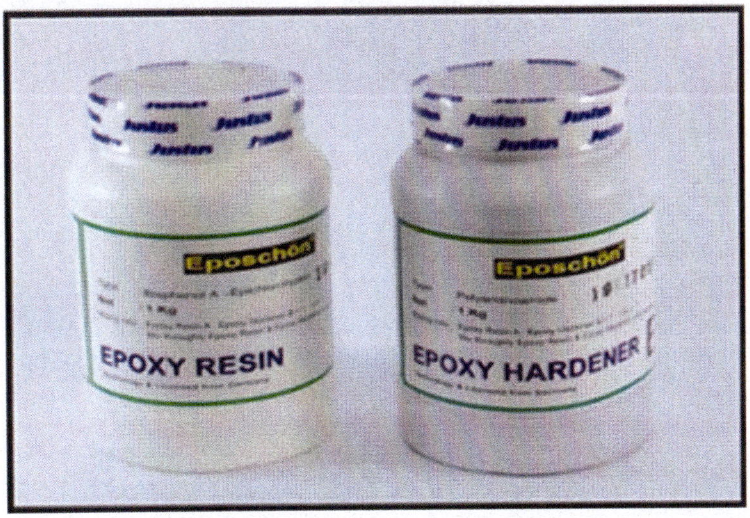

Fig. 4 Epoxy resin bisphenol-A

The epoxy matrix that has been prepared according to the predetermined composition is stirred until well mixed. In accordance with the recommendations of the epoxy bisphenol-A resin product, the drying process is carried out for 24 h, and then, the composite can be removed from the mold [4]. The printing device using vacuum-assisted resin infusion (VARI) can be seen in Fig. 5.

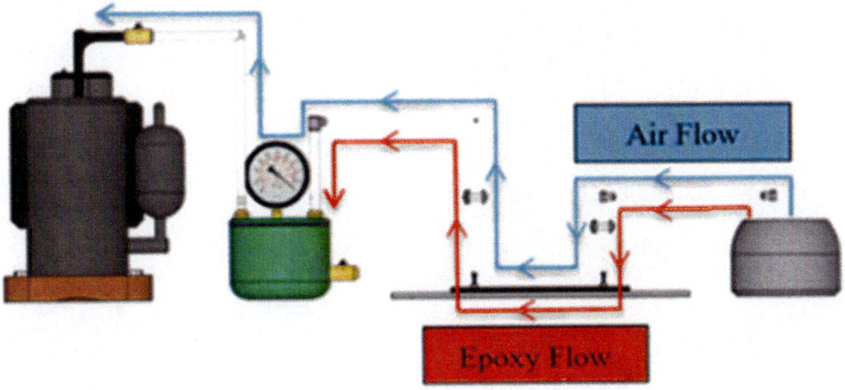

Fig. 5 Vacuum-assisted resin infusion (VARI) diagrams

Table 1 Natural fiber composite manufacturing parameters

Factor	Level 1	Level 2	Level 3
NaOH concentration (%)	4	6	8
Post-curing temperature (°C)	80	100	120

Table 2 Matrix orthogonal array $L_9 3^2$

Experiment	A	B
1	1	1
2	1	2
3	1	3
4	2	1
5	2	2
6	2	3
7	3	1
8	3	2
9	3	3

5 Taguchi Method

In this study, the Taguchi experimental design was used on the $L_9 3^2$ orthogonal array matrix using Minitab19 software. The Taguchi study design was used to reduce the number of studies to be conducted, reduce costs, and achieve optimal results in a shorter time [5]. The Taguchi method provides a systematic and efficient way to optimize the system parameter design, but its impact is much smaller than most of the optimization techniques required [6]. The Taguchi method also uses the signal-to-noise ratio to analyze the influence of factors that contribute to the response. SNR has three properties. In the parameter optimization process, "the lowest is the best", "the highest is the best", and "the highest nominal is the best" [7].

The parameters used as inputs in the use of the Taguchi method have been determined as experimental designs in this study as given in Table 1.

The factor that becomes the parameter in this study is the NaOH concentration factor (A) and then the post-curing temperature factor (B) (Table 2).

6 ANOVA Analysis

ANOVA was used to assess the significance of the regression model for the parameters and individual model coefficients [8]. Table 6 presents the ANOVA analysis of the experimental results for impact strength and flexural strength. The experimental design was evaluated at a 95% confidence level. The value of the contribution becomes a consideration in determining the level of significance of a variable.

Table 3 Strength of natural fiber composite material

Experiment	Uncode		Impact strength (J/mm^2)	Flexural strength (MPa)
	A	B		
1	1	1	0.24	77.4
2	1	2	0.25	78.5
3	1	3	0.26	80.1
4	2	1	0.23	75.8
5	2	2	0.22	76.8
6	2	3	0.24	77.3
7	3	1	0.23	74.6
8	3	2	0.23	76.9
9	3	3	0.24	79.2

Analysis of variance (ANOVA) is an analytical tool used in statistics that splits the observed aggregate variation found in a dataset into two parts: systematic and random factors. Systematic factors have a statistical impact on a given dataset, while random factors do not. Analysts use her ANOVA test to determine the effect of the independent variables on the dependent variable in the regression analysis.

7 Result and Discussion

S/N Ratio Analysis. The experimental response in the form of the composite material strength value was obtained from experiments with different manufacturing parameters as given in Table 3. The highest natural fiber composite material strength value was obtained from the optimal manufacturing composition. In this study, there are two conditions of optimal manufacturing parameters that produce maximum material strength which are simplified as follows: A_1B_3 (impact strength); A_1B_3 (flexural strength). Each optimal parameter setting has a tendency toward a maximum S/N ratio. The optimal material strength value is proportional to the maximum S/N ratio because the selected S/N ratio criterion which is larger is better [9]. This indication shows that the parameter quality is optimal with maximum material strength (Fig. 6).

8 Optimal Composition Selection Based on Taguchi Method

Prediction of the optimal manufacturing composition (Composition) of the impact strength can be expressed as follows:

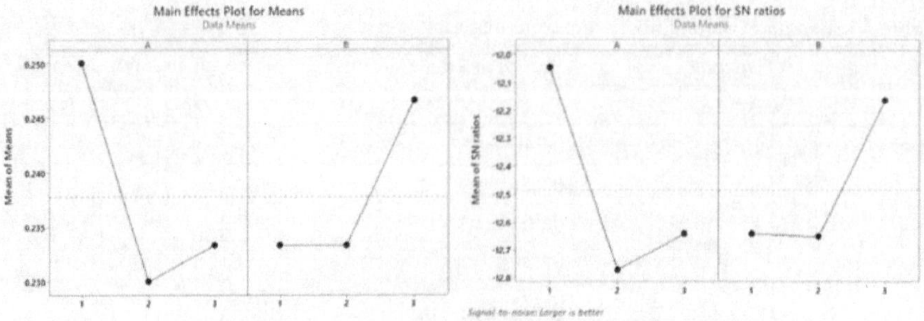

(a) Impact Strength

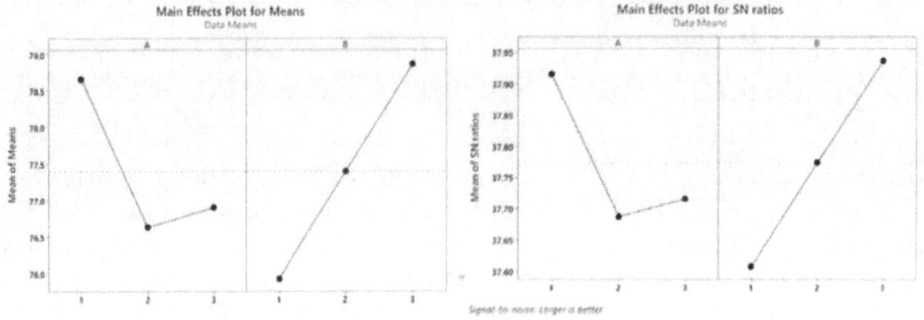

(b) Flexural Strength

Fig. 6 Plot mean and S/N ratio material strength

$$\text{Composition}_{pred} = T_{\text{Composition\_}exp} + \left(\overline{A_1} - T_{\text{Compositioni}_{exp}}\right) + \left(\overline{B_1} - T_{\text{Compositioni}_{exp}}\right) \tag{1}$$

where $T_{\text{Composition\_}exp}$= 0,237; $\overline{A_1}$=0,250; $\overline{B_1}$=0,246; then, Composition $_{\text{pred}}$ (impact strength) = 0,237 + (0,250–0,237) + (0,246–0,237) = 0,259 J/mm^2. As for the prediction of the optimal manufacturing, composition of the impact strength is $T_{\text{Composition\_}exp}$= 77,4; $\overline{A_1}$=78,66; $\overline{B_1}$=78,86; then, Composition $_{\text{pred}}$ (flexural strength) = 77,4 + (78,66–77,4) + (78,86–77,4) = 80,12 MPa. Prediction of optimal composition can also be determined by 3D plots, and this plot can make it easier to predict optimal parameters (see Fig. 7).

Confidence interval (CI) considered is to be able to predict the optimal value, and it can be calculated as follows [10]:

$$CI = \sqrt{F_{\alpha,dof\,V_{error}}\,V_{error} x \left(\frac{1}{n_{eff}}\right)} \tag{2}$$

$$n_{eff} = \frac{\text{Number of experiment}}{1 + \text{total dof in items used in estimate}} \tag{3}$$

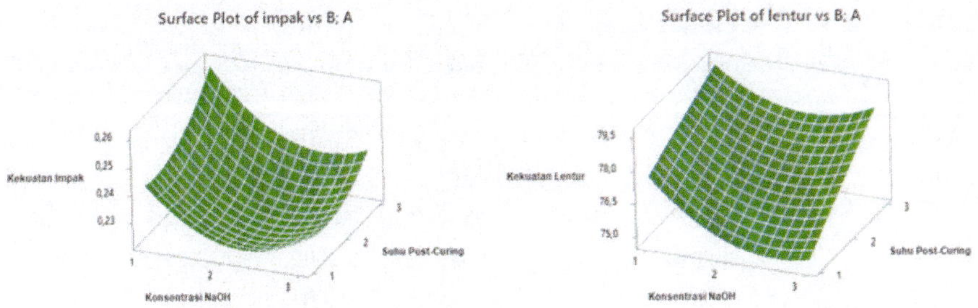

(a) Impact Strength (b) Flexural Strength

Fig. 7 Material strength 3D plot

Confidence interval for impact strength in Table 4 shows a value of 0.22854 where it is believed that 0.22854 is the parameter value of the population, while the confidence interval for flexural strength is 76.143 (Table 5).

ANOVA in RSM. ANOVA was used to evaluate the significance of the regression model and the individual model coefficients of the parameters that contribute to the strength of the material. ANOVA results for the strength of natural fiber composite materials are given in Tables 6 and 7. The contribution of P-value provides confidence in the level of significance for all variables. If the P-value is lower than 0.05, it means that both models are statistically significant at the 95% confidence level. These results are in accordance with previous research [11].

The contribution values for the response of the impact and flexural strength of the material are 92.55 and 92.86% with errors of 7.45 and 7.14%, respectively. Variations in this value indicate the level of influence on the value of the strength of the material. For the impact and flexural strength of the material, the combination of parameters A, B, A*A, B*B and A, B, A*A interacting with the parameters A*B provides the optimal setting for the material strength value. The percentage contribution of impact and flexural strength values is A (36.06%), B (23.08%) and A (20.57), B (56.71). Factor A (NaOH concentration) gave a more significant contribution to the impact strength value, and factor B (post-curing temperature) gave a more significant contribution to the flexural strength value.

Table 4 Confidence interval of impact strength

N	Mean	StDev	SE mean	95% CI for μ
9	0.23778	0.01202	0.00401	(0.22854; 0.24702)

Table 5 Confidence interval of flexural strength

N	Mean	StDev	SE mean	95% CI for μ
9	77.440	1.687	0.562	(76.143; 78.737)

Table 6 ANOVA on impact strength.

Source	DF	Seq SS	Contribution (%)	Adj SS	Adj MS	*F*-value	*P*-value
Model	5	0.001069	**92.55**	0.001069	0.000214	7.45	0.064
Linear	2	0.000683	59.13	0.000683	0.000342	11.90	0.037
A	**1**	**0.000417**	**36.06**	**0.000417**	**0.000417**	**14.52**	**0.032**
B	**1**	**0.000267**	**23.08**	**0.000267**	**0.000267**	**9.29**	**0.056**
Square	2	0.000361	31.25	0.000361	0.000181	629	0084
A*A	**1**	**0.000272**	**23.56**	**0.000272**	**0.000272**	**9.48**	**0.054**
B*B	**1**	**0.000089**	**7.69**	**0.000089**	**0.000089**	**3.10**	**0.177**
2-way interaction	1	0.000025	2.16	0.000025	0.000025	0.87	0.420
A*B	**1**	**0.000025**	**2.16**	**0.000025**	**0.000025**	**0.87**	**0.420**
Error	3	0.000086	7.45	0.000086	0.000029		
Total	8	0.001156	10,000				

The importance of giving bold letters to show that the sources (A; B; A*A; B*B; A*B) shown in the table really have a significant influence or the biggest contribution to the response measured in this paper

Table 7 ANOVA on flexural strength

Source	DF	Seq SS	Contribution (%)	Adj SS	Adj MS	*F*-value	*P*-value
Model	5	21.1358	**92.86**	21.1358	4.2272	7.81	0.061
Linear	2	17.5883	77.28	17.5883	8.7942	16.24	0.025
A	**1**	**4.6817**	**20.57**	**4.6817**	**4.6817**	**8.65**	**0.060**
B	**1**	**12.9067**	**56.71**	**12.9067**	**12.9067**	**23.84**	**0.016**
Square	2	2.6450	11.62	2.6450	1.3225	2.44	0.235
A*A	**1**	**2.6450**	**11.62**	**2.6450**	**2.6450**	**4.89**	**0.114**
B*B	1	0.0000	0.00	0.0000	0.0000	0.00	1.000
2-way interaction	1	0.9025	3.97	0.9025	0.9025	1.67	0.287
A*B	**1**	**0.9025**	**3.97**	**0.9025**	**0.9025**	**1.67**	**0.287**
Error	3	1.6242	7.14	1.6242	0.5414		
Total	8	22.7600	100.00				

The importance of giving bold letters to show that the sources (A; B; A*A; B*B; A*B) shown in the table really have a significant influence or the biggest contribution to the response measured in this paper

RSM-Based Modeling for Material Strength. RSM is used for modeling and analysis of the dependent and independent variables within a certain range by considering the experimental results. The material strength model can be expressed as follows:

$$\begin{aligned}\text{Impact strength} = {} & 0,2922 - 0,0500\text{A} - 0,0150\text{B} + 0,01167\text{A*A} \\ & + 0,00667\text{B*B} - 0,00250\text{A*B}\end{aligned} \tag{4}$$

$$\begin{aligned}\text{Flexural strength} = {} & 81,97 - 6,43A + 0,52B + 1,150A * A \\ & - 0,000B * B + 0,475A * B\end{aligned} \tag{5}$$

Equations (4) and (5) are used for the determination of R2. The response for the impact and flexural strength values resulted in R2 values of 92.55% and 92.86%, respectively. The above model can be used to predict the response of the material strength to the design parameters. The parameter interactions between manufacturing variables that contribute to the material strength values can be predicted from the 3D plots (Fig. 7). For manufacturing natural fiber composites at impact strength, maximum material strength can be achieved when selecting the lowest level of NaOH concentration and highest level of post-curing temperature, as well as flexural strength. This shows that a high concentration of NaOH will reduce the mechanical properties of the material as a result of this treatment the fibers will be brittle [12, 13].

Figure 8 shows that the maximum material strength for impact strength is 0.261 J/mm^2 and the maximum material strength for flexural strength is 79.658 MPa. These results are in accordance with previous research [8].

9 Conclusion

The Taguchi method approach and surface response in this paper are used to obtain the optimum composition of the kenaf/ramie composite material. The results showed that using the RSM method, the optimal combination of manufacturing parameters was obtained by selecting a low concentration of NaOH (4%) and a high post-curing temperature (120 °C). With these optimal parameters, the optimal flexural and impact strengths that can be achieved are 0.261 J/mm^2 and 79.658 MPa.

This research can be used as a reference for engineers to determine the optimum composite material for vehicle body applications made from natural fiber composites.

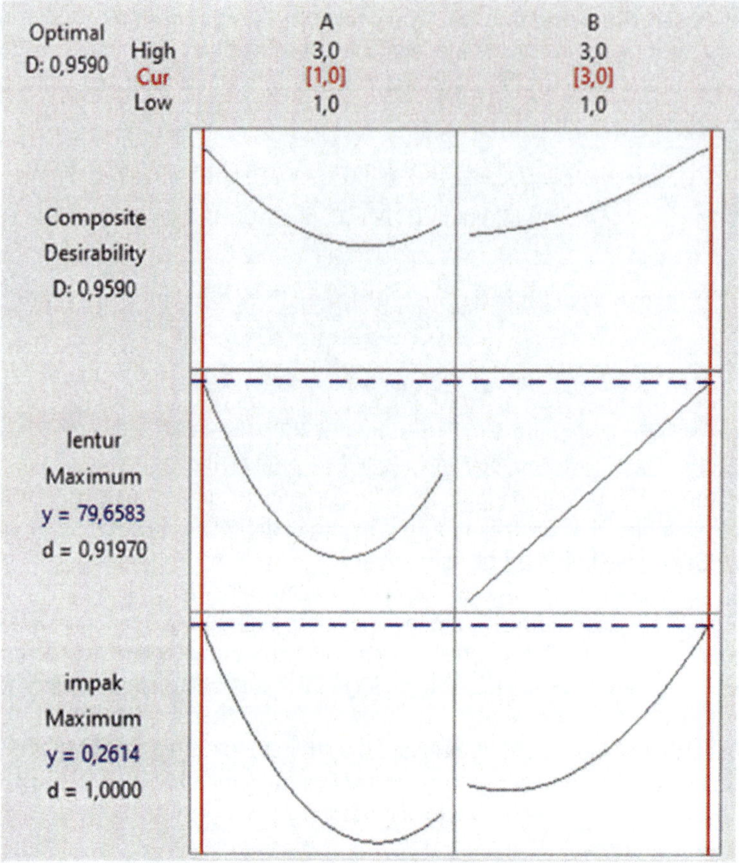

Fig. 8 Optimized results obtained with RSM

References

1. Kemenperin Home Page, https://www.kemenperin.go.id/artikel/22304/Upaya-Pemerintah-pada-Pertumbuhan-Industri-Kendaraan-Listrik.
2. Alzebdeh KI, Nassar MM, Arunacha-lam R (2019) Effect of fabrication parameters on strength of natural fiber polypro-pylene composites: statistical assessment. Measurement 146:195–207
3. Sanjay MR, Siengchin S, Parameswaranpillai J, Jawaid M, Pruncu CI, Khan A (2019) A comprehensive review of tecniques for natural fibers as reinforcement in composites: preparation, processing and characterization. Carbohyd Polym 207:108–121
4. Sudarisman S, Hidayat A, Haniel H, Tiopan M, Himarosa RA, Muflikhun MA (2022) The characteristic of CFRP laminates related to water absorption subjected to tensile, bending, and compression. Preprints. https://doi.org/10.20944/preprints202203.0382.v1
5. Terzioğlu H (2020) Analysis of effect factors on thermoelectric generator using Taguchi method. Measurement 149(106992):1–9
6. Montgomery DC, Runger GC (2014) Applied statistics and probability for engineers. Wiley, USA, pp 575–662

7. Anggoro PW, Anthony AA, Bawono B, Jamari J, Bayuseno AP, Tauviqirrahman M, Nugroho A (2019) CNC milling of EVA foam with varying hardness for custom orthotic shoe insoles and process parameter optimization. J Mech Eng Sci 13(3):5347–5370
8. Anggoro PW, Anthony AA, Tauviqirrahman M, Jamari J, Bayusena AP, Aylie H (2020) Machining parameter optimization of EVA foam orthotic shoe insoles. Int J Eng Technol Innov 10(3):179–190
9. Maazinejad B, Mohammadnia O, Ali GA, Makhlouf AS, Nadagouda MN, Sillanpää M, Sadegh H (2020) Taguchi L9 (34) orthogonal array study based on methylene blue removal by single-walled carbon nanotubes-amine: adsorption optimization using the experimental design method, kinetics, equilibrium and thermodynamics. J Mol Liq 298:112001
10. Roy RK (1990) A Primer on the Taguchi Method. Van Nostrand Reinhold, New York, USA
11. Bawono B, Anggoro PW, Bayuseno AP, Jamari J, Tauviqirrahman M (2019) Milling strategy optimized for orthotics insole to enhance surface roughness and machining time by Taguchi and response surface methodology. J Ind Prod Eng 36(4):237–247
12. Hamidon MH, Sultan MT, Ariffin AH, Shah AU (2019) Effects of fibre treatment on mechanical properties of kenaf fibre reinforced composites: a review. J Market Res 8(3):3327–3337
13. Mahjoub R, Yatim JM, Sam ARM, Hashemi SH (2014) Tensile properties of Kenaf fiber due to various conditions of chemical fiber surface modifications. Constr Build Mater 55:103–113

Identification of Administrative Worker Postures to Reduce Musculoskeletal Disorders with Ergonomic Approach

Gisya Amanda Yudhistira, Chancard Basumerda, Qurtubi, and Muhammad Ragil Suryoputro

Abstract Musculoskeletal disorders (MSDs) are complaints of pain in body parts due to poor posture, repetitive movements, and movements with too large an angle. This study examines operators who work in the administration department in a learning industry in Yogyakarta. This study aims to determine the level of risk that exists in static workers using the LS-CMDQ questionnaire and the ROSA method. The results of the LS-CMDQ questionnaire show that the back has the highest score of 14, and workers experience pain in that area 3–4 times a week. The ROSA method shows a value of 5 for administrative activities, which is very dangerous in the long run. The recommendations are redesigning an ergonomic work chair, adjusting the work position with minimal movement, and tilting the wrist not more than 15°.

Keywords LS-CMDQ · ROSA · Score · Work posture

1 Introduction

In 2000, as many as 80% of daily workers were required to use computers [1]. Jobs that involve computers are completed with static work postures. The length of time working in a static position can cause a variety of work complaints. A study by da Rosa et al. showed no significant difference between the right and left side views on a person's work posture. It only affects the angle formed and produced in the head and knee positions [2]. Identification of complaints of pain or pain in the body points is often interpreted as musculoskeletal disorders (MSDs). The research results by Besharati et al. explain that MSDs can result from physical movements such as poor static posture, repetitive movements, less ergonomic workplace, worker psychology, and organizational design (work cycle) [3]. It is potential in all industrial sectors with the movement and work risks experienced in posture positions.

G. A. Yudhistira · C. Basumerda · Qurtubi (✉) · M. R. Suryoputro
Departement of Industrial Engineering, Universitas Islam Indonesia, Yogyakarta, Indonesia
e-mail: qurtubi@uii.ac.id

Irwansyah et al. (eds.), *Proceedings of the 4th International Conference on Experimental and Computational Mechanics in Engineering*, Lecture Notes in Mechanical Engineering, https://doi.org/10.1007/978-981-99-7495-5_16

Musculoskeletal disorders (MSDs) are complaints of static work with movements in the hands that have the potential for injury and musculoskeletal problems. Furthermore, it could lead to chronic health problems [4]. In fact, repetitive movements can cause damage to ligaments, tendons and joints with prolonged work risk [5]. If there is no improvement, it will result in prolonged complaints of pain due to tense muscles. One of the activities for static workers is administrative workers by working every day staring at a computer screen. Therefore, occupational risk considerations for MSDs are commonplace for workers with computers. So it is necessary to measure work posture for ergonomic conditions to improve work [6]. Risk factors for postural discomfort are caused by posture, duration, and type of computer used [7]. If a job activity is not based on comfort and safety, its make awkward posture which will hinder work process, and direct biomechanical risks in work postures until decrease in productivity [8]. Thus, the most important thing to identify the operator's work posture to know the categories of work risks experienced that can be evaluated and provide improvements in the future.

There is a need to identify static workers, one of which is in the administrative field. Using the method (Rosa Rapid Office Strain Assessment) to identify the work posture of workers by considering the movement of the mouse, keyboard, monitor, and other work equipment. With the addition of the Cornell questionnaire, which was used to determine the total level of worker discomfort with static work [9].

2 Method

The subject of this study is female administrative workers, 23 years old, one of the learning industries located in Yogyakarta. Work 9 h a day with a 1-h break. The task of administrative work is to administer the participant's learning system by becoming customer service to clients who will register learning in this tutoring.

Direct observation data were collected by filling out the ROSA observation sheet and the CMDQ questionnaire to identify work postures. ROSA (Rapid Office Strain Assessment) is an assessment method in office ergonomics, where an assessment is carried out to measure computer-related risks to determine the level of change actions and ergonomics interventions [10]. The Cornell Musculoskeletal Discomfort Questionnaire (CMDQ) is a subjective measuring tool for mapping body parts that are felt sick at work. CMDQ is a combination questionnaire that is essentially taken through a similar questionnaire, namely the Nordic Body Map (NBM), with additional questions about the prevalence of musculoskeletal pain, severity, and whether it interferes with the respondent's performance at work [11].

3 Result and Discussion

Initial identification of the operator is using direct observation of the operator's position. The following are the results of the observations made (Table 1):

A. LS-CMDQ Result

LSCDQ is applied to determine which body parts are experiencing pain complaints. Pain can be identified from body parts with the highest score from 12 categories: neck, shoulder, upper back, upper arm, lower back, forearm, wrist, hand/finger, hip/butt, thigh, knee, and lower leg. The following is the recapitulation of each body point part (Table 2).

The results show the highest level values with dominant pain, namely in the lower back. The most significant value is 14 because the seat design is not ergonomic, which results in a stooped posture. The operator felt pain 3 to 4 times in the last week. Other body parts do not feel pain because they move a lot. From subjective observations, the researchers saw that the operator experienced pain because he often looked at the computer where the operator's right side was not parallel to the body.

B. ROSA Result

The highest category is section A, with a score of 5. The position between the back of the knee and the chair's seat is too high, causing the foot position to look unnatural, so the backrest has a value of 3. The seat design position is not ergonomic, and the operator does not carry out activities on the back of the chair. The presence of back support on the seat makes the operator less tired and does not cause back pain or muscle tension in the back of the body. Then, the score of part B is 3, which belongs to the low category on the monitor display and the neck position is more than 30o. The position of the computer is to the right of the operator. So, make the operator see the tilted head position. The score of part C is 3, which the category is the mouse grip parallel to the shoulder, or the mouse position is in front of the body in parallel and can work efficiently to hold or move, but the mouse size is a bit, so that it does

Table 1 Preliminary observation results

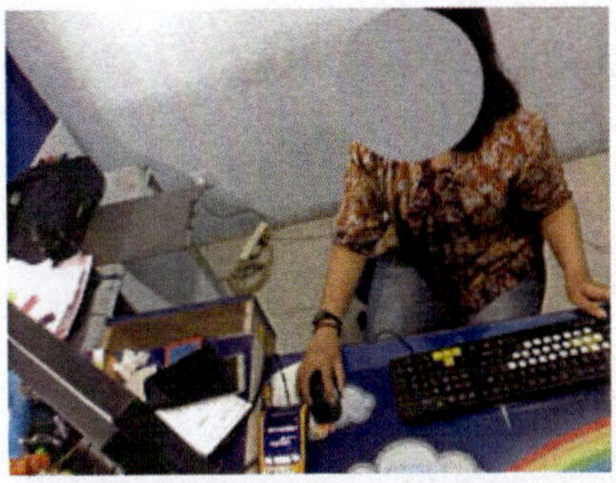	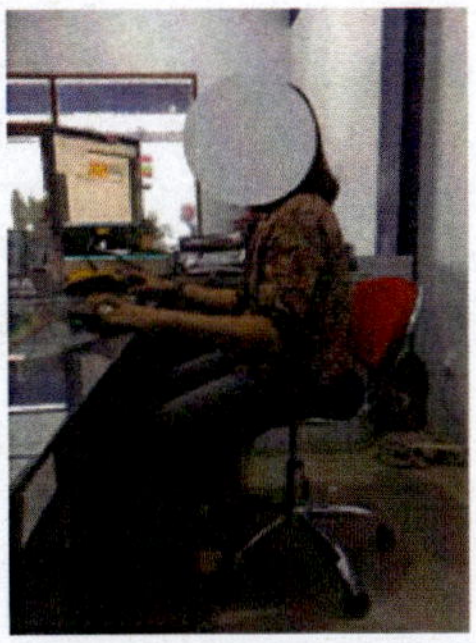	

Table 2 LS-CMDQ recapitulation

No	Part of body	Score1	Score2	Score3	Output
1	Neck	1.5	2	2	6
2	Shoulder right	0	1	1	0
	Shoulder left		1	1	
3	Upper back	0	1	1	0
4	Upper arm right	0	1	1	0
	Upper arm left	0	1	1	0
5	Lower back	3.5	2	2	14
6	Forearm right	0	1	1	0
	Forearm left	0	1	1	0
7	Wrist right	1.5	2	1	3
	Wrist left	0	1	1	0
8	Hand/finger right	0	1	1	0
	Hand/finger left	0	1	1	0
9	Hip/buttocks	0	1	1	0
10	Thigh right	0	1	1	0
	Thigh left	0	1	1	0
11	Knee right	1.5	2	1	3
	Knee left	1.5	1	1	1.5
12	Lower leg	0	1	1	0
	Lower left	0	1	1	0

not match the size of the finger and the keyboard factor is typical or can relax the shoulders (Table 3).

The highest ROSA score is 14. This number is over the limit and can cause pain in the lower back. Thus, using ROSA on the LS-CMDQ provides accurate information for researchers. Therefore, this study suggests that operators follow the ergonomics rules while working, making operators feel safe and comfortable while working, thus increasing productivity.

C. Recommendation

1. The chair's design is according to the ergonomic concept. The office chair has the right size for the human body, and the office chair has the right shape to use.
2. Making the chair after the height of the table and the position of the operator's body, there is no excessive movement when viewing the computer.
3. Using a tilt-down keyboard (no more than 15° tilt):

   a. Significantly reduces wrist extension while typing and at rest,
   b. Greatly improves both lower back pain and shoulder pain,
   c. Wrist and hand maintain a more neutral position > 60% of typing time [9].

Table 3 Recapitulation ROSA score

Section	Category	Classification	Score	Area score
Section A	Chair height	Too high–knee angle	2	3
		Non-adjustable	+1	
	Pan depth	Too long–Less than 3″ of space	2	3
		Non-adjustable	+1	
	Armrests back support	Too high/low (Arms unsupported)	2	2
		No back support	2	3
		Non-adjustable	+1	
	Chair height–pan depth			5
	Arm rest and black support			6
	Section A + Duration			5 + 1 = 6
Section B	Monitor	Too low (below 30°)	1	3
		Neck twist greater than 30°	+1	
		Duration	+1	
	Telephone	Heat set/one hand on phone and natural neck posture	1	2
		Duration	+1	
	Section B			3
Section C	Mouse	Mouse in line with shoulder	1	3
		Pinch grip on mouse	+1	
		Duration	+1	
	Keyboard	Wrists straight, shoulders relaxed	1	3
		Platform non-adjustable	1	
		Duration	+1	
Monitor and peripheral score	Section B			3
	Section C			3
ROSA FINAL SCORE	5	Monitor and peripheral score		3
		Chair		5

References

1. Ayoub Ghanbary EH (2015) Evaluation of Musculoskeletal disorders among computer users in Isfahan Reprovado No Artigo. Iranian J Health Safety Environ 2(3):330–334
2. da Rosa BN, Secrieru J, Candotti CT (2022) Analysis of postural asymmetry on sagittal plane between right and left side views using photogrammetry. J Bodyw Mov Ther 29:251–256
3. Besharati A, Daneshmandi H, Zareh K, Fakherpour A, Zoaktafi M (2020) Work-related musculoskeletal problems and associated factors among office workers. Int J Occup Saf Ergon 26(3):632–638
4. Mohammadipour F, Pourranjbar M, Naderi S, Rafie F (2018) Work-related musculoskeletal disorders in Iranian office workers: prevalence and risk factors. J Med Life 11(4):328
5. Widodo L, Ruslie C (2020) Ergonomic analysis by using REBA, WERA and biomechanics method in the production process of women's bags in small industry (SME). In: IOP conference series: materials science and engineering, vol 1007. IOP Publishing, p 012088
6. Lotfollahzadeh A, Feiz Arefi M, Ebadi Gurjan H, Razagari N, Ebadi B, Babaei-Pouya A (2019) Musculoskeletal disorders among healthcare network staff using rapid office strain assessment. Int J Musculoskelet Pain Prevent 4(4):270–276
7. Osama M, Ali S, Malik RJ (2018) Posture related musculoskeletal discomfort and its association with computer use among university students. J Pak Med Assoc 68(4):639–641
8. Ranavolo A, Draicchio F, Varrecchia T, Silvetti A, Iavicoli S (2018) Wearable monitoring devices for biomechanical risk assessment at work: current status and future challenges—a systematic review. Int J Environ Res Public Health 15(9):2001
9. Shariat A, Cleland JA, Danaee M, Kargarfard M, Moradi V, Tamrin SBM (2018) Relationships between cornell musculoskeletal discomfort questionnaire and online rapid office strain assessment questionnaire. Iran J Public Health 47(11):1756
10. Rodrigues MSA, Leite RDV, Lelis CM, Chaves TC (2017) Differences in ergonomic and workstation factors between computer office workers with and without reported musculoskeletal pain. Work 57(4):563–572
11. Hedge A (2006) Ergonomic strategies. In: 10 Tips for using a computer mouse, p 18

Aerodynamic Analysis of the Rear Spoiler of Hatchback Vehicles Using SST K-ω Turbulence Model

Andre Kuwito, Steven Darmawan, and Harto Tanujaya

Abstract In the development of the times and technology, demands for forms and characteristics have a good aerodynamic value. As a result, various types of new models are emerging rapidly in the automotive market. In addition, many drivers modify their cars. One of them is by adding a body kit or just adding a spoiler to make it more esthetic. As time goes by, many car manufacturers pay attention to several aspects in designing their cars. One of them is aerodynamic. The aerodynamics performance of a car is affected by the drag coefficient (Cd) or drag when the car is in motion. Furthermore, to get this Cd value, a study will be carried out that focuses on the effect of adding spoilers with NACA 2412 on vehicles. The research can be carried out using the CFD simulation. In the study, a simulation was carried out using a Hatchback car and two spoiler models with a NACA 2412 airfoil profile. The result obtained is that a car without a spoiler produces a Cd of 0.562, and for a car model with a spoiler 1, it produces a Cd of 0.383, while for a car with a spoiler of 2 it produces a Cd of 0.409.

Keywords Aerodynamics · Car · CFD · K-ω turbulence model · Rear spoiler

1 Introduction

Aerodynamics is defined as a change in the movement of an object due to air resistance when moving at high speeds. The aerodynamics of a car is also called the drag coefficient (Cd) or drag when the car is in motion. Aerodynamics is inseparable from

A. Kuwito (✉) · S. Darmawan · H. Tanujaya
Universitas Tarumanagara, Jakarta, Indonesia
e-mail: andre.515180030@stu.untar.ac.id

S. Darmawan
e-mail: stevend@ft.untar.ac.id

H. Tanujaya
e-mail: hartotan@ft.untar.ac.id

Irwansyah et al. (eds.), *Proceedings of the 4th International Conference on Experimental and Computational Mechanics in Engineering*, Lecture Notes in Mechanical Engineering, https://doi.org/10.1007/978-981-99-7495-5_17

the science regarding the profile of the Naca airfoil. Spoiler is a car component that affects the aerodynamics of a car, Rear spoiler is responsible to the drag and downforce experienced by the car. Spoilers work by applying downforce or downward pressure force on the car to reduces the lift generated by high speed [1].

In particular, rear spoiler of a car come in many geometries based on the function. One could come from the dealership, or as accessories. The application of rear spoilers on a car is also often related to the esthetics of the vehicle, especially in hatchback type. Application of inappropriate rear spoiler geometry may relate to safety. Rear spoiler effect on a car can be analyzed with experimental methods using wind tunnel, which usually complicated, limited only for some geometries and costly. Other way to analyze the effect of rear spoiler is by using CFD method [2]. As the next step of this Cd analysis research, a study will be conducted that focuses on the effect of adding spoilers with NACA 2412 on a hatchback type vehicle, the vehicle to be used is the third-generation VW Scirocco [3–5]. The choice of this vehicle is because the model of the vehicle is attractive and quite in demand in the market. The formulation of the problem in this study is how the value of Cd with the use of spoilers with the Naca 2412 airfoil profile [6–8]. The formulation of other problems is what factors affect Cd in vehicles and how to analyze the fluid flow rate in vehicles.

In this study, the limitation of the problem used was a car of the hatchback type and the speed of the vehicle used started from 60 km/h. To obtain the results of this study, it will be compared the Cd value produced between the car with the addition of spoilers with NACA 2412 and with those that are not. The purpose of this study is to analyze the influence of aerodynamics of vehicle performance by using rear spoilers and to determine the effect of using spoilers with NACA 2412 on the downforce of hatchback type vehicles [9].

1.1 Problem Formulation

Looking at the problems that arise, it can be known what to do, namely:

1. How does the use of spoilers with NACA 2412 airfoil profile affect the Cd value?
2. How is the analysis of fluid flow rate in vehicles?
3. What factors affect Cd in cars?

2 Research Method

Analysis of the study of Cd on steel plate with airfoil NACA 2412 profile was carried out by experimental method. Experiments were carried out to determine the aerodynamic performance of the plate and also the best mesh configuration. This research uses computational fluid dynamics (CFD) simulation. This simulation uses the ANSYS Student 2022 R1 and Autodesk Fusion 360 software which is carried out on an Acer Aspire E14 laptop. The simulation is divided into three stages, starting with the geometric model process, then the CFD model, and the last one is post-processing [10, 11].

2.1 Geometry Model

In this study, it used a hatchback type vehicle. An example car used is the third-generation VW Scirocco car. The VW Scirocco has dimensions of 4256 × 1810 × 1404 mm. The creation of this geometry is done using the Autodesk fusion 360 application (Figs. 1, 2).

Next is the spoiler design process which is also carried out on the Autodesk fusion 360. The spoiler is designed by utilizing the shape of the NACA 2412 airfoil profile, and the dimensions of the spoiler are 1800 × 150 mm as seen on Fig. 3.

Fig. 1 VW Scirocco [12]

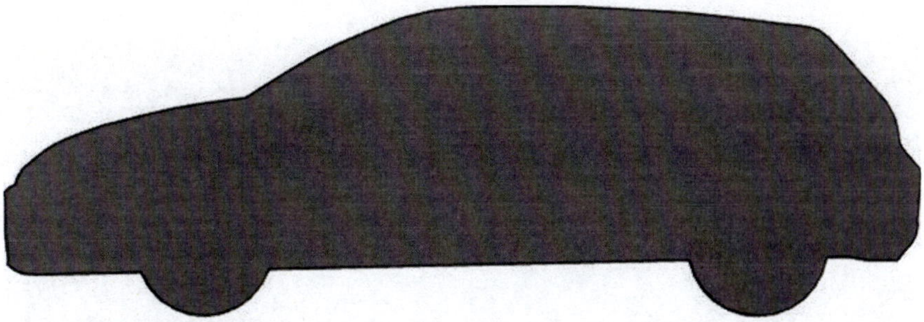

Fig. 2 Car geometry design

Fig. 3 Spoiler model

The computational domain used for simulation has a length of 1418.64 mm, a width of 784.73 mm, and a height of 463.16 mm with details of the length of the front wall, right side, left side, and top which is two times the size of the car length. For the rear length, it is 5 times the length of the car. And the bottom wall is 50 mm.

2.2 CFD Models and Configuration

CFD simulation is conducted by using Ansys Fluent Student Edition. In the simulation process of this vehicle, a 0.005 m mesh is used and with a type of mesh, namely tetrahedral SST *k-ω turbulence model is used* [13]. For the vehicle simulation process, the velocity used is 16.66 m/s or 60 km/h, the density of air is 1.225 kg/m^3, and the temperature is 288.16 K. After carrying out the feeding meshing process, the number of nodes was obtained by 87,382 and the number of cells was 469,049 (Fig. 4).

The car used for the simulation process will be reduced by a scale of 1:24 or reduced by 24 times its original size. This is intended to simplify the CFD simulation process that will be carried out. After the vehicle and spoiler design process is carried out, the process continued to assembling the two component. The spoiler is placed on the back of the car with a slope level of 25°, and for the second spoiler, it is placed 150 mm from the vehicle body. CFD simulation conducted three-dimensionally with single-phase air. In order to simplify the simulation process, the geometry model is

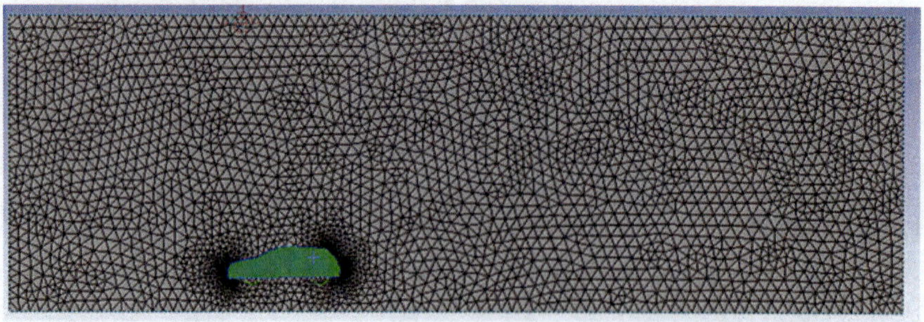

Fig. 4 Meshing results

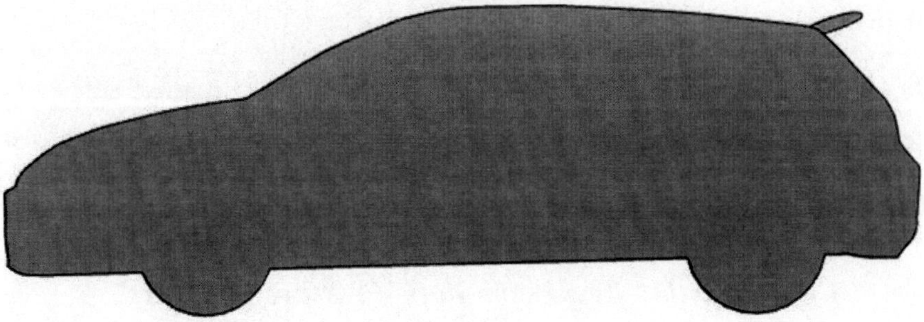

Fig. 5 Car configuration 1

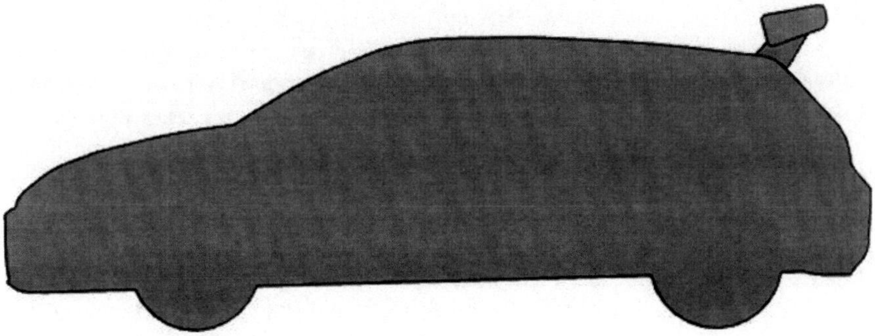

Fig. 6 Car configuration 2

reduced 24 times than its original size. Since the right and left side geometry of the model is identic, the simulation is assumed to be symmetric (Figs. 5, 6).

2.3 Turbulence Model

This k-omega turbulent SST model is a k-omega turbulent model that has been refined several times due to its weakness in sensitivity to free stream boundary conditions. In the model refined by Menter, the k-omega model is combined with the k-epsilon model, so that it has good capabilities in the area around the wall and a low Reynold number as an advantage of k-omega, and flows with adverse pressure gradients and insensitivity in the free stream area which is the advantage of k-epsilon.

$$\frac{\partial}{\partial_t}(\rho k) + \frac{\partial}{\partial_{x_i}}(\rho k u_i) = \frac{\partial}{\partial_{x_j}}\left[\tau_\omega \frac{\partial k}{\partial_{x_j}}\right] + G_k - y_k + S_k \quad (1)$$

$$\frac{\partial}{\partial_t}(\rho \underline{\omega}) + \frac{\partial}{\partial_{x_j}}(\rho \underline{\omega} u_j) = \frac{\partial}{\partial_{x_j}}\left[\tau_\omega \frac{\partial \varepsilon}{\partial_{x_j}}\right] + G_\omega - Y_\omega + D_\omega + S_\omega \quad (2)$$

3 Result

The final result of the simulation will later compare the best Cd values of cars without spoilers, cars with spoilers 1, and cars with spoilers 2. The simulation will be carried out by providing a fluid flow inlet on the X axis.

3.1 CFD Simulation Results on Static Pressure Contours

In the simulation results carried out on the three car models, it can be seen that Fig. 8 which is the result of static pressure simulation for cars without spoilers has a fairly small static pressure with a static pressure value of 1616 Pa compared to two car models with other spoilers. Figure 9 illustrates the greatest spread of static pressure on the front bumper, windshield, and in the spoiler section of the car with a value of 1942 Pa. Meanwhile, Fig. 10 which depicts a simulation of a car with spoiler 2 shows the largest spread of static pressure points in the front bumper, windshield, and legs that connect the spoiler and car body with a static pressure value of 1720 Pa.

By looking at the contours of static pressure in Figs. 7, 8, and 9, it can be seen that with the addition of a spoiler, it can affect the static pressure on the car. The greater the static pressure, the greater the downforce on the car. This happens because static pressure occurs in the spoiler and the direction of the static pressure is toward the bottom. So the greater the static pressure that exists, the greater the downforce value, with an average value increase of 0.63%.

3.2 CFD Simulation Results on Velocity Contours

In the simulation with a car model without a spoiler depicted in Fig. 11, the speed produced after the separation process is equal to the inlet speed of 16.66 m/s. This event also affects the resulting wake being less massive. This happens because the data collection point chosen is the point that will be affected by the presence of a spoiler, so that in a car simulation without a spoiler, the resulting speed does not have an impact on body separation.

In the second simulation, namely a car with spoiler 1, it can be seen in Fig. 12, the speed produced after the flow separation process is 2.55 m/s. The speed is reduced due to the presence of a large Wake, so that it has an impact on reducing the flow speed.

In the third simulation using a car model with spoiler 2 whose results can be seen in Fig. 13, the resulting speed is 12.85 m/s. This happens because the Wake produced after separation of the flow is not too large, so that it has an impact on the speed on the back of the car.

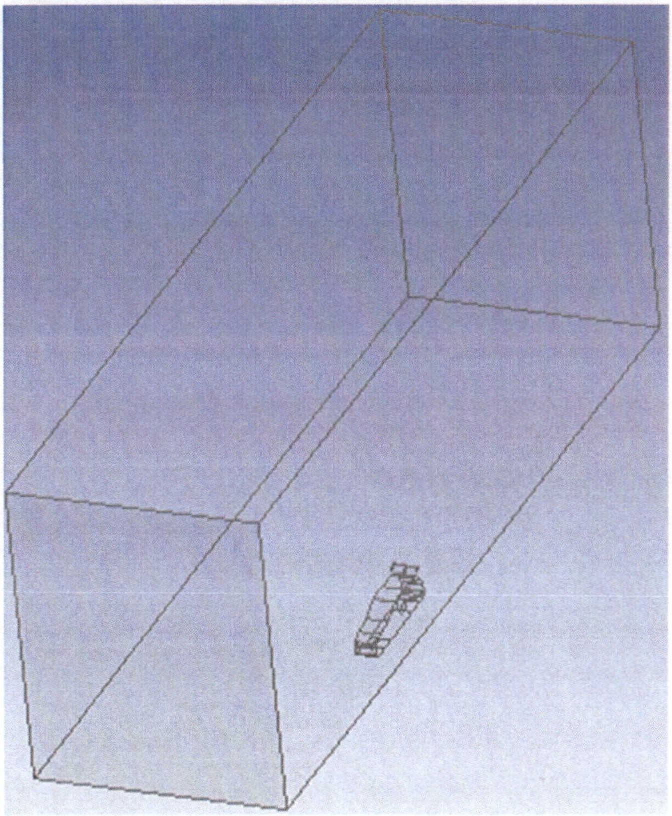

Fig. 7 Symmetry domain

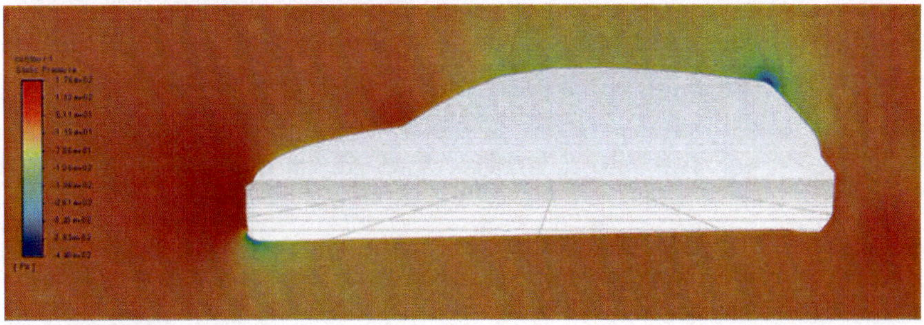

Fig. 8 Static pressure car without spoiler

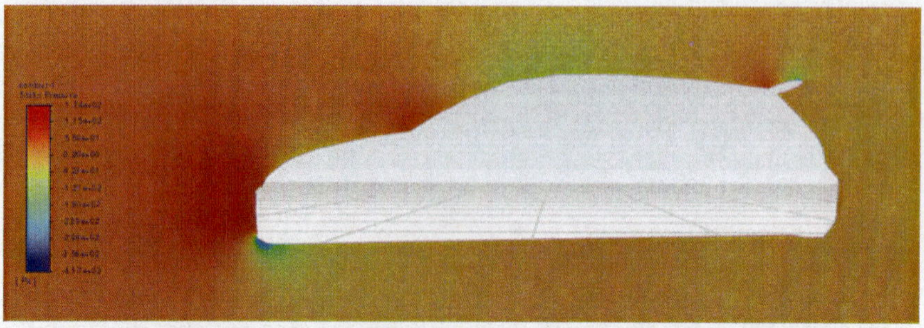

Fig. 9 Static pressure generated by spoiler model 1

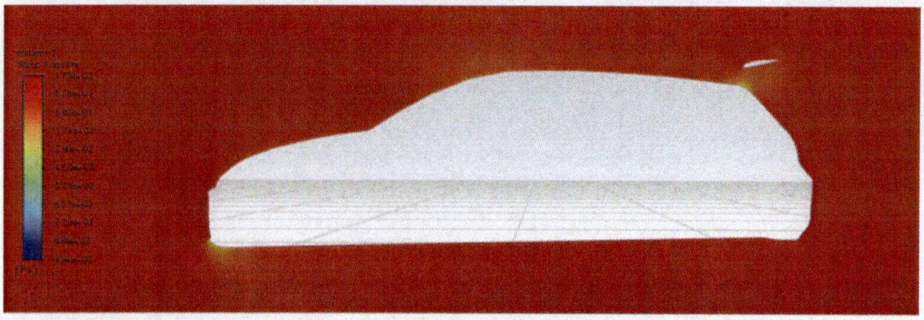

Fig. 10 Static pressure generated by spoiler model 2

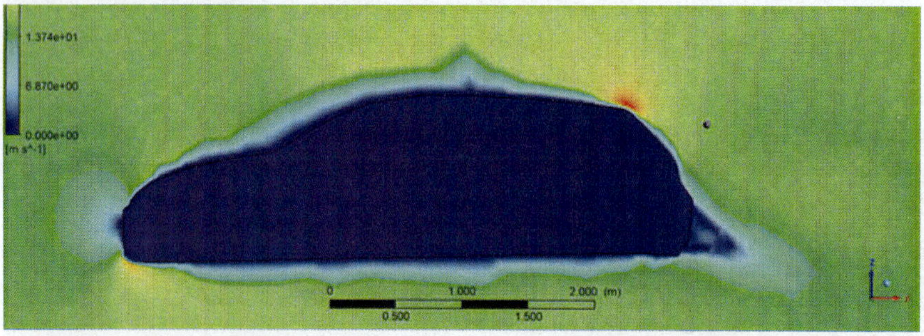

Fig. 11 Car velocity contours without spoiler

3.3 CFD Simulation Results on Aerodynamic Force

Based on the presentation of simulation data and calculations that have been carried out above, it can be seen that the Cd value generated by cars with spoiler 1 has a smaller value of 0.383 while cars without spoilers have a Cd value of 0.562; this indicates that the Cd on cars with spoiler 1 depreciated by 31.73%, while in cars with spoiler 2 which have a Cd value of 0.409, it depreciated from cars without spoilers

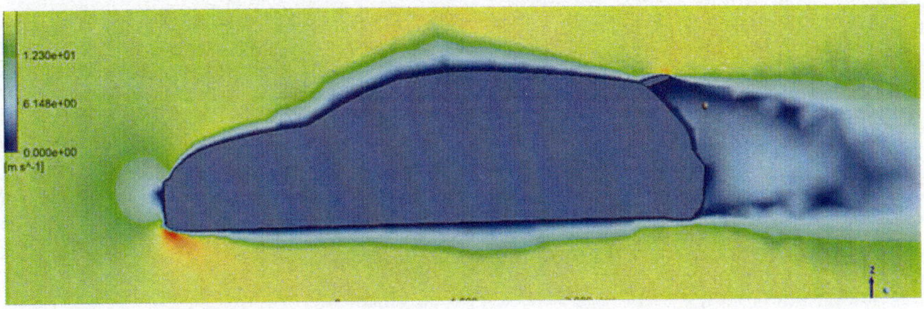

Fig. 12 Car velocity contour with spoiler 1

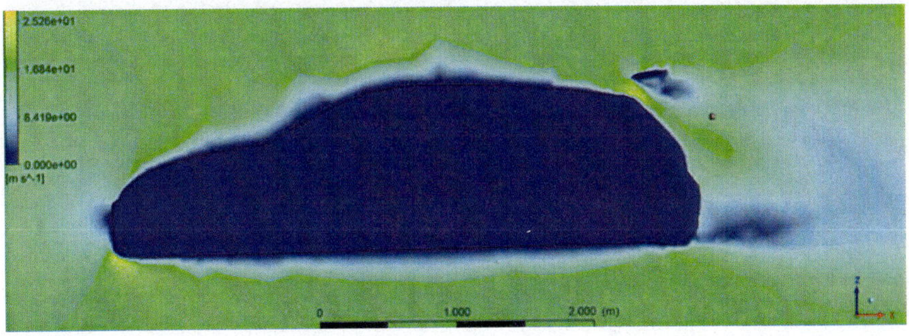

Fig. 13 Car velocity contour with spoiler 2

as much as 27.13%. This CFD simulation results related on Cd and Cl provided in Table 1.

From this, it can be said that the addition of spoilers has proven to be able to reduce the value of Cd which makes drag on vehicles improve. This is also evidenced by the European Journal of Computational Mechanics with its journal entitled "Reduce drag Coefficient of a Hatchback car utilizing fractional factorial design algorithm", suggesting that cars with spoilers can produce Cd that are less valuable than cars that do not have a spoiler [14]. The journal simulates a car with a hatchback type and produces a Cd on a 0.282-sized car and a 0.411 non-spoiler car.

In addition to Cd, there are simulation results in the form of lift or cl coefficients, the value of this coefficient will decrease as static pressure increases. This can be proven by the cl value in cars with spoilers that have been reduced by 5.47% compared

Table 1 Drag coefficient and lift coefficient data

No	Variations	Cd	Decrease value of Cd	Cl	Decrease value of Cl
1	Without spoiler	0.562	–	0.737	–
2	Spoiler 1	0.383	31.73%	0.696	5.47%
3	Spoiler 2	0.409	27.13%	0.698	5.21%

to cars without spoilers with a static pressure value of 1942 Pa. As for the cl value on cars with spoiler 2, it was reduced by 5.21% with a static pressure value of 1.720.

4 Conclusion

Based on the simulation results with the CFD method and the presentation of data that has been previously provided, it can be concluded:

1. Based on the simulation results on three car models with a speed of 16.66 m/s or 60 km/h, it can be concluded that adding a spoiler can make drag on the car improve, and this can be seen in Fig. 11 describing the results of the car velocity simulation with spoiler 1 in addition to that in Fig. 8 explaining that the vehicle with spoiler 1 has the most static pressure among 2 other car models, namely 1.942 Pa which will make the compressive force down the better.
2. Wake created by the separation of the flow of the vehicle body can cause a drag effect on the vehicle the greater and longer the Wake created, the better the drag on the vehicle and vice versa the smaller and shorter the Wake created by the separation of the vehicle body flow, the worse the drag that exists on the vehicle.
3. Taking into account the aerodynamic forces in vehicles, the NACA 2412 airfoil chosen as the basic form of spoiler manufacturing was shown to increase drag on the vehicle. This can be viewed from the Cd value generated by each vehicle model. Cars without spoilers have a Cd value of 0.562, for cars with spoiler 1 have a Cd value of 0.383, and for cars with spoiler 2 have a Cd value of 0.409. This shows that the use of spoilers with a NACA 2412 airfoil profile can reduce the Cd produced by cars without spoilers.

References

1. Cakir M (2012) CFD study on aerodynamic effects of a rear wing/spoiler on a passenger vehicle. St Cl Univ Sch Commons pp 1–72. Retrieved from http://scholarcommons.scu.edu/mech_mstr
2. Kesuma P, Darmawan S, Halim A (2020) Aerodynamics analysis of mobil irit tarumanagara using CFD method. IOP Conf Ser Mater Sci Eng 1007(1). https://doi.org/10.1088/1757-899X/1007/1/012032
3. Eken S (2019) Free vibration analysis of composite aircraft wings modeled as thin-walled beams with NACA airfoil sections. Thin-Walled Struct 139:362–371. https://doi.org/10.1016/j.tws.2019.01.042
4. Yossri W, Ben Ayed S, Abdelkefi A (2021) Airfoil type and blade size effects on the aerodynamic performance of small-scale wind turbines: Computational fluid dynamics investigation. Energy 229. https://doi.org/10.1016/j.energy.2021.120739
5. Palanivendhan M, Chandradass J, Bannaravuri PK, Philip J, Shubham K (2021) Aerodynamic simulation of optimized vortex generators and rear spoiler for performance vehicles. Mater Today Proc 45:7228–7238. https://doi.org/10.1016/j.matpr.2021.02.537

6. Dey S, Saha R (2020) Performance of NACA cambered and symmetrical airfoils as rear spoilers on race car aerodynamic drag & lift: computational fluid dynamics study. Proc Int Exch Innov Conf Eng Sci 6:292–298. https://doi.org/10.5109/4102505
7. Chodagudi KBR (2012) Analysis Of Naca 2412 for automobile rear spoiler using composite material certain applications. Int J Eng Res Technol 1(7):1–5
8. No R (1931) The national advisory committee for aeronautics. Science 74(1923):451. https://doi.org/10.1126/science.74.1923.451
9. Dey S, Saha R (2018) CFD Study on aerodynamic effects of NACA 2412 air foil as rear wing on a sports car CFD study on aerodynamic effects of NACA 2412 airfoil as rear wing on a sports car. no. December, pp 0–6. Retrieved from https://www.researchgate.net/project/CFD-Study-on-Aerodynamic-Effect-of-NACA-2412-Airfoil-as-Rear-Wing-on-a-Sports-Car
10. Taha TR (2005) An introduction to parallel computational fluid dynamics. 6(4). https://doi.org/10.1109/mcc.1998.736434
11. Fluent A, "Ansys Fluent Theory Guide," *ANSYS Inc., USA*, vol. 15317, no. November, pp. 724–746, 2013.
12. "Volkswagen Scirocco GTS teased." https://www.motortrend.com/news/1504-volkswagen-scirocco-gts-teased/ (accessed Jul. 17, 2022).
13. K.-J. Bathe, *Computational fluid and solid mechanics 2003[Recurso electrónico] :] proceedings, Second MIT Conference on Computational Fluid and Solid Mechanics, June 17–20, 2003*. 2003. [Online]. Available: http://0-www.sciencedirect.com.jabega.uma.es/science/book/9780080440460
14. Vahdati M, Beigmoradi S, Batooei A (2018) Minimising drag coefficient of a hatchback car utilising fractional factorial design algorithm. Eur J Comput Mech 27(4):322–341. https://doi.org/10.1080/17797179.2018.1550962

A Fatigue Life Assessment of Automotive Coil Spring and Lower Arm Due to Road Surface Contours

Teuku Edisah Putra, Husaini, Muhammad Ikbal, Muhammad Rizal, and Amir Zaki Mubarak

Abstract This study aims to determine the effect of road surface contours on the life of car suspension components. In this study, strain gauges were fixed on the critical points of a coil spring and a lower arm. The car went through smooth and rough road surfaces with speeds over 30 and below 20 km/h, respectively, for 60 s. The strain signals obtained were analyzed according to the Coffin-Manson, Morrow and Smith–Watson–Topper models. According to the strain amplitudes measured, the coil spring fatigue life of the rough road surface was 16% lower compared with the smooth surface, whereas the lower arm of the rough road surface was 27% lower than the smooth surface. It was also found that the coil spring failed 36% faster than the lower arm when driving on the rough road surface. This is because the contour of the surfaces provides a vertical load affecting the coil spring at higher rates than the lower arm. It is in accordance with the function of the coil spring which works to reduce the load vertically.

Keywords Vibration · Stress · Strain · Failure

T. E. Putra (✉) · Husaini · M. Ikbal · M. Rizal · A. Z. Mubarak
Department of Mechanical Engineering, Universitas Syiah Kuala, Darussalam, Banda Aceh 23111, Indonesia
e-mail: edi@usk.ac.id

Husaini
e-mail: husainiftm@usk.ac.id

M. Rizal
e-mail: muh.rizal@usk.ac.id

A. Z. Mubarak
e-mail: amir_zm@usk.ac.id

Irwansyah et al. (eds.), *Proceedings of the 4th International Conference on Experimental and Computational Mechanics in Engineering*, Lecture Notes in Mechanical Engineering, https://doi.org/10.1007/978-981-99-7495-5_18

1 Introduction

There are many fatigue failures occurring in the vehicle components. These can be caused by different physical conditions of roads, because the stability of the vehicle is absolutely dependent on the friction of tires with the road surface [1]. The friction produces vibrations which increase during the time the vehicle passes a rough road surface at high speed. If the vibration occurs repeatedly, it causes fatigue damage to the components.

The suspension system is a mechanical structure with the role of reducing displacement vertically and horizontally [2–5] and provides maximum results for tire friction and roads to stabilize the vehicle [6, 7]. Two vital suspension components are the coil spring and lower arm. The coil spring functions to hold, reduce, and absorb impact, torsional and cyclic loads [8, 9], while the lower arm acts as a movement controller both back and forth and during the vehicle turning [10].

Components often fail because residual stresses occur repeatedly and can affect their durability [11–13]. In previous studies, fatigue life assessments of a coil spring [14–18] and a lower arm [19] were carried out. However, the components were analyzed separately. Therefore, the purpose of this study is to predict the fatigue life for both components simultaneously to study the reaction of each component when running on the same road surface.

2 Material and Methods

Chemical Composition Test. The coil spring and lower arm of a passenger car were used as the case study. The research started with a chemical composition test aiming to determine the chemical elements contained in the components. The testing samples (Fig. 1) first had to be grinded in order for the surface to be clean of contaminants that might disturb the testing results. The sample was inserted into a spectrometer and then injected with argon fluidized electrode.

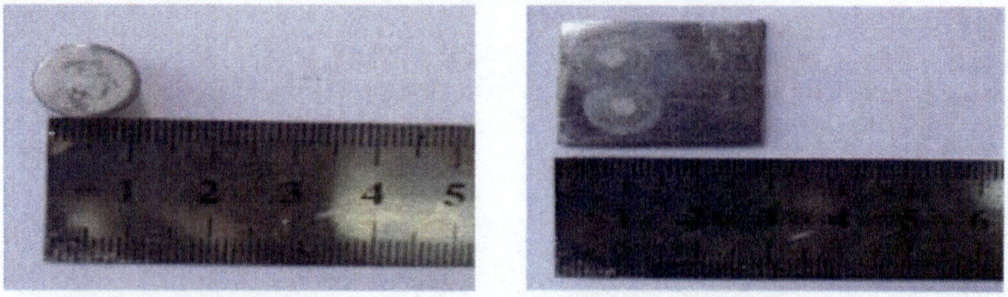

(a) (b)

Fig. 1 Specimens for the chemical composition tests: **a** coil spring, **b** lower arm

Stress Analysis. An engineering problem that aims to deteermine the critical point in three-dimensional components can be solved by the finite element analysis (FEA) which analyzes by dividing components into several small elements. The accuracy of the simulation results is very dependent on the preprocessing stage where the value and direction of the load and the boundary conditions are included. The meshing used for the simulation was a tetrahedron. This type of meshing produces better results than other meshing [20]. The total nodes and elements for the coil spring were 10,999 nodes and 3710 elements, while for the lower arm there were 36,562 nodes and 18,530 elements.

Considering that the weight of the car was 1045 kg and the weight of the passenger(s) and the load carried was 420 kg, the total weight was divided by four by assuming the load was distributed to all wheels. This way, the load received by the coil spring and lower arm was 366.25 kg or 3591.7 N. The load was given at the bottom of the coil spring (B), while the upper part was given fixed support (A). For the lower arm, the load was given to the ball joint (C), while the bushing sections (A and B) were given roller supports. The mechanism for giving loads and supports can be seen in Fig. 2.

von Mises [21] is derived from the three-dimensional main stress relationship, namely hoop, longitudinal, and radial directions. According to the relationship, the yield criterion occurs when σ_e is equal to yield stress σ_{ys}, which can be defined by:

$$\sigma_e = \frac{1}{\sqrt{2}}\left[\left(\sigma_x - \sigma_y\right)^2 + \left(\sigma_y - \sigma_z\right)^2 + \left(\sigma_z - \sigma_x\right)^2 + 6\left(\tau_{xy}^2 + \tau_{yz}^2 + \tau_{zx}^2\right)\right]^{1/2} \quad (1)$$

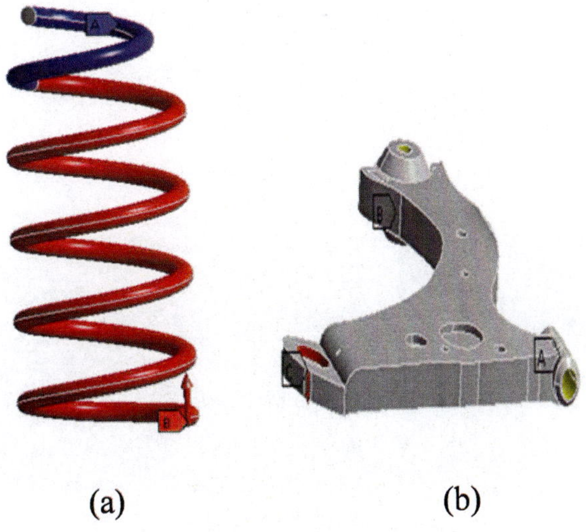

Fig. 2 Boundary conditions: **a** coil spring, **b** lower arm

where σ_x, σ_y, and σ_z are the normal stresses in the directions of the x, y, and z-axes, while τ_{xy}, τ_{yz}, and τ_{zx} are shear stresses.

Strain Signal Measurement. Strain signal measurements were carried out by attaching 3-mm strain gauges with a resistance of 350 Ω on the coil spring and the lower arm. The strain gauge mounting locations were chosen according to the highest stress distribution area. The frequency of the measurement must be greater than 400 Hz [22]. Therefore, a frequency of 500 Hz was suitable to detect and store all destructive load cycles in both components.

After the installation of the equipment, the passenger car was driven on smooth and rough road surfaces. In order to determine the road surface profiles, international roughness index (IRI) [23] was used as follows: in order for a road to be categorized as smooth or rough, the IRI value must be below 4 and over 12, respectively. The IRI is defined as [24, 25]:

$$IRI = \frac{1}{L}\int_{0}^{T} |\dot{z}|\mathrm{d}t \tag{2}$$

where $\dot{z}$ is the suspension velocity due to the road excitation, L is travel distance, and T is the time. The vehicle was driven at a speed over 30 km/h on the smooth road surface and below 20 km/h for the rough surface. Figure 3 shows how to install the strain gauges, and Fig. 4 shows the road surfaces passed by the car.

Fatigue Life Assessments. In the study, fatigue life assessments were performed according to the strain-life approaches, namely the Coffin-Manson, Morrow, and Smith–Watson–Topper (SWT) models. The Coffin-Manson model [26, 27] can be expressed as:

$$\varepsilon = \frac{\sigma\prime_f}{E}(2N_f)^b + \varepsilon\prime_f(2N_f)^c \tag{3}$$

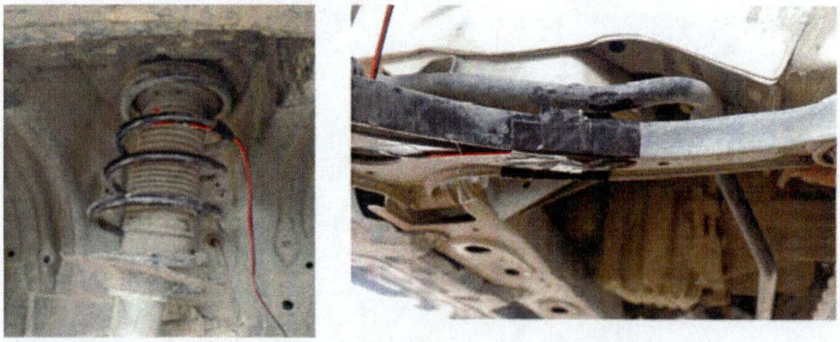

(a) (b)

Fig. 3 Installation of strain gauges: **a** coil spring, **b** lower arm

(a) (b)

Fig. 4 Road surfaces: **a** Smooth, **b** Rough

where ε is the strain amplitude, σ'_f is the fatigue strength coefficient, E is the material modulus of elasticity, N_f is the number of cycles, b is the fatigue strength exponent, ε'_f is the fatigue ductile coefficient, and c is the fatigue ductile exponent.

Two more models, Morrow [28] and SWT [29], consider the mean stress effect [30, 31] expressed, respectively, by:

$$\varepsilon = \frac{\sigma\prime_f - \sigma_{\text{mean}}}{E}\left(2N_f\right)^b + \varepsilon\prime_f\left(2N_f\right)^c \tag{4}$$

$$\sigma_{\max}\varepsilon = \frac{\sigma\prime_f^2}{E}\left(2N_f\right)^{2b} + \sigma\prime_f\varepsilon\prime_f\left(2N_f\right)^{b+c} \tag{5}$$

where σ_{mean} is the mean stress and $\sigma_{\max}$ is the maximum stress amplitude.

Fatigue damage for each loading cycle D_i is:

$$D_i = \frac{1}{N_f} \tag{6}$$

Then, the cumulative fatigue damage was calculated utilizing the Palmgren–Miner rule [32, 33]:

$$D = \sum\left(\frac{n_i}{N_f}\right) \tag{7}$$

where n_i is the number of applied cycles. The process of this research is summarized in Fig. 5.

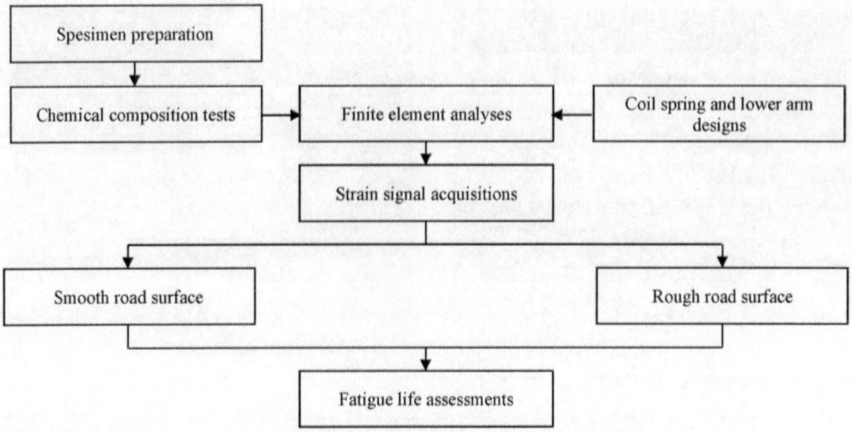

Fig. 5 Process flow of the research

3 Results and Discussion

Material Composition. The results of the chemical composition tests are listed in Table 1. The chemical composition of the coil spring is similar to the SAE 5160 carbon steel, a material often used in the automotive industries to make coil springs, while the chemical composition of the lower arm is the AISI 1513 carbon steel. The mechanical properties of the materials are given in Table 2.

Stress Distribution. The stress concentrations for the coil spring and the lower arm are presented in Fig. 6. The area with the highest stress concentration is denoted by red, followed by orange, yellow, green, and blue. The red area of the coil spring has a maximum von Mises stress of 997.51 MPa. Referring to Table 2, the maximum stress is low compared with the yield strength of the SAE 5160 carbon steel which is 1,487 MPa. The maximum von Mises stress for the lower arm is 358.82 MPa, which is still below the yield strength of the AISI 1513 carbon steel of 450 MPa in Table 2.

Table 1 Results of the chemical composition tests

Chemical elements	Coil spring (%)		Lower arm (%)	
	Testing results	SAE 5160 [34]	Testing results	AISI 1513 [35]
Carbon (C)	0.6415	0.56–0.64	0.09496	0.100–0.160
Silicon (Si)	0.1631	0.15–0.3	0	0
Manganese (Mn)	0.7780	0.75–1	1.33170	1.10–1.40
Chromium (Cr)	0.7682	0.7–0.9	0.03094	0
Molybdenum (Mo)	0.1838	0.15–0.25	0	0
Vanadium (V)	0.15171	0.15	0.00285	0

Table 2 Material mechanical properties of the SAE 5160 carbon steel and the AISI 1513 carbon steel [36]

Mechanical properties	SAE 5160	AISI 1513
Ultimate tensile strength (MPa)	1584	585
Modulus of elasticity (GPa)	207	210
Yield strength (MPa)	1487	450
Fatigue strength coefficient (MPa)	2063	1,089.4
Fatigue strength exponent	−0.08	−0.0787
Fatigue ductility exponent	−1.05	−0.5425
Fatigue ductility coefficient	9.56	0.4109
Cyclic strain hardening exponent	0.05	0.1169
Cyclic strength coefficient (MPa)	1940	978.68
Poisson ratio	0.27	0.27

This indicates that the load does not cause a failure of the components. The surface area of the inner spring is smaller than the outside, therefore, the critical point occurs inside [37]. For the lower arm, the highest stress occurs in the area near the roller support on the right bushing. This is due to the fact that the component has many holes that cause the surface area to become narrow.

Strain Signals. The smooth road surface had an IRI value of 2.4, while the one for the rough road surface was 15. The total length of the strain signals recorded were 60 s for 30,000 data. As shown in Fig. 7 for the coil spring strain signals, the smooth road strain signal (S1) had an amplitude range between −769 με and −569 με. A higher amplitude range was provided by the rough road strain signal (S2), which was between −769 με and −237 με. For the lower arm strain signals (Fig. 8), the smooth road strain signal (S3) was distributed between −24 με and 14 με. Meanwhile, the

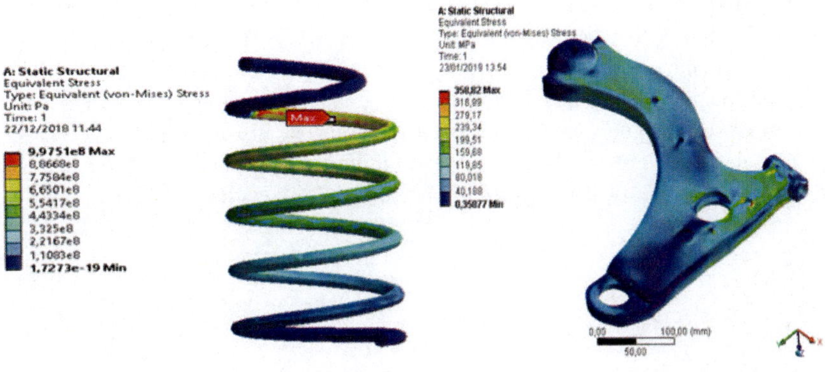

(a) (b)

Fig. 6 Stress distributions: **a** coil spring, **b** lower arm

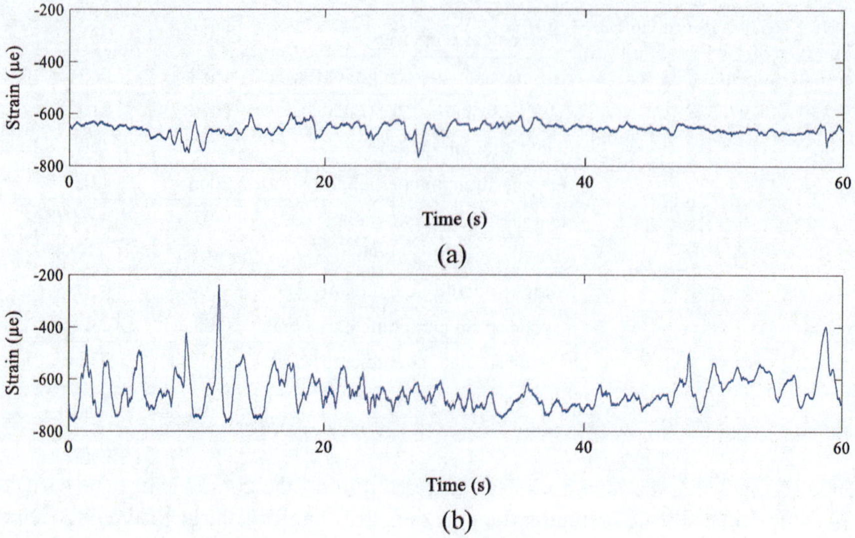

Fig. 7 Coil spring strain signals: **a** Smooth-S1, **b** Rough-S2

rough road strain signal (S4) was distributed between −35 and 16 με. All the strain signals have a nonzero mean value, namely −658.319, −649.771, −5.067, and −11.570 με for S1, S2, S3, and S4, respectively. A negative mean value shows that the components received a compressive load.

Fatigue Life. Based on the Coffin-Manson model, the fatigue damages of S1, S2, S3, and S4 were 4.888E−2 damage per block, 2.273E−1 damage per block, 1.773E−2 damage per block, and 1.002E−1 damage per block, respectively. The Morrow model gave the fatigue damages of 4.067E−4 damage per block, 2.113E−3 damage per block, 1.701E−2 damage per block, and 9.513E−2 damage per block, respectively. The SWT model provided the fatigue damages of 2.489E−2 damage per block, 1.557E−1 damage per block, 1.533E−2 damage per block, and 8.095E−2 damage per block, respectively. Since they were the one that provided lower fatigue damage, the smooth strain signals gave a longer fatigue life. The life spans of S1 and S3 were 8,036 cycles to failure and 13,050 cycles to failure, respectively. For S2 and S4, the life spans were of only 1284 cycles to failure and 3580 cycles to failure.

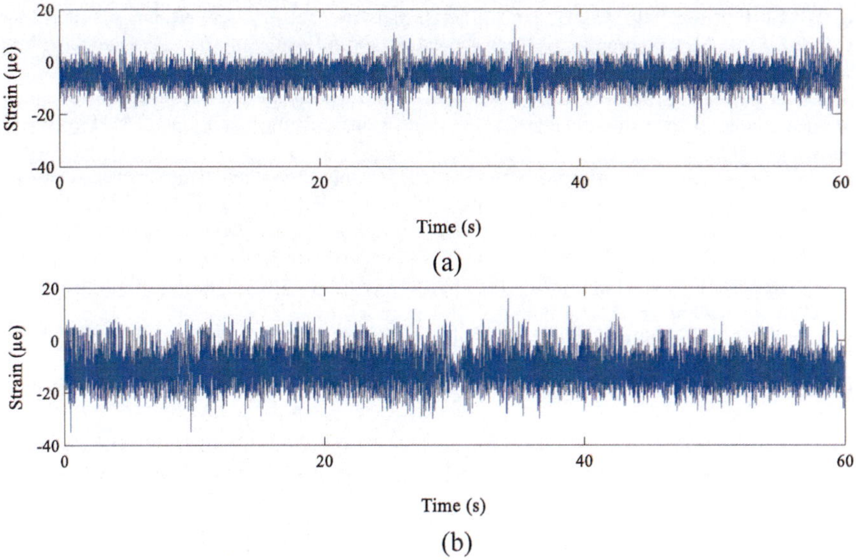

Fig. 8 Lower arm strain signals: **a** Smooth-S3, **b** Rough-S4

4 Conclusion

The objective of the study is to identify the effect of road surface contour on the fatigue life of the coil spring and lower arm of a passenger car. From the fatigue life, S2 was 16% lower than S1, whereas S4 was 27% lower than S3. It was due to higher amplitudes contained in the rough surface strain signals. When the car goes over the rough road surface, the coil spring fails 36% faster than the lower arm since the road surface profiles give a load vertically, which is suitable with the coil spring function. Meanwhile, the lower arm function holds the load when turning, so the component is inactive when receiving a vertical load.

Acknowledgements The authors would like to express their gratitude to Universitas Syiah Kuala for financial support for this research through grant no. 13/UN11.2.1/PT.01.03/DPRM/2022.

References

1. Jin LQ, Ling M, Yue W (2017) Tire-road friction estimation and traction control strategy for motorized electric vehicle. PLoS ONE 12:e0179526
2. Anubi OM, Crane C (2014) A new semiactive variable stiffness suspension system using combined skyhook and nonlinear energy sink-based controllers. IEEE Trans Control Syst Technol
3. Banitalebidehkordi H (2014) Vibration and force analysis of lower arm of suspension system. Université Du Québec

4. Tandel A, Deshpande AR, Deshmukh SP, Jagtap KR (2014) Modeling, analysis and PID controller implementation on double wishbone suspension using SimMechanics and Simulink. Procedia Eng 97:1274–1281
5. Ka'ka S, Himran S, Renreng I, Sutresman O (2018) The pneumatic actuators as vertical dynamic load simulators on medium weighted wheel suspension mechanism. IOP Conf Ser: J Phys: Conf Ser 962:012022
6. Huihui P, Weichao S, Huijun G, Jinyong Y (2015) Finite-time stabilization for vehicle active suspension systems with hard constraints. IEEE Trans Intell Transp Syst 16:2663–2672
7. Wang J, Zhang J (2015) Model-free tracking control for vehicle active suspension systems. Proceedings of the 34th Chinese control conference, pp 8067–8072
8. Ansari AA, Jain KK (2016) Review of design and numerical analysis of helical spring used in automobile. Glob J Eng Sci Res 3
9. Kumhar V (2016) Structure analysis of shock absorber spring using finite element analysis. VSRD Int J Mech Civ Autom Prod Eng 6
10. Do-Hyoung K, Dong-Hoon C, Hak-Sung K (2014) Design optimization of a carbon fiber reinforced composite automotive lower arm. Composites: Part B 58:400–407
11. Cruz MLS, Carrillo J, de Almeida SFM (2016) Effect of thermal residual stresses on buckling and post-buckling properties of laminated composites perimetrally reinforced. Latin Am J Solids Struct 13:435–455
12. Mahalov MS, Blumenstein VY (2016) Finite element surface layer inheritable condition residual stresses model in surface plastic deformation processes. IOP Conf Ser: Mater Sci Eng 126:012004
13. Yanxun X, Da T, Mingxi D, Yunze L, Changjun L, Fuzhen X (2018) Characterization of local residual stress at blade surfaces by the V(z) curve technique. Metals 8:651
14. Husaini, Putra TE, Ali N (2018) The Morlet wavelet transform for reducing fatigue testing time of an automotive suspension signal. AIP Conf Proceed 1983. 030003-1–030003-6
15. Putra TE, Husaini (2018) Identifying strain signal characteristics of automotive suspension system subjected to road surface vibrations. AIP Conf Proc 1983. 030004-1–030004-6
16. Putra TE, Husaini (2018) Investigating the road surface effect to the fatigue life of an automotive coil spring. IOP Conf Ser: Mater Sci Eng 352:012016
17. Nasir NNM, Abdullah S, Singh SSK, Haris SM (2018) Extreme condition strain signal reliability assessment using empirical mode decomposition. In: Proceedings of mechanical engineering research day, pp 23–24
18. Rahim AAA, Abdullah S, Singh SSK, Nuawi MZ (2018) Discrete wavelet transform method for fatigue analysis on car suspension system. In: Proceedings of mechanical engineering research day, pp 22–26
19. Ijagbemi CO, Oladapo BI, Campbell HM, Ijagbemi CO (2016) Design and simulation of fatigue analysis for a vehicle suspension system (VSS) and its effect on global warming. Procedia Eng 159:124–132
20. Mulla TM, Kadam SJ, Kengar VS (2012) Finite element analysis of helical coil compression spring for three wheeler automotive front suspension. Int J Mech Ind Eng 2:74–77
21. von Mises R (1913) Mechanik der festen Körper im plastisch-deformablen Zustand. Nachrichten von der Gesellschaft der Wissenschaften zu Göttingen, Mathematisch-Physikalische Klasse 1:582–592
22. Ilic S (2006) Methodology of evaluation of in-service load applied to the output shafts of automatic transmissions. The University of New South Wales
23. ASTM E1926-08 (2021) Standard practice for computing international roughness index of roads from longitudinal profile measurements. ASTM International, West Conshohocken
24. Chen S-L, Lin C-H, Tang C-W, Chu L-P, Cheng C-K (2020) Research on the international roughness index threshold of road rehabilitation in metropolitan areas: a case study in Taipei city. Sustainability 12:10536
25. Šroubek F, Šorel M, Žák J (2021) Precise international roughness index calculation. Int J Pavement Res Technol

26. Coffin LF Jr (1954) A study of the effects of cyclic thermal stresses on a ductile metal. Trans ASME 76:931–950
27. Manson SS (1965) Fatigue: a complex subject—some simple approximation. Exp Mech 5:193–226
28. Morrow J (1968) Fatigue design handbook. Society of Automotive Engineers, Warrendale
29. Smith KN, Watson P, Topper TH (1970) A stress-strain function for the fatigue of materials. J Mater JMLSA 5:767–778
30. Ince A (2017) A mean stress correction model for tensile and compressive mean stress fatigue loadings. Fatigue Fract Eng Mater Struct 40:939–948
31. Shan L, Yunlai S, Mao Y, Yan L (2018) A modified Walker model dealing with mean stress effect in fatigue life prediction for aeroengine disks. Math Prob Eng
32. Palmgren A (1924) Die Lebensdauer von Kugellagern. Zeitschrift VDI 68:339–341
33. Miner MA (1945) Cumulative damage in fatigue. J Appl Mech 67:A159–A164
34. ASTM A322-91 (2001) Standard specification for steel bars, alloy, standard grades. ASTM International, West Conshohocken
35. ASTM A29/A29M-05 (2006) Standard specification for steel bars, carbon and alloy, hot-wrought, for general requirements. ASTM International, West Conshohocken
36. nCode (2018) GlyphWorks, nCode International, Ltd. Sheffield
37. Putra TE, Husaini, Machmud MN (2020) Predicting the fatigue life of an automotive coil spring considering road surface roughness. Eng Fail Anal 116:104722

Talang Salted Fish Processing Packaging Model Using Value Engineering Methods

A. Amri and Saifuddin Muhammad Jalil

Abstract One product that needs attention, especially regarding packaging, is the talang salted fish product, which has been trading talang salted fish without sacrificing packaging. This study aims to organize the packaging of talang salted fish so that it has a selling value that can attract consumers to shop. The strategy used is value engineering and the analytical hierarchy method (AHP). First, the engineering method is feasible to be applied to expand the value, performance, and value of packaging design, which consists of brand name criteria, Internet weight measurement, product identity, attractiveness, color suitability, typography, and font size. Then, the analytical hierarchy method is applied at the analysis stage to verify the priority criteria that follow consumer desires. The results showed that the chosen design with the best performance was an alternative (A-B-D-E), to be precise, the display was made of clear plastic packaging material, the colors used were bright, the packaging was rectangular, and each capital letter was used. This elite alternative design includes a value of 0.039. This value is obtained from the performance of 38.9 and the price of IDR thousand per pack. The novelty of this research is that the packaging model is easier to recognize by customers because the packaging is more attractive, more hygienic and does not cause odor from talang foreign fish. The method is more innovative because of the combination of product design and the selection of product alternatives if the product design uses value engineering and if the alternative design uses AHP.

Keywords Packaging design · Talang salted fish · Value engineering · Analytical hierarchy process

A. Amri · S. M. Jalil (✉)
Department of Industrial Engineering, Universitas Malikussaleh, Lhokseumawe 24351, Indonesia
e-mail: saifuddin@unimal.ac.id

A. Amri
e-mail: iramri@unimal.ac.id

Irwansyah et al. (eds.), *Proceedings of the 4th International Conference on Experimental and Computational Mechanics in Engineering*, Lecture Notes in Mechanical Engineering, https://doi.org/10.1007/978-981-99-7495-5_19

1 Introduction

Pusong village, Lhoseumawe town, could be a sizeable fish-producing space within the Lhokseumawe city area; however, the individuals in Pusong village, particularly the fishers tribe, lack the data to initiate the packaging of processed salt-cured fish merchandise into modern salted fish packaging. The absence of contemporary packaging forms and solely exploiting clear and unlabeled plastic luggage as containers have risks that affect the standard and sturdiness of food. So the shortage of product attractiveness and considering the massive range of talang salted fish in the village of Pusong, Lhokseumawe city makes researchers inquisitive about adding to the commerce worth of the packaging design of processed talang salt-cured fish, which might later become an attraction for customers to shop for. This study aims to style talang salted fish packaging to extend the commerce worth, which can later be an attraction for consumers to buy it. With this attractive design, tourists will also become consumers of souvenirs within the type of talang salted fish in Pusong village. It is necessary to design a gorgeous talang salted fish packaging, correct information regarding color, and a style maintained once traveling, and so on other. A product's packaging can be crucial in maintaining or increasing sales. For food products, the packaging is formed to safeguard the contents from external contamination akin to substances that will harm the taste, shape, and suitableness of consumption and increase the product's commercial value. Packaging is essential in influencing customers directly or indirectly in crucial the selection of merchandise be purchased [1, 2]. The tactic used is value engineering. The worthy engineering method is applied to extend the value, performance, and value of packaging style consisting of name criteria, web weight measurement, product identity, attractiveness, color suitability, typography, and font size [3, 4]. Previous research which can be carefully associated with this study is as follows: Wardah and Mardiana [5] Department of Industrial Engineering, Indragiri Islamic University, her studies is entitled "Proposed Packaging of Salted Lome Fish (Harpodon Nehereus) Using the Value Engineering Method." The variables used are packaging and fee functions. This look concludes that there may be a boom with inside the price of packaging layout with the use of the price engineering method, i.e., a boom of 17.284%, and there may be an extensive distinction between the antique packaging layout and the primary packaging layout which has the very best price, via way of means of 86.81%. IDR compares the fee of various designs among the antique packaging layout and the brand new packaging.

2 Methodology

2.1 Types and Sources of Data

To get the facts needed, researchers got here without delay to the Pusong market, Lhokseumawe city, mainly at UD. Wirna, Jl. KP3 Lhokseumawe city to peer stay fish sales. Talang salted fish layout is covered withinside the joint examination due to the fact that this layout makes use of qualitative and quantitative facts. The facts used for the evaluation become acquired through gathering number one and secondary facts in keeping with the wishes of facts evaluation.

2.2 Data Collection Techniques

There are numerous techniques utilized in gathering records in this packaging layout proposal, specifically:

1. Literature studies, specifically by studying references, books, journals, works, the net, or literature related to the theoretical basis.
2. Observation is gathering records within the area by immediately seeing and staring at the item of studies via recording records and associated topics.
3. Documentation is a way of gathering records owned through the supplier concerned.
4. Interviews, specifically records withinside the shape of evaluations acquired from actors and events with a hobby in operations concerning salted fish packaging, capital, market, capital, and the essential data.
5. Questionnaire is fixed of statements submitted to be submitted to respondents. This questionnaire aims to acquire written data from respondents associated with the study objectives. The questionnaire on this has a look that consisted of:
   A. This open-ended questionnaire became used as an initial look at the reports to give the attributes of the salted fish packaging layout consistent with the purchaser's wishes. The wide variety of respondents for this open questionnaire distribution is forty.
   B. Closed questionnaire, specifically brainstorming the outcomes of the look for purchaser wishes on an open questionnaire that diagnosed variables and benchmarked merchandise with respondents. This closed questionnaire was allotted to forty respondents. The evaluation on a closed questionnaire makes use of a Likert scale, i.e., to look at the extent of respondents' approval of a question. The Likert scale has a gradation from very fantastic to very negative, inclusive of strongly agree (SS): 5, agree (S): 4, uncertain (R): 3, disagree (TS): 2, strongly disagree (STS):1

2.3 Analysis Method

The analytical technique utilized in value engineering (VE) or fee engineering with inside the layout of talang salted fish packaging is to apply the standard job plan, which includes numerous tiers or tiers as follows [4–15].

1. **Information Phase**

The statistics degree objectives to accumulate statistics wished in product improvement, including customer needs, prices, product functions, materials, advantages and downsides of current merchandise, and so on, in addition to using FAST diagrams. Information changed received by way of interviews, observations, and questionnaires. This section collects all forms of statistics about merchandise, shoppers, and matters wished for fee engineering techniques. Then examined the validity and reliability of the use of the software [16, 17].

2. **Creative Phase**

This section is on producing creativity that is achieved through creativity techniques. To this degree, it may begin while the preferred feature has been determined [18, 19].

3. **Analysis Phase (Judgment Phase)**

An evaluation degree is supposed to assess all options that seem in order that they may be used as a foundation for selecting the first-rate opportunity. Calculating the hierarchical weight component of all standards is achieved to determine the extent of the significance of a criterion in opposition to different standards. This calculation makes use of Expert Choice eleven software. Writers are an assessment section of the options generated from the innovative section to determine which opportunity can be used for similar research [20–23]. In this study, the technique used is to grow the feature two times and the value to growth as soon as with the aid of using the formula:

In this manner, to boost the promotion fee of talang salted fish, it is essential to enhance the packaging characteristic by growing the fee of packaging design.

4. **Development Phase**

This section is a deeper evaluation of the options decided on via the evaluation section that may be completed by thinking about technical and financial factors.

5. **Presentation and Recommendation Stage**

This section is a presentation to selection makers of what has been advanced and encouraged within the improvement section.

3 Results and Discussion

In this study, to reap the favored talang salted fish packaging, purchaser voice information changed into wanted via the distribution of open questionnaires, closed questionnaires, and coupled questionnaires (AHP).

Table 1 Open questionnaire design data for talang

Attribute			
Primary		Secondary	Tertiary
Design	1	Square packaging	Shape
	2	Packaging colors	– Light colors – Simple
	5	Packaging dimensions	– Net weight 500 gr. Length and width are adjusted according to net weight
	6	Ornaments	– Own a trademark – Sewer fish pictures – Net weight information
Material	7	Packaging	Transparent plastic
Multi-function	9	Main function	Packaging
	10	Additional functions	– Attractiveness and aesthetics – Product durability – Minimize odor when carried

Source Data collection 2021

1. **Open Questionnaire Results**

The consequences of the open questionnaire dispensed to forty respondents are summarized in Table 1.

After an open questionnaire was carried out, a closed questionnaire was carried out, which was distributed to forty respondents. Data obtained from the paired questionnaire method (AHP) can be seen in Table 1. The geometric calculation to get a paired matrix is to fill withinside the clean boxes, certainly considered one among that is divided with the aid of using a couple of matrices in order that the consequences of the recapitulation of the chance or geometric of every criterion may be visible in Table 2.

Analysis Model

After all the essential records, each number one and secondary records had been collected, then records processing and evaluation were carried out. The evaluation is through a cost engineering technique, which includes five stages, precisely the five-stage work plan.

Information Stage

At this stage, the outcomes of the statistics acquired concerning the preliminary product can be defined so one can be used as a reference in making the layout of the talang salted fish packaging, e.g., talang salted fish may be visible in discerning two and the plastic field for talang salted fish is in discern three.

Figure 1 is talang salted fish and Fig. 2 is plastic container for talang salted fish. Writers are bought on the Pusong marketplace even as parent three is a plastic field;

Table 2 Matrix of probability comparison between criteria

Criteria	Dadang brand	Measures net weight product	Identity attractive	Interesting	Appropriate color	Typography	Font size
Trademark	1.000	1.170	0.381	1.426	1.919	1.486	1.104
Net weight measure	0.855	1.000	0.295	1.219	1.952	1.219	2.246
Product identity	2.627	3.394	1.000	2.627	3.536	2.521	2.950
Interesting	0.701	0.820	0.381	1.000	1.104	1.219	1.346
Color match	0.521	0.512	0.283	0.906	1.000	0.635	1.000
Typography	0.673	0.820	0.397	0.820	1.575	1.000	1.919
Font size	0.906	0.445	0.339	0.743	1.000	0.521	1.000

Source Data collection 2021

this is used as a talang salted fish field if there are purchasers who purchase it. Moreover, from the consequences of the distribution of open questionnaires, which have been accomplished on forty respondents, it became that selling salted fish without packaging has subsequent shortcomings.

- Less attractive
- Does now no longer have a product identity
- Does now no longer have a trademark
- Less hygienic
- Less durable

Fig. 1 Talang salted fish figure

Fig. 2 Plastic container for talang salted fish

A. **Validation and Reliability Test**

From the consequences of the evaluation of the rating cost of the announcement with the whole rating, we evaluate it with the cost of r desk. The cost of r desk at 5% is widespread with a two-sided check and n = forty, then the r desk maybe 0.3120. Table 2 gives that each statement within the closed questionnaire is valid (r count > r desk). The consequences of the reliability evaluation acquired an alpha cost of 0.716, even as the crucial r cost at 5% importance with a two-sided check and n = 98 (df = n − 2 = 38) of 0.320. Writers are evidenced using the crucial r cost > the crucial desk r cost (0.712 > 0.320).

B. **Function Analysis System Technique Method**

The function analysis system technique (FAST) technique is a scientific diagramming approach for figuring out capabilities and describing the relationships among those capabilities. The effects of brainstorming patron desires and product benchmarking can become aware of the capabilities wished withinside the talang salted fish packaging design, which may be visible in Fig. 3.

FAST diagram results in Fig. 3, it is acknowledged that the primary characteristic of the talang salted fish packaging layout is to bundle the product. Moreover, the primary characteristic received nine by-product functions. From the derived characteristic received, its miles are geared up with a goal characteristic. The goal characteristic describes the necessities of the second characteristic.

Creative Phase

This degree increases seven standards with one goal. This degree describes numerous options that can be prioritized primarily based on a settlement among researchers and customers at the options that can be raised. Numerous standards and the burden of every criterion are made using the AHP technique as a decision-making system.

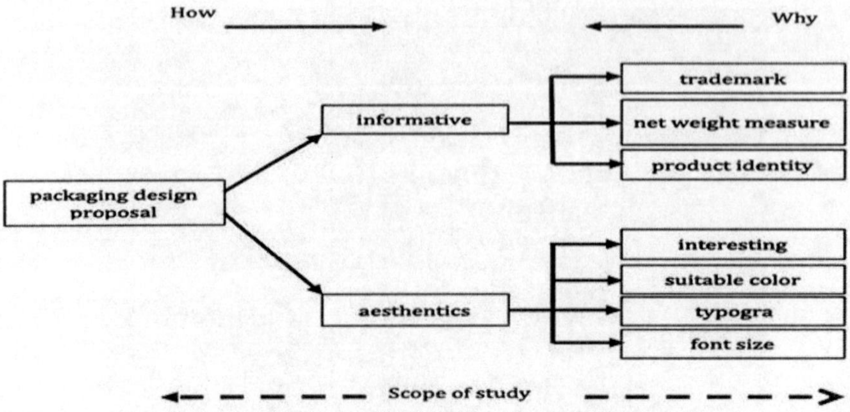

Fig. 3 FAST diagram of the proposed talang salted fish packaging design

Writers are taken into consideration to be assessed. The hierarchical shape of the talang salted fish packaging layout may be visible in Fig. 4.

C. **Alternative Generator Layout**

The opportunity layout generator is primarily based on the effects of brainstorming toward the standards that purchasers need. It is a bodily circumstance that represents the feature and standards of the packaging layout itself. The effects of the morphology chart packaging layout may be visible in Table 3.

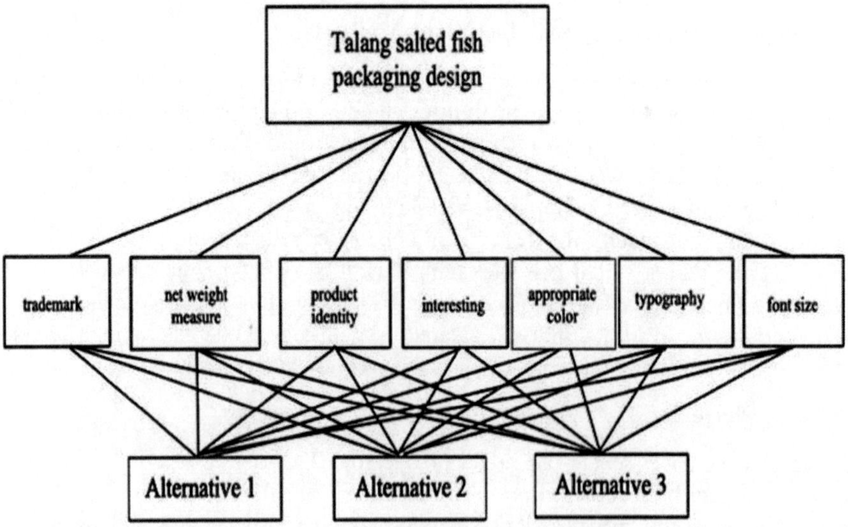

Fig. 4 Hierarchical structure of talang salted fish packaging. *Source* Processing results

Table 3 Packaging design morphology chart

No.	Element	Alternative	
1	Packaging material	A	
2	Color of packaging design	B	C
3	Packaging shape	D	
4	The typeface used in the packaging design	E	F

Source Data processing 2021

Three Phases of Analysis

At this stage, layout options might be evaluated and analyzed, so the chosen layout is acquired. From the layout options which have been acquired on the innovative stage, three opportunity designs might be designed, namely:

1. Alternative 1 (A-B-D-E)
    - The packaging cloth is fabricated from obvious plastic (A)
    - The color used is a mild shade (B)
    - Rectangular packaging (D)
    - Use all capital letters in product identity (E)
2. Alternative 2 (A-C-D-E)
    - The packaging cloth is fabricated from obvious plastic (A)
    - The color used is darkish (C)
    - Rectangular packaging (D)
    - Use all capital letters in product identity (E)
3. Alternative three (A-B-D-F)
    - The packaging cloth is fabricated from obvious plastic (A)
    - The color used is a mild shade (B)
    - Rectangular packaging (D)
    - Using capital letters and lowercase letters on product identity (F)

Furthermore, an evaluation of the cost of diverse options and standards is accomplished withinside the following Table 4.

After getting the fee of the evaluation of every criterion in opposition to the alternative, normalization is carried out. Finally, an evaluation of the consistency index (CI) and the inconsistency ratio (RI) is calculated using Expert Choice eleven software. To test the information inconsistency ratio, the inconsistency ratio fee in every matrix evaluation is in Table 5.

It may be concluded that the pairwise comparisons given via means of professional respondents have a minor inconsistency ratio cost of 0.1 because of the most restricted inconsistency ratio cost. Thus the outcomes of the blended geometric calculation of the respondent's records are pretty consistent. The outcomes of the recapitulation of the overall performance of the three options may be visible on Table 6.

Table 4 Assessment of standards and opportunity scores

Criteria/Alternative	Alternative 1	Alternative 2	Alternative 3	Total
Trademark	0.421	0.301	0.278	1.000
Net weight measure	0.378	0.301	0.321	1.000
Product identity	0.367	0.322	0.311	1.000
Interesting	0.431	0.206	0.363	1.000
Color match	0.417	0.294	0.290	1.000
Typography	0.401	0.327	0.272	1.000
Font size	0.356	0.333	0.312	1.000

Source Data processing 2021

Table 5 Ratio of the inconsistency of evaluation among standards in opposition to alternatives

No.	Element comparison matrix	CR value
1	Comparison of trademark criteria	0.00036
2	Comparison of the criteria for the size of the net weight	0.03
3	Product identity criteria comparison	0.00104
4	Interesting criteria comparison	0.00072
5	Comparison of appropriate color criteria	0.10
6	Typographic criteria comparison	0.09
7	Comparison of font size criteria	0.00104

Source Data processing 2021

Table 6 Performance calculation

Alternative	Pn	Rank
Alternative 1	38,9	1
Alternative 2	30,4	3
Alternative 3	30,7	2

Source Data processing 2021

The calculation outcomes show that the opportunity with the best overall performance is opportunity 1 (ABDE) with the following specifications: packaging fabric manufactured from clear plastic, brilliant colors used, rectangular packaging, and all capital letters. This opportunity is then taken into consideration for manufacturing compared with different alternatives.

Development Phase

Calculation of font length is in order that the brand new packaging layout is simpler to examine via way of means of consumers, and on this look at the usage of a viewing distance of three meters or three hundred cm from the item of research. The outcomes of the calculation of the font length may be visible in Table 7 below:

Table 7 Calculation of font size

No.	Type	Size (cm)
1	Letter height (H)	1.5
2	Font bold	0.25
3	Distance of two letters/numbers	0.3
4	Lowercase height	1
5	Letter width	1
6	Two words distance	1

Source Data processing 2021

Cost Analysis

The following info on the expenses that want to be incurred for manufacturing talang salted fish packaging may be visible within the Table 8. Selected opportunity (alternative 1).

Alternative considerations (opportunity 2 and opportunity 3).

Calculation of Fee

The remaining degree of fee engineering decides the fee acquired from the overall performance rating divided by the price consistent with the product at every opportunity (Table 9).

Based on the outcomes of figuring out the price of every one of the options above, it could be analyzed that the highest rate is possibility 1 with a rate of 0.039, just so possibility 1 is a possibility that wants to be generated because of the reality it is by that rate. Engineering strategies and by the requirements of consumer desires.

Table 8 Design costs of selected alternative designs (alternative 1)

Material and labor cost	Price per unit (IDR)
Plastic screen printing price	300
Price transparent plastic size 25 × 38	400
Price of the packing fee	300
Total	1000

Source Market research 2021

Table 9 Results of determining the value (value)

Alternative	Cost per package (IDR)	Performance score	Value	Rank
1	1000	38.9	0.039	1
2	1000	30.4	0.030	3
3	1000	30.7	0.031	2

Source Data processing 2021

Function Value

Based on Table 9, the output of calculating the characteristic rate of selected possibilities, main possibility 1 (A-B-D-E) is 0.039, and the possibility of selected format (A-B-D-E). Pictures of the selected talang salted fish packaging format proposals can be seen in Figs. 5, 6, and 7.

Fig. 5 Proposed packaging design 1

Fig. 6 Proposed alternative design 2

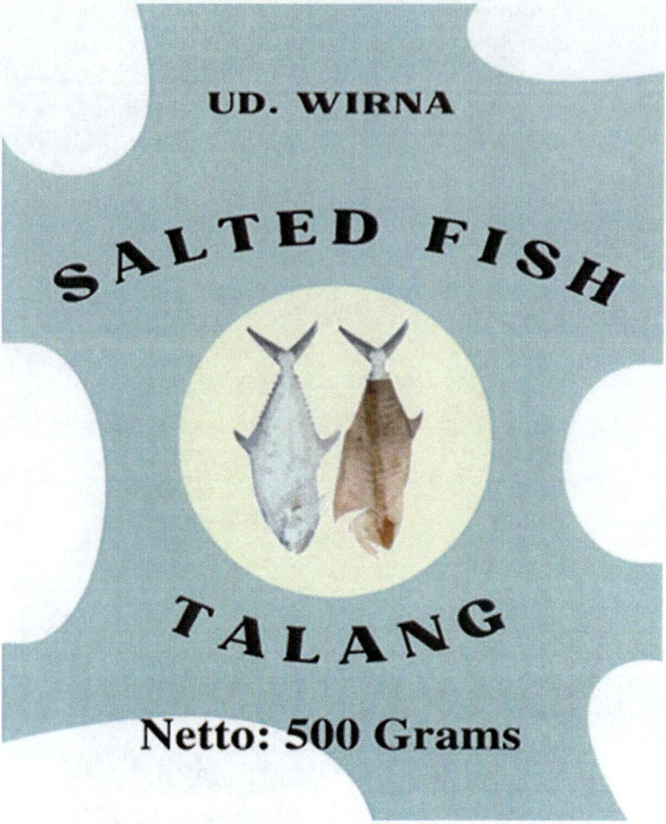

Fig. 7 Suggested alternative design 3 packaging of salted fish talang

4 Conclusion

Based on the outcomes of studies and dialogue of packaging layout standards consistent with patron desires, packaging with a hallmark has an Internet weight measurement, clear, appealing product identification, suitable color gradation, typography, and suitable font length. Therefore, the maximum precedence on the current standards is to have a product identification with a concern component of 0.323, trademark 0.142, Internet weight length 0.138, typography 0.118, appealing 0.112, font length 0.086, and suitable color 0.081. Therefore, the overall performance of every opportunity is as follows: opportunity 1 is 38.9, opportunity 2 is 30.4, and opportunity 3 is 30.7.

Based on the cost calculation outcomes, the salted fish packaging layout talang alter-local 1 (ABDE) is a packaging layout with packaging substances product of clear plastic, the colors used are vibrant patterns, and the packaging is square. This decided-on opportunity layout has a cost of 0.039. This cost is received from the overall performance of 38.9 and the value of IDR, one thousand packs.

Acknowledgements This research is fully supported by PNBP research grant. The authors fully acknowledged Ministry of Higher Education (DIKTI) and Universitas Malikussaleh for the approved fund which makes this important research viable and effective.

References

1. Bedward DA, Brinston RM, Kotler J (1995) Converting from ETO to radiation sterilization: educating the medical supply industry. Radiat Phys Chem 46(4 Part 1):443–448. https://doi.org/10.1016/0969-806X(95)00190-9
2. Ang SH, Leong SM, Kotler P (2000) The Asian Apocalypse: crisis marketing for consumers and businesses. Long Range Plann 33(1):97–119. https://doi.org/10.1016/S0024-6301(99)00100-4
3. Wang Q, Wu D, Li G, Gao W (2021) A virtual model architecture for engineering structures with twin extended support vector regression (T-X-SVR) method. Comput Methods Appl Mech Eng 386:114121. https://doi.org/10.1016/j.cma.2021.114121
4. Galbraith J (2021) Values in early-stage climate engineering: The ethical implications of 'doing the research.' Stud Hist Philos Sci Part A 86:103–113. https://doi.org/10.1016/j.shpsa.2021.01.009
5. Mardiana, Wardah S (2017) Proposed lome (Harpodon Nehereus) salted fish packaging design using the value engineering method. J Teknik Ind UNISI 1(1):11–19
6. Hidayat R, Arendra A, Akhmad S (2014) Bread fermentation machine development using value engineering (Ve) approach. Spektrum Ind 12(1):1. https://doi.org/10.12928/si.v12i1.1645
7. Rondini A, Bertoni M, Pezzotta G (2020) At the origins of product service systems: supporting the concept assessment with the engineering value assessment method. CIRP J Manuf Sci Technol 29:157–175. https://doi.org/10.1016/j.cirpj.2018.08.002
8. Setti PHP, Canciglieri Junior O, Estorilio CCA (2021) Integrated product development method based on value engineering and design for assembly concepts. J Ind Inf Integr 21:100199. https://doi.org/10.1016/j.jii.2020.100199
9. Tao J, Yu S (2018) Product life cycle design for sustainable value creation: methods of sustainable product development in the context of high-value engineering. Procedia CIRP 69:25–30. https://doi.org/10.1016/j.procir.2017.11.099
10. Pelzer K et al (2017) Penerapan Value Engineering (VE) Sebagai Pemilihan Alternatif Pembuatan Kantong Tas Belanja Wanita Dengan Konsep Green Product. Solid State Ionics 2(1):1–10 [Online]. Available: http://linkinghub.elsevier.com/retrieve/pii/S0167273817305726; https://doi.org/10.1038/s41467-017-01772-1; http://www.ing.unitn.it/~luttero/laboratoriomateriali/RietveldRefinements.pdf; http://www.intechopen.com/books/spectroscopic-analyses-developme
11. Haryanto LT (2015) The redesign of walker for Pasca-Stroke patient use a value engineering method. Available: https://adoc.pub/oleh-lulus-tri-haryanto-nim-i.html
12. Nugroho S, Pujotomo D, Gitakusuma A (2018) Aplikasi Value Engineering Untuk Mengatasi Value Problem Pada Produk Foodcart Studi Kasus Di Master Gerobak. Ind Eng 7(3):1–9. Available: https://ejournal3.undip.ac.id/index.php/ieoj/article/view/22318/20471
13. Kawakami H, Katai O, Sawaragi T, Konishi T, Iwai S (1996) Knowledge acquisition method for conceptual design based on value engineering and axiomatic design theory. Artif Intell Eng 10(3):187–202. https://doi.org/10.1016/0954-1810(95)00028-3
14. Umaya BI (2017) Penerapan value engineering (VE) sebagai pemilihan alternatif pembuatan kantong tas belanja wanita dengan konsep green product. Univ. Nusant. PGRI Kediri, 01:1–7 [Online]. Available: http://www.albayan.ae

15. Meliala S, Saifuddin, Rosdiana (2019) Drying design model for dried-anchovy using solar collector and solar cell panel position control. Int J Recent Technol Eng 7(6S3):1430–1435 [Online]. Available: http://www.scopus.com/inward/record.url?eid=2-s2.0-5067196861&partnerID=MN8TOARS
16. Heinzel J et al (2020) Use of the CatWalk gait analysis system to assess functional recovery in rodent models of peripheral nerve injury—a systematic review. J Neurosci Methods 345:108889. https://doi.org/10.1016/j.jneumeth.2020.108889
17. Tenaglia G, Romanelli F, La Rovere S, Polli GM, Gabellieri L, Valisa M (2021) Functional analysis for the diagnostic systems to support the exploitation of the Divertor Tokamak test facility. Fusion Eng Des 170:112692. https://doi.org/10.1016/j.fusengdes.2021.112692
18. Martí-García S, Relinque-Medina F, Fernández-Borrero MÁ, Vázquez-Aguado O (2021) Structural and functional analysis of cases of family treatment treated in the public social services system. Heliyon 7(5):e06825. https://doi.org/10.1016/j.heliyon.2021.e06825
19. Hornum ST, Bolwig S (2021) A functional analysis of the role of input suppliers in an agricultural innovation system: the case of small-scale irrigation in Kenya. Agric Syst 193:103219. https://doi.org/10.1016/j.agsy.2021.103219
20. Hendratmoko C, Istiyanto B, Ayu I, Rachmawati K (2015) Development of empowerment models for fish processors to increase income (case study on fish processors in Cilacap Regency). J Paradig 12(02):158–178. Available: https://www.neliti.com/id/publications/116481/pengembangan-model-pemberdayaan-bagi-pengolah-ikan-guna-meningkatkan-pendapatan
21. Saaty TL, Rokou E (2017) How to prioritize inventions. World Pat Inf 48:78–95. https://doi.org/10.1016/j.wpi.2017.02.003
22. Saaty T (2017) Neurons the decision makers, Part I: the firing function of a single neuron. Neural Netw 86:102–114. https://doi.org/10.1016/j.neunet.2016.04.003
23. Saaty T (2017) Part 2—The firings of many neurons and their density; the neural network its connections and field of firings. Neural Netw 86:115–122. https://doi.org/10.1016/j.neunet.2016.04.004

Utilization of Citronella Oil Refining Industry Solid Waste: Feasibility Study

Bakthiar, Khairul Anshar, Adi Setiawan, and Agam Muarif

Abstract The amount of solid waste from the distillation of citronella oil is mostly still not utilized optimally so that it only accumulates and pollutes the environment. Especially in Aceh Province, waste from citronella oil distillation on average is only burned when it has accumulated. The potential for the amount of waste is relatively large, with the area of citronella in Aceh Province being 17,445 ha, the waste generated from the distillation process can reach an average of 6,959 tons/year. When managed properly, the waste still has a high economic value, one of which is by turning the waste into briquettes. This study aims to analyze the feasibility of processing citronella solid waste into briquettes with a focus on the financial aspect. Processing refining citronella oil solid waste into briquettes is through five stages starting from the drying process, raw material chopping, pyrolysis process, refining, to briquetting. Testing was conducted by looking at the value of the break even point and net present value. It costs IDR 19,652,000 as investment cost to make briquettes from citronella oil refining solid waste. Based on the BEP value, briquette products must at least be able to be sold at a price of IDR13,525/kg to cover the cost of production, while the production unit should be able to produce 8.95 kg briquettes every day. Based on the calculation of the net present value which is worth IDR 5,380,817, it indicates that the business or investment design is economically feasible to be implemented. However, the by-products of liquid smoke and gas from pyrolysis have great potential to be utilized.

Bakthiar · K. Anshar (✉)
Department of Industrial Engineering, Universitas Malikussaleh, 24351 Aceh, Indonesia
e-mail: khairul.anshar@unimal.ac.id

Bakthiar
e-mail: bakhtiar@unimal.ac.id

A. Setiawan
Department of Renewable Energy Engineering, Universitas Malikussaleh, 24351 Aceh, Indonesia
e-mail: adis@unimal.ac.id

A. Muarif
Department of Chemical Engineering, Universitas Malikussaleh, 24351 Aceh, Indonesia
e-mail: amuarif@unimal.ac.id

Irwansyah et al. (eds.), *Proceedings of the 4th International Conference on Experimental and Computational Mechanics in Engineering*, Lecture Notes in Mechanical Engineering, https://doi.org/10.1007/978-981-99-7495-5_20

Keywords BEP · NPV · Briquette · Citronella oil · Pyrolysis

1 Introduction

Indonesia is the ninth largest essential oil producing country in the world. Based on data from the International Trade Center (ITC), the export value of Indonesian essential oils in 2009 was USD 91 million, and the value even jumped to USD 161 million in 2011. Therefore, in recent years, essential oils have received considerable attention from the government. Indonesia has already produced nine types of essential oils: clove oil, ylang oil, patchouli oil, vetiver oil, nutmeg oil, eucalyptus oil and citronella oil until now. Citronella oil export growth is currently around 9–10%. BPS data shows that the contribution of citronella oil to the export income of essential oils is around 6.89%, the third largest after patchouli oil at around 60% and vetiver oil around 12.47%. *Cymbopogon nardus* are spread in almost all parts of Indonesia. However, the main producers of citronella oil are the provinces of Aceh, West Java and Central Java with production of more than 95% from Indonesia's total production [1].

The global use of citronella oil was estimated at more than 2000 tons every year in 2007, with the largest area recorded in the province of Aceh at 17,445 ha. However, productivity is still low (135 kg/ha) compared with West Java Province which has an area of 1102 ha and reached a productivity of 202 kg/ha [2].

The main product of *Cymbopogon nardus* is essential oil. Essential oil content of *Cymbopogon nardus* is 0.5–1.5%, while the rest is solid waste (dregs of raw materials) and distilled water. Citronella distillation waste is generally used as an insecticide, air freshener, source of ruminant feed and raw material for organic fertilizer, but in Aceh Province the solid waste is still not utilized optimally. Solid waste from citronella oil distillation in Aceh Province is estimated to reach 256,500 tons/year, and if used properly, the waste still has high economic value.

Another use of citronella solid waste, which is still very rare, is to be used as raw material for briquettes. Briquettes sourced from biomass are an alternative in producing energy with materials that are abundantly available. There have been many researches related to the production of briquettes with biomass as raw materials. Among the biomass studied were coconut shells, peanut shells, coffee grounds and rice husks [3]. The manufacture of briquettes is done by pressing or compacting the process which aims to increase the calorific value per unit area of a biomass that will be used as alternative energy.

Currently, the citronella oil distillery in the village of KM 6, Simpang Keuramat subdistrict, still relies on firewood as a source of heat energy to produce steam in the distillation process, for a one-time distillation process that lasts for three working days. The cost needed for firewood for three working days can reach one million rupiah, so there is demand that the cost of firewood needs can be reduced to increase the profit. An alternative that can be chosen with the potential for low costs is the utilization of citronella dregs waste by processing it as briquettes which is expected to be a fuel additive to reduce the overall use of firewood. In addition to the economic benefits of using briquettes, it also has a positive impact on the environment by reducing tree cutting as firewood for industrial needs. Citronella distillation waste is in the form of citronella leaf pulp which has been extracted for its oil but still has a fairly high energy content, gross energy (GE) of 3353 kcal/GE/Kg [4]. In order to ensure that the utilization of this citronella waste can reduce the cost of burning raw materials, a study related to the economic analysis of the use of citronella plant refining waste into briquettes was carried out as a tool for firewood in the citronella distillation process.

The method for economic analysis is currently the most widely used method for evaluating projects or investment options. NPV is considered to be the most popular economic valuation technique. This is because the NPV considers all future cash flows (both inflows and outflows) generated from the project and then sums them up [5]. Parameters in business management, such as price as well as product, number of working days in one year and interest rate [6] have been considered in calculating business feasibility with economic analysis.

Break Even Point (BEP) is a business calculation used to determine the amount of sales needed to break even or return capital, at which point the business begins to expect profits [7]. BEP always shows the equation of the total cost and product price. One of the functions of BEP is to predict the impact of changes in cost and efficiency on profitability. Break Even Point is used to envision how many products or services must be sold to prevent losses [8, 9]. Break Even Point analysis can also be done for economic analysis by considering environmental aspects [10]. Based on previous studies, the utilization of citronella solid waste has not been widely carried out and not led to biochar or briquettes. Therefore, in this study, the analysis of the economic feasibility (business feasibility) in the utilization of citronella leaf waste into briquettes will be performed.

2 Methodology

The study was conducted on an area of 120 m^2 (12 m × 10 m). A roofed area was required when it is raining or hot but not an enclosed space. The main need for land was for drying raw materials. The production process and observations were carried out using 8 kg of citronella solid waste in one production cycle and produced an average of 3.2 kg of biochar in 2.5 h. Three production cycles could be carried out every day.

Briquetting. The process of making briquettes consists of five main processes starting from the processing of the main raw materials, leaves from the distillation of citronella oil to the briquette printing process.

1. **Drying Process**. The first step in a series of briquette-making processes was the drying. The drying of raw materials for leaves from citronella distillation products aimed to reduce the water content of the raw materials, so that during

the pyrolysis process did not require excessive heat and time to reduce the water content of the raw materials [11, 12]. The drying process of raw materials was carried out by drying directly in the sun for three to four days depending on weather conditions until the raw materials are dry enough.

2. **Raw Material Chopping**. The chopping process used a chopping machine (spes), aimed to reduce the size of the raw materials to be processed by pyrolysis, thus speed up the reaction rate in the composition of the raw materials, and also to increase the amount of raw materials to be entered into the reactor for further processing, slow pyrolysis [13, 14].
3. **Pyrolysis Process**. The pyrolysis process was started with the preparation of tools and raw materials for pyrolysis. Raw materials that were ready to be processed were weighed as much as 8 kg and put into the reactor. The designed reactor had a maximum capacity of 8 kg. Furthermore, the reactor was heated to a temperature of 300°C–330°C using LPG gas as fuel. When the reactor was heated, the water in the condenser began to circulate with the help of a pump. In this pyrolysis process, the time used was not determined, because the process was carried out until the pyrolysis process ends, the end of the pyrolysis process was marked by the pressure in the reactor dropping or no longer showing an increase [15]. The temperature rise will be observed and recorded at intervals of one minute during the process.

   The pyrolysis process produces three types of products, namely charcoal, liquid smoke, and gas. In order to obtain charcoal products in the reactor, the reactor needs to be at ambient temperature to prevent burned charcoal product because of reaction with oxygen. The liquid product is taken immediately after the pyrolysis process is complete and after the gas product is removed to avoid clumping or sticking of tar in the separator the gas product would be removed during the pyrolysis process gradually, due to control of the pressure during the pyrolysis process and will also be removed entirely after the pyrolysis process is complete. Gas products will be accommodated using truck tires or will be burned directly.
4. **Refining**. The refining process aims to reduce the size of the charcoal product that will be used as a briquette product, so as to increase the mass density and energy density of the resulting briquette product [13]. The refining process is accompanied by sifting which aims to homogenize the results obtained from the refining process using a sieve size or a 60 mesh sieve. The charcoal refining process begins with the preparation of the refining machine along with the charcoal to be mashed. The refining device uses an electric motor as a drive with a capacity (1.5 kW).
5. **Briquetting**. The briquetting process begins by providing charcoal that has been refined with a size of 60 mesh, preparing an adhesive medium made of starch with a ratio of 500 ml of water to 20 g of starch which will be mixed with charcoal [14]. The adhesive manufacturing process begins by heating 500 ml of water together with 20 g of starch at an electric stove temperature of 400 °C for 20 min and stirring slowly until it's finished [16]. After the manufacture of the adhesive is complete, the next process is mixing between the adhesive and the charcoal to

be bonded, the mixing ratio for making the bond is 70%:30% with the provisions of 70% charcoal and 30% adhesive from the total weight of the bonding [17, 18]. After the charcoal and adhesive are evenly mixed, prepare the mold and press that will be used in the compaction process where, in this process a press pressure of 300 kg/cm^2 is used with a mold diameter of 2.5 cm and a height of 8.2 cm in the form of a cylindrical tube.

3 Break Even Point (BEP)

1. **Calculation of Total Cost (TC)**. The total cost shows all the costs incurred to carry out production activities which in this case is the manufacture of briquettes with the raw material of citronella solid waste. In calculating BEP, costs can be grouped into four cost components, namely fixed costs (FC), variable costs (VC), mixed costs (MC) and cost of goods manufactured (HPP). In this study, the total cost consists of fixed costs and variable costs. The total costs incurred to process citronella solid waste into briquettes can be calculated using Eq. 1.

$$\mathrm{TC} = \mathrm{FC} + \mathrm{VC}, \tag{1}$$

   where
   FC = Fixed cost (IDR/month).
   VC = Variable cost (IDR/month).
2. **Calculating Total Revenue (TR)**. Revenue is derived from product sales. The calculation of the total cost is obtained from the number of goods that can be produced multiplied by the selling price of the product. Product sales are accumulated in one month. The calculation of total income can be written as follows.

$$\mathrm{TR} = Q \times P, \tag{2}$$

   where
   Q = Amount of production every month (kg/month).
   P = Selling price of briquettes (IDR/kg).
3. **Calculate the BEP Production ($\mathbf{BEP_{op}}$)**. BEP production is calculated to find out how much briquettes should be produced to cover all costs incurred for the production process. BEP output can be calculated by Eq. 3.

$$\mathrm{BEP_{op}} = \mathrm{FC}/[(P - \mathrm{VC}/Q) \times N], \tag{3}$$

   where
   N = Number of working days in a month.

4. **Calculate the BEP Price (BEP_{pr}).** The calculation of the BEP value is carried out with the aim of knowing what the selling price of the product is, which is with the current situation and condition of the business unit to cover the total cost of production. The value of the price BEP can be calculated using the following Eq. 4.

$$BEP_{pr} = P = (FC + VC)/Q. \quad (4)$$

4 Net Present Value (NPV)

1. **Calculating Gross Income (GI)**. The amount of income is calculated from the number of products sold in one month. The basic assumption in calculating income is that all products that can be produced are sold. Another variable to consider is productivity. The amount of gross income can be calculated by Eq. 5.

$$GI = Q \times P \times PI \times 12, \quad (5)$$

where
PI = Productivity index.

2. **Calculating Expenses (E)**. The cost used in this method is the annual value in the month period of all costs incurred. Included in the cost components are fixed costs and variable costs.
3. **Calculating Net Present Value (NPV)**. Calculating NPV has two steps. The first step is to calculate the present value (PV) of total expenses per year and total profit per year. The second step is to add up each present value (PV) of total profits and total expenses, then calculate the difference between the sum of the two. Assuming there is no change in production capacity and interest rate, the value of NPV can be calculated by Eq. 7.

$$NCF_t = GI_t - E_t \quad (6)$$

$$NPV = IC - NCF_t(P/F, i, t), \quad (7)$$

where
NCF_t = net cash flow in year (t).
IC = Investment cost.
i = Interest rate.

Table 1 Distribution of pyrolysis results, energy consumption and processing time

Test	Energy consumption (kg)	Results (gram)			Time (hours)
		Solid	Liquid	Gas	
1	1.62	3.204	2.368	2.428	2.05
2	1.53	2.752	2.360	2.888	2.00
–	–	–	–	–	–
–	–	–	–	–	–
10	1.52	3.208	2.752	2.040	1.83
Average	1.5	3.204	2.782	2.014	2

5 Results and Discussion

5.1 Pyrolysis Experiments

The core process in the manufacture of briquettes is the pyrolysis process. This process converts the raw material of citronella waste into biochar. The pyrolysis process is the main focus because the quality and quantity of charcoal produced from this process have a direct impact on the yield of briquettes. Also, this process is the process that takes the longest of a series of briquette manufacturing processes (bottle neck). One cycle of the pyrolysis process is with a device designed to process 8 kg of citronella solid waste. This process will produce three outputs consisting of solid (biochar), liquid (liquid smoke) and gas. The results of the pyrolysis process are summarized in Table 1.

From Table 1 above, it can be concluded that the pyrolysis process with citronella raw materials with a capacity of 8 kg, took 2 h and consumed 1.5 kg of gas until the pyrolysis process was completed. The pyrolysis process produced 40.1% biochar, 34.8% liquid smoke and 25.2% gas. This data was then used as a reference in the calculation of financial analysis using BEP and NPV.

5.2 Break Even Point (BEP) Analysis

1. **Calculation of Total Cost (TC)**. The total costs spent on producing briquettes are grouped into fixed costs and variable costs. The total cost of each month is calculated by Eq. 1.

   $\text{TC} = \text{FC} + \text{VC}$

   $\text{TC} = 2,327,250 - 1,365,000$

   $\text{TC} = \text{IDR } 3,692,250/\text{month}$

   Based on the calculation of the total cost above, it can be concluded that the cost of IDR is 3,692,250 per month to produce 273 kg of briquettes.

2. **Calculating Total Revenue (TR)**. Income is obtained from 273 kg of briquettes which are sold at a price of IDR 15,000 per kg. By using Eq. 2, the total monthly income is as follows.

$$TR = Q \times P$$

$$TR = 273 \times 15,000$$

$$TR = \text{IDR } 4,095,000/\text{month}$$

3. **Calculate the BEP production**. By using Eq. 4, output BEP value is.

$$BEP_{op} = FC/[(P - VC/Q) \times N]$$

$$BEP_{op} = 2,327,250/[(15,000 - 1,365,000/273) \times 26]$$

$$BEP_{op} = 8.95 \text{ kg}$$

The business unit must produce at least 8.95 kg every day so that all costs incurred during the period are covered. Every gram of briquettes that exceeds BEP_{op} will provide benefits for the business unit.

4. **Calculating the BEP Price**. Selling price is an important factor in product marketing. The determination of the selling price is based on many things, from production costs, raw material prices, profit targets, market mechanisms and others. Based on Eq. 7, the BEP value of the resulting briquette product is.

$$BEP_{pr} = P = (FC + VC)/Q$$

$$BEP_{pr} = 3,692,250/273$$

$$BEP_{pr} = \text{IDR } 13,525 \text{ kg}$$

The price of BEP is an important value that must be known by business units. This value can be used as the basis for adjusting the selling price of the product. With the current daily production capacity of 9.6 kg of biochar, at least the product can be sold at a price of IDR 13,525 per kg briquettes to return all costs spent each month (Fig. 1).

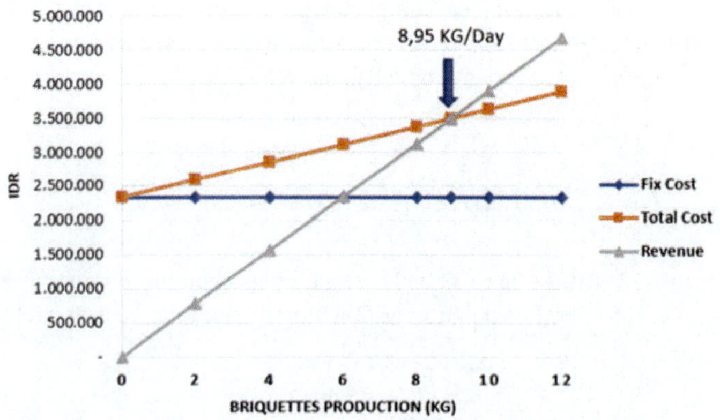

Fig. 1 BEP production

Table 2 Parameters considered for NPV

Parameter	Value	Unit
Annual operating days	26	days/month
Production capacity	273	kg/month
Briquettes sales price	15,000	IDR/kg
Firing cost	10,000	IDR/kg
Maintenance cost	150,000	IDR/month
Investment cost	19,652	IDR
Plant life	4	Year
Salvage value	1,850,000	IDR
Interest rate	8	%

5.3 Net Present Value (NPV)

When calculating NPV, there are two main variables that must be calculated first, namely gross income and expenses. The parameters used to calculate gross income, expenses and NPV are given in Table 2.

1. **Calculating Gross Income (GI)**. The amount of income is obtained from the sale of briquettes that can be produced. Factors that affect the level of production is the level of productivity. In this case, the productivity in the first year is assumed to be 0.9 and 1 in the second year until the end of the economic life. A recapitulation of gross income for each year will be presented in Fig. 2. The calculation of gross income in the first year using Eq. 5 is as follows.
   $\mathrm{GI}_1 = 273 \times 15,000 \times 0.9 \times 12$
   $\mathrm{GI}_1 = 44,226,000$
2. **Calculating Expenses (E)**. Expenses are costs required to produce briquettes. These costs consist of fixed costs, namely labor costs and maintenance costs, and

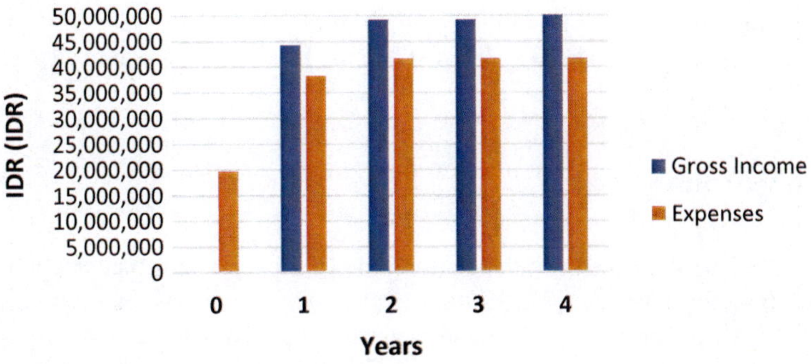

Fig. 2 Gross income and expenses throughout the economic life of the investment

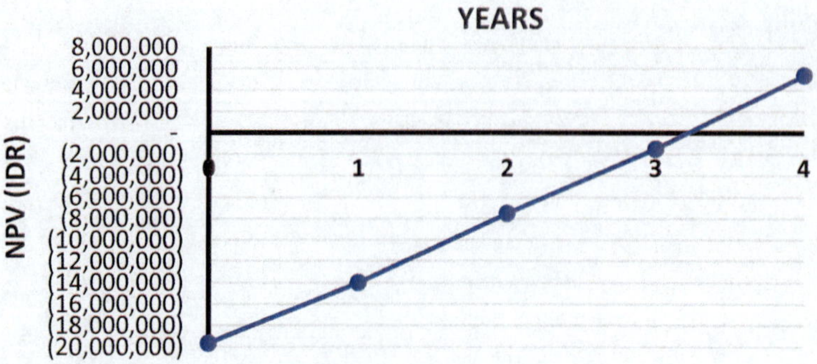

Fig. 3 NPV throughout the economic life of the investment

variable costs, namely process materials and raw materials. The recapitulation of expenses and gross income is presented in Fig. 2 above.

3. **Calculating Net Present Value (NPV)**. After calculating the gross income and expenses in Sects. (1) and (2), the net income will be calculated. Net income begins to be obtained in the first year because in year 0 it is assumed the period of business establishment. After the NCF value in each year is obtained, then the value will be converted into present value which can then be calculated net present value by adding up the total investment cost. The following is the calculation for the first year.

$$\mathrm{NCF}_1 = \mathrm{GI}_1 - \mathrm{E}_1$$
$$\mathrm{NCF}_1 = 44,226,000 - 38,142,000$$
$$\mathrm{NCF}_1 = 6,084,000$$
$$\mathrm{NPV} = 19,652,000 - 6,084,000(P/F, 8\%, 1)$$
$$\mathrm{NPV} = 19,652,000 - 6,084,000(0.92593)$$
$$\mathrm{NPV} = 14,018,667$$

Figure 3 illustrates the NPV curve during economic life of investment. The income calculation is derived from the sale of briquettes where quantity and production prices of the product is mentioned in Table 2. Based on the calculation of NPV using the parameters in Table 2, it can be concluded that the business is economically feasible.

6 Conclusion

The solid waste of leaves from the distillation of citronella has great potential to be utilized. In order to produce 273 kg of briquettes made of citronella leaves, it costed IDR 3,692,250, with an initial investment value of IDR 19,652,000. The current condition and production capacity unit, the business unit has the potential to incur a monthly profit of IDR 402,750. Based on the calculation of the net present value which is worth IDR 5,380,817, it can be concluded that the business is economically feasible. The business unit must at least be able to produce 8.95 kg of briquettes every day, so that the costs incurred can be covered. In accordance with the current production capacity, briquettes should be sold at least IDR 13,525/kg so that the business unit does not suffer losses. Based on the BEP criteria, because the production capacity of the business is greater than the BEP production (10.5 > 8.95) and the current selling price exceeds the BEP price (15,000 > 13,525), it can be concluded that the business is economically viable.

Acknowledgements The authors are grateful to the Institute for Research and Community Services (LPPM), Malikussaleh University, for the research funds, under contract number 43/PPK-2/SPK-JL/2021.

References

1. Suryanti Z, Rahmat N (2020) Produksi Minyak Sereh Wangi Sebagai Parfum Lokal Anti Bakteri Masyarakat Desa Tambon Tunong Kecamatan Dewantara Kabupaten Aceh Utara, vol 4, no 1, pp 97–101
2. Sulaswatty A, Rusli MS, Abimanyu H, Tursiloadi S (2019) Quo Vadis: Minyak Serai Wangi dan Produk Turunannya, vol 9, no 2. Lembaga Penelitian Indonesia Press
3. Magnago RF et al (2020) Briquettes of citrus peel and rice husk. J Clean Prod 276. https://doi.org/10.1016/j.jclepro.2020.123820
4. Djazuli M, Suheryadi D, Penelitian B, Obat T, Tentara J, No P (2011) Seraiwangi (Cymbopogon nardus L) Sebagai Penghasil Minyak Atsiri, Tanaman Konservasi Dan Pakan Ternak. Inovasi Perkebunan 2011:174–180
5. Žižlavský O (2014) Net present value approach: method for economic assessment of innovation projects. Procedia Soc Behav Sci 156:506–512. https://doi.org/10.1016/j.sbspro.2014.11.230
6. Patel S et al (2019) Transformation of biosolids to biochar: a case study. Environ Prog Sustain Energy 38(4):1–11. https://doi.org/10.1002/ep.13113
7. Sayuti M (2008) Analisis Kelayakan Pabrik. Graha Ilmu, Yogyakarta
8. Sayuti M, Puspasari C, Anshar K, Zeki M (2019) Potensial use of backyard for oyster mushroom (Pleurotus Ostreatus) cultivation to increase family income; studies on break-event point analysis. IOP Conf Ser Mater Sci Eng 536(1). https://doi.org/10.1088/1757-899X/536/1/012132
9. Potkany M, Krajcirova L (2015) Quantification of the volume of products to achieve the break-even point and desired profit in non-homogeneous production. Procedia Econ Financ 26(15):194–201. https://doi.org/10.1016/s2212-5671(15)00811-4
10. Cottafava D, Costamagna M, Baricco M, Corazza L, Miceli D, Riccardo LE (2021) Assessment of the environmental break-even point for deposit return systems through an LCA analysis of single-use and reusable cups. Sustain Prod Consum 27:228–241. https://doi.org/10.1016/J.SPC.2020.11.002
11. Cai W, Luo Z, Zhou J, Wang Q (2021) A review on the selection of raw materials and reactors for biomass fast pyrolysis in China. Fuel Process Technol 221(106919):1–14. https://doi.org/10.1016/j.fuproc.2021.106919
12. Genieva S, Gonsalvesh L, Georgieva V, Tavlieva M, Vlaev L (2021) Kinetic analysis and pyrolysis mechanism of raw and impregnated almond shells. Thermochim Acta 698(178877):1–11. https://doi.org/10.1016/j.tca.2021.178877
13. Bello RS, Onilude MA (2021) Effects of critical extrusion factors on quality of high-density briquettes produced from sawdust admixture. Mater Today Proc 38:949–957. https://doi.org/10.1016/j.matpr.2020.05.468

14. Nagarajan J, Prakash L (2021) Preparation and characterization of biomass briquettes using sugarcane bagasse, corncob and rice husk. Mater Today Proc 47:4194–4198. https://doi.org/10.1016/j.matpr.2021.04.457
15. Aryati T, Williansyah A, Zulnazri, Setiawan A (2021) Slow pyrolysis of areca-nut fibres in a-pilot scale batch reactor. Lecture Notes Mech Eng 263–270. https://doi.org/10.1007/978-981-16-0736-3_26
16. Alchalil, Setiawan A, Juwaini, Bin Nur T (2021) Effect of densification pressure on physical and combustion properties of binderless briquettes made from rice-husk and coffee-pulp. Lect Notes Mech Eng 1–8. https://doi.org/10.1007/978-981-16-0736-3_1
17. Aransiola EF, Oyewusi TF, Osunbitan JA, Ogunjimi LAO (2019) Effect of binder type, binder concentration and compacting pressure on some physical properties of carbonized corncob briquette. Energy Rep 5:909–918. https://doi.org/10.1016/j.egyr.2019.07.011
18. Espuelas S, Marcelino S, Echeverría AM, del Castillo JM, Seco A (2020) Low energy spent coffee grounds briquetting with organic binders for biomass fuel manufacturing. Fuel 278:1–8. https://doi.org/10.1016/j.fuel.2020.118310

The Finite Element Method Simulation of Rear Driveshaft Failure in Trucks

Fauzan, Husaini, and Nurdin Ali

Abstract There was a failure in the truck's rear driveshaft in the spline during the transportation of rock and sand in Aceh Besar, in Indonesia. Therefore, this research aims to analyze the causes of failures in the rear driveshaft of the truck by performing a finite element analysis (FEA) simulation. It was discovered that the standard for the shaft material is AISI 4150. Based on the results, the maximum shear stress was s 233.21 MPa, this has exceeded the allowable value of 80.7 MPa and shear stress Tresca which is 190 MPa. Furthermore, the Von misses shear stress analysis obtained is 426.2 MPa, which is above 179.13 MPa. The analysis of the stress intensity factor also obtained an average value of 27.7 MPa$\sqrt{}$m, which is more than the 10 MPa$\sqrt{}$m ($K_I > K_{IC}$) fracture toughness value, thereby triggering crack propagation to fracture.

Keywords Failure analysis · Driveshaft · Finite element method · Stress intensity factor

1 Introduction

The driveshaft transmits the driving force to the wheel, which is mounted on the axle shaft [1, 2]. It includes the main shaft and the pinion gear which connects the wheels and gears in the differential [3, 4]. The driveshaft is affected by torsion and shear stress by resisting large stresses [4, 5] that require special attention to design. Based on the shaft housing, the driveshaft is divided into three types, namely full, three-quarter, and semi-floatings. Trucks with large loads use the full floating types that are designed to accept torsional loads [6–8]. One method of failure analysis is

Fauzan · Husaini (✉) · N. Ali
Department of Mechanical Engineering, Laboratory of Computational Mechanics, Universitas Syiah Kuala, Darussalam, Banda Aceh 23111, Indonesia
e-mail: husainiftm@usk.ac.id

N. Ali
e-mail: nurdin.ali@usk.ac.id

Irwansyah et al. (eds.), *Proceedings of the 4th International Conference on Experimental and Computational Mechanics in Engineering*, Lecture Notes in Mechanical Engineering, https://doi.org/10.1007/978-981-99-7495-5_21

the finite element method. Mesh size will affect the accuracy of the results, with small mesh size yields high accuracy when compared with the large mesh size [9].

Shaft failure commonly occurs because it is an element in every machine construction. Several investigations have been carried out on shaft failure analysis [10–12], but the differences include dimensions, loads, and fractures on the shaft, which make the results applicable for future consideration.

The truck's rear driveshaft broke before nine months of use in the spline of the shaft. Previous research has been carried out on failed shafts using only experimental methods [13]. Therefore, this research aims to determine the factors causing the failure to avoid similar cases in the future using the finite element method.

2 Research Object

This research used a failed rear driveshaft of a truck which occurred when it was crossing extreme roads and carrying heavy loads. The failed shaft is shown in Fig. 1, while the initial crack in the shaft is manually rendered in a semicircle with a radius of 0.25 mm and a width of 0.25 mm and the results of the CAD design are given in Fig. 2. The driveshaft has a spline part at the end of the shaft as a torque receiver from the differential.

3 Research Method

Shaft Design. The simulation process begins with designing the shaft using a design application. The first step is to measure the failed shaft and design the CAD of the shaft as shown in Fig. 2 with the dimensions presented in Fig. 3.

Meshing. The meshing process is the method of dividing elements into smaller parts to obtain detailed simulation values. Therefore, the mesh used is tetrahedral with a size of 0.01 as shown in Fig. 4.

Mechanical Properties. The mechanical properties determined in previous research [13] discovered that the shaft material is AISI 4150, as given in Table 1.

Load and Constraint. Loads and constraints significantly affected the simulation results. The load was given in the spline area with a torque load of 372.4 Nm, while the constraints given were in two types, namely bolt and bearing as shown in Fig. 4.

Maximum Shear Stress Analysis. Maximum shear stress analysis is the process of determining whether the highest value that occurs in the shaft does not exceed the allowable shear stress. When the maximum shear stress on the shaft exceeds the allowable value, it can trigger the shaft to break.

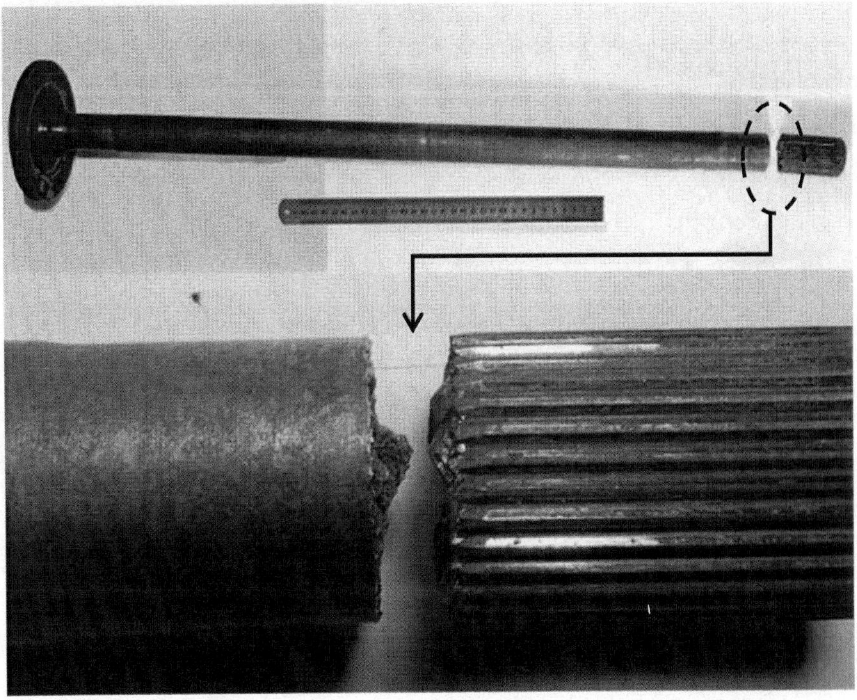

Fig. 1 Shaft fail

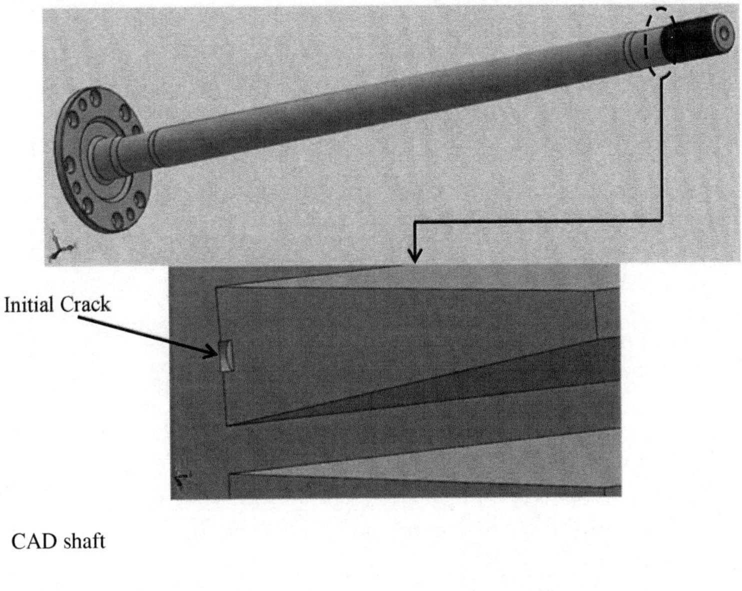

Fig. 2 CAD shaft

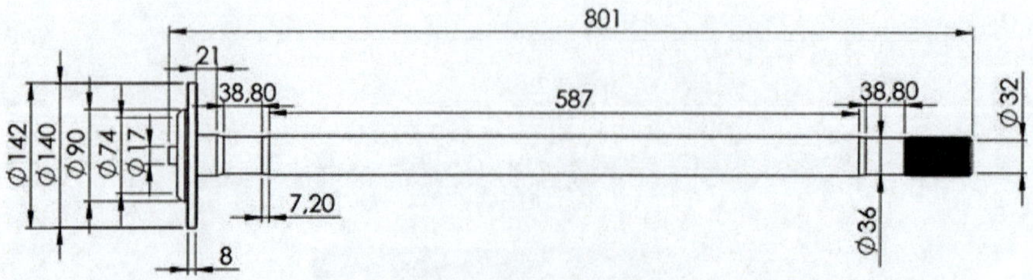

Fig. 3 Shaft dimension

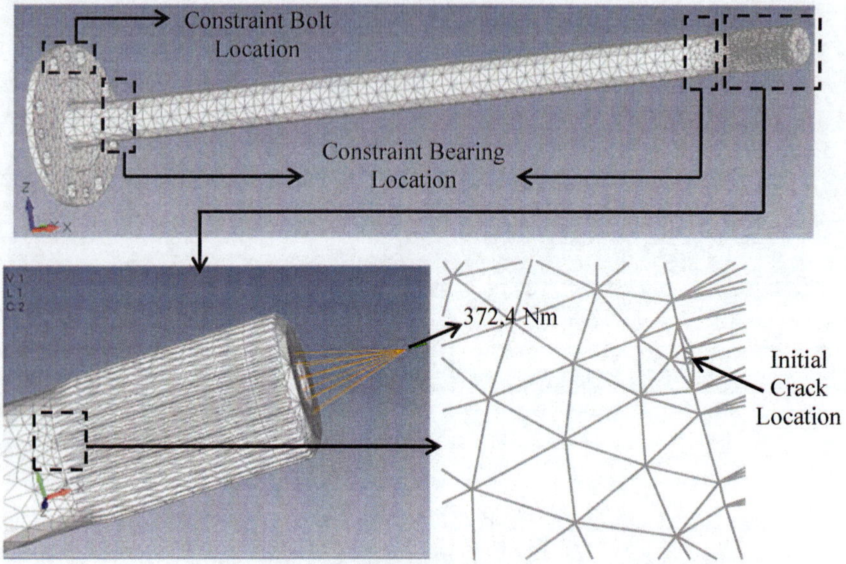

Fig. 4 Meshing, constraints and loads on shaft

Table 1 Mechanical properties of AISI 4150 [13–15]

No	Properties	Metric
1	Tensile strength, σ_B	731 MPa
2	Yield strength, S_y	380 MPa
3	Shear modulus, G	80 GPa
4	Poisson's ratio, v	0.27–0.30
5	Shear stress Von Misses, τ_{oct} [13]	179.13 MPa
6	Shear stress Tresca, τ_{Tresca} [13]	190 MPa
7	Allowable shear stress	80.7 MPa
8	Fracture toughness, K_{IC} [15]	10 Mpa$\sqrt{}$m

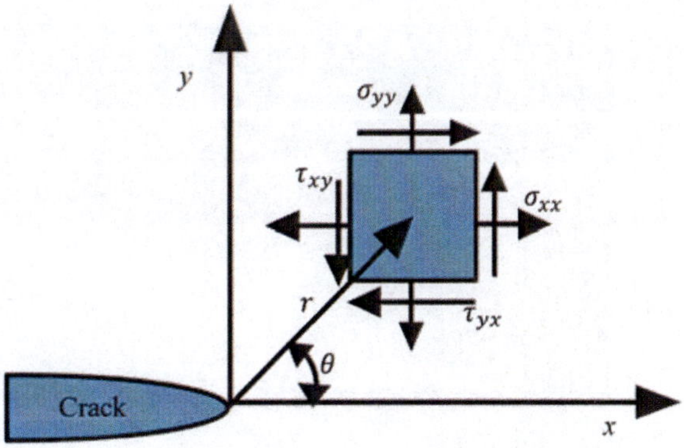

Fig. 5 Definition of the coordinate axis near the crack tip [17]

Von Misses Theory Analysis. This process aims to analyze the shear stress on the shaft based on the Von misses failure theory. This failure will occur when the shear stress is equal to or exceeds the value of the Von misses.

Stress Intensity Factor. The results of the analysis using the FEMAP software showed the values of the shear stresses that occur on the shaft, therefore, values of the stress intensity factor can be concluded with calculation manually. The stress intensity factor was carried out at four nodal points and calculated using Eq. 1 [16]:

$$K_I = \frac{\tau\sqrt{2 \times \pi \times r}}{\cos\left(\frac{\theta}{2}\right) \times \sin\left(\frac{\theta}{2}\right) \times \sin\left(\frac{3\times\theta}{2}\right)}, \tag{1}$$

where σ is the stress, r is the distance from the crack tip to the element point, and θ is the angle formed. Moreover, the coordinate axes near the crack tip [17] were defined as shown in Fig. 5.

4 Result and Discussion

Maximum Shear Stress Analysis. From Fig. 6, the shaft simulation results showed that the shear stress value of the initial area of the crack on the spline shaft has exceeded the allowable value, which is marked in red. The results of the shear stress analysis showed that the maximum value occurred in the initial crack area with a value of 233.21 MPa as shown in the figure with arrows. This value has exceeded the allowable shear stress which is 80.7 MPa and shear stress Tresca which is 190 MPa of the shaft.

Von Misses Theory Analysis. Shaft simulation with Von misses theory showed that the value in the initial area of the crack on the spline shaft has also exceeded the allowable limit. The simulation results of the driveshaft with the Von misses theory are shown in Fig. 7.

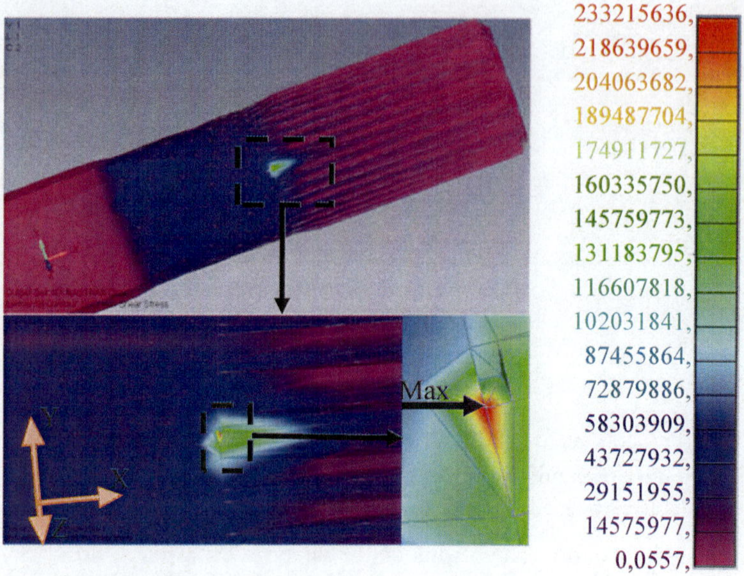

Fig. 6 Maximum shear stress

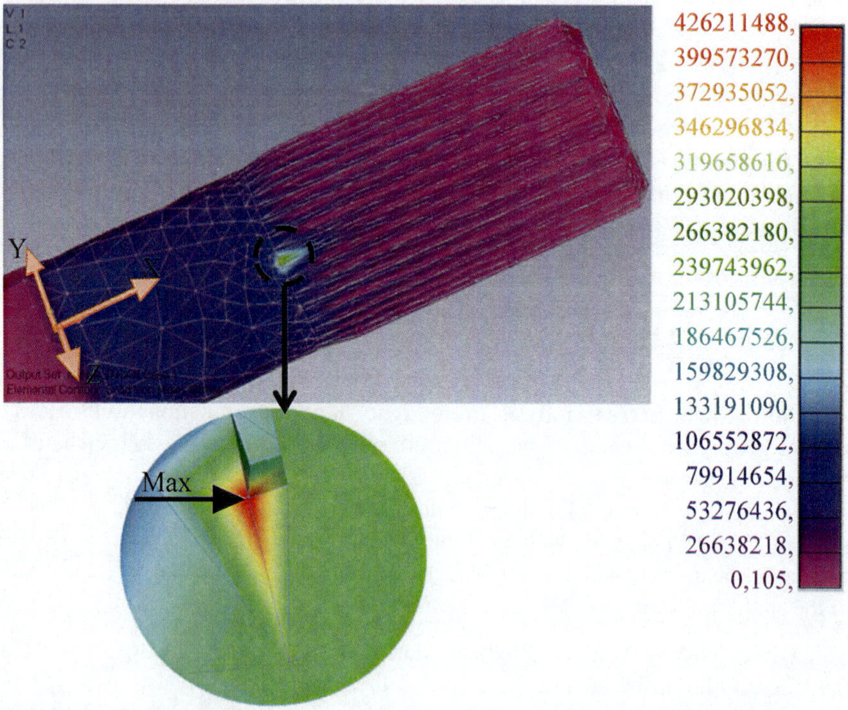

Fig. 7 Von misses shear stress

The results of the analysis obtained the shear stress that occurred in the initial crack area with a value of 426.2 MPa. This value has exceeded the Von misses shear stress of the shaft material, which is 179.13 MPa.

Stress Intensity Factor. The value of the stress intensity factor is taken from four points as shown in Fig. 8, which is a magnification from Fig. 7.

The stress intensity factor value at the initial crack area was calculated using Eq. (1) [16] and the results are given in Table 2.

The results of the stress intensity factor in Table 2 showed that the average value is 27.7 MPa$\sqrt{m}$. Meanwhile, the closest value to the crack tip is the highest with the largest stress. The stress intensity factor that occurred on the shaft has exceeded the fracture toughness value of 10 MPa$\sqrt{m}$ [15], indicating the occurrence of crack propagation [16–18].

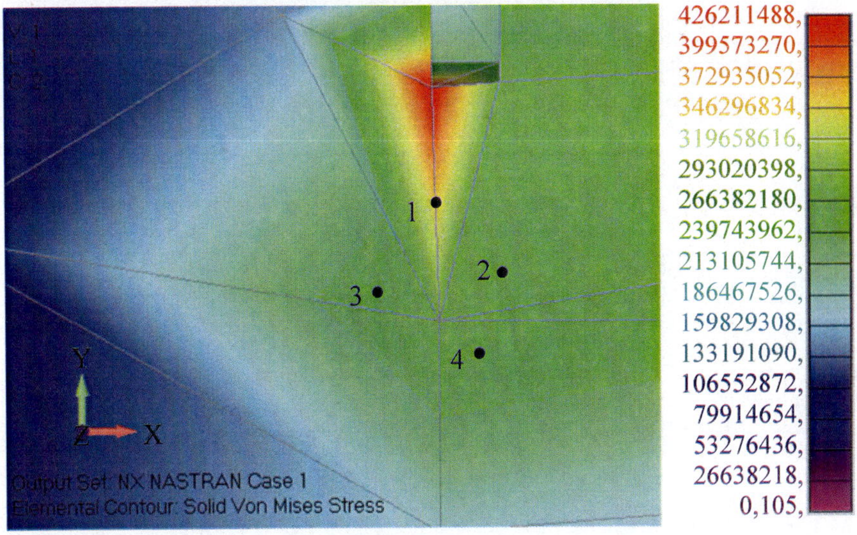

Fig. 8 Calculation point of stress intensity factor value

Table 2 Results of the calculation of the value of the stress intensity factor, K_I

Point	τ(MPa)	$r(m)$	θ(°)	K_I(MPa$\sqrt{m}$)
1	226.9	0.00041	92.1	34.6
2	154.3	0.00065	55.3	24.6
3	139.9	0.00076	72.2	21.3
4	79.6	0.00101	103	30.5
Average				27.7

5 Conclusion

Based on the results, the following conclusions are obtained:

a. The maximum shear stress simulation of 233.21 MPa occurred in the area where the initial crack has exceeded the allowable value which is 80.7 MPa and shear stress Tresca which is 190 MPa. This showed that the shaft is unable to withstand the stress.
b. The Von misses shear stress simulation obtained a value of 426.2 MPa, which has exceeded the allowable value of 179.13 MPa, causing failure to occur.
c. The analysis of the stress intensity factor obtained an average value of 27.7 MPa$\sqrt{m}$, while the fracture toughness of AISI 4150 was 10 MPa$\sqrt{m}$ [15]. This indicated that the stress intensity factor has exceeded the fracture toughness, causing crack propagation to fracture.

Acknowledgements The authors would like to thank Universitas Syiah Kuala for providing partial financial support to this research through Research Grant No. 001/UN11.2.1/PT.01.03/ PNBP/2022.

References

1. Samlawi K, Firdaus SR (2018) Fracture analysis of rear axle shaft vehicles truck HINO 500. Prosiding SNTTM XVII 28:6–12
2. Bhagoria P, John ST, Patangia P, Purohit R (2017) Failure analysis of the axle shaft of an automobile. Mater Today: Proc 4:5398–5407
3. Pang H, Ngaile G (2019) Development of a non-isothermal forging process for hollow power transmission shafts. J Manuf Process 47:22–31
4. Baig A, Langde AM (2016) Failure analysis and optimization of a propeller shaft for heavy duty vehicle. Int J Adv Eng Manage Sci 2:1189–1196
5. Tawancy HM, Al-Hadhrami LM (2013) Failure of a rear axle shaft of an automobile due to improper heat treatment. J Fail Anal Prev 13:353–358
6. Kumar L, Singh C, Dewangan BK, Sen PK, Bohidar SK (2014) Study on the front axle and rear axle attached to differential system. Int J Innov Res Sci Technol (IJIRST) 1:9–12
7. Getu M (2018) Design of rear axle shaft for light weight green vehicle. Thesis Mechanical Engineering, Addis Ababa University, Ethiopia
8. Thecartech, Axle shafts design. Available at http://thecartech.com/subjects/design/axle_shafts_design.htm
9. Liu W, Geni M, Yu L (2011) Effect of mesh size of finite element method in modal analysis for periodic symmetric struts support. Key Eng Mater 462–463:1008–1012
10. Husaini, Saputra E, Husni, Putra TE, Rachman M (2019) Failure analysis of a centrifugal pump shaft experiencing plastic deformation using the exact and finite element method. IOP Conf Ser Mater Sci Eng 523:012073
11. Siczek K, Wozniak M, Ozuna G (2020) The study on the damage of the rear axle shaft in a KIA truck. Adv Sci Technol Res J 14:115–124
12. Guimaraes AV, Brasileiro PC, Giovanni GC, Costa LRO, Araujo LS (2016) Failure analysis of a half-shaft of a formula SAE racing car. Case Stud Eng Fail Anal 7:17–23
13. Fauzan, Husaini, Ali N (2021) Fracture analysis on the rear driveshaft experienced failure which applied in the truck 130 PS. In: The 2nd Virtual conference on computational and experimental mechanics (VCCEM 2021), vol 1, pp 97–98

14. Azom (2012) AISI 4150 alloy steel (UNS G41500). Available at https://www.azom.com/article.aspx?ArticleID=6770
15. Nagarajan VR, Putatunda SK, Boileau J (2017) Fatigue crack growth behavior of austempered AISI 4140 steel with dissolved hydrogen. Metal 7:466
16. Wathan S, Husaini, Ali N (2021) Failure analysis on the fracture shaft of a centrifugal pump used for diesel engine cooling system. Trans Tech Publications Ltd, Switzerland, vol 892, pp 107–114
17. Husaini, Handayani LR, Ali N, Sadrawi M (2021) Failure analysis of a fractured leaf spring as the suspension system applied on the dump truck. Key Eng Mater 892:89–98
18. Husaini, Ali N, Riantoni R, Putra TE, Husin H (2019) Study of leaf spring fracture behavior used in the suspension systems in the diesel truck vehicles. IOP Conf Ser Mater Sci Eng 541:012046

Promoting Smart Industry with a Smart Monodzukuri Support Team in Thailand: A Case Study

Noppadol Sriputtha

Abstract This work aimed to promote improvement of manufacturing through adopting smart manufacturing in Thailand. As the manufacturing industry in Thailand moves toward Industry 4.0, many SMEs are struggling to start up and invest. To help them, the Thai and Japanese governments are collaborating to provide a Smart Monodzukuri support team to assist SMEs to have more skills and be more competitive. This team will learn techniques to implement Japan's Industrial Value Chain Initiative (IVI), an organization that uses cyber-physical manufacturing, and encourages all companies to take part in a "small start" initiative. This work provides a case study using an auto parts manufacturing company. We describe how a Smart Monodzukuri team could collaborate with this company's staff following the IVI guidelines. The results show that by using the IVI idea, the team could incorporate cyber-physical manufacturing with few difficulties and low investment. This is the first simple and feasible step toward Industry 4.0 for SMEs.

Keywords Monodzukuri · Smart industry · Industry 4.0 · IVI · Cyber-physical manufacturing

1 Introduction

The planet is evolving at a faster rate than ever, and has become much more interconnected, leading to increased financial competitiveness and disruptions in society, politics and culture. Progress in information and communication technology (ICT) is changing how people think, live, learn, and interact [1]. This progress has had far-reaching effects for all aspects of national development and economic progress. The effective use of digital technology can help match the demands of the planet's increasing population. Many businesses, particularly large corporations, have begun

N. Sriputtha (✉)
Robotics and Lean Automation Engineering, Faculty of Engineering, Thai-Nichi Institute of Technology, Suanluang, Bangkok 10250, Thailand
e-mail: noppadol@tni.ac.th

Irwansyah et al. (eds.), *Proceedings of the 4th International Conference on Experimental and Computational Mechanics in Engineering*, Lecture Notes in Mechanical Engineering, https://doi.org/10.1007/978-981-99-7495-5_22

to adapt their factories to use modern technology, to increase production efficiency [2].

"Industry 4.0" refers to the use of developments in automation and data technology in production and service industries. In Germany, the phrase has been used to describe the digital shift of manufacturing [3]. In Thailand, Industry 4.0 first appeared in the country's 20-year national policy (2017–2036). Thailand is an ASEAN industrial hub with supply chains for automotive and electric/electronic sectors. However, a serious labor shortage and rising labor costs [1, 4] have challenged Thailand's strongly competitive position. Consequently, there is a dire need for industrial upgrade via computerization and the use of robotics [5]. Many small and medium-sized enterprises are finding it difficult to obtain capital for startup and investment. For this reason, the Thai government is collaborating with the Japanese government to establish a Smart Monodzukuri support team to help small and medium-sized businesses to become more skillful and competitive. The Department of Industrial Promotion (Thailand), the Technology Promotion Association (Thailand-Japan), the Ministry of Economy, Trade and Industry (Japan), the Association for Overseas Technical Cooperation and Sustainable Partnerships (AOTS), the Japan-Thailand Economic Cooperation Society (JTECS), the Industrial Value Chain Initiative (IVI), and others have all provided support for this collaboration.

Monozukuri, sometimes written Monodzukuri, means "production" or "creation of things" in Japanese and is the Japanese word for "manufacturing." The wider definition includes a combination of technological ability and in-depth knowledge, and the Japanese industrial spirit. This comprises an honest attitude toward production, with pride, skill, and commitment, as well as a craving for originality and superiority. Monodzukuri is a buzzword in Japan, and many Japanese people believe it has brought Japan to a leading position in the global market [6–8].

The first Smart Monodzukuri support team visited Japan to study the use of smart technologies, IoT, and using robots in small and medium-sized businesses, as well as applications in large organizations. All the team were studying the concepts of IVI, an organization that integrates design, production, and information technology, and wants to promote company collaboration in a "small start" initiative. The Industrial Value Chain Initiative (IVI) is a platform for companies to collaborate in the design of our future society, through integrating manufacturing and digital technology. After completing the theoretical learning and study trip, the support team was required to practice with SMEs in Thailand in order to develop each company's support team.

An auto parts manufacturing firm was the subject of our case study. Together with the company's staff, the support team went through seven phases of on-the-job training, beginning with a viewing of the present situation. The second stage was to promote the theory of IVI and then brainstorm using the KJ technique. After forming an AS-IS model and a TO-BE model, the next stage was ranking issues in order to choose themes. The final step was to choose appropriate actions and activities. The team was able to integrate cyber-physical conditioning successfully, which resulted in fewer obstacles and lower investment. It serves as a departure point for the transition of SMEs to Industry 4.0 in Thailand.

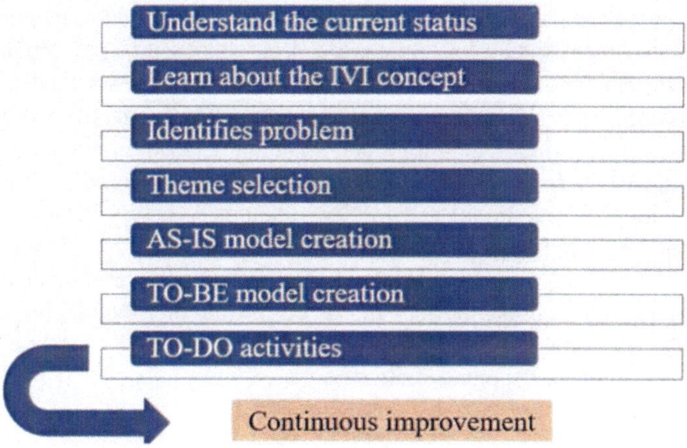

Fig. 1 Steps of case study implementation

2 Case Study

The case study focused on an auto parts manufacturing company. The auto parts manufacturing firm chose its own support staff to take part in this exercise before they could begin hands-on training. The smart Monodzukuri support team's approach was based on the stages of implementation of IVI ideas. The tour of the production area was the first stage of this approach. This was to understand what was currently going on [9]. This allowed the Monodzukuri support team to understand the operations involved in the production. Once the process was understood, the Monodzukuri support staff cooperated with Japanese IVI specialists to demonstrate the implementation of the IVI idea. During the shop floor visit, the smart Monodzukuri support team visited the manufacturing area with the company's team. They asked them to explain the process of making their products, and raised some concerns and problems about their production. Once the smart Monodzukuri support team had gathered all the information provided by the company support team, it was then easier to understand the problems and other concerns. Figure 1 shows the actions taken for the case study.

3 IVI's Methodological Approach

The smart Monodzukuri support team must help the IVI Japanese specialist in teaching the IVI methodology to the business support team. At this point, Japanese specialists explain to the company's participants about the IVI process activities, with the smart Monodzukuri support team clarifying the concepts.

IVI Concept, IVI has opened an active debate about how smart technology such as IoT will transform human-based manufacturing and aims to develop a mutually connected system architecture based on zones of collaboration between enterprises. IVI does not stress a firm's competitive advantage, but analyzes how companies can interact naturally, forming a growing understanding about basic models of internal connections, without forcing them into a single standard model [10]. In complicated industries with both traditional and modern features, a new system unwilling to change would face challenges. A strategy based on the real situation, but involving the best current solutions, should be able to make the best progress. Smart manufacturing systems are based on collecting data and using it to predict future needs [11]. IVI aims to boost the value of each enterprise using cyber-physical production systems based on the "small start" connection model.

4 Brainstorming

The IVI concept, in contrast to top-down management, aims to develop the shop floor with bottom-up management. So the smart Monodzukuri support team begins by brainstorming with the firm's team to collect information from shop floor personnel. The KJ approach was used to gather problems or the worries people have [12]. Even though the people concerned work in different sections and at different levels, brainstorming must be carried through. At the end it will be found that there are common threads or connections. Figure 2 shows a result of the brainstorming technique using the KJ method.

Fig. 2 Problems discovered through brainstorming

In Fig. 2, members of the company's support team collaborated with the smart Monodzukuri support team members serving as advisors and guides. Each team member wrote down as many problems or concerns as they could on post-it notes. After members had written down their problems and worries, the post-it notes were placed on a board in a location where everyone could see them. Then, using post-it notes of different colors for the topic of each group, similar problems were arranged into categories.

5 Theme Selection

After brainstorming and assessing the challenges, these key problem areas were ranked into the matrix table along with their defining performance criteria [13]. In Table 1, the concerns have been placed into 14 groups, and then a company team member wrote down the primary issues gathered into the topic selection matrix and prioritized the set of recognized themes. In the theme matrix selection, performance indicators were prioritized based on team consensus covering four dimensions: importance level, IoT applicability, can be executed by themselves, and on-time achievement. A rating of 1–4 was assigned to all dimensions. Members of both support teams voted depending on their opinions. The votes in each dimension were counted by multiplying the votes with the specified score and adding the scores received together to give the total score for each topic.

The 12th and 11th problem topics, communications and report system were ranked first and second, respectively, among the fourteen issue topics. As a result, they were selected for future improvement by the company's support team. In this work, both are implemented and reported; however, the communication problems were complicated, impacting many people, which might have resulted in the project timeline being extended.

6 AS-IS Model

At this stage, the two groups collaborate to create the current scenario. This is referred to as the AS-IS model and uses several symbols, as given in Table 2.

As shown in Fig. 3, a die casting process was selected to create an AS-IS model of the reporting system problems. This AS-IS model is then used to simulate on-site situations, determining the characters, their jobs, responsibilities, and the tools they use. After that, the team will identify the difficult issues and write them down on a red post-it.

The model in Fig. 3 shows the interaction between all of the people involved in this procedure. In this scenario, there are 15 persons. Each person displays the work description, tools or equipment, and information that was important to them. The red tags in the scenario show problems or worries they have.

Table 1 Theme matrix selection for prioritization of all issues

No	Topic	Important level				IoT applicability				Execution by themselves				On-time achievement				Total Score	Rank
		1	2	3	4	1	2	3	4	1	2	3	4	1	2	3	4		
1	Waiting		4	8	4			5	11			12	4		6	10	1	205	3
2	Defects				16		8	8		1		13	2	9		6		179	8
3	Mold spare parts			12	3		3	4	8		2	4	8		6	7	1	183	7
4	M/C breakdown			10	6		4	6	5		1	9	5	2	4	9		186	6
5	Packaging	4	7	4		2	3	7	3	2	4	5	4	1	4	8	2	153	13
6	Manpower		6	8	1	1	5	7	2			10	4	1		7	7	176	9
7	Layout		5	10		2	5	7		1	6	7	1	5	4	5	1	143	14
8	Setup time		3	8	4	2	4	7				13	2	2	3	6	4	166	12
9	Maintenance		1	7	7		5	8	1		2	11	2	1	6	8		171	11
10	QC inspection			8	7		5	7	3	1	2	10	2	1	7	7		174	10
11	Report system		1	5	10				16	1		7	8			3	13	236	*2
12	Communications				16				15		1	8	7		1	2	13	238	*1
13	WIP		2	6	7	1	2	1	11	1	3	8	3	1		9	5	193	5
14	Planning		6	9			1	2	12		5	5	5			6	9	194	4

* The 1st rank and 2nd rank were selected for improvement

Table 2 Symbols used to construct the present scenario

Symbol	Name of symbol	Description
	Actor	The individual in this circumstance is?
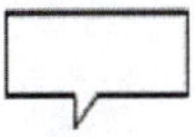	Activities	Tasks performed by the actors in the situation
	Things	Physical things used by actors
	Data/Information	Information about what happened during the activity
	Digital mark	The use of IoT in this data

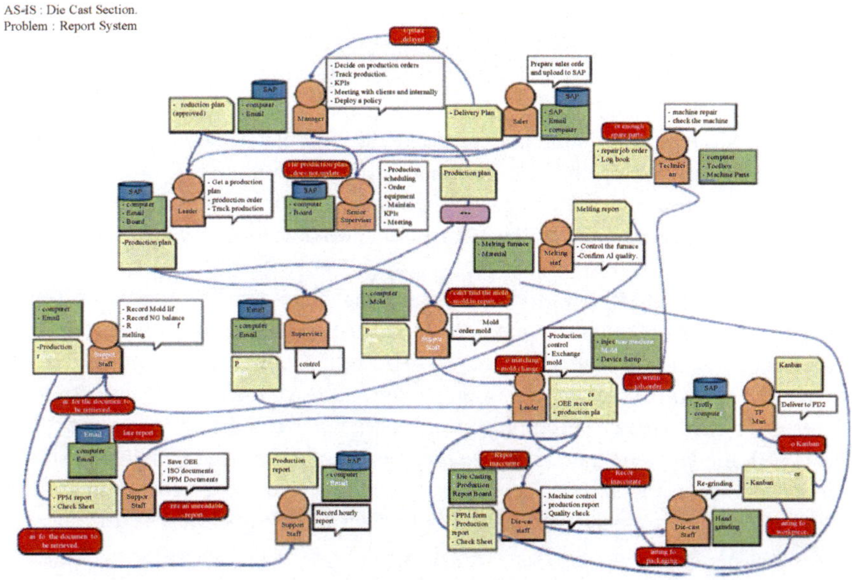

Fig. 3 AS-IS model of reporting system for the die-casting process

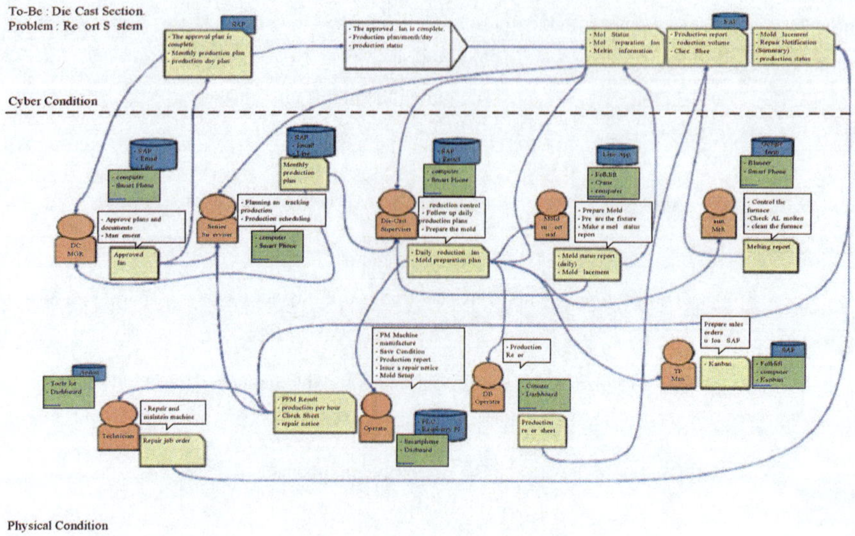

Fig. 4 TO-BE model of reporting system for the die-casting process

7 TO-BE Model

The smart Monodzukuri support team and the company team design the TO-BE model for improving the present manufacturing operation. They try to eliminate the red issues and introduce cyber-physical manufacturing into the TO-BE scenario model. Figure 4 shows the TO-BE scenario developed by the two teams in collaboration.

Figure 4 depicts an improved scenario using lean manufacturing as its basis. Using a lean manufacturing system to implement the TO-BE scenario model demonstrates the relevance of this production system [14]. In this setting, they used the ECRS approach to minimize the number of employees used for the die-cast process from 15 to 9, and the activities were combined and restructured to simplify the process [15]. The TO-BE model is separated into two sections: physical and cyber. It demonstrates the connection between physical and cyber data so that data may be conveyed without stuttering. The main goal of the model is to offer a real-time view of the state of production.

8 TO-DO Activities

Following the demonstration of the TO-BE scenario, the teams create an action plan for implementation and establishment of the project's goal. Table 3 gives the TO-DO activities and plan. The project's three major aims were to eliminate duplicate work,

Table 3 TO-DO activities and execution

No.	Activities	Estimated Budget (THB)	In charge	Timeline (FY2021) Jan	Feb	Mar	Apr	May	Jun
1	Obtaining data from the diecast machine		Team						
	Raspberry PI and I/O	7000							
	Power supply	2000							
	Noise filter	1500							
	Lan cable	200							
	Programming (Node Red)	-	Team						
2	Display design (dashboard)		Team						
	Display (smart TV)	10,000							
	TV stand	1500							
	HDMI cable	500							
3	Procurement of equipment		PU						
4	Installation		MN						
5	System test		Team						
6	Monitoring and improvement		Team						
7	Project summary		Team						
	Total	22,700							

eliminate paper notes, and gather real-time production data. To achieve the goals, the team reviewed and decided how to proceed with the small start concept.

The Raspberry Pi is a microcomputer that can be accessed by anybody using simple and inexpensive IoT devices [16]. The Raspberry Pi was chosen for this project because of its powerful processing capabilities as a single-board embedded computer, as well as its ability to connect to the Internet wirelessly via Wi-Fi or Bluetooth [17]. It cost all together about 22,700 THB, which is inexpensive and easy to present to executives.

Configuration system, based on the components chosen, we decided on the layout of the system configuration shown in Fig. 5. The signal flow and interconnections between various components are proposed. The Raspberry Pi serves as the primary controller in the developed system, collecting data from the die-cast machine's PCL unit. Node-RED is a programming tool for connecting Raspberry PI devices and processing data to dashboards and online services in mobile applications.

9 Result and Discussion

The support staff and the company team can use IVI ideas to construct a cyber-physical production environment, giving the team a way to accomplish its objectives as above. Furthermore, because of the improvement, the number of unnecessary employees was cut from 15 to 9, representing a 40% reduction. The time spent entering data to the whiteboard was reduced from 24 to 0 min per day, the time spent writing the report sheet was reduced from 20 to 0 min per day, the time spent entering data into the Excel file was reduced from 5 to 0 min per day, and the amount of A4

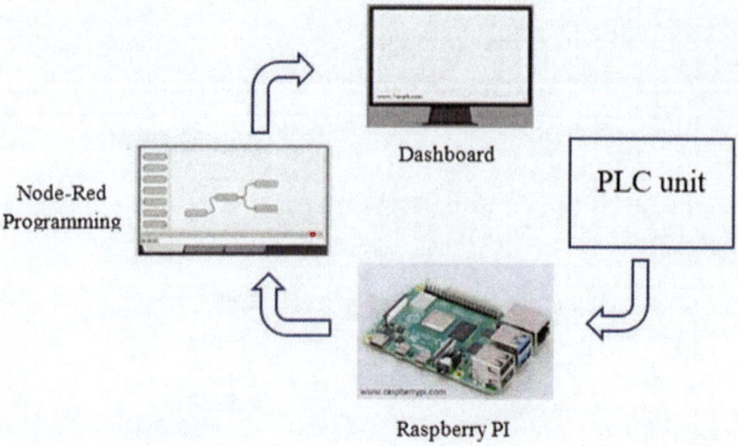

Fig. 5 System configuration

paper used was reduced from 2 to 0 sheets per day. However, even if it achieves this aim, there are additional issues that arise when it is deployed, e.g., the staff find the program inconvenient, it will take some time to learn how to utilize it and there is the issue of delayed ordering, which results in shorter test times. Figure 6 depicts the Node-red application program developed by the team, and Fig. 7 depicts the dashboard provided by the Node-red program.

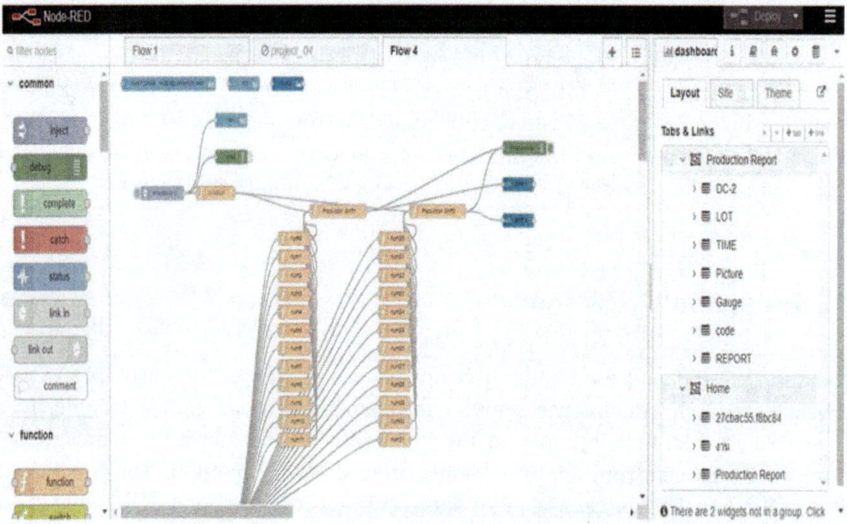

Fig. 6 Node-red program

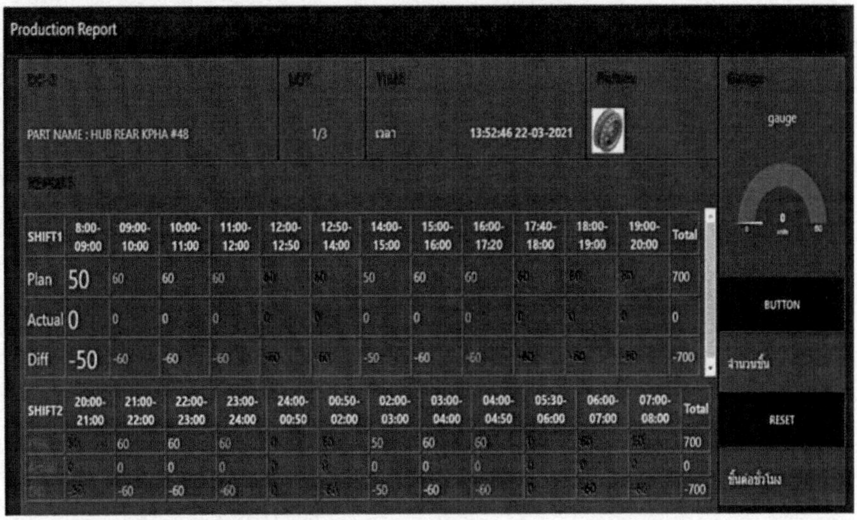

Fig. 7 Dashboard showing the production in real time

10 Conclusion

The project Promoting the Smart Industry in Thailand with the Smart Monodzukuri Support Team: A Case Study of Auto Parts Manufacturers with a Small Start and Bottom-Up Management was tested with a company. The support staff who gained expertise from IVI Japan were able to assist SMEs in adjusting to Industry 4.0. The results show that the team could integrate cyber-physical manufacturing with only a few issues and at low cost, with help from the IVI concept. This is the first basic and doable step toward attaining Industry 4.0 for SMEs. However, before basic IoT can be used in the production process, SMEs must improve lean manufacturing. Additionally, if it is not lean enough, it will result in waste from the use of technology, which would be counterproductive.

References

1. Buasuwan P (2018) Rethinking Thai higher education for Thailand 4.0. Asian Educ Dev Stud 7(2):157–173. https://doi.org/10.1108/AEDS-07-2017-0072
2. Kazuyuki N (2015) Support for Monozukuri in Southeast Asia. IHI Eng Rev 48(2):38–43 [Online]. Available: https://www.ihi.co.jp/en/technology/review_library/review_en/2015/48_02.html
3. Kohpaiboon A (2020) Industry 4.0 policies in Thailand. ISEAS Economic Working Papers, no. 2020, p. 1018 [Online]. Available: https://www.iseas.edu.sg/category/articles-commentaries/iseas-economics-working-papers/

4. Puriwat W, Tripopsakul S (2020) Preparing for industry 4.0-will youths have enough essential skills? An evidence from Thailand. Int J Instr 13(3):89–104. https://doi.org/10.29333/iji.2020.1337a
5. Charlie J, Paitoon P (2017) Innovative ideas: Thailand 4.0 and the fourth industrial revolution. Asian Int J Soc Sci 17(1):4–32
6. Wikipedia (2021) Monozukuri. Wikipedia. https://en.wikipedia.org/w/index.php?title=Monozukuri&oldid=913237347. Accessed 26 Oct 2021
7. Tanaka H (2017) Where has the Japanese core of 'Monodzukuri' come from, and where is it going? Rom Econ Bus Rev 12(2):57–71
8. Kovacic M (2018) The making of national robot history in Japan: monozukuri, enculturation and cultural lineage of robots. Crit Asian Stud 50(4):572–590. https://doi.org/10.1080/14672715.2018.1512003
9. Pokuaa-Duah G (2020) Value stream mapping in enhancing operational performance of Ghanaian food processing industry: mediating role of supply chain collaboration Gloria Pokuaa-Duah level 3 office building of the deputy vice chancellor (research and innovation) University Pu. J Int Vol Soc Sci 11(4):49–60. https://doi.org/10.30845/ijbss.v11n4p6
10. The Industrial Value Chain Initiative (IVI) (2021) What is IVI? Online [Online]. Available: https://iv-i.org/en/about-us/whatsivi/
11. Martin NL, Dér A, Herrmann C, Thiede S (2020) Assessment of smart manufacturing solutions based on extended value stream mapping. Procedia CIRP 93(1):371–376. https://doi.org/10.1016/j.procir.2020.04.019
12. Kunifuji S (2016) A Japanese problem-solving approach: the KJ Ho method. Adv Intell Syst Comput 364:165–170. https://doi.org/10.1007/978-3-319-19090-7_13
13. Izudi J, Akot A, Kisitu GP, Amuge P, Kekitiinwa A (2016) Quality improvement interventions for early HIV infant diagnosis in Northeastern Uganda. Biomed Res Int 2016. https://doi.org/10.1155/2016/5625364
14. Dos Santos DMC, Dos Santos BK, Dos Santos CG (2021) Implementation of a standard work routine using lean manufacturing tools: a case study. Gest Prod 28(1). https://doi.org/10.1590/0104-530X4823-20
15. Sriputtha N, Kositwat J, Kaewdook D (2020) Reducing waste in detergent packaging process by using ECRS technique. In: The 5 Rajamangala manufacturing and management technology conference, pp 190–196
16. Karunamoorthi R et al (2020) Design and development of IoT based home computerization using Raspberry pi. Mater Today Proc. https://doi.org/10.1016/j.matpr.2020.10.673
17. Jabbar WA, Wei CW, Azmi NAAM, Haironnazli NA (2021) An IoT Raspberry Pi-based parking management system for smart campus [Formula presented]. Internet Things (Netherlands) 14:100387. https://doi.org/10.1016/j.iot.2021.100387

Design of a Control System on the HVAC Damper to Regulate Temperature and Humidity in the Isolation Room

M. Salamul Fajar Sabri, Samsul Rizal, Rudi Kurniawan, Rachmad Almi Putra, and Suheri Suheri

Abstract Most infectious diseases are caused by viruses that are spread from one person to another through the air. Isolation room is a room that keeps virus in infected people from spreading to other people by maintaining temperature and humidity. Therefore, a system is needed that regulates the isolation room to remain sterile, where this system must control and maintain the stability of air circulation in the room. For this purpose, this study proposes the design of an automated control system to regulate the temperature and humidity in an isolated room by controlling the supply air with damper. The method that is used in making this control system is experimental and analysis from sensor readings that will be used as a sign of decision-making for the damper output that will control the air flow to the isolation room with closed loop algorithm for the system control, the controller will send digital signal to the damper relay based on sensor reading, so that the data results of the experiment and analysis can see whether the control system can work well or not. The successful performance of the control system is seen from results of sensor readings; when the system changed the damper position it can maintain the isolation room situation around 22 °C to 24 °C for temperature and 50% until 60% for the humidity.

M. Salamul Fajar Sabri
Master of Mechanical Engineering Program, Faculty of Engineering, Universitas Syiah Kuala, Banda Aceh 23111, Indonesia
e-mail: salamul@mhs.unsyiah.ac.id

S. Rizal (✉) · R. Kurniawan
Department of Mechanical and Industrial Engineering, FacultyofEngineering, Universitas Syiah Kuala, Banda Aceh 23111, Indonesia
e-mail: samsul.rizal@usk.ac.id

R. Kurniawan
e-mail: kurniawan@usk.ac.id

R. A. Putra · S. Suheri
Doktoral Program, Faculty ofEngineering, Universitas Syiah Kuala, Banda Aceh 23111, Indonesia
e-mail: Rachmad.almi@unsam.ac.id

S. Suheri
e-mail: suheri@unsam.ac.id

Irwansyah et al. (eds.), *Proceedings of the 4th International Conference on Experimental and Computational Mechanics in Engineering*, Lecture Notes in Mechanical Engineering, https://doi.org/10.1007/978-981-99-7495-5_23

Keywords Damper control · Control system · Isolation room · HVAC

1 Introduction

Infectious diseases are the result of various factors, such as viruses and other parasites, that can spread from one person to another through the air. Some of the infectious diseases that are transmitted through the air are pneumonia, acute respiratory infection (ARI), tuberculosis (TB), influenza, varicella, severe acute respiratory syndrome (SARS), and COVID-19. [1–4]. In general, sufferers of the disease are advised to avoid contact with other people, or better known as self-isolation. Only some sufferers of the disease can receive intensive medical treatment in health facilities such as hospitals. Patients with the disease are treated in an infectious disease isolation room. Isolation rooms are generally used for the treatment of patients with airborne infectious diseases, these isolation rooms have a temperature of 18 °C–26 °C and have a lower pressure (negative pressure) to keep the room sterile [5, 6]. The air flow in the isolation room is generally controlled so that it does not mix with the air in the surrounding room. Some of the parameters that are set in the isolation room are temperature, pressure, ventilation, and humidity, even though these parameters have been set with optimal settings in the isolation room. Therefore, some system is needed that controls the isolation room to remain sterile, where this system must keep humidity and temperature by maintaining stable air pressure, temperature, air velocity, and circulation in the room that is called the HVAC system (heating ventilating and air conditioning) [7]. HVAC is a system that functions to regulate air flow so that the temperature and humidity conditions in a room can be maintained and the air in the room becomes comfortable [8, 9]. By adjusting the HVAC system both on return air (RA), exhaust air (EA), and supply air (SA), it will affect the risk of spreading the virus indoors by changes in the movement and direction of airflow [10–12].

Several previous studies, simulations using CFD were carried out to analyze the air flow in the isolation room, it helped to get the design of laying the air supply and exhaust. In this study [13, 14], there were two models CFD simulation of air flow conditions and air circulation in the HVAC system that were suitable for isolation rooms. By placing an optimal air supply, it can improve the performance of the air ventilation. To create air flow in the isolation room, it is necessary to place an appropriate HVAC system but determining the placement of HVAC is not enough to maximize the air condition in the isolation room, so it is necessary to do a control system to regulate the amount of air that enters the isolation room. In this research a control system will be designed to regulate the temperature and humidity in an isolated room by using damper in the ducting as an air intermediary to the room. Control systems are now starting to shift from manual control to automated systems that require the use of computers, so that human intervention in controlling is very small. When compared with manual work, computer-controlled equipment systems will provide advantages in terms of efficiency, safety, and accuracy. Computer capabilities, both hardware and software, can be utilized for various control applications, such as temperature control. This research used an experimental method to automate the control system that regulates the situation in the isolation room. It analyzed the temperature change by controlling the damper in the ducting that supplied air to the isolation room. For the system model that is used in this research is closed loop, this model will work by looping the system for reading the sensor value and makes decision continuously. With the closed loop system control, we can see the successful performance of the control system from the results of sensor readings when the system changed the damper position.

2 System Overview

To control the HVAC system for the isolation room, this study uses an air supply setting by utilizing the damper to regulate the air flow in the duct and the output temperature in the isolation room, where the optimal temperature environment in the isolation room is between 18 °C and 26 °C [15–17], and for humidity among 40–60% [15, 17, 18]. With setting the damper, it can reduce the air flow into the patient room so that the air temperature in the room can be controlled. Isolation rooms usually consist of the patient room, bathroom, anteroom, and waiting room, as shown in Fig. 1. In testing the HVAC system control in the isolation room, this research was only focused on the patient room where there are 2 air supplies located in the middle of the room and 4 exhausts with 2 of them located near the patient's head and the other at the edge of the room. Damper changes that affect the isolation room are damper no. 2, so this study will focus on the effect of setting control damper no. 2 on the situation of the isolation room.

To find out the control system is running well, some sensors are needed to see the air output is change properly. We use temperature and humidity sensors to make some parameters that needed to prove whether the system logic is working properly to change the position of the damper in the duct. Temperature measurements were carried out using a thermocouple type K, two of them placed at the diffuser that the air supply comes out to the isolation room and another one located at the middle of the isolation room to see the general temperature in the room. Thermocouple type K is the most commonly used temperature sensor in industry, because it is quite resistant to oxidation. Besides that, this thermocouple can also avoid low oxygen or reduce the atmosphere (similar to vacuum condition). This thermocouple can withstand temperatures from −270 °C to 1300 °C [19, 20] so the sensors can comply the criteria for reading the temperature in the isolation room. For humidity sensor we use HTM 2500, this sensor can read relative humidity from 10% until 99% with accuracy ± 3% and this sensor can communicate with analog signal from 0 to 5 VDC [21]. The actuator motor that used for the damper is Honeywell CN4605, the motor runtime is from 0 to 90° about 110 s with motor torque 5Nm [22]. To adjust all

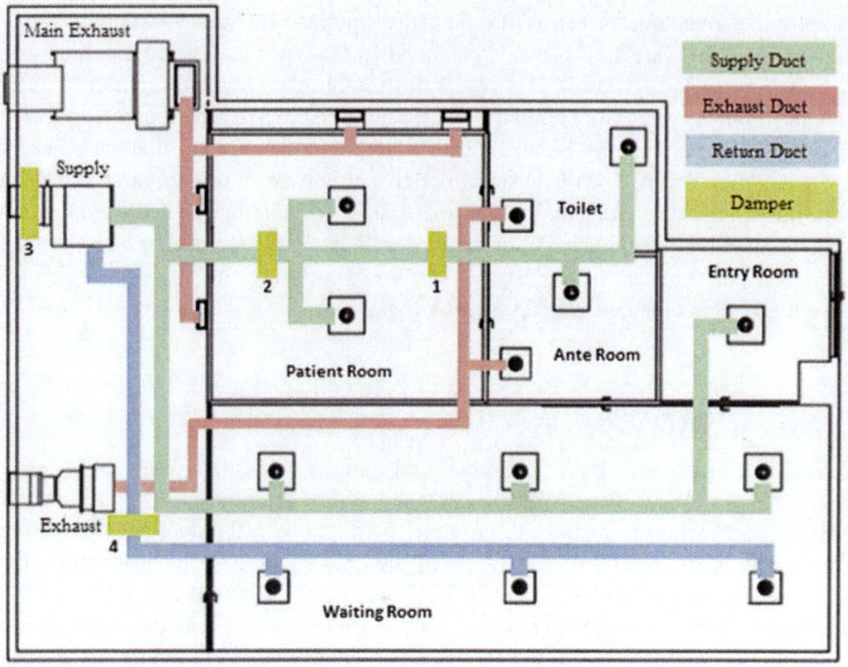

Fig. 1 Isolation room sketch

dampers also read temperature and humidity sensors, a control system is designed which basically can be seen in Fig. 2.

The controller is the center of the system which is where data processing is carried out. From the diagram, the input data to the controller can be in the form of a command from the user or an input data from humidity sensor or temperature sensor. Direct commands from the user are an absolute communication for the controller reprogramming, program updates, and direct decision-making commands. Data received

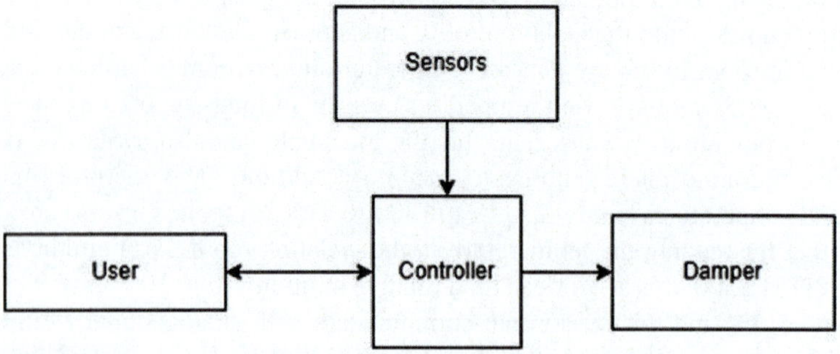

Fig. 2 Basic system diagram

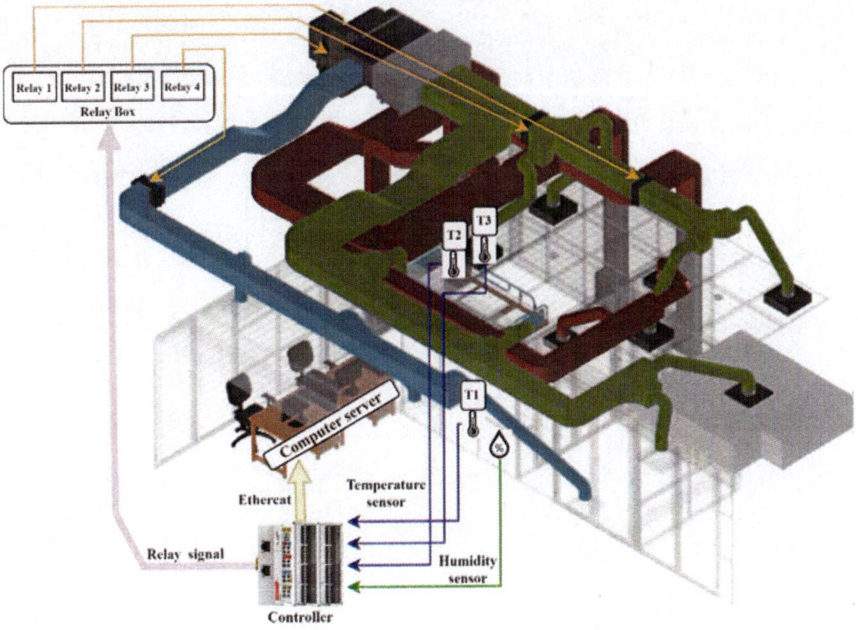

Fig. 3 Experimental setup for system testing

from sensor readings will be processed by the controller as a reference for making decision when the system runs automatically. The results of the processing decisions will be continued to regulate the HVAC system in the isolation room in the form of damper movement to change the air supply capacity.

The experimental setup of this study is shown in Fig. 3 where there are four dampers that regulate air flow, damper 1 is a damper that goes to the toilet room and entry room, damper no. 2 goes to the isolation room, damper 3 functions as a regulator of the entry of fresh air, and the last damper is a damper that regulates air circulation in the waiting room. Place the humidity sensor in the middle of the room. For temperature sensors, there are three sensors, two of the temperature sensors are placed on the diffuser, and another sensor is placed in the middle of the isolation room to read the temperature of the entire room. In this experiment, the setup involves the exhaust being turned off, and for the temperature at remote is 24 °C.

3 Control Algorithm

In designing a control system, the communication flow is important in the delivery of data both as input and as output. In general, the signal on the controller is divided into four, analog signal, discrete signal, sample data signal, and digital signal. In this study, the signals used are analog signals and digital signals as shown in the flow

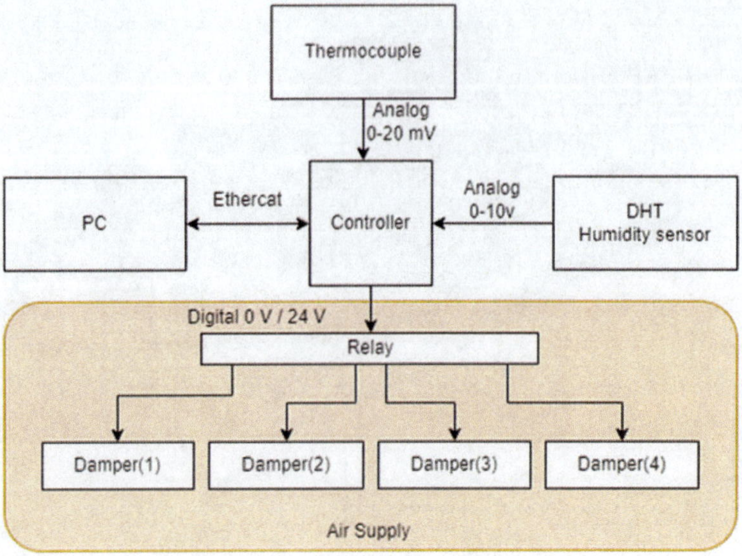

Fig. 4 Signal flow communication

diagram in Fig. 4, where the analog signals are used as sensor readings and digital signals as HVAC regulator outputs.

The computer and the controller communicate through an Ethercat cable, which enables fast and stable data transmission for programming and troubleshooting while the system is running [23, 24]. The command digital signal is given to the damper, where at the relay a voltage of 0 V is used to turn off/close the damper and 24 V to activate the damper. The placement of the damper in the duct is adjusted to regulate the air flow into the patient room, the placement of the damper can be seen in Fig. 2. Meanwhile the thermocouple sensor sends information with an analog signal from 0 to 20 mV to the controller for reading the temperature in the isolation room. For humidity sensor use analog signal from 0 to10 V for communication with controller.

In this research, the algorithm that used to adjust the damper is a closed loop control system, where in this closed loop control makes the system continue to run until the system is turned off. The close loop method of control system that has been designed in the study can be seen in Fig. 5. When the initial declaration of the relay that was given a 24 V signal is to the relay no. 2 and relay no. 3, from Fig. 2 it can be seen that relay no. 2 is a relay damper that regulates air entry into the isolation room, while relay no. 3 is a damper that regulates the entry of air from the outside environment. Damper 1 and damper 4 are set in the closed position because this study will focus on setting the damper for air entering the isolation room.

The decision-making on this system depends on the temperature value, so the system will continuously read the temperature. In this control system, the cycle time used for temperature readings continuously is at 10 ms/cycle so that the decisions on the system can be made quickly. The temperature value in this study is set at 24 °C.

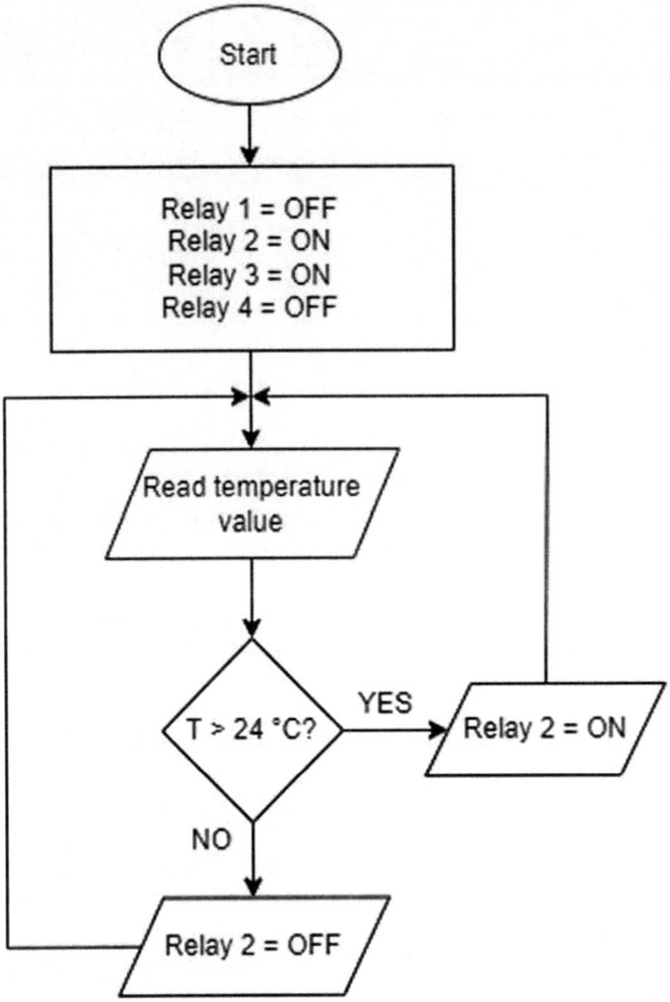

Fig. 5 System flowchart

The decision for the damper signal will be made if the temperature is more or less than the setup temperature.

4 Result and Discussion

In this section, we can see the condition of the isolation room when the control system is working. The condition that can be adjusted in this control system is by opening or closing the damper, so that the situation of the isolation room is divided into two

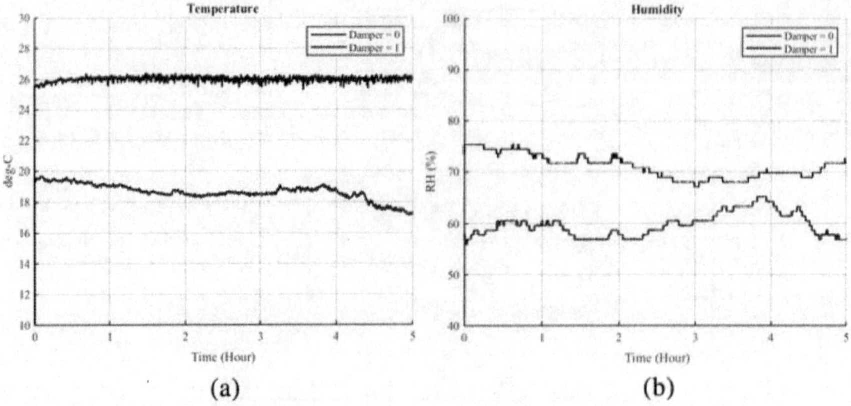

Fig. 6 Isolation room situation: **a** temperature, **b** humidity

conditions as shown in Fig. 6, where the condition is when the damper is opened or when the damper is closed where with the results of the temperature and humidity sensor readings we can see the difference between the two conditions of the isolation room. When the controller gives signal 0 to the damper, the damper will be closed so that the temperature condition in the room exceeds the desired temperature and the humidity in the room also goes over 60% which is the value that has been set for the isolation room. Meanwhile, when the damper is given a signal 1, the damper is open which causes the room temperature to drop very significantly until 18 °C, but in this condition the humidity value obtained is sufficient in accordance with the criteria for the isolation room. The drop of the temperature when the damper is open is caused by an error in the variable refrigerant flow (VRF) in the air condition. These VRF faults contain temperature sensor faults, the system faults, and indoor unit faults [25]. This can also be caused by thermostat overriding during demand response events targeting the temperature in their air conditioners [26] (Fig. 7).

By using an automatic system control, the movement of the damper will change according to the sensor readings so that the temperature and humidity conditions will be maintained by the system with the parameters that have been made in the system algorithm. The temperature and humidity ranges are in accordance with the needs with temperatures below 24 °C and humidity between 50 and 60%. However, during the transition at the time of decision-making, there are fluctuations that cause the condition of the room to be unstable. This is due to the damper setting that just can be done only by opening and closing for controlling the air flow, with these limitations the results obtained are not stable.

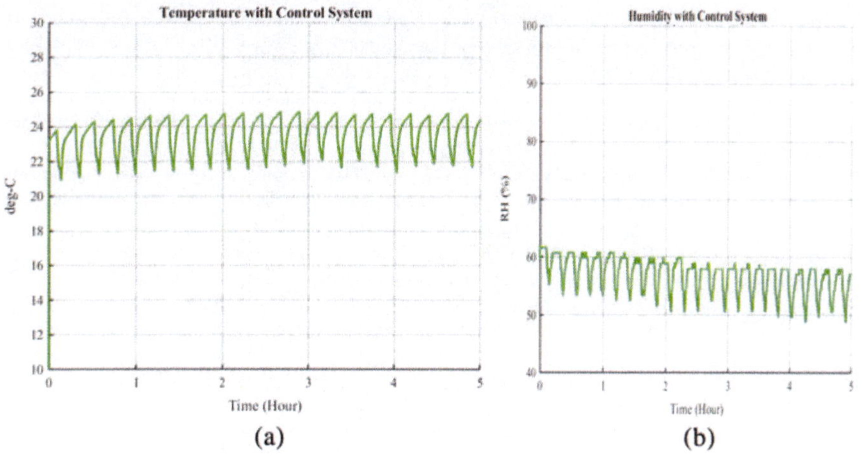

Fig. 7 Isolation room situation with control system: **a** temperature, **b** humidity

5 Conclusion

In this paper, we presented control system design that used to control the situation of air flow in the isolation room by setting the air supply on the damper in the HVAC system. The controller can receive data well from sensors to determine the condition of the isolation room and for decision-making parameters on the system. Therefore, the system control strategy for the isolation room that regulates changes in air flow has been successful with evidence of changes in temperature and humidity sensors according to the parameters needed when the system is running. Even though the situation of isolation room is still unstable, this control system can maintain the situation according to the needs required by the isolation room.

Funding This research was funded by Lembaga Pengelola Dana Pendidikan (LPDP), managed by Indonesian Science Fund (ISF): RISPRO/KI/B1/TKL/5/15448/2020; Syiah Kuala University: 370/UN11.2.1/PT.01.03/PNBP/2021).

References

1. Moghadami M (2017) A narrative review of influenza: a seasonal and pandemic disease. Iran J Med Sci 42(1):2–13
2. Pai M et al (2016) Tuberculosis. Nat Rev Dis Prim 2. https://doi.org/10.1038/nrdp.2016.76
3. Schneider E (2012) Severe acute respiratory syndrome (SARS). Netter's Infect Dis 537– 543. https://doi.org/10.1016/B978-1-4377-0126-5.00089-6
4. Gershon AA et al (2015) Varicella zoster virus infection. Nat Rev Dis Prim 1:1–41. https://doi.org/10.1038/nrdp.2015.16
5. Borro L, Mazzei L, Raponi M, Piscitelli P, Miani A, Secinaro A (2021) The role of air conditioning in the diffusion of Sars-CoV-2 in indoor environments: a first computational fluid dynamic model, based on investigations performed at the Vatican State Children's hospital. Environ Res 193. https://doi.org/10.1016/j.envres.2020.110343

6. Bahl P, Doolan C, De Silva C, Chughtai AA, Bourouiba L, Macintyre CR (2020) Airborne or droplet precautions for health workers treating coronavirus disease 2019? pp 1–8. https://doi.org/10.1093/infdis/jiaa189
7. Valenzuela PE, Ebadat A, Everitt N, Parisio A (2020) Closed-loop identification for model predictive control of HVAC systems: from input design to controller synthesis. IEEE Trans Control Syst Technol 28(5):1681–1695. https://doi.org/10.1109/TCST.2019.2917675
8. Sporr A, Zucker G, Hofmann R (2019) Automated HVAC control creation based on building information modeling (BIM): ventilation system. IEEE Access 7:74747–74758. https://doi.org/10.1109/ACCESS.2019.2919262
9. Anuntasethakul C, Banjerdpongchai D (2021) Design of supervisory model predictive control for building HVAC system with consideration of peak-load shaving and thermal comfort. IEEE Access 9:41066–41081. https://doi.org/10.1109/ACCESS.2021.3065083
10. Jacob S, Yadav SS, Sikarwar BS (2019) Design and simulation of isolation room for a hospital. Springer Singapore
11. European Centre for Disease Prevention and Controls (2020) Heating, ventilation and air-conditioning systems in the context of COVID-19. Eur Cent Dis Prev Control 1–5 [Online]. Available: https://www.ecdc.europa.eu/sites/default/files/documents/Ventilation-in-the-context-of-COVID-19.pdf
12. Khankari K (2016) Patient room HVAC. ASHRAE J 58(6):16–26
13. Cho J (2019) Investigation on the contaminant distribution with improved ventilation system in hospital isolation rooms: effect of supply and exhaust air diffuser configurations. Appl Therm Eng 148:208–218. https://doi.org/10.1016/j.applthermaleng.2018.11.023
14. Wang F, Chaerasari C, Rakshit D, Permana I, Kusnandar (2021) Performance improvement of a negative-pressurized isolation room for infection control. Healthcare 9(8):1–11. https://doi.org/10.3390/healthcare9081081
15. Weng CL, Kau LJ (2021) Planning and design of a full-outer-air-intake natural air-conditioning system for medical negative pressure isolation wards. J Healthc Eng 2021. https://doi.org/10.1155/2021/8872167
16. Saran S et al (2020) Heating, ventilation and air conditioning (HVAC) in intensive care unit. Crit Care 24(1):1–11. https://doi.org/10.1186/s13054-020-02907-5
17. Zheng W et al (2021) COVID-19 impact on operation and energy consumption of heating, ventilation and air-conditioning (HVAC) systems. Adv Appl Energy 3:100040. https://doi.org/10.1016/j.adapen.2021.100040
18. Santos AF, Gaspar PD, Hamandosh A, de Aguiar EB, Filho ACG, de Souza HJL (2020) Best practices on HVAC design to minimize the risk of COVID-19 infection within indoor environments. Braz Arch Biol Technol 63:1–11. https://doi.org/10.1590/1678-4324-2020200335
19. Thermometry T, Usage G, Guide to secondary thermometry, pp 1–62
20. Webster E (2021) A critical review of the common thermocouple reference functions. Metrologia 58(2). https://doi.org/10.1088/1681-7575/abdd9a
21. Measurement Specialities (2012) HTM2500LF—Temperature and relative humidity module, p 9
22. Control F, Non-spring return direct-coupled damper actuators for contents of package, pp 1–8
23. Prytz G (2008) A performance analysis of EtherCAT and PROFINET IRT. In: IEEE International conference on emerging technologies and factory automation (ETFA), no 1396, pp 408–415. https://doi.org/10.1109/ETFA.2008.4638425
24. Cena G, Bertolotti IC, Scanzio S, Valenzano A, Zunino C (2012) Evaluation of EtherCAT distributed clock performance. IEEE Trans Ind Informatics 8(1):20–29. https://doi.org/10.1109/TII.2011.2172434
25. Guo Y et al (2019) An expert rule-based fault diagnosis strategy for variable refrigerant flow air conditioning systems. Appl Therm Eng 149:1223–1235. https://doi.org/10.1016/j.applthermaleng.2018.12.132
26. Sarran L, Gunay HB, O'Brien W, Hviid CA, Rode C (2021) A data-driven study of thermostat overrides during demand response events. Energy Policy 153:112290. https://doi.org/10.1016/j.enpol.2021.112290

Assessment of Hole Quality When Drilling Particleboard Made of Oil Palm Empty Fruit Bunch (OPEFB)

Ismet Hari Mulyadi and Ilham Sumardi

Abstract Assembling particleboard made of palm oil empty fruit bunches requires good quality holes so that the assembled product can meet quality requirements. Machinability is critical in producing high-quality holes, especially for manufacturing products from particleboard. It is owing to low level of density and structural strength of particleboards. Accordingly, the machinability of the particleboards in terms of hole quality is a primary concern in this study. The hole quality was assessed in terms of the diameter deviations, the hole's perpendicularity and the surface roughness. The results showed that the response deviations were relatively large for all observation aspects. In addition, the variations in cutting conditions and the appropriate selection of the tool geometry failed to render their defined effects on the hole quality. In other words, the effect of cutting conditions and tool geometry on the hole quality is difficult to define accurately.

Keywords Particleboards · Oil palm empty fruit bunch · Drilling processes

1 Introduction

Along with the rapid growth of the furniture industries, the limited availability and the rise of wood prices as the primary raw material for making furniture, the use of bio-composite materials is a potential alternative. Besides, the bio-composite material would also be beneficial in increasing added value for natural waste materials utilized as the bio-composite main composition. For furniture applications, the bio-composites are usually manufactured in the form of particleboards such as low-density fibreboards (LDF) and medium-density fibreboards (MDF). Grand View Research reported in 2019 that the sales volume of MDF had been valued at $61.3 billion, and it is expected to increase on average per year from 2020 to 2027 by

I. H. Mulyadi (✉) · I. Sumardi
Mechanical Engineering Department, Andalas University, Kampus Unand, Limau Manis, Padang, Indonesia
e-mail: ismet@eng.unand.ac.id

Irwansyah et al. (eds.), *Proceedings of the 4th International Conference on Experimental and Computational Mechanics in Engineering*, Lecture Notes in Mechanical Engineering, https://doi.org/10.1007/978-981-99-7495-5_24

6.6% [1]. Meanwhile, the demand for particleboard, as reported by Mordor Intelligence, is also expected to increase by 5.5% in 2022–2027 [2]. The demand for the above two types of bio-composite is dominated by the furniture, accents, flooring and speakers industries [3, 4]. Nowadays, for furniture and cabinets production, it takes approximately 85% of particleboard utility [5].

Oil palm empty fruit bunches are a waste by-product of palm oil into edible oil. If further processed, many parts of this waste can produce various product groups with significant added value [6]. One of the product groups that can be developed from palm oil empty fruit bunches is bio-composites for manufacturing furniture such as MDF and particleboards [7, 8]. Both particleboards and MDF have the disadvantage of not being able to hold screws [9]. Whereas as is known, particleboard can be utilized to develop the structure of a furniture product. To make it happen, the assembly process through a mechanical joining process using bolts, screws, or pegs requires. Therefore, drilling quality in terms of cut smoothness and dimensional accuracy of the hole is essential [10] to maintain the strength of the joint of the bolts and pins used in the joining processes. Therefore, parameter analysis of cutting conditions and tool geometry for various types of particleboard needs to be carried out. This is to maintain the strength of the connection of the bolts and pins used in the connection process. Therefore, parameter analysis of cutting speed, cutting tool diameter, cutting angle, feeding motion, and cutting tool geometry for various types of particleboard forming materials needs to be carried out [11].

Several studies have been carried out to optimize the cutting parameters to improve the quality of the particleboard in the drilling process [12–18]. However, the type of particleboards studied is predominantly commercial particleboard. Meanwhile, research on the quality of the machining process, especially the results of the drilling process on particleboard from empty palm oil bunches, is still limited. In fact, as an environmentally friendly material, particleboard made of empty palm oil bunches has potential for the furniture industry in the future. It is highly valued and marketed at competitive prices compared to other bio-composite materials such as plywood [19].

The research related to the machining process of particleboard made of empty fruit bunches of oil palm is limited to several processes, such as the turning process [20] and wood milling using a router [21]. Ratnasingam et al. [20] investigated the wear of a tungsten carbide cutting tool while turning particleboard made of empty palm oil bunches. Meanwhile, Ratnasingam et al. [21] studied the wear of tungsten carbide chisels using a router machine to cut profiles on the particleboard made of empty palm oil bunches. Both of these studies were conducted with open machining processes where the cutting temperature, which is the leading cause of the low quality of the process, can be better controlled. In contrast, some furniture features are in the form of holes used for the joining process. For this reason, research on closed machining processes, such as drilling processes needs to be performed.

In this study, the quality of the holes resulting from drilling the particleboard made of empty oil palm fruit bunches was carried out. The quality of the holes evaluated was perpendicularity, diameter deviation and surface roughness.

2 Materials and Methods

The workpiece selected in this study was particleboard made from palm oil empty fruit bunches that have a size of 300 × 300 × 25 mm. This workpiece has a composition of 80% oil palm empty fruit bunches and 20% adhesive. The adhesive used was a mixture of Paraformaldehyde, Gambier and Aquadest that is capable of producing particleboard with the structural strength of 3.5 kg/cm^2 [22, 23], while the cutting tool used was a high-speed steel (HSS) with a diameter of 10 mm. This cutting tool material was chosen because it is commercially available and widely used for wood machining processes.

The cutting tool was mounted on an Edu-Mill CNC machine type V2M-2825P with a maximum spindle rotation of 3000 RPM. Furthermore, the drilling process was performed to produce a blind hole using the cutting conditions as depicted in Table 1.

The quality of the holes produced by the drilling process was assessed in terms of diameter deviations, perpendicularity, and surface roughness. The perpendicularity and diameter deviations were measured qualitatively using a telescope bore gauge that was positioned at the centre of the hole by utilizing the machine's spindle. The diameter indicated by the gauge would be quantified utilizing a vernier caliper. Sixteen points radially and three points along the hole depth were taken as measurement points as shown in Fig. 1.

Table 1 Variation of cutting conditions used in this study

No	V_c (m/min)	f (mm/rev)	κ_r (^{o})
1	10	0.05	45
2	30	0.06	60

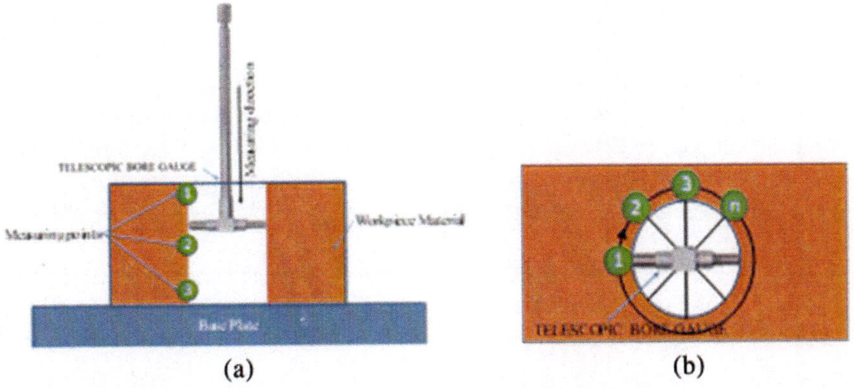

Fig. 1 Measurement method of hole diameter deviatian and perpendicularity. **a** Axial direction, **b** Radial direction

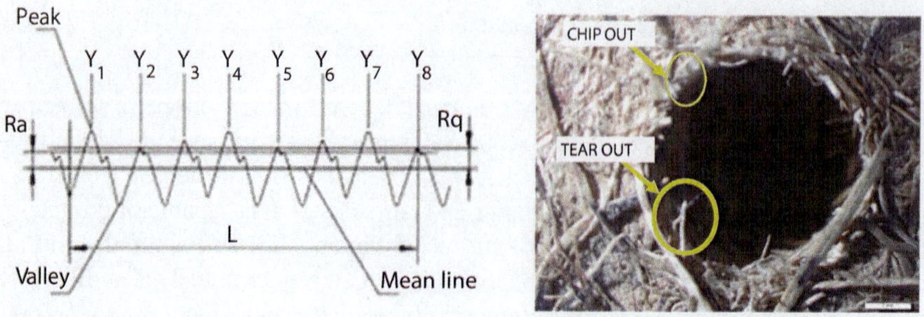

Fig. 2 Arithmetic surface roughness measurement method. **a** Measurement concept, **b** Real measurement

Furthermore, the image of the surface roughness of the hole's surface that had been previously cut was captured by an Olympus optical microscope type SZX10. The captured images were then calibrated to equalize so that they could be measured using AutoCAD software. The use of AutoCAD software is due to the rough surface profile of the workpiece materials, making it difficult to measure using a surface roughness measuring tool. The surface roughness indicator chosen was the arithmetic surface roughness number (Ra) calculated by surface roughness parameters application belonging to OriginLab© based on the concept shown in Fig. 2. 15 points along hole depth were taken as surface profile observation.

3 Results and Discussions

Assessment of Hole Diameter Deviations. In this research, the hole quality is assessed in terms of diameter deviation, hole perpendicularity and the hole's surface roughness. Figure 3 and Fig. 4 show the effect of feeding movement (f) on diameter deviation for various cutting speed choices and tool geometry (in this case, the tool point angle). The diameter deviations were measured at three locations, namely near the top surface (entry), in the middle (middle) and near the bottom surface (exits). In general, the deviation of holes on particleboard made from a mixture of palm oil bunches fibre and Gambier shows inconsistency with changes in cutting parameters and the point angle. This indicates the lower machinability of palm oil-based particleboards [24].

In addition, as can be seen in Figs. 3 and 4, the hole diameter deviation tends to be negative. In other words, the hole is shrinking compared with the nominal diameter of the tool used (d = 10 mm). This can be identified from the type of hole failure, namely chip out and tear out, as shown in Fig. 5 [25]. This type of failure occurs due to differences in the density of the fibres in the particleboard structure [26].

Meanwhile, hole dilation (positive diameter deviation) indicates that delamination occurs in the hole. This is especially the case in the holes produced by the drilling process using a cutting tool with a combination of a sharp point angle and

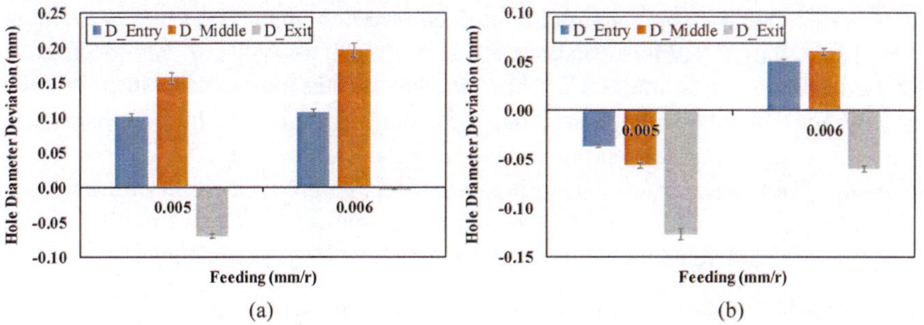

Fig. 3 Effect of feed movement on hole diameter deviation (entry, middle and exit) for cutting tool with point angle of 90° (($\kappa_r = 45°$) at cutting speed (Vc) of **a** 10 m/min and **b** 30 m/min

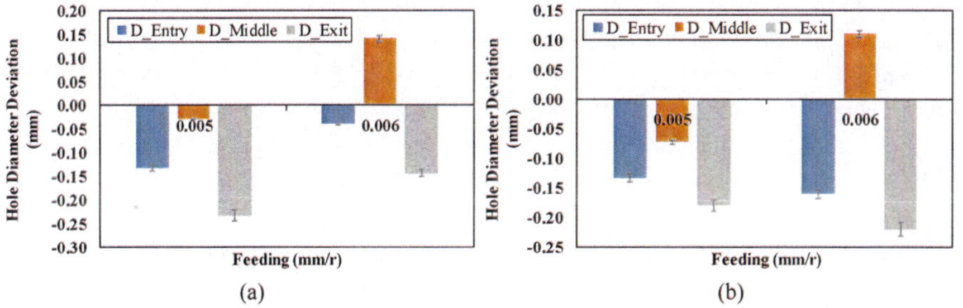

Fig. 4 Effect of feed movement on hole diameter deviation (entry, middle and exit) for cutting tool with point angle of 120° ($\kappa_r = 60°$) at cutting speed (Vc) of **a** 10 m/min and **b** 30 m/min

Fig. 5 Hole image captured at the top of particleboard surface

low speeds. Delamination occurs due to localized bending, which makes the fibre unbroken completely. Fibres that are not entirely cut off will also spin during the rotation of the cutting tools. The rotating fibre will be pushed and flow through the cutting tool's helical angle to enlarge the cutting diameter. On the top surface, it would widen the hole to produce peel-up delamination, while at the bottom of the hole would cause push-out delamination. Meanwhile, a nasty cut of the fibre at the middle of the hole would also enlarge the diameter. This is especially true for materials with low inter-fibre density and strength, such as particleboard.

Assessment of Hole Perpendicularity. The diameter deviation at the hole entry, middle and exit positions also causes a deviation of the centre of the hole so that the axis of the hole is no longer perpendicular to the reference surfaces. In this study, assessment results of the axis hole perpendicularity are depicted in Fig. 6. The magnitude of the hole axis perpendicularity was obtained using a linear regression approach. The observed centre deviation of the holes at the hole's entry, middle and exit was plotted to create linear regression models. The models were then differentiated to obtain the slope value. By using the application, the slope was converted into the angle, thus obtaining the magnitude of the perpendicularity in degrees [27].

From Fig. 6, as is hole diameter deviation, inconsistency of axis misalignment also occurs. There is a hole axis tilted to the left and the right. The sharp point angle of the cutting tool coupled with the high cutting speed and feed movement results in minimum deviation. Meanwhile, the cutting speed for a blunt point angle should be chosen low to produce slight deviations. This is to minimize the high compressive force [28]. However, the direction of the deviation varies between the blunt and the sharp point angle.

Assessment Hole Surface Quality. With low density and bond strength between fibres, in addition to producing holes with low geometric quality, it also affects the quality of the surface of the hole itself. Theoretically, the surface quality of the drilled hole surface would depend on appropriate cutting conditions and tool geometry selection. A higher surface finish would be gained if high cutting speed, low feed

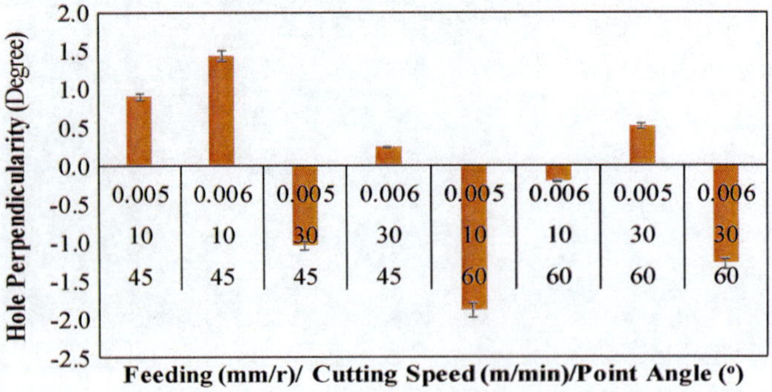

Fig. 6 Effect of cutting conditions and point angle on the hole perpendicularity

movements and sharp tool point angle are opted [29]. However, in several respects, the results of this study, as shown in Fig. 7, slightly deviate from the existing theoretical concept. For the same point angle and cutting speed, the surface roughness of the drilled hole on the particleboard made of empty bunches of oil palm decreased with increasing feeding movement. The results indicate only high cutting speeds and blunt tool point angles, which align with existing theoretical concepts. It can be argued that the variations in the selected cutting conditions are not suitable or the range is too short. This argument can be strengthened by the results of the analysis of variances (ANOVA) test to seek the difference between each response, as given in Table 2. All the experimental factors have a p-value lower than 0.05 with a 95% confidence level (CF-confidential factor). It means the observed factors give an insignificant contribution to the observed response.

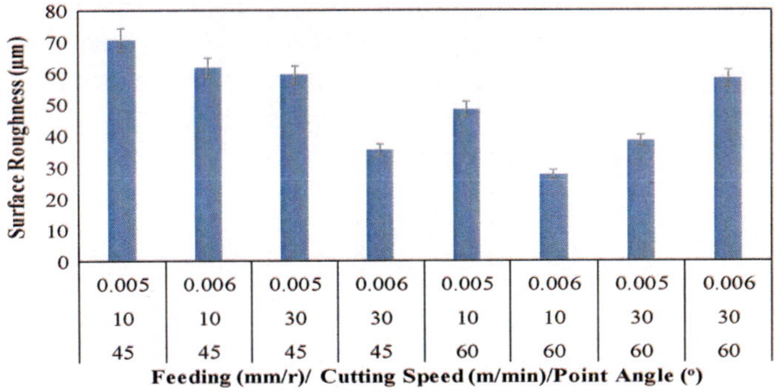

Fig. 7 Effect of cutting conditions and point angle on the hole surface roughness

Table 2 Analysis of variances (ANOVA) test for surface roughness

Source	DF	Seq SS	Contribution (%)	Adj SS	Adj MS	F-Value	ρ-Value
Vc	1	72.16	0.74	72.16	72.16	0.10	0.758
f	1	283.51	2.90	283.51	283.51	0.39	0.543
Kr	1	754.76	7.71	754.76	754.76	1.04	0.327
Error	12	8680.53	88.66	8680.53	723.38		
Lack-of-fit	4	2041.03	20.85	2041.03	510.26	0.61	0.664
Pure error	8	6639.49	67.81	6639.49	829.94		
Total	15	9790.96	100.00				

4 Conclusions

The limited sources of raw materials from natural wood make the furniture industry turn to agricultural waste, such as empty fruit bunches of oil palm as a substitute. Appropriate processing of empty palm oil bunches with other materials to form bio-composite particleboard can be suitable for replacing natural wood in the furniture industry. However, particleboard might have to go through a secondary manufacturing process, especially one that supports the joining process, such as the drilling process. In order to produce high structural strength furniture products, the joining process had to be supported by the higher quality of the holes. However, the quality of the holes on particleboard made of empty palm oil bunches is still relatively low. The hole diameter deviation and the hole axis perpendicularity prove this. In addition, the surface roughness of the hole is also relatively high (i.e. rough).

Furthermore, the effect of cutting conditions and tool geometry on hole quality shows inconsistencies. This is the impact of the low density and bond strength between fibres. For this reason so that the potential use of bio-composites from empty oil palm fruit bunches can be used as the primary material in the manufacture of furniture, it is necessary to increase the mechanical properties.

References

1. Grand View Research (2019) Medium density fiberboard market size, share & trends analysis report by product (Standard MDF, Moisture Resistant MDF), by application (Furniture, Construction), by region, and segment forecast, 2020–2027. Grand View Research. https://www.grandviewresearch.com/industry-analysis/medium-density-fiberboard-market. Accessed 10 Jun 2022
2. Mordor Intelligence (2021) Particle board market—growth, trends, covid-19 impact, and forecasts (2022–2027). Mordor Intelligence. https://www.mordorintelligence.com/industry-reports/particle-board-market. Accessed 10 Jun 2022
3. Wu Q, Vlosky RP (2000) Panel products: a perspective from furniture and cabinet manufacturers in the southern Unites States. For Prod J 50(9):45–50
4. Narangba Timbers (2022) Top uses for particle board. Narangba Timbers. https://narangbatimbers.com.au/top-uses-particle-board/. Accessed 13 Jun 2022
5. Kowaluk G (2012) Machining processes for wood-based composite materials. In: Hocheng H (ed) Machining technology for composite materials. Woodhead Publishing, pp 412–425
6. Abdulrazik A, Elsholkami M, Elkamel A, Simon L (2017) Multi-products productions from Malaysian oil palm empty fruit bunch (EFB): analyzing economic potentials from the optimal biomass supply chain. J Clean Prod 168:131–148
7. Izani MAN, Paridah MT, Astimar AA, Nor MYM, Anwar UMK (2012) Mechanical and dimensional stability properties of medium-density fibreboard produced from treated oil palm empty fruit bunch. J Appl Sci 12:561–567
8. Kasim J, Ahmad AJ, Yamani SAK, Nasir NM (1997) Particleboard from fibres of oil palm empty fruit bunches. J Trop For Prod 3(1):81–87
9. Klpomart (2018) Particle boards, MDF and Plywood. Klopmart. https://www.klopmart.com/article/detail/particle-board-mdf-plywood. Accessed 10 Jun 2022
10. Davis EM (1957) Machining test for particle board; some factors involved

11. Kumar RS, Bhowmik, Jayasval R (2019) Influence of drilling parameters and mechanical properties of biodegradable particleboard composite panel. In: Kumar K, Davim JP (eds) Biodegradable composites: materials, manufacturing and engineering. De Gruyter, pp 167–182. https://doi.org/10.1515/9783110603699-008
12. Mercy JL, Prakash S (2012) Adaptive neuro fuzzy inference system (ANFIS) modeling of thrust force in drilling particle board (PB) composites. Int J Des Manuf Technol 6(1):22–30
13. Kumar AM, Jayakumar K (2018) Drilling studies on particle board composite using HSS twist drill and spade drill. IOP Conf Ser: Mater Sci Eng 402:1–7
14. Valarmathi TN, Palanikumar K, Latha B (2013) Measurement and analysis of thrust force in drilling of particle board (PB) composite panels. Measurement 46:1220–1230
15. Szwajka K (2015) Analysis of thrust force and torque in drilling process of particle board. Mater Sci Forum 818:233–238
16. Mercy JL, Prakash S, Goswami K, Vijayalakshmi P (2014) Mathematical modelling and statistical analysis of delamination in the drilling of particle board wood composite. In: 5th International All India manufacturing technology, design and research conference, pp 1–6
17. Mercy JL, Prakash S, Vijayalakshmi P (2014) Multi response optimization of drilling parameters during drilling of particle board using Grey relational analysis. Appl Mech Mater 592–594:530–533
18. Sofuogle SD, Kacal A (2013) Investigating of machining performance in terms of tool wear and hole accuracy in drilling of particleboard surfaced with synthetic resin sheet. In: Proceedings of 21st International wood machining seminar, pp 233–240
19. Enters T (1997) Technology scenarios in the Asia-Pacific forestry sector. Kuala Lumpur
20. Ratnasingam J, Ma TP, Ramasamy G, Manikam M (2009) The wear characteristics of cemented tungsten carbide tools in machining oil palm empty fruit bunch particle board. J Appl Sci 9(18):3397–3401
21. Ratnasingam J, Tek TC, Farrokhpayam SR, Ioras F (2013) Tool wear characteristics of oil palm empty fruit bunch (OPEFB) particleboard. J Inst Wood Sci 19(2):91–94
22. Junaidi J, Kasim A, Budiman D (2016) Pengaruh Jenis Serat Tandan Kosong Sawit (TKS) Hasil Defiberasi Secara Mekanis dan Kadar Perekat Gambir Terhadap Kualitas Papan Komposit. In: Prosiding Seminar Nasional Rekayasa Teknologi Industri dan Informasi (ReTII) ke-10, pp 889–895
23. Junaidi J (2018) Pengaruh Suhu dan Waktu Kempa Terhadap Sifat Fisik dan Mekanik Papan Komposit dari Serat Tkks Berperakat Gambir berlapis Anyaman Bambu. J Ilm Poli Rekayasa 13(2):54–63
24. Ratnasingam J, Scholz F (2006) An assessment of the machining characteristics of oil palm empty fruit bunch particleboard. Germany
25. Ratnasingam J, Meulty T, Manikam M (2008) The machining characteristics of oil palm empty-fruit bunches particleboard and its sustainability for furniture. Asian J Appl Sci 1(3):253–258
26. Klamecki BE (1980) Interaction of tool wear mechanisms in wood cutting. Wood Sci 13:122–126
27. Engineering Toolbox (2009) Slope-degree, gradient and grade calculator. https://www.engineeringtoolbox.com/slope-degrees-gradient-grade-d_1562.html. Accessed 12 Jul 2022
28. Valarmathi TN, Palanikumar K, Latha B (2013) Measurement and analysis of thrust force in drilling of particleboard (PB) composite panels. Measurement 46:1220–1230
29. Valarmathi TN, Palanikumar K, Sekar S, Latha B (2020) Investigation of the effect of process parameters on surface roughness in drilling of particleboard composite panels using adaptive neuro fuzzy inferences system. Mater Manuf Process 35(4):469–477

Characteristics of Heat Transfer of Cooling Tower Flat Type Flat Storage

Afdhalul Rajai, Ahmad Syuhada, and Razali

Abstract Cooling tower is a heat transfer device that works through water media, along with air used to cool the temperature using water media with direct contact with air which causes some of the water to evaporate. To improve the performance of the cooling tower, many trials and research have been carried out, one of which is by adjusting the hole in the cooling tower area as a liquid medium distribution. From this phenomenon, this study examines the effect that occurs on the number of holes on the rate of movement of the flat plate cooling tower with the overall height of the cooling tower 2.4 m, the number of distribution media for the cooling tower water flow is 5 plates in succession, and the length of each plate is 0.7 m with a width of 0.5 m. Plate variations are given to the cooling tower, namely plate without holes, plate 25 plate holes, 35 holes, 45 holes zig-zag hole model on the cooling tower. The diameter of each plate hole is 0.02 m with the slope ratio of each plate arrangement which is 15°. The results showed the highest heat transfer rate occurred in the cooling tower water inlet temperature of 75 °C. The plate variation used was the plate without the inlet hole model. And the lowest heat transfer rate occurs at the inlet water temperature of 55 °C with a plate variation of 45 holes with a zig-zag hole model. So that the variation of the hole pattern affects the temperature change with the heat transfer rate on the tower performance.

Keywords Cooling tower study · Heat transfer · Number of holes

A. Rajai
Master of Mechanical Engineering Program, Faculty of Engineering, Universitas Syiah Kuala, Banda Aceh 23111, Indonesia
e-mail: afdhalul.2@mhs.usk.ac.id

A. Syuhada (✉) · Razali
Department of Mechanical Engineering, Universitas Syiah Kuala, Darussalam, Banda Aceh 23111, Indonesia
e-mail: syuhada_mech@yahoo.com

Razali
e-mail: razalithaib@unsyiah.ac.id

Irwansyah et al. (eds.), *Proceedings of the 4th International Conference on Experimental and Computational Mechanics in Engineering*, Lecture Notes in Mechanical Engineering, https://doi.org/10.1007/978-981-99-7495-5_25

1 Introduction

Industries currently use thermal equipment for factory activities. This thermal equipment in operation will emit excessive heat, as a result of this excessive heat will result in thermal fatigue of the equipment. To overcome the above, it is necessary to dissipate heat from thermal equipment, so a cooling system is needed. This cooling system is usually a water cooling medium.

Cooling towers can be used in the water cooling process in the industrial world [1]. Cooling towers are used to lower the temperature of the water flow by extracting heat from the water and emitting heat to the atmosphere [2]. The cooling tower is used as a fluid cooling device with air as the cooling medium. A cooling tower is a device used to lower the temperature of a water stream by extracting heat from the water and emitting heat into the atmosphere. Cooling tower can also be defined as a heat exchanger whose working fluid is water and air which functions to cool water by direct contact with air which causes a small part of the water to evaporate [3].

Many researches on cooling towers have been carried out to improve their performance. Comparative research on cooling tower performance was carried out for the induced counter flow type [4]. Comparison of the characteristics of the cooling tower type of pipe arrangement as a liquid distribution [5]. Characteristics of cooling towers with fibers filled with fibers [6]. In analysis of the cooling tower induced draft counterflow heat load with wulung bamboo as filler [7], water is made to drip drop by drop, or form a thin layer on a flat surface, so that it comes into contact with air moving upwards in the opposite direction [8].

The entire surface of the filler is expected to be wetted by water, so that the heat transfer that occurs is optimal. The main function of the filling material in the cooling tower is to break up the flow and slow down the falling motion of water so as to increase the contact time and contact area of hot water and the water–air mixture [9].

In this study, the planned cooling tower is a flat plate type where the plates are arranged horizontally and the fluid flows over the plate, this flow pattern is to slow down the mass flow rate of the hot fluid (water), so that the contact area of the heat transfer surface between the hot fluid and air can be reduced and Enlarged, and the contact time of hot fluid with air can be extended.

By increasing the touch surface and extending the contact time between the hot fluid and air, it can increase the rate of heat transfer, so that the cooling tower's heat energy dissipation into the air can be increased. The cooling tower housing that collects the water cooled by all the cells is a homogeneous tank, and consequently, the temperature is uniform throughout the basin to take into account that due to turbulence[10].

2 Methods

The equipment used in this study is the cooling tower which is shown in Fig. 1.

Test Materials and Materials. The overall dimensions of the planned cooling tower are shown in Fig. 1. The cooling tower frame uses angled iron and a reservoir, and the cooling tower plate uses aluminum with a thickness of 0.2 mm. The cooling tower manufacturing process takes approximately one month. Figure 2 describes the dimensions of the cooling tower plate used. The test material in this study is water which is heated at a cold temperature, and then, the water cooling process is carried out.

Research Tools and Measuring Tools. *Heater* is used to heat water to a temperature of 75 °C. The water pump used consists of two types of *pumps*. Based on the device specification data, it shows two pumps with a capacity of 100 Watts. The first pump is used to raise water from the main reservoir to the upper reservoir, and the second pump is used to circulate or move water from the lower reservoir to the main reservoir. The measuring instrument used consists of two measuring instruments, namely measuring water temperature and measuring time. The water temperature measuring instrument used is a digital thermometer. The time measurement tool used is a stopwatch. A thermometer is a temperature measuring device. In order to measure quantitatively, the *thermometer* needs to be equipped with a scale. How do

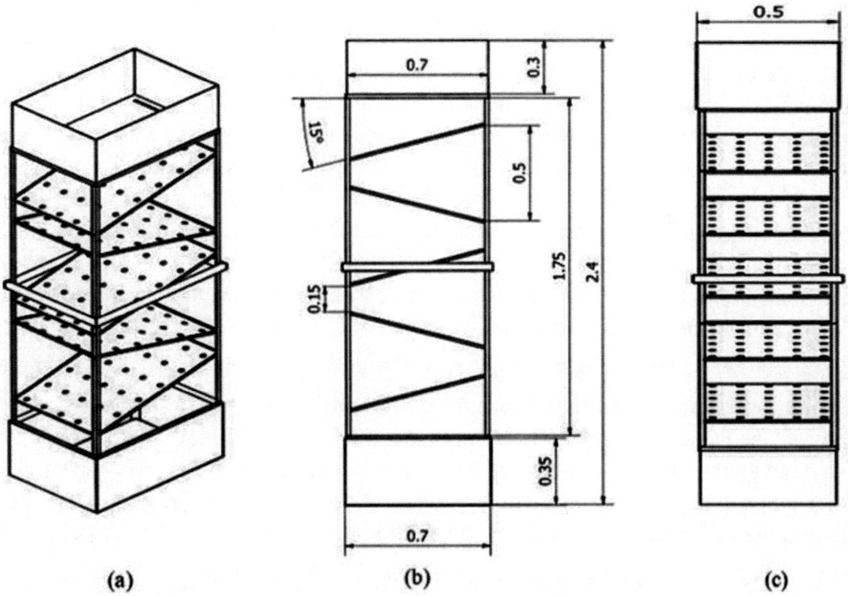

Fig. 1 **a** Three-dimensional cooling tower, **b** side view of cooling tower and **c** front view of cooling tower

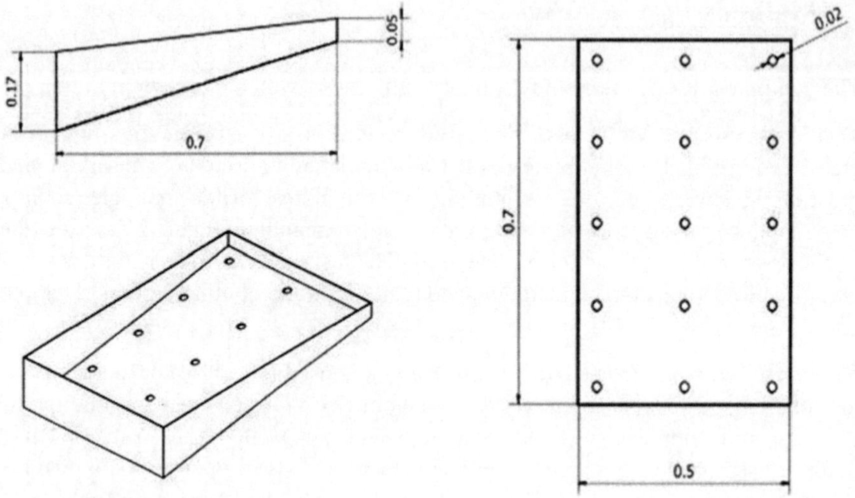

Fig. 2 Dimensions of flat plate

you put a scale on a thermometer? What are the physical considerations? All types and types of thermometers are based on natural phenomena related to changes in the physical properties of a quantity due to heat entering or leaving the quantity. Certain physical quantities whose properties can change because the temperature changes or is changed are called thermometric quantities (Thermometric Property) [11].

Equipment Schematic. The research was carried out by flowing water from the main reservoir to the upper reservoir of the cooling tower, so that the water from the upper reservoir fell and flowed throughout the plate to the lower reservoir. In the main water reservoir, the temperature is maintained according to the desired temperature variation by using a heater. Then, the water temperature in the upper reservoir is also stabilized using a heater to match the cool inlet temperature. The temperature of the inlet water, the temperature of the water on the plate, and the temperature of the water when it falls from one plate to the other until the temperature of the water is out are measured using a digital thermometer. Data collection for each variation is carried out in four stages with a difference of 2 h.

Testing and Data Collection Process. In this study, the value of the heat transfer rate and the value of the convection heat transfer coefficient will be calculated on the flat plate inlet and outlet zig-zag on the cooling tower, as well as the test tower with plate type zig-zag outlet hole. This activity will also look at the rate of heat transfer that occurs in the research media on the cooling tower.

This study used several assumptions. The heat transfer when water falls from the upper reservoir to the first plate has not been completely exposed to environmental air, the ambient air will absorb the heat of the water completely starting from the

Table 1 Data on water temperature measurement results in cooling towers flat plate

No hole			25 holes		
Measuring point	55	75	Measuring point	55	75
1	54.90	74.88	1	55.13	75.12
2	54.76	74.72	2	54.86	74.67
3	54.48	74.36	3	54.91	73.48
4	54.20	74.00	4	54.41	72.96
5	53.94	73.66	5	53.67	72.46
6	53.68	73.32	6	52.84	71.71
7	53.43	72.99	7	52.45	70.89
8	53.20	72.69	8	51.79	70.35
9	52.98	72.40	9	51.82	69.67
10	52.77	72.15	10	51.38	69.52
11	52.57	71.91	11	50.64	68.75
12	52.38	71.67	12	50.41	68.31

second plate of the cooling tower to the lower reservoir, and the air tin and air tout use the ambient air temperature.

3 Results and Discussion

Temperature in Tower Water. The results of the tests that have been carried out is presented, and the parameters measured in this study are the temperature when the water enters the cooling tower, the temperature when the water falls from each plate of the cooling tower, the temperature at the cooling tower plate, and the temperature of the water out. Tables 1 and 2 is the result data from the cooling tower.

4 Graph from Table

Figure 3 shows the distribution of the cooling tower water temperature drop for each point when the water inlet temperature is 55 °C, where the cooling tower plate without holes experienced a decrease in average temperature from 55 °C to 52.38 °C. In the cooling tower plate with 25 holes, the temperature drops up to 50.41 °C. For the cooling tower plate 35 holes, the temperature decreased up to 49.44 °C. On the cooling tower plate with 45 holes, the temperature of the water out reaches 47.65 °C. While on the cooling tower plate with55 holes, the temperature of the water out reaches 48.95 °C. Then, the distribution of the highest water temperature drop occurs in the cooling tower with a plate of 45 holes.

Table 2 Data on water temperature measurement results in cooling towers flat plate

35 holes			45 holes			55 holes		
Measuring point	55	75	Measuring point	55	75	Measuring point	55	75
1	55.17	75.21	1	55.31	75.26	1	55.11	75.29
2	54.81	74.68	2	54.86	74.67	2	54.73	75.00
3	54.44	73.98	3	54.17	73.8	3	54.27	74.79
4	53.96	73.56	4	53.47	73.35	4	53.67	74.27
5	53.34	72.86	5	52.62	72.55	5	52.92	73.86
6	52.76	72.14	6	51.74	71.66	6	52.37	73.34
7	52.5	71.48	7	51.27	70.52	7	51.83	72.82
8	51.88	70.84	8	50.94	69.29	8	51.48	72.36
9	51.32	70.28	9	50.28	68.73	9	50.98	71.76
10	50.78	69.76	10	49.62	68.23	10	50.43	71.25
11	50.44	69.24	11	48.76	67.96	11	49.66	70.85
12	49.44	68.81	12	47.65	67.2	12	48.95	70.37

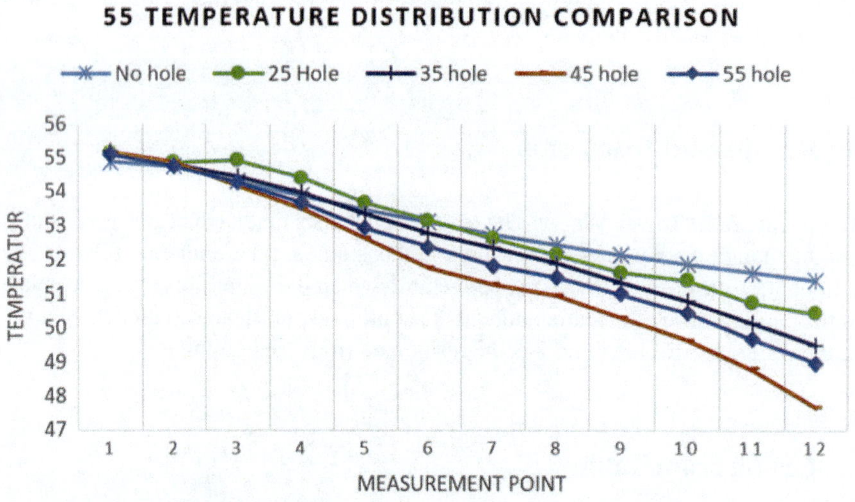

Fig. 3 Graph 55 temperature distribution

Figure 4 shows the distribution of the cooling tower water temperature drop for each point when the inlet water temperature is 75 °C, where the cooling tower plate without holes experienced a decrease in average temperature from 75 °C to 71.67 °C. On the 25 holes cooling tower plate, the temperature decreased to 68.31 °C. For the cooling tower plate with 35 holes, the temperature decreased to 67.81 °C. On the cooling tower plate with 45 holes, the outlet water temperature reaches 67.20 °C.

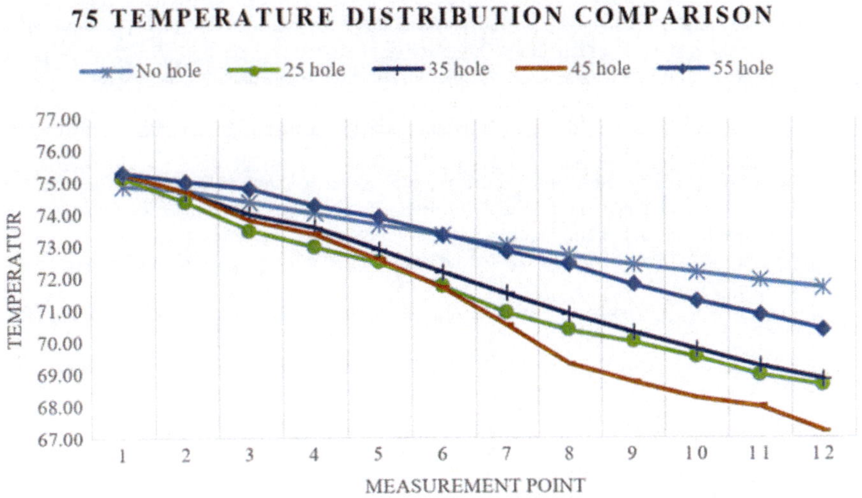

Fig. 4 Graph 75 temperature distribution

While on the cooling tower plate with 55 holes, the temperature of the water out reaches 70.31 °C. Then, the distribution of the highest water temperature drop occurs in the cooling tower with 45 holes plate.

5 Conclusion

The heat transfer rate with the lowest temperature occurs in the cooling tower with an inlet water temperature of 55 °C with variations hole 45 in the zig-zag hole at night with a decrease in temperature of 10.21 °C. The configuration of the type of hole pattern on the cooling tower plate affects the convection coefficient and the rate of heat transfer that occurs in the cooling tower. The convection value is influenced by the value of the heat transfer rate, surface area, and the difference in water temperature in the cooling tower.

References

1. Hamid ADKK (2017) Pengaruh Bentuk dan Konfigurasi Alur Sekat Terhadap Unjuk Kerja Menara Pendingin (cooling tower). J Rotor 10(2):1–5
2. Ardani ADKK (2018) Pengaruh Sudut Alur Sekat Terhadap Unjuk Kerja Menara Pendingin (cooling tower). Dinamika Teknik Mesin 88(1):21–29
3. Pratiwi NP, Nugroho G, Hamidah NL (2014) Analisa Kinerja cooling tower induced draft tipe IBC W-300 Terhadap Pengaruh Temperatur Lingkungan. J Teknik Parnits 7(7):1–6

4. Wibisiono Y (2005) Perbandingan unjuk kerja antar bahan pengisi pada menara pendingin tipe induced counter flow. J Teknologi Pertanian 6(3):152–162
5. Johanes S (2011) Komparasi Karakteristik menara pendingin menggunakan beberapa tipe susunan pipa-pipa sebagai pendistribusi cairan. Forum Teknik 34(1):67–75
6. Johanes S (2010) Karakteristik menara pendingin dengan bahan isian tali ijuk. Forum Teknik 33(3):188–194
7. Taufik HA, Listyadi SD, Sutjahjono H (2014) Analisis beban kalor cooling tower induced draft counterflow dengan bahan pengisi bambu wulung. Artikel ilmiah hasil penelitian mahasiswa, Jurusan teknik Mesin, Fakultas Teknik, Universitas Jember
8. Kumar VS (2018) Performance analysis and optimization of cooling tower. Int J Sci Res Publ 8(4):225–237
9. Mulyono PDKK (2000) Karakteristik Menara Pendingin dengan Bahan isian Plastik Bergelombang. Media Teknik 22(1):98–103
10. Munson Bruce R, Young Donald F (2009). Fundamental of fluid mechanics. Wiley, New Jersey, p 13
11. Hamid AA (2017) Kalor Dan Termodinamika. Universitas Negeri Yogyakarta Journal Termodinamika 1:20–21

Review of Research on the Application of CNG Fuel in Vehicle Engines: Research Bibliography 2017–2021

Suroto Munahar, Muji Setiyo, Fadhlurrachman Masykur, and Bagiyo Condro Purnomo

Abstract Compressed Natural Gas (CNG) is an alternative fuel used to solve cleaner exhaust emissions, global warming, and limited world fossil oil production problems. It emits cleaner gases during combustion, is economical, can be applied to light/heavy vehicles, and has good adaptability to diesel/petrol engines. This research reviews the application of CNG in vehicle engine articles published from 2017 to 2021. There were 82 articles on CNG, while 38 were on CNG as vehicle fuel. The articles reviewed were classified in several focus areas of research, such as the development of the ignition system on the CNG engine, the injection method, the application of Exhaust Gas Recirculating (EGR) technology, variations in the compression ratio of the CNG engine combustion chamber, the application of Reactivity Controlled Compression Ignition Engine (RCCI) on CNG machine, system modeling, mixer and method. The results showed a lack of research on the development of CNG that integrates intelligent control systems. Therefore, further research is suggested to explore the potential effect of the next period.

Keywords CNG · Vehicle · Article · Engine

S. Munahar (✉) · M. Setiyo · F. Masykur · B. C. Purnomo
Laboratorium of Automotive Engineering, Universitas Muhammadiyah Magelang, Megelang, Indonesia
e-mail: munahar@unimma.ac.id

M. Setiyo
e-mail: mujisetiyo@unimma.ac.id

B. C. Purnomo
e-mail: bagiyo_condro@ummgl.ac.id

S. Munahar · M. Setiyo · B. C. Purnomo
Center of Research, Development, and Community Services, Universitas Muhammadiyah Magelang, Megelang, Indonesia

Irwansyah et al. (eds.), *Proceedings of the 4th International Conference on Experimental and Computational Mechanics in Engineering*, Lecture Notes in Mechanical Engineering, https://doi.org/10.1007/978-981-99-7495-5_26

1 Introduction

Compressed Natural Gas (CNG) is an alternative fuel significantly utilized due to its ability to produce cleaner emissions, economical price, adaptability to gasoline/diesel engines, and application to light/heavy vehicles. It also helps solve the problem of increasing energy consumption due to the rise in the number of vehicles [1] and the limited fossil energy production globally [2, 3]. Therefore, the application of CNG as an alternative fuel is essential to produce vehicles with cleaner emissions and high economic value.

The application of CNG continues to be developed sustainably with various fields of research, especially regarding its engine performance. This alternative fuel improves engine performance by several aspects, such as injection, setting the ignition system, applying EGR, modifying the combustion system, and utilizing various mixing methods, hence it more homogeneous. Meanwhile, multiple studies have been carried out to determine the CNG modeling in terms of pressure in the reducer/fuel line, flow regulation approach, and a mixer.

The CNG setting with injection is designed to regulate the fuel entering the engine (4,5). Yuvenda [4] developed an injection timing system, which improved the CNG entering the engine. Meanwhile, Duc [5] designed an injection system by adopting the single-hole injector technology. However, both studies focused on developing an injection technology without creating an Engine Control Unit (ECU) to control the Air-to-Fuel Ratio (AFR) between CNG and air. Muhssen [6] and Kar (7) then achieved AFR stoichiometry values, while Muhsessen [6] and Kar [7] developed a CNG engine with an AFR stoichiometry approach oriented toward a range value of 17.2–17.4. This research, somehow, was unable to set AFR, which is oriented toward achieving high fuel efficiency/savings.

Ignition system settings are developed by determining the most suitable method, which leads to the use of lasers to reduce the particle content in exhaust emissions [8]. Intelligent Dual Sequential (IDS) is applied by igniting the spark plug 2 times in 1 engine cycle, therefore, the fuel content entering the exhaust manifold due to the engine flushing step can be burned to achieve cleaner emissions [9]. Correct calibration of the CNG engine ignition system is the key to power production, using the multiplexed extremum method [10]. However, the studies above have not developed an ignition system to improve fuel economy.

Another increase in CNG engine performance is by applying Exhaust Gas Recirculation (EGR). According to Mahla [11], this technique helps to reduce NOx gas in the exhaust manifold of CNG engines and focuses on achieving cleaner emissions. Besides the content of Nox gas, Kontses et al. [12] stated that the Particle Number (PN) in the exhaust gas is a concern after comparing CNG with other alternative fuels. A unique way to reduce emissions is to observe the driving style, such as Gha [13] research using 60 CNG vehicles. Modification of the CNG engine combustion with RCCI [14] and variations in the compression ratio has been developed to increase engine performance [15]. This research focused on the search for the most optimal combustion modification by analyzing the power generated by the CNG engine without discussing efforts to increase vehicle savings.

CNG engine modeling is the best method for predicting its engine capabilities. Computational Fluid Dynamics (CFD) software can be used to simulate the mixer design [16]. Sadah [16] observed the process of mixing CNG with air to obtain the most optimal mixture. Meanwhile, Alper and Do (9) used CFD to achieve variable torque generated by CNG engines. Dual sequential ignition was observed on the change in torque of the CNG engine. However, these two modeling research have not discussed the ECU development modeling to save fuel.

The development of CNG study has been developed with several variables. Still, several variables have not been discussed, including the development of AFR, which is oriented toward increasing fuel economy appropriately. CNG ignition system calibration that considers the fuel consumption factor. The application of EGR considers fuel consumption. CNG engine modeling considers the fuel use factor. Therefore, this article provides an overview of research trends to determine opportunities for developing and improving the application of CNG engines in vehicles.

2 Method

Reviews from studies on the application of CNG fuels were traced from 2017 to 2021 through "Science Direct". A total of 82 articles were found, with only 38 on CNG application to vehicle engines. Improved engine performance was the most discussed theme. The research reviewed was divided into several themes, namely ignition system, EGR application, mixer, injection method, compression ratio variation, RCCI application, CNG modeling, and other applications. The distribution of research themes on the application of CNG engines to vehicles and per year is shown in Figs. 1 and 2, respectively. Furthermore, approximately 5, 2, 2, 15, 3, 4, 1, and 6 articles were found on ignition system research, EGR application, mixer, injection method, compression ratio variation, RCCI application, CNG modeling, and other applications, respectively. Imran [17] previously conducted a review on CNG application in vehicles from 1991 to 2016. The themes discussed include its safety systems, the environment, the addition of hydrogen, vehicles with bi-fuel, fuel injection/combustion characteristics, engine performance, control system design, and economic aspects. The distribution of the research themes reviewed is presented in Figs. 1 and 2.

Environmental friendly and energy-efficient vehicles motivate the development of spark ignition engines [18, 19] since global energy availability is a major concern [20]. Even worse, there is a continuous rise in energy demand in the transportation sector along with an increase in the number of vehicles. The development of injection technology is one of the efforts to realize energy-efficient vehicles [21]. Lee [21] developed dual port injection technology, which led to a fair good fuel economy. Furthermore, Hydraulic Variable Valve Actuation (HVVA) was introduced to increase fuel efficiency [22]. The element of flexibility and application of genetic

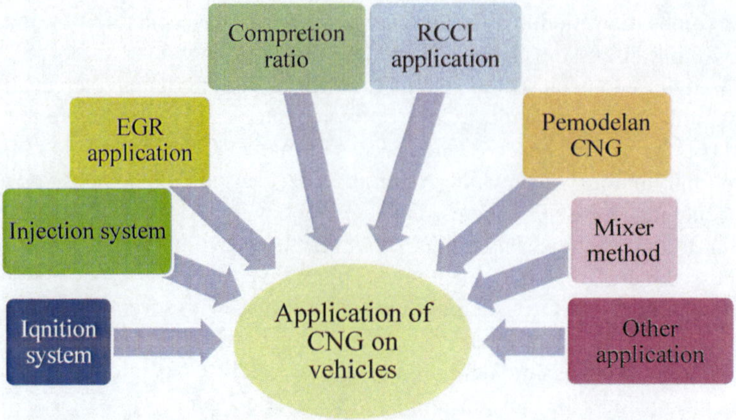

Fig. 1 Distribution of research themes on the application of CNG engines to vehicles

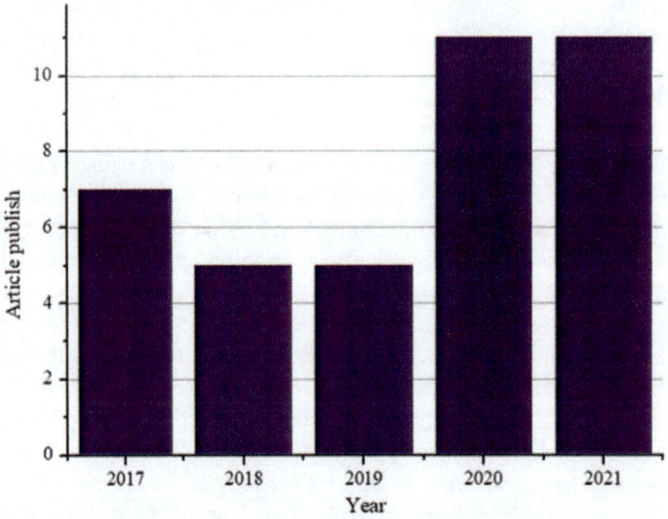

Fig. 2 Number of articles published on the application of the CNG engine

algorithms were also considered in this study. The results were quite interesting, with about 13% achievement of fuel economy, but both studies did not include steering operation as a control element.

Electronic Control Unit (ECU) is a fuel system control module in vehicles, continuously being developed to improve internal combustion engines. Meanwhile, Air-to-Fuel Ratio (AFR) is the main element to increase fuel economy, which is a driver's need [23–25]. It has three ranges of values, including lean AFR above 14.67, stoichiometry of 14.67, and rich below of 14.67 [26]. Currently, the realization of stoichiometry values is the focus of the AFR controller. The stoichiometry is generally maintained at the ideal value of 14.67.

Recently, artificial intelligence (AI) is being applied as a controller in the fuel control system [10, 11]; an example is fuzzy logic [27–29]. Furthermore, this study used fuzzy logic to control AFR but is yet to consider steering operation. Therefore, there is a need for further studies using the intake manifold pressure variable as an AFR clustering with neuro-fuzzy. The output would be in the form of a fuel injection control system. Another study used oxygen sensors to control AFR [30]. Besides fuzzy logic, the neural network is an AI that can be used for control systems [31–33]. Therefore, there is a continuously increasing number of studies on neural networks as a fuel control system [34]. Arsei [34] utilized the Recurrent Neural Network (RNN) to achieve the AFR value in the stoichiometry range. However, the study has not considered steering operation.

2.1 Literature Review

a. **Ignition System**

Research to improve the performance of CNG engines with variable ignition systems was carried out with several variations. Singh [8] stated that the laser method has the ability to reduce the particle content in exhaust emissions, thereby leading to cleaning emissions. Meanwhile, the Intelligent Dual Sequential (IDS) method is able to provide information stating that the performance of CNG is better in terms of gas emission. It allows the spark generated by the plug to be ignited 2 times in 1 cycle, including the compression and exhaust cycle [9]. Furthermore, the multiplexed extremum model is used to calibrate the CNG ignition system, which showed the ability to use proper adjustment to improve the fuel efficiency of CNG engines [10].

b. **Injection System**

The injection system developed with several varied methods, such as dual injection (Direct injection-DI, Port Fuel Injection–PFI), pilot injection, Start of Injection (SOI), timing, and CNG jets, is used to improve engine performance. According to Ramasamy [35], the dual injection method combines CNG and other fuel engines to produce an optimal mixture of 65% CNG and 25% gasoline. Lee [36] utilized EGR combined with a dual injection system, where the CNG rate was controlled by 70–80% flow, thereby indicating thermal efficiency. Furthermore, dual injection added with hydrogen gas by experimental DI technology on CNG engines showed an increase in brake thermal and engine power efficiency with a decrease in its fuel consumption [37]. Although these three studies discussed the dual injection method of the CNG fuel system, none analyzed the development based on an intelligent control system.

Song [38] compared the volumetric efficiency between CNG-DI and CNG-PFI to determine the injection system using DI/PFI type technology. Subsequently, CNG-DI can produce higher volumetric efficiency and fuel conversion efficiency than CNG-PFI. Furthermore, [36] researched the development of the DI type integrated with the turbocharger through experimental DI technology, which was compared between gasoline and CNG engines. The outcomes indicated that DI-CNG has the ability to improve thermal efficiency at full load by increasing the spark time due to its higher octane number. Research conducted by Melaika [39] compared the efficiency of CNG engines with gasoline/gasoline-DI(GDI), CNG-PFI, and CNG-DI types, where CNG-DI was set at stoichiometry conditions. The rate of heat release of CNG obtained is lower than gasoline fuel, however, the stable combustion process of DI-CNG provides additional benefits, and the efficiency of the CNG engine is quite high. Kar [7] compared GDI and CNG-DI to determine how the engine works with charge dilution. The results showed that the highest efficiency was obtained on CNG-DI with the lean burning operation and AFR stoichiometry with charge dilution. Although these four studies compared the volumetric efficiency of the CNG fuel system, none analyzed the development based on an intelligent control system.

Verma [40] designed a pilot injection to spray CNG with low load injection timings set at 32, 29, and 26°BTDC showing energy efficiencies of 8.5%, 11.1%, and 11.9% for biogas, CNG, and hydrogen operations, respectively. According to Meng [41], this method is also used to determine the throttling effect by applying it to the diesel-type CNG engine. Meng stated that dual fuel (diesel–CNG) engines are more advantageous for increased efficiency and emissions at low loads. However, both studies led to good engine performance, not developed a pilot injection capable of increasing fuel saving based on the engine's external control system.

Yuvenda [5] sets injection time from 70° to 150° ATDC and CNG from 70° to 150°CA with intervals of 20° at low load. Subsequently, the CNG engine is able to produce the highest cylinder pressure and Heat Release Rate (HRR) values by slowing down the CNG injection time (130° ATDC). Injection Timing (TI) settings measure Brake Thermal Efficiency (BTE) on CNG machines with different loadings. The results showed that advancing the injection time increases BTE earlier in time [42]. Channappagoudr [43] experimented with the injection time on a BIO-CNG engine and concluded an enhanced performance at 29° BTDC. The three research have tried to use the injection time variable to improve CNG performance but have not linked the injection time to the external engine variable.

Other variations of CNG injection methods include testing engine brake power, pilot fuel Start of Injection (SOI), and the CNG jet method. Duc et al. (2019) Duc [4] increased the engine performance by improving the injection method tested for the engine's brake power. This method decreases up to 19%, increasing brake-specific energy consumption (BSEC) by 14% (4). SOI is controlled for dual fuel engines, namely diesel-CNG, with a pressure of 7.5 bar from the Indicated Mean Effective Pressure (IMEP) and a mixture of B10CNG60 and B20CNG60 [41]. Presently, there is a massive application of the jet/CNG jet method globally. A Schlieren photograph was used to observe the characteristics of CNG injection, and the results showed that the increase in axial jet penetration is strongly influenced by room pressure [44].

These three studies have explored various improvements to the CNG engine with different variations of methods without discussing the development of the ECU to control CNG with variables from environmental aspects.

c. **EGR Application**

EGR as a method of reducing NOx gas in CNG engines has numerous advantages, therefore, it continues to be pursued. Mahla [11] added Bio-Diesel/B20 (castor oil) to the CNG engine, integrating EGR and B20 used as pilot fuel. This method is able to reduce NOx at engine load without damaging the combustion characteristics [11]. Another method is to apply EGR to a CNG-diesel engine with a certain mixture used to decrease the movement of BTE and BFSC under full loading conditions. Although Mahla [11] and Kumar [45] applied EGR to CNG engines, they have not yet developed CNG EGR controlled by an intelligent control system to provide better regulatory effects.

d. **Compression Comparison Variations**

The performance of the CNG engine is carried out by varying its compression ratio. Srivastava and Agarwal [15] observed a compression ratio from 9.9 to 11.8 with laser ignition on a CNG-air mixture. This method is used to determine lean mixtures and find a suitable compression ratio for CNG engines. The result showed the lean flammability limit from 1.62 to 1.76 and also sheds light on laser ignition that can be used with CNG-air lean mixtures [15]. Research conducted by Sahoo [46] controlled the compression ratio to reduce detonation in CNG engines. The compression ratio of 12 produces the highest engine beats. Meanwhile, Bhasker and Porpatham [47] observed the impact on the compression ratio of CNG engines 12.5:1, which increased brake power output and thermal efficiency.

e. **RCCI Application**

RCCI is a strategy used to reduce emissions with a control system by promoting cleaner combustion. Sunmeet [14] used biodiesel (B100) to inject pilot fuel into the engine with the addition of CNG and 18% HCNG to reduce emissions. Therefore, HC, CO, and smoke emissions in all parts of the HCNG energy are less than that of basic CNG. However, NOx emissions increased slightly with HCNG and less than basic biodiesel.

Another research used RCCI in diesel engines with CNG as the main fuel and a mixture of diesel and Karanja biodiesel (BD20) as pilot fuel. Furthermore, 10 ms CNG induction with 90% Energy Share (ES) and BTHE (Brake Thermal Efficiency) experienced the highest rate of 24.2% and the smallest BSFC 0.3 kg/kW-hr. According to Wategave [48], low Reactive Fuel is better than diesel at 60% ES. RCCI is also applied with injection timing settings at pressures of 500, 750, and 1000 bar. The integration of loading, injection time, and pressure can improve engine performance with optimal conditions at a 750 bar [49].

2.2 Synthesis

A summary of several research articles that focus on discussing the application of CNG in vehicle engines is given in Tables 1, 2, 3, 4, 5, 6, 7, 8, 9 and 10.

3 Conclusion

The literature review on the application of CNG as a fuel in vehicles has two periods, including 1991–2016 and 2017–2021. The first period focused on the research of regional experiences with CNG vehicles, Economic Aspects, Engine Design, Control and Performance, Combustion and Fuel Injection Characteristics, Diesel Dual Fuel Operation, Hydrogen Enriched CNG Vehicles, Environmental Aspects, and Safety Aspects. Meanwhile, for the second period, the focus was on improving the performance of CNG engines. The most widely carried out research themes with more than 70% were a modification of the ignition system, injection adjustment, RCCI application, CR investigation, EGR design, etc. However, research on CNG mixer modeling using CFD has started to be developed.

In the first and second periods, the literature review on the application of CNG as a vehicle fuel did not imply the development of control systems associated with intelligent control systems. Therefore, this literature review becomes a potential for future research development because this field has not been widely examined previously. Moreover, it is interesting to develop an external engine system and CNG control with smart sensors in a control system considering environmental aspects.

Table 1 Research articles about CNG fuel system development part one

No	Author	Research focus	Method	‘Analysis of results	Finding
1	[50]	Improvement of engine performance	Hydrogen + CNG mixing investigation with a dynamic system	This research examines the mixing of hydrogen + CNG (HCNG). The 30HCNG mixture has the most optimal conditions, which provides information on the most optimal blending method but does not cover how to mix HCNG with air	Mixture of HCNG with 30% yields better performance than basic CNG
2	[47]	Improvement of engine performance	Observing the impact on the compression ratio of CNG engines	Bhasker is used to set the compression ratio for CNG engines. The result signified that the CNG engine is capable of working on ultra-lean combustion. However, it has not considered the engine loading aspect	The ratio of 12.5:1 is proven to increase brake power output and thermal efficiency
3	[35]	Improvement of engine performance	Using dual injection CNG and gasoline	Ramasamy applied a dual injection system between CNG and gasoline to enhance engine performance. However, no research has been conducted on the stoichiometry mixing process	The most optimal dual injection blend is at 65% CNG and 35% gasoline

Table 2 Research articles on CNG fuel system development part two

No	Author	Research focus	Method	‘Analysis of results	Finding
4	[44]	Improvement of engine performance	Experimenting with jet characteristics on CNG	Erfan used Schlieren photography to observe CNG injection with the jet method. Although the results were the most optimal conditions with the jet method, the research has not been applied to vehicle engines	The results provided information on the strategies needed to increase the penetration of the axial jet, which is strongly influenced by room pressure
5	[38]	Improvement of engine performance	Developing DI and PFI injection system and timing	The results of comparing the volumetric efficiency between CNG-DI and CNG-PFI had numerous advantages. However, the research has only been carried out at 1700 rpm engine speed yet with various rotations	CNG-DI has the ability to have higher volumetric efficiency and fuel conversion efficiency than CNG-PFI
6	[14]	Improvement of engine performance	Developing a clean combustion strategy to reduce emissions	Sunmeet used biodiesel (B100) as pilot fuel in its CNG engines and 18% HCNG to reduce emissions. RCCI was also integrated into the research. The results tend to reduce emissions. However, studies on mixing with the AFR method have not been considered to enhance fuel economy	HC, CO, and smoke emissions in all parts of HCNG energy are less than basic CNG. However, NOx emissions increased slightly with HCNG and are still less than basic biodiesel
7	[40]	Improvement of engine performance	Using pilot injection on the CNG-Biogas-Hydrogen engine	Verma compared the efficiency of dual fuel (DF) CNG–biogas–hydrogen in diesel engines. This research provided recommendations on energy efficiency by mixing CNG–biogas–hydrogen. However, studies on mixing the air entering the engine have not been discussed	At low load, injection timings of 32, 29, and 26°BTDC demonstrated the highest energy efficiency of 8.5%, 11.1%, and 11.9%, respectively, for biogas, CNG, and hydrogen DF operations

Table 3 Research articles on CNG fuel system development part three

No	Author	Research focus	Method	‘Analysis of results	Finding
8	[15]	Improvement of engine performance	Implementing variable compression and laser ignition system–plasma	Srivastava observed the laser ignition of the CNG-air mixture to determine the lean mixture and find a suitable compression ratio for the engine. The result indicated that this method can increase the flammability limit of CNG machines. However, the uses of this flammability limit have not been discussed in detail	It obtained a lean flammability limit from = 1.62 to = 1.76 and increased the compression ratio from 9.9 to 11.8. Therefore, the CNG-air mixed laser ignition can be developed through a certain mixture
9	[51]	CNG model design	Creating a multi-zone thermodynamics model on a CNG engine	Baratta reproduced / interpreted data to gain insight into engine combustion processes and pollutant formation. Although the research considered kinematic chemistry, it has not considered the molecular type	The model-developed results were able to predict greater engine capability improvements
10	[10]	Improvement of engine performance	Using the multiplexed extremum model to calibrate the CNG ignition system	In this research, multiplexed extremum was used to determine the appropriate model to improve CNG performance. The designed system focused on the application of calibration accuracy without developing self-calibration	Experimental results exhibited that under proper tuning, the proposed controller can improve the fuel efficiency of CNG engines
11	[11]	Improvement of engine performance	Using EGR on CNG engines with added Bio-Diesel (castor oil)	Mahla used (EGR) on a diesel engine (CNG) with a blend of Jatropha biodiesel (B20) as pilot fuel. However, the results were cleaner emission reductions, and no research has been conducted on the power generated	The EGR method can reduce NOx in the engine load without changing the engine combustion characteristics

Table 4 Research articles on CNG fuel system development part four

No	Author	Research focus	Method	'Analysis of results	Finding
12	[9]	Improvement of engine performance	Using i DSI (intelligent dual sequential ignition) engine on bi-fuel (gasoline-CNG)	Alper and Do compared the performance of CNG-Gasoline engines with various engine loads. The i-DSI ignition system was applied to a bi-fuel engine (gasoline-CNG). Furthermore, the power and volumetric efficiency results were discussed in detail. However, the AFR between fuel and air has not been discussed	This method is able to state that the performance of gasoline is better than CNG in terms of power. However, in terms of CNG emissions, it is better
13	[4]	Improvement of engine performance	To improve the performance of CNG engines and the injection method	Experimental research on the performance of gasoline-CNG engines with injection ports denoted an increase in BSEC and a decrease in emissions. However, it has not measured the AFR between fuel and air	When fueled with CNG, the brake power of the test engine was reduced by 19%, while the brake-specific energy consumption (BSEC) was increased by 14%
14	[52]	Improvement of engine performance	Using pilot fuel Start of Injection (SOI) on the engine (diesel-CNG)	Meng used SOI to increase Indicated Thermal Efficient (ITE) and reduce emissions optimally. However, it has not included the aspect of power generated by the engine. This condition would be better variable power discussed	At a pressure of 7.5 bar, the Indicated Mean Effective Pressure, B10CNG60, and B20CNG60 ITE increase with a decrease in emissions
15	[53]	Improvement of engine performance	Addition of hydrogen to reduce emissions in CNG engines	This research assessed the emission of gasoline, CNG, and HCNG engines. The most optimal mixture of HCNG acquired is approximately 18%, with the ability to reduce emissions drastically. However, it has not used artificial intelligence to regulate the mixing method	The lowest emission was achieved in HCNG with a mixture of 18%

Table 5 Research articles on CNG fuel system development part five

No	Author	Research focus	Method	'Analysis of results	Finding
16	[54]	Improvement of engine performance	Developing CNG engine modeling with CFD	Garg designed a CNG engine modeling with CFD. The results showed that the overall engine efficiency increases when the case of pre-mix stoichiometry is compared with the lean stratified condition. However, the power generated has not been revealed	The coated casing exhibits 30% lower NOx emissions when compared to pre-mixed casings at the same engine load
17	[5]	Improvement of engine performance	The time setting of the CNG injection that enters the engine	Yumenda optimized the CNG injection time. The results revealed that a late injection time leads to a high heat release rate and low emissions. However, it has not discussed the increase in CNG savings by setting the injection time	By slowing down the injection time of CNG (130° ATDC), the highest cylinder pressure and heat release rate (HRR) is produced
18	[55]	Improvement of engine performance	Using a pilot injection system to determine the effect of throttling	Meng examined the effectiveness of increasing throttling on efficiency and emissions in dual fuel combustion modes numerically. As an outcome, the pin reduction numerical calculation can significantly improve combustion efficiency and emissions. However, the research discussed on combustion efficiency has not discussed fuel efficiency	Engines with dual fuel (diesel–CNG) with the pilot injection method are more profitable for increased efficiency and emissions at low loads
19	[56]	Improvement of engine performance	Optimization of diesel injector injection with CNG substitution	Lee used a CNG rate at 70–80% flow to enhance thermal efficiency with varying EGR. The computational indicated that a 25% smaller diameter nozzle (150 m) with continued ROI increased the mean effective pressure by 3.4%. However, this research has failed to reveal the power generated by the engine and fuel efficiency	This research used a CNG rate of up to 80%. The results obtained can increase the thermal efficiency of the engine and reduce CO_2 in the exhaust gas

Table 6 Research articles on CNG fuel system development part six

No	Author	Research focus	Method	'Analysis of results	Finding
20	[43]	Improvement of engine performance	Modifying the injection timing setting on the BIO-CNG engine	This research modified the diesel engine into BIO-CNG. The accomplishment showed an increase in engine performance and emissions. However, the set injection time has not been able to increase fuel efficiency	The injection time can improve the performance of the Bio-CNG engine at 29° BTDC
21	[37]	Improvement of engine performance	Experimenting with DI technology and the addition of hydrogen to a CNG engine	Hydrogen was added to the CNG engine with DI technology to observe the engine performance. The results have the ability to improve thermal brake efficiency and engine power, reducing specific fuel consumption. However, it has not discussed the AFR setting on the addition of hydrogen	This method is able to increase the efficiency of brake thermal and engine power while decreasing its specific fuel consumption
22	[36]	Improvement of engine performance	Conducting experimental DI technology on gasoline and CNG engines	Lee compared DI technology in gasoline and CNG engines with the addition of a turbocharge. Subsequently, CNG increases the thermal efficiency under maximum load conditions by raising the spark time. However, it has not used AFR to measure engine performance	DI-CNG improves thermal efficiency at full load by increasing the spark time, with a rise in the octane number
23	[8]	Improvement of engine performance	Developing laser ignition system	Sigh used a laser ignition system on a CNG–hydrogen engine. Consequently, using CNG–hydrogen with a laser ignition system reduces the particle content in the exhaust gas. However, it has not enhanced fuel economy	The combination of CNG–hydrogen and increasing ignition timing improves CNG engine performance

Table 7 Research articles on CNG fuel system development part seven

No	Author	Research focus	Method	‘Analysis of results	Finding
24	[12]	Improvement of engine performance	Observing the Particle Number (PN) produced by gasoline-PFI, diesel-DPF, hybrid, LPG-PFI, and CNG-PFI fuels	In this research, the effect of PN on vehicles with gasoline-PFI, diesel-DPF, hybrid, LPG-PFI, and CNG-PFI fuels was compared. Subsequently, PN can be significantly reduced on LPG and CNG engines. However, this research has not linked the resulting PN to the engine during acceleration and deceleration	Significant PN reduction potential was observed with LPG and CNG versus gasoline by up to 93% and 90%, respectively
25	[13]	Improvement of engine performance	Analyzing changes in emissions from CNG-fueled vehicles on driver behavior	Gha tested private vehicles/ taxis with CNG engines based on driving patterns. Consequently, emissions increase with the rise in vehicle speed. However, changes in emissions to environmental changes have not been observed	For CO, gas increases depending on vehicle speed, while for HC, it decreases
26	[57]	AFR control	Observing the robust stochastic system for the AFR setting	This research proposed setting AFR with robust stochastic on CNG machine. The result showed the AFR control in the closed loops, whereof type CNG machine area, while the changes based on machine learning technology have not been observed	AFR application with robust stochastic can regulate AFR in the stoichiometry area, thereby improving engine performance
27	[6]	AFR control	Designing a mixer on a CNG machine	Muhssen designed the CNG Mixer previously customized in CFD. The results were in the form of a mixer design used to mix CNG with air to obtain a more homogeneous mixture. It has not been tested with various engine load variations	The mixer design results can help achieve AFR CNG in the stoichiometry area

Table 8 Research articles on CNG fuel system development part eight

No	Author	Research focus	Method	'Analysis of results	Finding
28	[46]	Improvement of CNG engine performance	Observing the Compression Ratio (CR) aspect to knock on CNG engines	Sahoo used CR to determine the knock effect on a CNG engine. The results exhibited that the CR 12 gasoline engine produced the highest knocking, while the CNG engine can work on CR 16 and is more resistant to knocking symptoms. However, it has not added CNG to other fuels	On the CR 16, the CNG engine is still able to operate without knock-out effects
29	[39]	Improving CNG engine performance	Injection method control	This research compared the efficiency of GDI, PFI-CNG, and DI-CNG systems using stoichiometry conditions. Although the results implied that DI-CNG has the highest efficiency and reduced emissions, the applied injection method has not considered the application of an intelligent control system to improve the engine performance further	The application of DI-CNG provides better engine performance
30	[42]	Improving CNG engine performance	Injection method control	Rai used timing injection (TI) settings to determine the performance of CNG engines with different loadings. Subsequently, IT applications are evidenced to reduce NOx and increase smoke. However, the intelligent control system has not been considered for IT control	The results showed that increasing the injection time leads to a rise in BTE over time
31	[7]	Improving CNG engine performance	Injection method control	Kar compared GDI and DI-CNG to determine how the engine works with charge dilution. The results indicated that the lean burn strategy led to higher brake thermal efficiency than stoichiometric operation. However, research has not been carried out on environmental conditions	EGR/lean burn application increases the total CO_2 equivalent emissions relative to stoichiometric operations without dilution

Table 9 Research articles on CNG fuel system development part nine

No	Author	Research focus	Method	‘Analysis of results	Finding
32	[49]	Improving CNG engine performance	Injection method control	This research used an RCCI set with injection time at 500, 750, and 1000 bar pressures. The results showed that 750 (Bar) injection pressure and the injection time (BTDC) of 12.5 led to the most optimal engine performance. However, this research has not considered the input from the environment/intelligent control system	By integrating loading, injection time, pressure and increasing engine performance under optimal conditions at an injection pressure of 750 bar
33	[45]	Improving CNG engine performance	Use of EGR	Kumar used dual fuel to increase the performance of the CNG engine. The results showed that BTE and BSFC decreased at a higher load of 2–3% under all loading conditions compared to diesel. However, no research has been carried out on applying an intelligent control system to control mixing in a dual fuel system	In dual fuel mode, NOx is decreased by 21%–18%, and under all loading conditions, BTE reduces by 1–2%, and BSFC decreases at higher loads by 2–3% compared to diesel
34	[48]	Improvement of engine performance	Using reactivity controlled compression ignition (RCCI)	Watagave set fuel operating limits to reduce CNG engine noise and provided cleaner combustion. The results showed that 10 ms CNG induction with 90% ES was the operating limit of the CNG engine, while the 60% CNG content produced the cleanest combustion. However, it has not observed AFR CNG restrictions related to rough sound and combustion quality	In 10 ms CNG induction with 90% energy share (ES), and BTHE experienced the highest 24.2% and the lowest BSFC 0.3 kg/kW-hr. Low Reactive Fuel is better than diesel at 60% ES

Table 10 Research articles on CNG fuel system development part ten

No	Author	Research focus	Method	'Analysis of results	Finding
35	[58]	Improvement of engine performance	Using RCCI and CR	Aydin used a variation of CR to enhance the performance of the CNG engine. The analysis proved that CNG with biodiesel increases BTE and BSEC, while BSFC experienced a decrease in biodiesel/CNG RCCI and various CR operations	With this variation of the CR method, BTE tends to control BTE, BSEC, and BSFC
36	[59]	Improvement of engine performance	Using RCCI and CBG	Harari applied RCCI with a combination of CNG and Compressed Bio Gas (CBG). The analysis indicated that D + CNG gave better results in terms of performance and emission characteristics. However, in AFR, CNG with CBG has not been observed	This method is able to increase (BTE) at a loading flow of 75%
37	[60]	Improvement of engine performance	Using laser system ignition on CNG engines	Alper compared CNG engines to conventional laser and ignition systems	The results obtained with the laser system reduced the propagation time and kernel flame formation as well as NOx gas
38	[16]	CNG Simulation	Simulating the mixer on the CNG reducer design with CFD	Laser ignition is confirmed to reduce propagation time, Kernel flame formation, and NOx. However, studies on the relationship between AFR and CNG engine laser ignition have not been conducted	The CFD simulation better predicts the CNG engine capability at 10 injection holes in the mixer

Acknowledgements The authors are grateful to the BRIN Research and Technology as well as LPPM (Research institutions and community service) of Muhammadiyah University for supporting and funding this research.

References

1. EIA (2017) Global transportation energy consumption: examination of scenarios to 2040 using ITEDD. Washington DC

2. Al-fattah SM (2020) Non-OPEC conventional oil : Production decline, supply outlook and key implications. J Pet Sci Eng 189:107049. Available from: https://doi.org/10.1016/j.petrol.-2020.107049
3. Kutlu O (2020) Global oil production declines in June 2020. Energy
4. Duc Nguyen K, Duy Nguyen V, Hoang-dinh L, Viet Nguyen T (2019) Performance and emission characteristics of a port fuel injected, spark ignition engine fueled by compressed natural gas. Sustain Energy Technol Assessments 31:383–9. Available from: https://doi.org/10.1016/j.seta.2018.12.018
5. Yuvenda D, Sudarmanta B, Wahjudi A, Muraza O (2020) Improved combustion performances and lowered emissions of CNG-diesel dual fuel engine under low load by optimizing CNG injection parameters. Fuel 269:117202. Available from: 10.1016-/j.fuel.2020.117202
6. Muhssen HS, Masuri SU, Sahari B, Bin Hairuddin AA (2020) Design improvement of compressed natural gas (CNG)-Air mixer for diesel dual-fuel engines using computational fluid dynamics. Energy 118957. Available from: https://doi.org/10.1016/j.energy.2020.118957
7. Kar T, Zhou Z, Brear M, Yang Y, Khosravi M, Lacey JA (2021) Comparative study of directly injected, spark ignition engine performance and emissions with natural gas , gasoline and charge dilution. Fuel 304:121438. Available from: https://doi.org/10.1016/j.fuel.-2021.121438
8. Singh AP, Kumar D, Agarwal AK (2020) Particulate characteristics of laser ignited hydrogen enriched compressed natural gas engine. Int J Hydrogen Energy 45(35):18021–31. Available from: https://doi.org/10.1016/j.ijhydene.2020.05.005
9. Alper A, Do Y (2017) Investigation of the effects of gasoline and CNG fuels on a dual sequential ignition engine at low and high load conditions. Fuel 2018(232):114–123
10. Sharafi J, Moase WH, Manzie C (2017) Multiplexed extremum seeking for calibration of spark timing in a CNG-fuelled engine. Control Eng Pract 72:42–52. Available from: https://doi.org/10.1016/j.conengprac.2017.11.005
11. Mahla SK, Dhir A, Gill KJS, Muk H, Chang H, Singh B (2018) Influence of EGR on the simultaneous reduction of NO x -smoke emissions trade-off under CNG-biodiesel dual fuel engine. Energy 152(x):303–12. Available from: https://doi.org/10.1016/j.energy.-2018.03.072
12. Kontses A, Triantafyllopoulos G, Ntziachristos L, Samaras Z (2020) Particle number (PN) emissions from gasoline, diesel, LPG, CNG and hybrid-electric light-duty vehicles under real-world driving conditions. Atmos Environ J 222(2019)
13. Gha O, Reza M, Ahmadikia H (2020) On-road performance and emission characteristics of CNG-gasoline bi-fuel taxis / private cars at the roadside environment. Atmos Pollut Res 11(July):1743–1753
14. Sunmeet SK, Subramanian KA (2017) Experimental investigations of effects of hydrogen blended CNG on performance, combustion and emissions characteristics of a biodiesel fueled (RCCI).pdf. Int J Hydrogen Energy 42(7):4548–60
15. Srivastava DK, Agarwal AK (2018) Combustion characteristics of a variable compression ratio laser-plasma ignited compressed natural gas engine. Fuel 214:322–9. Available from: https://doi.org/10.1016/j.fuel.2017.10.012
16. Sadah H, Ujila S, Bin B, Aziz A (2021) Design improvement of compressed natural gas (CNG) -Air mixer for diesel dual-fuel engines using computational fluid dynamics. Energy 216:118957. Available from: https://doi.org/10.1016/j.energy.2020.118957.
17. Imran M, Yasmeen T, Ijaz M, Farooq M, Wakeel M (2016) Research progress in the development of natural gas as fuel for road vehicles : a bibliographic review (1991–2016). Renew Sustain Energy Rev 66:702–41. Available from: https://doi.org/10.1016/j.rser.-2016.08.041
18. Alkidas AC (2007) Combustion advancements in gasoline engines. Energy Convers Manag 48(11):2751–2761
19. Kiencke U, Nielsen L (2005) Automotive control system for engine, driveline and vehicle, 2nd edn. Springer, Berlin Heidelberg, Second
20. Al-Fattah SMAl (2020) Non-OPEC conventional oil production decline, supply outlook and key implications. J Pet Sci Eng 189:107059. Available from: http://www.elsevier.com/locate/petrol

21. Lee Y, Oh S, Kim C, Lee J, Lee K, Kim J (2018) The dual-port fuel injection system for fuel economy improvement in an automotive spark-ignition gasoline engine. Appl Therm Eng 138:300–6. Available from: https://doi.org/10.1016/j.applthermaleng.2018.04.-027
22. Li Y, Khajepour A, Devaud C, Liu K (2017) Power and fuel economy optimizations of gasoline engines using hydraulic variable valve actuation system. Appl Energy 206:577–93. Available from: https://doi.org/10.1016/j.apenergy.2017.08.208
23. Rajagopalan SSV, Midlam-Mohler S, Yurkovich S, Dudek KP, Guezennec YG, Meyer J (2014) A control design and calibration reduction methodology for AFR control in gasoline engines. Control Eng Pract 27(1):42–53. Available from: https://doi.org/10.1016/j.conengprac.2014.02.020
24. Triwiyatno A, Sinuraya EW, Munahar S, Setiyawan JD (2015) Smart controller design of air to fuel ratio (AFR) and brake control system on gasoline engine. In: International conference on information technology, computer, and electrical engineering (ICITACEE)
25. Khajorntraidet C, Shen T (2018) Adaptive lean air-fuel ratio control and analysis of commercial gasoline engines. IFAC-PapersOnLine 51(31):423–8. Available from: https://doi.org/10.1016/j.ifacol.2018.10.094
26. Stone R, Ball JK (2018) Automotive engineering fundamentals
27. Robertson D, Prucka R (2020) Evaluation of autoignition models for production control of a spark-assisted compression ignition engine. Int J Engine Res 1–13. https://doi.org/10.1177/1468087420934555
28. Wang Y, Shi Y, Cai M, Xu W (2020) Predictive control of air-fuel ratio in aircraft engine on fuel-powered unmanned aerial vehicle using fuzzy-RBF neural network. J Franklin Inst 357:8342–8363. https://doi.org/10.1016/j.jfranklin.202
29. Sardarmehni T, Aghili-Ashtiani A, Menhaj MB (2019) Fuzzy model predictive control of normalized air-to-fuel ratio in internal combustion engines. Soft Comput 23(15):6169–6182. Available from: https://doi.org/10.1007/s00500-018-3270-2
30. Barghi F, Safavi AA (2011) Experimental validation of recurrent neuro-fuzzy networks for AFR estimation and control in SI engines. IEEE Int Conf Comput Intell Meas Syst Appl Proc 107–12
31. Du KL (2010) Clustering: a neural network approach. Neural Netw 23(1):89–107
32. Hebboul A, Hachouf F, Boulemnadjel A (2015) A new incremental neural network for simultaneous clustering and classification. Neurocomputing 169:89–99. Available from: https://doi.org/10.1016/j.neucom.2015.02.084
33. Xu R, Wunsch D (2005) Survey of clustering algorithms. IEEE Trans Neural Netw 16(3)
34. Arsie I, Pianese C, Sorrentino M (2008) Development and real-time implementation of recurrent neural networks for AFR prediction and control. SAE Tech Pap 1(1):403–412
35. Ramasamy D, Goh CY, Kadirgama K, Benedict F, Noor MM, Najafi G (2017) Engine performance , exhaust emission and combustion analysis of a 4-stroke spark ignited engine using dual fuel injection. Fuel 207:719–28. Available from: https://doi.org/10.1016/j.fuel.-2017.06.065
36. Lee J, Park C, Bae J, Kim Y, Lee S, Kim C (2020) Comparison between gasoline direct injection and compressed natural gas port fuel injection under maximum load condition. Energy 197:117173. Available from: https://doi.org/10.1016/j.energy.2020.117173
37. Zareei J, Haseeb M, Ghadamkheir K, Farkhondeh SA (2020) The effect of hydrogen addition to compressed natural gas on performance and emissions of a DI diesel engine by a numerical study. Int J Hydrogen Energy 45(58):34241–53. Available from: https://doi.org/10.1016/j.ijhydene.2020.09.027
38. Song, J., Choi, M., Park, S. : Comparisons of the volumetric efficiency and combustion characteristics between CNG-DI and CNG-PFI engines. Appl Therm Eng. 2017;121:595–603. Available from: https://doi.org/10.1016/j.applthermaleng.2017.04.110.
39. Melaika M, Herbillon G, Dahlander P (2021) Spark ignition engine performance, standard emissions and particulates using GDI, PFI-CNG and DI-CNG systems. Fuel 293:120454. Available from: https://doi.org/10.1016/j.fuel.2021.120454
40. Verma S, Das LM, Bhatti SS, Kaushik SC (2017) A comparative exergetic performance and emission analysis of pilot diesel dual-fuel engine with biogas, CNG and hydrogen as

main fuels. Energy Convers Manag 151:764–77. Available from: https://doi.org/10.1016/j.enconman.2017.09.035
41. Meng X, Tian H, Long W, Zhou Y, Bi M, Tian J (2019) Experimental study of using additive in the pilot fuel on the performance and emission trade-offs in the diesel/CNG (methane emulated) dual-fuel combustion mode. Appl Therm Eng 157:113718. Available from: https://doi.org/10.1016/j.applthermaleng.2019.113718
42. Rai AA, Bailkeri NK, Br SR (2021) Materials today : proceedings effect of injection timings on performance and emission characteristics of CNG diesel dual fuel engine. Mater Today Proc 46:2758–63. Available from: https://doi.org/10.1016/j.matpr.2021.02.509
43. Channappagoudra M, Ramesh K, Manavendra G (2020) Effect of injection timing on modi fi ed direct injection diesel engine performance operated with dairy scum biodiesel and Bio-CNG. Renew Energy 147:1019–32. Available from: https://doi.org/10.1016/j.renene.-2019.09.070
44. Erfan I, Hajialimohammadi A, Chitsaz I, Ziabasharhagh M, Martinuzzi RJ (2017) Influence of chamber pressure on CNG jet characteristics of a multi-hole high pressure injector. Fuel 197:186–93. Available from: https://doi.org/10.1016/j.fuel.2017.02.018
45. Kumar S, Nayyar A, Goel V (2020) Optimization of EGR effects on performance and emission parameters of a dual fuel (Diesel + CNG) CI engine : an experimental investigation. Fuel 291:120183. Available from:https://doi.org/10.1016/j.fuel.2021.120183
46. Sahoo S, Kumar D (2021) Effect of compression ratio on engine knock, performance , combustion and emission characteristics of a bi-fuel CNG engine. Energy 233:121144. Available from: https://doi.org/10.1016/j.energy.2021.121144
47. Bhasker JP, Porpatham E (2017) Effects of compression ratio and hydrogen addition on lean combustion characteristics and emission formation in a compressed natural gas fuelled spark ignition engine. Fuel 208:260–270
48. Wategave SP, Banapurmath NR, Sawant MS, Elahi M, Soudagar M, Reza M (2021) Clean combustion and emissions strategy using reactivity controlled compression ignition (RCCI) mode engine powered with CNG-Karanja biodiesel. J Taiwan Inst Chem Eng 000(2):1–16. Available from: https://doi.org/10.1016/j.jtice.2021.04.055
49. Ramachander J, Gugulothu SK, Sastry GR, Surya MS (2021) Statistical and experimental investigation of the influence of fuel injection strategies on CRDI engine assisted CNG dual fuel diesel engine.pdf. Hydrog Energy 46:22149–64
50. Sagar SMV, Agarwal AK (2017) Experimental investigation of varying composition of HCNG on performance and combustion characteristics of a SI engine. Int J Hydrogen Energy 42(18):13234–44. Available from: https://doi.org/10.1016/j.ijhydene.2017.03.063
51. Baratta M, Ferrari A, Zhang Q (2017) Multi-zone thermodynamic modeling of combustion and emission formation in CNG engines using detailed chemical kinetics. Fuel 231:396–403. Available from: https://doi.org/10.1016/j.fuel.2018.05.088
52. Meng X, Tian H, Zhou Y, Tian J, Long W, Bi M (2019) Comparative study of pilot fuel property and intake air boost on combustion and performance in the CNG dual-fuel engine. Fuel 256:116003. Available from: https://doi.org/10.1016/j.fuel.2019.116003
53. Lather RS, Das LM (2019) Performance and emission assessment of a multi- cylinder S . I engine using CNG & HCNG as fuels. Int J Hydrogen Energy 44(38):21181–92. Available from: https://doi.org/10.1016/j.ijhydene.2019.03.137
54. Garg M, Ravikrishna RV (2018) In-cylinder flow and combustion modeling of a CNG-fuelled stratified charge engine. Appl Therm Eng 2019(149):425–438
55. Meng X, Meng S, Cui J, Zhou Y, Long W, Bi M (2020) Throttling effect study in the CDF/RCCI combustion with CNG ignited by diesel and diesel/biofuel blends. Fuel 279:118454. Available from: https://doi.org/10.1016/j.fuel.2020.118454
56. Lee S, Kim C, Lee S, Lee J, Kim J (2020) Diesel injector nozzle optimization for high CNG substitution in a dual-fuel heavy-duty diesel engine. Fuel 262(2019):116607. Available fromhttps://doi.org/10.1016/j.fuel.2019.116607
57. Yang J, Wang J, Zhou X, Li Y (2020) Stochastic air-fuel ratio control of compressed natural gas engines using state observer. Hindawi 1–8

58. Aydin H (2021) An innovative research on variable compression ratio in RCCI strategy on a power generator diesel engine using CNG-saf flower biodiesel. Energy 231:121002
59. Harari PA, Yaliwal VS, Banapurmath NR (2021) Materials today : proceedings effect of CNG and CBG as low reactivity fuels along with diesel and TPME as high reactivity fuels in RCCI mode of combustion by varying different loads. Mater Today Proc 47:2491–4. Available from: https://doi.org/10.1016/j.matpr.2021.04.557
60. Alper A, Wong V (2021) Influence of laser ignition on characteristics of an engine for hydrogen enriched CNG and iso-octane usage. Int J Hydrogen Energy 46(74):37071–82. Available from: https://doi.org/10.1016/j.ijhydene.2021.08.206

Deformation and Strain Analysis in Calcaneus Plate Using Digital Image Correlation Method

Irwansyah, Muhammad Dirhamsyah, Edwar Iswardy, T. Nanta Aulia, Marzuki Alkindi, and S. N. Diah Fitriani

Abstract Analysis of strain distribution is a key topic in mechanical engineering. The two traditional techniques for measuring strain are making grids on test samples and using a strain gauge. Both of these methods require a lot of time and work. Digital image correlation (DIC), which is based on cutting-edge camera and computer technologies, is now an essential tool for measuring strain fields for a variety of test specimen scales. The fundamental idea behind DIC is to capture the entire process as images and then process those images so that they can be compared with deformed image sequences. This comparison can be used to determine the displacement and strain distribution down to the sub-pixel level. Several studies have highlighted the DIC method in solving bio-mechanic cases. However, limited knowledge is available regarding the deformation and strain distribution of bone plate implants. The goal of this study is to use the digital image correlation method to measure calcaneal bone plate deformation and strain during compression loading. A three-point bending test was performed on a calcaneal plate of 3.5 mm and screws that were fixed on the Sander's IIB fractured platform. The method of DIC measurement was discovered to determine the location and magnitude of strains at the bone-screw interface as well as the magnitude and location of plate stress.

Keywords Digital image correlation · Strain · Bone plates

Irwansyah (✉) · M. Dirhamsyah · E. Iswardy · M. Alkindi · S. N. D. Fitriani
Department of Mechanical and Industrial Engineering, Universitas Syiah Kuala, Banda Aceh 23111, Indonesia
e-mail: irwansyah@unsyiah.ac.id

M. Dirhamsyah
e-mail: mdirhamsyah@unsyiah.ac.id

E. Iswardy
e-mail: edwar@unsyiah.ac.id

T. N. Aulia
Department of Medicine, Universitas Syiah Kuala, Banda Aceh 23111, Indonesia
e-mail: tnantaaulia@unsyiah.ac.id

Irwansyah et al. (eds.), *Proceedings of the 4th International Conference on Experimental and Computational Mechanics in Engineering*, Lecture Notes in Mechanical Engineering, https://doi.org/10.1007/978-981-99-7495-5_27

1 Introduction

With modern advancements in computational methods, optics, and graphics computing, 3D scanning is rapidly becoming more prevalently adopted in society. High-density 3D scanning can be performed at rates of real time or faster, thus broadening the scope of applications to which these technologies can be applied. A structured light scanning the system projects different light patterns, or structures, and captures the light as it falls onto the scene. It then uses the information about how the patterns appear after being distorted by the scene to eventually recover the 3D geometry.

The potential speed of data acquisition, noncontact nature, the availability of necessary hardware, and the high precision of measurement offered by modern 3D structured light scanning technologies are what make them highly adoptable into industries such as medicine, biology, manufacturing, security, communications, remote environment reconstruction, and consumer electronics.

As the number of applications in which structured light techniques are employed increases, more interesting and challenging problems arise. It should be noted that there is not one 3D sensing technology that solves each issue and works as a general solution. The handbook by Zhang (1), which discusses many of the major 3D acquisition techsnologies, can be used to identify advantages and disadvantages of each approach. This article, however, focuses on discussing structured light scanning techniques.

Structured light techniques have benefited greatly from recent advancements in digital technology. To achieve real-time 3D data acquisition and reconstruction, much computational power is required, yet it can now be matched by today's modern computers; even some modern tablet computers can be used for these purposes. Given this, it is clear that the barrier to entry for some applications to use these technologies is quite low. However, in the past; if a manufacturing operation, for example, wanted to use these techniques, it may have been quite difficult to obtain the minimally required hardware, let alone to deploy the actual system itself. Modern advancements have ensured that the required hardware is relatively easily available.

Software techniques required to reconstruct 3D data have improved greatly as well and these will also be discussed.

To be useful for a wide range of applications, a system that can capture and reconstruct 3D information in real time (online), instead of retroactively (offline), is necessary.

Real time in this context typically refers to acquiring, processing, and visualizing the captured 3D data at speeds of at least 24 Hz (2). As mentioned earlier, advancements in hardware have made this a possibility, yet it is not a mere trivial task; many problems had to be solved before the technology reached this point.

As mentioned earlier, structured light scanning is only one method of 3D data acquisition. Within the area of structured light scanning, there are many different techniques, as well. One method stands out that has shown many advantages and derived from laser interferometers:digital fringe projection (DFP). The DFP technique involves varying sinusoidal patterns; using these methods, speeds of up to 120 Hz have been realized (3). However, the DFP approach has limits when it comes to its actual implementation, due to the sinusoidal pattern itself.

Eight-bit grayscale images are required to display the sinusoidal patterns, yet modern projectors can only display 8-bit patterns at certain rates. Due to these speed limitations, among other reasons (such as nonlinearity calibration and correction), other methods have been developed.

These other methods allow for increased speed while limiting the increase of system's complexity.

Such improvements to DFP include the squared binary defocusing technique (4). Whereas speed barriers were approached on today's digital image projection units with 8-bit images, 1-bit structured images can be projected much faster. By displaying 1-bit patterns through a defocused projector, the bits are blended together to reproduce a natural sinusoidal pattern. Another benefit, other than speed, of the binary defocusing technique is its lack of additional complexity due to calibration. As only two intensity values are being projected, nonlinearity calibration can be ignored. The smaller data transfer rate, increased rate of projection, and lack of complexity make this technique very advantageous and 3D shape measurement speeds greater than 120 Hz can be achieved. Using binary defocusing and digital light processing (DLP) platforms, Zhang et al. Have been able to successfully develop a system that performed 3D shape measurement at tens of kHz (5).

This article aims to provide a review of the different principles used in structured light scanning technologies, 3D data acquisition, and, by extension, a summary of its many different fields of application. By no means does this article position itself as a comprehensive body of work covering all structured light technologies and their finer details; several other surveys have been written regarding structured light techniques that may be useful, as well (6,7). It should be noted that the content relies on published work, either in journals or in conference proceedings, done by us based on our own experiments or by others in the field.

Section 2 will cover different structured pattern encodings. Section 3 will detail the steps required to properly calibrate a structured light scanning system. Section 4 will discuss structured light scanning at real-time speeds including DLP technology, 3D data acquisition, processing, and visualization. Section 5 will outline several of the many fields in which 3D structured light scanning technologies are proving themselves to be useful, including communications, human–computer interaction (HCI), entertainment, medicine and biology, security, and remote environment reconstruction. Finally, Sect. 6 will conclude this article along with discussing future directions for 3D scanning and, specifically, structured light scanning.

Non-contact optical imaging methods have replaced traditional strain gauges in many material testing laboratories in recent years. Standard strain gauges cannot be utilized in some applications due to a number of factors, including the usage of delicate or thin materials or the necessity to measure an object's whole area of operation without mounting several gauges. The most widely used optical technique is digital image correlation (DIC), which measures strain over the test sample using stereoscopic imaging and analysis. In order to monitor surface displacement and provide in-plane 2D measurements of strain, the digital image correlation method was originally employed in a single camera configuration in the 1980 [1]. Measurements of the mechanical properties such as deformation and strain present numerous difficulties in the field of bio-mechanics. Particularly, specimens of bone plates are not uniformly shaped and difficult to quantify using conventional strain gauges. Moreover, it complicates data interpretation of the stress and strain connection. The DIC approach is well-suited for biomechanical assessments because it can precisely detect fullfield, non-contact, three-dimensional stresses and displacements on biomedical components at both the macro and microscales. Additionally, DIC may be utilized for measurements of a variety of bio-mechanics, including as tension, compression, torsion, bending, and combined of those loading. The DIC method not only promptly produces measurement findings, but also gives the researcher access to these results as a complete dataset. Such information is essential for understanding complicated materials and structures, especially those with anisotropic properties, and for comparing experimental findings to computational models. Measurements of biomechanic components deformation and strain frequently confront particular difficulties due to complicated geometries and interactions of materials with different compositions, such as bone with implants.

Calcaneal fractures are a frequent foot fracture brought on by high energy trauma, such as a fall or motorcycle accident. About 1–2% of all fractures in the human body are calcaneal fractures, the most common tarsal bone to be injured. Approximately 75% of calcaneal fractures are intra-articular, affecting the subtalar joint, and are typically brought on by a fall from a height with the heel striking the ground directly [2, 3]. Even the most skilled trauma surgeons find it difficult to treat calcaneal fractures. The treatment option of plate fixation is available, but reports of technical issues and fracture reduction failure have been made. These designs have only recently been made available for calcaneal fractures, following the successful application of contoured plates and locking screws for other complex fractures. Deformation and strain in the surface of loading materials can be measured using DIC, according to a number of previous experimental studies [4–6]. In this experimental study the aim is to investigate to use the digital image correlation method to measure calcaneal bone plate deformation and strain during compression loading. Our null hypothesis was that the different locking plates do not differ in mechanical stability during experimental fixation of calcaneal fractures.

2 Digital Image Correlation Method

DIC is an non-contact measurement technique. Comparing the grayscale relationship between the images before and after distortion is the basic idea behind the digital image correlation technique. Before a load was applied, it was assumed in two-dimensional DIC that the coordinates that would be tracked could be written as

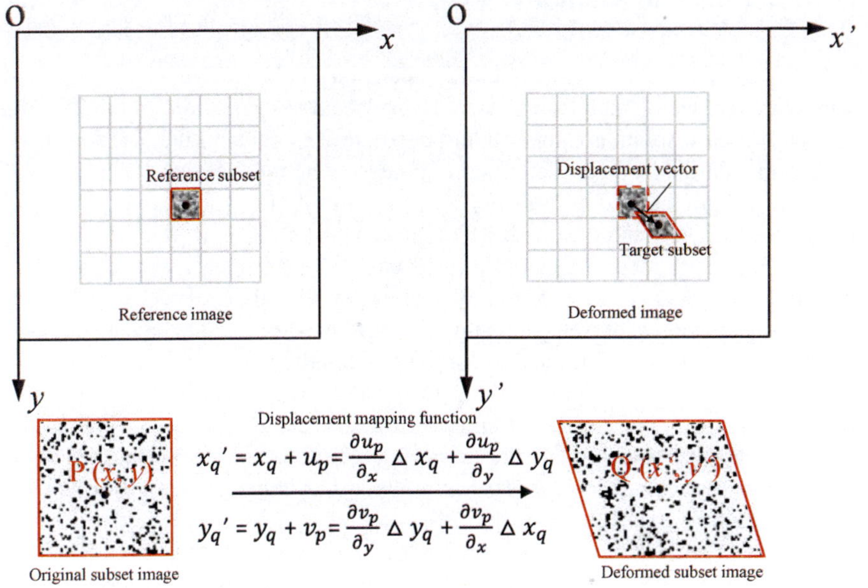

Fig. 1 Basic principle of digital image correlation method and the coordinate correspondences for the same point's displacement before and after loading

(x_p, y_p). The updated positions are represented as $(x_p{'}, y_p{'})$ after loading. On the image after deformation, the location of the corresponding point from the original, as depicted in Fig. 1, will be determined [7].

The points (x_p, y_p) can be seen as the center of the subset from which other points, like (x_q, y_q) can be referenced. The reference image can be used to correlate the deformed image to the original by using subset matching because it shows where the coordinates were before being loaded. The equations below govern the correlation algorithms that compute deformed positions relative to initial coordinates [7, 8]:

$$x_q\prime = x_q + u_p = \frac{\partial u_p}{\partial_x}\Delta x_q + \frac{\partial u_p}{\partial_y}\Delta y_q, \tag{1}$$

$$y_q\prime = y_q + v_p = \frac{\partial v_p}{\partial_y}\Delta y_q + \frac{\partial v_p}{\partial_x}\Delta x_q, \tag{2}$$

where u_p and v_p represent the x and y parts of the displacement vector at point p, respectively. As a result, x_q and y_q are q's pre-loading coordinates.

3 Materials and Methods

In the experiments, 2D DIC method rather than conventional strain gauge was used to measure elongation and distributions of displacement and strain. In the 2D DIC method, a high-resolution camera is mounted perpendicular to the surface of the object and records the entire deformation process. The measurement accuracy and precision are gradually improved by using high-resolution and frame-per-second (fps) cameras. The Canon D250 DSLR camera with 18–55 mm (f/3.5–5.6) macro lens and 58 mm diameter was used in this study. All experiments were captured at a resolution of 1280 × 720 pixels at 60 frames per second. The mechanical three-points test platform was developed in this to measure the strength of calcaneus plate. For the test specimen, a calcaneal locking plate (Synthes, Solothurn, Switzerland), screws of 3.5 mm were employed [9]. The installed 2D DIC measurement setup is shown schematically in Fig. 2. Digital images will be captured under various loads, and DIC will be used to measure the strains that developed in the bonded region.

A high-resolution camera is used to follow the surface's distinctive speckle pattern. Before applying the speckles, the specimen was prepared by spraying either a white or a black background onto the surface. It's crucial to get a sufficient speckle pattern for the digital image correlation in order to get accurate results (black speckles with different shapes and sizes). The pattern will then be captured under lighting conditions that maximized the distribution of gray levels across the dynamic range, as shown in Fig. 3.

After the camera and other tools have been set up, images have been acquired, and images have been extracted from video, it is time for the crucial stage of how to analyze the strain field that was discovered through the correlation technique on images. The Ncorr v1.2.2 program, an open source 2D digital image correlation MATLAB program (Matlab R2019a + toolboxes; Image Processing and Statistics),

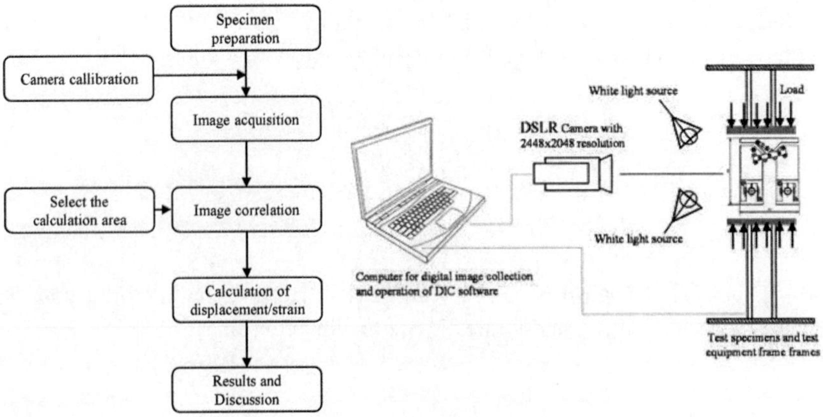

Fig. 2 Workflow of DIC displacement and strain measurement

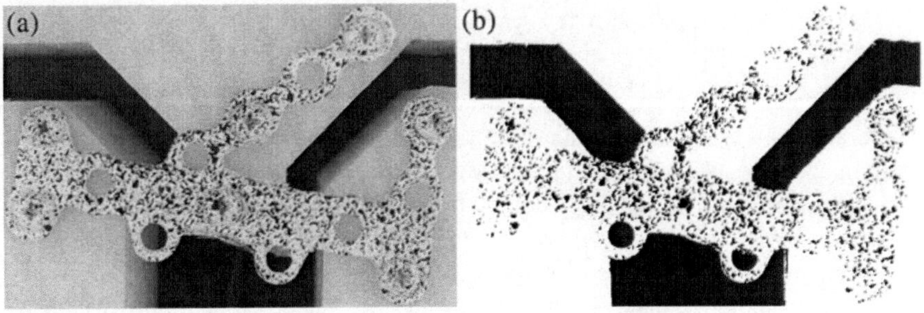

Fig. 3 Spraying the speckle pattern on surface specimen. **a** Black-white speckle pattern, **b** Gray-level distribution

was used to determine the amount of deformation and strain on the specimen surface [10, 11].

A region of interest (ROI) based on the gauge area and strain area is created as part of setting the DIC parameters. As shown in Fig. 4, the application, including the loading conditions, anticipated failure mode, and deformation is what determines the normal situation. Subset location appears to be taken into account by several parameters. The spacing between the dots determines how big the subsets are. In essence, reducing the computational load is done by increasing the distance between the dots. Options for multiple threads are designed to accelerate computation. More threads mean a more precise result, so the more threads, the better.

The DIC analysis's initial hypotheses are provided by the seed placement process. The seed is then verified in order to avoid processing errors. To convert the displacements from pixels to real units during the calibration and scaling phase of the process, it is necessary to specify the corresponding pixels that were used.

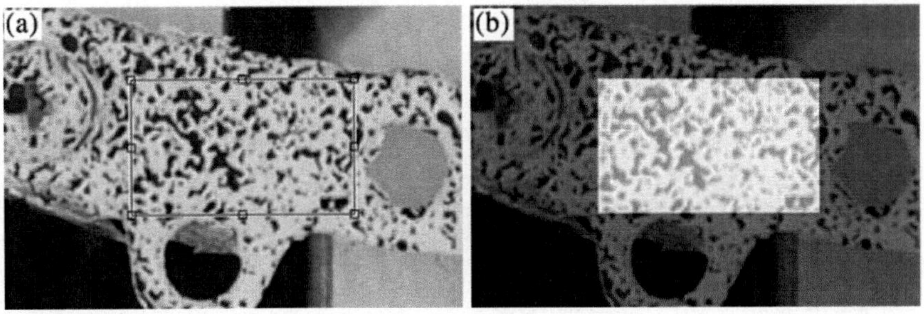

Fig. 4 Subset option and region of interest in Ncorr v1.2.2 **a** Region of interest, **b** Region seed

4 Results and Discussion

The current calcaneal bone plate was loaded as the DIC system was setup using the maximum 500 N force loading that was used in the experiment [12, 13] in order to set up images for correlation and processing to create a displacement or strain field. After the experiment was fully set up, the reference image was loaded at zero loads, as shown in Fig. 5. The sample was photographed for the first time before it was deformed, creating the reference image. All of the additional images that need to be correlated are gathered in the current images interface and prepared for processing by the correlation algorithm. The current image's format (TIF/Tag Image File Format) must be the same for this to work.

The strain option includes a parameter for the strain subset area when calculating strain, as shown in Fig. 6. To compute the fitting of a plane, this chosen area has a collection of points. In order to avoid noisy strain data, the ideal strain area should be chosen similarly to the ideal subset area.

It is assumed either the displacement has been formatted or the strain has been calculated [14]. The value of the strain corresponding to the load applied can be determined from the contour of strain at 500 N, the value of median strain εxy is 0.0014, as presented in Fig. 7.

5 Conclusion

This study has demonstrated measurement of the strain and deformation of the calcaneus plate using digital image correlation. This technique offers an alternative to the common method for measuring elongation, such as a strain gauge. The findings showed that DIC can be used effectively to study the displacement and strain field caused by deformation of the calcaneal plate. Accuracy and practicality can be increased by utilizing a system that combines load measurement and image matching at the same time, as well as high resolution and magnification lenses, a high speed camera, and light.

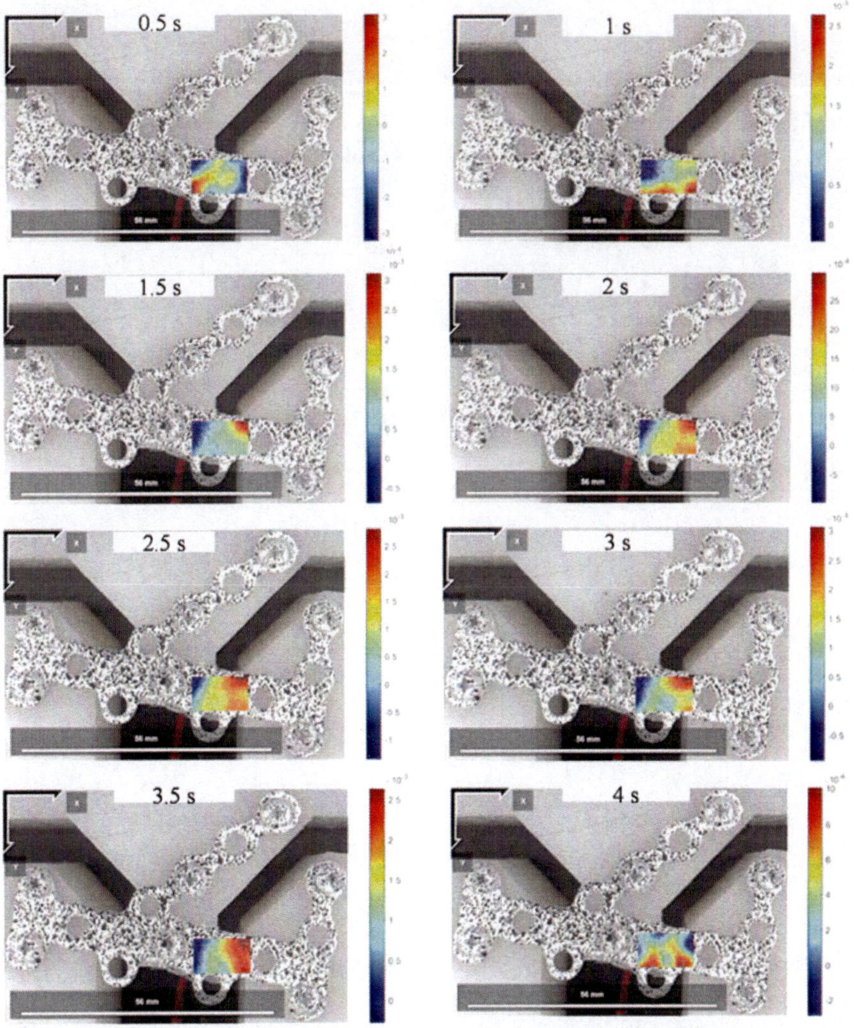

Fig. 5 Results of deformation on the surface of calcaneal plate by current loading

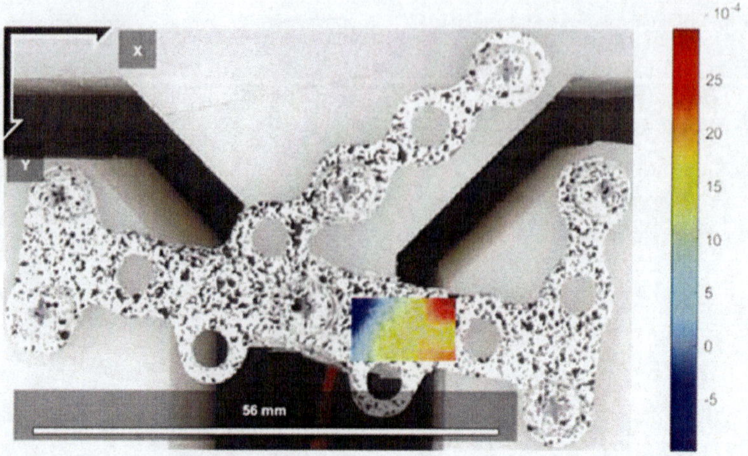

Fig. 6 Strain Ɛxy contour and median value

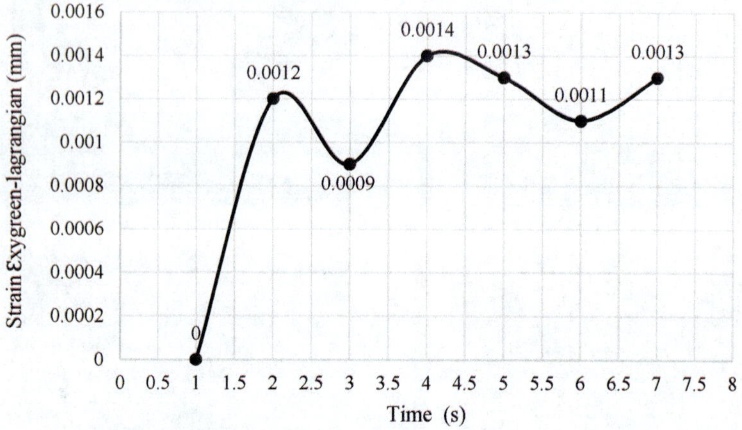

Fig. 7 Strain Ɛxy deformation of calcaneus plate

Acknowledgements The authors would like to express their gratitude to Universitas Syiah Kuala, Ministry of Education, Culture, Research and Technology, Republic of Indonesia for financial support of this research through the Grant Nomor: 145/UN11/SPK/PNBP/2022 Tanggal 11 Februari 2022.

References

1. Rankin K, Browne M, Dickinson A (2017) Digital image correlation for strain analysis of whole bones and implants. In: Zdero R (ed) Experimental methods in orthopaedic biomechanics, Academic Press, pp 65–83
2. Davis D, Seaman TJ, Newton EJ (2022) Calcaneus fractures. In: StatPearls treasure Island: StatPearls Publishing
3. Galluzzo M et al. (2018) Calcaneal fractures: radiological and CT evaluation and classification systems. Acta Biomed 89(1-S):138–150. https://doi.org/10.23750/abm.v89i1-S.7017
4. Rusin T, Weissman C (2018) Non-contact 3D deformation and strain measurements using digital image correlation for biomedical applications. In: Conference: biomechanics, Zielona Gora, Poland
5. Palanca M, Tozzi G, Cristofolini L (2016) The use of digital image correlation in the biomechanical area: a review. Int Biomech 3(1):1–21
6. Sutton MA, Wolters WJ, Peters WH, Ranson WF, McNeill SR (1983) Determination of displacements using an improved digital correlation method. Image Vis Comput 1:133–139
7. Ab Ghani AF, Ali MB, DharMalingam S, Mahmud J (2016) Digital image correlation (DIC) technique in measuring strain using opensource platform Ncorr. J Adv Res Appl Mech 26(1):10–21
8. Blaber J, Adair B, Antoniou A (2015) Ncorr: open-source 2D digital image correlation matlab software. Exp Mech 55(6):1105–1122
9. Walter B et al. (2015) Calcaneal plating with small diameter screws: mechanical performance and anatomical considerations. DePuy Synthes, Corpus ID: 51791356
10. Hung PC, Voloshin A (2003) In-plane strain measurement by digital image correlation. J Braz Soc Mech Sci Eng 25:215–221
11. Pan B, Qian K, Xie H, Asundi A (2009) Two-dimensional digital image correlation for in-plane displacement and strain measurement: a review. Meas Sci Technol 20(6):062001
12. Coombs DJ, Wykosky S, Bushelow M (2014) Calcaneal fixation plate test method development. SIMULIA Community Conference
13. Ouyang H et al (2017) Biomechanical comparison of conventional and optimised locking plates for the fixation of intraarticular calcaneal fractures: a finite element analysis. Comput Methods Biomech Biomed Eng 20(12):1339–1349
14. Yu B, Chen WC, Lee PY et al (2016) Biomechanical comparison of conventional and anatomical calcaneal plates for the treatment of intraarticular calcaneal fractures—a finite element study. Comput Methods Biomech Biomed Eng 19:1363–1370

Maximum Power Point Tracker for Dual-Energy Dryer Harvesting Systems Using Internet of Things

Saifuddin Muhammad Jalil, Husaini, Rizal Munadi, and Ira Devi Sara

Abstract In our modern environment, electrical energy has become the most important resource. However, it has access to many areas of the world, particularly agriculture. As a result, there is a huge need not only to identify alternative energy sources, but also to make them available to each sector of the world. Solar energy could entirely meet the energy needs, yet it is currently the most untapped energy resource. To achieve an alternate method of agriculture product drying system, a dual-energy dryer design employing maximum power points tracker was designed in this research. Solar panels will generate electricity to charge the battery throughout the day, which will subsequently be utilized to power the system, increasing system efficiency. The results of real-time data using Internet of Things show the results of battery performance and battery output loads as well as significant load devices by applying maximum power point tracker as the main mainframe in increasing the performance of the double dryer system. Better output results from this dual drying system at maximum power control. The novelty of the research in this activity shows that the maximum power capture value is more efficient in the dual-energy dryer

S. M. Jalil
School of Engineering, Universitas Syiah Kuala, Banda Aceh, 23111 Aceh, Indonesia

Department of Electrical, Universitas Malikussaleh, Lhokseumawe, 24351 Aceh, Indonesia

S. M. Jalil
e-mail: saifuddin@unimal.ac.id

Husaini (✉)
Department of Mechanical Engineering, Universitas Syiah Kuala Darussalam, Banda Aceh, 23111 Aceh, Indonesia
e-mail: husainiftm@usk.ac.id

R. Munadi · I. D. Sara
Department of Electrical and Computer Engineering, Universitas Syiah Kuala, Banda Aceh, 23111 Aceh, Indonesia
e-mail: rizal.munadi@usk.ac.id

I. D. Sara
e-mail: ira.sara@usk.ac.id

Irwansyah et al. (eds.), *Proceedings of the 4th International Conference on Experimental and Computational Mechanics in Engineering*, Lecture Notes in Mechanical Engineering, https://doi.org/10.1007/978-981-99-7495-5_28

system and the system performance is higher and more accurate with data obtained through the Internet of Things.

Keywords Energy sources · Dual-energy dryer · Maximum power points tracker · IoT · Solar panel

1 Introduction

Electricity consumption is continually expanding in our fast-growing planet. The scarcity of traditional resources, as well as its negative consequences, prompted humanity to consider an alternative and sustainable source of energy [1, 2]. Renewable energy sources are the way of the future for powering the entire country [3]. Solar, wind, biomass, and geothermal are just a few examples. Solar energy is one of the most promising alternatives. The key benefits of such a prospective energy source includes simple availability, eco-friendliness, and low operational and maintenance costs [4].

To optimize the energy received from solar radiation, solar collectors must be oriented at the optimal angle. Appropriate design, high thermal performance, and collector sizing are critical factors in determining the thermal performance of solar collectors [5]. The most critical characteristic that directly impacts the drying rate is the temperature within the drying chamber.

Nonetheless, the changeable weather conditions and the intermittent nature of solar energy have a significant impact on the temperature threshold of drying air, the quality of the dried items stands greatly, solar dryers remain restricted and ineffective in the absence of another energy source and complete control over the whole drying parameters [6]. The energy generated by the solar collector is determined by both the solar radiation in the collector's position and the collector's position in reference to the sun. The fact that the sun's rays turn up the collector diagonally through the day and allows for optimal energy output from a solar collector. Solar collector thermal performance, an important element of solar-powered systems, varies depending on absorber design, solar radiation value, air velocity, and location [7].

Studies on solar tracking systems have been undertaken in the literature to make better use of solar energy, Maximum Power Point Tracking (MPPT) is used to precisely track the maximum available solar panel power [8, 9]. The fundamental objective of the MPPT is to acquire maximum power output from photovoltaic panels regardless of climatic conditions such as sun irradiation and temperature, these controllers provide the proper quantity of power, reducing costs and increasing system dependability [10]. The purpose of this research is to improve the performance and real-time data (RTD) of the prototype in the drying process in order to shorten the drying time, update the data and the problem of uniform distribution of drying heat to humidity with controlled and stable drying. parameters by implementing IoT-based innovations. The Internet of Things (IoT)-based solar dryer prototype will easily and in more detail get real-time data so that it can provide an overview for researchers,

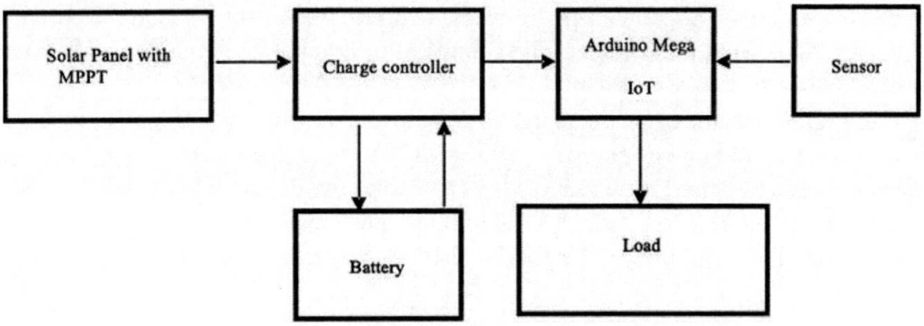

Fig. 1 Schematic of dual dryer with MPPT based Arduino Mega

practitioners, and policy makers about the many positive benefits of implementing IoT innovations that have been applied to solar-powered dryers prototype [11–13].

Typically, MPPT approaches operate on a relatively basic perturb and observe procedure. The converter first causes a disturbance in the panel voltage, the PV panel's power output changes as a result of this disturbance [14, 15]. Based on the measured transformation in power, a further evaluation was made on whether to further disrupt the panel's voltage and to what extent. The power tracker is compatible with several devices [16]. Component reduction and effective resource usage are among the principal advantages of the maximum power tracking approach. The integration of solar cells with smart grids is unachievable without the utilization of MPPT [17]. However, due to the speed characteristics, oscillation is likely near the tracking point. As a result, PV cell faults influence the efficacy of the maximum power point tracking method [18–20].

The current work describes the design and implementation of an Arduino Mega system-based MPPT for a dual-dryer system. Figure 1 depicts a block schematic of the proposed system. The system comprises a solar panel with a MPPT module connected to a charge controller that feeds an Arduino Mega and a battery to power the dryer system as a whole.

2 Methods

In this study, solar panels, Arduino Mega, a load control circuit, a buck-boost converter, a DC motor, a dual-channel 5 V relay, a temperature and humidity sensor, a heating element, a DC fan, and a conveyor belt are among the components utilized to overcome the drier process. The easy-to-use dual-energy solar-powered automated drier is designed to dry agriculture product rapidly and effectively. Figure 1 depicts the functioning block diagram. Solar panel with MPPT is has the main function in triggering the whole system to input the charge to battery and feed the system of

Arduino Mega to ensure the drying process occurred, the sensor also fed the parameter input to the Arduino Mega as the control of drying parameter as the main goal of the dual-energy drying system.

Figure 2 depicts the experimental device setup; the device consists of a solar panel and a cylindrical chamber equipped with a DC motor as the rotating device and a control system centered in Arduino Mega to ensure the drying process occurs efficiently; the dryer chamber is made of aluminum mesh material to provide flexibility and circulate the air to prevent heat imbalance inside the dryer chamber.

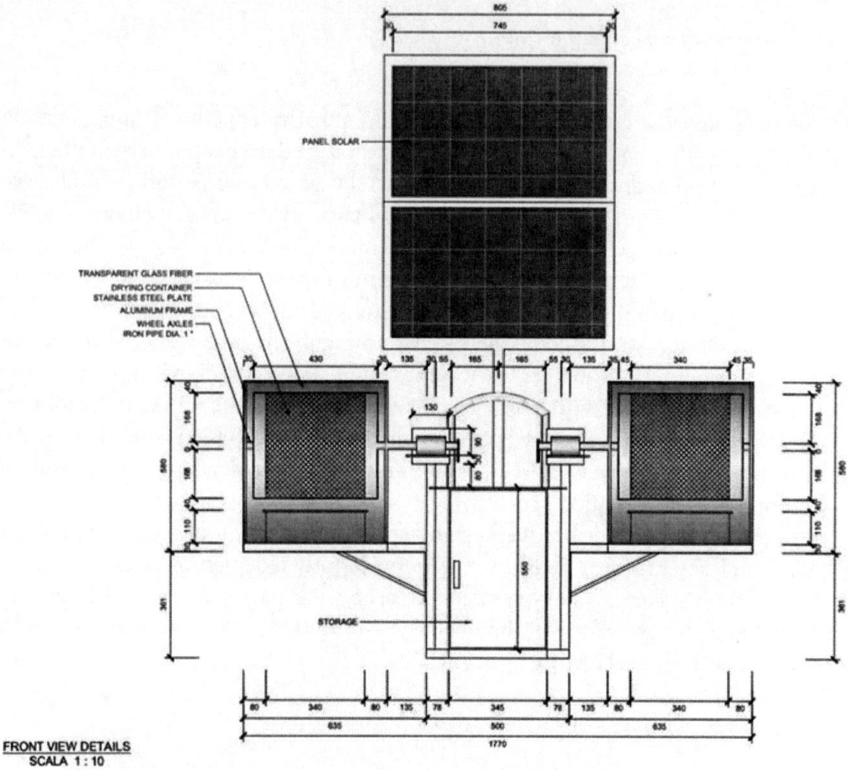

Fig. 2 Setup design for dual-energy dryer

3 Results and Discussion

3.1 Battery Performance of Dual-Dryer System

The usage of MPPT to power electrical equipment in this dual-dryer based on solar panel comes as a means to increase the efficiency of a solar panel. With MPPT, the system will start at Maximum Power Point (MPP) and create its maximum power output by measuring the maximum solar radiation that falls on the PV module. The MPPT solar charge controller transfers electricity from higher voltage levels to lower voltage levels by acting as a DC-DC transformer. The output current will be greater than the input current if the output voltage is less than the input voltage.

The measurement result shown in Fig. 3 is the process of battery charge during the test from morning time to afternoon as shown in graph the highest voltage measured 14,2 V using the MPPT.

As seen in Fig. 4, the phenomenon of current charging of a battery from morning to afternoon revealed 5 A. This shows that the MPPT effectively records the peak power point.

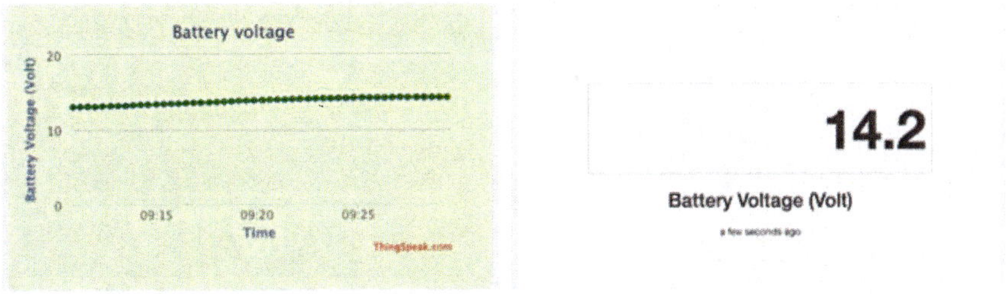

Fig. 3 Battery charge voltage measurement chart

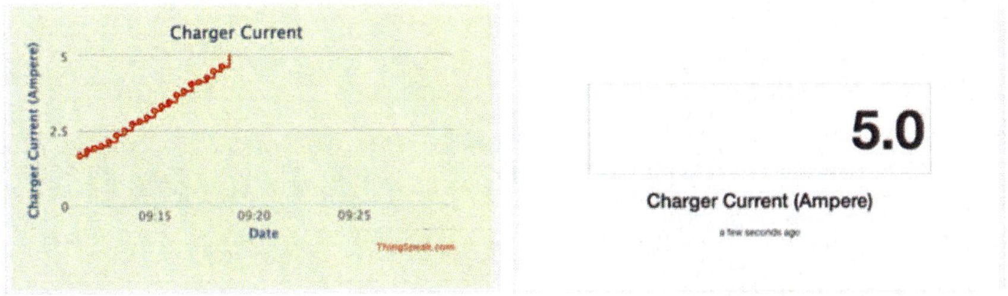

Fig. 4 Battery charge current measurement chart

4 Load Performance of Dual-Dryer System

Increasing incident solar radiation on the system and monitoring the maximum power point of the system have the ability to enhance the output power of a solar panel system. Figure 5 indicates that using MPPT has a significant influence on the performance of the load device, as evidenced by the tested dual-dryer system.

With high and moderate radiation levels, the performance of the dual-dryer system with the MPPT is investigated. Figure 5 depicts the observed power output measurement of solar panel system. The module power is clearly measuring the maximum power throughout the day. The MPPT keeps the load functioning at the maximum power point.

Figure 6 shows the measurement result. The decrease in humidity during the drying system test shows the main effectiveness of using MPPT.

The maximum power point tracking function is activated by a constant temperature control, resulting from the design parameters of light intensity and characteristics of the dual-dryer system, as depicted in Fig. 7.

Figure 8 shows the highest light intensity on drying system measurement indicates the maximum power rise rate as shown in the measurement results during the device's dual-dryer system test.

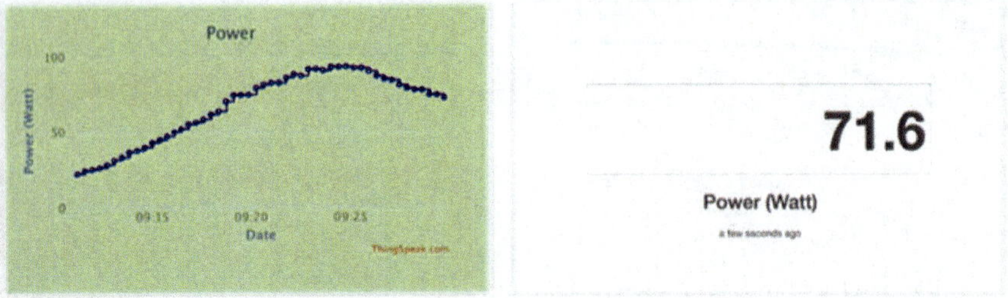

Fig. 5 Power output battery measurement of solar panel system

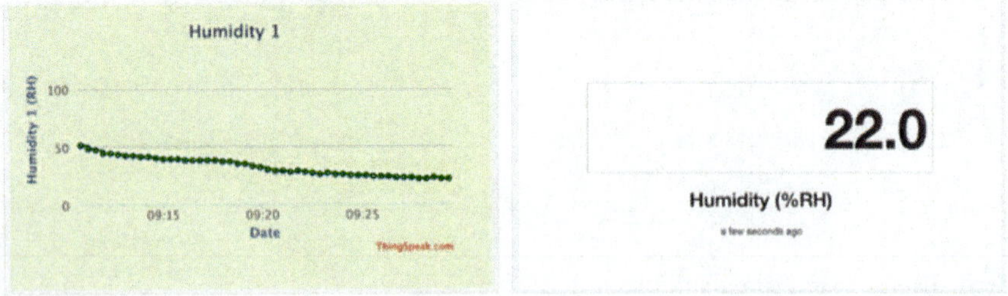

Fig. 6 Humidity measurement of solar panel system

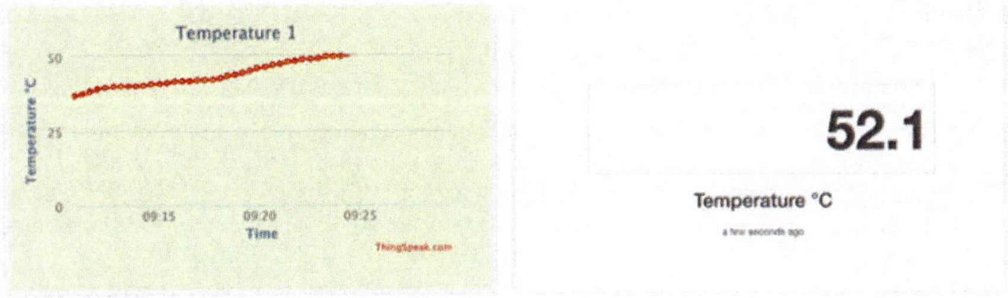

Fig. 7 Temperature measurement of solar panel system

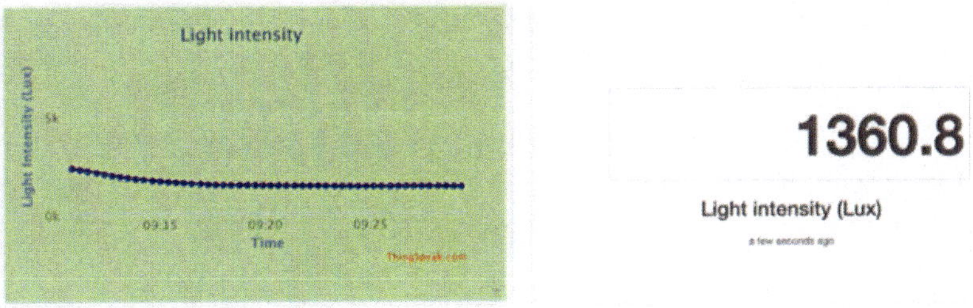

Fig. 8 Light intensity measurement

5 Conclusions

Solar energy is an important energy source for the future because it is the most predictable source of renewable energy and provides large amounts of energy to many sectors of the world, especially agriculture. Therefore, the development of MPPT technology in IoT-based drying systems is very important to increase the efficiency and output of solar panels while increasing dependence. This study proposes an Arduino Mega-based maximum power point tracking controller to determine the MPP and then adjust the solar panel to operate at the right operating voltage. Research output data for normal operating power, photovoltaic, and maximum power in terms of maximum power tracking performance show a significant improvement in the drying process.

Acknowledgements The authors are grateful to the Universitas Syiah Kuala for their financial support through Professor Research Grant No. 001/UN11.2.1/PT.01.03/ PNBP/2022 and the Ministry of Research Technology and the Higher Education Republic of Indonesia for the research grant program under the grant of Penelitian Disertasi Doktor No: 33/UN11.2.1/PT.01.03/DRPM/ 2021 The authors are also grateful to the Department of Mechanical Engineering, the Departement Electrical and Computer Engineering Universitas Syiah Kuala, Banda Aceh, for making the facilities accessible.

References

1. Saraswat SK, Digalwar AK (2021) Evaluation of energy alternatives for sustainable development of energy sector in India: an integrated Shannon's entropy fuzzy multi-criteria decision approach. Renew Energy 171:58–74. https://doi.org/10.1016/J.RENENE.2021.02.068
2. Bilgili F, Zarali F, Ilgün MF, Dumrul C, Dumrul Y (2022) The evaluation of renewable energy alternatives for sustainable development in Turkey using intuitionistic fuzzy-TOPSIS method. Renew Energy 189:1443–1458. https://doi.org/10.1016/J.RENENE.2022.03.058
3. Maka AOM, Salem S, Mehmood M (2021) Solar photovoltaic (PV) applications in Libya: challenges, potential, opportunities and future perspectives. Clean Eng Technol 5:100267. https://doi.org/10.1016/J.CLET.2021.100267
4. Shuba ES, Kifle D (2018) Microalgae to biofuels: 'Promising' alternative and renewable energy, review. Renew Sustain Energy Rev 81:743–755. https://doi.org/10.1016/J.RSER.2017.08.042
5. Ben Mansour R, Mateen Khan MA, Alsulaiman FA, Ben Mansour R (2021) Optimizing the solar PV tilt angle to maximize the power output: a case study for Saudi Arabia. IEEE Access 9:15914–15928. https://doi.org/10.1109/ACCESS.2021.3052933
6. Mofijur M et al. (2019) Phase change materials (PCM) for solar energy usages and storage: an overview. Energies 12(16):3167. https://doi.org/10.3390/EN12163167
7. Garcia RP, del R. Oliveira S, Scalon VL (2019) Thermal efficiency experimental evaluation of solar flat plate collectors when introducing convective barriers. Sol Energy 182:278–285. https://doi.org/10.1016/J.SOLENER.2019.02.048
8. Khodair D, Salem MS, Shaker A, El Munim HEA, Abouelatta M (2021) Application of modified MPPT algorithms: a comparative study between different types of solar cells. Appl Sol Energy 56(5):309–323. https://doi.org/10.3103/S0003701X20050084
9. Jalil SM, Abdurrahman F, Meliala S, Rosdiana R (2018) Design of maximum power point tracking for solar collector drying system: an experimental study. Int J Power Electron Drive Syst 9(4):1799. https://doi.org/10.11591/ijpeds.v9.i4.pp1799-1803
10. Abdallatif H, Abdel-Qader I, Harb A (2019) A fast MPPT algorithm for smart grid-PV connected system based on multiple sparse-aware time-varying stepsize adaptation technique. In: 2019 10th International renewable energy congress IREC 2019, March 2019. https://doi.org/10.1109/IREC.2019.8754547
11. Cui T, Li S (2020) System movement space and system mapping theory for reliability of IoT. Futur Gener Comput Syst 107:70–81. https://doi.org/10.1016/j.future.2020.01.040
12. Wilkins R, Brusey J, Gaura E (2018) Modelling uncontrolled solar drying of mango waste. J Food Eng 237:44–51. https://doi.org/10.1016/j.jfoodeng.2018.05.012
13. Satria DI, Jalil SM, Yusibani E (2020) The disaster notification system based sms to students at the school vulnerable flood disaster using single board computer of raspberry pi. Int J Adv Trends Comput Sci Eng 9(2):1063–1067. https://doi.org/10.30534/ijatcse/2020/25922020
14. Bahrami M et al. (2019) Hybridmaximum power point tracking algorithm with improved dynamic performance. Renew Energy 130:982–991. https://doi.org/10.1016/J.RENENE.2018.07.020
15. Venkatramanan D, John V (2019) Dynamic modeling and analysis of buck converter based solar PV charge controller for improved MPPT performance. IEEE Trans Ind Appl 55(6):6234–6246. https://doi.org/10.1109/TIA.2019.2937856
16. Zhu Y, Kim MK, Wen H (2018) Simulation and analysis of perturbation and observation-based self- adaptable step size maximum power point tracking strategy with low power loss for photovoltaics. Energies 12(1):92. https://doi.org/10.3390/EN12010092
17. Szczepaniak M, Otręba P, Otręba P, Sikora T (2021) Use of the maximum power point tracking method in a portable lithium-ion solar battery charger. Energies 15(1):26. https://doi.org/10.3390/EN15010026
18. Deshpande AS, Patil SL, Pandey SK (2020) maximum power point tracking using disturbance observer-based sliding mode control for estimation of solar array voltage. 48(1–2):148–161. https://doi.org/10.1080/15325008.2020.1731863

19. Meliala S, Saifuddin, Rosdiana (2019) Drying design model for dried-anchovy using solar collector and solar cell panel position control. Int J Recent Technol Eng 7(6S3):1430–1435. [Online]. Available: http://www.scopus.com/inward/record.url?eid=2-s2.0-85067196861&partnerID=MN8TOARS
20. Jalil SM, Abdullah L, Ahmad I, Abdullah H (2008) The effect of surface texturing on GaAs solar cell using TCAD tools. In: 2008 IEEE international conference on semiconductor electronics, pp 280–283.https://doi.org/10.1109/SMELEC.2008.4770323

Implementation Phase Change Material at Cold Side of Thermoelectric Cooler Box as Thermal Energy Storage

Adi Winarta, INyoman Suamir, IMade Rasta, and Achmad Wibolo

Abstract In a particular application, thermoelectric refrigeration has an unrivaled excellence which other refrigeration systems may fail to accomplish it. Especially in application related in medical purposes. An accurate temperature stability is essential for this application. This study examined the performance of a thermoelectric cooler box with a cold sink that fills with Phase Change Material (PCM). Two different PCM were tested and analyzed for their energy efficiencies related to the thermoelectric cooler box. The cooler box has 240 × 260 × 265 mm dimension and is supplied with two Peltier modules. The power supply to the module was controlled by a thermo-controller and set to interrupt the supply if a temperature condition was achieved during the test. The purpose is to exchange the cooling supplied from thermoelectric with cold PCM inside the container. Thus it can reduce the power consumption of thermoelectric. The result shows that PCM 2 which mixed of 95% of tap water and 5% soya ester could hold the temperature of the cooler box more stable and longer than tap water as PCM 1. Furthermore, the result also shows that using PCM 2 has potential energy storage in thermoelectric application devices compared with PCM 1.

Keywords Thermoelectric cooler box · Phase change material · Thermal energy storage

A. Winarta · I. Suamir · I. Rasta · A. Wibolo (✉)
Department of Mechanical Engineering, Politeknik Negeri Bali, Jln. Kampus Bukit Jimbaran, Kuta Selatan, Badung Bali 80364, Indonesia
e-mail: wibolo@pnb.ac.id

A. Winarta
e-mail: adi.winarta@pnb.ac.id

I. Suamir
e-mail: nyomansuamir@pnb.ac.id

I. Rasta
e-mail: imade.rasta@pnb.ac.id

Irwansyah et al. (eds.), *Proceedings of the 4th International Conference on Experimental and Computational Mechanics in Engineering*, Lecture Notes in Mechanical Engineering, https://doi.org/10.1007/978-981-99-7495-5_29

1 Introduction

Nowadays, refrigeration is an essential aspect of human life, exclusively in the medicine and food process industry. The vapor compression refrigeration system (VCRS) is the most applied in those industries. The system uses refrigerant as a working fluid; most of the refrigerant is not environmentally friendly. Therefore, researchers worldwide have to work hard to find and optimize systems that are more energy efficient and free from environmental issues. Fortunately, thermoelectric cooling (TEC) is one promising technology to answer the above problem. In addition, technological developments today in renewable energy (PV cells) and thermoelectric materials increase the feasibility of using thermoelectricity in refrigeration systems [1, 2]

Thermoelectric cooling (TEC) is a latter refrigeration system whose principal work is based on the Peltier, Seeback and Thomson effect. Some people also call thermoelectric with a Peltier module. Operation of thermoelectric module begin when a DC current passes through a junction made from couple of p-n type material of semiconductor then heat is carried out from the cold to the hot surface. The quantity of heat dissipated by the hot side related to the cooling method and the current supply.

Peltier module has several advantages such as no moving part, no friction and noise, light and compact structure, no refrigerant use, and almost free maintenance. Unfortunately, its energy performance is still very low compared with VCRS. The low performance of thermoelectric makes its development in many applications relatively slow compared with the VCRS.

Thermoelectric find their best use in medical, aviation and spaceship which requiring accurate and precise temperature control for specific refrigerated areas. Indeed, the researchers have made enormous progress in the field of cooling by thermoelectricity. Recently, low-cost thermoelectric cells have been rapidly developed by the industry and are produced in several sizes [3]. The good things is this module will obtain better parameter, such as higher cooling capacities, more compact dimension, varied shapes and sizes and lower prices in the next future [4, 5]. Therefore, this work then become more important and increase its possibility.

Many studies have been performed regarding to evaluating the performances of thermoelectric refrigeration. Enescu and Virjoghe [6] stated that the COP of thermoelectric refrigeration cannot more than 0,5 for the 20 K temperature difference if its power by conventional electricity. Only if uses renewable energy to supplied the electricity, it can reach more than 0,5. Gokçek and Shahin [7] successfully increase the performance of thermoelectric refrigerators using a water cooled heat sink with mini channel for the cooling block. They claim that the performance of their system around 0,41. Riffat and Qiu [8] numerically studied about cylindrical water cooled heat sink. They found that cylindrical water cooled has overall heat transfer 20 time compared with air cooled heat sink. Mirmanto et al. [9] proposed a study of thermoelectric cooler box using different type of heatsink on the hot side. Two type of heat sink such as a Heat Sink Fan (HSF) and Double Fan Heat Pipe (DFHP) were used to absorb the heat dissipation from the hot side. Their result shows that the HSF has lower energy consumption. Cuce et al. [10] conduct an experimental about improving COP of cooler box using water cooled block system and nanofluid working fluid. Al_2O_3, TiO_2 and SiO_2 were used as nanoparticle to water-based working fluid for water cooling block. They claim that using nanofluid for water block system has significant improvement for the temperature of cooler box cabin. Winarta el al. [11] conduct an experimental using vapor chamber as the cooling method of thermoelectric hot side. The highest COP obtained when the current supply to the Peltier module is 4 amperes.

Few researchers use phase change material (PCM) to increase the performance of thermoelectric cooling. Riffat et al. [12] were the researcher who investigated the employment of PCM (ClimSel C7) within the heat sink unit. The PCM inside the enclosure contacted with the cool side of the thermoelectric system through the aluminum block. Their results shows that encapsulated PCM improved the performance of thermoelectric cooler box. Omer et al. [13] then continued his study by adding thermosyphon as the thermal diode between Peltier module and the cold sink. In addition, they also compared the performance of their proposed system with the similar one but without the thermosyphon. They conclude that the proposed system could prevent heat leakage from PCM when the supply power off. Zhao and Tan [14] performed a study of integration PCM with thermoelectric for space cooling. Similar cooling effect also could achieve using the thermal diode as though the system without it. However, there is still a lack of study about implementing PCM as the thermal energy store for thermoelectric refrigeration.

The objective of this present study is to determine whether the effect of the implementation of PCM on the cold side of the thermoelectric module could save power consumption. Therefore, the thermoelectric cooler box is expected to have a more efficient and effective use of energy in the future (Fig. 1).

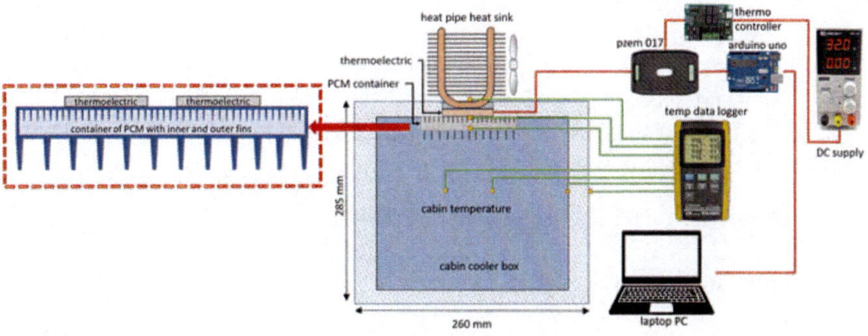

Fig. 1 Schematic diagram of experimental setup of the thermoelectric cooler box. The PCM container detail schematic picture shown at left side

2 Methodology

The experimental test consists of a cooler box, two thermoelectric modules, a heat sink for cold and hot sides of the module, and a data logger unit for temperatures and power input. The box dimension is 240 × 260 × 265 mm with a thickness of 40 mm and is made from polyurethane material. The top of the cooler box is partly wood and framed with polyurethane material. It is also composed of two TEC2-19,006 with 40 × 40 × 6.3 mm of dimension. The thermoelectric is attached to the cold sink, which also functions as a PCM container. The detail of the cooler box is shown in Fig. 2. Tap water and mixture of water and soya ester are used in this experiment. The liquid volume is about 360 ml. The PCM container is made from aluminum and attached with fins to extend the heat transfer area. In Fig. 2, the detailed schematic picture of PCM container is also presented for the detailed construction. Two heat pipe heat sinks with fans were used to dissipate the heat from the hot side. The gaps between the cold sink, thermoelectric and the heat sink heat pipe are minimized with thermal paste. Temperature measurements are carried out at 11 points such as hot (T_{hot}TEC) and cold side (T_{cold}TEC) thermoelectric, PCM liquid (T PCM), PCM container (T Cold Sink), cabin (T Cabin), cabin wall (T wall Cabin) and ambient (T ambient). Type K of the thermocouple and Lutron BTM-4208SD 12 channels data logger was employed to measure the temperature history of this experiment. Power consumed by the thermoelectric was captured by a measurement system using PZEM-017 DC meter coupling with Arduino Uno. The energy data is then stored on a laptop for later analysis.

Figure 2 shows the picture of experimental test. The power supply to the thermoelectric was regulated using a thermo-controller W1219 dual display. The temperature of liquid PCM was used as the setting value for controlling the power supply to the thermoelectric. The setting value was −7 °C with 2.5 °C differential. Liquid PCM inside the container was assumed already solid with this setting value. After that, the supply current of the thermoelectric will be cut by the thermo-controller. The solid phase of PCM will cool down the temperature of the cooler box cabin. On this occasion, no heat load was located inside the cooler box. The DC supply was set at 13 V and 4.5 A to power the Peltier module. The ambient temperature is set at 25 °C using air conditioning.

3 Results and Discussion

Figure 3 shows evolution temperature versus time for the test using tap water as PCM 1. Initially, all the temperature measurement were display around 25 °C near the ambient temperature. As soon as the current flows to the Peltier module, the cold and hot sides temperatures separate each other. The temperature of T_{cold}TEC falls rapidly to −3 °C in 2070s (34 min). After that, its curve slowly decreases until the 14750th second (245 min). At the same time, T_{hot}TEC increases sharply and then

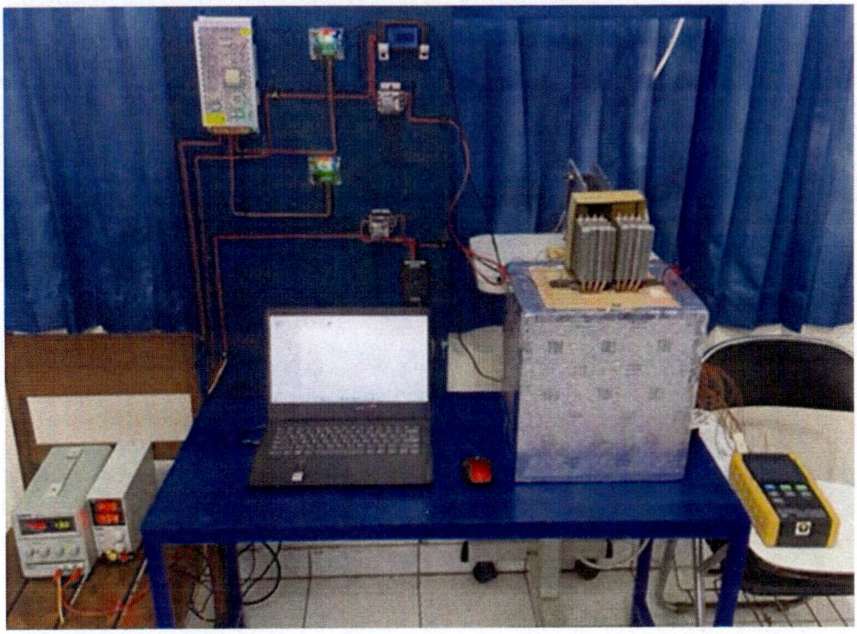

Fig. 2 Actual experimental work

gets stable around 42–43 °C. This result validates our experimental work because it shows a similar trend with previous publications [7, 9, 11, 15, 16]. Furthermore, T PCM trend follows the pattern of cold junction temperature with few degrees higher. The slope of T_{cold}TEC temperature increases from 14,750 to 20,550 s, dropping to −8,4 °C. The changing slope of temperature might be the effect of the transition process of PCM from liquid to solid. Due to the low cooling capacity of thermoelectric, the phase change process was time-consuming. At 20,560 s, the temperature of T_{cold} TEC and PCM suddenly jump to 0.3 °C due to the response of thermo-controller cut-off. The electricity supplied to the thermoelectric was shut off for around 150 s due to the thermo-controller setting value. Three sets of cut-offs and cut-in are shown clearly in the graph, which makes the cabin temperatures fluctuate.

Figure 4 presents the temperature data using PCM 2, which contains 95% of tap water and 5% soya ester. As seen on the graph, the temperature of TcoldTEC, during the cooling phase, falls to −301 °C in 3070 s and then drops slowly and almost stably. After being stable for nearly 180 min, thermo-controller cut-off the power to the Peltier module, which automatically increases the temperature of the cold side TEC. The PCM temperature also follows this increment. On the other hand, T_{hot}TEC temperature drops below the ambient temperature due to no heat dissipating. Surprisingly, the cabin temperature is almost stable, around 7–8 °C, despite a temperature change of PCM due to a power supply interruption to the Peltier module.

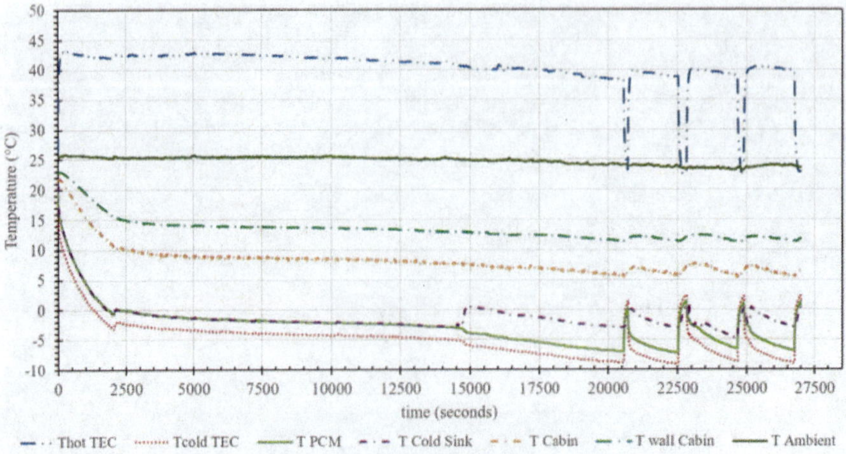

Fig. 3 Experimental data of variation of temperatures TEC cooler box using PCM1 (tap water)

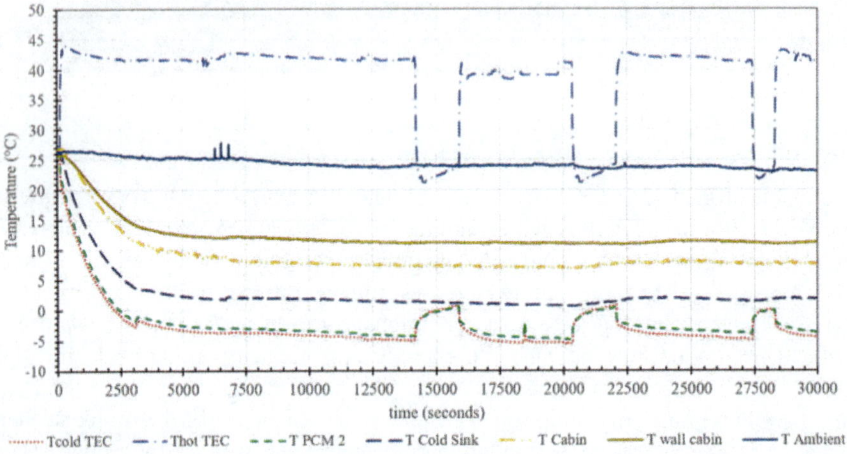

Fig. 4 Experimental data of variation of temperatures TEC cooler box using PCM 2

The charging process of PCM, regarding of cooling process, is related to the transition phase from liquid to solid material. The storage of cold energy in the phase change materials causes the solidification of the PCM material. Based on temperature history analysis, PCM 1 shows lower temperature and longer cooling period than PCM 2. This might be due to delay of solidification process of PCM 1, which composed from 100% tap water. Hence, using PCM 2, implemented on a cold surface of the Peltier module, is more feasible for saving energy than PCM 1.

The on–off cycle of profile temperatures from both tests is observed into a detailed graph, as shown in Fig. 5, to examine the phenomena in more detail. One complete cycle consisting of an on-and-off period could be interpreted as a charging and discharge process, as cold energy is absorbed and released to maintain the cabin temperature. Fast increase of PCM 1 temperature (a) indicating rate of heat transfer from cabin to the water tap as PCM. The temperature of PCM 1 rises with a higher Delta T (dT), from −9 to 2.7 (11,7 K) compared with PCM 1, when the thermo-controller cuts off the current supply to the thermoelectric. As shown in the graph, fluctuations in PCM 1 temperature and the cold side of TEC cause disturbances in cabin temperature. These phenomena may occur due to an incomplete phase change process in PCM 1 becoming solid. The phase change process of tap water requires supercooling almost 40 degrees below the phase change temperature (0 °C) [17, 18]. The data also shows that PCM 2 has a longer off-time than PCM 1. The off-time of thermoelectric energy supplied is related to discharge process of phase change material. The discharge process on the PCM 2 occurs slower and longer due to higher energy which has been storage during the charging process, compared with PCM 1. Or, we could say that, PCM 2 has been completely frozen before the thermo-controller stop the energy supply, and then giving out their energy storage to absorb the heat from the cabin. Much cold energy can be absorbed by the PCM 2 compared with PCM 1.

The off-time of thermoelectric supply power is related to discharge time in phase change material, which shows the material's ability to maintain the cabin temperature without an electricity supply. PCM 2 is more capable of maintaining a more stable cabin temperature than PCM 1. Our results find that using PCM 2, the mix of tap water and soya ester can maintain better temperature stability in the cabin for this cooler box application.

Figure 6 shows the comparison of power consumption of thermoelectric in the experimental test using PCM 1 and PCM 2. The graph shows that PCM 2 has lower energy consumption due to a longer off-time cycle corresponding with longer discharging time, as shown in Fig. 4. As indicated by the increased distance between power off and on to the Peltier module. Using tap water mixed with soya ester for thermal energy storage at the cold side of the thermoelectric can reduce the Peltier module's energy consumption. On the other hand, the capability of reduced energy consumption in PCM 1 failed to show, which is depicted in a shorter off cycle compared with PCM 2, as described in Fig. 6.

Tables 1 and 2 compile the data on the thermoelectric cycling process, such as on–off condition time, TEC and PCM temperature, thermo-controller time, and charging-discharging period which already shown at Fig. 5a and b. From the data, the average ratio of discharge and charge is calculated for each PCM. This ratio may be indicating the effectiveness of PCM for the thermal energy storage. The average ratio of discharge and charge for PCM 2 is 0.36 meanwhile, 0.121 for PCM 1. PCM 2 has three times higher than PCM 1. The result shows that PCM 2 has the three time of discharge and charge ratio than PCM 1, which means that the capability of thermal energy storage for PCM 2 is better than PCM 1.

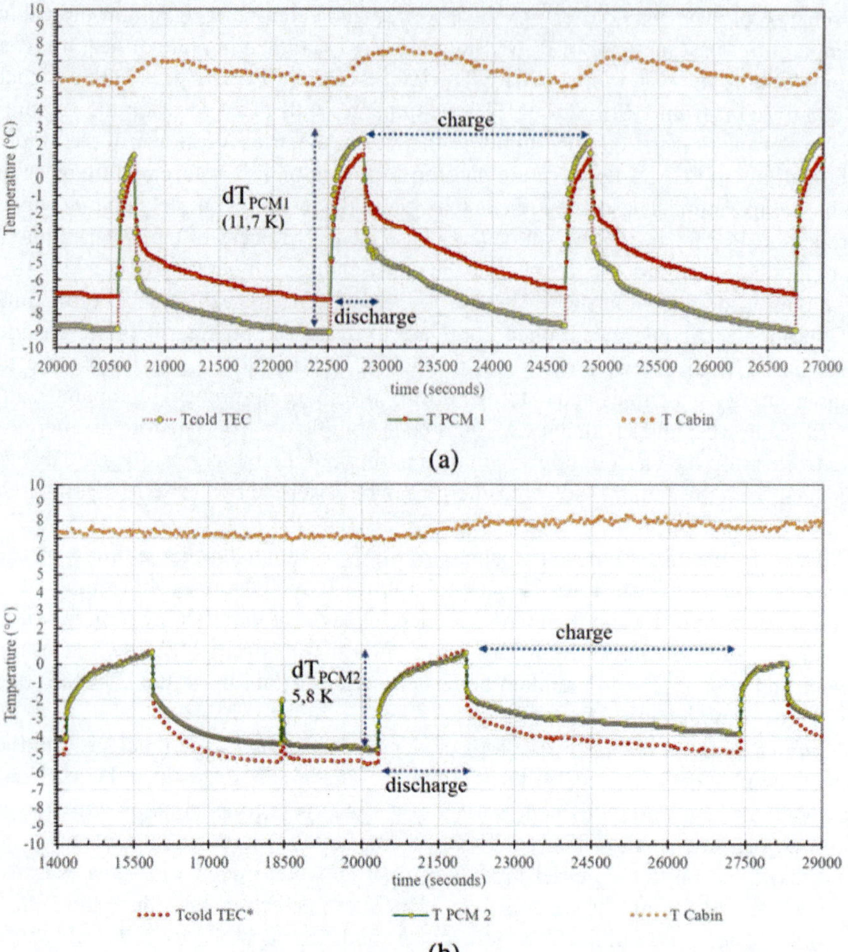

Fig. 5 Temperature evolution during cycling process of Tcold TEC and PCM in PCM 1 **a** and PCM 2 **b** experimental result

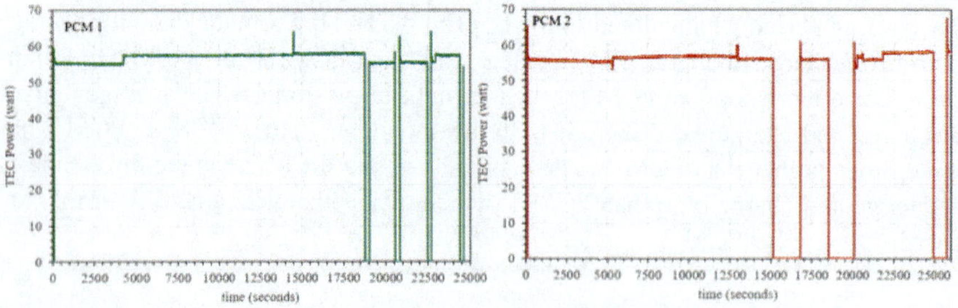

Fig. 6 Power consumption comparison between PCM 1 and PCM 2

Table 1 Time and temperature of cycling on–off, charging and discharging period using PCM1

Stages	Thermo-controller	Time	TEC temperature	PCM temperature	Charging period	Discharge period	Ratio discharge and charge
		[second]	[C]	[C]	[minutes]	[minutes]	
I	Off	20,560	−8.4	−7.05		2.50	
	On	20,710	1.05	0.3			
II	Off	22,520	−9.1	−7.25	30.17	5	0.083
	On	22,820	2.3	1.35			
III	Off	24,660	−7.3	−6.6	30.67	3.67	0.163
	On	24,880	2.2	1.1			
IV	Off	26,750	−7.8	−6.95	31.17	3.83	0.118
	On	26,980	2.3	1.2			
Average ratio of discharge and charge							0.121

Table 2 Time and temperature of cycling on–off, charging and discharging period using PCM2

Stages	Thermo-controller	Time	TEC temperature	PCM temperature	Charging period	Discharge period	Ratio discharge and charge
		[second]	[C]	[C]	[minutes]	[minutes]	
I	Off	14,170	−5.5	−4.2		28.67	
	On	15,890	1.1	0			
II	Off	20,340	−5.5	−4.8	74.17	28.7	0.39
	On	22,060	1.1	0.6			
III	Off	27,400	−5	−3.9	89.0	15	0.32
	On	28,300	0.4	0			
Average ratio of discharge and charge							0.36

4 Conclusion

The implementation of PCM as thermal energy storage in the cold side of thermoelectric cooler was tested in this study. The PCM container was tested using two materials (tap water as PCM 1 and mix of tap water and soya ester PCM 2). The results showed that the use of PCM 2 produced better thermal energy storage capabilities than PCM 1, indicating greater savings in thermoelectric power supply. The cycle on–off temperature at PCM 2 is longer than PCM 1. The cabin temperature also more stable at 7–8 °C. Better power reduction also achieved at PCM 2 and this led to higher ratio of discharging-charging ratio.

Acknowledgements The authors would like to acknowledge Mechanical Engineering Department Politeknik Negeri Bali for its research facility. The research is supported by the Politeknik Negeri Bali through Penelitian Unggulan DIPA PNB with the contract number: **SP DIPA-023.18.2.677608/ 2022** Revisi ke 03 tanggal 15 Februari 2022.

References

1. Zhao L-D et al. (2014) BiCuSeO oxyselenides: new promising thermoelectric materials. Energy Environ Sci 7(9):2900–2924. https://doi.org/10.1039/C4EE00997E
2. Kim JY, Lee J-W, Jung HS, Shin H, Park N-G (2020) High-efficiency perovskite solar cells. Chem Rev 120(15):7867–7918. 12 Aug 2020
3. Hadjiat MM, Salhi K, Ouali S, Ait-Ouali A, Hazmoune M, Imessad K (2020) Experimental study of a thermoelectric refrigerator. In: ICREEC 2019, Singapore, 2020, Springer, Singapore, pp 27–34
4. Salsabila E, Ajiwiguna T, Suhendi A (2019) Performance evaluation of TEC1–12706 thermoelectric cooler module at low temperature experimentally. In: 2019 International conference on information and communications technology (ICOIACT), 2019, IEEE, pp 835–838
5. Enescu D, Ciocia A, Mazza A, Russo A and assessments (2017) Solutions based on thermoelectric refrigerators in humanitarian contexts. 22:134–149
6. Enescu D, Virjoghe EO (2014) A review on thermoelectric cooling parameters and performance. Renew Sustain Energy Rev 38:903–916. 01 Oct 2014
7. Gökçek M, Şahin F (2017) Experimental performance investigation of minichannel water cooled-thermoelectric refrigerator. Case Stud Therm Eng 10:54–62. 01 Sep 2017
8. Riffat SB, Qiu GQ (2006) Design and characterization of a cylindrical, water-cooled heat sink for thermoelectric air-conditioners. 30(2):67–80
9. Mirmanto M, Sayoga I, Sutanto R, Alit I, Nurchayati N, Mulyanto A (2018) Experimental cooler box performance using two different heat removal units: a heat sink fin-fan, and a double fan heat pipe. 10
10. Cuce E, Guclu T, Cuce PM (2020) Improving thermal performance of thermoelectric coolers (TECs) through a nanofluid driven water to air heat exchanger design: an experimental research. Energy Convers Managem 214:112893. 15 June 2020
11. Winarta A, Rasta IM, Suamir IN, Puja IGK (2021) Experimental Study of thermoelectric cooler box using heat sink with vapor chamber as hot side cooling device. In: Proceedings of the 2nd international conference on experimental and computational mechanics in engineering: ICECME 2020, Banda Aceh, October 13–14, Akhyar, Ed. Singapore, Springer Singapore, pp 389–399
12. Riffat SB, Omer SA, Ma X (2001) A novel thermoelectric refrigeration system employing heat pipes and a phase change material: an experimental investigation. Renew Energy 23(2):313–323. 01 June 2001
13. Omer SA, Riffat SB, Ma X (2001) Experimental investigation of a thermoelectric refrigeration system employing a phase change material integrated with thermal diode (thermosyphons). Appl Therm Eng 21(12):1265–1271. 01 Aug 2001
14. Zhao D, Tan G (2014) Experimental evaluation of a prototype thermoelectric system integrated with PCM (phase change material) for space cooling. Energy 68:658–666. 15 Apr 2014
15. Midiani LPI, Subagia IWA, Suastawa IW, Sapteka AANG, Winarta A (2020) Preliminary investigation of performance and temperature distribution of thermoelectric cooler box with and without internal fan. J Phys: Conf Ser 1450(1), pp 012088. 01 Feb 2020
16. Winarta A, Made Rasta I, Ike Midiani LP, Wayan Adi Subagia I, Gde Sapteka AAN (2021) Experimental study of thermoelectric cooler box using heat sink with u-shape heat pipe and methanol working fluid. In: IOP conference series: materials science and engineering, vol 1034(1). pp 012033. 01 Feb 2021

17. Safari A, Saidur R, Sulaiman FA, Xu Y, Dong J (2017) A review on supercooling of phase change materials in thermal energy storage systems. Renew Sustain Energy Rev 70:905–919. 01 Apr 2017
18. Rasta IM, Suamir IN (2018) The role of vegetable oil in water based phase change materials for medium temperature refrigeration. J Energy Storage 15:368–378. 01 Feb 2018

Supercooling Behavior of Water-Based Solution with Propylene Glycol

Adi Winarta, Achmad Wibolo, INyoman Suamir, and IMade Rasta

Abstract Thermal energy storage using phase change material (PCM) is considered a promising technology for future energy needs due to its low-cost and high storage capacity under isothermal conditions. Supercooling of phase change materials (PCM) during the transition process from liquid to solid is a major problem in thermal energy storage, which reduces energy efficiency and exacerbates energy wastage. This study focuses on the supercooling characteristics of PCM under heterogeneous nucleation, which provides a new idea to investigate the effect of nucleating agents on the supercooling rate of aqueous solutions. Understanding, predicting, and, where possible, preventing, or at least reducing, supercooling is particularly important for latent heat thermal energy storage systems, as the temperature differences within them must be small to achieve higher efficiencies. Water is the most popular PCM today because it has good characteristics despite high supercooling. Various PCM samples were prepared and tested in this study to reduce water supercooling. The sample comprises water and propylene glycol solution in various percentages (10/90, 20/80, and 30/70). The t-history method is applied to test the samples. The results showed that the water supercooling decreased with the increase in the percentage of propylene glycol solution in water, from 10 degrees to 6, 4, and 2 degrees, respectively, making it suitable for cooling applications.

Keywords Supercooling · Nucleated agent · Phase change material · Thermal energy storage

A. Winarta · A. Wibolo · I. Suamir · I. Rasta (✉)
Department of Mechanical Engineering, Politeknik Negeri Bali, Jln. Kampus Bukit Jimbaran, Kuta Selatan, Badung Bali 80364, Indonesia
e-mail: maderasta@pnb.ac.id

A. Winarta
e-mail: adi.winarta@pnb.ac.id

A. Wibolo
e-mail: wibolo@pnb.ac.id

I. Suamir
e-mail: nyomansuamir@pnb.ac.id

Irwansyah et al. (eds.), *Proceedings of the 4th International Conference on Experimental and Computational Mechanics in Engineering*, Lecture Notes in Mechanical Engineering, https://doi.org/10.1007/978-981-99-7495-5_30

1 Introduction

The technology of latent heat thermal energy storage has become one a popular research topic due to its huge benefit and massive industrial applications, including aerospace, air conditioning and refrigeration, solar thermal storage systems, and building heating. A large amount of literature on experimental and numerical studies has been published in the last few decades [1–3].

The thermal storage concept has been investigated since the 1970s [4]. After that, the phase change material (PCM) was introduced to improve the performance of thermal energy storage due to the massive energy density and minimum temperature changes during the phase transition [5].

PCM is divided into two categories: organic and inorganic. Inorganic PCM has the characteristics of a significant phase transition temperature range, large enthalpy value, and unstable energy output. In contrast, organic PCM has the characteristics of low heat storage density, does not corrode, and does not undergo supercooling [6]. The enthalpy of the phase transition of inorganic salts is large. Although its transition temperature is stable, the supercooling rate during the phase transition is significant. This inorganic salt is also corrosive to the storage material. Furthermore, phase separation occurs with increased phase change cycles [7].

There are several methods to hinder the PCM supercooling, such as nucleating agents [8–10], ultrasonic vibration [11–14], and mechanical vibration [15]. Adding nucleating agents and thickening agents to phase change materials are mostly adopted methods by researchers to reduce the rate of supercooling and phase separation [16–20].

Supercooling is a phenomenon in which a liquid freezes below its usual freezing point. The PCM is still liquid below its melting point, which can cause the material to fail to function as thermal energy storage. Hence, controlling supercooling becomes crucial in PCM technology to ensure that latent heat energy storage can be performed at isothermal temperatures.

The importance of supercooling phenomenon is reflected in the increasing number of published articles and patents which discuss supercooling in their topics over the last few decades. Figure 1 illustrates this trend based on Google Scholar data.

Depending on the given application, supercooling can be helpful or destructive. However, treating this phenomenon requires a good knowledge of its material behavior under different conditions. Supercooling is a prevalent phenomenon in nature and technological processes, but it is still difficult for researchers and developers to control.

Supercooling was introduced through its emergence in nature or everyday applications such as food preservation, water transition from liquid to solid, and many natural phenomena. Water is the most well-known PCM when studying supercooling in nature. The importance of considering this phenomenon in experimental and research work is explained by presenting its direct effect on application performance and efficiency.

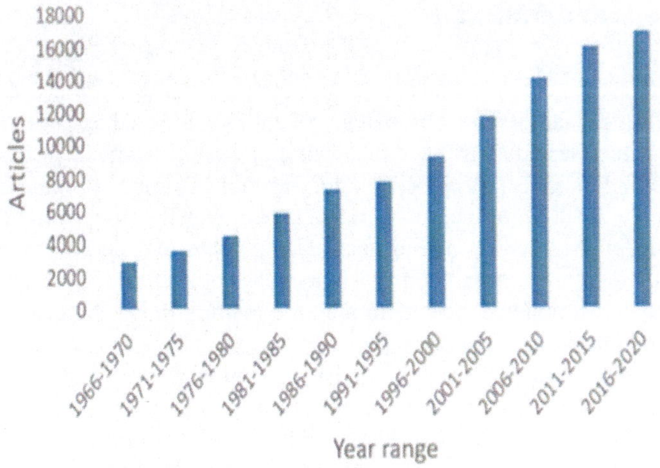

Fig. 1 Trend of supercooling handling in published article until 2020 [21]

Searching for low-cost materials that can be used for proper latent thermal energy storage has led researchers to study new materials or maximize the potential of known materials. The new phase change material should have better performance or thermal and chemical stability regarding PCM performance.

Thus, proper nucleation ability while releasing and absorbing latent heat is essential. In addition, the supercooled material must have enormous activation energy for nucleation to avoid spontaneous crystallization under supercooled conditions. Although several developers have provided well-established theories of meta-stable fluids, calculating thermophysical properties remains an open issue. Controlling the behavior of supercooled or liquids is a new challenge in technical applications. Therefore, understanding the crystallization process for supercooling control is critical to PCM research and technology advances.

Water is currently a well-known PCM because it has good characteristics. However, water has a high supercooling capacity, limiting its practical application as a thermal energy store. The scope of this paper is to select the best nucleating agent for use with water to optimize its performance as a phase change material by reducing its supercooling. The choice of nucleating agent was considered and compared with the results obtained to strengthen their effect.

This study aims to reduce the degree of supercooling of water by adding propylene glycol solution. Due to supercooling being a crucial parameter and critical issue from a practical point of view of thermal energy storage, methods for controlling supercooling are the basis for advanced thermal energy technology.

2 Methods and Materials

The material used as PCM in this study is water-based with various percentages (5, 10, 15, 20, 25 and 30%) of propylene glycol solutions. Those mixtures were prepared and then tested as PCM. After that, their result compared with common water tap to find the effect of glycol solution regarding to supercooling perceptive. There are a few methods of measuring the degree of supercooling, which basically measures supercooled liquids melting and freezing points with acceptable accuracy. One common measurement method is T-history, as depicted in Fig. 2. This simple method provided simplicity for determining the melting point, the heat of fusion, specific heat, and thermal conductivity of PCM with reasonable accuracy. Those mixtures were PCM samples (each of 10°C) were put inside the glass tube and then immersed in the water bath chamber, as shown in Fig. 2. The water chamber was cooled down from 25 to −20 °C with refrigeration until below the transition phase temperature. The temperature versus time curves of PCM samples were taken, and their thermophysical properties were obtained by comparing them with the other known reference materials (tap water) [22]. Zhang and Jiang measured the thermo-physical properties of several PCMs using the T-history method and found sufficient concordance between the results and the data available in the literature [23]. The detail of T-history method used in this work is explained clearly in [24]. Temperature measurement of the PCM sample performed using type K thermocouple and data acquisition with personal computer.

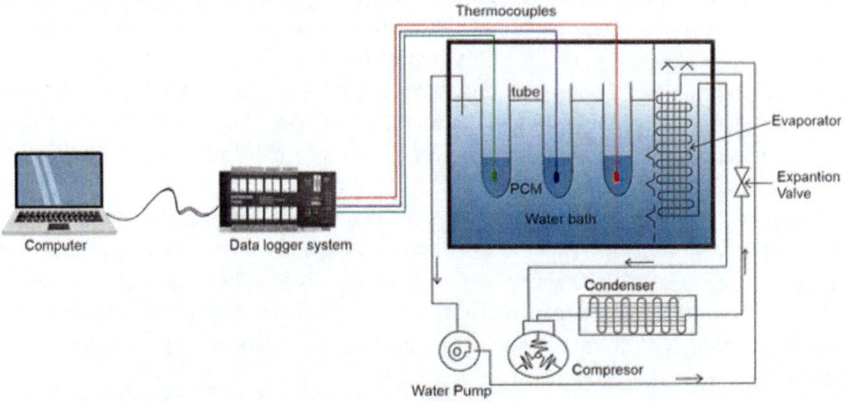

Fig. 2 T-history method schematic diagram [23, 24]

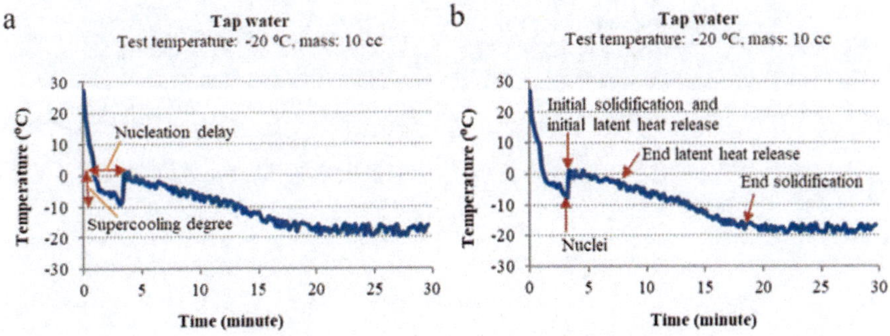

Fig. 3 **a** Temperature curve of liquid in different phase, **b** Freezing process of tap water with supercooling

3 Results and Discussion

3.1 *Supercooled Liquid Thermal Energy Storage*

Figure 3a shows the supercooling curve and the thermal behavior of the supercooled liquid for tap water before solidification begins (change from liquid to solid). Poor crystal growth or nucleation makes a deep supercooling process, which takes a lot of energy to perform the process. The graph proves that it takes longer for ice cores to form a bulk solid due to nucleation delay. In addition, the temperature of water falls below its melting point. The temperature of tap water rises and stabilizes at the melting point (T_m) immediately after crystallization. Figure 3b shows the detail explanation for transitions of water tap from liquid to solid. It shows that the temperature of liquid fall below the melting point before the nuclei begin to form. After that, the temperature of water tap increases suddenly for initial solidification and latent heat release. The water temperature is almost stable until the end of latent heat release when solidification occurs.

Figure 4 shows a supercooled or very cold liquid and its solid state. Liquids in which they remain in the liquid phase when cooled below their melting point temperature. Supercooled liquids require additional energy to release the stored latent heat. In addition, the supercooled material must have a large activation energy for nucleation and avoid spontaneous crystallization under supercooled conditions. For example, pure water can be supercooled to −41 °C, at atmospheric pressure in the laboratory without transitioning to the solid phase [24].

3.2 *Degree Supercooling*

The level of supercooling is the difference between the theoretical melting temperature of a substance and the lowest temperature reached by the liquid phase before

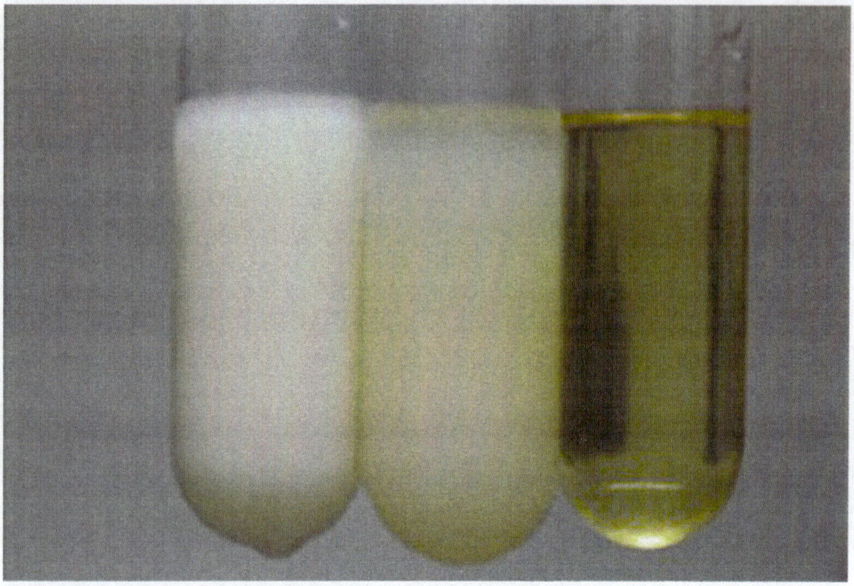

Fig. 4 Picture of supercoled liquid in different phase

the solidification process begins. In this period, the nucleation process or ice core formation occurs. Supercooling degree expressed as a hysteresis equation below [25]:

$$\Delta T_s = T_m - T_n \tag{1}$$

where ΔT_s refers to the degree of supercooling, T_m is the melting temperature, and T_n is the nucleation temperature. Reduced the difference between T_m and T_n would increase the effective heat capacity of supercooled liquids.

Figure 5 shows a graph of the temperature in the solidification process of water with the addition of propylene glycol percentages of 5, 10, 15 and 20% variations to tap water. Adding glycol solution with different percentages to tap water results in different temperature evolution, as shown in Fig. 5a–d. In general, the PCM solutions freezes after the cooling temperature are well below its typical freezing point temperature. The crystallization process goes through several phases: the induction phase, the crystal growth phase, and the crystal regrowth phase. During the induction phase, the nucleus is formed and grows to a sufficient size to become stable (a nucleation center is formed). Furthermore, PCM crystals diffuse toward the nucleus to be adsorbed on its surface and grow the nucleus. The adsorbed material migrates along the surface and is incorporated into a crystalline form. These tiny crystals continue to grow and eventually become large enough and large enough to sustain the rapid rate of crystal growth. The rate of crystallization slows down as the freezing process is almost complete. Even after the material is entirely frozen, the redistribution process continues to modify the particle shape and size distribution.

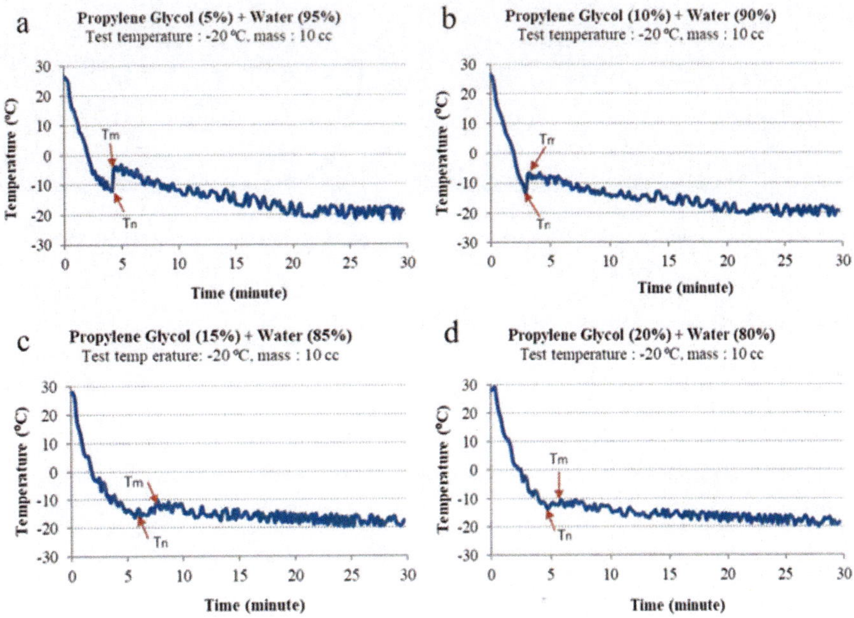

Fig. 5 Crystallization behavior of PCM samples 5, 10, 15 and 20% of propylene glycol in water

Adding propylene glycol to water causes the freezing point of water to decrease, and the degree of supercooling also decreases. Adding various percentages of propylene glycol solution (5, 10, 15 and 20%) into water causes the freezing point of water to decrease to 5, 8, 11 and 13 °C. The same thing also happened to the degree of supercooling, where there was a decrease in the degree of supercooling to 8 K, 7 K, 4 K, and 1 K, respectively.

Meanwhile, with 25 and 30% propylene glycol addition to the water, there is no state change from a liquid to a solid (solidification), as shown in Fig. 6. The low degree of supercooling is probably due to the solution's freezing point being lower than the test temperature. No particles can trigger nuclei formation for crystallization growth. Furthermore, the testing time is not enough to achieve solidification. Therefore, the material only stores sensible heat.

Figure 7a shows the freezing temperature of water and solutions of various percentages of propylene glycol in water. Figure 7b shows the degree of supercooling of water and some aqueous propylene glycol solutions. In detail, the freezing temperature, degree of supercooling of water, and various percentages of propylene glycol in water are given in Table 1.

Experimental results show that adding propylene glycol to water causes the freezing temperature and degree of supercooling of water to decrease. The most exciting result is that the addition of 20% propylene glycol in water causes the supercooling of water to decrease drastically, even minimal (1 K), making it very suitable for low-temperature thermal energy storage applications (refrigeration).

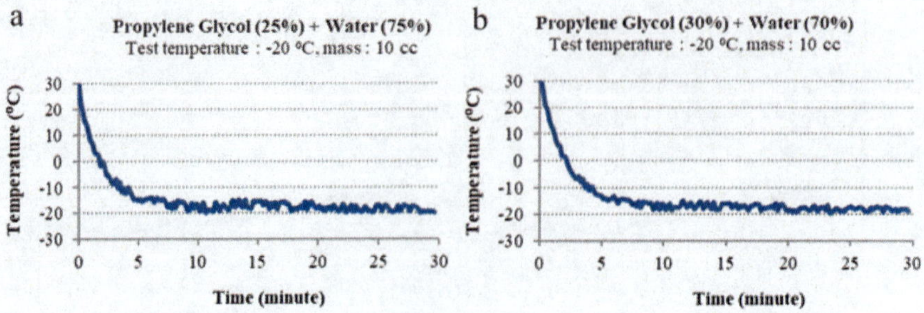

Fig. 6 Cooling curve 25 and 30% propylene glycol in water

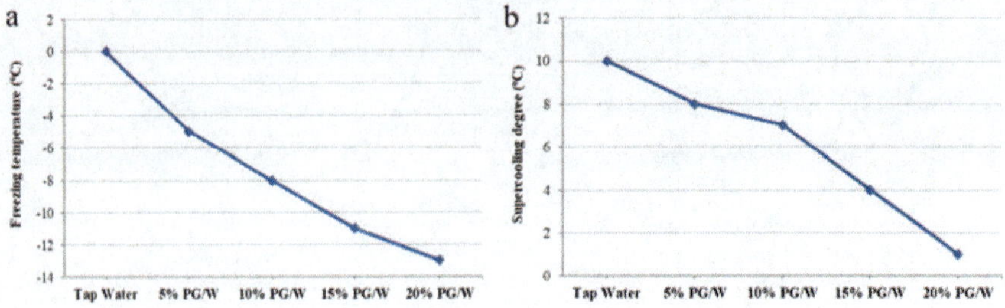

Fig. 7 **a** Freezing temperature of tap water and various percentages of propylene glycol in water. **b** Supercooling degree of tap water and various percentages of propylene glycol in water

Table 1 PCM thermal properties of tap water and propylene glycol mixtures in water

Sample (Vol.%)	T-history	
	Cooling process	
	Freezing temperature (Tm, °C)	Supercooling degree (K)
Tap water	0	10
5/95 (PG/W)	5	8
10/90 (PG/W)	8	7
15/85 (PG/W)	11	4
20/80 (PG/W)	13	1
25/75 (PG/W)	No solidification	-
30/70 (PG/W)	No solidification	-

PG/A = Propylene glycol in water

4 Conclusion

The efficiency of a phase change thermal energy storage device is directly related to its thermophysical properties. New blends of pure PCM and additives are being developed for thermal energy storage. These materials solutions are important areas of research today. Most methods for improving PCM properties, production, and cost-effectiveness are still in the testing phase. However, the findings show they are very promising for the thermal energy storage industry.

This study is focused on increasing the thermal storage capacity of PCM. The thermal storage capacity of PCM can be increased by using suitable additives in appropriate proportions. However, improper concentration can lead to a reduction in heat storage capacity.

Adding suitable additives or nucleating agents to PCM is one of the most frequently used methods to reduce the degree of supercooling because it can induce nucleation. The test results of several water-based PCM samples developed with the addition of propylene glycol additives can reduce and even eliminate the degree of supercooling of water and lower the freezing point of water, making it suitable for cold storage applications. The degree of supercooling depends on the container's shape and size, the PCM's composition, heating and cooling rates, heat transfer area, and other related areas.

Acknowledgements The authors acknowledge the financial support from Politeknik Negeri Bali through DIPA PNB contract number **SP DIPA-023.18.2.677608/2022.**

References

1. Farid MM, Khudhair AM, Razack S, Al-Hallaj S (2004) A review on phase change energy storage: materials and applications. Energy Conver Managem 45(9):1597–1615. https://doi.org/10.1016/j.enconman.2003.09.015
2. Rehman TU, Ali HM, Janjua MM, Sajjad U, Yan WM (2019) A critical review on heat transfer augmentation of phase change materials embedded with porous materials/foams. Int J Heat Mass Transf 135:649–673. https://doi.org/10.1016/j.ijheatmasstransfer.2019.02.001
3. Sharma A, Chauhan R, Kalliolu MA, Chinnasamy V, Singh T (2020) A review of phase change materials (PCMs) for thermal storage in solar air heating systems. Mater Today Proc 44:4357–4363. https://doi.org/10.1016/j.matpr.2020.10.560
4. Xu J, Wang RZ, Li Y (2014) A review of available technologies for seasonal thermal energy storage. Sol Energy 103:610–638
5. Zhang YP, Ding JH, Wang X, Yang R, Lin KP (2006) Influence of additives on thermal conductivity of shape stabilized phase change material. Solar Energy Mater Solar Cells 90(11):1692–1702
6. Swa B, Tya B, Zka B, Wpa B (2020) Thermal conductivity enhancement on phase change materials for thermal energy storage: a review. Energy Storage Mater 25:251–295
7. Wang Q, Wang J, Chen Y, Zhao CY (2019) Experimental investigation of barium hydroxide octahydrate as latent heat storage materials. Sol Energy 177:99–107
8. Hu P, Lu DJ, Fan XY, Zhou X, Chen ZS (2011) Phase change performance of sodium acetate trihydrate with AlN nanoparticles and CMC. Solar Energy Mater Solar Cells 95:2645–2649

9. Garay Ramirez BML, Glorieux C, San Martin Martinez E, Flores Cuautle JJA (2014) Tuning of thermal properties of sodium acetate trihydrate by blending with polymer and silver nanoparticles. Appl Therm Eng 62:838–844
10. Wei LL, Ohsasa K (2010) Supercooling and solidification behavior of phase change material. ISIJ Int 50:1265–1269
11. Inada T, Zhang X, Yabe A, Kozawa Y (2001) Active control of phase change from supercooled water to ice by ultrasonic vibration 1. Control of freezing temperature. Int J Heat and Mass Transf 44:4523–4531
12. Zhang X, Inada T, Yabe A, Kozawa Y (2001) Active control of phase change from supercooled water to ice by ultrasonic vibration 2. Generation of ice slurries and effect of bubble nuclei. Int J Heat and Mass Transf 44: 4533–4539
13. Miyasaka E, Takai M, Hidaka H, Kakimoto Y, Hirasawa I (2006) Effect of ultrasonic irradiation on nucleation phenomena in a $Na_2HPO_4 \cdot 12H_2O$ melt being used as a heat storage material. Ultrasonic Sonochem 13:308–312
14. Zhang X, Cai L, Zhongjie SU, Ping C, Zhong Y (2010) Effects of ultrasound on phase separation and crystallization of sodium acetate trihydrate. CIESC J 61:104–108
15. Pan LH, Huang LW, Qiao Y, Zhao J, Jiang M (2008) Influence of vibration on the supercooling relex of inorganic salt solution as a phase change material. J Zhejiang Univer Technol 36:655–658
16. Fang Y, Jin C, Liang X, Gao X, Zhang Z (2015) Preparation and performance of sodium acetate trihydrate/formamide composite phase change material. CIESC J 66:5142–5148
17. Fu W, Zou T, Liang X, Wang S, Gao X, Zhang Z, Fang Y (2018) Thermal properties and thermal conductivity enhancement of composite phase change material using sodium acetate trihydrate–urea/expanded graphite for radiant floor heating system. Appl Therm Eng 138:618–626
18. Dongling WU, Tingxian LI, Feng HE, Wang R (2018) Preparation and performance of modified sodium acetate trihydrate composite phase change material for thermal energy storage. CIESC J 69:2860–2868
19. Jintian L, Jinfeng M, Weihua L, Jing L, Xiaoyan X (2009) Supercooling mechanism and experimental study of sodium acetate trihydrate. J Refrigeration 30:32–35
20. Jintian L, Jinfeng M, Jing L, Shiguang YA (2011) A selection and optimization experimental study of additives to thermal energy storage material sodium acetate trihydrate. J Function Mater 42:144–147 (In Chinese)
21. Pennec SF, Biwole P, Fardoun P (2022) Supercooling of phase change materials: a review. Renew Sustain Energy Rev 158:112172
22. Zhang Y, Zhou G, Lin K, Zhang Q, Di H (2007) Application of latent heat thermal energy storage in buildings: state of the art and outlook. Build Environ 42:2197–2209
23. Yinping Z, Yi J (1999) A simple method, the-history method, of determining the heat of fusion, specific heat and thermal conductivity of phase-change materials. Meas Sci Technol 10(3):201
24. Rasta IM, Suamir IN (2018) The role of vegetable oil in water phase change materials for medium temperature refrigeration. J Energy Storage 15:368–378
25. Awasthi A, Ali Shah N, Oh Kyung Kwon YJ, Chung JD (2022) Supercooling effects and solidification of water inside a horizontal cylinder with a rough, sinusoidal surface shape. J Energy Storage 51:104442
26. Beaupere N, Soupremanien U, Zalewski L (2020) Experimental measurement of the residual solidification duration of supercooled sodium acetate trihydrate. Int J Therm Sci 158:106544

Simulation on Compressive Testing Softwood Materials Using Finite Element Analysis

Z. Salleh, M. F. M. Nor, and M.A.A.Zullastri

Abstract Nowadays, in many industries such as marine, aerospace and civil, the utilization of several materials must undergone a product testing and one of it is a compression test. The purpose of the compression test is to clarify whether the material possesses excellent mechanical properties or vice versa peculiarly with regard to the compressive behaviour. The compression test will determine the compressive strength, elastic modulus as well as the specific compression of the material. In this research, the simulation of the compression test of the activated carbon composite will be executed via Finite Element Analysis. The activated carbon that will be used in this research is softwood materials such as Mangrove Bark Activated Carbon (MBAC), an eco-friendly type activated carbon since it is derived from the mangrove bark of the mangrove tree. The model of the activated carbon composite with different diameter will be tested on the simulation software to compare the compressive strength of each models.

Keywords Finite element analysis · Activated carbon · Mangrove Bark

1 Introduction

According to (Taylor and Francis 2005), activated carbon, occasionally to be known as activated charcoal, is a kind of unique adsorbent due to its extremely permeable structure that permits it to effectually lay hold of the materials [1]. In addition, the activated carbon had been widely used throughout a number of industries. Activated carbon is a carbon-based material that can be procured from carbon-rich organic materials including [2]:

- Bamboo stem

Z. Salleh (✉) · M. F. M. Nor · M.A.A.Zullastri
Universiti Kuala Lumpur, Malaysian Institute of Marine Engineering Technology, Bandar Teknologi Maritim Jalan Pantai Remis, 32200 Lumut Perak, Malaysia
e-mail: zulzamri@unikl.edu.mymail

Irwansyah et al. (eds.), *Proceedings of the 4th International Conference on Experimental and Computational Mechanics in Engineering*, Lecture Notes in Mechanical Engineering, https://doi.org/10.1007/978-981-99-7495-5_31

- Mangrove bark
- Oil palm empty fruit bunch (EFB)
- Coconut shells

The physical properties as well as the chemical properties of the activated carbon can differ significantly depending on the source of the material and also the processing techniques implemented to produce the activated carbon [3]. Due to this, commercially produced activated carbons are exceptionally particular to accomplish the best outcomes for a given application. MBAC stands for Mangrove Bark Activated Carbon is an activated carbon that is procured from the mangrove bark of the mangrove tree. Mangroves are one kind of tropical trees that survive in climate most timber could never bear; salty, coastal waters, and the tide flow. Mangrove trees also has the ability to store enormous amounts of carbon (Fig. 1).

Mangroves are usually make a habitat along sheltered coastlines whether in the tropics or sub- tropics as they fulfil important socio-economic as well as environmental responsibility. In 1980, as a part of FAO/UNEP Tropical Forest Resources Assessment, the total mangrove population were approximated for the first time and estimated that 15.6 million hectares are concluded as a world total of the mangrove area [4] (Fig. 2).

There are many advantages of mangrove tree. From the roots, leaves, bark, all have their beneficial function whether towards the human or environment. One of the popular utilisations of the mangrove is the benefit of the mangrove bark. From the mangrove bark, the activated carbon is procured through a string of heating times and also the suitable temperatures [6]. In this research, the material is softwood

Fig. 1 Mangrove forests

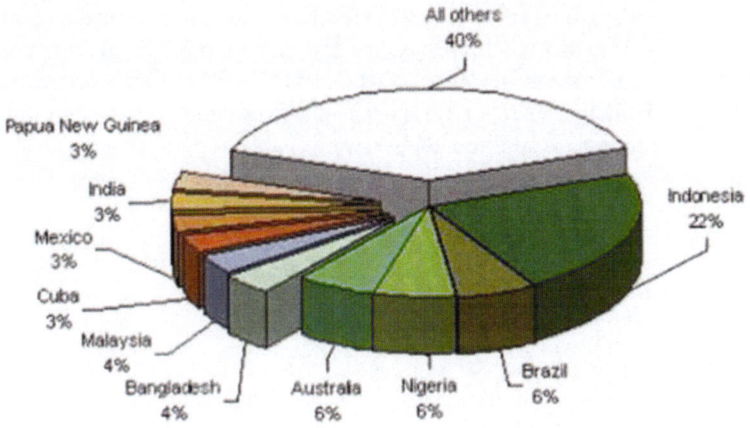

Fig. 2 Mangrove area extent worldwide [5]

which is similar properties with MBAC properties. Hence, the material properties data are also available in CREO simulation database. The softwood properties such as Tensile strength: 104 MPa, Bending strength: 87 MPa, Shear strength: 10 MPa, Impact strength: 70 kJ/m^2, Hardness: 4Brinell and Modulus Elasticity: 12000 MPa. The similar materials also show that MBAC with different weight percentage (wt.%) can reduce the barnacle growth when applied for marine applications [7].

In this research, the finite element analysis software will be used to conduct the simulation. Finite Element Analysis or also recognized as Finite Element Method (FEM) is a popular method used in science and engineering. As highlighted by (Rao 2004), the FEM has been applied widely as a solution to variety of problems that arise in applied science and engineering [8]. Since the experimental testing required a lot of time investment and resources, FEM is an alternative and better option since FEM delivered much more detailed set of results and furthermore it is less time consuming and less expensive. This is very useful when we had global disease issues such as COVID-19 pandemics has been attacked around the world since 2020. While we are using this simulation and modelling, it will reduce our attended to make the experimental work physically but can be alternate wit simulation and modelling research work. There are a lot of FEM advantages for simulation and modelling particularly related with structural analysis for engineering applications.

The list of the advantages as bellows:

(a) Comprehensive result sets, producing the physical reaction of the framework at any area, counting a few which might have been ignored in an expository approach.
(b) Secure simulation of possibly unsafe, damaging or illogical stack conditions and disappointment modes.
(c) The concurrent calculation and visual representation of a wide assortment of physical parameters like stress or temperature, empowering the architect to quickly examine execution and conceivable adjustments.

There are several software's are available for the Finite Element Analysis software such as ANSYS, Strand 7, CREO, Maxsurf, and many mores. In this research, the simulation for models will be used PTC CREO. PTC CREO has been widely used as a modelling tool whether in 2D or 3D shape and also provide every design-through-manufacturing process. In addition, the software is able to perform an analysis with ease.

2 Methodology

The actual size for compression samples has been used in this research and converted into modelling for FEA. After the design of the model has been imported into CREO Software, the design need to be set according to the below procedures before running the full analysis. This is benefit for any simulation for any mechanical testing could be performed before it is compared with actual axial load experiment such as flexural testing [9]. The procedures are repeated with remaining designs of the model shows in Figure 3 for details.

There are five models to be design and simulated and it is followed according to the ASTM (D695) standard with differences diameter is shown in Fig. 4:

The samples have been develop based on the different diameter ranging from 6 to 102 mm. The details for the specification can be referred in Table 1. The process set-up for the samples preparation are shows bellows:

(a) Step 1: Set the load and constraint
(b) Set the load to 10kN and the constraint opposite to the load.
(c) Step 2: Assign the materials into the model
(d) The materials that being assigned from the database into the model is a softwood pine which has similar mechanical properties to MBAC.
(e) Step 3: Refine the model by using AutoGEM shows in Fig. 5.

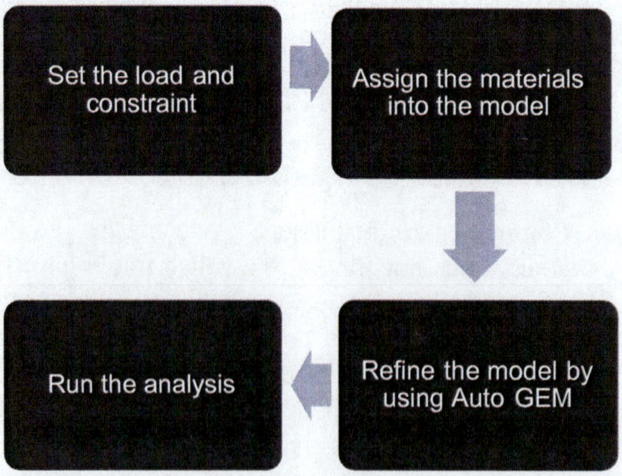

Fig. 3 Simulation procedures before running

Fig. 4 Cylinder model

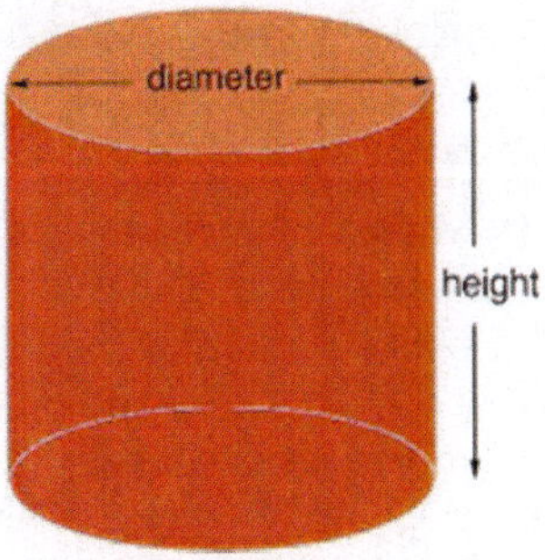

Table 1 Specification for compression samples

Model	Height (mm)	Diameter (mm)
1	50.8	6.35
2		12.70
3		25.40
4		50.80
5		101.60

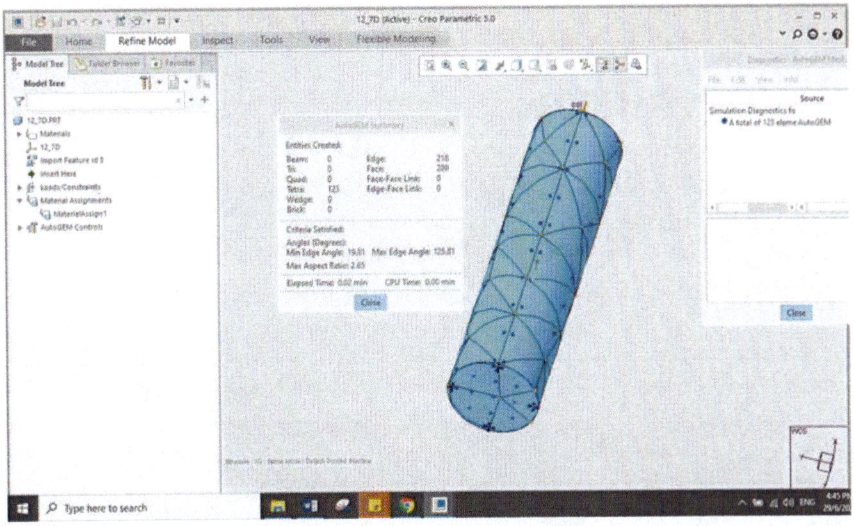

Fig. 5 Overview the AutoGEM summary

(f) Step 4: Run the analysis
(g) Steps 1 until 4 were repeated for all five models. The results are summarised in Results and Discussion part.

3 Result and Discussion

In this research, there are five models of data or result that can obtained which is stress, strain and displacement according to the Table 1. All the data or result for each specimen was collected and tabulate. The results from the Table 1 clarify that the stress experienced by the model of larger diameter which is model 5 has lesser compression stress than the other remaining models. The maximum compression von misses stress that the model 1 experienced is only 435 MPa. Furthermore, model 5 also has the lowest maximum principal strain compared to the other models which is 202×10^{-6} MPa. The details of the all models showed in Fig. 6.

From Figs. 7 and 8, there are differences of value of stress, strain, and displacement between all the five specimens. When the diameter changes with percentage of the mangrove wood activated carbon increase, all the value of maximum stress, maximum strain, and maximum displacement also show decreased when the diameter is larger size given in Table 2 for details. The results from the Table 2 clarify that the stress experienced by the model of larger diameter which is model 5 is lesser than the other remaining models. The maximum stress that the model 5 experienced is only 4.648 MPa. Furthermore, model 5 also has the lowest maximum principal strain compared to the other models which is 202×10^{-6}. The displacement of the model 5 also is at the lowest which 5.24×10^{-6} mm. The larger diameter of the model deliver the larger compressive strength compared to the other models since it experienced a small stress, strain, and displacement, making the model higher resistance

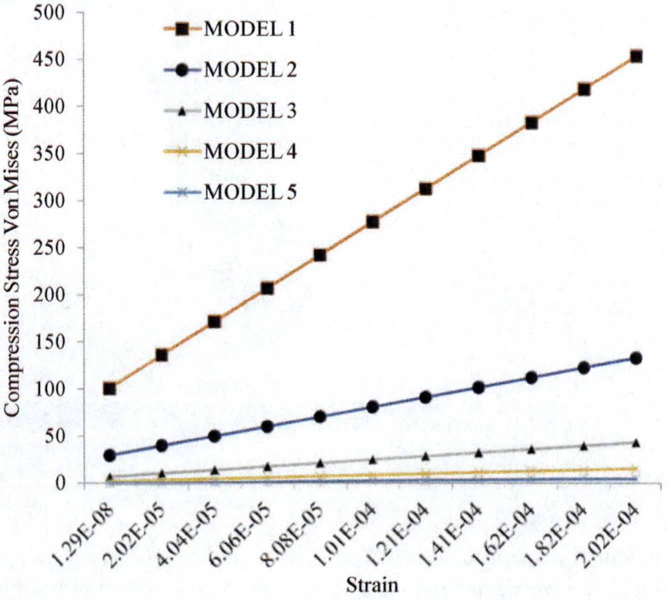

Fig. 6 Graph for compression stress–strain

to breaking under compression. The result also shows the isotropic materials such as softwood could similar behaviour when compressive testing has been performing [10].

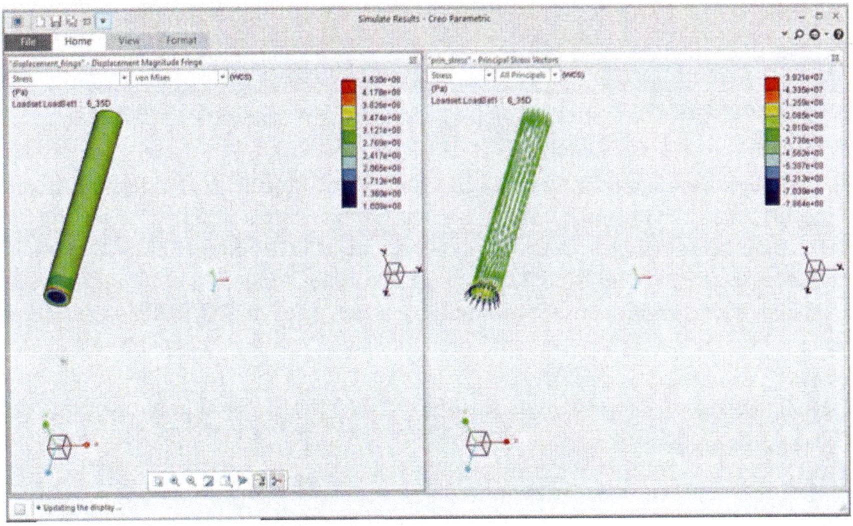

Fig. 7 Specimen with H: 50.8 mm, D:6.35 mm

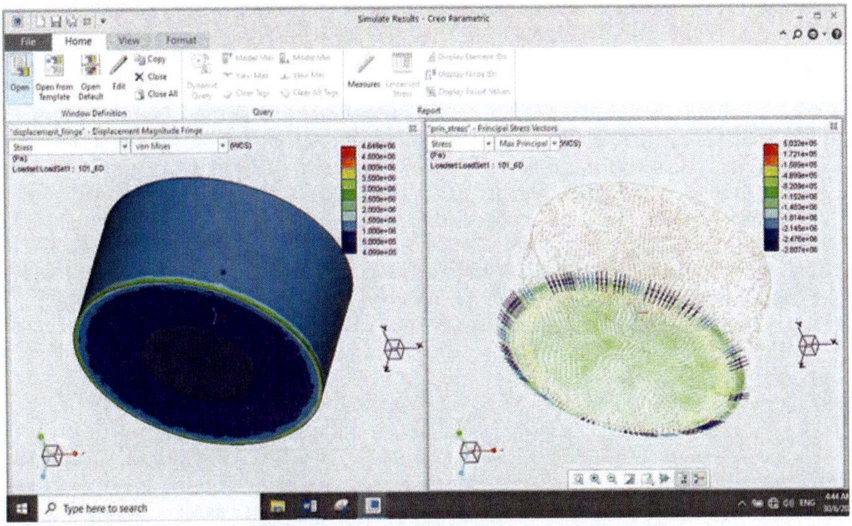

Fig. 8 Specimen with H:50.8 mm, D: 101.6 mm

Table 2 Comparison data for all specimens

Model	Max compression stress (MPa)	Max compression strain	Max displacement (mm)
1	2.77×10^{2}	9.85×10^{-2}	9.46×10^{-04}
2	8.12×10^{1}	2.98×10^{-03}	2.35×10^{-04}
3	2.51×10^{1}	9.86×10^{-04}	5.83×10^{-05}
4	8.40	6.00×10^{-03}	1.45×10^{-05}
5	2.51	1.01×10^{-04}	5.24×10^{-06}

4 Conclusion

This research had successfully satisfied all the three objectives. Firstly, the model of the specimens had been developed and design by using a design software which one of models following the ASTM D695 Standard. The simulation test also had been successfully executed and the comparison of the compressive strength of each models also had been done in an orderly manner. The MBAC model with larger diameter possess larger compressive strength since the stress it experienced lesser compared to the other MBAC models. In overall, although the magnitude of the load is same, the area of force applied are different. The more area of the force, the less stress experienced to the subject.

Acknowledgements The author would like to thank the Ministry of Higher Education (MoHE), Malaysia, for providing a grant (FRGS/1/2021/TK0/UNIKL/02/20) to the first author in carrying out and completing this work.

References

1. Bansal RC, Goya M (2016) Activated carbon adsorption. Boca Raton, Taylor & Francis Group, 2005. Google Books, Web, September 19
2. Abdul Khalil HPS, Jawaid M, Firoozian P, Alothman OY, Paridah MT, Zainudin ES (2014) Flexural properties of activated carbon filled epoxy nanocomposites. The Malaysian J Anal Sci 18:391–397
3. Ebbenis A, Carlson C (2019) Activated carbon: a growing market of opportunity, September 6 Retrieved from https://feeco.com/activated-carbon-a-growing-market-of-opportunity/
4. FAO (2003) Status and trends in mangrove area extent worldwide. In: Wilkie ML, Fortuna S (eds) Forest resources assessment working paper no. 63. Forest Resources Division, FAO, Rome
5. Clough BF (1993) The Economic and environmental values of mangrove forests and their present state of conservation in the South-East Asia/Pacific Region. Mangrove Ecosystems Technical Reports vol.3 ITTO/ISME/JIAM Project PD71/89. Rev. 1(F) Okinawa, Japan, ISME. pp 202
6. Zulkarnain Z, Hussein MZ, Badri M (1993) Activated carbon from mangrove wood (Rizophora apiculata): preparation and characterization. Pertanika. J Sci Technol 1(2):169–177

7. Salleh Z, Mazlan TMIAT (2022) Study on mangrove barks activated carbon (MBAC) for fibre reinforce plastic (FRP) rehabilitation pile structure. Springer book chapter: technological advancement in instrumentation and human engineering, vol 882. pp 327–338
8. Rao A, Khanna R, Gangopadhyay M (2012) Applications of finite element methods (FEM)—overview. In: International conference on mathematical sciences, Dec 28–31, 2012, Nagpur, India
9. Salleh Z, Kee GW (2021). Simulation of three-point bending sandwich composite panels through finite element analysis. Springer book chapter: design in maritime engineering contributions from the ICMAT 202, vol 167. pp 35–44
10. Salleh Z, Islam MM, Epaarachchi JA (2015) Compressive behavior of low density polymeric syntactic foams. Appl Mech Mater 799–800, 135–139. Submitted: 2015–06–26 © (2015) Trans Tech Publications, Switzerland

Object Classification System Using Convolutional Neural Network for Several Environmental Conditions

Udink Aulia, Iskandar Hasanuddin, Muhammad Dirhamsyah, and Nasaruddin Nasaruddin

Abstract At this time, autonomous mobile robots are used to move groceries. Therefore, object classification is required to move groceries to avoid collisions with other objects. Object classification using a convolutional neural network (CNN) can significantly increase mobile robot autonomy. Therefore, this paper investigates object classification performance using CNN for several environmental conditions. The research method is computer simulation using several steps: image collection, image annotation, and training. SSD Mobilenetv2FPNLite was used as a CNN model for object classification. The objects for the classification using the pre-trained CNN are cars, motorcycles, persons, and goods rickshaws. The results enabled us to estimate the effectiveness of using pre-trained CNN for classifying different objects. The result also shows that objects can be recognized with a classification accuracy of 100%.

Keywords Autonomous mobile robot · Convolutional neural network · Object classification · Conditions

U. Aulia (✉)
School of Engineering, Universitas Syiah Kuala, Banda Aceh 23111, Indonesia
e-mail: uaulia2@unsyiah.ac.id

U. Aulia · I. Hasanuddin · M. Dirhamsyah
Department of Mechanical and Industrial Engineering, Universitas Syiah Kuala, Banda Aceh 23111, Indonesia
e-mail: iskandarhasanuddin@unsyiah.ac.id

M. Dirhamsyah
e-mail: mdirhamsyah@unsyiah.ac.id

N. Nasaruddin
Department of Electrical and Computer Engineering, Universitas Syiah Kuala, Banda Aceh 23111, Indonesia
e-mail: nasaruddin@unsyiah.ac.id

Irwansyah et al. (eds.), *Proceedings of the 4th International Conference on Experimental and Computational Mechanics in Engineering*, Lecture Notes in Mechanical Engineering, https://doi.org/10.1007/978-981-99-7495-5_32

1 Introduction

The increased need for service robots has motivated academics to enhance technology for automating these platforms. Their navigation system is one of the most crucial technologies for automating these platforms [1]. Autonomous mobile robot (AMS) navigation is an active research topic that has piqued many researchers' interest in recent years [2]. AMRs are developed and programmed to replace people in various jobs. Humans are typically unable to undertake such duties for a variety of reasons, including poor environmental circumstances, harmful and dangerous surroundings, and repeated work that can be tedious. Due to these constraints and the necessity for robots under challenging situations, there has been a great study on mobile robots in recent years. The goal is to have such devices that can do most jobs automatically and without human interaction. Various hurdles accompany all of the benefits that can be obtained by AMR use. The key issues with these robots' self-controlling systems are localization and navigation.

Hence, emerging AMRs employ cameras and deep learning methods for object detection and classification [3]. These techniques identify probability linked to each bounding box that encloses identified objects and predicts their bounding boxes. Convolutional neural network (CNN)-based techniques, particularly, have demonstrated promising outcomes in detecting pedestrians, cars, and other objects [4]. The method of object detection resembles that of picture classification. Only that image classification seeks to determine a class for the entire image. Object detection, on the other hand, seeks to identify a particular position of each class and mark it with a bounding box and name. Detection by categorization is the common name for this process.

In poor lighting situations, especially early in the day, during the day, in the evening, and at night, the quality of the visual signals that autonomous cars acquire might be compromised and distorted. The capacity of the autonomous mobile robot to recognize important things in the surroundings may suffer significantly under such circumstances, which reduces scene contrast and visibility. An object's geometrical representation can be used to identify it to the human eye. Our capacity to discern distinctions in hue and light intensity is used to comprehend geometry. Similar to this, a sudden shift in hue in computer vision may point to a particular geometry of the object being seen. How well the geometry can be deciphered will be greatly influenced by the quantity of light reflected by an item and detected by the camera sensors.

Deep learning is a subclass of machine learning that has shown to be quite effective in various applications such as robotics. This success and advancement in machine learning have inspired academics to examine AMR localization in outdoor contexts using deep learning algorithms. Most of these systems are based on vision-based localization, which is often implemented as image discovery, which discovers the picture in the database with the most similar characteristics to the query image [5]. Many approaches have been successfully implemented to tackle the localization problem using the deep learning methodology; nevertheless, they have limitations due to the restricted size of the private datasets. Deep learning models have millions of parameters, and training them with minimal datasets might result in overfitting. Even though transfer learning and learned weights from previously trained models can help to alleviate this problem, the model still has finite accuracy due to insufficient generalization capabilities.

In this paper, we focus on determining the influence of light consistency on the performance of CNNs in classifying objects on the road in Banda Aceh City. We collected images under three lighting conditions. We also studied transfer learning from datasets into CNN frameworks used in classifying the object.

2 Related Work

Recently, many models have been put up that employ these layers to carry out the classification task. The Rectified Linear Unit (ReLU), a pioneer model that was proposed to increase accuracy and performance, is the AlexNet [6]. ReLU has the benefit of significantly accelerating convergence during network training. Later, factorized convolutions were included in the VGG model [7], which enhanced CNN-based models. The convolutional layers of VGG employ smaller receptive windows, such as kernels or filters, which results in a smaller network and better performance. However, embedded systems cannot benefit from the huge parameter size. Google researchers responded by publishing a MobileNet version of their design [8]. MobileNet was specifically created for embedded or mobile devices with more stringent hardware requirements. With this methodology, training time was cut, while performance was elevated. The accuracy of MobileNet on the ImageNet dataset was 87%, and it had a total of 28 layers.

The introduction of one-stage detectors established the next major milestone in object detection. The key idea of a one-stage detector is to avoid region proposals that consume time and perform detection in one step. You Only Look Once (YOLO), proposed by Joseph et al. [9] in 2015, improved detection speed. YOLO achieved 45 fps at an accuracy of 63.4% mAP on the VOC07 dataset. By reducing the convolutional layers, the fastest version of YOLO achieved 155 fps at an accuracy of 52.7% mAP evaluated on the same dataset.

The second one-stage detector that appeared was the Single Shot MultiBox Detector (SSD). W. Liu et al. proposed an SSD approach in 2015[10]. The main contribution is related to introducing multi-reference and multiresolution detection techniques. These techniques significantly improved the detection accuracy of small objects. SSD achieved a speed of 59 fps at an accuracy of 76.8% mAP evaluated on the VOC07 dataset. However, one-stage detectors could not compete in accuracy with two-stage detectors until Y. Lin et al. discovered that the foreground–background class imbalance experienced during training was a reason for low accuracy. Researchers proposed RetinaNet as a solution in 2017 [11]. RetinaNet utilizes focal loss that focuses more on misclassified background examples during training. The suggested technique allows one-stage detectors to retain very fast detection speeds while achieving accuracy equivalent to two-stage detectors.

The fact that training data are often obtained in dry weather with high visibility conflicts with the necessity that autonomous mobile robot systems perform dependably in various lighting conditions. As a result, algorithms for classifying objects perform worse under difficult lighting circumstances. A straightforward approach to address this problem is to train a given CNN to classify objects using images captured in some environmental lighting conditions. Unfortunately, as we highlighted earlier, sufficient annotated datasets captured in realistic urban environments in Banda Aceh City in natural lighting conditions are not readily available. Moreover, small datasets inherently make them inadequate to train deep learning architectures for object classification reliably.

3 Method

3.1 Image Collection

We used a camera for image collection. The camera provides a 12-megapixel resolution. This image collection configuration captured detailed visual information about the vehicles and persons while allowing image collection to occur under natural lighting conditions without much influence from the shadow of the image collection system. Image collection took place in Banda Aceh City. The light factor is measured for each condition (morning, afternoon, and evening) and expressed in light units.

3.2 Image Annotation

Since images have a dimension of 4160×3120, which is too large for neural network training, the images were divided into 800×600 pixels. The object types in the image set were grouped into four categories: car, motorcycle, person, and goods rickshaw. The last category contains a new object not pre-trained on the CNN model.

3.3 Training

A large dataset and training hardware are needed to create a CNN model from scratch. The use of transfer learning and augmentation is one method for overcoming this weakness. Transfer learning involves retraining a model for a comparable job after it has been trained on a huge dataset like ImageNet or COCO. The TensorFlow 2 Detection Model Zoo is a collection of several contemporary models and modeling approaches for TensorFlow users. Code implementations and pre-trained models from disseminated research articles may be found in the Tensorflow Detection Model Zoo's Object Detection directory. These models, which were trained using the COCO dataset, may be adjusted and utilized for transfer learning. In this work, the input model was changed to 800 × 600 to avoid image degradation. Moreover, it also avoided losing the information within the images. The architecture training was undertaken in Google Colab, providing free GPU access for speeding up the training time, with time restrictions on use.

4 Result and Discussion

This section presents the results obtained by training selected CNN models. A set of experiments described in the previous section has been prepared and evaluated to understand the impact of several lighting environmental conditions on object classification. The test was carried out to obtain qualitative data on the effect of light conditions on the ability of the CNN SSD Mobilenetv2 FPN lite model for object classification. The tested images have a resolution of 800 × 600 pixels and are in color. The parameters tested are object names and classification accuracy for four objects, including cars, motorcycles, people, and goods rickshaws. Environmental conditions that will be tested on the ability to classify objects are in the morning, afternoon, and evening.

4.1 Morning Environmental Condition

The image was taken at the Lam Dingin location in the morning at 9.45 am. The objects that can be seen in Fig. 1 consist of 2 car objects and one goods rickshaw object. The weather conditions are clear, and the shadows have appeared. The position of the car object is from the rear right side, and the goods rickshaw object is from the left side.

Image testing on the model's ability to classify objects for morning conditions shows that the CNN model can recognize objects such as cars and goods rickshaws. In Fig. 1, the car object, which consists of a light-colored car and a dark-colored car from the rear right side, can be classified as two pieces with a classification accuracy of 81 and 98%. The vehicle's color affects detection accuracy; for light colors, the classification accuracy reaches 98%, and for dark cars, the detection accuracy is 81%. The car's size shows the car's distance to the camera, and the CNN model can classify objects correctly even though the pixel size of the light color car is 200 × 100 pixels and the dark color is 250 × 150. On a dark car, a small part of it is covered by a board on the back; objects can still be recognized with a classification accuracy of 98%. The goods rickshaw is a new object that has never been trained before. With a large number of datasets, it can be classified with 100% accuracy in the left-side

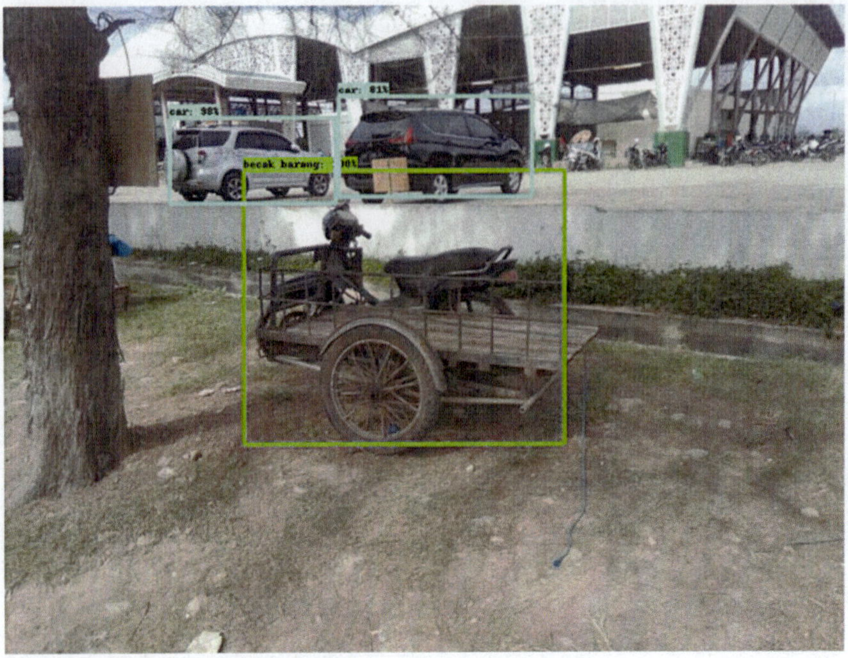

Fig. 1 Classification of objects in morning conditions

Table 1 Object name and classification accuracy in morning condition

No.	Object name	Classification accuracy (%)
1	Car	81
2	Car	93
3	Goods rickshaw	100

position. It produces a classification accuracy of up to 100%. The results of object classification and classification accuracy are displayed in Table 1.

4.2 Afternoon Environmental Condition

In afternoon conditions, the image was taken at the Batoh location at 11.16 am. The objects seen in Fig. 2 are one car object, two-person objects, and two motorcycle objects. Sunny weather conditions and shadows dominate the road. The object's position is the car from the rear, the person, and the motorcycle from the rear.

Figure 2 depicts the car parts as seen from behind. Even if the shadow on the bottom of the car is nearly indistinguishable from the black color of the wheels, the simulation test can be classified with 100% accuracy. Motorcycles are recognized

Fig. 2 Classification of objects in afternoon conditions

with 100% accuracy, and 99% on the rear left position. The shadow on the underside of the motorcycle does not reduce detection accuracy. Furthermore, the person object is recognized with an accuracy of 73 and 97%; the difference in the accuracy value of the person object classification is caused by the lighter blue color of the clothes with the red motorcycle color, so the resulting value is 97%. The results of object classification and classification accuracy are given in Table 2.

Table 2 Object name and classification accuracy in afternoon condition

No.	Object name	Classification accuracy (%)
1	Car	81
2	Person	73
3	Person	97
4	Motorcycle	100
5	Motorcycle	99

4.3 *Evening Environmental Condition*

The image was taken at the Darussalam location in the afternoon at 17.25 pm. The objects seen in Fig. 3 are cars, motorcycles, and people with a large number of varying scales and orientations. The weather conditions are less sunny, and the shadows are almost non-existent. The goods rickshaw is in the left-side position, while the car is in the back right position.

With less harsh lighting and shadows on the automobile's undercarriage, automotive objects seen from behind can be categorized with 100% accuracy. First, motorcycles are classified with 100% accuracy in the left rear position and front position, as shown in Fig. 3. Then, the person object is classified with an accuracy of 75% with a slightly dark orange clothing color. The results of object classification and classification accuracy are given in Table 3.

Fig. 3 Classification of objects in evening conditions

Table 3 Object name and classification accuracy in evening condition

No.	Object name	Classification accuracy (%)
1	Car	100
2	Person	75
3	Motorcycle	100

5 Conclusion

We have investigated object classification performance using CNN for several environmental conditions. A computer simulation has been conducted to classify the objects. The CNN SSDMobilenetv2FPNLite model can classify cars, motorcycles, and person objects according to test findings on images taken in the morning, afternoon, and evening. An image resolution of 800 $\times$ 600 pixels is sufficient for object classification to keep inference time to a minimum. Tests on images that have shadows show that there is no effect on object classification abilities. Accuracy of object classification on car, motorcycle, and goods rickshaw objects can be achieved up to 100%. Meanwhile, for person objects, the maximum classification accuracy reaches 97% and a minimum of 73%. For car objects, the maximum classification accuracy reaches 100% and a minimum of 81%. In contrast, the motorcycle object's maximum classification accuracy reaches 100% and a minimum of 99%. From testing the object classification model, SSDMobilenetv2FPNLite allows it to be used for object classification on autonomous mobile robots.

Acknowledgements This work is funded by Doctoral dissertation research, the Ministry of Research, Technology, and Higher Education of the Republic of Indonesia, under grant no. 67/UN11.2.1/PT.01.03/DPRM/2022.

References

1. Cao J et al (2020) Front vehicle detection algorithm for a smart car based on improved SSD model. Sensors (Switzerland) 20(16):1–21. https://doi.org/10.3390/s20164646
2. Nilwong S, Hossain D, Kaneko SI, Capi G (2019) Deep learning-based landmark detection for mobile robot outdoor localization. Machines 7(2). https://doi.org/10.3390/machines7020025
3. Klančar G (2022) Shop floor automation
4. Mahaur B, Singh N, Mishra KK (2022) Road object detection: a comparative study of deep learning-based algorithms. Multimed Tools Appl 81(10):14247–14282. https://doi.org/10.1007/s11042-022-12447-5
5. Aslan SN, Uçar A, Güzeliş C (2022) New convolutional neural network models for efficient object recognition with humanoid robots. J Inf Telecommun 6(1):63–82. https://doi.org/10.1080/24751839.2021.1983331
6. Sharma V, Mir RN (2020) A comprehensive and systematic look up into deep learning based object detection techniques: a review. Comput Sci Rev 38:100301. https://doi.org/10.1016/j.cosrev.2020.100301
7. Wu X, Sahoo D, Hoi SCH (2020) Recent advances in deep learning for object detection. Neurocomputing 396:39–64. https://doi.org/10.1016/j.neucom.2020.01.085
8. Sandler M, Howard A, Zhu M, Zhmoginov A, Chen LC (2018) MobileNetV2: inverted residuals and linear bottlenecks. Proc IEEE Comput Soc Conf Comput Vis Pattern Recognit, pp 4510–4520. https://doi.org/10.1109/CVPR.2018.00474
9. Morera Á, Sánchez Á, Moreno AB, Sappa ÁD, Vélez JF (2020) Ssd versus Yolo for detection of outdoor urban advertising panels under multiple variabilities. Sensors (Switzerland) 20(16):1–23. https://doi.org/10.3390/s20164587

10. Liu W et al (2015) SSD: single shot multibox detector Wei 794:185–192 [Online]. Available: http://arxiv.org/abs/1512.02325
11. Lin TY, Goyal P, Girshick R, He K, Dollar P (2020) Focal loss for dense object detection. IEEE Trans Pattern Anal Mach Intell 42(2):318–327. https://doi.org/10.1109/TPAMI.2018.2858826

Investigation of Muscle Contraction on Motorcycle Handlebar Using Electromyography

Hartomo Soewardi and Khoirul Anwar Hasibuan

Abstract A motorcycle is commonly used by people in Indonesia as a transportation device. It is because this device is flexible and affordable as well as easy to maneuver. However, the risk of driving a motorcycle is an emergence of the Carpal Tunnel Syndrome (CTS) on the wrist aside from the accident. This syndrome occurs when tissues surrounding the flexor tendons swell, providing pressure on the median nerve. It is caused by the vibration on the motorcycle handlebar that results from the engine and it will impair the muscle work. The objective of this study is to investigate the correlation between the *Flexor Carpi Ulnaris* muscle contraction and vibration occurred in the motorcycle handlebar and how much the muscle contraction comes about. An experimental study was conducted in an ergonomic laboratory in which six respondents have participated. Electromyography was used to identify Maximum Voluntary Contraction (MVC) on *Flexor Carpi Ulnaris* muscle which is functioned as controlling the throttle, clutch, and brake levers. Manual and automatic motorcycle with 110 and 125 machine capacity, and 1000 and 5000 speed level for a case study. Statistical analysis was conducted to test the hypothesis. The result of this study shows that the percentage of maximum voluntary contraction on *flexor carpi ulnaris* is higher on a manual motorcycle than the automatic motorcycle in 5000 rpm for 110 and 125 machine capacity, respectively, in which it will be higher as long as vibration increases on a motorcycle handlebar. An important finding of this study was that the machine capacity, the type of transmission, and the speed of engine will generate the vibration on hand's muscle contraction. Thus, it is required to design the personal protective equipment to mitigate the affliction on hand for next study.

Keywords Vibration · Muscle contraction · Electromyography · *Flexor Carpi Ulnaris* · Motorcycle

H. Soewardi (✉) · K. A. Hasibuan
Department of Industrial Engineering, Industrial Technology Faculty, Universitas Islam Indonesia, Jalan Kaliurang Km. 14.5, Yogyakarta 55584, Indonesia
e-mail: hartomo@uii.ac.id

Irwansyah et al. (eds.), *Proceedings of the 4th International Conference on Experimental and Computational Mechanics in Engineering*, Lecture Notes in Mechanical Engineering, https://doi.org/10.1007/978-981-99-7495-5_33

1 Introduction

The motorcycle is the vehicle commonly used by people. It is affordable, easy to drive, and does not consume a lot of petrol compared to another vehicle [1]. In Indonesia, the number of motorcycles increased by 6.3% from the year 2015 to 2016, and in the following year, it increased by 7.4% [2]. The fact indicates that people prefer driving a motorcycle as the main transportation device to other means. Thus, a rider should have the ability and skills in driving, which is proven by driving license. However, an accident may occur because of the uncomfortable and instability of the rider in driving [3]. One of the root causes is motorcyclists are exposed to the risk of fatigue, pollution from other vehicles, and physical damage due to noise, movement, and vibration [4].

The vibration occurred in the motorcycle comes up from engines exposes continuously along it is on, whether the motorcycle was moved or stopped [5]. This vibration will expose directly to the whole body especially on hand in which it is called human arm vibration (HAV) [6]. HAV will produce a large variety of effects on hand such as discomfort, numbness, and tingling that annoy the muscle contraction [7, 8] as well as another type of neurological disorder that is carpal tunnel syndrome (CTS) [9]. National Institute of Neurological disorders explained that the CTS is one of the most common neuropathies occurred when the median nerve was depressed or squeezed at the wrist. This syndrome or disorder occurs often when the excessive amount of vibration from the engine was exposed on the handlebar that giving pressure on the wrist.

Electromyography (EMG) is a tool to identify the quantitative signal amplitude of muscle activity and has been used in several studies related to the activity of hand. The EMG signal can describe hand muscle contraction when riding a motorcycle.

This study presents an Electromyography investigation to evaluate the relationship between hand muscle contraction and vibration when riding a motorcycle.

2 Method

Six subjects were participated in this study and taken randomly. Their age is in range from 21 to 23 years old. All of them have been experienced in riding a motorcycle for more than 2 years and none never getting accident [10, 11].

An experimental study was conducted on all subjects by riding a motorcycle in static condition with standard handlebar at two different speeds that are 1000 rpm and 5000 rpm for manual and automatic engine type, respectively, as well as at different machine capacity, 110 cc and 125 cc, during thirty minutes.

The vibration on the handlebar was measured by sticking an accelerometer 25 g. The speed of the engine is controlled by using a tachometer in order to maintain a fixed rotation per minute. Raw signal of vibration is filtered by using Fast Fourier Transform (FFT) to get rid of the noise.

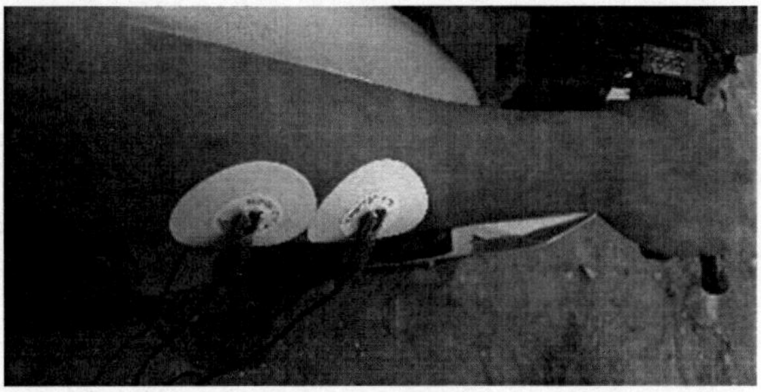

Fig. 1 Electrode attachment

Muscle contraction on hand is also measured by using the electromyography by attaching an electrode at the *flexor carpi ulnaris* muscle to record and normalize raw signal of the maximum voluntary contraction (MVC). Location of attached surface is illustrated in Fig. 1. The surface of electrodes attached should be parallel to muscle fibers between one electrode to another on the sterilized skin area, whereas one reference electrode is attached on nearby unaffected area [12]. Electromyography signal was set at 500 Hz with high-pass and low-pass amplifier, band pass filter was set to 10–500 Hz with 100 ms Root Mean Square for smoothing the signal [12].

A personal computer is used to present the raw signal of EMG and vibration using logger pro version 3.8.7 (Vernier software and technology). Parametric statistical analysis by using independent T-test was conducted to test the hypothesis about difference among some variables examined.

3 Result and Discussion

The result of experiment for the automatic transmission is presented in Fig. 2. It shows that the muscle contraction increases along with the faster of rotation per minute (rpm) whether for the machine capacity of 110 cubical centimeter (cc) or 125 cubical centimeter (cc). This because of there is difference the vibration of machine between both capacities with speed increase. Thus, this condition will affect the comfort of hand in steering motorcycle such that it can lead to instability, even accident. The same state of affairs will also occur on the manual transmission as shown in Fig. 3.

The result of maximum voluntary contraction (MVC) on automatic transmission with machine capacity 110 cubical centimeter is 15.86 and 35.79% with vibration 0.65 and 2.41 m/s^2 and for 125 cubical centimeter (cc), 16.13 and 34.02% with vibration 0.76 and 1.64 m/s^2 for 1000 and 5000 rotations per minute (rpm). Whereas in manual transmission, the maximum voluntary contraction (MVC) in 110 cubical centimeters is 21.01 and 42.79% with vibration 1.57 and 6.02 m/s^2 and in 125 cubical centimeters, 22.37 and 41.89% with vibration 1.66 and 3.09 m/s^2. Reference [13] stated that sign of muscle fatigue begins at 11% of the MVC value due to continuous and prolonged muscle contractions. During the contraction, the internal pressure will

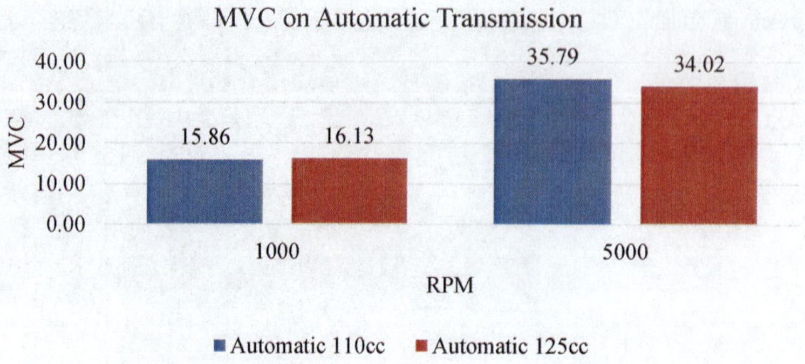

Fig. 2 MVC on automatic transmission

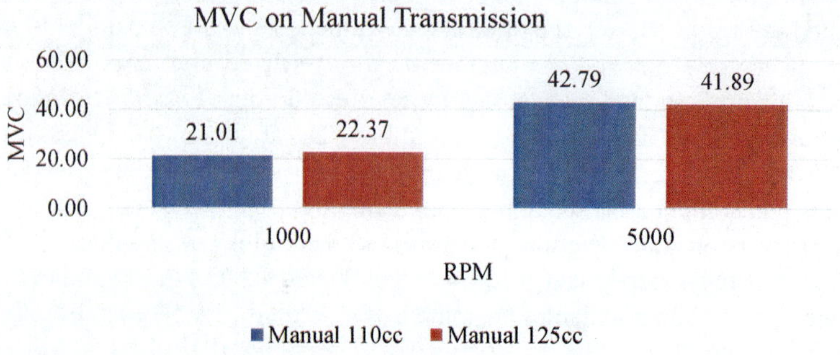

Fig. 3 MVC on manual transmission

shorten the muscle fibers at 10% to 50% of the maximum voluntary contraction. The higher of MVC value will endanger the muscle and reducing endurance of muscle made the muscle easier to be fatigue.

Table 1 shows result of the independent T-test, in which there is a significant difference in MVC value between 1000 and 5000 rpm for both automatic and manual transmission on 110 cc of machine capacity. 125 cc of machine capacity also had the same result as the test. This difference is caused by increased vibration in the engine due to increase engine speed such that the contraction on muscle in the hand of the greater. Likewise, the MVC testing between automatic and manual transmission also produced significant differences in engine rotation for both 110 and 125 cc. Thus, this study found a fact that automatic transmission is better than manual transmission and also 110 cc of machine capacity is better than 125 cc because of lower contraction occurred.

Reference [14] stated that there is strong relationship among muscle activity, interval time, and vibration level. This result is in line with [14] found that increasing of vibration level will increase heart rate and RMS value. The reason of MVC value as the vibration increase because the muscle is sensitive to change in posture and respond the muscle activity when hand is exposed to vibration [15].

Table 1 Independent T-test result of MVC

MVC				
	110 cc		125 cc	
	Automatic	Manual	Automatic	Manual
1000 rpm	0.0000051	0.000068		
5000 rpm	0.0000026	0.00019		
Independent T-test	0.000013	0.000000090	0.0021	0.000025

4 Conclusion

The vibration effects are hazardous for motorcyclist, so it needs to be minimized. Based on the analysis, it can be concluded that the engine vibration increase will work up the muscle voluntary contraction (MVC) and automatic transmission, the low machine capacity, and the lower speed have the lower contraction of muscle. An important finding of this study was that the machine capacity, the type of transmission, and speed of engine will generate the vibration on hand's muscle contraction. Thus, this study recommends to involve all relevant muscles of human body part in further investigation and to design the personal protective equipment to mitigate the affliction on hand for the future study.

Acknowledgements We wish to acknowledge the Directorate of Student Development Islamic University of Indonesia Yogyakarta Indonesia for supporting this project.

References

1. Rezali KA, As'arry A (2017) Transmission of vibration from motorcycle handlebar to the hand. J Soc Automot Eng Malays, 191–197
2. Statistik BP (2017) Perkembangan Jumlah Kendaraan Bermotor Menurut Jenis, 1949–2016
3. Nugroho, Wahyu P (2018) Analisis Kenyamanan dan Kestabilan Kendaraan Roda Dua Dalam Tinjauan Satu Derajat Kebebasan, Jurnal Energi dan Teknologi Manufaktur 01–06
4. Jaiyesinmi A, Areoye O, Olagbegi O, Bolarinde S, Uduonu E (2018) Work-related musculoskeletal disorders and predisposing factors among commercial motorcyclist in Ibadan North Local Goverment Area, Nigeria, Occupational Health Southern Africa
5. Baad SM, Patil RJ, Qaimi MG (2017) Hand arm vibration alleviation of motorcycle handlebar using particle damper. I.J. Eng Manuf
6. Griffin MJ (1990) Handbook of human vibration. Academic Press Inc., San Diego
7. Margareta N, Andersson G, Pope MH (1997) Musculoskeletal disorder in the workplace: principles and practice. Mosby, Missouri
8. Bovenzi M (2011) Hand-transmitted vibration. Retrieved from http://iloencyclopaedia.org/part-vi-16255/vibration/84-50vibration/hand-transmitted-vibration. January 20
9. National Institute of Neurological Disorders and Stroke (2017) Carpal tunnel syndrome. United States of America, Bethesda, Maryland
10. Balasubramanian V, Jagannath M (2014) Detecting motorcycle rider local physical fatigue and discomfort using surface electromyography and seat interface pressure. Transp Res, 150–158

11. Rashida H, Ma'arofa MI, Haronb R, Fauzia WM, Omara AR, Jaafara R (2015) Selection of muscle groups for surface electromyography (Semg) measurement in analyzing motorcycling activity. J Teknologi
12. Konrad P (2005) The ABC of EMG, Noraxon, Inc., USA
13. Criswell E (2011) Surface electromyography, 2nd edn, Jones and Barlett, Canada
14. Dawal S, Mirta W, Ling Y, Zadry H (2008) The effect of electric drill laboratory hand tool on muscle activity and heart rate measurement. In: IFMBE Proceedings, 4th Kuala Lumpur international conference on biomedical engineering, ISSN: 1680-0737, 21
15. Dawal M, Widia Md SZ (2011) The effect of hand-held vibrating tools on muscle activity and grip strength. Aust J Basic Appl Sci 5:198–211

Parameter Study of Wing-Type Structure Subjected to Impact Loading

Awangku Muhammad Asyraf bin Pengiran Haji Mat Rais and Yulfian Aminanda

Abstract This research scopes to the modelling and simulating of one section of wing-type structure subjected to a bird strike. Modelling and simulations are performed through finite element (FE) approach by using ANSYS Workbench (Explicit Dynamic). The bird is modelled as a rigid body, having a similar weight to an actual bird. Varying desired parameters of the leading edge—curvature radius (R), skin thickness (T), and number of ribs (N) behind the leading edge—provides a better understanding of the extent of each design modification. Six different values of R and T are studied, with none ($N0$), one ($N1$), and two ribs ($N2$) behind the leading edge's skin considered. By impacting the model with 120 m/s of initial velocity, the effect of T is seen to exhibit linear relationship with energy absorbed (EA) and specific energy absorbed (SEA). For R, a peak EA and SEA has been identified. Adding ribs behind the leading edge's skin was also proven to provide more strength. Spacing of ribs is expected to vary the performance. The effect of impact location has been analysed as well.

Keywords Aircraft wing · Structural analysis · Bird strike · Finite element analysis · Impact engineering

1 Introduction

Air accidents have been recorded ever since the introduction of air travel which causes damage to the corresponding aircraft as well as to the extent of casualties [1]. One reason includes the occurrence of bird strike during travel in which intersection between aircraft and the bird typically causes a high energy impact [2]. The general aviation accidents had damage due to foreign object debris for as much as 90% [3, 4]. Forefacing external parts of the aircraft are prone to such damage, by which wing's

A. M. A. P. H. M. Rais · Y. Aminanda (✉)
Mechanical Engineering Department, Universiti Teknologi Brunei, Tungku Highway, Gadong 1410, Brunei Darussalam
e-mail: yulfian.aminanda@utb.edu.bn

Irwansyah et al. (eds.), *Proceedings of the 4th International Conference on Experimental and Computational Mechanics in Engineering*, Lecture Notes in Mechanical Engineering, https://doi.org/10.1007/978-981-99-7495-5_34

leading edge is considered to be the one of the most vulnerable [4, 5]. Therefore, there is a great opportunity to study within this area to suggest further improvement. Numerous researches have been carried out to replicate bird strike phenomena on wing structure to study the effect and subsequently propose new designs [6]. The use of conventional and FE methods can be performed, which the latter are widely utilized.

Aircraft's wings are generally comprised of the skin, ribs, and spars. For the wing's leading edge, effect of reconfiguring the parts is typically studied. This includes altering its radius *R*, its thickness *T*, and number of ribs *N*. Promising improvement to the wing's impact capabilities has been presented by numerous authors by increasing either the *R*, *T*, and *N*. Increasing *R*, especially by incorporating additional reinforcement behind the skin, can induce higher force requirement to fail the skin [7–9]. More dispersion of bird during impact on models with greater *T* can be achieved which discourages perforation [10]. It as such also constitutes to higher effective impact kinetic energy, which translates to higher EA [11]. Introducing less space between two ribs, or using more *N*, also promotes greater stiffness to the structure, hence, requiring higher energy for perforation to occur [12, 13]. This work deal with analysing the effect of *R*, *T*, and *N* extensively using FEA in order to draw the trend of energy absorption capability of the structure and its deformation mechanism.

2 Materials and Method

2.1 Validation

Validation was initially performed by referring to an existing study of impact test by Aisyah [14]. The test comprised of semicircular aluminium plate and a spherical impactor of diameter 13 mm with mass of 28.45 kg. An illustration for the test is shown in Fig. 1, with impact test to proceed with initial velocity of 10 m/s. *R* is set to 70 mm with a length of 150 mm.

Fig. 1 Illustration for validation set-up

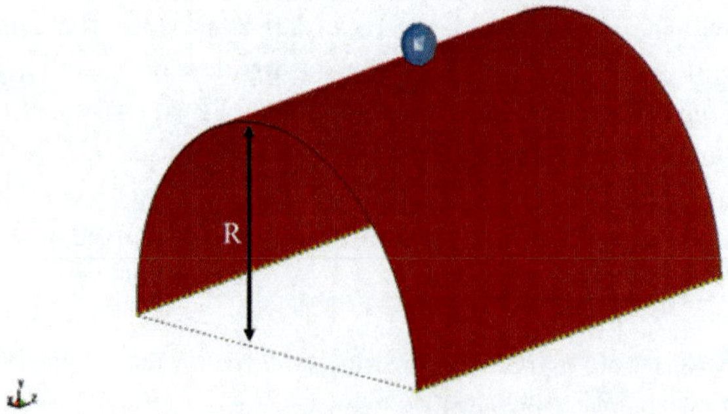

Different values of skin thickness T (0.40, 1.00, 1.25, and 1.50 mm) for the plate are accounted. Respective results obtained will be used for comparative study for validation. Also, different mesh sizes are used for the plate for mesh study—5, 7.5, 10, and 15 mm. The impactor is simply meshed using 'body sizing' of size 2.5 mm.

2.2 Actual Research

Simple sketch of wing model is shown in Fig. 2. Wing's leading edge is considered to be part A, C, and D with fuel tank being the box behind it. Parts B and E are the front and rear wall, respectively. F is the fuselage body with G to be considered as the side wall of fuel tank or the end rib. All parts are applied using material AA 2024-T35. Fixed support is applied to part F and the connected circumferential edge of part A. Varying wing configurations for parametric study is provided in Table 1.

Variation of R affects the profile of part A, and variation of T affects the wing's leading edge, whilst N affects the void between parts A and B. Parts unaffected by varying T have constant thickness—7 mm for part B and 2 mm for other parts.

Impactor (Fig. 3) is considered as a rigid structure and take account of bird's mass (approximately 3 kg) which has a density of 2300 kg/m^3. It is placed 10 mm away from the wing's leading edge with initial velocity of 120 m/s. Impactor is placed at central-fore position (Fig. 4a) except for models with N1 and N2—placed 200 mm away from the end rib (Fig. 4b, c).

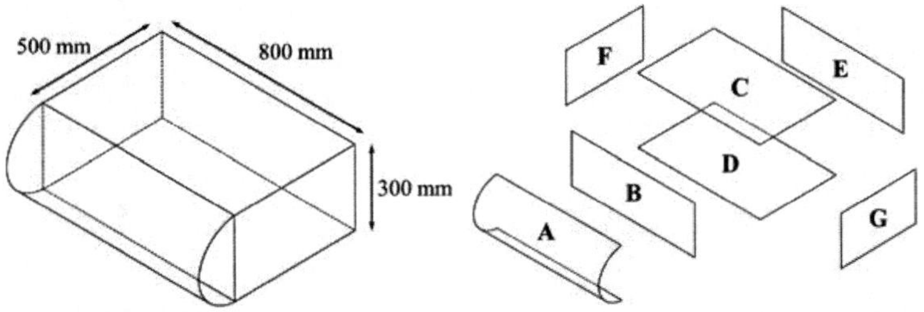

Fig. 2 Illustration for wing model

Table 1 Variations of R, T, and N

R1 [mm]	R2 [mm]	R3 [mm]	R4 [mm]	R5 [mm]	R6 [mm]
37.50	75.00	150.00	350.00	750.00	900.00
T1 [mm]	T2 [mm]	T3 [mm]	T4 [mm]	T5 [mm]	T6 [mm]
0.50	1.00	1.25	1.50	1.75	2.00
N0		N1		N2	
No ribs		1 additional rib		2 additional ribs	

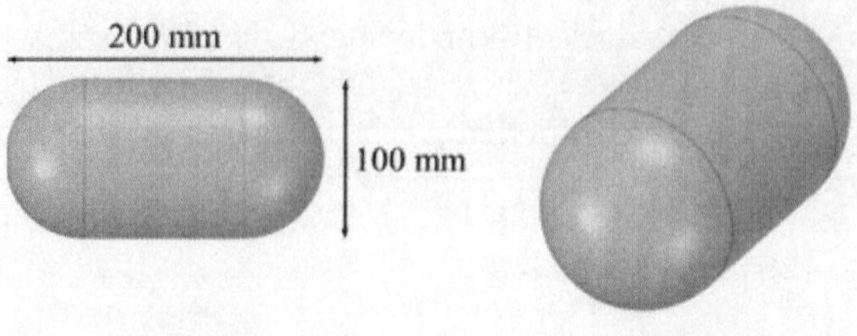

Fig. 3 Dimension of impactor (bird model)

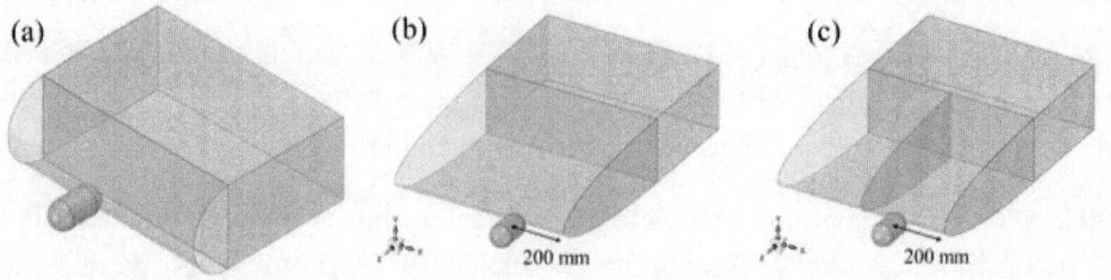

Fig. 4 Initial position of impactor for wing models with **a** N0, **b** N1, and **c** N2

In order to systematically assess the viability of each parametric variation, the impact is modelled through the steps listed below:

Step A1—Vary *R* and *T* with N0 to determine the best performing *R* and *T*.

Step A2—Determine the effect of varying N by considering the best performing *R* and *T*.

Observing the impact successions aids understanding of wing's behaviour to withstand the impact through different failure modes. The use of numerical analyses would support any claims from the observations. EA and SEA are determined by using Eqs. 1 and 2. m, V_0, V_i, and m_s denote the impactor's mass, impactor's initial velocity, impactor's prompt velocity, and mass of leading edge's skin (parts A, C, and D), respectively.

$$\mathrm{EA} = 0.5 \times m \times \left(V_0^2 - V_i^2\right) \tag{1}$$

$$\mathrm{SEA} = \mathrm{EA}/m_s \tag{2}$$

3 Results and Discussion

3.1 Validation and Mesh Study

Energy absorption (EA) at each thickness *T* and different mesh sizes are given in Fig. 5.

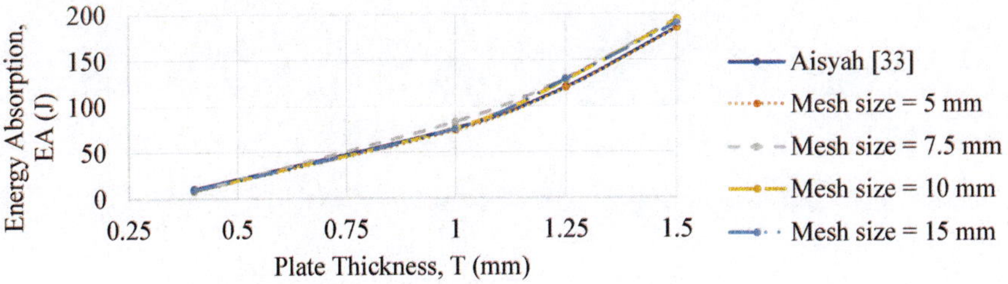

Fig. 5 EA at t = 0.007 s against different *T* and mesh sizes

Based on Fig. 5, a similar trend has been achieved—gradual increase in EA, and T increases. As part of mesh study, the use of 5 mm mesh size can be observed to be much closer Aisyah [14] in comparison to the use other mesh sizes.

3.2 *Failure Mechanisms (Step A1)*

By varying *T* and *R* at N0, four different failure mechanisms have been identified via simulation, as briefly described in Table 2. Respective wing models affected with these types of failure mechanism are provided in Table 3.

Table 2 Different types of failure mechanisms for A1

	FM1	FM2	FM3	FM4
Brief description	Impactor reaches front wall and causes perforation, without fully penetrating the leading edge's skin	Impactor reaches front wall and causes perforation, after fully penetrating the leading edge's skin	Impactor reaches front wall creating crack and bounces back, after fully penetrating the leading edge's skin	Impactor reaches front wall and bounces back, after fully penetrating the leading edge's skin

Table 3 Wing models associated with each failure mechanism

	R1	R2	R3	R4	R5	R6
T1	FM1	FM1	FM2	FM2	FM2	FM2
T2	FM1	FM1	FM2	FM2	FM2	FM2
T3	FM1	FM1	FM2	FM2	FM2	FM2
T4	FM1	FM1	FM2	FM2	FM2	FM2
T5	FM1	FM1	FM2	FM2	FM3	FM3
T6	FM1	FM1	FM2	FM3	FM4	FM3

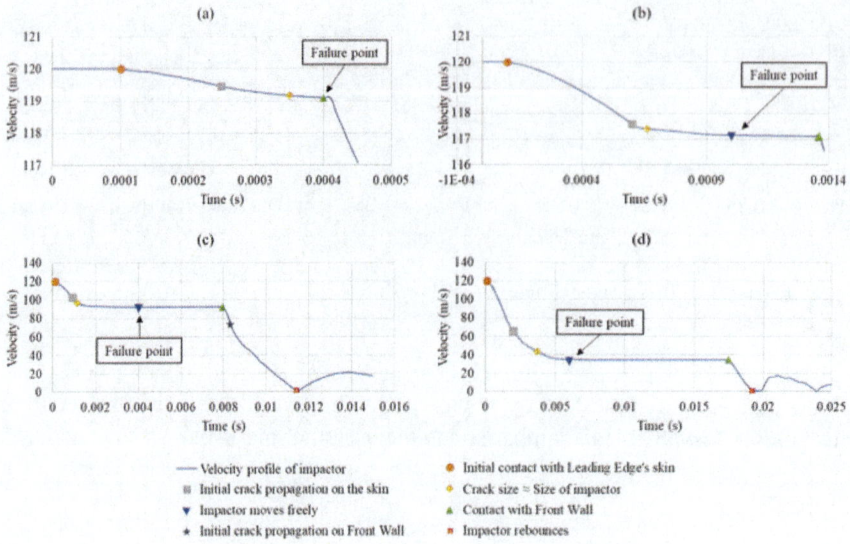

Fig. 6 Velocity–time plot for **a** R1T1 for FM1, **b** R3T1 for FM2, **c** R5T5 for FM3, and **d** R5T6 for FM4

For all failure mechanisms, all respective models have similar trend in its velocity–time plot which only vary in magnitude. As example, the velocity profile for wing model with R1T1, R3T1, R5T5, and R5T6 is shown in Fig. 6. This corresponds to failure mechanism FM1, FM2, FM3, and FM4, respectively.

Generally, constituting models with FM2, FM3, and FM4 are considered to fail prompt to complete perforation by the impactor through the wing skin (Fig. 7b–d, respectively). Dissimilarly, failure for FM1 occurs prompt to impactor reaching the front wall (Fig. 7a). As such is due to the nature of R1 and R2 which are considerably small. For all models, crack initiates on the wing skin around the proximity of impact—arises from higher high stress concentration which exceeds the material's yield point.

For FM1, impact is considered the most disastrous—not highly affected by the wing. Crack continues to develop on the skin as impactor travels against it. As it proceeds, it eventually reaches the front wall (Fig. 7a) and causes the structure to bulge which then cause crack to initiate and propagate. The crack on front wall is large enough to allow impactor to puncture through and exposing the internal part of the fuel tank. These progressions of deformations are similar with that of FM2. The only difference identified is the lack of obstruction by the wing skin due to large crack size, allowing impactor to easily travel further (Fig. 7b). This can also be reflected by the sudden constant acceleration after such failure point in Fig. 6b.

For both FM3 and FM4, sufficient energy from impact is able to be absorbed by the wing skin since it disallows impactor from entering the fuel tank. The progression of deformation initially starts with crack propagation on the skin as well as observation of buckling (Fig. 8a, c), which is dissimilar from FM1 and FM2. Despite such, the

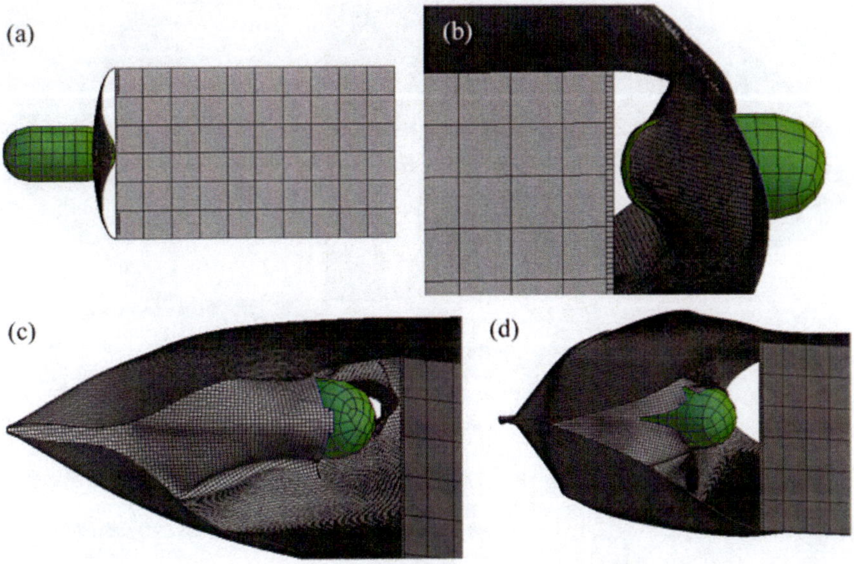

Fig. 7 Failure point of wing model with **a** R1T1 for FM1, **b** R3T1 for FM2, **c** R5T5 for FM3, and **d** R5T6 for FM4

impactor was still able to eventually perforate through the skin (Fig. 7c, d). Following then, impact reaches front wall and causes it to bulge inwards. Crack on front wall only occurs for models with FM3. Eventually, surplus impact energy is absorbed entirely by the front wall and causes the impactor to bounce back and travel to the opposite direction. Maximum deflection suffered by models with FM3 and FM4 is approximately 126 mm and 28 mm (Fig. 8c, d), respectively.

3.3 Energy Absorption Capabilities (Step A1)

Energy absorption (EA) and specific energy absorption (SEA) at each model's failure point are compared, as shown in Fig. 9a, b, respectively. A similar distribution can be observed for the case of EA plots. EA increases with R and reaches a peak value at a point. Peak value is observed to occur for wing models with R5, except for the case of T1 which occurs at R4. Although comparing each trendline's peak value, a linear relationship is obtained with highest EA for wing model with T6 at 19,819.4 J and lowest EA for T1 at 1567.5 J. For SEA comparison, each plot at different T is not entirely similar. However, it is observed that for each R, the dominant T which allows greater is wing model with T6. Overall, by considering both EA and SEA comparison, wing model with R5T6 is the best as it has inherent greatest magnitude of EA and SEA.

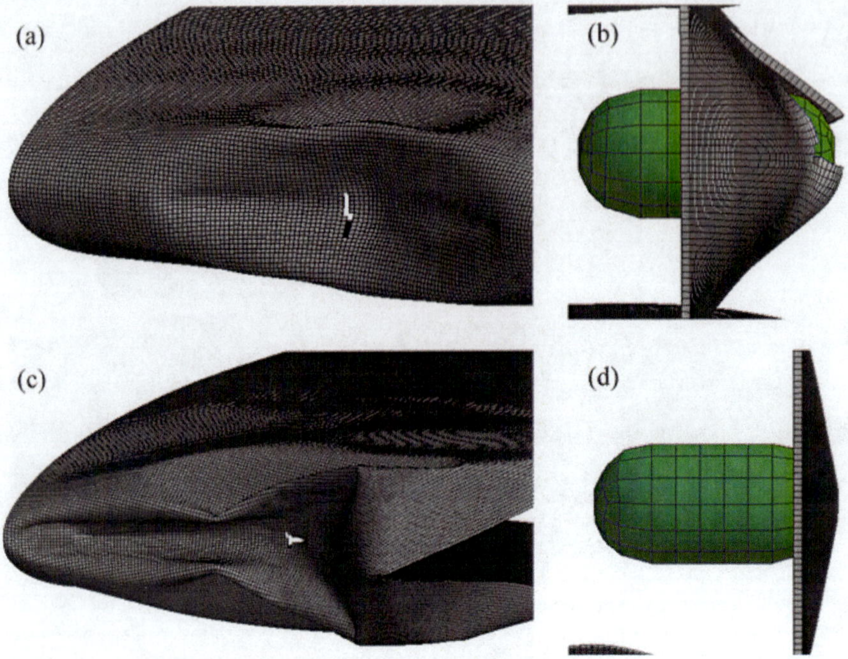

Fig. 8 **a** crack initiation and buckling of R5T5 for FM3, **b** maximum deflection of front wall of R5T5 for FM3, **c** crack initiation and buckling of R5T6 for FM4, and **d** maximum deflection of front wall of R5T6 for FM4

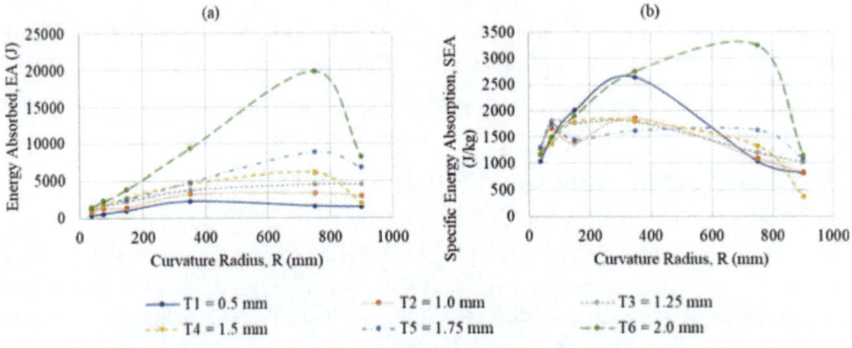

Fig. 9 **a** Energy absorbed (EA) comparison, **b** specific energy absorbed (SEA) comparison

3.4 Failure Mechanisms (Step A2)

As wing model with R5T6 has been identified to be the best from step A1, effect of additional ribs (N1 and N2) is analysed. The deformation progressions are almost similar with failure mechanism FM4 from step A1. Failure point still occurs prompt

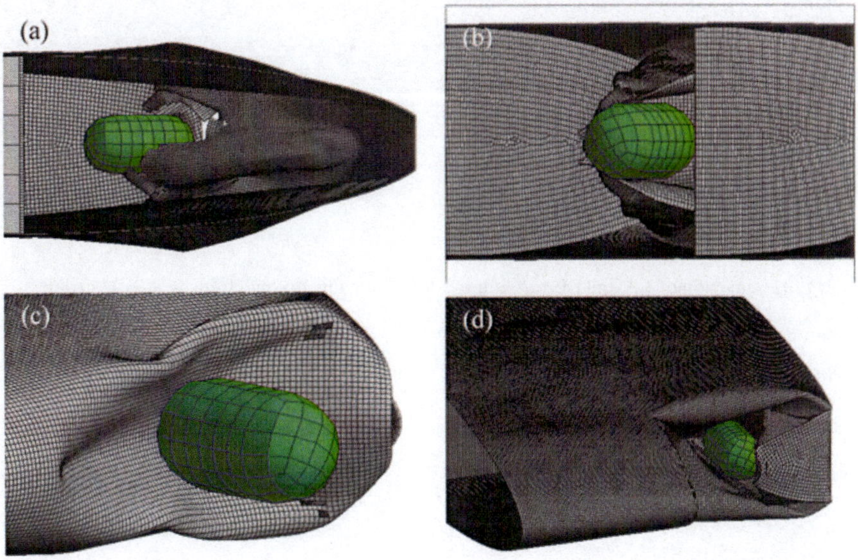

Fig. 10 **a** Failure point for R5T6N1, **b** failure point for R5T6N2, **c** initial crack propagation, buckling and skin detachment for R5T6N1, and **d** skin tear and end rib deflection for R5T6N2

to impactor perforating through the wing skin (Fig. 10b). For R5T6N1, crack propagation and buckling initially occur with skin detached from the end rib (Fig. 10c). For R5T6N2, the initial deformation is similar with R5T6N2. However, an additional skin tear has been observed due to the central rib. This also caused the end rib to deflect inwards throughout the impact succession (Fig. 10d). Following the progression of deformations, both models eventually reach a failure point and cause the impactor to reach the front wall. As more deformation occurs for R5T6N2 than R5T6N1, the latter results in greater deflection of front wall with approximately 38 mm whilst the former deflects by 22 mm.

3.5 Energy Absorption Capabilities (Step A2)

EA and SEA at failure point for R5T6 with different N are provided in Fig. 11. For R5T6N0 and R5T6N1, SEA is obtained by considering the entire mass of the wing's leading edge. Dissimilarly, R5T6N2 only considers wing's leading edge from the end rib to the central rib.

Comparing each model's EA and SEA, it is definite that R5T6N2 is the best performing model (EA = 20,696 J; SEA = 4176 J/kg) with R5T6N1 being the worst (EA = 18,427 J; SEA = 2614 J/kg). The effect R5T6N2 proves that having well-designed support behind the wing's leading edge is important. Stiffness offered by the central with wing skin's distortion behaviour during impact allows more energy to be

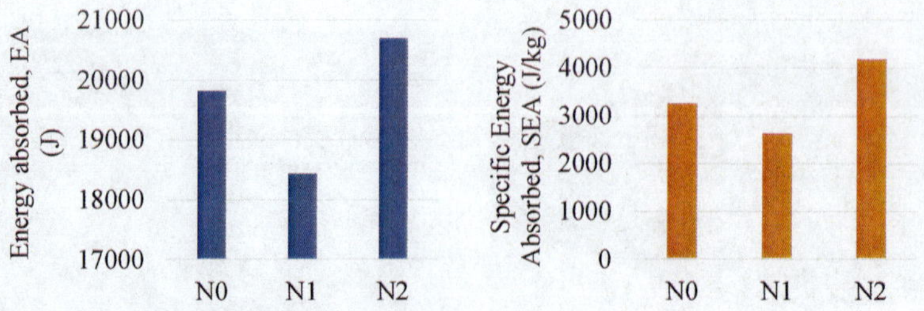

Fig. 11 Comparison of EA and SEA at failure for R5T6 with varying N

dissipated or absorbed, hence, causing less destruction to constituting components behind the leading edge.

4 Summary

Different parametric variations have shown great potential in altering the impact strength performance, especially for the case of bird strike. Generally, a linear relation has been found for the case of varying *T*. Increasing *T* was observed to potentially increase the energy absorption capabilities. For the case of *R*, non-linear trend has been observed. Between six different values of *R* tested, R5 (750 mm) was found to exhibit the best performance for almost all values of *T*. The effect of rib(s) was not extensively studied, although promising results have been obtained. It has been found that incorporating additional ribs are capable in enforcing the structural integrity of the wing's leading edge by providing more support and stiffness, which improves EA and SEA capabilities. Overall, wing model with R5T6N2 configuration would offer the best design in absorbing impact energy.

References

1. Kumer PD, Asim KD (2016) Quantifying the risk of extreme aviation accidents. Phys A: Stat Mech Its Appl 463:345–355
2. Dennis N, Lyle D (2009) Bird strike damage and windshield bird strike. Atkins Limited, Food and Environment Research Agency, Farnham
3. Xue Y, Fu G (2018) A modified accident analysis and investigation model for the general aviation industry: emphasizing on human and organizational factors. J Saf Res 67(1):1–15
4. Anjan Kumar D, Girija V (2018) A review on bird strike analysis on leading edge of an aircraft wing structure using a SPH formulation. Int Res J Eng Technol (IRJET) 5(10):330–334
5. Ravi K (2013) Comprehensive bird strike simulation approach for aircraft structure certification. Cyient
6. Zhu Y, Huang X, Fang W, Li S (2010) Trajectory planning algorithm based on quaternion for 6-DOF aircraft wing automatic position and pose adjustment method. Chin J Aeronaut 23(1):707–714

7. Long S, Mu X, Liu Y, Wang H, Zhang X, Yao X (2021) Failure modeling of composite wing leading edge under bird strike. Compos Struct 255(1):1–14
8. Arachchige B, Ghasemnejad H, Yasaee M (2020) Effect of bird-strike on sandwich composite aircraft wing leading edge. Adv Eng Softw 148(1):1–11
9. Di Caprio F, Cristillo D, Saputo S, Guida M, Riccio A (2019) Crashworthiness of wing leading edges under bird impact event. Compos Struct 216(1):39–52
10. Yijing G, Pinghui J, Guanxin H (2012) Research on bird strike simulation of composite leading edge. AASRI Procedia 3(1):674–679
11. Yu Z, Xue P, Yao P, ZMS (2019) Analytical determination of the critical impact location for wing leading edge under birdstrik. Latin Am J Solids Struct 16(1):1–17
12. Michael M, Sonja A-K, Volker L, Dominik L, Conchin CA, Mathieu I (2021) Bird strike resistance of a CFRP morphing leading edge. Compos Part C: Open Access 4(1):1–7
13. Salem SC, Gaddikeri KM, Gowda M, Sundaram R, Dhayanidhi J, Nagarajappa B, Sathiyanaarayan S, Chandra S, Nayak A, Petley V, Ramachandra S (2011) Design and testing of wing leading edge of a light transport aircraft. In: NASAS 2011, Bangalore, Karnataka
14. Nur Aisyah MK (2011) Impact on wing leading edge (study on the effect of various skin radius of curvatures). International Islamic University Malaysia

Experimental of the Distribution and Direction of Droplet Movement Using a High-Speed Camera

Rachmad Almi Putra, Hamdani Umar, Samsul Rizal, Maimun Syukri, and M. Salamul Fajar

Abstract Droplets are liquids or splashes of water that are expelled through the respiratory tract when a person talks, coughs, or sneezes. The average size of the droplets released during sneezing is > 5 μm with a distance of < 1 m. Meanwhile, when speaking, it is 5 μm at a distance of 1 m, which is often called nuclei or respiratory droplets. Due to their smaller sizes and the inability to observe with the naked eye, the direct spread of infections through these droplets is a major factor in the increase in disease cases. Therefore, this research is focused on the set-up of droplet observation instrumentation in the room, where the velocity of airflow is neglected. The novelty lies in determining the size distribution of droplets produced and examining the transmission speed when someone sneezes. The results showed that the set-up successfully identifies the droplets expelled during sneezing. The

R. A. Putra
School of Engineering, Universitas Syiah Kuala, Banda Aceh, Indonesia

Faculty of Engineering, Universitas Samudra, Langsa, Indonesia

R. A. Putra
e-mail: rachmad.almi@unsam.ac.id

H. Umar (✉) · S. Rizal
Department of Mechanical Engineering, Faculty of Engineering, Universitas Syiah Kuala, Banda Aceh, Indonesia
e-mail: hamdani@unsyiah.ac.id

S. Rizal
e-mail: samsul.rizal@unsyiah.ac.id

H. Umar
Department of Mechanical Engineering, Faculty of Engineering, Universitas Samudra, Langsa, Indonesia

M. Syukri
Medical Faculty, Universitas Syiah Kuala, Banda Aceh, Indonesia
e-mail: maimun_62@unsyiah.ac.id

M. S. Fajar
Faculty of Engineering, Universitas Syiah Kuala, Banda Aceh 23111, Indonesia
e-mail: salamul@mhs.unsyiah.ac.id

Irwansyah et al. (eds.), *Proceedings of the 4th International Conference on Experimental and Computational Mechanics in Engineering*, Lecture Notes in Mechanical Engineering, https://doi.org/10.1007/978-981-99-7495-5_35

mucosal sac expands and is dispersed and retracted, which makes the ligaments to be visible when pulled or detached from the mother's pouch, causing the sac to disintegrate and form droplets. Based on observations, the number of droplets produced when sneezing is approximately 109 particles, where 39% are 5 μm and 61% are above 5 μm, with a speed of 2.08 m/s.

Keywords HSC · Droplets · PIV · Sneezing · Size

1 Introduction

Infectious diseases such as Acute Respiratory Infections (ARI), Pneumonia, Tuberculosis (TB), and SARS are caused by several biological factors, namely viruses, bacteria, and parasites that are transmitted through the air. In 2020, the world was faced with the emergence of coronavirus, often called COVID-19. It is a new variant of the SARS-Cov 2 virus and the worst outbreak from 2002 to 2003 [1, 2]. This pandemic is experienced in almost all countries and one of the main spreads is through droplets [3]–[5]. Meanwhile, droplets are liquids or splashes of water that are expelled through the respiratory tract when a person talks, coughs, or sneezes. The average size of droplets expelled by a person when sneezing is > 5 μm with a distance of < 1 m. When a person speaks, the size is 5 μm at a distance of 1 m [6], which is often referred to as nuclei or respiratory droplets. Due to the smaller size of droplets and the inability to observe with the naked eye, the direct spread of a disease is a major factor in the increase in COVID-19 cases [5]–[7]. Patients diagnosed with respiratory infectious diseases are generally recommended to self-isolate by staying in a room, at home, and avoiding contact with other people, including those living together.

Isolation measures for people with infectious diseases are also carried out in health care facilities such as hospitals, which are called isolation rooms [6–8]. An isolation room is a place with lower air (negative) pressure compared to the surrounding space for patients with airborne infectious diseases. The design is aimed to direct the airflow to the isolation room and prevent air contaminated with pathogens from entering other rooms [9, 10]. It also aims to make the airflow in the isolation room comes out into free air, therefore, adequate pressure, humidity, temperature, and ventilation settings are needed [12]. One form of air disinfection in the isolation room is the droplets released by a patient, which are mixed with the airflow. This makes the release of air from the isolation room into free space, the risk of infecting other patients. These include visitors and medical personnel, who are around the isolation room and even aggravate the illness suffered by the isolated patient. Based on these conditions, an analysis is needed to calculate the velocity of air or droplet flow and evaluate the movement of air contaminated with pathogens in the isolation room.

A previous related investigation stated that the droplet velocity can be observed using a High-Speed Camera (HSC). Based on the observation, it was discovered that the average droplet velocity was 5–10 m/s when sneezing [13]. Another report was also conducted on the safe distance of people suffering from COVID-19 in public places by considering the level of aerosol concentration that varies and accumulates through inhalation using the Monte Carlo method and obtained quantitative data about the length of time exposure in different public spaces [14][15]. Therefore, this research focuses on the set-up of droplet observation instrumentation in the room, where the velocity of airflow is neglected. The novelty lies in determining the size distribution of droplets produced and examining the transmission speed when sneezing. The results are expected to be a benchmark as the basis for further investigations.

2 Method

2.1 Experiment Settings

The droplets released by the subject were recorded using a High-Speed Camera (HSC) Phantom brand model T-1340. The lighting during the shooting process was from the GSVITEC Multi-led G8 brand LED. This type of LED was selected because it is equipped with a choice of lighting dispersion angles in form of sheet illumination. The droplet recording process used sheet illumination with a beam angle of 15° to focus the light directly on the droplet and produce a clear image. The lamp power during the recording process was 85% and was considered the main light, while the HSC position, light source, and subject were at the same height and placed in front of a black background. The subject position was directly opposite to the LED as a light source and in a straight line. Similarly, the HSC position to the subject was parallel in a straight line and perpendicular to the light source. Since the camera focus was directed to the droplet (object), it forms an angle of 10° from the subject position (Fig. 1).

2.2 HSC Configuration

Since the duration of droplets released by humans through sneezing ranges from 150 to 200 ms [16], HSC is needed to record the movement. Droplets were recorded at 1024 × 768 resolution at a frame rate of 1000 frames-per-sec (fps) and an exposure time of 1000 μs. The camera position was faced 10° from the subject direction and 80° toward the light source to reduce perspective distortion and clarify the recorded object in form of droplets. This configuration was also carried out in previous investigations [13, 17] with a distance of 60 cm from the camera to the subject and 90 cm to the light source.

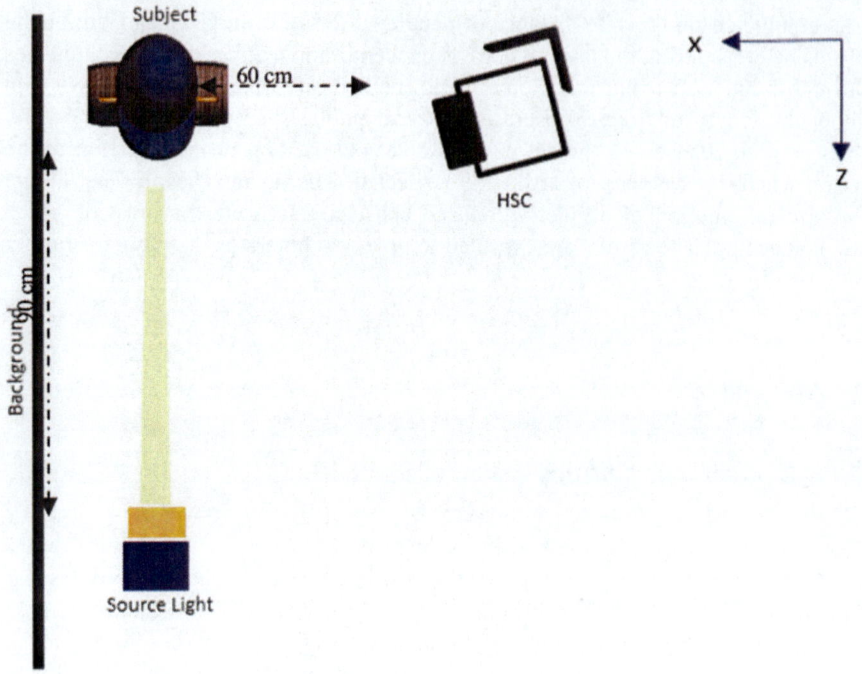

Fig. 1 Set-up of the position of the droplet retrieval instrument

2.3 Image Processing

The data were analyzed using the Particle Image Velocimetry (PIV) method with the Particle Tracking Velocimetry (PTV) algorithm. Before the application of the algorithm, PTV samples in form of HSC recordings went through several stages, namely Sampling Rate, Pre-Processing, and Post-Processing.

2.3.1 A. Sampling Rate

The sampling rate is the number of shots taken sequentially with a maximum distance of 1/20 s between the images. For measuring the speed of particle motion, a minimum of two images is required to analyze the movement in the frame divided by the time interval. The image footage is saved in Tagged Image File Format (.tiff) and the processing is carried out starting from pre- to post-processing.

2.3.2 B. Pre-Processing Image

Pre-processing is the initial processing of image data from video footage to improve image quality before analysis. This includes increasing image intensity for each pixel, lighting and capping intensity, image acquisition, feature extraction, and image evaluation.

2.3.3 C. Post-Processing Image

Post-processing image is a step to filter outliers by selecting a speed limit that is determined manually. This is followed by data interpolation techniques, commonly the nearest neighbor-interpolation. Subsequently, data exploration, which is the appearance of the velocity vector direction of the particle motion, is carried out.

3 Result and Discussion

Observations were made on the sneezing of a male with an age of 30 years, to confirm that the instrumentation set-up described can be used to observe the droplets released by a person when sneezing. The result showed that the duration of sneezing is about 120 ms, which is faster than the values in previous reports [17]–[19]. Figure 2 shows the formation of a mucosal sac when a person sneezes. The sac is formed from a fluid with a specific level of viscosity that comes from the secretions of the respiratory cavity mixed with saliva. The impact of air pressure makes the sac to expand and promote dispersion. The edges of the mucosal sac retract and the ligaments become visible when pulled or detached from the mother pouch, which occurs due to shear stress in the fluid. The mucosal sac is assumed to be fluid in motion (dynamic), while the surrounding air is considered to be fluid at rest (static). This is illustrated in Fig. 2.

The orange line shows the boundary layer or between the mucosal sac and the air. It was formed because the effect of fluid viscosity still occurs. Therefore, the shear stress that occurs in the fluid is systematically described in the equation below:

$\tau = \frac{F}{A} = \frac{du}{dy}$ …(10).

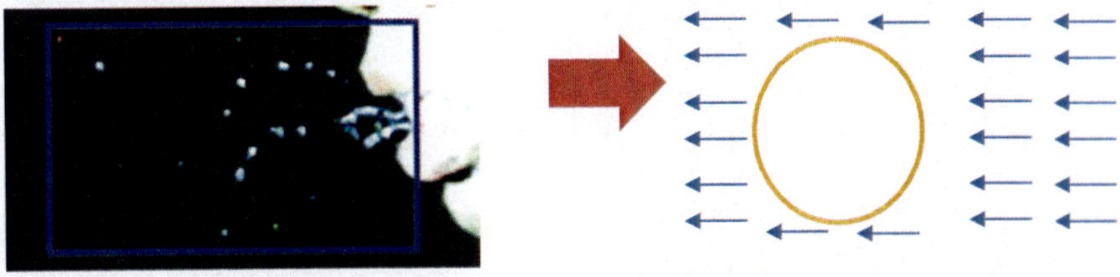

Fig. 2 Illustration of the position of the mucosal sac when sneezing

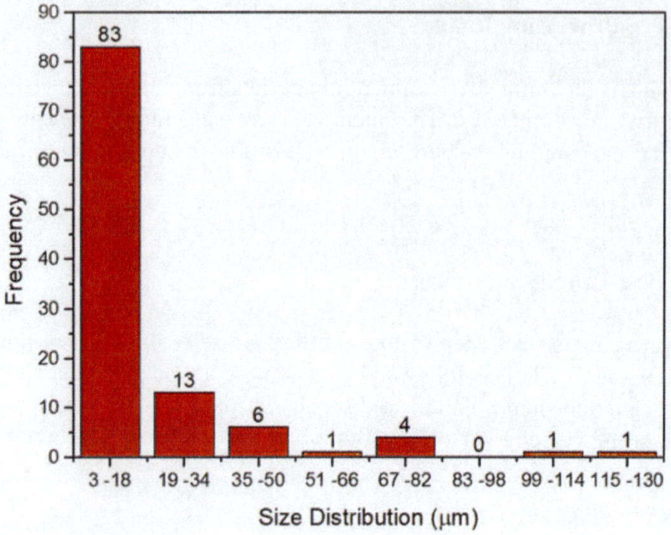

Fig. 3 Droplet size distribution

$= \mu \frac{du}{dy}$...(11).

Where the value of du/dy is the rate of deformation of the fluid and μ is the proportionality factor between τ and du/dy, which depends on the viscosity. Equations (10) and (11) indicated that the shear stress experienced by a fluid with a high viscosity will have a higher deformation than the value of a fluid with a lower viscosity. This causes the mucosal sac to disintegrate and form droplets that contain pathogens or viruses with the potential to spread in the air [20].

The distribution of droplets released by a person ranges from 109 particles, where 39% is ≤ 5 μm in size and 61% more than 5 μm, which was distributed as shown in the graph (Fig. 3). The average, smallest, and maximum droplet sizes obtained are 17 μm, 3 μm, and 127 μm, respectively.

The maximum speed of the droplet is 2.08 m/s, which is close to the value obtained in a previous report [21], which is graphically shown in Fig. 4.

The speed is influenced by the pressure coming from the nasopharynx, assuming the amount of pressure when everyone sneezes are the same, namely 2.19 kPa [22]. The direction of droplet flow is shown in Fig. 5.

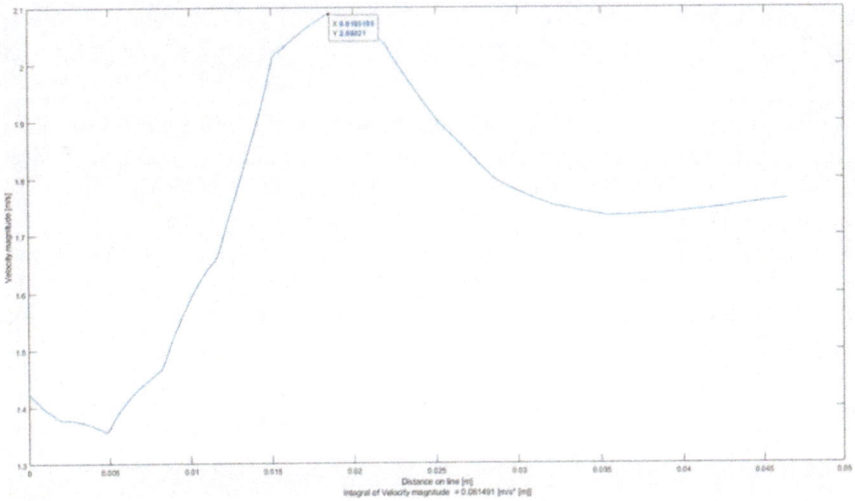

Fig. 4 Graph of droplet movement speed

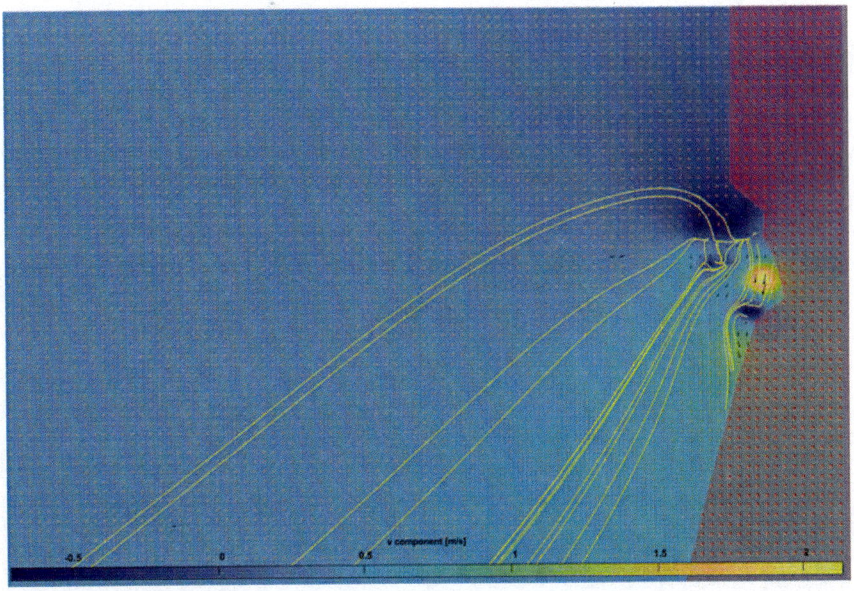

Fig. 5 Droplet Flow Direction

4 Conclusion

The experimental set-up of droplet measurement and distribution using a High-Speed Camera (HSC) was examined. The results showed that the design can observe the droplets released from the spray and when sneezing. It was discovered that droplets from a spray are produced by a thin layer of fluid that is centrifuged, while those generated when sneezing are through the mucosal sac. When someone sneezes, the sac expands and is dispersed and retracted. The effect of this deformation causes the sac to disintegrate and form droplets. The droplet size obtained varies from 3 to 127 μm and the average size from a sneeze is approximately 17 μm, where 39% was ≤ 5 μm and 61% was more than 5 μm.

Acknowledgements This research was funded by Lembaga Pengelola Dana Pendidikan (LPDP), managed by Indonesian Science Fund (ISF): RISPRO/KI/B1/TKL/5/15448/2020; Syiah Kuala University: 370/UN11.2.1/PT.01.03/PNBP/2021).

References

1. S. Bhattacharyya, K. Dey, A. R. Paul, and R. Biswas, "A novel CFD analysis to minimize the spread of COVID-19 virus in hospital isolation room," *Chaos, Solitons and Fractals*, vol. 139, Oct. 2020, doi: https://doi.org/10.1016/j.chaos.2020.110294.
2. Y. Yang *et al.*, "The deadly coronaviruses: The 2003 SARS pandemic and the 2020 novel coronavirus epidemic in China," *Journal of Autoimmunity*, vol. 109. Academic Press, May 01, 2020. doi: https://doi.org/10.1016/j.jaut.2020.102434.
3. World Health Organization, "Coronavirus disease 2019 Situation Report 76 05th April 2020," *World Heal. Organ.*, vol. 2019, no. April, p. 2633, 2020, [Online]. Available: https://www.who.int/emergencies/diseases/novel-coronavirus-2019
4. Maehata T et al (2021) A novel mask to prevent aerosolized droplet dispersion in endoscopic procedures during the coronavirus disease pandemic. Medicine (Baltimore) 100(26):e26048. https://doi.org/10.1097/MD.0000000000026048
5. I. M. Viola *et al.*, "Face Coverings, Aerosol Dispersion and Mitigation of Virus Transmission Risk," *IEEE Open J. Eng. Med. Biol.*, vol. 2, no. November 2020, pp. 26–35, 2021, doi: https://doi.org/10.1109/ojemb.2021.3053215.
6. B. Chow, "Experimental study on specialized ventilation system of airborne infection isolation room".
7. Kao PH, Yang RJ (2006) Virus diffusion in isolation rooms. J Hosp Infect 62(3):338–345. https://doi.org/10.1016/j.jhin.2005.07.019
8. Li Y et al (2007) Role of ventilation in airborne transmission of infectious agents in the built environment - A multidisciplinary systematic review. Indoor Air 17(1):2–18. https://doi.org/10.1111/j.1600-0668.2006.00445.x
9. Cho J (2019) Investigation on the contaminant distribution with improved ventilation system in hospital isolation rooms: Effect of supply and exhaust air diffuser configurations. Appl Therm Eng 148:208–218. https://doi.org/10.1016/j.applthermaleng.2018.11.023
10. Kalliomäki P, Hagström K, Itkonen H, Grönvall I, Koskela H (2019) Effectiveness of directional airflow in reducing containment failures in hospital isolation rooms generated by door opening. Build Environ 158:83–93. https://doi.org/10.1016/j.buildenv.2019.04.034

11. F. Wang, C. Chaerasari, D. Rakshit, I. Permana, and Kusnandar, "Performance improvement of a negative-pressurized isolation room for infection control," *Healthc.*, vol. 9, no. 8, Aug. 2021, doi: https://doi.org/10.3390/healthcare9081081.
12. Lind MC, Koskela H, Venås B, Vikan AW, Kalliomäki P, Harsem TT (2019) Designing simplified airborne infection isolation rooms to reduce infection rate in future pandemics. ASHRAE Trans 125:280–287
13. Bahl P, de Silva CM, Chughtai AA, MacIntyre CR, Doolan C (2020) An experimental framework to capture the flow dynamics of droplets expelled by a sneeze. Exp Fluids 61(8):1–9. https://doi.org/10.1007/s00348-020-03008-3
14. Vuorinen V et al (2020) Modelling aerosol transport and virus exposure with numerical simulations in relation to SARS-CoV-2 transmission by inhalation indoors. Saf Sci 130(100):1–49. https://doi.org/10.1016/j.ssci.2020.104866
15. N. S. Punn, S. K. Sonbhadra, S. Agarwal, and G. Rai, "Monitoring COVID-19 social distancing with person detection and tracking via fine-tuned YOLO v3 and Deepsort techniques," pp. 1–10, 2020, [Online]. Available: http://arxiv.org/abs/2005.01385
16. Busco G, Yang SR, Seo J, Hassan YA (2020) Sneezing and asymptomatic virus transmission. Phys Fluids 32(7):1–18. https://doi.org/10.1063/5.0019090
17. P. Bahl, C. De Silva, C. R. MacIntyre, S. Bhattacharjee, A. A. Chughtai, and C. Doolan, "Flow dynamics of droplets expelled during sneezing," *Phys. Fluids*, vol. 33, no. 11, 2021, doi: https://doi.org/10.1063/5.0067609.
18. Bourouiba L (2020) Turbulent Gas Clouds and Respiratory Pathogen Emissions: Potential Implications for Reducing Transmission of COVID-19. JAMA - J. Am. Med. Assoc. 323(18):1837–1838. https://doi.org/10.1001/jama.2020.4756
19. Scharfman BE, Techet AH, Bush JWM, Bourouiba L (2016) Visualization of sneeze ejecta: steps of fluid fragmentation leading to respiratory droplets. Exp Fluids 57(2):1–9. https://doi.org/10.1007/s00348-015-2078-4
20. D. Verreault, S. Moineau, and C. Duchaine, "Methods for Sampling of Airborne Viruses," vol. 72, no. 3, pp. 413–444, 2008, doi: https://doi.org/10.1128/MMBR.00002-08.
21. Li Y, Guan D, Wu C, Guo C (2017) An Experimental Study on the Transmission of Respiratory Droplet Aerosol Using the Available Agent in an Enclosed Chamber. Procedia Engineering 205:2365–2369. https://doi.org/10.1016/j.proeng.2017.10.321
22. P. Zhenxing Wu, PhD1, John R Craig, MD2, Guillermo Maza, MD1, 3, Chengyu Li, PhD1, 4, Bradley A Otto, MD1, Alexander A Farag, MD1, Ricardo L. Carrau, MD, FACS1, Kai Zhao, "Peak sinus pressures during sneezing in healthy controls and post-skull base surgery patients," *Physiol. Behav.*, vol. 130, no. 9, pp. 1–12, 2021, doi: https://doi.org/10.1002/lary.28400.Peak.

An Investigation of Friction and Wear Behavior of Hybrid Polymer Composites

M. Iqbal, L. B. Abhang, and P. M. karandikar

Abstract In automobile and consumer product industries, companies require long-life reliable, lightweight, and high performance self-lubricating composite materials, properties of these materials are promising materials for guide-ways, bearings, bushes, and piston rings that demand low friction and wear in the absence of wet lubrication. The present work investigates the tribological properties of hybrid composites. PEEK, PTFE reinforced with some new additives. Tribological tests are conducted at different temperatures on tribometer under dry sliding conditions and different environmental temperatures (room temperature) to elevated temperature. The experimental tests were conducted as per standard duration according to work conditions for all specimens. The hybrid composite exhibited lower friction and lower wear rate than those of without hybrid composite materials at both low and elevated temperatures. The combined effect of solid-state lubrication of hybrid PTFE composite was found satisfactory at higher temperature. Dominating friction and wear mechanisms in the hybrid polymer composites at the different temperatures were exhibited to get PTFE adhesion. Formation of transfer films on the steel surface was also analyzed at various temperatures.

Keywords Hybrid composites · Piston rings · Friction · Wear · Temperature

M. Iqbal (✉)
Department of Mechanical and Industrial Engineering, Universitas Syiah Kuala, Banda Aceh, Indonesia
e-mail: mohd.iqbal@unsk.ac.id

L. B. Abhang · P. M. karandikar
Mechanical Engineering Department, Pravara Rural Engineering College, Loni, SPPU, Pune, Maharashtra, India
e-mail: karandikarpm@pravaraengg.org.in

Irwansyah et al. (eds.), *Proceedings of the 4th International Conference on Experimental and Computational Mechanics in Engineering*, Lecture Notes in Mechanical Engineering, https://doi.org/10.1007/978-981-99-7495-5_36

1 Introduction

PEEK is a thermoplastic polymer with a glass transition temperature (Tg) of 143 °C whose composites are widely used in many tribological applications. For elevated service temperatures, a maximum operating temperature of 250 °C was reported [1]. Tribological performance of Nylon composites with nano additives was studied for self-lubrication purposes. A systematic study comparing the wear behavior of composites with nylon matrix (PA 66, 46, 12) and different nano additives and reinforcing additives like graphite, graphene, Mos_2, and Zro_2 has been carried out experimentally. In order to achieve a proper self-lubricant material for bearing cages [2], performance analysis of PTFE as journal bearing material. Among all the PTFE composites under experimental investigation due to friction and wear mechanism, the lowest wear loss is observed for the composite of PTFE reinforced with 15% Glass fiber + 5% MoS_2 particles by weight for all the values of loads and sliding velocity [3]. Hybrid PTFE glass hierarchical braided epoxy composites with self-lubricating and high tribological performance were investigated. In this study, hybrid PTFE/ glass fibers were employed in hierarchical braided structure as a composite reinforcement. Tibological experiments were performed on the hybrid composites. The results showed self-coefficient due to the PTFE transfer film formation, SEM micrographs confirmed the transfer film formation. Friction coefficient of 0.112, 0.105, and 0.096 were obtained under loading of 20, 30, and 40 N forces, respectively [4]. Studied comparison of tribological performance of PEEK, UHMWPE, glass fiber reinforced PTFE, and PTFE reinforced PEI composite materials was obtained under dry and lubricated conditions. The tribological performance of PEI + 10% PTFE, PEEK, UHMWPE polymers, and glass fiber reinforced PTFE composites under dry and lubricated environment conditions was obtained at room temperature [5]. The friction and wear response of neat and reinforced PEEK polymers were investigated by means of a reciprocating wear test with a 100Cr6 steel as a counter surface. From the study, it is concluded that the tribological behavior of filled polymers is unique for different materials and test parameters. Solid lubricants are fairly effective in reducing the coefficient of friction but not the wear rate in tribosystem [6]. Study the comparative wear behavior of two PEEK and PBIat high temperatures done experimentally. During the experiment, the polymer exhibits superior wear resistance due to formation of transfer films along with PTFE polymer. PBI has observed lower wear rate at low temperature because of its good thermomechanical characteristics [7]. The author Iqbal et al. [8], investigated the hole elimination in drilling Kevlar composite materials panel. Author [9] has designed and developed asymmetric gear, fabricated by 3-D printing method, and results were analyzed and compared to experiment as well as simulation by ANSYS method. Author [10] has studied the bending strength of the FML with different fiber content. Five panels of abaca FRP were prepared using hand-lay-up methods. Authors [11] have recently studied the friction and wear properties of hybrid polymer with different percentage of glass fiber rotating under dry and lubricated conditions by using Tribometer (TR-20). A Taguchi approach is used to design the experiment for scientific study for experimentation. Abhang

[12] has demonstrated the application of a simple multi-objective optimization on the basis of ratio analysis (MOORA) method to solve the multi-criteria (objective) optimization problem in engineering field. Author [13, 14] has studied effect of bio-lubricant in tribological systems of piston ring under the concentration of bio green (Jatrophaoil) by using response regression analysis and fuzzy-logic technique. Objective of this research work is mainly to investigate the tribological properties (friction and wear) of lightweight materials like peek and peek hybrid composites subjected to various ageing conditions. The lightweight material is used commonly in aerospace, automotive, oil, and gas industries etc.

2 Experimental Work

2.1 Materials and Compounding

The composite was prepared by compression as well as injection molding. First, PEEK, PTFE, Glass fibers, Bronze, and Mos_2 were mixed with different proportion for various batches with batch size 100 gm for compression molding and 15 gm for injunction molding. Preparations of samples are prepared and given their nominations as follows, as shown in Table 1 and Fig. 1a.

The mixtures of the PEEK with the fillers were prepared by twin screw-extruders and injection molding. The counterpart that is disc is selected with consideration of application like air compressor. So, the disc material was selected steel with grade EN 8 as picture shown in Fig. 1b. Disc hardness was checked by Brinell hardness tester and surface roughness 0.4 μm was checked at Vishal Engineering Nashik (Maharashtra, India).

Wear behavior of composites is studied using a pin-on-disc (Ducom madetribometer) apparatus under dry sliding condition which is available in the mechanical engineering laboratory, at PREC, Loni as shown in Fig. 2. Wear monitoring setup was supplied by DUCOM and the sliding wear test is performed according to ASTM G99. The coefficient of friction is determined on the same experimental setup. The amount of wear is determined by weighing specimens before and after the test.

Table 1 Prepared samples

S. No	Polymer materials	Name of sample
1	Pure PEEK	Pin1
2	PEEK (70 %wt) + PTFE (15 %wt) + glass fibers (15%)	Pin 2
3	PEEK(65 %wt) + PTFE(15 %wt) + Glass fibers(15 %wt) + MoS_2(5 %wt)	Pin 3
4	PEEK(50 %wt) + PTFE(15 %wt) + Glass fibers(15 %wt) + MoS_2(5 %wt) + Bronze (15 %wt)	Pin 4

Fig. 1 **a** Manufactured specimens for testing on Tribometer. **b** Manufactured polish steel discs with size Ø 165 × 8 mm

Specifications and testing parameters are MODEL: BD-NL, working pressure = 0.3471 mpa, speed = 450 to 850 rpm, stroke length = 120 mm, cylinder diameter = 100 mm. Operating parameters are load 15 N, sliding velocity = 2.5 m/s, temperature = 25 to 180 °C, with each test having a duration of 30 min.

During the experimental test, the friction force and the normal force were measured. The coefficient of friction is calculated from these measured forces according to formula, ratio of the friction force FF, and the normal force FN. The amount of wear is determined by weighing both specimens before and after the experimental test. The amount of wear depends upon the number of factors such as the applied load, machine characteristics, sliding speed, sliding distance, the environment, and the material properties like material behavior etc.

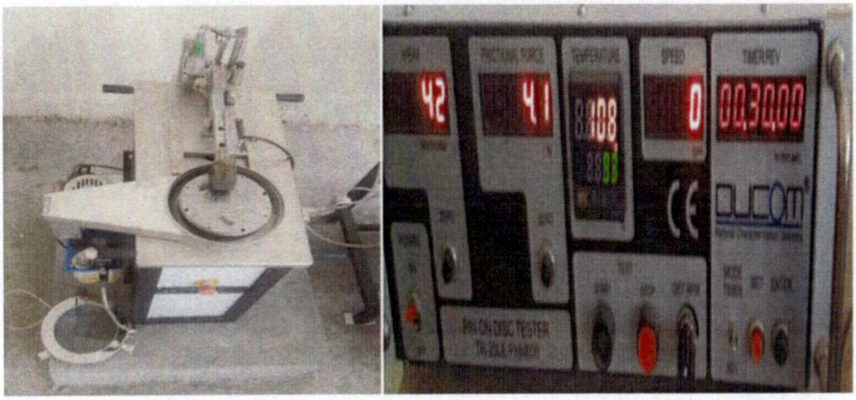

Fig. 2 Experimental setup of pin-on-disc for measurement of wear and force (frictional)

The specimen pin, with 4 mm diameter and 30 mm long, was run against the polished steel disc of grade EN-8 with surface roughness of 0.4μ. The contact pressure ranges from 0.25 to 0.5N/mm^2. The values of contact pressure were selected for dry lubricating air compressor applications in actual practice, like working pressure is 10mpa and sliding velocity was selected in the range from 1.8 to 3.4 m/s. To evaluate the durability of material at elevated temperature, the pin was kept in environment. During the test, first specimen p1 was held in an elevated temperature, load 15N, and sliding velocity 2.5 m/s as input parameter for wear and frictional force measurement. The calculated specific wear rate is also used to validate the reliability of the recorded height loss of specimens. The wear results were performed at the elevated temperature (150°). Each result is the average value of last three experimental data, and frictional coefficient and temperature of pin in the tables are the mean values during the steady state of the sliding process.

$$\text{Coefficient of Friction}(\mu) = \text{Frictional force/Normal load} \tag{1}$$

The wear rates have been determined as wear volume loss (mm^3) per unit sliding distance (m). The Archard's Law of wear expressed as,

$$Wr = V/FNL \tag{2}$$

where, Wr—wear rate (mm^3/m), V = wear volume (mm^3), L = sliding distance (m), FN = Normal load (N)

Wear rate in pure PEEK (Pin-p1) at elevated temperature was found higher and increased continuously up to the end of test as shown in Fig. 3. This indicates that the pure PEEK has less lubricating properties. Though PEEK has good mechanical property, sudden friction with counterpart at time of starting released heat and starts to deteriorate.

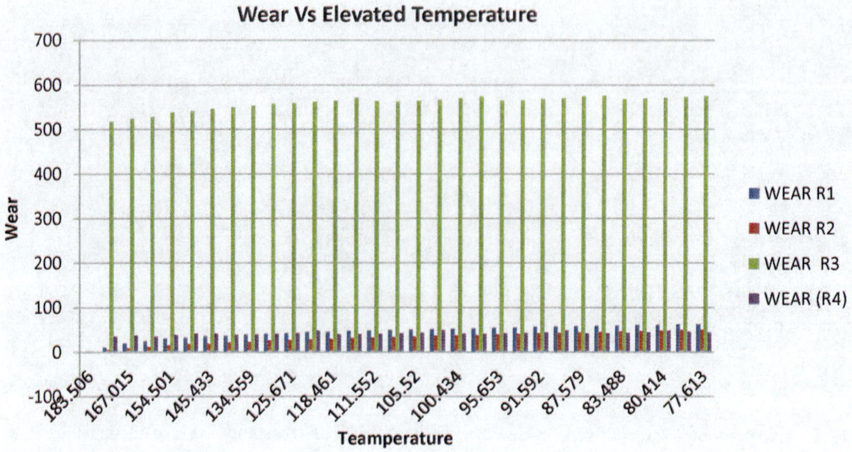

Fig. 3 Graph shows the Wear Vs Temperature after 30 min at elevated temperature

Figures 3 and 4—wear vs. time at room temperature and elevated temperature—indicate that initially, wear of pure PEEK (Pin P1) has tremendously increased because of sudden friction due to sudden application of normal load on the PIN -1 this is due to the asperities from counterpart and pin both are rubbed because of relative sliding motion between them and show abrasive effects on pin as well as counterpart resulting high wear loss. That's why the graph initially shows the high wear value. As the time increased, the asperities removed out from the pin have to be settled on the counterpart, and formed a solid lubricating film that causes the less wear. It has been found that wear again increased with respect to the time and became stable for value 49 micron after 30 min. The specific wear rate was found to be $K_0 = 2.65 \times 10^{-6} \text{mm}^3/\text{Nm}$.

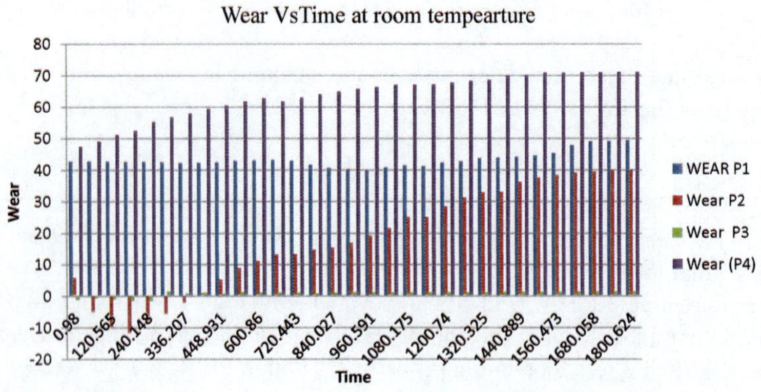

Fig. 4 Graph shows the wear vs time after 30 min at room temperature

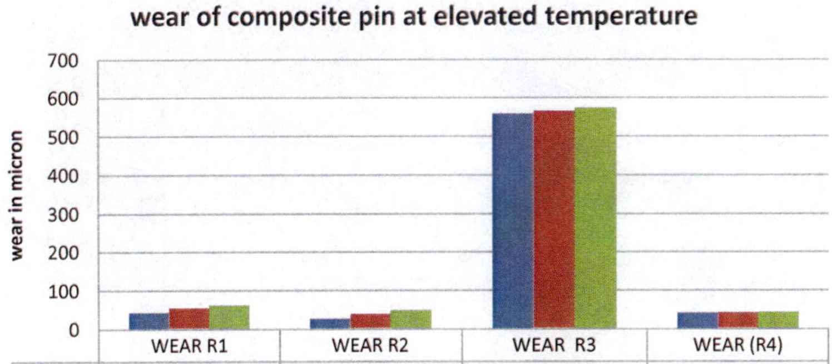

	WEAR R1	WEAR R2	WEAR R3	WEAR (R4)
10 min	42.5	25.72	558.49	41.64
20 min	54.21	38.42	564.27	41.3
30 min	62.02	49.6	574.22	43.22

Fig. 5 Wear of hybrid composite pin at elevated temperature

Figure 4 shows the effect of time on specific wear rate of PEEK at standard operating conditions. Initially, the wear rate was negative, because of sudden thermal expansion between disc and pin. The expansion deteriorates the mechanical properties due to which, base material initially loses some pin material in a vicinity of counter parts and adheres on the surface of pin. This transfer film creates resistance against wear properties. The addition of MoS_2 with glass fiber acts as self-lubricating agent, which helps to form a thin black gray color transfer film on disk along with thermal expansion of pin material. It was observed that with respective time, the wear rate increases and stabilizes after 6 mints. The specific wear rate is 3.31×10^{-11} mm^3/Nm after 30 min.

Composition of PEEK/PTFE reinforced with glass fiber (Pin R^2) shows (Figs. 5 and 6) less wear as compared to pure peek. This is because glass fiber has good mechanical property which acts as resistance to wear also overcome the thermal expansion of pin material.

The effect of time on wear rate of PEEK composites at standard operating conditions. It has been observed that specific wear for pin-p2 was initially somewhat high because of sudden normal and tangential force acted on the pin during the starting of test rig. As the time increases, the wear decreases. For first 20 min, the wear for pin p2 was very less and also showed the negative value of wear, that is -10 μm. This is due to the sudden start-up of the test rig and some particles were removed and settled on the wear track through which pin was moved. A very thin layer of material transfer is called transfer film on counter face. Initially, we get the high wear value but because of formation of transfer film, wear rate decreased. The specific wear rate is found to be 9.44×10^{-11} mm^3/Nm after 30 min.

It was observed that from Figs. 5 and 6 with addition of 5% MOS_2 and reducing 5% Peek from PEEK/PTFE enhanced the wear performance. The wear properties drop down vigorously as compared to pure PEEK and PEEK/PTFE/GF. The addition of inorganic filler material like Mos2, which is used as solid lubricant and has greater

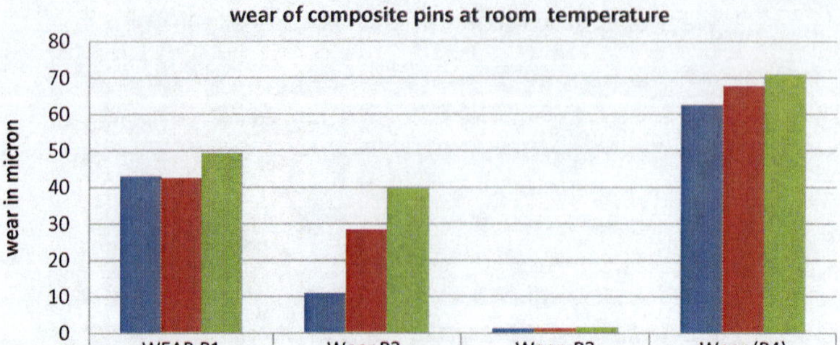

	WEAR P1	Wear P2	Wear P3	Wear (P4)
10 min	43	10.87	1.2	62.64
20 min	42.36	28.33	1.23	67.65
30 min	49.3	39.86	1.48	71.02

Fig. 6 Wear of hybrid composite pin at room temperature

lubricating properties, helps in formation of transfer film as compared to other filler material. The presence of GF also helps to provide resistance to wear because of its toughness and simultaneously, the formation of Mos2 transfer film around the pin surfaces occurs rapidly, leading to a decrease in the wear rate and coefficient of friction. Addition of glass fiber with Mos2 increased wear resistance, compressive strength, reduced cold flow, and increased rigidity. The percentage of MoS_2 with glass fiber should be 5%. above this it loses its mechanical properties show the greater wear rate.

The PIN-R4 has a composition of Bronze, PEEK/PTFE/GF/MoS_2. The significant result was found as compared to PIN-R3. During analysis, it was found that the Pin-4 shows low wear rate than PIN-R1, PIN-R2, and PIN-R3. And wear rate of PIN-4 is less as compared to PIN-R1 and PIN-R3. It was 2.66×10^{-3} mm^3 /Nm, because the percentage of bronze shows in Figs. 5 and 6. It improves the thermal conductivity of the composite materials. So it is used as additive materials. The specimen made by PEEK (50 %wt) + PTFE (15 %wt) + Glassfibers (15 %wt) + MoS_2 (5 %wt) + Bronze (15 %wt) has designated PIN-P4. The graph of pin-P4 shows the effect of time on wear rate or specific wear rate of PEEK at standard operating conditions. During the test, the thin brown color transfers film form on disc. The specific wear rate is found to be 2.98×10^{-9} mm^3/Nm after completion of test.

3 Comparative Study with Frictional Force and Coefficient of Friction

Coefficient of friction (μ) is a dimensionless number that is defined as the ratio between friction force and normal force. COF depends on the nature of the materials and surface roughness. Figures 7, 8 and 9 represent effect of time on frictional force for PIN-R4 at same operating conditions. It has been found that initially, the frictional force was 15 N at load 1 kg and contact pressure 0.347 Mpa and velocity 2.5 m/s. And coefficient of friction for this value is found to be $\mu = 0.433$. The value of frictional force of PIN-3 is very less as compared to all four pins but with respective time, this value increases. This indicates that there is significant relation between wear and frictional force. It has been found that the coefficient of friction for PIN-P4 is very less as compared to all other three pins as shown in Figs. 7 and 9. The least values of coefficient of friction are found to be 0.65. Figures 10 and 11 show coefficient of friction at elevated temperature and room temperature, respectively.

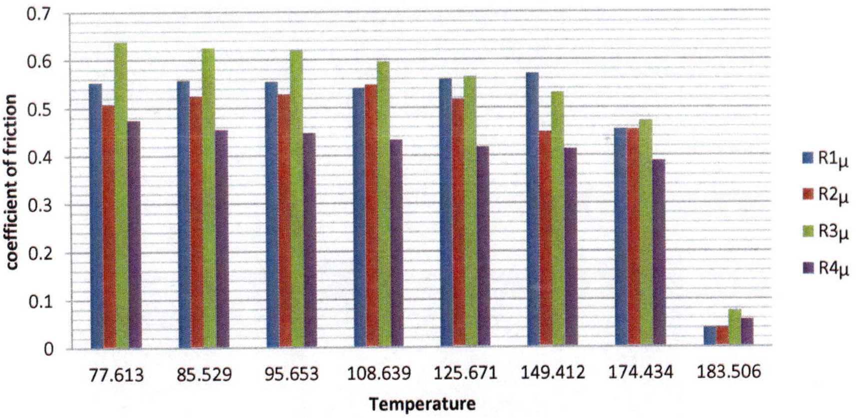

Fig. 7 Graph shows coefficient of friction versus elevated temperature after 30 min

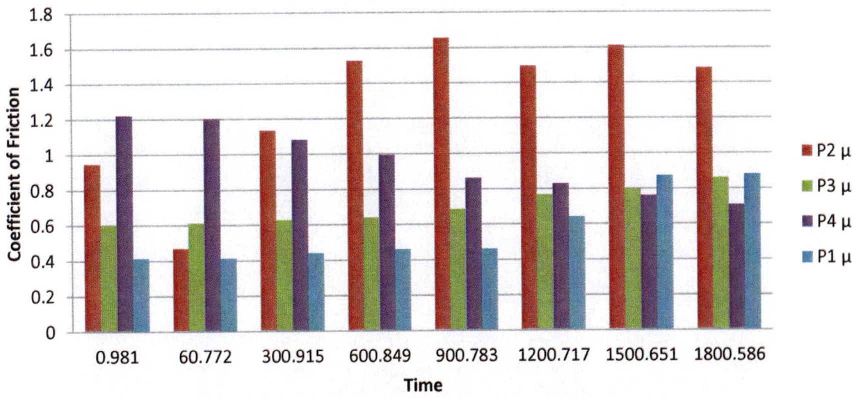

Fig. 8 Graph shows the result of coefficient of friction versus time after 30 min at room temperature

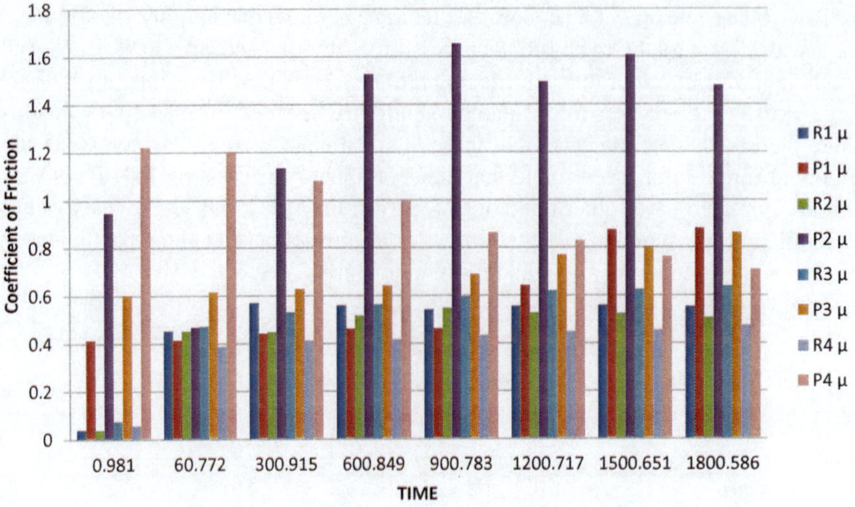

Fig. 9 Graph shows the comparison of coefficient of Friction after 30 min at room temperature and elevated temperature

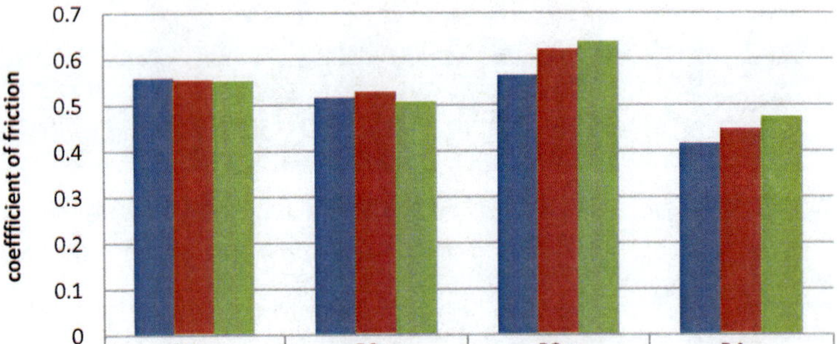

	R1 μ	R2 μ	R3 μ	R4 μ
10 min	0.558666667	0.516	0.564	0.416
20 min	0.554	0.527333333	0.62	0.446666667
30 min	0.552666667	0.507333333	0.637333333	0.474

Fig. 10 Coefficient of friction at elevated temperature

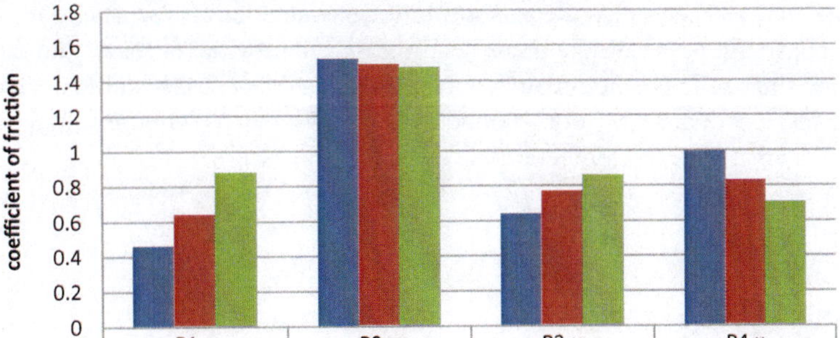

	P1 μ	P2 μ	P3 μ	P4 μ
10 min	0.46	1.526666667	0.64	0.993333333
20 min	0.64	1.493333333	0.766666667	0.826666667
30 min	0.88	1.48	0.86	0.706666667

Fig. 11 Coefficient of friction at room temperature

4 Conclusion

- Pure PEEK shows high wear rate, $k_0 = 0.355 \times 10^{-6}$mm^3/Nm at elevated temperature 180 °C and Pure PEEK shows high wear rate, $K_0 = 2.65 \times 10^{-6}$mm^3/Nm at room temperature
- The addition with glass fiber to PEEK/ PTFE improves its wear resistance properties at both conditions (room temperature and elevated temperature).
- It was found that the PIN –R2, 15% GF reinforced with PEEK / PTFE at elevated temperature, improves the friction & Wear behavior of polymer Composite. Also, Fiber Filler improves the creep resistance & composite strength of the PEEK composite and result enhance wear resistance $K_0 = 0{\cdot}108$ mm^3/N.m. At room temperature the pin-P2, it is $K_0 = 9.44 \times 10^{-11}$ mm^3/Nm.
- In pin-R3 Composite 5% Mos_2 with 15% GF in PEEK / PTFE exhibited at elevated temp. high wear resistance. $K_0 = 4.16 \times 10^{-3}$ mm^3 / N.m as compared to PIN – P3
- Mos_2 and Bronze work as a solid lubricant material in polymer composite. It shows low coefficient of friction and high wear resistance. In dry sliding condition, the lowest coefficient of friction is measured of normal load 15N. This is helpful to reduce the wear and increase wear life of element.
- By adding various filler materials in polymer composite improve the tribological properties. Pin-R4 shows greater wear resistance at same load condition and elevated temperature of 180 °C. Wear rate = 2.66×10^{-3} mm^3/N·m
- It was also found that at room temperature, the composite pin-P3 composition 5% MoS_2 with 15% glass fibers in PEEK/PTFE exhibited low coefficient of friction and high wear resistance. It shows very less wear rate than that of PEEK/PTFE reinforced with glass fibers. It is 3.331×10^{-11} mm^3/NM. This is because of proportion of solid lubricant.

- MoS_2 and Bronze are used as solid lubricant material. These materials easily enter the roughness valley and stably stay on disk. It provides necessary lubrication during sliding. This is helpful to reduce the wear and increase wear life of element.
- The test conducted at room temperature shows much better result than test conducted at elevated temperature.

References

1. Claveria I (2020) Tribological performance of Nylon composites with nano additives for self lubricationpupose
2. Desale DD, Pawar HB (2018) Performance analysis of polytetrafluoroethylene as journal bearing material. Procedia Manuf 20:414–419
3. Zahabi SR (2020) Hybrid PTFE -glass hierarchical braided epoxy composites with self-Lubricating and high tribological performance
4. Unal H, Mimaroglu A (2012) Comparison of tribological performance of PEEK, UHMWPE, glass fiber reinforced PTFE and PTFE reinforced PEI composite materials under dry and lubricated conditions. Polym Eng 32:349–354
5. Rodriguez V, Sukumaran J, Schlarb AK, DeBaets P (2016) Influence of solid lubricants on tribological properties of polyetheretherketone (PEEK). Tribol Int 103
6. Qu S, Penaranda J, Wang S-S (2016) Tribological behavior of PTFE/PEEK composite. In: Proceedings of the American society for composites: thirty-first technical conference
7. Chang L, Zhang G, Wang H, Fu K (2017) Comparative study on the wear behavior of two high-temperature-resistant polymers. Tribol Lett 65(2):34
8. Iqbal M, Firmansyah F, Tadjuddin M, Abhang LB (2020) The investigation of hole delamination in drilling Kevlar composite panel using HSS drill tool. Defect Diffus Forum 402:108–114. https://doi.org/10.4028/www.scientific.net/ddf.402.108
9. Abhang LB, Dekhane AG (2018) Use of Polylactic Acid (PLA) material for asymmetric spur gear manufactured by 3-D printing. In: Antony K, Davim J (eds) Advanced manufacturing and materials science. Lecture notes on multidisciplinary industrial engineering. Springer, Cham. https://doi.org/10.1007/978-3-319-76276-0_11
10. Iqbal M, Satrianda MS, Firsa T, Azan SA, Abhang LB (2021) Bending strength of fiber metal laminate based on abaca fiber reinforced polyester and aluminum alloy metal sheet. KEM 892:134–141. https://doi.org/10.4028/www.scientific.net/kem.892.134
11. Pawar DS, Abhang LB (2022) Experimental investigation and prediction of wear behavior of polymeric composites. Mater Today Proc. https://doi.org/10.1016/j.matpr.2022.04.100,Vol ume62,Part13,2022,Pages7255-7260
12. Abhang LB, Lqbal M, Hameedullah M (2020) Optimization of machining process parameters using moora method. Defect and Diffusion Forum 402(July 2020):81–89. https://doi.org/10.4028/www.scientific.net/ddf.402.81
13. Sanjay MR, Abhang LB (2021) Experimental analysis of effect of bio-lubricant between tribological systems of piston ring under the Jatropha Oil. In: Techno-Societal 2020, 2021. Springer, Cham, pp. 863–872. ISBN: 978-3-030-69920-8
14. Abhang LB, Hameedullah M (2021) Modeling and analysis of surface roughness with statistical and soft computing approach. AST 106:109–115. https://doi.org/10.4028/www.scientific.net/ast.106.109

Optimization of Production Surface Facilities in Mature Gas Compression System in APO and Point B with Economic Valuation Strategy

Edy Kurniawan, Helmi, Supriadi, Hasan Yudie Sastra, and Syifaul Huzni

Abstract Mature gas field optimization for new local company for operatorship shall begin with main production problems identification and cost readiness, for Gas Compression Area Cluster IV (APO) and Unit 26 (Point B) operation, the main problem is the production facility equipment needs to modify surface production facility to adapt the current operating conditions. Excessive fuel own use has more challenging for new operator, in additional, several joint agreements in Facilities Sharing Agreement also need to be considered. First and foremost, generating surface production facility resizing programs that directly contribute to reduce fuel own use should be put on priority with business scheme toward Operation Expenses i.e., rental equipment models should be put in first consideration as the company needs cash to do exploration in the area. Gas Compression is a vital system in the distribution of Gas. The existing gas booster compressor cluster IV (K/KGT-4920) and gas booster compressor unit 26 (K/KGT-2601) are 2 (two) stage LP/HP Centrifugal machine driven by Gas Turbine. Fuel gas consumption of K/KGT-4920 in the amount of 3.8 MMSCFD and fuel gas consumption of K/KGT-2601 in the total amount of 6.2 MMSCFD. There is one project were in Cluster IV Gas Compression area that will be executed to reduce fuel own use in gas booster compressor package for Cluster IV and Unit 26 as well as increasing sales gas. Resizing of new gas booster compressor package cluster IV (K/KGT-4920) is adjusted to the gas booster compressor package unit 26 (K/KGT-2601) output by means directly delivered to the Inlet treating unit of Point B. The new size of gas booster compressor will reduce total amount of fuel gas consumption ± 6.9 MMSCFD, therefore, the equal gas sales will increase in the same amount that will contribute extra income of 37,464 USD on daily basis. Based on the project economic valuation from Q4 2023 up to Q4 2028, the expected Net Present Value (NPV) worth 27,800,000 USD after replacement of the existing compressor looks profitable.

E. Kurniawan (✉) · H. Y. Sastra · S. Huzni
Magister of Industrial Engineering Program, Syiah Kuala University, Banda Aceh, Indonesia
e-mail: edy.kurniawan@outlook.com

E. Kurniawan · Helmi · Supriadi
Aceh Upstream Oil and Gas Management Agency (BPMA), Banda Aceh, Indonesia

Irwansyah et al. (eds.), *Proceedings of the 4th International Conference on Experimental and Computational Mechanics in Engineering*, Lecture Notes in Mechanical Engineering, https://doi.org/10.1007/978-981-99-7495-5_37

Keywords Optimization · Gas compression · Production facility · Fuel own use · Economic valuation

1 Introduction

Arun field is well-known for its remarkable hydrocarbon resource in Indonesia. Discovered in 1971, the field has supported Liquefied Natural Gas (LNG) sales to Asian market [1]. The Arun field has been producing since 1977. Currently, more than 99% of expected ultimate gas recovery has been produced. Surface facilities were designed for a peak capacity of 3500 MMSCFD with multiple well clusters and production trains [2].

The first gas was produced in July 1977, and the first condensate was exported in October 1977, and the first LNG tanker was loaded in October 1978 [1].

Since 17 May, 2021, this field has been operated by PT Pema Global Energi (PGE) as a new operator (Production Sharing Contractor) to Aceh Upstream Oil and Gas Management Agency (hereinafter called "BPMA"). Previously, the field had been operated by PHE-NSB.

The production field area consists of Arun, South Lhoksukon A (SLS-A), and South Lhoksukon D (SLS-D), [3]. Gas production from the fields will be transported via 42-inch pipeline 32 km to Point B treating unit ex.PT Arun NGL and PHE-NSO Onshore for blending and further process with gas from PHE-NSO and from Triangle Pase Indonesia.

This study describes changing operation mode and resizing surface production facilities to reduce fuel own use consumption. The effective implementation of the concept to reduce cost and fuel own use shall be analyzed comprehensively. In the next stage, several surface facilities changing and resizing will be selected to additionally offer utilized innovations in various units, such as power generation system and gas compressor in Aceh Production Operation (APO) field and other ageing facility.

The issue of ageing facilities leads to an increased risk of loss of containment and other failures due to plant and equipment deterioration. The significance of deterioration and damage relates to the potential effect on the equipment's availability and reliability. Ageing facilities are no longer considered fully fit for purpose due to obsolescence in its integrity or functional performance [4].

The traditional approach to reduce the cost production and fuel consumption is to change the existing surface production facility which appropriates with the current operating condition (Fig. 1).

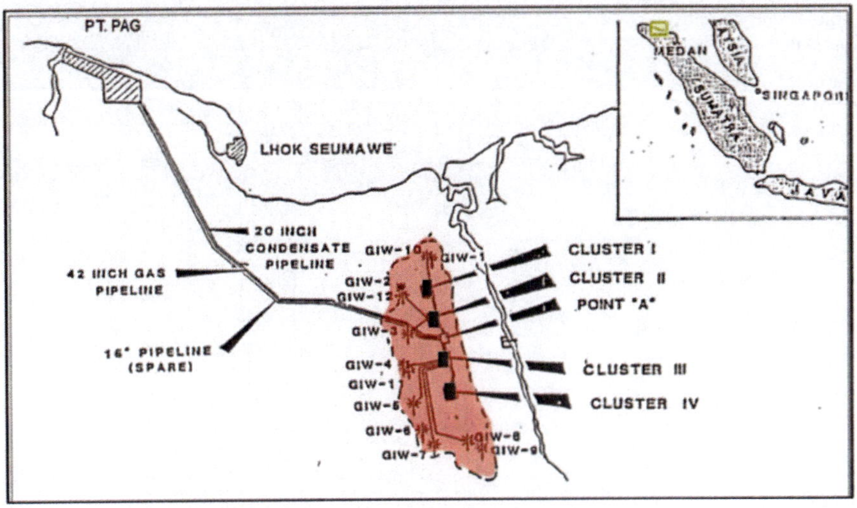

Fig. 1 Arun field map [5]

2 Data and Method

Currently, the production from Pema Global Energi (APO) approximately is 48 MMSFD and 1000 barrels of stabilized condensate per day. The APO field information is shown in Table 1

The Arun field consists of Cluster I, II, III, and IV and each cluster consisted of production wells and other surface facilities. Gas from SLS-A, SLS-D delivered to Cluster IV as well as gas from Cluster I, II, and III. In cluster IV, stream gas undergoes a three-phase separation and then refrigerated dehydration unit. Dry gas and separated condensate are then delivered to metering station in Point A by Booster Compressor K-4920 (located in Cluster-IV) before transfer to the Point B [2].

Feed gas from the PGE (APO field) still contains impurities. Feed gas from PGE and PHE NSO (outlet SRU unit) will be commingled shared facilities booster gas

Table 1 APO field information

Field information	Arun	SLS-A	SLS-D
Discovery	1971	1972	1991
No. of Wells	141	14	16
No. Active Wells	43	2	8
IGIP (Bscf)	14,112	306	696
Gp (Bscf)	13,267	286	616
Rem. Reserve (BCF)	56	1	3
Gas Prod. (MMscf/d)	36.62	2.18	6.19
RF (%)	94	94	89

Table 2 Gas composition from APO

No	Component	Composition Mol fraction
1	N_2	0.0045
2	CH_4	0.6845
3	H_2S	0.000
4	CO_2	0.2156
5	C_2H_6	0.0514
6	C_3H_8	0.0203
7	iC_4H_{10}	0.0051
8	nC_4H_{10}	0.0063
9	iC_5H_{12}	0.0033
10	nC_5H_{12}	0.0021
11	C_6H14+	0.0069
Total		10,000

compressor (KGT-2601), the feed gas will require further purification and processing to remove impurities and condensate stabilization before delivered to the buyer. Gas purification and processing in this step causes a large consumption of fuel for its own use (FOU).

The operation scheme at Point B using Facility Sharing Agreement (FSA) and the FSA is operated by PHE-NSO. In addition, there are cost for utilities service provided by PT. PAG in this area. The utilities services are intended for two main areas, namely Sulfur Recovery Unit (dedicated for PHE-NSO) and the other Gas Treating and Condensate Treating Unit (share between PGE and PHE NSO). For Gas Treating unit, the allocation of cost sharing will be based on ratio of the volume of feed gas rate. And for condensate treating unit, the allocation cost will be based on the ratio of the rate of volume of the condensate (Table 2).

After purified, gas is then delivered to buyer according to gas nomination. This gas is also used for operational needs at Point B. On the other hand, to support operating gas purification and processing, both parties need additional items including services and consumable such as manpower, chemical and catalyst, lubricant, and other supporting utilities.

Different strategies were created by previous operator (ExxonMobil) to maintain profitability of mature field. Arun Cluster Consolidation Project (ACCP) was started in 2004 to centralize gas dehydration and booster compression facilities into one cluster. The benefit from this project is reduced active assets by 60%, reducing operating expense by $350 K annually, and saving fuel gas consumption by 9 MMSCFD by deactivated 2-unit gas turbine booster compressor. The ACCP was followed by the Liquid Handling Project (LHP) to consolidate liquid separation and transfer facilities from four clusters into two. Deactivation of two condensate pumps is enabled until 20% reduction of power consumption. In addition to the condensate pumps, the LHP also eliminated several other types of equipment, including condensate-water separators, produced water treatment tanks, and produced water transfer pumps [2].

3 Cost Sharing, Feed Gas, and Gas Supply

The treated gas that is flowed from the shared facility to all buyers is a commingled gas sale of the parties. Sales gas for each party is measured based on feed gas after deduction with fuel consumption. Gas for fuel is taken from treated gas from shared facilities, the fuel gas is used for power plants, steam generators, gas turbine for booster compressor and NSO facilities.

Measurement of fuel consumption is carried out at the measuring point of each equipment. Fuel consumption for NSO treating unit will be borne by NSO, and the fuel consumption used for PGE treating unit (APO field) is also fully borne by PGE. Gas from the parties entering the sharing facilities is measured at the Gas Transition Point in volume units (MMSCFD) using the following measuring tools:

PGE gas is measured using flow meter FQI-56001/2, which is as flow meter for PGE. Gas from NSO is measured by calculating the difference from the combined gas measured by flow meter FQI-2602A, which is located downstream of booster compressor unit-26 (Fig. 2).

Gas for fuel is taken from the gas that has entered the shared facility, which will be used as fuel for booster compressor (Unit-26). The fuel consumption for booster compressor will be charged to each party (PGE and NSO) proportionally based on the feed gas volume of each party. Fuel consumption for power plants and steam generators is charged on a prorated basis based on the use of electricity and use of steam by each party. Feed gas volume from each party is measured by flow meter (custody) as shown in Table 3.

The recording of the feed gas volume measurement is carried out daily at 00:00–24:00. Analysis of gas composition to determine the GHV value is carried out on a weekly basis.

The volume of gas sales to each buyer is commingle gas sales by PGE and NSO, measured by the flow meter also as shown in Table 3.

The recording of the sales gas volume measurement is carried out daily at 00:00–24:00. Analysis of gas composition to obtain the GHV value will be carried out periodically in accordance with the agreement with each buyer. Recording and reading of gas usage in the system is carried out each day, weekly, and monthly.

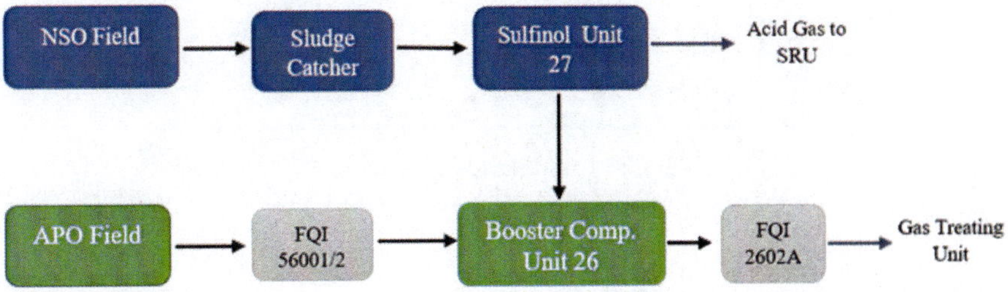

Fig. 2 APO field information

Table 3 Custody meter supply gas

Tag number	Services	Remark
FQI-6751	Total treated gas	
FQI-1905	Gas to Arbel	KIM and Jargas Lhoksukon
M-401/402	PT Pupuk Iskandar Muda	
M-403/404	PT Kertas Kraft Aceh (KKA)	PLN-KKA
CM	Jargas Aceh Utara	
CM	Jargas Lhokseumawe	

Sales gas for each party is calculated by prorating it based on the comparison of each party's feed gas after deducting for fuel consumption (PGE and NSO) as described below (Fig. 3).

$$\text{Sales Gas PGE} = \text{Total gas sales} \times \frac{\text{Feed Gas PGE} - \text{Fuel PGE}}{(\text{Feed Gas NSO} - \text{Fuel}) + (\text{Feed Gas PGE} - \text{Fuel PGE})} \quad (1)$$

$$\text{Sales Gas NSO} = \text{Total gas sales} \times \frac{\text{Feed Gas NSO} - \text{Fuel NSO}}{(\text{Feed Gas NSO} - \text{Fuel NSO}) + (\text{Feed Gas PGE} - \text{Fuel PGE})} \quad (2)$$

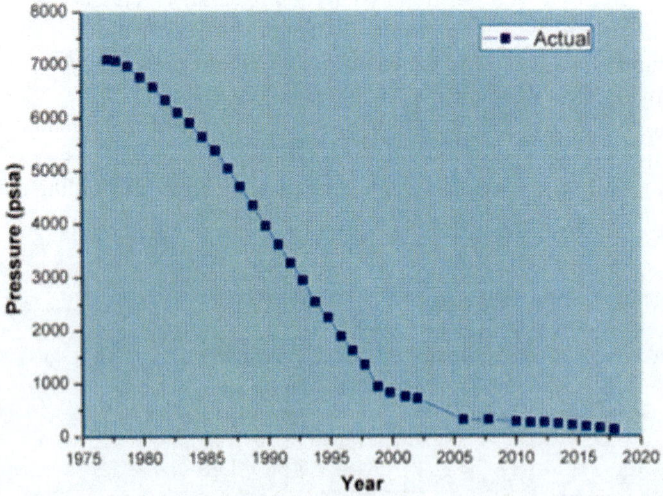

Fig. 3 Arun field reservoir pressure [3]

The calculation of fuel own use is carried out in energy units. The measurement of the volume of gas fuel own use and heating value at the measuring point is described in Table 4.

The measurement of gas volume is recorded daily at 00:00–24:00.

In many mature gas fields, production and fuel own use is a big issue. For APO field, optimal use of existing surface facilities or intervention of surface equipment, including installing the new facilities to adapt with current condition, is the key for reducing production costs and increasing economic value (Fig. 4).

Table 4 Measuring point of fuel own use

No	Equipment	Meter	Remark
1	Condensate stripper	FI-2760	Dedicated NSO
2	SRU plant	FI-6702	Dedicated NSO
3	HRSG	FI-9239	Share based on using of Steam
4	Power generation	FI-9001	Share based on electricity consumption after deduction with load of PAG
5	KGT-2601	FI-5731	Share based on feed gas volume
6	KGT-4501	FI-4562	Share based on feed gas volume

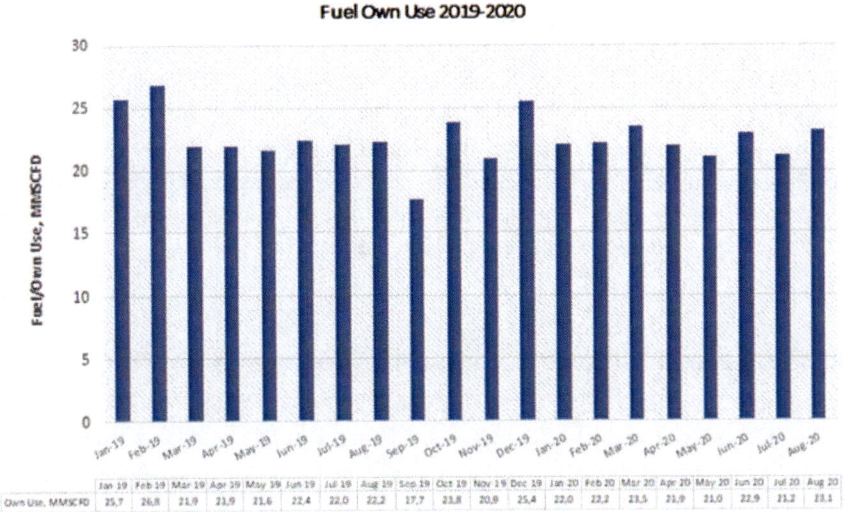

Fig. 4 Fuel own use 2019–2020

4 Result and Discussion

The Arun and SLS A/D fields are mature fields with the oil and gas depletion rate reaching 94%. The carbonate reservoir drain rate is generally below 80%. The rate of decline in natural production from year to year reaches 20%.

Apart from that, along with decline of gas production and excessive fuel own use in Block B, cause on production life becomes shorter and at certain times, all the production will be used for fuel gas only.

With the existing operating conditions, the economic value of the Arun field will be negative in 2022/2023. To prevent this event, PGE as new operator should define the objective to be achieved, including prioritizing and identifying effectiveness of production cost.

Average production of gas from APO field is 48 MMSCFD (March 2022). In this field, the biggest fuel own use consumption is for Gas Booster Compressor Cluster IV (K/KGT-4920) that consumes 3.8 MMSCFD. After deduction with fuel own use in APO field, feed gas should be delivered to Point B in amount of 42 MMSCFD. At point B, feed gas from APO field will commingle with gas from NSO (outlet of unit 27), with the amount 21 MMSCFD. The total feed gas commingled at Point B is 63 MMSCFD.

From the amount of feed gas, 20.24 MMSCFD is used for fuel own use, and this is heavily influencing the commercialization of sales gas. The biggest fuel own use in Point B is the treating unit that consumes + 10 MMSCFD, followed by power turbine generator, which consumes 7.5 MMSCFD. The amount of impurities gas is 18.44 MMSCFD. Finally, total amount sales gas is 21.3 MMSCFD with 15.05 MMSCFD for PGE portion, 5.25 MMSCFD. (This data is based on PGE daily report August 2021).

First and foremost, generating surface production facility intervention programs that directly contribute to reduce fuel own use (efficiency) should be put on priority. In this scenario, there is one project that located at Gas Compression Area Cluster IV that will be executed to reducing fuel own use and increasing sales gas.

5 Replacement Booster Compressor Package KGT-4920

The project is replacing the existing Booster Compressor Cluster IV (K/KGT-4920), which has a design capacity of 400 MMSCFD with a new booster compressor unit that adjusts to the feed in amount of 50 MMSCFD.

Initially from the existing flowline design from APO to Point B before the reservoir experienced decrease pressure, Booster Compressor Cluster IV (K/KGT-4920) sent feed gas directly to the Inlet Point B treating unit.

With the current operating condition, the suction pressure is ± 30 psig at the inlet flange of Booster Compressor Cluster IV (K/KGT-4920) and final stage discharge pressure at the outlet flange is ± 380 psig.

Unit 26 is a booster compressor (K/KGT-2601) which aims to increase feed gas pressure exit from Cluster IV to the Inlet treating unit of Point B. The feed gas comes from PGE gas and PHE NSO gas.

The inlet pressure of K/KGT-2601 is ± 265 psig and outlet pressure is ± 738 psig. To compress this amount of untreated gas, K/KGT-2601 requires treated gas fuel consumption of 6.2 MMSCFD, the amount of fuel is not much different from the design fuel requirement. This amount of total fuel gas consumption is the portion divided between PGE and PHE NSO based on throughput feed gas. The fuel gas consumption of K/KGT-4920 is 3.8 MMSCFD (Figs. 5 and 6).

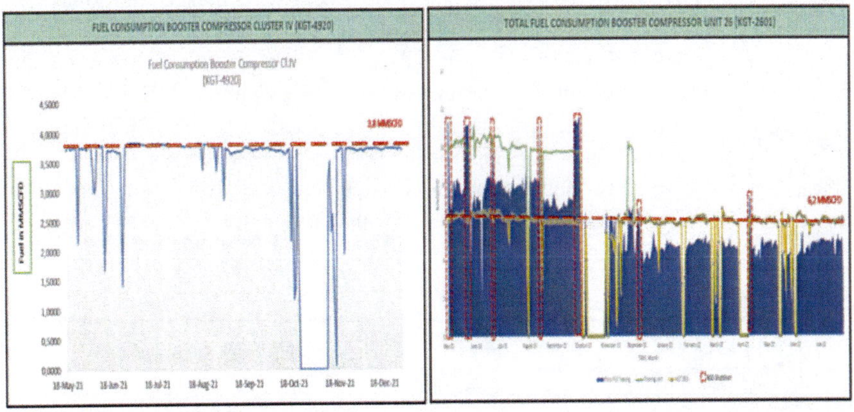

Fig. 5 Fuel consumption of KGT-4920 and KGT-2601

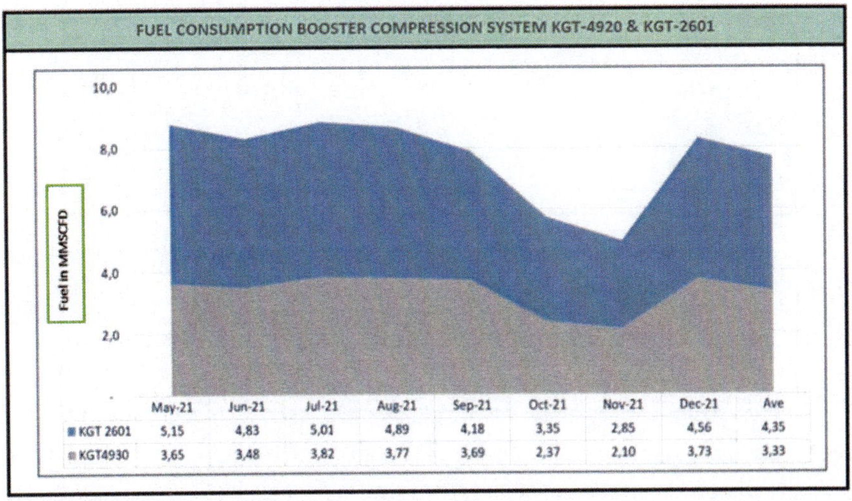

Fig. 6 PGE portion fuel consumption of KGT-4920 and KGT-2601

By reducing fuel gas consumption at the booster compressor unit by replacing the booster compressor with a smaller capacity with discharge pressure ± 800 psig, is adjusted to the gas booster compressor package unit 26 (K/KGT-2601) output by mean directly delivered to the Inlet treating unit of Point B. Decrease in fuel gas consumption can be used as additional gas sales to increase gas revenue.

Due to the high compression ratio between discharge and suction pressure, the most suitable type of Compressor in Cluster IV to meet the current operating condition should be a Reciprocating machine.

The new booster compressor in Cluster IV will be selected with fuel gas consumption maximum 1.8–2.0 MMSCFD, therefore will be a fuel saving from the unit Cluster IV dan unit 26 can reach ± 6.9 MMSCFD and this fuel saving can be additional sales gas for the portion PGE itself.

In addition, PHE NSO will also replace the existing booster compressor unit 26 with suitable capacity with fuel gas consumption maximum 0.5–0.7 MMSCFD, therefore will be a fuel saving from the unit 26 can reach ± 1.3 MMSCFD and this fuel saving can be additional sales gas for the portion PHE NSO (Fig. 7).

Raise in economically viable awareness of current operation mode has led to increased fuel efficiency at APO and point B Gas Compression area, almost ± 7.4 MMSCFD. This fuel saving provides economic advantage for PGE and PHE NSO.

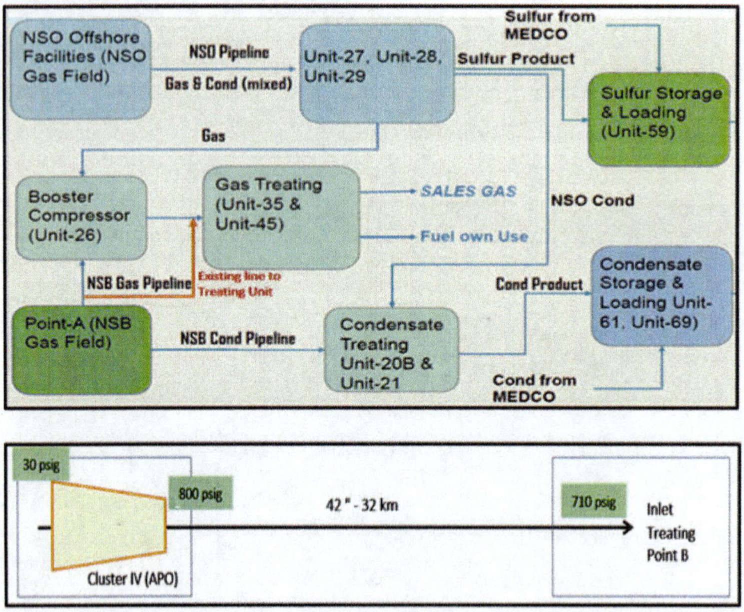

Fig. 7 Schematic gas compression to treating unit point-B

6 Project Economic Cost and Benefit Valuation

Fuel consumption for a reciprocating compressor is lower than centrifugal compressor unit (existing) for same condition on operation. The reciprocating compressors are widely used in many industries related to the compression and supply gas to consumers. Compressor units can be made on single frame design. With interstage devices and all necessary piping, placed on a single platform with a compressor. The modern automation system of the compressor units guarantees the safety and easy use of the equipment.

The project for replacement of the Booster Compressor Package Cluster IV is planned to be executed in the current year. The plan is to split the execution in 2 (two) separate contracts as follows:

1. Contract for Rental, Operation & Maintenance (O&M) On-skid Booster Compressor Package for the period of 60 (sixty) months.
2. Contract EPCI Off-skid Booster Compressor Package, which consists of Civil foundation, Piping interconnecting, Electrical and Instrumentation interface, etc.

The economic benefits of replacing the Booster Compressor Cluster IV, the savings in gas fuel consumption as well as operation and maintenance costs obtained for 60 (sixty) months period from the existing Booster Compressor Cluster IV and the existing Booster Compressor Unit 26 are shown in Table 5.

The best method for arriving at the decision for project investment is usually by applying methods like Net Present Value (NPV) and Internal Return Rate (IRR) calculation, which will provide financial and economic viability for the investment.

The estimated Cumulative Cash Portion as described in Table 6, also known as Net Present Value (NPV), earned for the duration of the project from Q4 2023 to Q4 2028, is to the tune of $27.8 million USD and the calculated value of the Internal Rate of Return (IRR) is 159%. Operating expenses are based on rental costs of the unit new booster compressor package and EPCI Off-skid booster compressor package work. Assuming the inflation rate is around 5.0% per year.

Detail breakdown calculation of NPV and IRR are shown in Table 6.

7 A Sustainable Efficiency Effort

Currently, the portion of PGE fuel at point B is 13 MMSCFD. In the next option, the new gas booster compressor KGT-4920, located in Cluster IV with a discharge pressure of 800 psig, will not be delivered to point B for blending with gas from NSO. And for this option, the gas purification and treating unit will be installed at Cluster IV.

As mentioned above, gas purification and treating unit will be installed at cluster IV. This project Cluster IV purification and treating unit is designed to treat raw gas from APO field and then directly delivered to the buyer and no further processing at

Table 5 Benefit of fuel and O&M cost saving

Fuel Cost Saving Wise (USD)									
Item description	Value	Unit	Daily saving	2023	2024	2025	2026	2027	2028
Benefit cost of fuel saving									
Total Saving Fuel (New Reciprocating Compressor Cluster IV)	1.8	MMSCFD	8190	–	2,907,450	2,907,450	2,907,450	2,907,450	2,907,450
GHV Fuel Gas	910	BTU/SCF		–					
Total Saving Fuel (By Passing Booster Compressor Unit 26-Point B)	5.1	MMSCFD	27,795	–	9,867,225	9,867,225	9,867,225	9,867,225	9,867,225
GHV Fuel Gas	1090	BTU/SCF		–					
1 Years (Reliability > 97%)	355	Day		–					
Gas prices	5.0	USD		–					
Operation & Maintenance (O&M) cost									
Gas Turbine Compressor KGT-4920 Preventive Maintenance & Service			–	–	200,000	200,000	100,000	200,000	200,000
Equipment certification			–	–	–	50,000	–	50,000	–
Two Years Spare Parts for Gas Turbine Compressor			–	–	–	–	400,000	–	400,000
Major Inspection & Services (KGT-4920)			–	–	–	–	900,000	–	–
Gas Turbine Compressor KGT-2601 Preventive Maintenance & Service	66.7	%	–	–	133,333	133,333	66,667	133,333	133,333
Equipment certification	66.7	%	–	–	–	33,333	–	33,333	–
Two Years Spare Parts for Gas Turbine Compressor	66.7	%	–	–	–	–	266,667	–	266,667

(continued)

Table 5 (continued)

Fuel Cost Saving Wise (USD)									
Item description	Value	Unit	Daily saving	2023	2024	2025	2026	2027	2028
Major Inspection & Services (KGT-2601)	66.7	%	–	–	–	–	600,000	–	–
Total Benefit Fuel and O&M Saving per Year:		USD	37.465	–	13,108,008	13,191,341	15,108,009	13,191,341	13,774,675

Table 6 Net present value and internal return rate project analysis

OPEX Scheme Installation New Booster Compressor Cluster IV			Annual Sales (Fuel O&M Saving)	Operating expenses	Gross profit	Net income (Tax 10%)	Present value (PV)	Cumulative cash portion
Interest or Inflation Rate:	**5**	%						
Breakdown in Year 2023 (Initial Investment)		USD	0	4,055,000	– 4,055,000	– 4,055,000	– 4,055,000	– 4,055,000
Breakdown in Year 2024		USD	13,108,008	5,474,560	7,633,448	6,870,104	6,542,956	2,487,956
Breakdown in Year 2025		USD	13,191,342	5,474,560	7,716,782	6,945,104	6,299,414	8,787,369
Breakdown in Year 2026		USD	15,108,008	5,474,560	9,633,448	8,670,104	7,489,561	16,276,931
Breakdown in Year 2027		USD	13,191,342	5,474,560	7,716,782	6.945.104	5,713,754	21,990,685
Breakdown in Year 2028		USD	13,774,675	5,474,560	8,300,115	7.470.104	5,853,022	27,843,706
Sub Total		USD	68,373,375	31,427,800				
Total Net Present Value (NPV)		USD	**27.843.706**					
Payback Period (PP)	Year		**1**					
Internal Rate of Return (IRR)		%	**159.75**					

NPV shall positive for good investment
PP is very short which is very good
IRR far more better that Minimum attractive rate of return

Point B is needed. The purification and treating unit at Cluster IV includes installation of CO2 removal unit, TEG Dehydration Unit, H_2S removal and sales point of gas from PGE is located at Point A (Custody Meter). Potential customers to sales gas directly from APO field are PT PIM or PTGN.

This scenario will provide economic benefits for PGE which can increase gas sales by 12 MMSCFD. This option should be seriously considered to prevent economical issue of APO field production and the application of new concept has generated numerous opportunities to extend field life.

8 Conclusions

1. PGE as a new operator has a lot of challenges to operate Arun mature field. The new operator should be seriously considered innovative method by generating surface facilities to reduce excessive fuel own use and intervention for mature equipment to expand life of production profile.
2. Optimization of existing surface facilities or intervention of surface equipment, including installing the new facilities to adapt with current condition, is the key to reducing production costs and increasing economic value.
3. Without facility intervention program to reduce excessive fuel own use, at certain time, all the production from this field will be used for fuel gas only.
4. In view of the excessive fuel consumption and aging facilities, which have operational and reliability challenges, replacement of existing Booster Compressor is one of the solutions to improve efficiency, reliability, and integrity of the gas compression area.
5. Reduce fuel consumption and cost reduction for maintenance at APO and Point B is a cost reduction strategy and this is a key to a new operator to deal with economic issue in mature gas compression area.
6. It can be concluded that, based on the project economic evaluation from Q4 2023 up to Q4 2028, Internal Rate of Return (IRR) is estimated to be 159% and Net Present Value (NPV) is worth $ 27,800,000 USD which makes this project economically viable and profitable.

Acknowledgements The author would like to thank the management Aceh Upstream Oil and Gas Management Agency (BPMA) for their permission to publish this paper. The author would also like to acknowledge all kind of support from PT PGE team.

References

1. Pathak P, Fidra Y, Avida H, Kahar Z, Agnew M, Hidayat (2004) The Arun gas field in Indonesia: Resource management of a mature field. SPE 87042
2. Avida H, Soedarmo AA, Utama DP (2013) Aceh operations mature field management: Arun Assets Consolidation and North Sumatera Offshore (NSO) Liquid Handling. SPE-165929
3. Luthfi M, Nugroho BA, Setiyadi S, Hidajat IT (2019) Rectification of propane dehydration system in Aceh operation mature surface facilities. SPE-196391-MS
4. Horrocks P, Mansfield D, Parker, J Thomson, Atkinson T, Worlsley J (2010) Managing ageing plant a summary guide. RR823. www.he.gov.uk
5. Suhendro (2017) Review of 20 years hydrocarbon gas cycling in the Arun gas field. SPE-186280-MS

Combustion Test of Densified Arabica Coffee Pulp Biochar in a Purposely Built Chamber

Siti Nurjannah, Yoandra Andika, Adi Setiawan, Faisal Faisal, Muhammad Muhammad, and Lukman Hakim

Abstract The study of solid fuel combustion characteristics is still continuing due to increases in demand of clean and renewable source of energy. Agricultural waste such as coffee pulp is an alternative source of biomass which has not optimally been utilized. The aim of this study is to examine the characteristics of the solid fuel combustion of densified coffee pulp biochar in a purposely built furnace. Initially, a simple mechanical design of combustion test furnace was prepared equipped with data logger for measuring weight lost and temperature changes. As main testing fuel, arabica coffee pulp was pyrolyzed and densified by adding starch binder. For comparison, locally available commercial coconut briquette and young brown coal were also tested under similar condition. The weight of each fuel sample was ≤ 50 g. Based on combustion test results, required time for ignition of coffee pulp and coconut briquettes was 2 min, and it took 5 min for coal. The combustion rate of densified coffee pulp biochar, coconut briquette, and coal was 0.5 g/min, 0.59 g/min, and 0.23 g/min, respectively. The maximum combustion temperature of 480 °C at 35 min, 375.9 °C at 45 min, and 487 °C coal at 140 min was observed during the combustion test of densified coffee pulp biochar, coconut briquette, and coal, respectively.

Keywords Solid fuel combustion · Coffee pulp · Rate of combustion · Combustion temperature · Coconut briquette · Brown coal

S. Nurjannah
Magister Program in Renewable Engineering, Faculty of Engineering, Universitas Malikussaleh, Bukit Indah, 24352 Lhokseumawe, Indonesia

Y. Andika · A. Setiawan (✉) · F. Faisal
Mechanical Engineering Departement, Faculty of Engineering, Universitas Malikussaleh, Jalan Batam, Bukit Indah, Muara Satu, 24352 Lhokseumawe, Indonesia
e-mail: adis@unimal.ac.id

M. Muhammad · L. Hakim
Chemical Engineering Departement, Faculty of Engineering, Universitas Malikussaleh, Jalan Batam, Bukit Indah, Muara Satu, 24352 Lhokseumawe, Indonesia

Irwansyah et al. (eds.), *Proceedings of the 4th International Conference on Experimental and Computational Mechanics in Engineering*, Lecture Notes in Mechanical Engineering, https://doi.org/10.1007/978-981-99-7495-5_38

1 Introduction

Biomass is organic material produced through photosynthesis, either in the form of products or waste. Agricultural waste, biomass, is a great alternative energy source containing relatively large energy content [1]. In Indonesia, the biomass which can be used as an energy source is abundant. Wastes originating from animals and plants have the potential to be processed and developed [2]. Apart from utilizing waste, biomass as a product for energy sources is currently being developed rapidly. Oil palm, jatropha seeds, soybeans are several types of plants mainly used as raw materials for making biodiesel while sawdust, corn cobs, coconut husks, and husks are products that are often used as materials for making biobriquettes [3].

Biobriquette can be a solid fuel made from biomass raw materials with a mixture of a little adhesive. There are several types of biobriquette, which have been developed such as biobriquette from sugar cane, straw, sawn waste, coconut husk, and husks [4]. There have been many biobriquette manufacturing, but it is necessary to conduct more in-depth experiments on the factors that affect the quality of the biobriquette. By carrying out in-depth experiments, it is expected that the factors influencing the quality of biobriquettes can be known as one of the factors is the combustion rate and the burning characteristics of briquette fuel.

Coffee growth and consumption were known to increase every year [5]. This has directly caused more and more waste to be generated, while the management is still very old-fashioned. There are several previous studies that have been carried out related to the characteristics of combustion from coffee waste biomass. The research conducted by [3] studied the combustion ability of spent coffee ground biobriquettes on variations in the use of adhesives, which gave the result of a maximum combustion temperature at 533.4 °C and a combustion rate 0.09–0.20 g/s. The highest peak temperature is obtained by biobriquettes with the addition of adhesive 10% and at the percentage of air supply in the combustion process as much as 850%. Kang et al. (2017) studied combustion of spent coffee ground in a small boiler. They reported that spent coffee ground can be used in a 6.5 kW boiler and the emissions of O_2, CO, and NOx were found to be better than commercial gas and wood pellets [6]. Alchalil et al. (2021) studied the combustion characteristics from rice husk and coffee pulp biobriquettes with different densified pressures. The results showed that the biobriquettes of rice husks and coffee skins produced had calorific values, combustion rates, and ash content of 4764 cal/g, 0.019 g/s, and 12 wt% ash, respectively [7].

Setiawan et al. (2019) reported the combustion characteristics of coffee pulp biobriquettes with different addition of starch adhesive. In this study, coffee pulp biobriquettes were produced by carbonization of raw coffee pulp through a slow pyrolysis method, then crushed the particle size of coffee pulp charcoal into a mesh of 20 and densified with a pressure of 150 kg/cm^2. The results showed that the coffee pulp biobriquette had a calorific value of 4427.96–4662.98 cal/g, the combustion rate was 0.41–0.97 g/s, the maximum temperature was 450–488 °C, and 8.41–9.86% ash [8]. Based on the literature, it can be seen that biobriquettes have the ability for using as a solid fuel. The aim of this study is to examine the characteristics of the solid fuel combustion rate of densified coffee pulp biochar, coconut shell briquette, and coal in a purposely built furnace. Combustion temperature, combustion rate, and % of ash were determined and used as criteria for the best combustion condition. The results of the study are expected to provide an alternative choice for using waste as energy.

2 Materials and Methods

2.1 Materials

The solid fuels used in this study were coffee pulp biobriquettes, coconut shell briquettes, and coal. Biobriquette coffee pulp is prepared according to the procedure in the previous research report [8]. Coconut shell briquettes were purchased from local market in Lhokseumawe.

2.2 Design and Fabrication of Combustion Test Apparatus

Figure 1 shows the design chamber for evaluating the combustion characteristics of solid fuels. This chamber was a cylinder made from A106 grade carbon steel with 160 mm inner diameter, 170 mm outer diameter, and 170 mm height in the combustion chamber. The height of the chimney is 95 mm with a diameter of 20 mm, and the pipe for air entry has a diameter of 30 mm. The design of the tool is shown in Fig. 1. This equipment was then manufactured and tested using three solid fuels: coffee skin charcoal biobriquette, commercial briquettes, and coal with a weight of 50 g per fuel.

2.3 Experimental Procedure

Densified coffee pulp biochar was prepared following procedure reported in previous publication [8]. The weight of sample was set $\leq$ 50 g and put into the combustion chamber. Temperature inside the chamber and at the flue gas outlet was recorded by using type-k thermocouple sensors. This set-up was equipped with data logger. To start ignition, a portable gas burner was used. During the combustion process, the sample weight loss with respect to time was recorded every 1 min until the fuel has completely burnt out. The remaining ash after the combustion was then recorded.

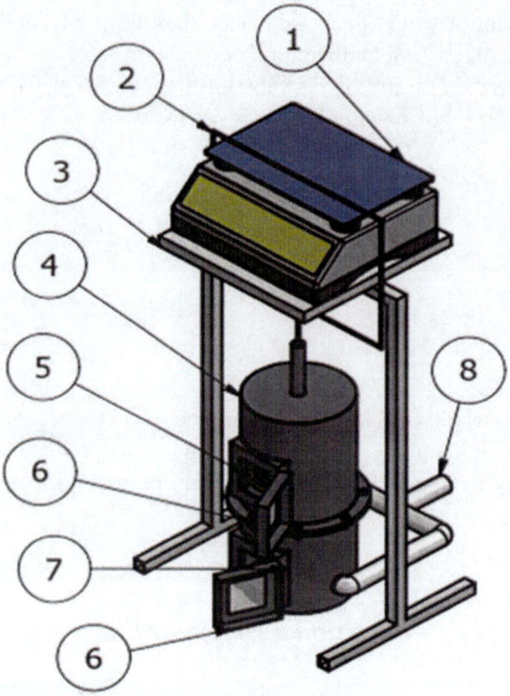

1. Digital scale
2. Tray hanger wire
3. Frame structure
4. Combustion chamber
5. Tray sample
6. Door of combustion chamber
7. Bottom part of combustion chamber
8. Primary air intake pipe

Fig. 1 Set-up of combustion test chamber

3 Result and Discussions

Table 1 details all parameters recorded during combustion experiments. Ignition of coffee pulp and coconut biobriquettes took 2 min while brown coal required 5 min to start combustion. During combustion test, the highest flame temperature was observed from combustion of coffee pulp briquette, i.e., 675 °C. Duration of combustion process of coffee pulp and coconut briquettes is 100 and 80 min, respectively, while burning duration of brown coal is much longer, i.e., 212 min.

Figure 2 shows the pictures of combustion test chamber after fabrication and installation. There are several measuring instruments used at the time of testing including digital scale, portable gas lighter, thermocouple, and data logger. The total volume of combustion chamber is 73.72 mm^3. Prior to combustion test experiment, the sensors were calibrated against standard analog instrument.

Table 1 Measurement results

Measurement	Densified coffee pulp biochar	Coconut briquettes	Brown coal
Fuel weight (g)	50.03	50.12	49.95
Ash weight (g)	16.25	17.01	4.99
Ignition time (minutes)	2	2	5
Duration of fuel burn (min)	100	84	212
T flame (°C)	675.2	651.7	556.7
Tmax fuel (°C)	480	375.9	489.4
Ti chimney (°C)	46.07	59.98	108.29
To chimney (°C)	28.66	29.53	36.06
Ti wall (°C)	78.92	81.81	120.04
To wall (°C)	54.53	59.74	105.2
Ti combustion chamber (°C)	88.56	96.36	198.63
To combustion chamber (°C)	29.2	31.07	31.38

Fig. 2 Combustion test apparatus

3.1 Combustion Rate Test

The combustion rate test was carried out by burning the fuel to find out the duration of flame while measuring the mass of the burning sample. This test was conducted to determine the quality of a fuel in term of the extent of the fuel burning. This data is necessary to be used later in its application. The rate of combustion is the time required per gram of fuel to be burnt. Table 2 provides the results of combustion tests of three kind of fuels.

Table 2 Results of combustion rate test

Solid fuel	Mass of fuel burned (g)	Fuel duration (minutes)	Combustion rate (g/min)
Densified coffee pulp biochar	50.03	100	0.5
Coconut briquette	50.12	84	0.59
Brown coal	49.95	212	0.23

Based on the data in Table 2, it can be seen that the combustion rate of coal is longer than the coffee pulp and coconut briquettes. This occurs due to the low volatile matter and density in coal. In addition, the duration of the burning solid fuel and temperature toward the ignition time can be seen in the Fig. 3. From this figure, it is observable that the temperature at the time of testing was fluctuated because the ash in the fuel remains attached when it burns. The coffee pulp biobriquette has an ignition temperature of 480 °C in 35 min, while the coconut briquette ignited at 375.9 °C in 45 min. The ignition temperature of brown coal was 487.4 °C in 135 min.

3.2 Ash Content

The ash content analysis was carried out by weighing the weight of ash and fuel sample. The purpose of this test is to determine the percentage of ash content left after combustion test, where the percentage of ash content in each fuel is different. Ash content from three different fuels is displayed in Fig. 4. Ash is the remaining part of the combustion product, and its main element is the mineral silica, which contains poor fuel quality. The higher the ash contents of a fuel, the lower the calorific value. Based on the test, it can be seen that the ash content of commercial briquettes (33.90%) is higher than coffee skin charcoal briquettes (32.30%). Meanwhile, coal has a relatively smaller ash content compared to the other two fuels, where the ash content in coal fuel is only 9.98%.

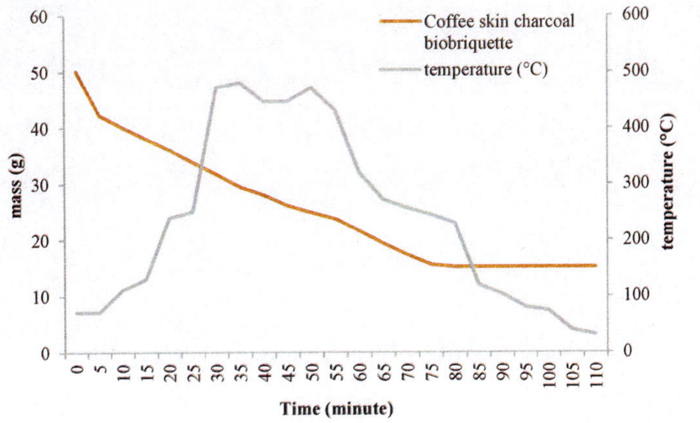

(a)

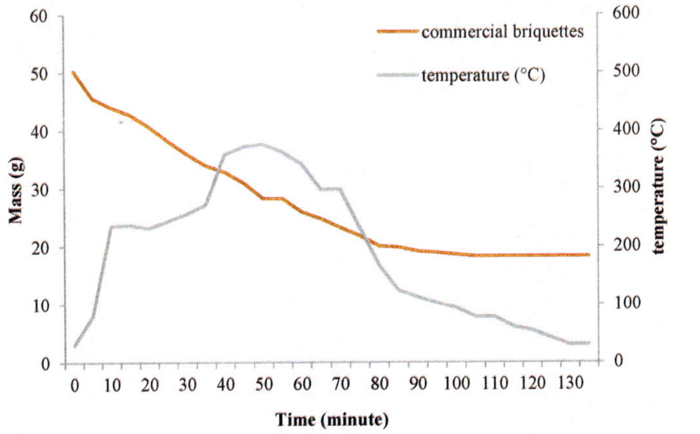

(b)

Fig. 3 The loss of weight and flame temperature as a function time **a** Coffee pulp briquette, **b** coconut briquette and **c** brown coal

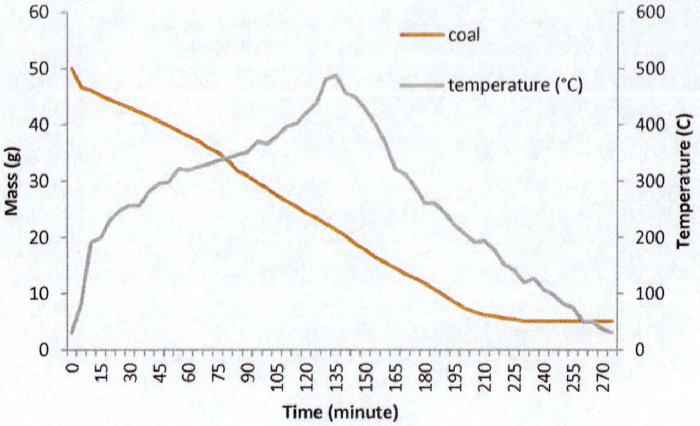

(c)

Fig. 3 (continued)

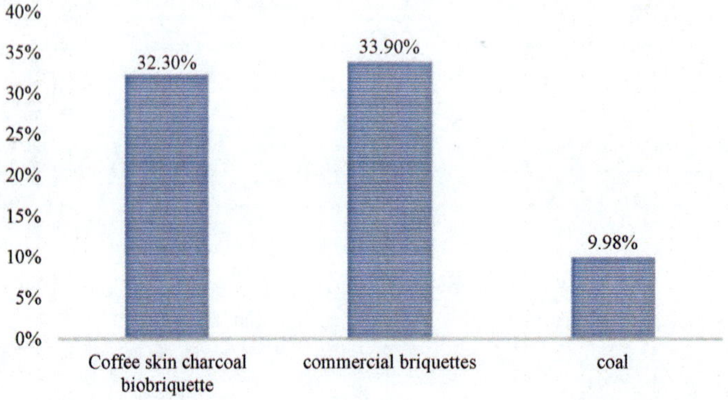

Fig. 4 Ash content in each fuel

4 Conclusion

The characteristics of densified coffee pulp biochar combustion has been studied in a purposely built combustion chamber and compared to coconut briquette and brown coal. The combustion rate of densified coffee pulp biochar is 0.5 g/min, comparable to coconut briquettes combustion rate. While, brown coal sample has the lowest combustion rate with only 0.23 g/min. In terms of ash content, coconut briquette has the highest percentage with 33.9% followed by coffee pulp briquette with ash content of 32.30%.

Acknowledgements We acknowledge the Directorate of Research, Technology, and Community Service, Ministry of Education, Culture, Research and Technology, Republic of Indonesia for their sponsorship under contract no. 028/E5/PG.02.00.PT/2022, sub-contract number 4/UN45.2.1/PT.01.03/III/2022.

References

1. Mitchell EJS, Gudka B, Whittaker C, Shield I, Price-Allison A, Maxwell D, Jones JM, Williams A (2020) The use of agricultural residues, wood briquettes and logs for small-scale domestic heating. Fuel Process Technol 210:106552. https://doi.org/10.1016/j.fuproc.2020.106552
2. Helwani Z, Ramli M, Rusyana A, Marlina M, Fatra W, Idroes GM, Suhendra R, Ashwie V, Mahlia TMI, Idroes R (2020) Alternative briquette material made from palm stem biomass mediated by glycerol crude of biodiesel byproducts as a natural adhesive. Processes. 8. https://doi.org/10.3390/pr8070777
3. Potip S, Wongwuttanasatian T (2018) Combustion characteristics of spent coffee ground mixed with crude glycerol briquette fuel. Combust Sci Technol 190:2030–2043. https://doi.org/10.1080/00102202.2018.1482888
4. Akolgo GA, Awafo EA, Essandoh EO, Owusu PA, Uba F, Adu-Poku KA (2021) Assessment of the potential of charred briquettes of sawdust, rice and coconut husks: Using water boiling and user acceptability tests. Sci African 12:e00789. https://doi.org/10.1016/j.sciaf.2021.e00789
5. USDA: Indonesia Coffee Annual Report 2020. USDA Foreign Agric Serv 9 (2020)
6. Kang SB, Oh HY, Kim JJ, Choi KS (2017) Characteristics of spent coffee ground as a fuel and combustion test in a small boiler (6.5 kW). Renew Energy 113:1208–1214. https://doi.org/10.1016/j.renene.2017.06.092
7. Alchalil, Setiawan A, Juwaini, Nur TB (2021) Effect of densification pressure on physical and combustion properties of binderless briquettes made from rice-husk and coffee-pulp. In: Lecture notes in mechanical engineering. Springer Science and Business Media Deutschland GmbH, pp 1–8
8. Setiawan, A., Hayat, F., Faisal, Nur, T.B.: Combustion characteristics of densified bio-char produced from Gayo Arabica coffee-pulp: Effect of binder. In: IOP Conference Series: Earth and Environmental Science. Institute of Physics Publishing (2019)

Design and Operation System of Small Robot for Water Spinach Planting

Arhami, Aulia Nanda Rizki, and Sabri

Abstract The innovation of agricultural equipment influences an increase in crop production. The process of planting vegetables is still a problem for the farmers, so it needs to be solved by designing tools in agriculture. This paper discusses the design and operation system of a small robot for planting spinach water. This study aims to design a small robot that can be used to plant spinach water seedlings quickly and with minimal labor. The planting operations for the spinach water seedlings include making holes, inserting seeds, and closing holes again. The mechanism for planting spinach water seeds is modeled as a robot that has a frame, electrical components, seed shelter, and soil hole punch. The design results of this robot can be the planting process approach of 25 kg of seed load. The operation of robot was tested using a control system on a Bluetooth-enabled cell phone. The robot control system operates on a standard electrical voltage range of 5–12 V and has a system performance of 16 min. With this workload, the robot can plant seeds on a 12 m × 0.5 m plot of soil.

Keywords Design · Agriculture robot · Operation system · Control system · Water spinach plant

Arhami (✉) · Sabri
Department of Mechanical Engineering and Industry, Syiah Kuala University, Banda Aceh, Aceh 23111, Indonesia
e-mail: arhami@usk.ac.id

Sabri
e-mail: sabri@unsyiah.ac.id

A. N. Rizki
Department of Mechanical Engineering, Buana Perjuangan Karawang University, Jawa Barat, Karawang 41361, Indonesia
e-mail: rizki.aulianandal@ubpkarawang.ac.id

Irwansyah et al. (eds.), *Proceedings of the 4th International Conference on Experimental and Computational Mechanics in Engineering*, Lecture Notes in Mechanical Engineering, https://doi.org/10.1007/978-981-99-7495-5_39

1 Introduction

The increase in population may outnumber the increase in food producers, one of which is vegetable producers. With these issues in mind, several engineers have developed the science of robotics and automation systems to create a robot that can work to increase the amount of food production, particularly in agriculture, to increase the amount of basic vegetable production that exists at this time.

Today, vegetables are a staple food that must be consumed daily to survive. According to the Indonesian Central Bureau of Statistics, the population grows by 1.2% per year. In this case, Indonesia must have adequate food supplies, including vegetables as a basic need. To provide smooth distribution of vegetables as staples, the increase in demand for vegetables must be balanced with the procurement of these vegetables.

However, as the population continues to grow and vegetable production continues to comply, vegetable production becomes less suitable for the needs of the Indonesian population. To address these issues, the researchers in this paper aim to design and develop a robot to carry out the process of planting plant seeds quickly according to procedures. This small robot is designed to plant water spinach seedlings quickly and without the need for additional labor. In general, the planting of water spinach seedlings has a planting procedure, which includes making a hole, inserting the seed, and closing it.

Robots are becoming less and less novel as time passes and technology becomes more sophisticated. Currently, robots can be found in almost every aspect of human life, including mobile robots. Because this robot can move freely, it is frequently used to transport goods. The mobile robot's weakness, however, is in navigating difficult and uncertain terrain. Complex robot movements require a navigation method that can handle problems and complete the given mission, such as observing its environment, interpreting information from sensors to improve knowledge of the robot's position, and environmental structure, and planning a route from the initial position to the destination position, avoiding obstacles, and controlling the angular speed and linear speed of the robot to reach the target [1]. The robot functions as follows. The weeding robot enters the plant beds with the assistance of the control and navigation system. The coordinates of the location of weeds in a row are determined automatically using a computer vision system [2].

The development of robots cannot be separated from the design and design to produce a robot shape by the mechanism of the planting process. Manufacturing includes design, material selection, assembly, and testing processes, but the design in question is design, design validation, design analysis, and production working drawings [3]. There are two general approaches to design a robotic for agricultural tasks. The first approach is to create and use a universal arm capable of performing a variety of tasks [4]. After comprehending the motion design, the motion must be adjusted to the sensor configuration to adjust the sensor and actuator navigation readings [5].

Depending on the model of the agricultural robot, there are two methods for acquiring turning angles. Navigational readings also affect the visual sensor of the robot, the robot will move according to the predetermined visual navigation. Robustness to changes in illumination, weather conditions, image background, and object appearance, for example, as plants grow, is required for agricultural robotic vision, as is sufficient accuracy and real-time performance to support onboard decision-making and vision-guided control of robotic systems. Active vision approaches that incorporate next-best-view planning may be required to ensure that all relevant information is available for robotic decision-making and control, for example, when the fruit or harvestable part of a crop is obscured by leaves or weeds [6].

The robot's motion will be perfect if the robot's support or chassis can support the load perfectly. The load-bearing strength also affects the resistance strength of the robot car chassis. A supporter with a less safety factor causes excessive loads and unstable motion to the speed of the robot car [7, 8].

The process of making and assembling electrical connections cannot be separated from the design of a robot car for planting water spinach seeds. Electricity is the primary component used to the power robot cars and control systems [9, 10].

This robot operates using a 12 V battery with a storage capacity of 7.5 Ah and the total load consumption of all the system is 20 W. So, it is necessary to take into account the length of time the robot will use the battery in every hour.

Following the calculation of the battery current is I = 20 W/12 V = 1.67 A. Hence, the usage time of the battery is 7.5 Ah/1.67 A = 4,5 h. Therefore, with battery deficiency assumption 20%, the usage time of the battery become to 4.5 h – (20% × 4.5 h) = 3.6 h. To build a robot car for growing water spinach seedlings, several aspects were previously described, including design construction, robot movement, visual navigation, control systems, and electrical system assembly. The operating system and testing by connecting to the mobile phone network. So, the motion system will automatically move forward, and there is a button to carry out the process of shedding seeds. Thus, it is hoped that this research will lead to technological advancements in agriculture.

2 Material and Method

The performance of the robot car for planting spinach water seedlings is comparable to that of conventional spinach water seedlings. The robot car's design must include a hole mechanism, a planter, and a manhole cover. There are several methods for producing a robot car suitable for planting spinach water seeds the design of a robot car for growing spinach water seed. Figure 1 depicts the procedure.

The method is explained by first reviewing several literature studies to get the desired robot shape. The process of analyzing how the process of planting water spinach seeds which will be converted into the form of a planting mechanism that will be carried out on the robot, will then be carried out. The planting process can be seen in Fig. 2.

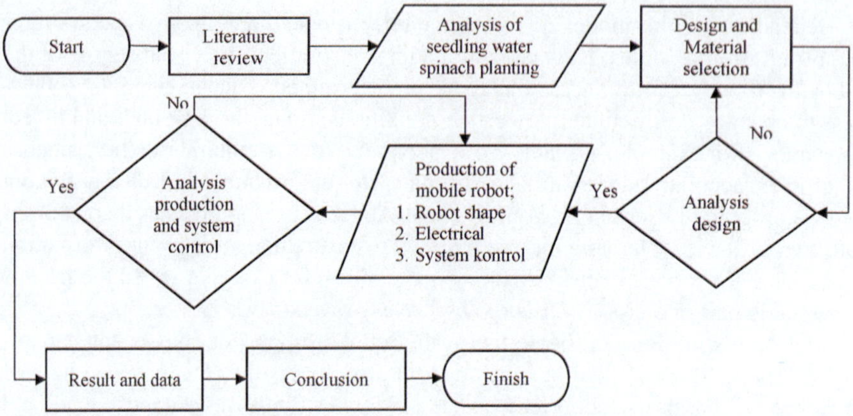

Fig. 1 Method of processing production mobile robot spinach water

Fig. 2 Process of water spinach planting

The process of planting spinach water seeds includes punching holes, inserting seeds, and closing holes. Based on this analysis, the design idea for designing a robotic mechanism for planting spinach water seedlings is developed, and the appearance of the design of the carrot robot for growing spinach water seedlings can be seen in Fig. 3.

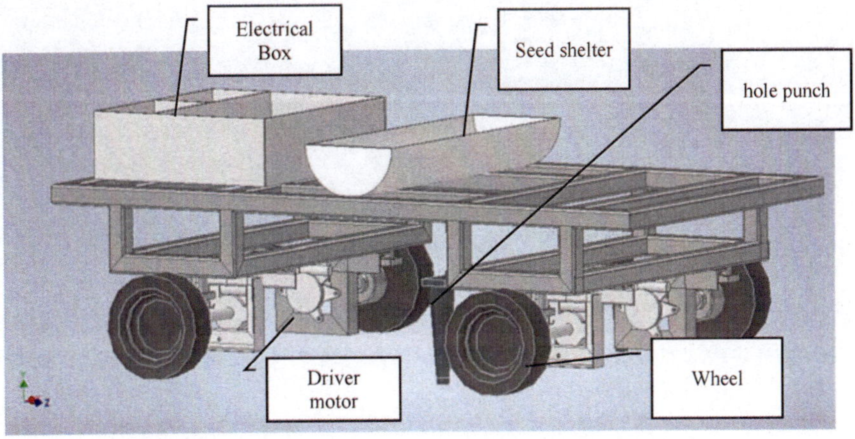

Fig. 3 Design of the mobile robot spinach water

The chassis, electrical housing, drive wheels, and seed storage are the main components of the water spinach seedling robot car, as shown in Fig. 3. The robot's motion system will operate by the control system commands that have already been connected and turned on; the control system will issue orders to the wheels, which will operate via the mobile control application. Figure 4 depicts the navigation procedure.

Several components are required to support the motion of the robot from the motion control navigation. Table 1 shows the components of the robot.

Following the design and material selection stages, the next step is to enter the results stage of the applied method, as shown in Fig. 3, analysis is carried out in the form of load analysis working on the robot car chassis, from the design, the process of making a robot car is carried out, motion analysis and control system analysis on robot cars.

3 Result

3.1 Design Model of Robot Frame

Loads acting on the chassis, such as load bearing loads and electrical system loads, are included in the design analysis. This design analysis only determines whether or not this chassis is appropriate for use on a water spinach seed growing robot car. Figure 5 depicts the placement of the load on the robot car for growing water spinach seedlings.

The total weight of the seed holder and the electrical load is 25 kg, resulting in a force of 196 N. To determine how much stress influences the chassis design of the robot car, a centralized load simulation is performed. By looking at the safety

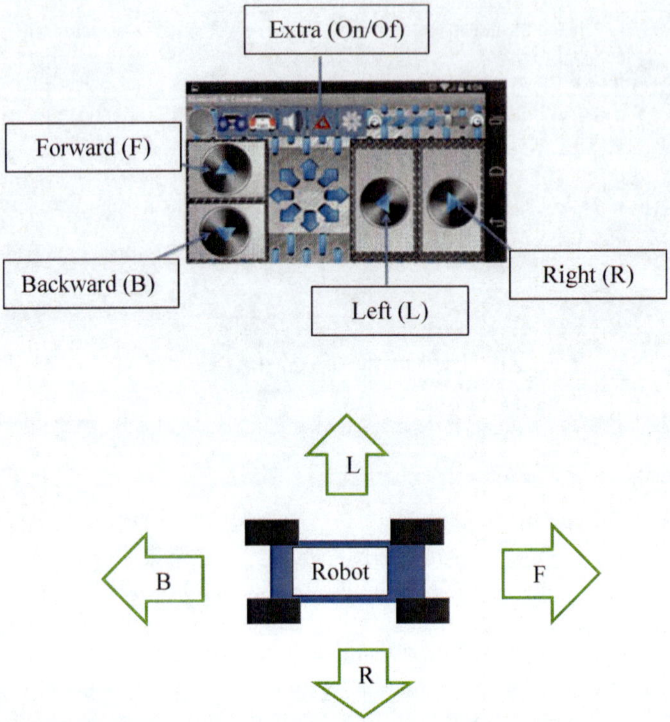

Fig. 4 Control Navigation of the Mobile Robot

factor, the normal load stress is a maximum of 5088 MPa for steel materials with dimensions of 30 × 30 × 2 mm. Figure 6 depicts the results of the stress from graph at a predetermined load.

Figure 6 shows that the influence of the chassis caused by the load causes the chassis in the middle position to deform and stress. But it is safe, and this design can be used because it has a maximum safety factor value of 3 on moving objects. The effect of stress in Fig. 6 is that it occurs only in 1 rod, namely the rod that is adjacent to the reservoir and the electrical box has a maximum force of 3.265 N on the FY symbol.

3.2 Mobile Robot and Electrical System Control

Figure 7 is an assembly form of each component consisting of a chassis, electrical box, wheels, and motor, and accompanied by operation test. Whereas, the Fig. 8 depicts the water spinach seed planting process.

Figure 8 depicts the robot's performance process, in which the robot's navigation is controlled by a cellular connected to Bluetooth, which sends commands to the

Table 1 Component mobile robot

Chassis	
Name	Function
Steel frame hollow 2.5 cm × 2.5 cm	As a support material for the entire component to be used
Motor DC 12 V	As the primary driver of the wheels, allowing the robot car to move by the control system
Wheel	As a mover
Box electrical	As storage of electrical components
Electrical	
Battery 12 V DC	To store current and distribute current to all electrical components so that they turn on
Arduino Mega	Robot car control center
Monster Moto Shield	DC rotation controller in CW or CCW
Step down	Reduce the excessive voltage on electrical components
Relay	Automatic circuit breaker controller
Bluetooth	Control network link between mobile and robotic systems
Motherboard	Electrical circuit main board
Cable jumper	Current connecting line

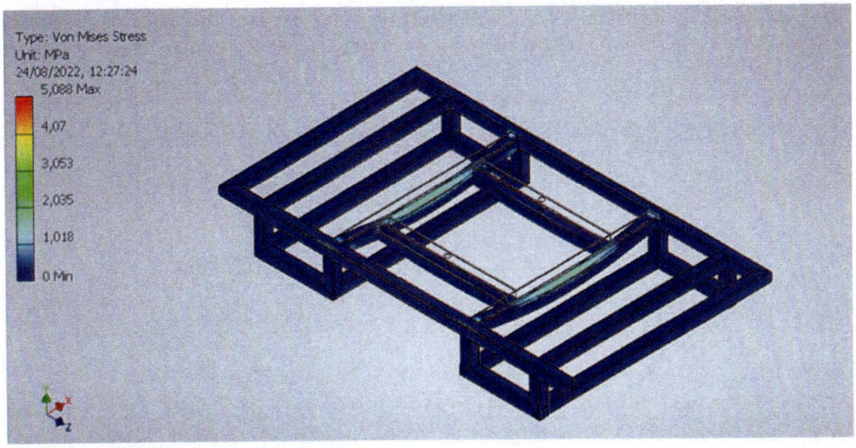

Fig. 5 Simulation of the chassis load

microprocessor or Arduino Uno, causing the robot to run by the commands on the cellular, but the navigation system or robot motion cannot be separated from the electrical system, which is always connected, so some electrical system analysis is also required. When the robot walks forward, the cellular gives the command to spill

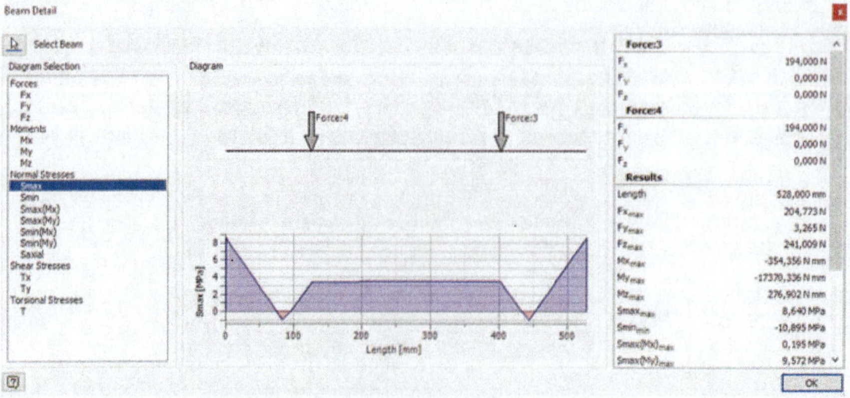

Fig. 6 Graph depicting the effect of load on the chassis

Fig. 7 Components assembly and operation test of small robot for water spinach planting

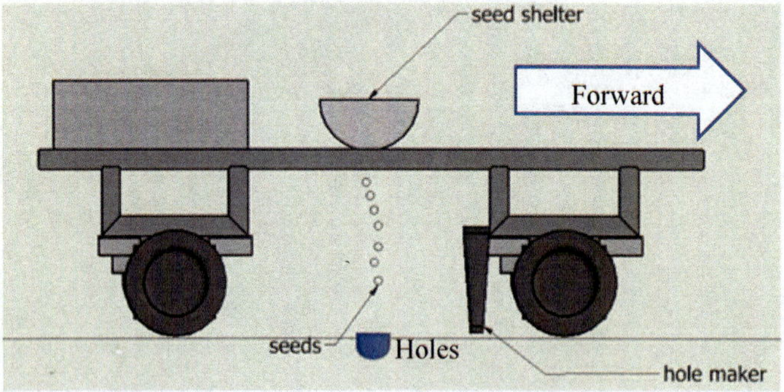

Fig. 8 Water spinach seed planting process

the seeds into the hole made by the hole punch. Using this mechanism, the seeds are guaranteed to enter the hole that has been formed.

The electrical display of the components described in Table 1 can be seen in Fig. 9, namely an analysis of the electrical system in the form of electric voltage and electric

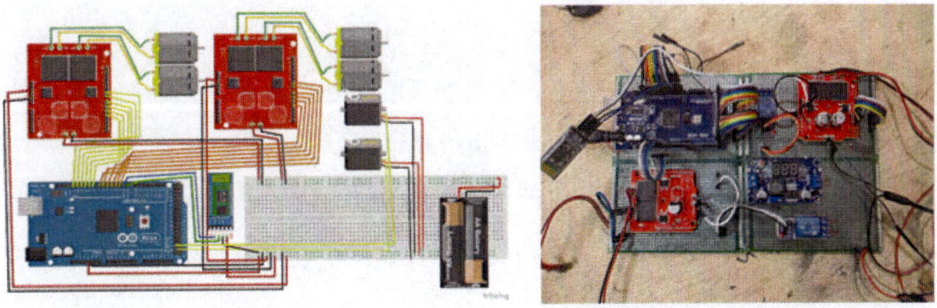

Fig. 9 Electrical system

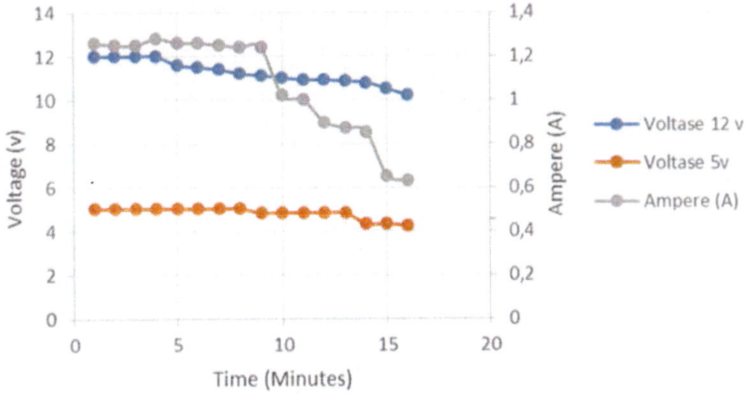

Fig. 10 The voltage changes during the operation

current, where a stable electric voltage has a stable electric current but if the electric current changes due to component division, the voltage also changes data. The graph in Fig. 10 is the result of this analysis.

Because this robot uses a 7.5 Ah battery, the electrical system is divided into two parts: one for the 5 V and one for the 12 V. Figure 10 shows the analysis of voltage changes as the robot moves forward, backward, and turns.

When the control command from the cellular cellphone uses the Bluetooth network, the robot car will drive according to the control system, when the robot car performs a maneuver, the electrical measurement of the system when the robot is working here. Figure 10 shows that the 12-V voltage is used for the 5-V propulsion for power to turn on the control system. If the working amperage continues to fall after 16 min, the voltage on the robot car also falls, but this does not affect the robot's performance. The robot can still operate at a voltage of 10 V and a current of 1 Amp. This system can plant seeds with a field size of 12 m × 0.5 m in about 16 min. Figure 11 depicts the results of the planting.

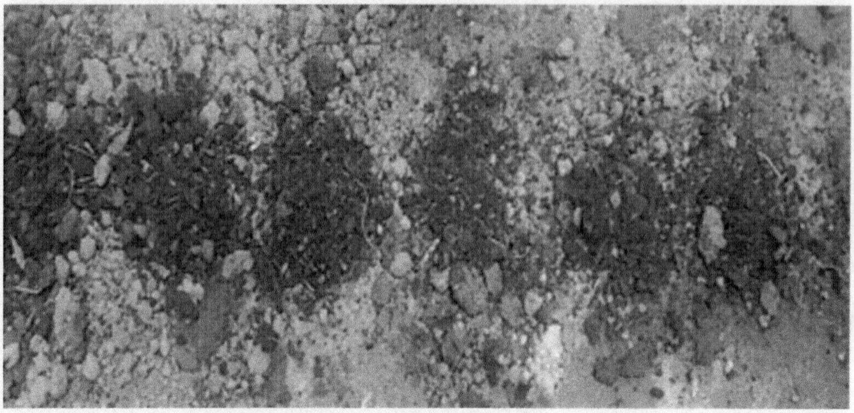

Fig. 11 The result of planting seeds

4 Conclusion

The design process and operation system of the small robot for water spinach were successfully done. Based on the results, several indicators of readiness in the image, namely the readiness of the results of the robot chassis analysis, the readiness of the electrical components, and the readiness to carry out the production process of the small robot. The operating system and testing were carried out. By connecting to the mobile phone network, the motion system will automatically move forward, and there is a button to carry out the process of shedding seeds. After the success of testing this robot, it will have an impact on farmers, allowing the process of planting seeds to be faster and more precise. The analysis of the robot's motion system is also considered successful because the electrical capacity operates by the specifications of the electrical components, ensuring that the robot is safe to use and operate. This robot car control system's voltage ranges from 12 to 5 V, with an average amperage of 2 Amps.

References

1. Usuman I, Prijodiprodjo W, Sejati PA (2019) "Navigasi robot mobile pada lingkungan tak pasti dengan pendekatan behavior based control. IJEIS Indonesian J Electron Instrum Syst 9(1):97. https://doi.org/10.22146/ijeis.44751
2. Ivanov AG, Zhoga VV, Vorob'yeva NS, Dyashkin-Titov VV (2022) A dynamic algorithm for stabilization of the working body of a mobile robot weeding for the future of agriculture. IOP Conf Ser Earth Environ Sci 965(1). https://doi.org/10.1088/1755-1315/965/1/012050
3. Ghafar ASA, Hajjaj SSH, Gsangaya KR, Sultan MTH, Mail MF, Hua LS (2021) Design and development of a robot for spraying fertilizers and pesticides for agriculture. Mater Today Proc XXXX. https://doi.org/10.1016/j.matpr.2021.03.174

4. Levin M, Degani A (2016) Design of a task-based modular re-configurable agricultural robot. IFAC-Papers Online 49(16):184–189. https://doi.org/10.1016/j.ifacol.2016.10.034
5. Zhao S, Zhang Z, Xiao D, Xiao K (2016) A turning model of agricultural robot based on acceleration sensor. IFAC-Papers Online 49(16):445–450. https://doi.org/10.1016/j.ifacol.2016.10.081
6. Duckett T et al (2018) Agricultural robotics: the future of robotic agriculture. (Online). Available: http://arxiv.org/abs/1806.06762
7. Finegan H et al (2019) Development of an autonomous agricultural vehicle to measure soil respiration. In: 2019 Systems information engineering design symposium SIEDS. https://doi.org/10.1109/SIEDS.2019.8735598
8. Majdoubi R, Masmoudi L (2021) Eco-design of a mobile agriculture robot based on classical approach and FEM criteria. In: Proceedings of IEEE 2021 international conference computer communication intelligent system ICCCIS 2021, pp 978–982. https://doi.org/10.1109/ICCCIS51004.2021.9397234
9. Nanda RA, Arhami A, Kurniawan R (2020) Perancangan Dan Pengujian Model Mobil Robot Penanam Bibit Kangkung. Rona Tek Pertan 13(2):14–28. https://doi.org/10.17969/rtp.v13i2.16982
10. Nanda RA, Dewadi FM, Rizkiyanto M (2022) "Mechanical Xplore (JTMMX) Archimedes" principle applied to buoy design for measuring purposes in offshore illumination conditions. 3(1):40–48

Commercial AC System Alternatives for a Case Study Hotel Building in Hot Climate Country: An Inclusive Comparison Analysis

I Nyoman Suamir, Adi Winarta, and I Made Rasta

Abstract Air conditioning (AC) systems are the main energy-significant services in hotel buildings. There is, therefore, remarkable potential to increase energy efficiency use of the hotel AC systems by optimization of both their system selection and operational characteristics. This paper presents an inclusive evaluation of commercial AC systems that can be appropriately functional for a case study star hotel. The hotel considered consists of resorts and villas of cooling load as high as 2400 tons of refrigeration. Site investigation, AC-specified characteristic review, and numerical approach were applied. The evaluation involved operational characteristics, coefficient of performance, and power and energy use per year. The evaluation results showed, for hotel building application, the central AC systems which generally have better diversity factor, could provide flexibility in installed system capacity. The results also showed that the water-cooled chiller system could perform with higher COP compared to the other AC systems. Additionally, the annual energy efficiency of the water-cooled chiller system was also found very superior with energy savings more than 50%. After central water-cooled chiller system, it was found that a modular water-cooled chiller system could also provide an excellent energy performance. Finally, the evaluation results can provide the most suitable AC system alternatives, specifically in hot climate country applications, to hotel facility consultants, hotel building contractors, and hotel owners.

Keywords Inclusive comparison analysis · Commercial AC system · Operational characteristics · A case study hotel building

I. Nyoman Suamir (✉) · A. Winarta · I. Made Rasta
Mechanical Engineering Department, Politeknik Negeri Bali, Campus Street, Kuta Selatan, Badung, Bali 80364, Indonesia
e-mail: nyomansuamir@pnb.ac.id

A. Winarta
e-mail: adi.winarta@pnb.ac.id

I. Made Rasta
e-mail: maderasta@pnb.ac.id

Irwansyah et al. (eds.), *Proceedings of the 4th International Conference on Experimental and Computational Mechanics in Engineering*, Lecture Notes in Mechanical Engineering, https://doi.org/10.1007/978-981-99-7495-5_40

1 Introduction

In Indonesia, the building sector, including hotel buildings, uses 50% of energy in general or 70% of total electricity consumption. This makes the sector the largest energy user and even exceeds the industrial and transportation sectors. In terms of energy use, the AC system in buildings consumes 65% of the total energy use in hotel buildings [1]. The hospitality sector itself is growing very rapidly in Indonesia, and tourist visits, especially to Bali, continue to increase.

AC plays a very important role in maintaining thermal comfort in the room, especially for hot and humid climates. The energy consumed by heating, ventilation, and air conditioning in tropical climates can exceed 50% of the total energy consumption of buildings [2, 3], 47% in the US [4] and 30–50% in China [5]. AC systems are the largest energy-significant utility of hotel buildings. Therefore, there is tremendous potential to increase the overall efficiency of the AC system in hotel buildings.

The largest electricity usage in hotel buildings is 70–80% for air- or water-cooled chiller operations, where the chiller is the main sub-system of the AC system. Hotel operational costs can be reduced by optimizing the use of energy in the chiller. Efficient management of multi-chiller systems can be very challenging as an alternative in optimizing their energy use [6]. The overall energy efficiency of an AC system with multi-chillers can have a major impact using the chiller operating sequence arrangement concept [7]. This operational method is a good practice in using multi-chillers with the best efficiency compared to chiller operations which consume more energy [8].

The energy demand for AC system can increase rapidly in the twenty-first century [9]. Higher outdoor temperatures impacting climate change affect cooling energy in terms of higher indoor temperatures and more stringent thermal comfort requirements. In addition, the impact of climate change on AC systems is expected to cause energy demand to soar up to 72%. The biggest consumers of energy related to AC systems come from developing countries. A large increase is expected in South Asia as energy demand for AC systems could increase by about 50% due to climate change [10].

To anticipate these changes, increasing energy efficiency in hotel buildings is currently the main target of energy policy at the regional, national, and international levels. Among building utilities, the increase in energy use of heating, ventilation, and air conditioning systems or better known as HVAC is very significant [11]. Several solutions have been proposed to promote energy conservation such as the development of an AC system equipped with a heat recovery system to heat water [12–16]; maintaining system performance through close monitoring of key chiller parameters including evaporator and condenser approach temperatures [17]; and optimizing the distribution of the cold water circulation system [18]. This could lead to wider adoption of energy efficiency, to support the Government of Indonesia's commitment to reduce greenhouse gas (GHG) emissions by 26% by 2020 [19]. Energy saving is one of the smart solutions to be applied to hotels. Therefore, actions from the building sector to mitigate and adapt to climate change early on from planning and selecting utilities including energy-significant AC systems become very important to reduce building energy consumption and GHG emissions [20].

This paper presents an alternative evaluation of various AC systems that can be applied to a case study five-star hotel. The hotel is a resort and villa with a peak cooling load based on a comprehensive design result of 2400 TR (tons of refrigeration). The main aspect that becomes the basis of evaluation in this paper is operational characteristics and the energy performance of the AC system including COP, power consumption, and energy consumption per year. Several other aspects were also discussed such as operational flexibility of the AC system, risk of damage, ease of maintenance and repair, and the economic life of the various AC systems evaluated.

2 Material and Methods

The commercial AC systems evaluated in this paper are AC systems that are commonly applied to hotel buildings, especially star hotels. The evaluated AC systems included five types of AC systems, namely: water-cooled chiller, air-cooled chiller, modular air-cooled chiller, water-cooled package (WCP), and variable refrigerant volume (VRV).

The main methods applied in this research include site investigation, AC-specified characteristic review, and numerical approach. Site investigation was conducted to the case study hotel in order to evaluate hotel building characteristics, the AC system used, and cooling demand. Various studies and reviews have also been conducted, among others: a study of the needs and characteristics of cooling loads in various commercial buildings, especially hotel buildings; study of technical specifications covering energy performance, cooling capacity, construction, installation, spatial requirements, and economic life. The study also involved operational characteristics as well as maintenance and repair of various types of commercial AC systems.

The results of the study were then evaluated comprehensively, and comparative analysis was carried out between the cooling load characteristics of the hotel building and the operational characteristics of the air conditioning system. Comparative analysis between various types of AC systems is also carried out to obtain the best AC system priority scale and the one that best suits the needs of hotel buildings, especially from the aspect of energy performance.

3 Results and Discussion

AC System Characteristics. Five commercial AC systems characteristics have been inclusively evaluated. They included water-cooled chiller, air-cooled chiller, modular air-cooled chiller, water-cooled package, and variable refrigerant volume AC systems. The results of the evaluation are systematically detailed below.

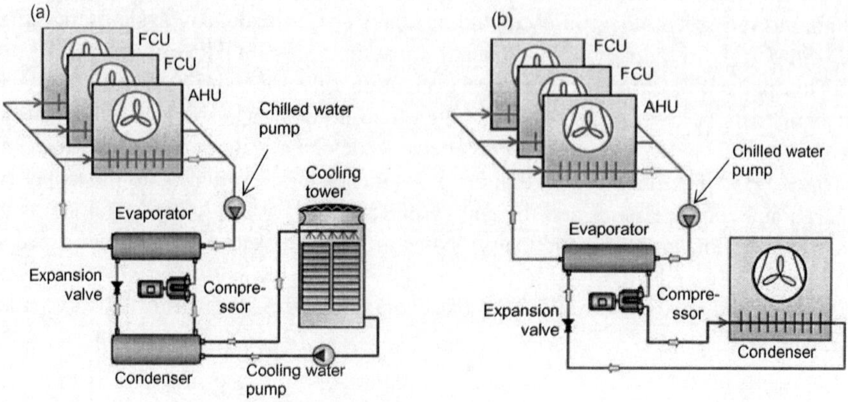

Fig. 1 Schematic diagram of **a** water-cooled chiller and **b** air-cooled chiller

Water-cooled Chiller. The water-cooled chiller is an AC system with a centralized or central system with a condenser cooled by water through a cooling tower system. For application in the case study hotel, the number of chiller units can reach 3 units. Each unit is with a capacity of 600 TR. The AC system produces chilled water with temperature in the range of 6–7 °C. The chilled water is distributed to the loading system in this case: the AHU (air handling unit) or FCU (fan coil unit) using a pump system. On the condenser side, the heat is absorbed by the cooling water and discharged to the environment through a cooling tower system. The schematic of the water-cooled chiller AC system is presented in Fig. 1a.

The characteristics of the water-cooled chiller AC system can be detailed: (i) The cooling capacity can be 20–30% lower than peak load due to diversity factor. For the case study hotel, the diversity factor can reach 70–80%; (ii) the compressor used is a centrifugal compressor with the required number 3 units of 600 TR each; (iii) a minimum enclosed refrigeration plant room is required with a floor area of 168 m^2 and a minimum height of 4.4 m [21]; (iv) the placement of the cooling tower in the open air is approximately 234 m^2 [21]; (v) required make up water for cooling tower of 13–15 L/hour per TR cooling capacity [21]; (vi) each room is serviced by an AHU or FCU loading system; (vii) AHU capacity with a range from 5 to 90 TR; (viii) small room serviced by FCU; (ix) the economic life of the water-cooled chiller AC system is generally up to 25 years assuming good maintenance; (x) relatively more flexible for buildings with varying room load conditions, or those equipped with rooms that require large load capacity or high latent loads such as ball rooms or convention halls; (xi) the system can meet the needs of large cooling load with 1 chiller unit; (xii) the refrigeration system is centralized and relatively simple; (xiii) the system has good annual efficiency; (xiv) the possibility of unit disturbance is relatively small because the refrigeration plant room only consists of 3 chiller units (3 × 600 TR).

Air-Cooled Chiller. Air-cooled chiller AC systems have the main difference in their condenser cooling systems compared to water-cooled chiller AC systems. The condenser of this AC system is cooled directly by the ambient air. The air-cooled chiller is also a centralized AC system. For the case study hotel, the required number of air-cooled chillers can reach 4 units each with a capacity of 450 TR. The AC system can also produce chilled water with temperatures in the range of 6–7 °C. The chilled water is also distributed to the AHU (air handling unit) or FCU (fan coil unit) systems using a chilled water pump system. On the condenser side, the condenser heat is cooled directly by the ambient air. The schematic of the air-cooled chiller AC system is presented in Fig. 1b.

The characteristics of the air-cooled chiller involve: (i) For the case study hotel building applications, the cooling capacity of the air-cooled chiller system can also be 20–30% lower than the peak cooling load due to the diversity factor; (ii) the compressor used is generally a screw type. The hotel requires 4 units of 450 TR each; (iii) a minimum open refrigeration plant room of area 420 m^2 [21] is required, usually located on the roof top; (iv) each conditioned room is serviced by AHU or FCU; (v) small rooms such as hotel rooms are serviced by FCUs; (vi) the economic life is the same as that of a water-cooled chiller, generally up to 25 years; (vii) relatively more flexible for buildings with various room conditions; (viii) the system can meet the needs of large cooling loads with 1 chiller unit; (ix) the refrigeration system is centralized and relatively simpler than a water-cooled chiller because there is no cooling tower and cooling water system; (x) the possibility of unit disturbance is relatively small because the refrigeration plant room only consists of 4 chillers (4×450 TR) completed with piping system, chilled water pump, and the control system.

Modular Air-Cooled Chiller. The installation of the modular air-cooled chiller must be in an open place, usually on the roof top so that the heat dissipation process occurs properly into the ambient air. The distance between the AHU and FCU loading systems and the refrigeration unit can be far because the system is equipped with a chilled water piping system. Chiller operation is very easy and more flexible. In addition, the equipment required is relatively less compared to the water-cooled chiller system. The operation of this modular air-cooled chiller is relatively similar to that of air-cooled chiller in that the capacity is smaller. So, the construction is simpler. The modular air-cooled chiller is an AC system with a non-centralized system, meaning the refrigeration system is spread throughout the building according to loading requirements. The capacity of the modular air-cooled chiller ranges from 14 TR (48 kW) to 66 TR (232 kW) [22]. The AC system also produces chilled water with temperatures in the range of 6–7 °C. The chilled water is also distributed to the AHU (air handling unit) or FCU (fan coil unit) using a pump system. The schematic of the modular air-cooled chiller is presented in Fig. 2a.

The characteristics of the modular air-cooled chiller include: (i) For hotel building applications, the cooling capacity of the modular air-cooled chiller can be 15% lower than the total peak load, because the system can operate semi-centrally with a diversity factor of up to 85%; (ii) the modular installation can be as many as 5 units to get the cooling capacity. Therefore, it is a semi-central AC system (centralized but spread throughout the building) [22]; (iii) for the case study hotel, 8 semi-central systems are required with each consisting of 5 units so that 40 units of modular

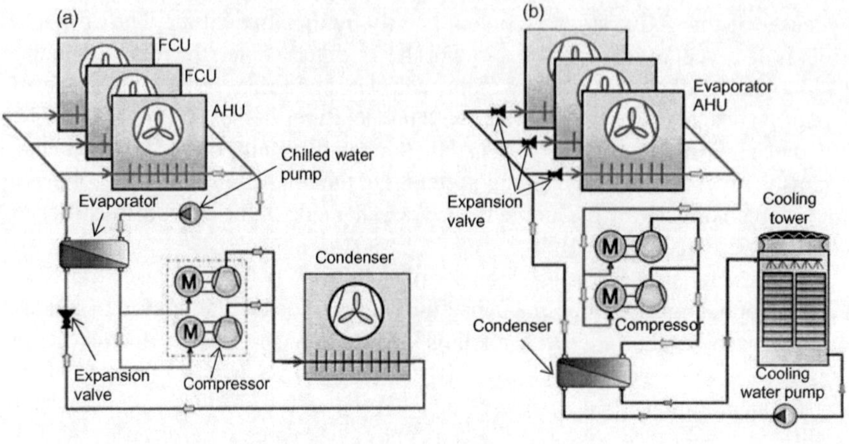

Fig. 2 Schematic diagram of **a** modular air-cooled chiller and **b** water-cooled package (WCP)

air-cooled chiller are required. Each unit is with a cooling capacity of 50 TR; (iv) the compressor used is a scroll inverter compressor; (v) each conditioned room is also serviced by the AHU or FCU system; (vi) the economic life is generally up to 15 years assuming good maintenance; (vii) the system is quite flexible for buildings with various room conditions; (viii) the system is also capable to meet the needs of large loads, and a combined chiller unit is required; (ix) the refrigeration system is semi-centralized or centralized but still spread throughout the building in the form of groups or combined chillers. So, the refrigeration system is relatively more complex; (x) the system has a very good annual efficiency, especially for hotel buildings that have a fluctuating load factor; (xiii) the possibility of unit disturbance is relatively high because the large number of systems required.

Water-Cooled Package. The installation of the water-cooled packaged (WCP) AC system can be on the floor or on the roof top. The distance between indoor and outdoor cannot be far. So, it should be installed close to the conditioned room. The operation of the WCP unit is relatively more complex, because before the refrigeration system is turned on, the cooling tower and cooling water pump must be run first. The schematic diagram of the WCP AC system can be seen in Fig. 2b.

The characteristics and operations of the WCP AC system can be explained as follows: (i) The WCP AC system is direct expansion with water cooling in the condenser; (ii) consists of hanging type WCP with cooling capacity in the range of 9 TR to 56 TR [23]; (iii) for the case study hotel, approximately 48 WCP units with a capacity of 50 TR each are required; (iv) the main components of the refrigeration system, such as the refrigerant piping system, compressor, and condenser, are already one packaged unit; (v) the compressor used is a scroll inverter compressor; (vi) the WCP system condenser is cooled by cooling water from the central cooling tower unit; (vii) required make up water for cooling tower of 13–15 L/hour per TR cooling capacity; (viii) the installed system capacity is the total peak load, because the units work individually for each conditioned room with a diversity factor of 100%; (ix) installation of the WCP system can be in spread out spaces; (x) each room or group of rooms is served by a WCP system with an AHU system; (xi) the economic life is generally up to 15 years assuming good maintenance; (xii) less flexible, because WCP is a standard unit with a limited number of coil rows; (xiii) the installations are scattered and very numerous and cannot yet be classified as semi-central; (xiv) maintenance costs become more expensive and time-consuming; (xv) less chance of damage than split-type AC system.

Variable Refrigerant Volume. Variable refrigerant volume (VRV) AC system is a type of heating, ventilation, and air conditioning (HVAC) system in a building with a multi-split type that uses refrigerant flow control according to the cooling load requirements. In-building VRV system can save space for installation. Space efficiency is enhanced by compact individual unit sizes, long maximum piping lengths. In addition, the system provides superior design flexibility especially in that layout changes can be made easily. The outdoor unit can be placed on the floor or on the roof top. The schematic diagram of the VRV AC system is presented in Fig. 3.

The characteristics of the VRV AC system can be described as follows: (i) VRV AC system is an AC system with direct expansion and not central. The installation is spread throughout the building according to the cooling load of the conditioned room; (ii) consists of indoor unit and outdoor unit, where the installation of outdoor unit must be provided on each floor; (iii) capacity range from 6 to 30 TR; (iv) for the case study hotel, approximately 80 VRV units with a capacity of 30 TR each are needed [24]; (v) cooling of the condenser with ambient air; (vi) the amount of refrigerant flow in the system is variable according to the cooling load; (vii) the installed capacity is the total peak load, because the units operate individually with a diversity factor of 100%; (viii) the number of outdoor units can be very large, and their installation can be a problem for esthetically pleasing hotel buildings; (ix) each conditioned room is served by an indoor unit (IU) with a capacity range from

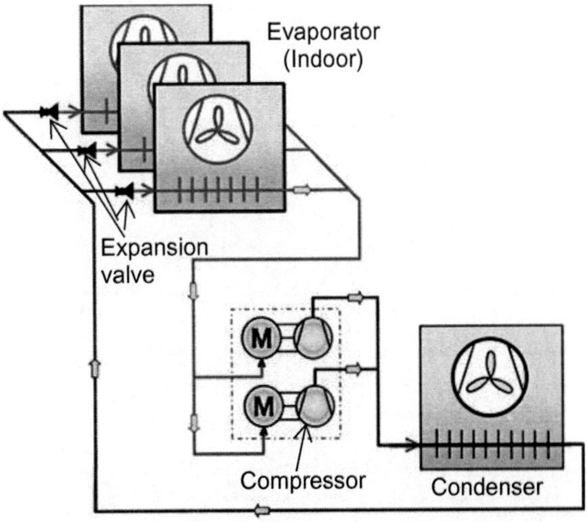

Fig. 3 Schematic diagram of a variable refrigerant volume AC system

0.5 TR to 20 TR; (x) the economic life is generally up to 15 years assuming good maintenance; (xi) less flexible, because the indoor unit is a standard unit with a limited number of row coils; (xii) for large spaces, many units are needed scattered around the room; (xiii) maintenance costs are more expensive and time-consuming compared to chiller systems; (xiv) there may be more refrigerant leaks, because pipe installation, refrigerant charging, and tests carried out at the installation site.

Performance and Energy Consumption Analyses. The results of the study of the performance of the investigated commercial AC system show various variations in performance. Their performances are influenced by heat rejection system in the condenser and the optimization of heat transfer in the heat exchanger components such as evaporator. Performance is also greatly influenced by the type of compressor and refrigerant used. The performance of the AC system also directly affects its energy consumption. Table 1 presents a comparison of performance and power consumption as well as annual energy consumption which was numerically determined based on the assumption of a use factor of 75% which means that the AC system is only used for 75% of the total hours in one year.

Based on the characteristics of the AC system and the results of the analysis presented in Table 1, the advantages and disadvantages of each AC system can be identified when applied to the case study hotel building with a large cooling load. The hotel building has characteristics with very fluctuating load factors and provides advantages for central AC systems with relatively better diversity factors. So that the installed system capacity can be smaller than the peak cooling load of the building.

Table 1 Result of investigation on the performance and energy consumption of the AC systems

AC system	Installed capacity (TR)	COP	Efficiency (kW/TR)	Power (kW)	Annual energy use (kWh)
Water-cooled chiller (centrifugal compressor) [25]	1800	5.79–6.29	0.61–0.56	1053	6,918,210
Air-cooled chiller (rotary crew compressor) [26]	1800	3.34–3.40	1.05–1.03	1872	12,299,040
Modular air-cooled chiller (inverter scroll compressor) [22]	2000	4.10–4.35	0.86–0.81	1670	10,971,900
Water-cooled packaged (WCP) (inverter scroll compressor) [23]	2400	3.55–4.29	0.99–0.82	2172	14,270,040
Variable refrigerant volume (VRV) (inverter scroll compressor) [24]	2400	3.43–3.75	1.03–0.94	2364	15,531,480

Assumption AC system operated at use factor of 75% (only 6570 h in a year). The variation of installed capacity due to diversity factor of AC system used

From the energy performance aspect, it can also be identified that the water-cooled chiller type AC system has the highest COP compared to other AC systems. Combined with the advantage that the installed capacity is also small, the annual energy consumption for water-cooled chiller is also very superior with energy savings of more than 50% compared to WCP and VRV type AC systems. Modular water-cooled chiller also has superior energy performance after water-cooled chiller. This AC system consumes 10.97 GWh of energy per year. With that amount of energy consumption, the modular air-cooled chiller has superior energy performance compared to the air-cooled chiller, WCP, and VRV AC systems.

The results of this study can provide alternative options for hotel owners, consultants, and building contractors in determining the most suitable AC system to be applied. It is also very useful for researchers to conduct further studies on various applications including the economic feasibility aspect of various AC systems for various commercial building applications.

4 Conclusions

Evaluation of various types of AC systems for a case study hotel building applications with large cooling capacity has been carried out. Based on the characteristics of the AC systems and the results of the analysis, it was found that the water-cooled chiller AC system could perform with the highest COP compared to other AC systems. Combined with the advantage that the installed capacity is also small, the annual energy consumption of the water-cooled chiller was also very superior with energy savings of more than 50% compared to the WCP and VRV type AC systems, and around 44% compared to the air-cooled chiller, and 37% compared to modular air-cooled chiller.

Modular air-cooled chiller also has superior energy performance after water-cooled chiller. The AC system consumes 10.97 GWh of energy per year. With that amount of energy consumption, the modular air-cooled chiller also has exceptional energy performance compared to the air-cooled chiller, WCP, and VRV AC systems.

Acknowledgements The authors would like to thank Bali State Polytechnic for the financial support on this study. The authors would also like to send our warm gratitude to technicians as well as students of the Bali State Polytechnic for their administrative support and data collection.

References

1. ICED (2015) Indonesia Clean Energy Development
2. Chua KJ, Chou SK, Yang WM, Yan J (2013) Achieving better energy-efficient air conditioning—a review of technologies and strategies. Appl Energy 104:87–104

3. Kurekci NA (2016) Determination of optimum insulation thickness for building walls by using heating and cooling degree-day values of all Turkey's provincial centers. Energy Build 118:197–213
4. US Department of Energy (2011) Building energy data book
5. Building Energy Conservation Research Center (2017) Annual report on China building energy efficiency
6. Yu FW, Chan KT (2012) Improved energy management of chiller systems by multivariate and data envelopment analyses. Appl Energy 92:168–174
7. Jayamaha L (2006) Energy-efficient building systems: green strategies for operation and maintenance, 1st edn. McGraw-Hill, US
8. Yu FW, Chan KT (2005) Experimental determination of the energy efficiency of an air cooled chiller under part load conditions. Energy 30:1747–1758
9. Morna I, Detlef PVV (2009) Modeling global residential sector energy demand for heating and air conditioning in the context of climate change. Energy Policy 37:507–521
10. Scott MJ, Huang YJ (2007) Annex A: technical note: methods for estimating energy consumption in buildings in effects of climate change on energy production and use in the United States a Report by the US climate change science program and the subcommittee on global change research, Washington DC
11. Deng J, He S, Wei Q, Liang M, Hao Z, Zhang H (2020) Research on systematic optimization methods for chilled water systems in a high-rise office building. Energy Build 209:109695
12. Suamir IN, Ardita IN, Wirajati IGAB (2017) Waste heat recovery from central AC system for hot water supply; a case study for hotel building application in Indonesia. Adv Sci Lett 23:12206–12210
13. Suamir IN, Ardita IN, Dewi NIK (2015) Integration of heat pump and heat recovery of central AC system for energy use reduction of hotel industry. Refrig Sci Technol 2015:3581–3588
14. Suamir IN, Baliarta ING, Arsana ME, Temaja IW (2017) The Role of condenser approach temperature on energy conservation of water cooled chiller. Adv Sci Lett 23:12202–12205
15. Suamir IN, Ardita IN, Rasta IM (2018) Effects of cooling tower performance to water cooled chiller energy use: a case study toward energy conservation of office building. Int Conf Appl Sci Technol (iCAST) 1:712–717
16. Suamir IN, Arsana ME, Subagia IWA, Rasta IM, Midiani LPI, Wibolo A (2019) Site investigation on water cooled chiller plant for energy conservation and environmental impact reduction of a large shopping mall. AIP Conf Proc 2187:020042
17. Suamir IN, Baliarta ING, Arsana ME, Sudirman, Sugina IM (2020) Field-based analyses on approach temperatures for performance evaluation of centralized air conditioning system in a shopping mall building. J Phys Conf Ser 1569:032041
18. Suamir IN, Sudirman, Ardita IN, Santanu G (2020) Experimental and numerical optimization on chilled water configuration for improving temperature performance and economic viability of a centralized chiller plant. J Phys Conf Ser 1450:012106
19. Building Energy Data United States (2010)
20. IPCC Climate Change (2014) Synthesis report: contribution of Working Groups I, II and III to the fifth assessment report of the Intergovernmental Panel on Climate Change IPCC, Geneva, Switzerland
21. Hawkins G (2011) Rules of thumb guidelines for building services, 5th edn, BSRIA
22. Trane, Modular air-cooled chillers: high seasonal efficiency (HSE) version cooling capacity 48–232 kW
23. Daikin (2021) Water cooled packaged unit. Daikin Industries, Ltd.
24. Daikin (2021) VRV III, Daikin Industries, Ltd.
25. Daikin (2021) Water cooled centrifugal chiller. Daikin Industries, Ltd.
26. Trane (2020) Air-cooled chillers: a full portfolio of solutions for comfort and process applications

The Effect of the Distribution of Air Temperature in the Isolation Room on the Condition of the Air Supply

Suheri Suheri, Razali Thaib, Hamdani Hamdani, Irwansyah Irwansyah, and M. S. Salamul Fazar

Abstract Isolation rooms for patients with infectious diseases in particular have a very high risk of being contaminated with various diseases such as Covid-19. Covid-19 is a severe respiratory disease caused by the coronavirus family that has become a worldwide pandemic. This type of disease spreads very quickly through the air. Isolation rooms for patients with infectious diseases must be prepared to prevent the spread of disease through the air. The air supply in the isolation room greatly affects the temperature and humidity of the air. This study aims to determine the air supply to the distribution of temperature and humidity in the isolation room. The research scheme was carried out in two conditions: Firstly, the isolation room was in normal condition without any air entering or leaving (diffuser–exhaust Off) and secondly, the condition of the isolation room being supplied with air (diffuser On–exhaust Off). The results of the first scheme show that there are slight differences between each of the measurement points ($T1$, $T2$, $T3$, and $T4$). The maximum temperature at point $T1$ is 27.85 °C, at positions $X = 0.2$ and $Y = 1.9$, while in other positions, there are no significant differences. This is because the positions $X = 0.2$ and $Y = 1.9$ are the locations of the measurement point on the patient's bed. The results of the second scheme in the diffuser area show that there is a significant change

S. Suheri
School of Engineering, Universitas Syiah Kuala, Darussalam, Banda Aceh 23111, Indonesia
e-mail: suheri@unsam.ac.id

R. Thaib (✉) · H. Hamdani · I. Irwansyah
Department of Mechanical and Industrial Engineering, Universitas Syiah Kuala, Darussalam, Banda Aceh 23111, Indonesia
e-mail: razalithaib@unsyiah.ac.id

I. Irwansyah
e-mail: irwansyah@unsyiah.ac.id

S. Suheri · H. Hamdani
Department of Mechanical Engineering, Universitas Samudra, Kampus Meurandeh, Langsa 24416, Indonesia

M. S. Salamul Fazar
Faculty of Engineering, Universitas Syiah Kuala, Darussalam, Banda Aceh 23111, Indonesia
e-mail: salamul@mhs.unsyiah.ac.id

Irwansyah et al. (eds.), *Proceedings of the 4th International Conference on Experimental and Computational Mechanics in Engineering*, Lecture Notes in Mechanical Engineering, https://doi.org/10.1007/978-981-99-7495-5_41

in temperature, recorded at the measurement point $T1$. The minimum temperature reaches 21.18 °C, while for the measurement points $T2$, $T3$ and $T4$, the minimum temperatures are 19.73 °C, 19.68 °C, and 19.63 °C, respectively. Other areas also experienced changes in temperature, and the highest temperature occurred at the measurement locations $X = 0.2$ and $Y = 1.9$, where these locations were at the top of the patient's bed and far from the diffuser position. Thus, the measurement results of each condition can be the initial basis for continuing further research.

Keywords Air temperature · Distribution temperature · Isolation room

1 Introduction

After the outbreak of COVID-19 cases that occurred in the first quarter of 2020, it is very important to provide isolation rooms for patients with infectious disease according to standards. The isolation room in the hospital has a special design and meets the standards for certain conditions. Ministry of Health Republic of Indonesia [1] has issued standards for isolation rooms and provides guidelines for carrying out procedures for handling patients in isolation rooms. The function of the isolation room is to meet the needs of patients suffering from infectious and very dangerous diseases [2].

Patients suffering from infectious diseases must receive special treatment in the isolation room to prevent transmission to medical personnel who care for them and their families who come to visit [3, 4]. The air inside can be conditioned by providing air inlet through the diffuser and air out through the exhaust. Research conducted by Liu et al. [2] has found that the position of the air inlet and outlet has an effect on the flow pattern that occurs in the room. The role of a good air conditioning system is needed to make the air cycle in the isolation room. According to ASHRAE [5], the use of HVAC (heating, ventilation, and air conditioning) can help condition the air in the isolation room. Standard air temperature ranges from 20 to 24 °C, and relative humidity of 50–60% with air changes of at least 12 times in 1 h. Types of infectious diseases such as bacteria, viruses and so on can quickly spread through the air. One of the keys to prevent the transmission of infectious diseases is to make isolation rooms according to standards and carry out air conditioning through several parameters [6, 7].

The air produced from the HVAC is channeled through the ducting and will then be distributed through 2 (two) diffusers in the ceiling of the isolation room. The diffuser ventilation system is designed to ensure the overall distribution of the air in the room [8]. The influence of the position of the diffuser and exhaust also determines the pattern of air flow that occurs in the isolation room. Airflow is highly dependent on the speed and dimensions of the diffuser [9].

Isolation rooms for patients with infectious disease are also highly dependent on ventilation systems. The design of the isolation room must also have adequate air intake and exhaust specifications [10]. The position of the inlet will affect the distribution of air in the room, while the exhaust duct will ensure that air can circulate properly. The configuration of the inlet and exhaust air ducts will provide comfort and at the same time prevent the spread of infectious diseases in the isolation room [9, 11].

The air supply in the isolation room greatly affects the temperature and humidity of the air. This study aims to determine the air supply to the distribution of temperature and humidity in the isolation room. Many researchers have carried out an isolation room model based on computational fluid dynamic (CFD) to obtain indoor air conditions, comfort levels, and the performance of the HVAC system [12–14].

The distribution of air in the isolation room is highly dependent on the position of the diffuser and exhaust. Proper positioning can increase the efficiency of the air cycle in the isolation room. Poor air quality can cause problems for both patients and medical personnel. Liu et al. [15] studied aerosol diffusion in ventilation chambers with downward air supply, and the results showed that lower air supply velocity is beneficial for aerosol diffusion control. Qian et al. [16] investigate the ventilation system down the hospital ward. It is recommended to further study the location of the air supply diffuser and exhaust outlet on the bottom ventilation system. Airflow structures that efficiently remove pollutants should be designed to reduce the risk of cross-infection in general hospital wards [17]. Jung et al. [18] believe that the air velocity in areas occupied by mixed ventilation is low and has no direction. The mixing of fresh and old air causes a wide spread of pollutants in the respiratory zone. This can lead to a higher risk of infection. Based on the above studies, serious efforts should be made to conduct experimental studies in isolation rooms. So that it can be determined with certainty the effect of the air supply entering the isolation room on temperature and humidity. This research involves many disciplines related to phenomena that occur in isolation rooms.

2 Methodology

2.1 Experiment Setup

This research was conducted experimentally in an isolation room laboratory using the HVAC (heat, ventilation, and air conditioning) system for one patient with infectious disease in the Faculty of Engineering building, Syiah Kuala University. The geometry of the isolation room has a length of 4 m, a width of 4 cm, and a height of 2.4 m, as shown in Fig. 1. This isolation room is specially made for one patient suffering from an infectious disease. This room has two diffusers located on the ceiling and has two exhausts behind the patient bed and two on the lower side position. The isolation room design has met the ASHRAE Standard which has dimensions of 4 m long, 4 m wide, and 2.4 m high.

The dimensions of the isolation room are distinguished in the direction of the *X*- and *Y*-axes to make it easier to pinpoint the location of the measurement point.

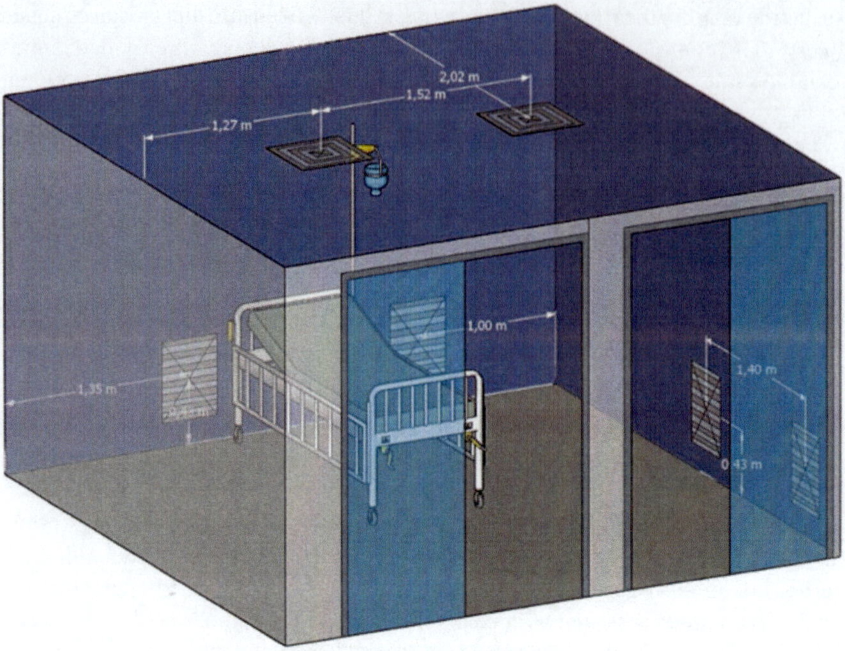

Fig. 1 Isolation room design for patients with infectious disease

This study measured the temperature distribution at eight points (*P*1, *P*2, *P*3, *P*4, *P*5, *P*6, *P*7, and *P*8) near the patient's bed. Figure 2 shows the location of the measurement point to determine the temperature distribution near the patient's bed. The measurement locations *P*3, *P*4, *P*5, *P*6, and *P*7 are located near the diffuser (Fig. 3).

Measurement of temperature distribution was carried out under two conditions: first, the condition of the isolation room without air supply (diffuser–exhaust Off) and second, the condition of the isolation room being supplied with air (diffuser On–exhaust Off). Air supply is carried out at 24 °C, and the light is on. The humidity results for each condition are only to provide additional information in the measurement.

2.2 Air Supply System

Measurements are made by varying the air supply in the isolation room. The first measurement scheme is to measure the temperature distribution when the isolation room is not supplied with air (diffuser and exhaust Off conditions). In this condition, the isolation room has been sterilized for 1 × 24 on the day before testing. The second measurement scheme is to measure the temperature distribution when the isolation

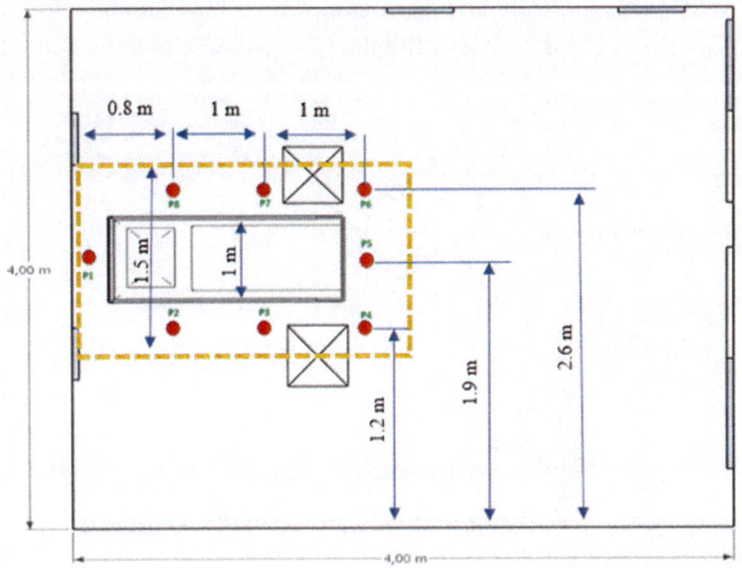

Fig. 2 Location of temperature distribution measurement points in isolation rooms for patients with infectious disease

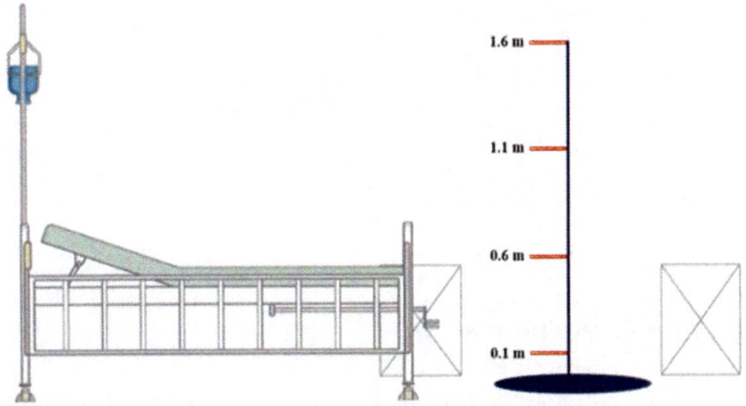

Fig. 3 Point of measurement of temperature distribution based on height

room is supplied with air (diffuser On and exhaust Off conditions). Air conditioning (AC) connected to the ducting duct supplies air through 2 (two) diffusers to circulate air in the isolation room for 12 h with a temperature of 24 °C (Table 1; Fig. 4).

Table 1 Conditions of air supply in the isolation room

Measurement Test	Measurement Time (h)	Air Supply	
		Diffuser	Exhaust
1	12	Off	Off
2	12	On	Off

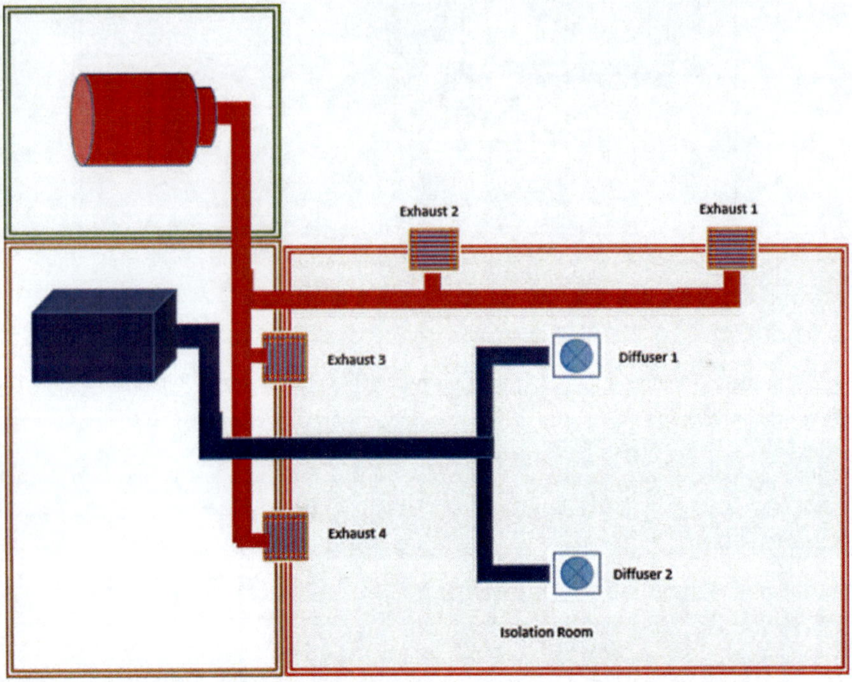

Fig. 4 Design of air supply for isolation room

3 Results and Discussion

The results of temperature distribution measurements with various measurement points have been carried out using experimental methods in the infectious disease isolation room laboratory and obtained results at each measurement point. There are two measurement conditions carried out in the isolation room: namely, the first condition where the diffuser and exhaust are in the Off state and the second condition where the diffuser is On and the exhaust is in an Off state. From the measurement results, the temperature distribution is distinguished based on the measurement position and altitude. The following is the result of measuring the temperature distribution in 2 (two) conditions:

A. **The results of the measurement of temperature distribution in the conditions of the diffuser and exhaust Off**

The results of the temperature distribution in the diffuser and exhaust Off conditions show that there are slight differences from each measurement point ($T1$, $T2$, $T3$, and $T4$). The maximum temperature at point $T1$ is 27.85 °C, at positions $X = 0.2$ and $Y = 1.9$, while in other positions, there are no significant differences. This is because the positions $X = 0.2$ and $Y = 1.9$ are the locations of the measurement point on the patient's bed (see Fig. 5). There was a change in the temperature of the measurement points $T2$, $T3$, and $T4$ but not significant, and the maximum temperature values for each measurement were 2.55 °C, 2.52 °C, and 2.27 °C.

B. **The results of the measurement of the temperature distribution in the conditions of the diffuser On and Exhaust Off**

The results of the temperature distribution at each measurement point change almost the same at each measurement point (see Fig. 6). The diffuser area experienced a significant temperature change, recorded at the measurement point $T1$, and the minimum temperature reached 21.18 °C, while for the measurement points $T2$, $T3$,

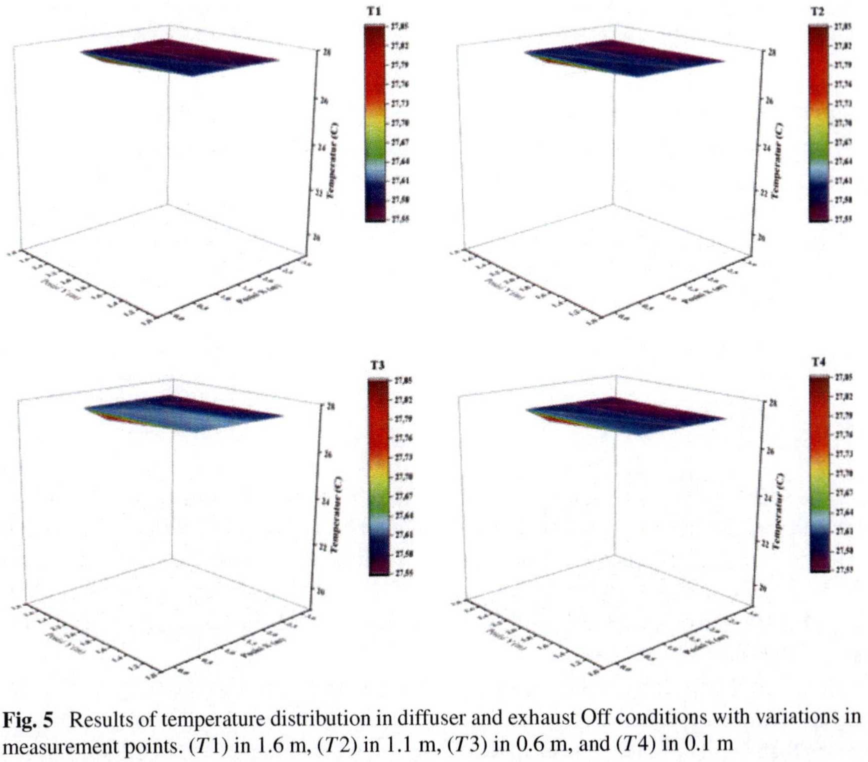

Fig. 5 Results of temperature distribution in diffuser and exhaust Off conditions with variations in measurement points. ($T1$) in 1.6 m, ($T2$) in 1.1 m, ($T3$) in 0.6 m, and ($T4$) in 0.1 m

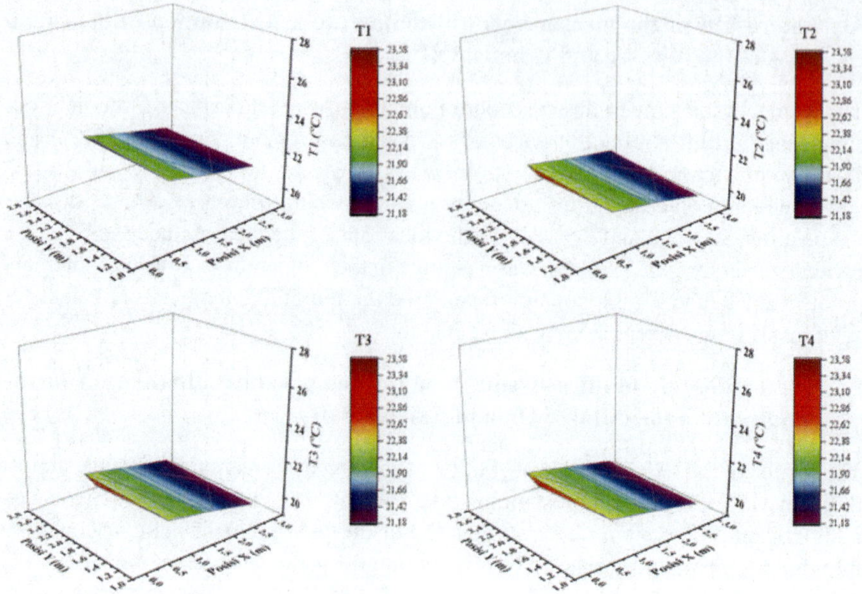

Fig. 6 Results of temperature distribution in diffuser On and exhaust Off conditions with variations in measurement points. ($T1$) in 1.6 m, ($T2$) in 1.1 m, ($T3$) in 0.6 m, and ($T4$) in 0.1 m

and $T4$, the minimum temperatures are 19.73 °C, 19.68 oC, and 19.63 °C, respectively. Other areas also experienced changes in temperature, and the highest temperature occurred at the measurement locations $X = 0.2$ and $Y = 1.9$, where these locations were at the top of the patient's bed and far from the diffuser position. The temperature obtained at this point has a significant difference from several other points. The temperature result for $T1$ is 23.57 °C, and the temperature results at points $T2$, $T3$, and $T4$ are 21.76 °C, 21.67 °C, and 21.64 °C, respectively.

4 Conclusion

In the two conditions of measurement carried out in the isolation room for infectious diseases, there are significant differences. The air supply provided through the diffuser has an effect on the temperature distribution at each measurement point and location ($T1$, $T2$, $T3$, and $T4$). The temperature value at each measurement point decreases based on the height of the measurement point. In both measurement conditions, the same phenomenon occurs for each measurement point. The temperature value at point $T1$ in all conditions shows the highest temperature value, then followed by the temperature value at points $T2$, $T3$, and $T4$. An isolation room with diffuser and exhaust Off conditions is needed to determine the temperature distribution when there is no air supply in the room, so that measurement data are obtained which will be the basis for further research development. In this condition, the average temperature value is 27.01 °C. As for the conditions of the diffuser On and exhaust Off, there are more significant differences from each measurement point and location. At positions $X = 0.2$ and $Y = 1.9$, the temperature values tend to increase at each measurement point. However, this does not happen at other measurement points and has almost the same temperature value. Thus, the measurement results of each condition can be the initial basis for continuing further research.

Acknowledgements This research was funded by Lembaga Pengelola Dana Pendidikan (LPDP), managed by Indonesian Science Fund (ISF): RISPRO/KI/B1/TKL/5/15448/2020; Syiah Kuala University: 370/UN11.2.1/PT.01.03/PNBP/2021).

References

1. Kesehatan DJBU (2012) Standar Akreditasi Rumah Sakit
2. Liu Z, Wang T, Wang Y, Liu H, Cao G, Tang S (2022) The influence of air supply inlet location on the spatial-temporal distribution of bioaerosol in isolation ward under three mixed ventilation modes. Energy Built Environ. https://doi.org/10.1016/j.enbenv.2022.03.002
3. K. Kesehatan -Ri, D. Bina, P. Penunjang, M. Dan, and S. Kesehatan, "Pedoman Teknis Prasarana Sistem Tata Udara Pada Bangunan Rumah Sakit."
4. Cho J, Woo K, Kim BS (2019) Removal of airborne contamination in airborne infectious isolation rooms. ASHRAE J 61(2):8–21
5. Sheerin MP et al (2020) Ventilation of health care facilities. 8400
6. Agirman A, Cetin YE, Avci M, Aydin O (2019) Influence of ceiling height on airflow and particle distribution in an operating room. E3S Web Conf 111:01032. https://doi.org/10.1051/E3SCONF/201911101032
7. Sadrizadeh S, Holmberg S, Tammelin A (2014) A numerical investigation of vertical and horizontal laminar airflow ventilation in an operating room. Build Environ 82:517–525. https://doi.org/10.1016/j.buildenv.2014.09.013
8. Alhamid MI, Budihardjo B, Raymond A (2018) Design of the ventilation system and the simulation of air flow in the negative isolation room using FloVent 8.2. AIP Conf Proc 1984(July 2018). https://doi.org/10.1063/1.5046600
9. Ahmed AQ, Gao S, Kareem AK (2016) A numerical study on the effects of exhaust locations on energy consumption and thermal environment in an office room served by displacement ventilation. Energy Convers Manag 117:74–85. https://doi.org/10.1016/j.enconman.2016.03.004
10. Yin Y et al (2009) Experimental study on displacement and mixing ventilation systems for a patient ward. HVAC R Res 15(6):1175–1191. https://doi.org/10.1080/10789669.2009.10390885
11. Vasic M, Stevanovic VD, Zivkovic B (2020) Uniformity of air flow from the ceiling diffuser by an advanced design of the equalizing element in the plenum box with side entry. Sci Technol Built Environ 26(5):676–686. https://doi.org/10.1080/23744731.2020.1730124
12. Memarzadeh F, Jiang J (2000) Methodology for minimizing risk from airborne organisms in hospital isolation rooms. ASHRAE Trans 106(January):2000
13. Cho J (2019) Investigation on the contaminant distribution with improved ventilation system in hospital isolation rooms: effect of supply and exhaust air diffuser configurations. Appl Therm Eng 148(November 2018):208–218. https://doi.org/10.1016/j.applthermaleng.2018.11.023
14. Wang Y, Liu Z, Liu H, Wu M, He J, Cao G (2022) Droplet aerosols transportation and deposition for three respiratory behaviors in a typical negative pressure isolation ward. Build Environ 219(April):109247. https://doi.org/10.1016/j.buildenv.2022.109247

15. Liu Z, Zhu H, Song Y, Cao G (2022) Quantitative distribution of human exhaled particles in a ventilation room. Build Simul 15(5):859–870. https://doi.org/10.1007/s12273-021-0836-1
16. Qian H, Li Y (2010) Removal of exhaled particles by ventilation and deposition in a multibed airborne infection isolation room. Indoor Air 20(4):284–297. https://doi.org/10.1111/j.1600-0668.2010.00653.x
17. Berlanga FA, de Adana MR, Olmedo I, Villafruela JM, San José JF, Castro F (2018) Experimental evaluation of thermal comfort, ventilation performance indices and exposure to airborne contaminant in an airborne infection isolation room equipped with a displacement air distribution system. Energy Build 158:209–221. https://doi.org/10.1016/j.enbuild.2017.09.100
18. Jung M, Chung WJ, Sung M, Jo S, Hong J (2022) Analysis of Infection transmission routes through exhaled breath and cough particle dispersion in a general hospital. Int J Environ Res Public Health 19(5). https://doi.org/10.3390/ijerph19052512

Charging–Discharging Characteristics of NaCl–KCl Eutectic Salt in a Thermal Box System

Zulmiardi Zulmiardi, Adi Setiawan, Meriatna Meriatna, Muhammad Sayuthi, Gunawati Gunawati, Shafira Riskina, and Dodi Lesmana Putra

Abstract Solar radiation is a varied source of energy, and the mismatch between the availability of these sources and domestic energy demand is a major challenge that must be faced. Energy storage system can be considered as effective solution to overcome the challenge in solar energy utilization. In this work, a thermal box was prepared as energy storage testing apparatus by using a 5W lamp as the heat source located at the center of the box. In the middle of thermal box, the sample of phase change material (PCM) was placed and monitored in order to find out the charging and discharging characteristics. The purpose of this study was to investigate the charging and discharging performance of NaCl–KCl eutectic salt at various ratios. The specific heat capacity and density of all PCM samples were also tested during this investigation. The changes in temperature were observed through type-k thermocouple sensors equipped with data logger. In general, it can be concluded that reducing KCl portion in PCM slightly decreases the specific heat capacity. Experiment results showed that PCM-1 (eutectic salt with 30 wt% NaCl–70 wt% KCl) has the highest specific heat capacity, i.e., 6.93 J/g°C. The highest rate of charging and discharging was exhibited by PCM-1, highlighting a positive effect of KCl content in eutectic salt.

Z. Zulmiardi · A. Setiawan (✉) · M. Sayuthi · D. L. Putra
Mechanical Engineering Departement, Faculty of Engineering, Universitas Malikussaleh, Jalan Batam, Bukit Indah, Muara Satu, 24352 Lhokseumawe, Indonesia
e-mail: adis@unimal.ac.id

M. Meriatna
Chemical Engineering Departement, Faculty of Engineering, Universitas Malikussaleh, Jalan Batam, Bukit Indah, Muara Satu, 24352 Lhokseumawe, Indonesia

G. Gunawati
Departement Physics, Faculty of Mathematics and Natural Sciences, Universitas Syiah Kuala, Darussalam, 23111 Banda Aceh, Indonesia

S. Riskina
Magister Program in Renewable Engineering, Faculty of Engineering, Universitas Malikussaleh, Bukit Indah, Muara Satu, 24352 Lhokseumawe, Indonesia

Irwansyah et al. (eds.), *Proceedings of the 4th International Conference on Experimental and Computational Mechanics in Engineering*, Lecture Notes in Mechanical Engineering, https://doi.org/10.1007/978-981-99-7495-5_42

Keywords Phase change materials · Eutectic salt · Thermal energy storage · NaCl · KCl

1 Introduction

The world's demand for energy continues to increase, while many economic and environmental problems are increasing year by year due to the use of fossil fuels. Concerns about rising global temperatures due to the consumption of fossil fuels are driving the development of environmentally friendly, sustainable, and safe renewable energy technologies [1]. One solution to this problem is to build energy storage technology. Energy storage technology consists of several methods, including mechanical energy storage, electrical energy storage, and thermal energy storage [2]. Currently, thermal energy storage is an energy that needs to be developed because it has several advantages compared to other energy storage. Thermal energy storage (TES), which can be called heat and cold storage, is a method of thermal energy storage that uses the principle of heat or cold storage for future use [3], and TES is one of the cheapest forms of energy storage [4]. There are two storage techniques in TES in the physical process, namely, by using the principle of sensible heat and latent heat [5].

In heat storage as sensible heat, the temperature will increase along with the stored energy, in contrast to heat storage as latent heat, where the storage process takes place at a constant temperature. Materials used as sensible/latent heat storage are referred to as phase change materials (PCMs). PCM has the ability to absorb or release energy quite well. Phase transitions (phase changes) are usually from solid to liquid, allowing these PCM materials to store thermal energy [6].

Eutectic is the minimum melting composition of two or more components, each of which melts and solidifies congruently. During the crystallization phase, a mixture of components is formed, hence acting as a single component. The components freeze into an intimate crystalline mixture and melt simultaneously without separation [7]. The eutectic can be a mixture of organic and/or inorganic compounds. Therefore, the eutectic can be made as a mixture of organic–organic, inorganic–inorganic, or organic–inorganic [8]. This leaves room for a wide variety of combinations that can be adapted for specific applications. Organic eutectic mixtures, the most commonly tested, consist of fatty acids, including capric acid/myristic acid [9], lauric acid/stearic acid, myristic acid/palmitic acid and palmitic acid/ stearic acid [10] and capric acid/ lauric acid [11].

The most common inorganic eutectic that has been studied consists of different hydrated salts. The benefit of a eutectic mixture is its ability to obtain more desirable properties such as a specific melting point or a higher heat storage capacity per unit volume. Organic PCM offers better chemical and thermal stability with congruent melting, and little or no sub-cooling. On the other hand, inorganic products are cyclically unstable, require nucleating agents and thickeners to minimize sub-cooling, and are highly reactive to metallic materials, as has been done by several researchers in Table 1.

Table 1 Eutectic salt that has been studied as thermal energy storage

Eutectic Salt	ΔT_m[a] (°C)	ΔH[b] (kJ/kg)	C_p[c] (kJ/kg.K)	k[d] (W/m.K)	Density (kg/m^3)	Refs.
$MgCl_2$–KCl–NaCl (60–20.4–19.6 wt%)	380	400	0.96	–	1800	[12]
KCl–NaCl (49.4–50.6 wt%)	657	360	–	–	–	[13]
KCl–$MgCl_2$ (69.8–30.2 wt%)	423	265	–	–	–	
KCl–$MgCl_2$ (65.3–34.7 wt%)	429	271	–	–	–	
KCl–$MgCl_2$ (41.6–58.4 wt%)	473	310	–	–	–	
$MgCl_2$–NaCl (43.1–56.9 wt%)	459	333	–	–	–	
$MgCl_2$–NaCl (52-48 wt%)	450	430	$0.92_{(s)}$*	$0.95_{(l)}$*	$2230_{(s)}$*	[14]
			$1.00_{(l)}$*	–	–	
$MgCl_2$–KCl (39–61 wt%)	435	351	$0.80_{(s)}$*	$0.80_{(l)}$*	$2110_{(s)}$*	[14]
			$0.96_{(l)}$*	–	–	
$MgCl_2$–KCl (64–36 wt%)	470	388	$0.84_{(s)}$*	$0.83_{(l)}$*	$2190_{(s)}$*	[14]
			$0.96_{(l)}$*	–	–	
NaCl–KCl (43.9–56.1 wt%)	659	313.0	–	–	–	[15]

[a] ΔTm = Temperature melting
[b] ΔH = Heat of fusion = enthalpy
[c] Cp = Specific heat capacity
[d] k = Thermal conductivity
* $_{(s)}$ = solid, $_{(l)}$ = liquid

Studies on NaCl–KCl that have been carried out are still lacking in terms of application, and how after mixing NaCl–KCl is applied directly for performance when used as an energy storage material. One of them is in the form of solar collectors, and thermal energy storage systems. Therefore, we need a tool that can observe the ability of this eutectic salt material to store thermal energy, to observe the charging and discharging process of the eutectic salt material.

Table 2 PCM sample ID

No.	Sample ID	Description
1	PCM-1	NaCl (30 wt%) + KCl (70 wt%)
2	PCM-2	NaCl (51.8 wt%) + KCl (49.4 wt%)
3	PCM-3	NaCl (80 wt%) + KCl (20 wt%)

2 Materials and Methods

2.1 *Eutectic Salt Preparation*

The eutectic chloride salt (NaCl–KCl) was prepared experimentally by the smelting method by mixing the concentrations of sodium chloride and potassium chloride, as shown in Table 2. The sodium chloride used was local salt taken from Tanoh Anoe, Bireun District, Province of Aceh. Production of NaCl used geomembrane system, which results in NaCl content of 96.4%. The two salts according to their respective concentrations were heated in a furnace at 680 °C for 50 min and then cooled in the room at room temperature. After that, the cold eutectic salt was then grounded using a mortar until smooth.

2.2 *Experimental Setup*

The thermal box was designed for a 5 L storage volume capacity with external dimensions of 452 × 299 × 237 mm. The box consists of a layer of wood 16 mm thick as a casing, a layer of styrofoam (22 mm thick), and the center of the box was installed a wall as a PCM retainer of styrofoam (3 mm thick) with a hole in the middle with a diameter of 73 mm. The heat source comes from a lamp with a power of 5W. Figure 1 shows a thermal box design with a PCM retaining wall.

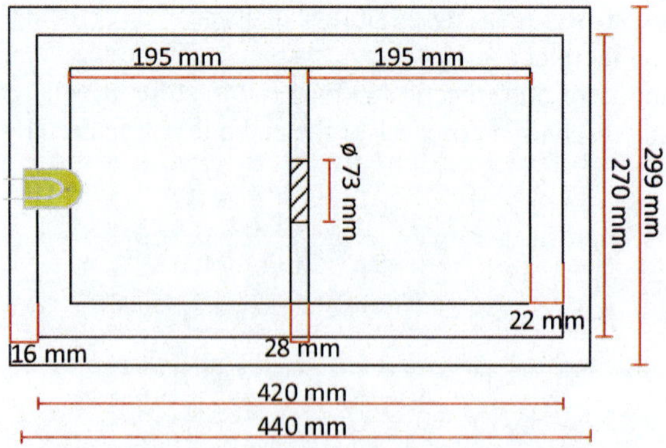

Fig. 1 Geometric model of thermal box

Experiments were carried out for three types of materials with varying concentrations of NaCl–KCl. The solid sample was pressed and inserted into the PCM retaining wall as much as 100 g. For charging, the wall containing PCM was installed into the box and tested by turned on the heat source (the lamp was turned on) for one hour. Then, after one hour, the lamp was turned off and the discharging process was continued until the temperature inside the box reached the same as the PCM temperature. The setup of this experiment are shown in Fig. 2.

Physical properties in the form of analysis of bulk density, apparent density, and specific heat capacity are shown in Fig. 3. Bulk density analysis was carried out using a 5-ml bottle. The PCM sample was poured into it until was full the mass of sample before and after was recorded, and then the density calculation was carried out. Apparent density was defined by measuring the total volume of pressed PCM sample and also its mass.

Specific heat capacity (Cp) is the amount of heat energy required to raise the temperature of one degree Celsius in one kilogram of solid [16]. Measurement of specific heat capacity in sensible heat conditions, the measurement technique, used

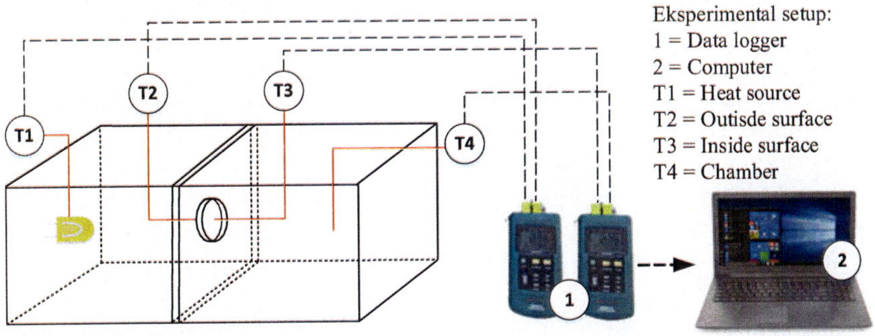

Fig. 2 Experimental setup of thermal box

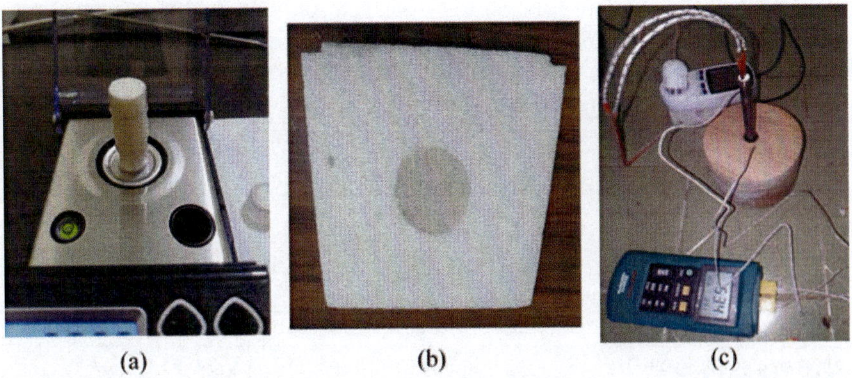

Fig. 3 **a** Bulk density test equipment, **b** apparent density, and **c** specific heat capacity test equipment

an insulator (wood) as a PCM container which was heat-treated using a cartridge heater with a voltage of 220 V and a power of 800 watts. An energy meter was used to determine the amount of energy needed to heat the PCM. The following equation calculates the specific heat capacity:

$$\frac{dH}{dt} = m.Cp.\frac{dT}{dt} \tag{1}$$

$$Cp = \frac{\text{Heat flow(mW)}}{(\text{massa(mg)}) \times (\text{Heat rate(°C)})} \times 60 \tag{2}$$

where, dH/dt is heat flow (mW), dT/dt is heat rate (°C/min), m is sample mass (mg), and Cp is specific heat capacity (J/mg°C).

3 Results and Discussions

3.1 Physical Properties

Density analysis was carried out using the bulk density analysis method, where the solid sample was inserted and compacted to the brim of a bottle with a volume of 4.915 ml. The density of PCM-1 sample is 1.613 gr/cm^3, PCM-2 sample is 1.591 gr/cm^3, and PCM-3 sample is 1.581 gr/cm^3. The density value obtained is higher than the density of local Aceh salt without treatment as reported by Gunawati et al., which is 1.047 gr/cm^3 and after being treated to a temperature of 800 °C, which is 1.133 gr/cm^3 [17].

Measurement of specific heat capacity in sensible heat conditions, the measurement technique, used an insulator (wood) as a PCM container which is heat-treated using a cartridge heater with a voltage of 220 V and a power of 800 watts. The results obtained are the PCM-1 sample of 6.93 J/gr °C, the PCM-2 sample of 6.67 J/gr°C, and the PCM-3 sample of 6.58 J/gr °C. The results obtained from each sample tend to be harder at the top (which is in direct contact with the heater), while the bottom was still in powder form (not hard). Table 3 shows the results of the physical characterization of the eutectic salt.

Table 3 Physical characteristics of the eutectic salt

Parameter	Description		
	PCM-1	PCM-2	PCM-3
Specific heat capacity (sensible heat) (J/g°C)	6.93	6.67	6.58
Bulk density (g/cm^3)	1.613	1.591	1.581
Apparent density (g/cm^3)	15.766	15.934	15.850

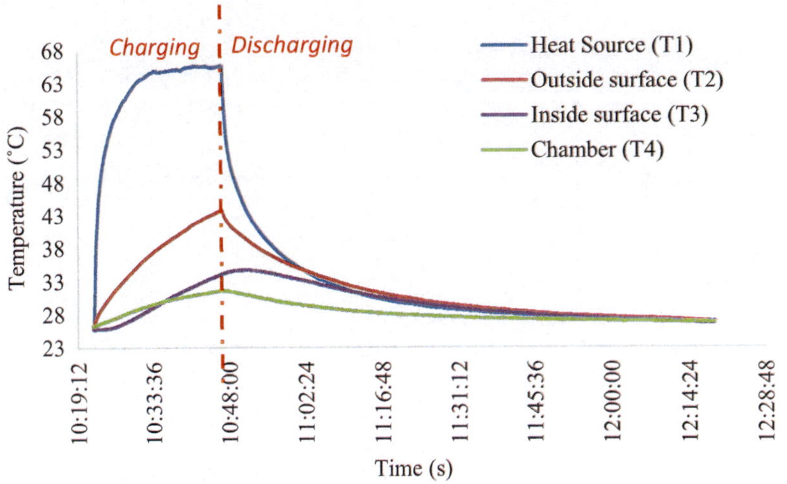

Fig. 4 Temperature profile of charging–discharging process PCM-1

3.2 *Charging–Discharging Experiment Results*

In this section, the results obtained through experimental tests are the process of charging and discharging of PCM samples of the eutectic salt material. The test was carried out with a mixture of two types of chloride salts which had different concentrations as shown in Fig. 4.

The performance test process using a thermal box was carried out for 2 h. The charging condition is carried out when the lamp as a heat source is turned on until it reaches its maximum point, which is at a temperature of 65°C, and the PCM outside surface temperature reaches 43 °C. The charging process is the process of absorbing energy from the heat source by the PCM outside surface and lasts for 30 min. The charging process is said to be complete when the PCM inside surface temperature is the same as the room temperature, which is at a temperature of 29.6 °C. After the temperature inside the surface and the room is the same, the heat source (light) is turned off and continued with the discharging process. The discharging process is a process of releasing temperature to ambient temperature, and this process lasts for 1 h 30 min (Fig. 4).

The performance of the thermal box with PCM-2 samples was carried out for 1 h. The charging condition is carried out when the lamp as a heat source is turned on until it reaches its maximum point at 80 °C, and the PCM outside surface temperature reaches 32°C. The charging process is the process of absorbing energy from the heat source by the PCM outside surface and lasts for 5 min. The charging process is said to be complete when the PCM inside surface temperature is the same as the room temperature, which is at a temperature of 29.6 °C. After the temperature inside the surface and the room is the same, the heat source (light) is turned off and continued with the discharging process. The discharging process is a process of

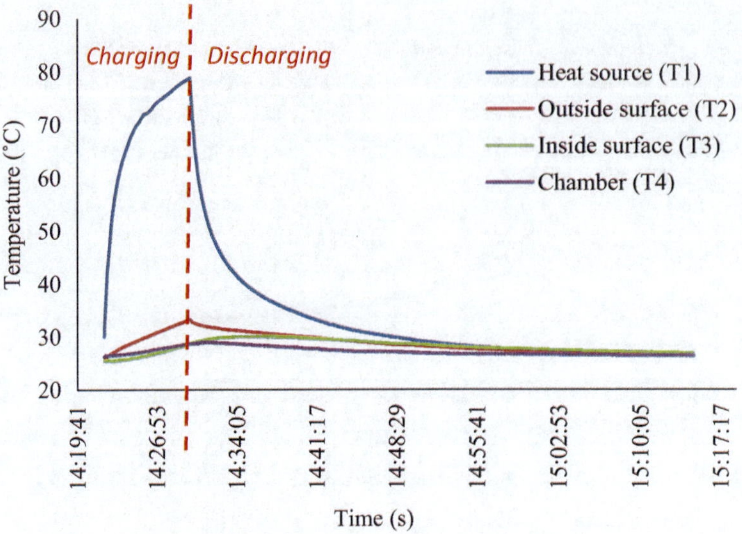

Fig. 5 Temperature profile of charging–discharging process PCM-2

releasing temperature to ambient temperature, and this process lasts for 1 h 30 min (Fig. 5).

The study on PCM-3 samples was carried out for 1 h, and the condition of the thermal box was closed. The temperature of the PMC outside surface increases until the heat collected by the PCM reaches a maximum point of 30°C; this process is called the charging process, which is marked when the inside surface temperature and room temperature condition are same, and then, the heat source (light) is turned on. After the PCM reaches its maximum point, the discharging process is continued where the heat source (lamp) is turned off, so that the temperature of the lamp, outside surface, and inside surface is close to room temperature, which is at a temperature of 24.9°C (Fig. 6).

The three samples measured using a thermal box, and it shows that the more concentration of KCl added, the higher the outer surface temperature of the PCM; this is in accordance with the results obtained from testing the specific heat capacity of latent heat conditions. This is in accordance with what has been described by Lutsenko and Fetsov [18] and Saranprabhu and Rajan [19] where the addition of a material increases the PCM side heat transfer coefficient and increases the specific heat capacity.

The charging process is a heating process, which is marked by an increase in the PCM temperature during performance tests using a thermal box; the heat stored by the PCM to increase its temperature is called sensible heat. The estimation of the amount of sensible heat can be known by Eq. 3, and the calculation of the value of sensible heat using the equation is obtained from the following data:

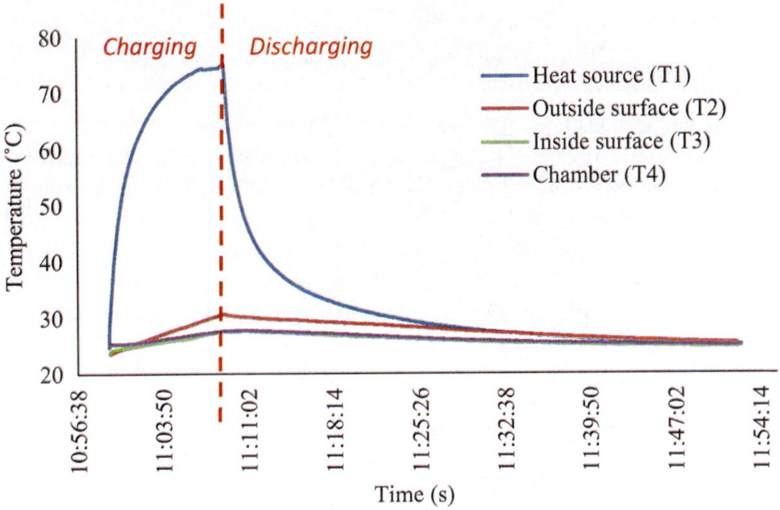

Fig. 6 Temperature profile of charging–discharging process of PCM-3

$$Q = \int_{T_i}^{T_f} mC_p dT = mC_p(T_f - T_i) \quad (3)$$

where, Q is the heat required or expended during the phase change (kJ), *Ti* is the initial temperature (°C), *Tf* is the final temperature (°C), m is the mass of the material used to store thermal energy (kg), and *Cp* is the heat type of material used to store heat energy (kJ/kg°C) (Table 4).

In the charging process, the eutectic salt in the heat exchanger on the solar collector receives heat from the sun's heat with different intensity and time intervals. However, because the value of specific heat capacity, initial temperature, and melting temperature are different, the amount of sensible heat that can be achieved is also different. The PCM-1 sample is able to store heat in the form of the sensible heat of 11.64 kJ, and after that, the remaining external heat is used to change the PCM phase, while in the PCM-2 and PCM-3 samples, the heat that can be stored in the form of sensible heat is 4.14 kJ and 4.28 kJ, respectively.

Table 4 Sensible heat that can be achieved from the measurement results of the thermal box

Sample ID	T_i(°C)	T_m(°C)	Cp solid (J/gr°C)	Q sensible (kJ)
PCM-1	26.2	43	6.93	11.64
PCM-2	25.8	32	6.67	4.14
PCM-3	23.5	30	6.58	44.28

4 Conclusion

In this work, a thermal box equipped with TES (thermal energy storage) based on a phase change material (PCM) was designed and fabricated to observe the charging–discharging characteristics of NaCl–KCl eutectic salts at various concentration ratios. Based on the characterization result of NaCl–KCl eutectic salt, it can be concluded that reducing KCl portion decreases slightly in specific heat capacity and melting temperature. Significant increase in sensible heat was observed when KCl content was added in eutectic salt. The highest rate of charging and discharging was exhibited by PCM-1 (eutectic salt with 30 wt% NaCl–70 wt% KCl) where its specific heat capacity is 6.93 J/g°C.

Acknowledgements The authors acknowledge the Institute of Research and Community Service (LPPM) Universitas Malikussaleh for their sponsorship under contract number 2/PPK-2/SPK-JL/2022.

References

1. Nazir H, Batool M, Bolivar Osorio FJ, Isaza-Ruiz M, Xu X, Vignarooban K, Phelan P, Inamuddin, Kannan AM (2019) Recent developments in phase change materials for energy storage applications: a review. Int J Heat Mass Transf **129**, 491–523
2. Sharma A, Tyagi VV, Chen CR, Buddhi D (2009) Review on thermal energy storage with phase change materials and applications. Renew Sustain Energy Rev 13:318–345
3. Mehling H, Cabeza F (2008) Heat and cold storage with PCM: an up to date introduction into basic and applications. Springer, Berlin
4. Rodríguez-Cumplido F, Pabón-Gelves E, Chejne-Jana F (2019) Recent developments in the synthesis of microencapsulated and nanoencapsulated phase change materials. J Energy Storage 24 (2019)
5. Gunawati, Dongoran AH, Setiawan A (2019) Evaluation on performance of cold storage box enveloped with phase change materials. J Phys Conf Ser 1242
6. Gunawati, Nasruddin MN, Setiawan A, Sebayang K, Makruf A, Zakaria (2021) Design and calibration of temperature monitoring device for a cold box with hydrated salts as phase change material (PCM). Lect Notes Mech Eng 199–210
7. Cabeza LF, Castell A, Barreneche C, de Garcia A, Ferandez AI (2011) Materials used as PCM in thermal energy storage in buildings: a review. Renew Sustain Energy Rev
8. Rodriguez-Ubinas E, Ruiz-Valero L, Vega S, Neila J (2012) Applications of phase change material in highly energy-efficient houses. Energy Build 50:49–62
9. Silva NSA (2013) Simulação numérica da influência da interface no fenómeno da hu-midade ascen- sional - Wufi 2D. niversidade do Porto (Portugal) ProQuest Dissertations Publishing
10. Hawlader MNA, Uddin MS (2003) MMK: microencapsulated PCM thermal-energy storage system. Appl Energy 74(1–2):195–202
11. Delgado JMPQ, Martinho JC, Vaz Sá A, Guimarães AS, Abrantes V (2019) Thermal energy storage with phase change materials In: A literature review of applications for buildings materials. Springer, Switzerland
12. Mohamed SA, Al-Sulaiman FA, Ibrahim NI, Zahir MH, Al-Ahmed A, Saidur R, Yılbaş BS, Sahin AZ (2017) A review on current status and challenges of inorganic phase change materials for thermal energy storage systems. Renew Sustain Energy Rev 70:1072–1089

13. Myers PD, Goswami DY (2016) Thermal energy storage using chloride salts and their eutectics. Appl Therm Eng 109:889–900
14. Liu M, Saman W, Bruno F (2012) Review on storage materials and thermal performance enhancement techniques for high temperature phase change thermal storage systems. Renew Sustain Energy Rev 16:2118–2132
15. Sergeev D, Kobertz D, Muller M (2015) Thermodynamics of the NaCl–KCl system. Thermochim Acta 606:25–33
16. O'Neill MJ (1966) Measurement of specific heat functions by differential scanning calorimetry. Anal Chem 38:1331–1336
17. Gunawati G, Humaidi S, Setiawan A, Sirait M, Jalil Z, Ramadhani N, Makruf A, Riskina S, Irhamni (2022) Physical characteristics of aceh traditional salt and its potential as raw material for thermal energy storage. J Ecol Eng 23:116–122
18. Lutsenko NA, Fetsov SS (2020) Effect of side walls shape on charging and discharging performance of thermal energy storages based on granular phase change materials. Renew Energy 162:1–12 (2020)
19. Saranprabhu MK, Rajan KS (2022) Thermal energy storage system with a high-temperature nanoparticle enhanced phase change material: charging and discharging characteristics upon integration with process preheating. J Energy Storage 55:1–10

The Failure Analysis of Crankshaft of Four Cylinder Diesel Engine Using Numerical Method

Akbar Rizqullah, Husaini, and Iskandar Hasanuddin

Abstract Crankshaft is one of the main components of a combustion engine, and it is designed to be used for continuous rotation. Therefore, this study aims to analyze the crankshaft failure of a jeep with a four-cylinder diesel engine which was used to transport 4 tons of logs on rough terrain and uneven roads for 12 months. The process involved modeling the crankshaft with the initial defect followed by the application of the finite element method to determine the value and distribution of stresses occurring on the crankshaft and around the location of the initial defect. The stress values obtained were later used to calculate the stress intensity factor (K_I), and the value obtained was compared with the fracture toughness (K_{IC}) value. The simulation results and stress intensity factor (K_I) calculated showed a concentration of stress in the area of the initial defect as indicated by the average value of the stress intensity factor recorded to be 293.13944 MPa $\sqrt{m}$ from the simulation, while the fracture toughness of the material is 91.73 ± 3.34 MPa $\sqrt{m}$. This means $K_I > K_{IC}$ at the crack tip, and this causes further propagation starting from the initial defect. The results simply showed that the crack propagation due to loading and the value of stress concentration at the crack tip area is the cause of the crankshaft failure. This is associated with the repeated loading characterized by beach mark patterns on the fractured surface.

Keywords Failure analysis · Crankshaft · Stress intensity factor · Finite element method

A. Rizqullah · Husaini (✉) · I. Hasanuddin
Laboratory of Computational Mechanics, Department of Mechanical Engineering, Universitas Syiah Kuala, Darussalam, Banda Aceh 23111, Indonesia
e-mail: husainiftm@usk.ac.id

I. Hasanuddin
e-mail: iskandarhasanuddin@usk.ac.id

Irwansyah et al. (eds.), *Proceedings of the 4th International Conference on Experimental and Computational Mechanics in Engineering*, Lecture Notes in Mechanical Engineering, https://doi.org/10.1007/978-981-99-7495-5_43

1 Introduction

Crankshaft is a major component in the piston engine which consists of a series of cranks and crankpins. Its connecting rods are usually attached to the clutch to convert linear to rotational motion [1]. Most crankshafts are made of medium carbon steel or alloys with chrome and nickel metals through forging or casting techniques to ensure they have outstanding mechanical strength. This component is required to be strong enough to withstand the thrust from the piston during the combustion process without applying much pressure. It also needs to be carefully balanced to avoid vibration caused by the center of mass. However, the force exerted by the piston is not uniform and this can lead to a decrease or increase in the crankshaft speed [2]. The presence of residual stress usually affects the durability of components, and this has the likelihood to cause failure due to fatigue after sustenance for a long time [3].

According to Husaini et al. [4], crack initiation and propagation have initially occurred despite the strong influence of stress concentration. This led to the analysis of the von Mises stress near the crack tip using the finite element method (FEM). Moreover, a counterweight was used to reduce the impulse effect based on the moment of inertia by making the flywheel absorb energy during the motor impulse cycle to ensure it returns to the crankshaft whenever the engine is stopped. This component rotates at high speed to perform billions of repeated loading cycles and is therefore usually designed for a certain service life [5]. Therefore, most of the crankshaft disturbances in engines are related to fatigue failure caused by bending and torsional loads mainly from high gas pressure transmitted to the crankpin through the connecting rod during the combustion process.

The other factors affecting the crankshaft life are high torque oscillations due to imbalance, insufficient oil between the main journal and crankpin bearings, high pressure, and the presence of cracks from combustion and manufacturing, especially in the cylinders. Meanwhile, surface cracks are associated with the formation of bends when there is insufficient oil between the main journal and the crankpin bearing [6, 7].

The errors created during disassembly and maintenance such as scratches which cause the initial defects leading to the initial disturbance need to be avoided. However, the failure usually starts from manufacturing defects at the critical points experiencing high-stress concentrations such as the fillet radius and crankpin installation errors [8]. Moreover, crack is mostly initiated on the surface of weak materials or highly stressed areas as indicated by scratches and pits due to repeated loading. Micro-cracks are initially formed and later developed into larger cracks until the occurrence of failure [9].

Husaini et al. [10] reported an initial defect in the fillet area of crankpin number 1, and visual observation showed the presence of beach marks on the fractured surface which is a characteristic of fatigue. Moreover, Husaini et al. [11] conducted another study on the leaf spring of a failed diesel turbine and reported that the fractures indicated by a beach mark pattern on the affected surface were caused by fatigue loads.

Engine components often experience failures such as fractures in areas with high-stress concentrations such as small fillet radii, sharp corners, grooves, splines, and scratches [12]. It is however important to note that the stress usually increases with an increment in the rotation of the crankshaft. Furthermore, preliminary studies reported the frequent occurrence of failures in the crankpin-web fillet region with high-stress concentrations.

The other factors affecting fatigue include high vibration due to crankshaft imbalance or lack of a counterweight, absence of lubrication between the bearing and main journal, high pressure caused by cylinder combustion, and the appearance of cracks during the manufacturing process, especially between the crankpin-web fillet areas [13].

The counterweights are useful in balancing the off-center weight of the crankpins and web. This is based on the application of moment of inertia to reduce the effects of impulses in the same manner a flywheel is used to store rotational energy in order to absorb the motorcycle impact which is subsequently felt at the crankshaft [14].

This research examines the failure of the crankshaft of a jeep which was used to carry logs for approximately 12 months. The vehicle operates on steep terrain and uneven roads, carrying a load of relatively 4 tons at a time, which led to the breakdown of the crankshaft. It was discovered that the failure occurred in the weight balancer portion.

2 Research Object

This study used a crankshaft that failed on the balancer weight section of a jeep-type vehicle with a four-cylinder diesel engine as shown in Fig. 1, while the CAD modeled drawing is presented in Fig. 2.

Research methods. The research was initiated by measuring the dimensions of the crankshaft followed by inputting data into the design made using CAD in Fig. 2 with the dimensions presented in Fig. 3.

Fig. 1 Crankshaft failure

Fig. 2 Crankshaft CAD

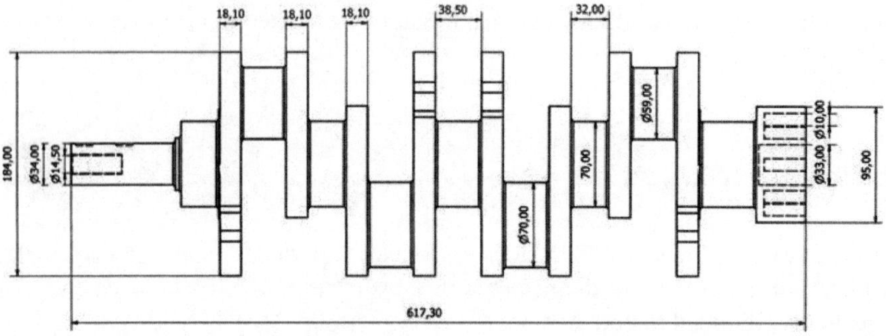

Fig. 3 Dimensions of the crankshaft (mm)

Meshing. The meshing process is a method of dividing elements into smaller parts to obtain detailed simulation values. The mesh used in this study is tetrahedral with a size of 0.2 as shown in Fig. 4.

Mechanical properties. The mechanical properties obtained from a previous study [15] showed that the shaft material is AISI 4340 as presented in Table 1.

Load and constrain. Loads and supports greatly affect the simulation results with the loading on the crankpin recorded to be $P_z = 16.7242$ MPa and $P_c = 5.855$ MPa as indicated in Fig. 5, while the support provided by each main journal of the crankshaft is presented in Fig. 6.

Von Mises Theory Analysis. This process was used to determine the von Mises stress to be used as input in the stress intensity factor formula.

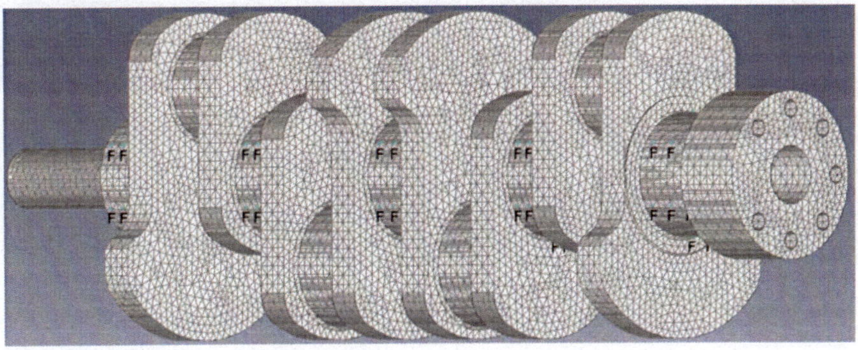

Fig. 4 Meshing on the crankshaft

Table 1 Mechanical properties of AISI 4340 [15, 16]

No.	Sifat Mekanik	Satuan
1	Tensile strength, (σ_B)	745 MPa
2	Poisons ratio, (υ)	0.27–0.30
3	Yield strength, (σ_{ys})	470 MPa
4	Young's modulus, E	190–210 GPa
5	Hardness Rockwell B	95 HRB
6	Shear modulus, (μ)	80 GPa
7	Bulk modulus	140 GPa
8	Fracture toughness, K_{IC}	91.73 ± 3.34 MPa $\sqrt{m}$

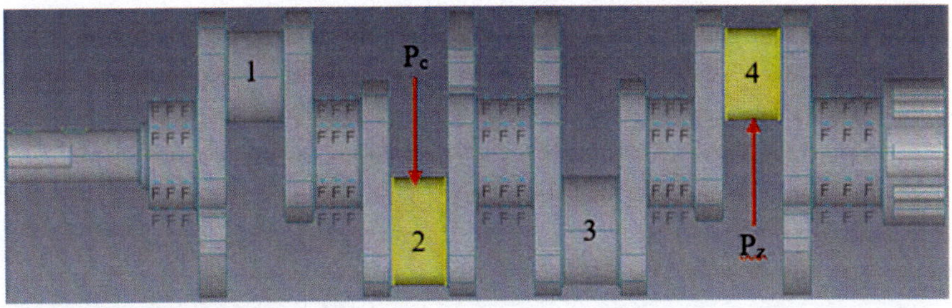

Fig. 5 Load on the crankpin

Stress Intensity Factor. FEMAP software was used to determine the value of the von Mises stress in the crankshaft, and the highest was found to have five nodes. Therefore, the stress intensity factor (KI) was calculated using the following Eq. 1 [12]:

$$K_I = \frac{\sigma_{xx}\sqrt{2\pi r}}{\cos\left(\frac{\theta}{2}\right)\left(1 - \sin\left(\frac{\theta}{2}\right)\sin\left(\frac{3\theta}{2}\right)\right)}. \tag{1}$$

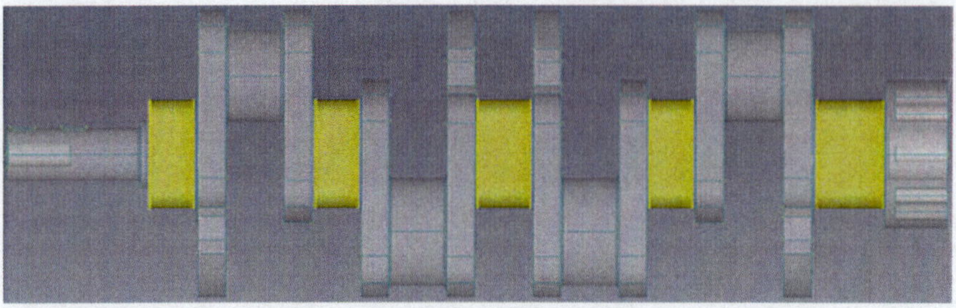

Fig. 6 Focus on each main journal

3 Results and Discussion

Von Mises Theory Analysis. The crankshaft was simulated using FEMAP and the highest von Mises value at crack initiation was recorded to be 351.1 MPA as indicated in Fig. 7. This value was used as a reference in calculating the stress intensity factor (K_I) value.

Stress intensity factor. The stress intensity factor value was obtained from five nodal points as shown in Fig. 8.

The stress intensity factor value of the initial crack to the nodal was calculated using Eq. (1), and the results are presented in Table 2.

The average stress intensity factor value was found to be 293.13944 MPa $\sqrt{m}$, and this exceeds the fracture toughness (K_{IC}) value of 91.73 ± 3.34 MPa $\sqrt{m}$ in the crankshaft [16].

Fig. 7 Simulation results of von Mises tegangan voltage

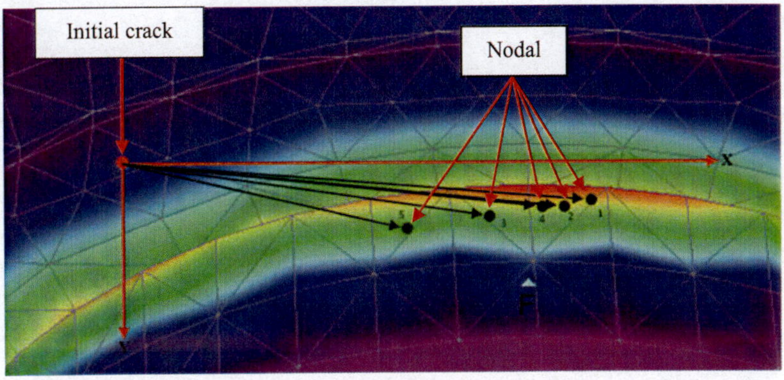

Fig. 8 Location of the nodes of the selected crack

Table 2 Stress intensity factor (K_I) value of the highest stress section

No.	Nodal ID	σ_{xx} (MPa $\sqrt{m}$)	r (m)	θ (°)	K_I (MPa $\sqrt{m}$)
1	102,679	177,1373	0.0252	84	201,7513
2	123,410	189,0826	0.0238	84	200,8252
3	34,719	424,578	0.0196	80	409,2266
4	117,193	462,8632	0.0227	84	480,1134
5	117,208	181,7146	0.0151	76	173,7807
Average					293,13,944

4 Conclusion

The numerical method applied showed that the failure of the four-cylinder diesel engine crankshaft was because the value of $K_I > K_{IC}$. The highest stress intensity factor (K_I) was found at node 117,193 with 480,1134 MPa $\sqrt{m}$, while the fracture toughness (K_{IC}) of AISI 4340 material is 91.73 MPa $\sqrt{m}$, and this means the crack propagation in the crankshaft was due to excessive load.

Acknowledgements The authors are grateful to the Universitas Syiah Kuala for partly their financial support through Professor Research Grant No. 141/UN11./SPK/ PNBP/2022. The authors are also grateful to the Department of Mechanical Engineering, Universitas Syiah Kuala, Banda Aceh, for making the facilities accessible.

References

1. Witek L, Stachowicz F, Załęski A (2017) Failure investigation of the crankshaft of diesel engine. In: 2nd International conference on structural integrity, vol 5. Funchal, Madeira, Portugal, pp 369–376
2. Aliakbari K (2018) The analysis of light-duty truck diesel engine crankshaft failure. J Stress Anal 2:11–17
3. Putra TE, Husaini MN (2020) Machmud, predicting the fatigue life of an automotive coil spring considering road surface roughness. Eng Fail Anal 116:1–28
4. Husaini, Kishimoto K, Notomi M (2004) Finite element analysis of rubber particles size distribution on fracture toughness of rubber-modified polymer. Key Eng Mater 261. Trans Tech Publications Ltd
5. Montazersadgh F, Fatemi A (2007) Stress analysis and optimization of crankshafts subject to dynamic loading. FIERF and American Iron and Steel Institute (AISI), Toledo
6. Fonte M, Freitas M (1997) Semi-elliptical fatigue crack growth under rotating or reversed bending combined with steady torsion. Fatigue Fract, Eng Mater Struct 20:895–906
7. Fonte M, Duarte P, Reis L, Freitas M, Infante V (2015) Failure mode analysis of two crankshafts of a single cylinder diesel engine. Eng Fail Anal 56:185–193
8. Fonte M, Duarte P, Reis L, Freitas M, Infante V (2015) Crankshaft failure analysis of a boxer diesel motor. Eng Fail Anal 56:109–105
9. Husaini, Zuhaimi (2016) Mixed mode fracture behavior of an aluminum alloy A6061 investigate by using compact tension shear specimens. Int J Technol (Ijtech) 3:456–462
10. Putra TE, Husaini, Ali N, Husin H, Zulfikar (2019) Failure analysis of the fracture surface of the crankshaft of a vehicle. IOP Conf Ser Mater Sci Eng 523:1–6
11. Husaini, Ali N, Riantoni R, Putra TE, Husin H (2019) Study of leaf spring fracture behavior used in the suspension systems in the diesel truck vehicles. IOP Conf Ser Mater Sci Eng 541:1–9
12. Husaini (2015) Fundamental of fracture mechanics (in Indonesia), Syiah Kuala University Press
13. Fonte M, Duarte P, Anes V, Freitas M, Reis L (2015) On the assessment of fatigue life of marine diesel engine crankshafts. Eng Fail Anal 56:51–57
14. Fonte M, Freitas M, Reis L (2019) Failure analysis of a demaged diesel motor crankshaft. Eng Fail Anal 102:1–6
15. Azom (2012) AISI 4340 Alloy Steel (UNS G43400). Available https://www.azom.com/article.aspx?ArticleID=6772
16. Li HF, Duan QQ, Zhang P, Zhou XH, Wang B, Zhang ZF (2019) The quantitative relationship between fracture toughness and impact toughness in high-strength steels. Eng Fract Mech 221:362–370

Failure Analysis of Crankshaft in Diesel Engine Vehicle Using Numerical Method

Osama, Husaini, and Teuku Edisah Putra

Abstract In this study, there has been a failure of the crankshaft on a pick up vehicle using a diesel engine. After 3 months of use, the crankshaft broke on the crank arm. The purpose of this study was to determine the cause of crankshaft failure in a diesel engine pick up numerically. The crankshaft is modeled by using CAD to locate the initial defects in the crank web, then applying the finite element method to obtain the value and distribution of stresses that occur in the crankshaft and around the location of the initial defects. The results of the stress values obtained are then applied to calculate the stress intensity factor (K_I) which will be compared with the fracture toughness value (K_{IC}). From the simulation results and the calculation of the stress intensity factor (K_I), it can be observed that there is a concentration of stress in the area of the initial defect. The amount of stress that occurs at the input in the value of the stress intensity factor is 91.827 MPa $\sqrt{}$m, while the fracture toughness of the material is 32.8934 MPa $\sqrt{}$m. The ratio of $K_I > K_{IC}$ at the crack tip causes crack propagation starting at the initial defect. Based on the results obtained, crack propagation due to loading and the concentration value at the crack tip area is the cause of failure of the crankshaft.

Keywords Failure analysis · Cast iron · *Crankshaft* · Stress intensity factor · Finite element method

Osama · Husaini (✉) · T. E. Putra
Laboratory of Computational Mechanics, Department of Mechanical Engineering, Universitas Syiah Kuala, Darussalam, Banda Aceh 23111, Indonesia
e-mail: husainiftm@usk.ac.id

T. E. Putra
e-mail: edi@usk.ac.id

Irwansyah et al. (eds.), *Proceedings of the 4th International Conference on Experimental and Computational Mechanics in Engineering*, Lecture Notes in Mechanical Engineering, https://doi.org/10.1007/978-981-99-7495-5_44

1 Introduction

One of the main components in internal combustion engine is a crankshaft, which is used to convert the up-and-down motion of the piston that becomes a circular motion when transmitted. By the design, crankshaft is designed in such a way with a material that is able to withstand a strong load, as a result of receiving a fairly high engine speed. Burden received by the crankshaft includes torsional loads, bending and centrifugal loads. The role of the crankshaft is quite important in optimizing transmission power, and if a failure occurs, it will result in an accident serious.

The crankshaft is a working engine component which changes the reciprocating motion of the piston by means of piston rod into rotary motion. To change it, a crankshaft requires a crank pin, an additional cushion placed at the end drive rods on each cylinder [1]. In real condition, crankshaft must be strong enough to withstand the bottom safe power stroke without excessive bending. Thus, the life of the combustion engine internally depends on the strength [2], the failure of a material or structure to represent multidisciplinary characteristics. There is no definition universally applicable to failure. In broad meaning, error represents a behavior where the material or structure is no longer capable of fulfilling the function [3].

In a simpler sense, failure is a condition that causes direct loss, usually caused by design errors, material defects and errors in maintenance [4]. Cracks generally start from the initial repetition usually occurs on the surface of weak or in areas of stress concentration, such as scratches, pits and others due to repeated loading. This early crack develops into micro-cracks or rapid propagation forming macro-cracks, culminating in failure [5]. The research by Kakade and Pasarkar can be used as a reference to state that the cause of fatigue failure of the crankshaft can be in the form of cracks, defects and vibrations triggered by variations in the load on the machine [6]. Another study was also conducted by Husaini et al., regarding broken screw spring front car suspension, and that research found the results of the analysis of the surface broken, indicating a failure that occurred on a spring screw which is fatigue failure in the mark early by the presence of cracks initiation and benchmark [7]. Steel crankshaft is usually produced by forging or die forging or by casting if the material is cast iron [8].

Failure usually occurs according to function, because this component will receive power or load which is very large, besides that the crankshaft rotates at a very high speed, due to voltage mechanics that appear as in small filets, sharp angles, grooves, splines and scratches. Shafts often break at the edges where there is a high level of stress concentration. In failure, there are supporting factors for the cause of failure, namely the selection design materials, heat treatment, fabrication, machining and assembling [9].

The crankshaft on a 2500 cc diesel engine pick up vehicle has been operating for 3 months and has failed. This usage period is still categorized as a short service life. Part of the crankshaft is broken on the crank arm. Therefore, from the observations obtained on failures that occur in a short period of time, further research is carried out. The purpose of this study was to determine the cause of crankshaft damage.

2 Research Object

This study used a 2500 cc diesel engine crankshaft pick up which failed at the crank web. The failed crankshaft section is shown in Fig. 1. The results of the drawing modeling with CAD are shown in Fig. 2.

Fig. 1 Crankshaft failure

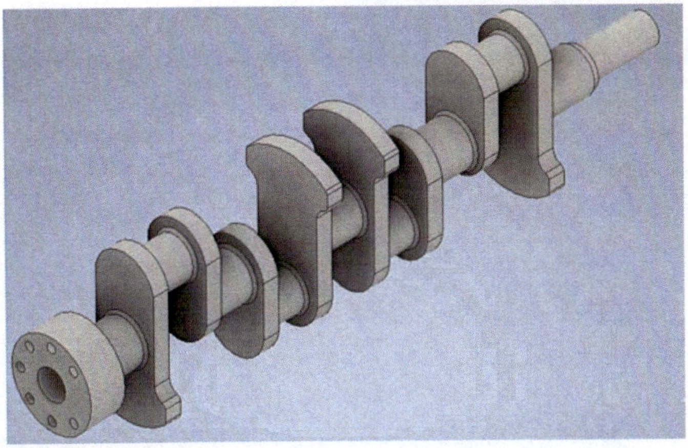

Fig. 2 CAD crankshaft

3 Research Method

Crankshaft Design. Starting with measuring the dimensions of the crankshaft then inputting data into the crankshaft design using CAD is shown in Fig. 2, and the dimensions of the crankshaft are shown in Fig. 3.

Meshing. This method is used for dividing elements into small parts in order to get a specific value for the division. In this study, the use of a mesh with a size of 0.02 tetrahedral is shown in Fig. 4.

Mechanical properties. Mechanical properties have been determined in previous studies [10], namely ASTM A536 Grade 80-55-06 material as shown in Table 1.

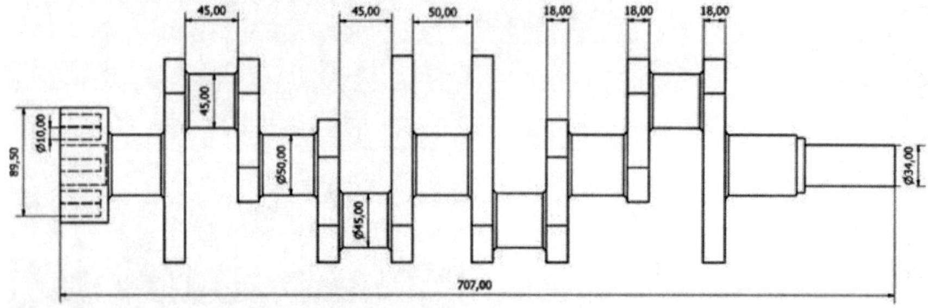

Fig. 3 Crankshaft dimension (mm)

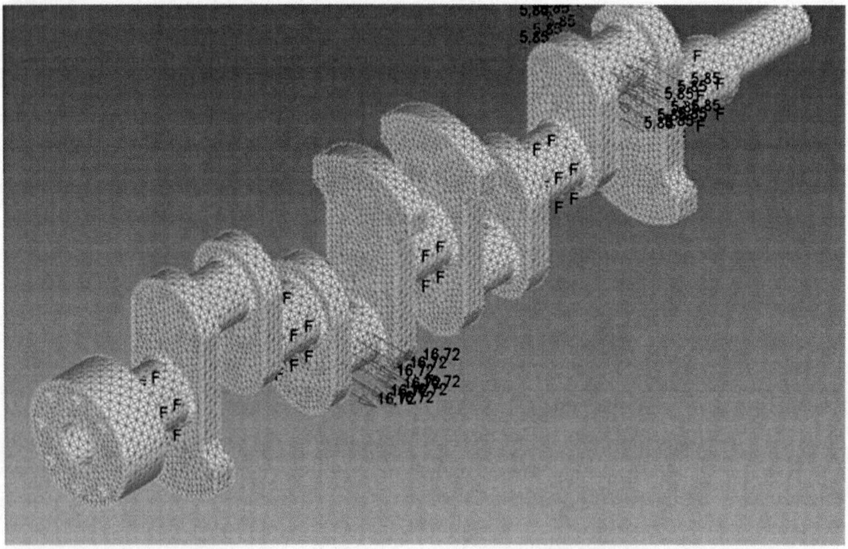

Fig. 4 Meshing of crankshaft

Table 1 Mechanical properties of ASTM A536 Grade 80-55-06

No.	Mechanical properties	Dimension
1	Tensile strength, (σ_B)	551 MPa
2	Poisson's ratio, (υ)	0.2925–0.2975
3	Yield strength, (σ_{ys})	470 MPa
4	Young's modulus, E	172,4 GPa
5	Hardness rockwell B	89,5 HRB
6	Fracture toughness, K_{IC}	32,8934 MPa $\sqrt{m}$ [11]

Load and Constraint. The load on the crankpin is carried out to find out how much load is given by the piston $P_z = 16.7242$ MPa and $P_c = 5.855$ MPa and can be seen in Fig. 5. Then, Fig. 6 shows the support given to each main journal on the crankshaft.

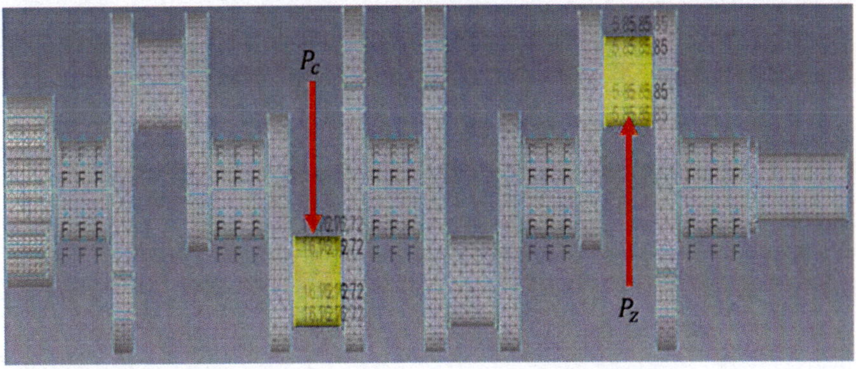

Fig. 5 Loading on the crankshaft

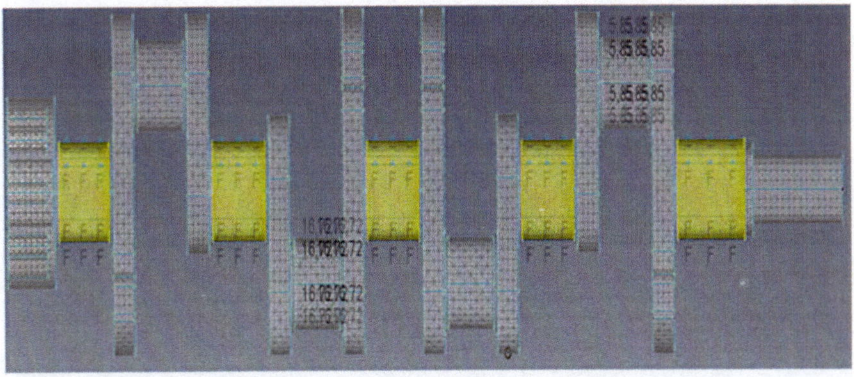

Fig. 6 Constraint on the crankshaft

Von Mises Theory Analysis. This process aims to analyze the von Mises stress, and the value obtained will be inputted into the stress intensity factor formula.

Stress Intensity Factor. Simulation using FEMAP software shows the value of the von Mises stress that occurs in the crankshaft so that the stress intensity factor value can be obtained through Eq. 1 [9]. The stress intensity factor is selected from the node with the highest von Mises value, namely 5 nodal points.

$$K_I = \frac{\sigma_{xx}\sqrt{2\pi r}}{\cos\left(\frac{\theta}{2}\right)\left(1 - \sin\left(\frac{\theta}{2}\right)\sin\left(\frac{3\theta}{2}\right)\right)} \quad (1)$$

4 Results and Discussion

Von Mises Theory Analysis. The results of the simulation using the FEMAP software obtained the highest von Mises value at crack initiation of 481.99 MPa as shown in Fig. 7. The results of the von Mises stress will be inputted as the basis for calculating the value stress intensity factor (K_I).

Stress intensity factor. The value of the stress intensity factor is taken from 5 nodal points as shown in Fig. 8.

The stress intensity factor value of the initial crack to the nodal has been calculated using Eq. (1). The results can be seen in Table 2.

The results of the stress intensity factor in Table 2 show that the average value of the stress intensity factor is 91.827 MPa $\sqrt{}$m. The stress intensity factor that occurs in the crankshaft exceeds the value of fracture toughness (K_{IC}) 32.8934 MPa $\sqrt{}$m [11].

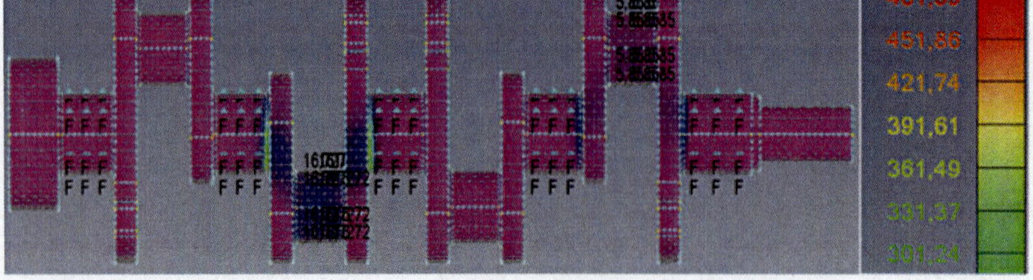

Fig. 7 Simulation results showing the value of the von Mises stress

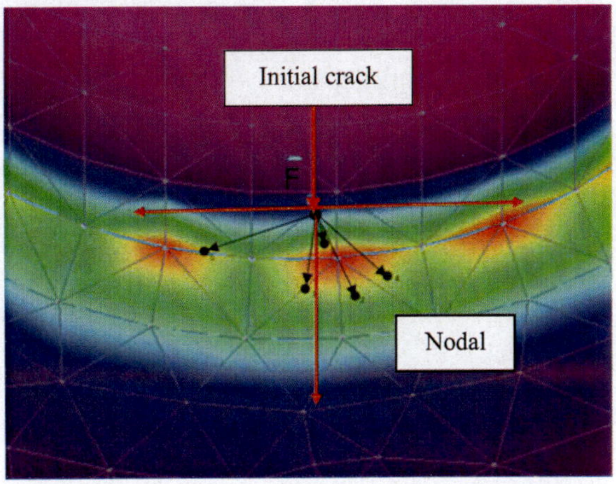

Fig. 8 Nodal point of initial crack

Table 2 Nodal selected around the crack tip along with the value of the stress intensity factor (K_I)

No.	Nodal ID	σ_{xx} (MPa m$^{1/2}$)	r (m)	θ (°)	K_I (MPa $\sqrt{m}$)
1	121,405	225,2713	0.007	80	129,7576
2	42,920	87,98,241	0.017	65	77,67
3	44,024	81,60,666	0.015	84	145,7576
4	126,079	82,96,323	0.02	45	49,52
5	120,775	118,705	0.022	28	56,43
Average					91,827

5 Conclusion

The results of the research from the case of crankshaft failure and the average stress intensity factor (K_I) was 91.827 MPa $\sqrt{}$m, and the fracture toughness (K_{IC}) of ASTM A536 grade 80-66-05 material only had a value of 32.8934 MPa $\sqrt{}$m. Therefore, the K_I value > K_{IC} value which causes crack propagation and then ends with failure of the crankshaft.

References

1. Prasetya T, Sarifuddin, dan Budi Joko Raharjo (2018) Crank pin journal crankshaft wear on diesel engine generators in. MV. Kartini Baruna, Jurnal Dinamika Bahari 9(1)
2. Shahane VC, Pawar RS (2018) Optimization of the crankshaft using finite element analysis approach. Automot Engine Technol
3. Huang Z-M, Sayed H (2019) Failure analysis, book of failure analysis. United Kingdom, London

4. Husaini, Putra TE, Zulfikar (2018) Analysis of cracks on a fractured surface of the vehicle crankshaft using the finite element method. Int J Eng Technol 7(4.36):1564–1568
5. Putra TE, Husaini, Ali N, Husin H, Zulfikar (2019) Failure analysis of the fracture surface of the crankshaft of a vehicle. IOP Conf Ser Mater Sci Eng 523(1):012067
6. Kakade, P., Pasarkar, M.D., Analyzing and identifying various approaches for crankshaft failures. State-Art Rev. J. Multidiscip. Eng. Sci. Technol. JMEST 2, 3159–0040. 2015.
7. Husaini, Ali N, Bakhtiar A (2016) Analysis on the surface of broken screw spring front car suspension. In: Proceedings of Seminar Nasional Teknologi Rekaysa (SNTR), vol 3
8. Myagkov LL, Mahkamov K, Chainov ND, Mahkamova I (2014) Advanced and conventional internal combustion engine materials, Chap 11. Elsevier
9. Husaini (2015) Fundamental of fracture mechanics (in Indonesia), Syiah Kuala University Press
10. Information on DuraBar, ASTM A536 grade 80-55-06. https://www.dura-bar.com/products/ductile-iron/80-55-06.cfm
11. Jen KP, Scardina JT, Smith DG (1985) Fracture behavior of as-cast pearlitic nodular iron. Elsevier

Study of Blades Number Influence on the Rotation of the Turbine Shaft on a Horizontal Axis Wind Turbine

Syuhada Ahmad, Sary Ratna, Afandi Dedi, and Fahriza Ikhsan

Abstract Wind energy is one of the energy sources that began to be developed to convert mechanical energy into an electrical need for the community. Indonesia is considered very worth developing wind turbines as power plants. The potential of wind energy in Indonesia reaches 9286 MW at an average wind speed of 5 m/s, based on data from the Indonesia Energy Outlook and Statistics 2004. In this study, what was discussed was related to the influence of the number of horizontal axis wind turbine blades on the rotation of the shaft. The model of the turbine used is a *propeller*-type wind turbine with variations in the number of blades 3, 4, 5, 6, and 7 made of meranti wood material (*Shorea multiflora*). Data collection on a wind turbine is carried out without using the load from the generator, then the data taken on measurements is only in the form of data on wind speed (m/s) and rotational speed (rpm). Data collection for 10 days, in the test of the wind turbine, was measured at an average wind speed of 1–6 m/s. The more blades on the wind turbine, the more rotation of the shaft will increase due to a large amount of wind that crosses the cross-sectional area of the blade. Horizontal axis wind turbine the number of 3–7 blades, getting the optimal number of blades, namely at the number of 5 blades produces a maximum shaft rotation of 1751 rpm when the wind speed is 6 m/s. In the wind turbine, the number of 5 blades obtained the optimal characteristic value of the wind turbine with a *tip speed ratio* of 6, and the power of the wind turbine (P λ *out*) is 166.78 watts.

Keywords Wind turbine · Wind speed · Rotation speed · Number of blades

S. Ahmad (✉) · S. Ratna · A. Dedi · F. Ikhsan
Mechanical and Industrial Engineering Department, Universitas Syiah Kuala, Jl.Tgk.Syech Abdurrauf No. 7 Darussalam, Banda Aceh 23111, Indonesia
e-mail: ahmadsyuhada@unsyiah.ac.id

S. Ratna
e-mail: ratnasary_ft@unsyiah.ac.id

A. Dedi
e-mail: dediafandi@unsyiah.ac.id

Irwansyah et al. (eds.), *Proceedings of the 4th International Conference on Experimental and Computational Mechanics in Engineering*, Lecture Notes in Mechanical Engineering, https://doi.org/10.1007/978-981-99-7495-5_45

1 Introduction

The wind is the air that moves due to the rotation of the earth and the difference in air pressure with the direction of wind flow from a place with a high level to a place with a low temperature or from an area that has a low temperature to a high-temperature area. The wind has a close relationship with sunlight because areas exposed to a lot of sunlight radiation will have higher temperatures as well as lower air pressure. This is proven when during the day, the wind will move faster than at night. Wind can also be caused by the movement of objects so that it can push the surrounding air to move to other places [1].

In Indonesia, wind speeds range from 2 m/s to 6 m/s at 6 h in 24 h/day. With such speed characteristics, Indonesia is considered very suitable for using small-scale (10kw) and medium-sized (10–100 kw) wind power plants, for energy use such as, lamps, pumps, electronic devices, and others. The center of wind power is still mostly in East Nusa Tenggara which has an average wind speed of up to more than 5 m/s. It is shown that the potential of wind energy in Indonesia reaches 9286 MW, where usage until 2004 was still less than 0.5 MW based on data from *Indonesia's Energy Outlook and Statistic 2004* [2].

Research on the potential of wind energy in Indonesia has been widely carried out, including the study of the effect of the height of the wind turbine installation site on the speed of wind energy, to optimize the energy produced by wind turbines [3]. Based on research conducted by Adhi Prasetya et al. [4], it shows that Indonesia has great potential in utilizing wind energy. Vertical axis wind turbines can be used in areas where the wind speed is less than 5 m/s so that the utilization of wind energy will be more evenly distributed. Research on energy potential in other regions of Indonesia has also been widely carried out, such as in Palu City [5], Gorontalo City [6], and Tasikmalaya Regency [7]. This also encourages researchers to conduct studies on the use of wind energy in the western region of Indonesia, namely Banda Aceh.

Research on the influence of other blades was conducted by Firman et al. with variations in the number of blades 3, 4, 5, and 6 [8]. Nobertus et al. conducted a study of the effect of the number of blades with variations of 2, 3, and 4 on vertical axis wind turbines [9]. The design characteristics of the blade also affect the energy output produced by the wind turbine [10, 11]. From the results of research conducted by Sayoga, et al. (2014). The effect of variations in the number of blades on performance in wind turbines. With a number of blades of 3, 4, and 5 at a wind speed of 3–5 m/s. Of the three variations, the number of blades 5 gives a higher average value of rpm compared to 3 and 4 blades. The highest rpm is obtained at a total of 5 blades, and a wind speed of 5 m/s which is an average of 576 rpm [12].

Based on the results of previous research by Rijal Mawi [12], comparison of the number of blades 3 and 4 made of wood material (meranti wood type) model *of propeller* wind turbine type *airfoil* Naca 4418 has shown a difference in rotational speed between wind turbines of 3 and 4 blades. Where in wind turbine of 3 blades, the rotation speed reaches 613.7 rpm, while in wind turbine of 4 blades, the speed of rotation only reaches 1143.8 rpm with an average wind speed of 5 m /s [13].

In this study, testing was carried out by adding the number of blades 5, 6, and 7 with a wind speed of 1–6 m/s in order to get the maximum power coefficient. The blade material remains the same as before, namely the type of meranti wood. The shape of the turbine blade is taken from the provisions of the Naca 4418-type airfoil on the horizontal shaft of the *propeller* type. So that it has a difference in rotational speed to the influence of many blades on the rotation (rpm) produced by the wind turbine.

The purpose of this study is to determine the most optimal number of blades to be used on *propeller* wind turbines in producing maximum wind turbine shaft rotation. This test only took data on wind speed and shaft rotation speed on a wind turbine. So this study limits the wind turbine to be tested for the shape of the blade of the turbine taken from the provisions of the Naca 4418-type airfoil, the blade material is made of wood material *(meranti wood type),* the test is carried out without using the generator load, and the test is carried out on a variation in the amount of 3–7 blades and measuring wind speed at 1–6 m / s.

2 Data Reduction

To calculate the value of the characteristic value on a wind turbine can use the equation below [14].

1. The energy contained in wind is kinetic energy, so it can be formulated as follows:

$$E_k = 0,5\ m\ v^2 \tag{1}$$

where

E_k kinetic energy (joule)
M mass (kg)
v wind speed (m/s)

2. Wind power (P_{in}) is the power generated by the wind at each blade area, which can be formulated.

$$P_{in} = 0,5\rho A v^3 \tag{2}$$

where

P_{in} power generated by the wind (Watt)
ρ density of air (1255 kg/m3).
A cross-sectional area of the blade (m^2)

3. The power generated by the turbine (P_{out}) is the power due to the wind crossing the blade of the turbine. The power of the turbine generated by the circular motion of the turbine can be formulated.

$$P_{out} = T\omega \tag{3}$$

where

P_{out} wind turbine power (watt)
T dynamic torque (Nm)
ω angular velocity (rad/s)

4. Torque (T) is a rotary moment acting on the shaft generated by the thrust force on the axis of the turbine, where this thrust force has a distance to the axis of the rotating shaft, so it can be formulated as follows:

$$Q = RFr \tag{4}$$

where

Q torque (Nm)
Fr resultant of *lift* and *drag force* (N)
R torque arm length (m)

5. The angular velocity of the turbine (ω) is the rotational speed of the turbine in units of rad/s. Angular velocity can be calculated using the formula.

$$\omega = \frac{2.\pi}{60}.n \tag{5}$$

6. *Tip speed ratio* (λ) is the ratio between the linear velocity of the blade end of the wind turbine to the wind speed. *The speed ratio tip* can be formulated as follows:

$$\lambda = \frac{2.\pi.r.n}{60.v} \tag{6}$$

where

r radius of the circle/cross-section of the blade of the turbine (m)
n rotation of the turbine (rpm)

3 Experimental Setup

The specifications of the turbines used in this study are shown in Table 1.

Table 1 Wind turbine specifications

Parameter	Description
Turbine type	Horizontal axis
Turbine diameter	200 cm
Number of blades	3, 4, 5, 6, and 7
Blade material	Meranti wood
Blade weight	1,2 kg
Blade length	100 cm
Blade width	20 cm
Blade thickness	3.5 cm
Airfoil type	Naca 4418
Pitch angle	10 °
Turbine height	300 cm
Hub diameter	27, 34.5, and 41 cm

Figure 1 shows the design of the wind turbine used in this study, where this study uses a horizontal axis-type turbine. The variations of turbines used in this study are turbines with a total of 3, 4, 5, 6, and 7 blades. The blade material is made using meranti wood with the type of airfoil NACA 4418. The length of the turbine blade is uniform, which is 1 m. In this study, it was carried out without using the generator load on the turbine.

Fig. 1 Horizontal wind turbine

4 Data Retrieval Procedure

This study measured the wind speed of 1–6 m/s and shaft rotation speed. The test was carried out for 10 days for testing wind turbines of 3, 4, 5, 6, and 7 blades. To measure wind speed (m/s), a digital anemometer is used, by directing the anemometer in the direction of coming wind with a height equivalent to the wind turbine shaft, which is 3 m from the ground level. To measure the rotating speed of the turbine shaft (rpm), a Tachometer (TASI-8740) is used. Directing the laser beam on the turbine shaft that has been affixed with sticker paper to be read by the sensor.

5 Result and Discussion

5.1 The Relationship of Wind Speed to the Rotation of the Turbine Shaft

Figure 2 indicates the speed of rotation of the shaft produced by the turbine by the number of blades 3, 4, 5, 6, and 7 at a wind speed of 1–6 m/s. From the chart, it can be seen that the more the wind speed of rotation of the shaft will also increase. However, in a 7-blade wind turbine when starting the initial rotation at a wind speed of 1 m/s, the turbine has not been able to rotate, the new turbine can rotate at a wind speed of 1.5 m/s, and at a wind speed of 1.5–4 m/s, the low rotation of the turbine shaft is produced compared to the number of blades 3, 4, 5, and 6. Whereas in turbines with 5 blades, the turbine can rotate at a wind speed of 1 m/s and experience a constant increase along with the increase in wind speed. This shows that turbines with 5 blades are more optimal because they are able to rotate at low wind speeds and experience a constant increase in speed to wind speeds.

5.2 The Relationship of Tip Speed Ratio *to Wind Speed*

Figure 3 shows the tip speed ratio produced by a wind turbine with a total of 3, 4, 5, 6, and 7 blades. The value of the tip speed ratio is influenced by the number of blades and the speed of the wind. At wind speeds of more than 5 m/s, turbines with 7 blades have a higher tip speed ratio value than turbines with blades of 3, 4, 5, and 6. However, at wind speeds below 5 m/s, the turbine with 7 blades has a low tip speed ratio value. Turbines with 5 blades have an almost constant tip speed ratio value at wind speeds of 1.5–6 m/s compared to turbines with a total of 3, 4, 6, and 7 blades. This shows that turbines with 5 blades are more optimally used at wind speeds of 1.5–6 m/s.

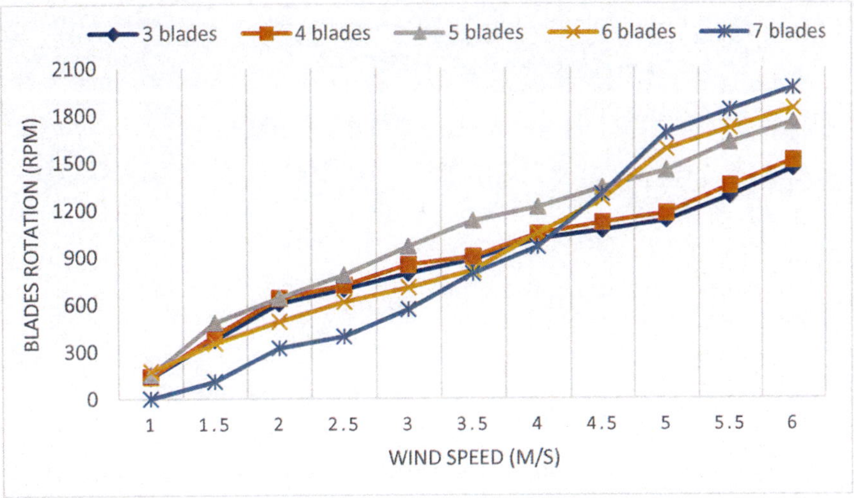

Fig. 2 Relationship of the rotational speed of the shaft to the wind speed

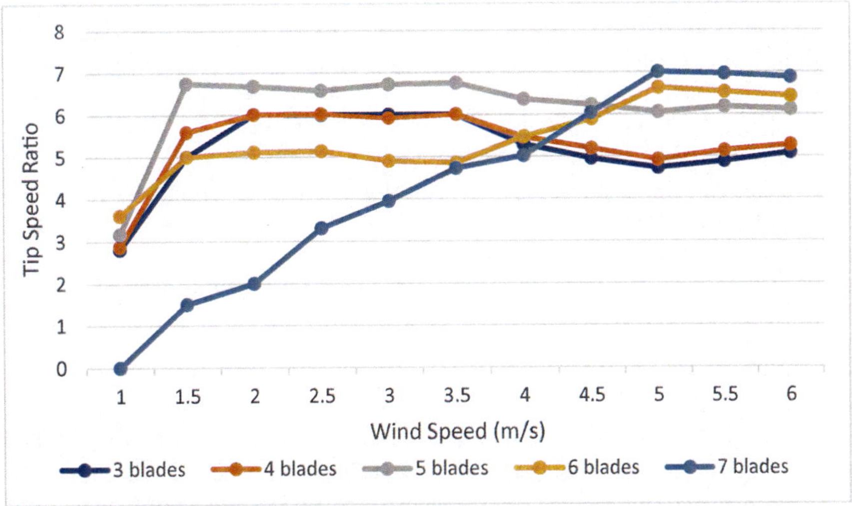

Fig. 3 Relationship of *tip speed ratio* to wind speed

5.3 The Power Relationship of Shaft to Wind Speed

Figure 4 shows the power relationship of the turbine shaft to the wind speed in the turbine by the number of blades 3, 4, 5, 6, and 7. On the graph, it can be seen that the higher the wind speed, the higher the shaft power generated. At wind speeds below 4 m/s, wind turbines with 7 blades produce the lowest shaft power compared

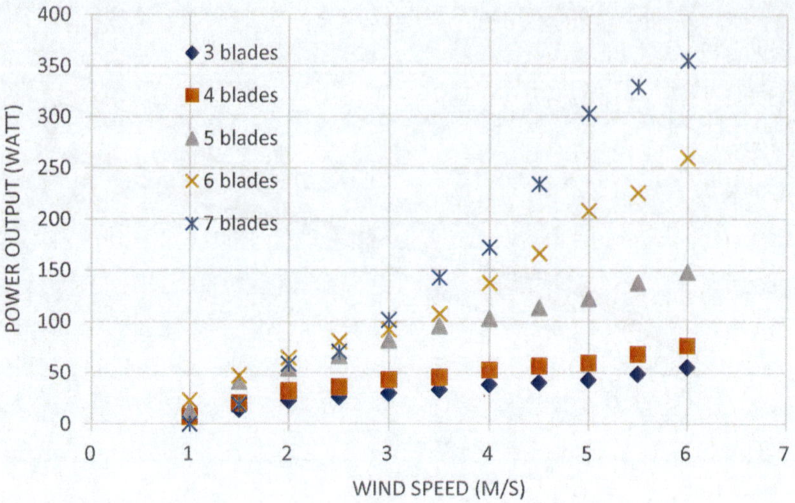

Fig. 4 Power relationship of the shaft to the wind speed

to turbines with a total of blades of 3, 4, 5, and 6. At wind speeds above 4.5 m/s, turbines with 7 blades produce the highest shaft power compared to turbines with blades of 3, 4, 5, and 6. Meanwhile, turbines with 5 blades produce the highest shaft power at wind speeds below 4.5 m/s and produce shaft power that is stable against increased wind speed. This shows a turbine with 5 blades is most optimal because it produces high power at wind speeds below 4.5 m/s and produces shaft power that is stable against increased wind speed.

6 Conclusion

From the analysis of research data, it can be concluded that a wind turbine with 5 blades produces the most optimal average shaft rotation, which is 1751 rpm. The most optimal average tip speed ratio is obtained in a wind turbine of 5 blades with a maximum torque of 7 λ. The most optimal average shaft power is generated in a wind turbine of 5 blades, which is 166.78 watts at a wind speed of 6 m/s. In this study, it showed that the most optimal wind turbine is a wind turbine with 5 blades.

References

1. Habibie MN, Sasmito A, Kurniawan R (2011) Kajian Potensi Energi Angin Di Wilayah Sulawesi Dan Maluku. J Meteorol dan Geofis 12(2):hal. 181–187. https://doi.org/10.31172/jmg.v12i2.99
2. Pusat Data dan informasi dan energy dan Sumber Daya Mineral KESDM, "Indonesia Energy Outlook 2010," hal. 1–198, 2010
3. Hardianto T, Supeno B, Saleh A, Setiawan DK, Gunawan, Indra S (2017) Potential of wind energy and design configuration of wind farm on Puger Beach at Jember Indonesia. Energy Procedia 143:579–584. https://doi.org/10.1016/j.egypro.2017.12.730
4. Prasetyo A, Notosudjono D, Soebagja H (2018) PEMBANGKIT LISTRIK TENAGA ANGIN Dalam penyusunan Tugas Akhir ini maksud dan tujuannya adalah : a. Mengkaji sistem penerapan dan pengembangan PLT Angin di Indonesia sebagai negara berkembang. b. Mengkaji potensi angin dan teknologi pada PLT Angin Sistem, hal. 1–12
5. Sam A, Daud P, Sam A, Patabang D (2005) Studi Potensi Energi Angin Di Kota Palu Untuk Membangkitkan Energi Listrik. J. SMARTek 3(1):21–26
6. Yunginger R, Nawir NS (2015) Analisis Energi Angin Sebagai Energi Alternatif Pembangkit Listrik Di Kota Di Gorontalo. Univ. Negeri Gorontalo 15:1–15
7. Simamora RP, Handarto H, Saukat M (2020) Analisis Potensi Energi Angin Dan Analisis Teknik Pembangkit Listrik Tenaga Bayu Untuk Membangkitkan Energi Listrik (Studi kasus di Gunung Kincir, Desa Ciheras Kecamatan Cipatujah Kabupaten Tasikmalaya). Pros Semin Nas Tek Elektro UIN Sunan Gunung Djati Bandung 0(0 SE):91–100
8. Aryanto F, Mara M, Nuarsa M (2013) Pengaruh Kecepatan Angin Dan Variasi Jumlah Sudu Terhadap Unjuk Kerja Turbin Angin Poros Horizontal. Din Tek Mesin 3(1):50–59. https://doi.org/10.29303/d.v3i1.88
9. Fachrudin AR (2018) Pengaruh Jumlah Sudu Terhadap Kinerja Turbin Angin Sumbu Vertikal Tipe Darrieus-H Naca 3412 Dengan Sudut Pitch 00. Info-Teknik 19(2):195. https://doi.org/10.20527/jit.v19i2.153
10. Kamal AM, Nawar MAA, Attai YA, Mohamed MH (2022) Blade design effect on Archimedes Spiral Wind Turbine performance: experimental and numerical evaluations. Energy 250:123892. https://doi.org/10.1016/j.energy.2022.123892
11. Pourrajabian A (2022) Effect of blade profile on the external/internal geometry of a small horizontal axis wind turbine solid/hollow blade. Sustain Energy Technol Assessments 51(December 2021):101918, 2022. https://doi.org/10.1016/j.seta.2021.101918
12. Rijal M (2017) Uji performa turbin angin propeller dengan bahan sudu dari kayu. Fakultas Teknik, Universitas Syiah Kuala, Jurusan Teknik Mesin dan Industri
13. Thoriq M (2016) Srudi eksperimen pengaruh jumlah sudu terhadap kerja turbin angin horisontal brbasis NACA 4415. Fakultas Teknik, Universitas Muhammadiyah Surakarta, Jurusan Mesin
14. Franciscus (2016) Unjuk Kerja Turbin Angin Propeler Tiga Sudu datar dari bahan Triplek dengan sudut Patahan 10° Lebar 10,5 Cm Dengan Empat Variasi Permukaan Sudu. Jurusan Mesin, Universitas Sanata Dharma

Analysis of the Stress Concentration on the Sport Utility Vehicle Camshaft at Different Conditions by Using the Finite Element Method

Azmi Ulul, Husaini, Edisah Putra Teuku, and Munandar Deski

Abstract This study was conducted to determine the stress in sports car camshaft using the finite element method. The simulation was based on three conditions which include flawless, wear defects, and wear with initial defects such as early cracks. The results showed that the maximum normal stress in the model without defects was 7.8 MPa at the fillet area, while the stress in the worn model was 22 MPa. These values were below the material permit limit of 29.8 MPa. Meanwhile, the worn model with initial defects such as cracks had 252 MPa, and this exceeds the limit. It was also observed that the stress intensity factor at the crack tip was 17.9 MPa $m^{1/2}$, and this indicates a propagation crack. Therefore, it was concluded that camshafts without defects or with wear and tear are safe to use, while those worn with defects such as cracks are not.

Keywords Camshaft · Finite element method · Stress intensity factor · Stress distribution

1 Introduction

A camshaft is an engine component that regulates the time interval for the opening and closing of the intake and exhaust valves on the internal combustion engine [1–5]. It is generally produced through casting or forging [6] and requires unavoidable friction between the cam and the follower component to perform its functions,

A. Ulul · Husaini (✉) · E. P. Teuku · M. Deski
Computational Mechanics Laboratory, Department of Mechanical Engineering, Universitas Syiah Kuala, Darussalam, Banda Aceh 23111, Indonesia
e-mail: husainiftm@usk.ac.id

E. P. Teuku
e-mail: edi@usk.ac.id

M. Deski
e-mail: deski.m@mhs.usk.ac.id

Irwansyah et al. (eds.), *Proceedings of the 4th International Conference on Experimental and Computational Mechanics in Engineering*, Lecture Notes in Mechanical Engineering, https://doi.org/10.1007/978-981-99-7495-5_46

thereby causing abnormal wear on the cam profile. Meanwhile, the bending cycle and torsional loads on the camshaft allow failure, starting from the initial defect [7].

The dimensions of camshafts often change during operation, and this can cause mechanical failure when the change in size and structure exceeds the permissible limits, thereby leading to ineffective functioning of the component [8]. The load or force exerted by the camshaft is known as working stress. Moreover, the maximum stress caused by the maximum load on the structure usually determines the ability of the shaft to withstand damage.

The failure of components usually starts with the stress concentrations originating from the geometry of the object such as holes, corners, and joints when subjected to repeated loading [9, 10]. This is the reason the estimation of the stress in geometries with certain defects is useful in designing a component or structure. Therefore, this study aims to analyze the stresses on these components with different defects using the finite element method.

2 Methodology

Material. The material used is the camshaft of a sport utility vehicle (SUV) which failed in the sliding bearing area as shown in Fig. 1. Previous studies showed that the camshaft is made of cast iron. Previous research stated that the crack toughness value for cast iron materials ranged from14 MPa $\sqrt{m}$ [11].

The 3D camshaft was modeled in this study using Autodesk Inventor as shown in Fig. 2 and simulated using FEMAP software. Previous studies showed that the type of material used is ASTM A48 standard gray cast iron [14], and the mechanical properties are presented in Table 1.

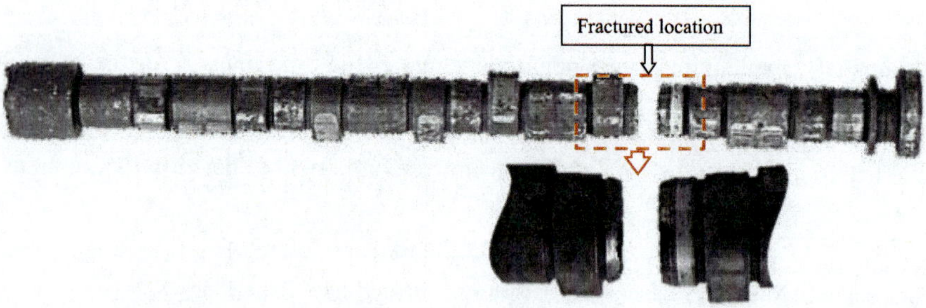

Fig. 1 Camshaft used as a case study

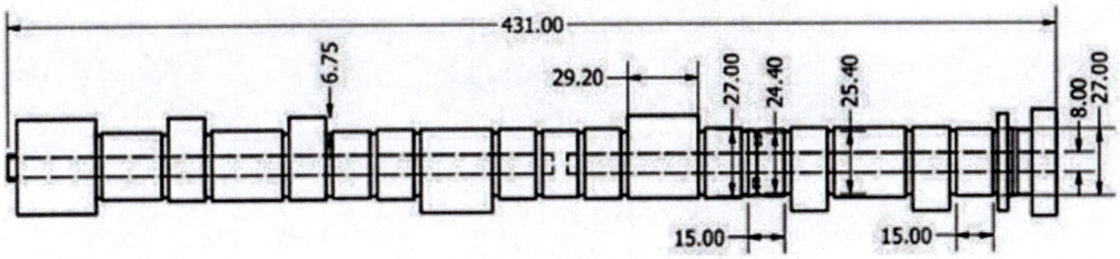

Fig. 2 Camshaft design

Table 1 Mechanical properties of gray cast iron [12, 13]

Mechanical properties	Value
Tensile strength, ultimate	179 MPa
Compression stress	669 MPa
Modulus of elasticity	114 GPa
Poisson ratio	0.29

3 Camshaft Design with Autodesk Inventor Software

Geometry modeling was conducted using Autodesk Inventor based on three conditions applied as the independent variables which include flawless, wear defects, and wear with initial defects as shown in the following Fig. 3.

Preprocessing consists of two stages with the first focused on ensuring the input material properties are in line with the ASTM A48 and the second was used to determine the boundary conditions such as the application of a normal force of 1085 N accompanied by a torque of 11.1 Nm as shown in Fig. 4.

The constraint to the area in direct contact with the sliding bearing was also provided as shown in Fig. 5.

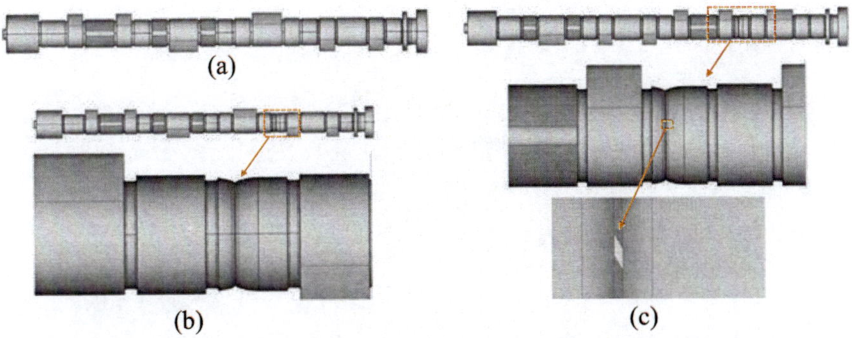

Fig. 3 **a** Flawless condition, **b** wear condition, and **c** wear condition and initial defects

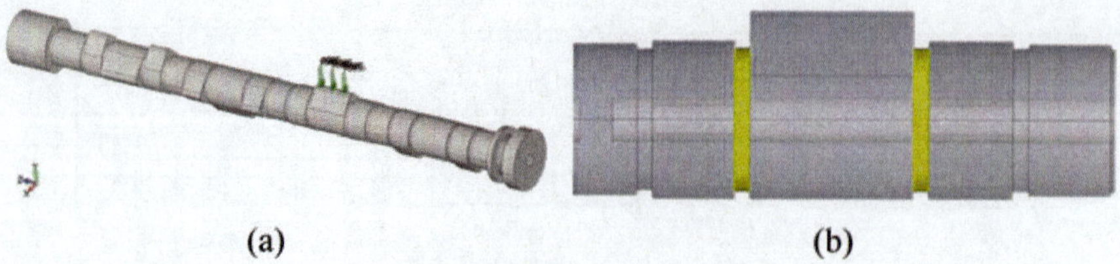

Fig. 4 **a** Normal force application; **b** torsion

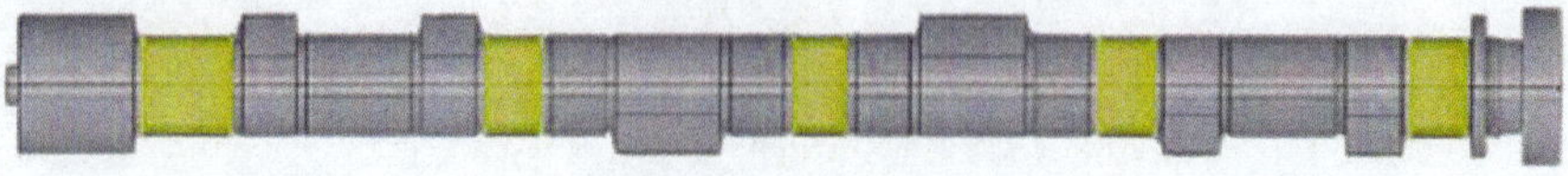

Fig. 5 Constraint area

Meshing and Analysis setup. Meshing was used to divide the camshaft model into smaller elements. This process involves subdividing a 3D model into discrete domains, and this is necessary due to the adoption of the finite element method (FEM) which requires solving the equation in each discrete domain. The meshing greatly affects the accuracy of the simulation results due to the fact that a smaller meshing setting in the 3D model usually ensures the final simulation result is closer to the real value.

The elements applied in the simulation are in the form of a triangle (Tet mesh) with a size of 2 mm in all parts of the model, and the results produced are presented in Fig. 6.

The setup analysis process was the final process in using the FEMAP application to determine the numerical solution of the model (camshaft) previously set [15, 16]. The running process takes a long time depending on the mesh size used after the results were displayed in the form of color contours according to the stress value recorded in the camshaft.

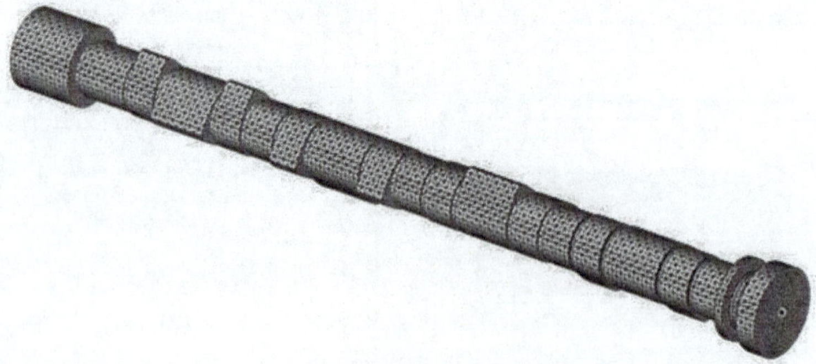

Fig. 6 Model in meshing

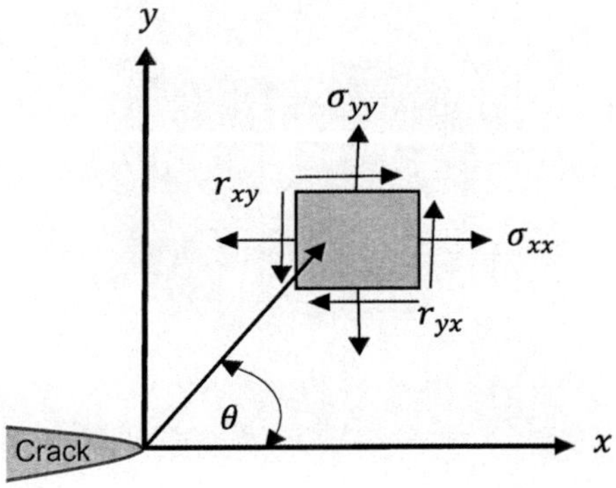

Fig. 7 Stress intensity factor analysis model [17]

Stress Intensity Factor Analysis (K_I). A material can be damaged due to different activities such as the manufacturing process, defective structural design, or during operation. This can further lead to cracks which subsequently produce high stress concentrations at the crack tip. The factor influencing the cracking phenomenon is the stress intensity factor which is symbolized by (K_I) and required to be analyzed only in the area around the crack tip because its value determines the rate of crack propagation and the length of time the material or component fails such that when the value of $K_I \geq K_{IC}$, crack propagation, and failure are expected to occur but when $K_I \leq K_{IC}$, there is no crack propagation [17].

The stress intensity factor in the failed camshaft was, therefore, calculated in this study using the following Eq. 1 (Fig. 7).

$$K_I = \frac{\sigma_{xx} \cdot \sqrt{2\pi r}}{\left(\cos\frac{\theta}{2}\right)\left[1 - \sin\left(\frac{\theta}{2}\right)\sin\left(\frac{3\theta}{2}\right)\right]} \tag{1}$$

where

σ_{xx} Stress
r Distance from the tip of the crack to the node
θ The angle formed from the tip of the crack to the node
K_I Stress intensity factor

4 Results and Discussion

Stress Analysis on Flawless Camshaft. The simulation of the camshaft without defects produced a maximum normal stress value of 11 MPa at the fillet area as shown in Fig. 8. This value is within the material permit limit of 29.8 MPa [13], and this means the initial design of the component is adequate.

Flawless Stress Intensity Factor. The simulation conducted without defects in the model was used to calculate the stress intensity factor (K_I) on the camshaft in order

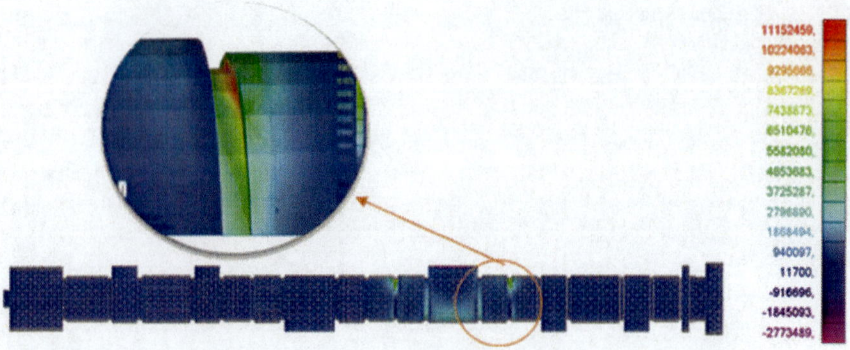

Fig. 8 Areas of stress concentration without defects

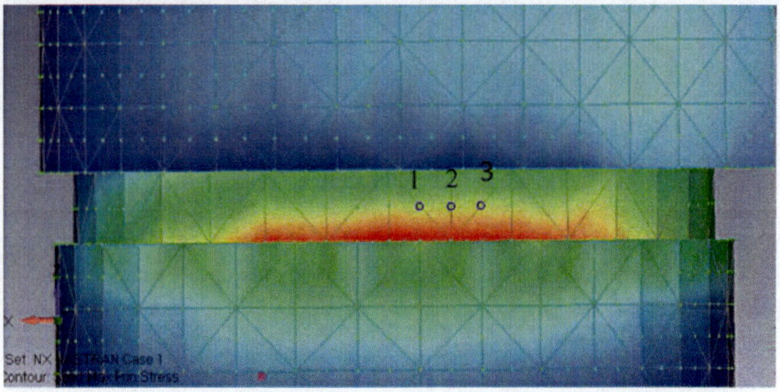

Fig. 9 Points reviewed on a flawless model

to determine the possible propagation of cracks [14] at 3 node points as shown in Fig. 9.

The values of the camshaft stress intensity without defects calculated using Eq. 1 are listed in Table 2, and it was discovered that the stress intensity value was smaller than the fracture toughness value at the load applied, and this means there was no crack propagation. The findings showed that the stress intensity factor around the fillet area ranged from 1.3 to 1.5 MPa $m^{1/2}$. This means the stress concentration does not exceed the fracture toughness for cast iron material which ranges from 14 MPa $\sqrt{m}$ [11], and this indicates there is no crack potential.

Stress Analysis on Worn Camshaft. The model with wear defects was observed to have a maximum normal stress of 22 MPa concentrated in the worn area as indicated in Fig. 10. This value is within the safe limits of the material [13] but higher than the 15.6 MPa recorded in previous studies. This is associated with the stress concentration in a certain part which makes the local stress to be greater than the applied stress [15], while the analytical calculations ignore the effect of geometry on stress.

Table 2 Stress intensity factor around the fillet area

Point	σ_{xx} (MPa)	r (m)	θ (°)	K_I (MPa m$^{1/2}$
1	7.6	0.0018	50	1.5
2	7.8	0.0028	29	1.3
3	7.8	0.0039	21	1.4
Average				1.4

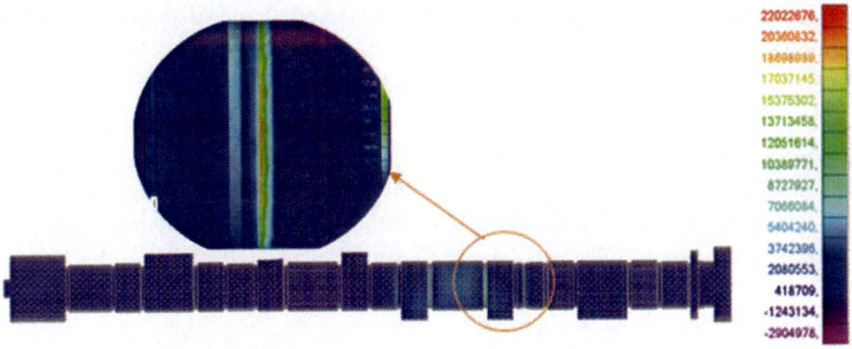

Fig. 10 Areas of stress concentration in wear conditions

Stress Intensity Factor with Wear Defect. The simulation of the model with defects was used to calculate the stress intensity factor *(K_I)* on the camshaft in order to determine the possible propagation of cracks [14] at 3 node points as shown in Fig. 11.

The stress intensity factor of the camshaft with wear defects was calculated using Eq. 1, and the value is listed in Table 3. It was discovered that the value does not

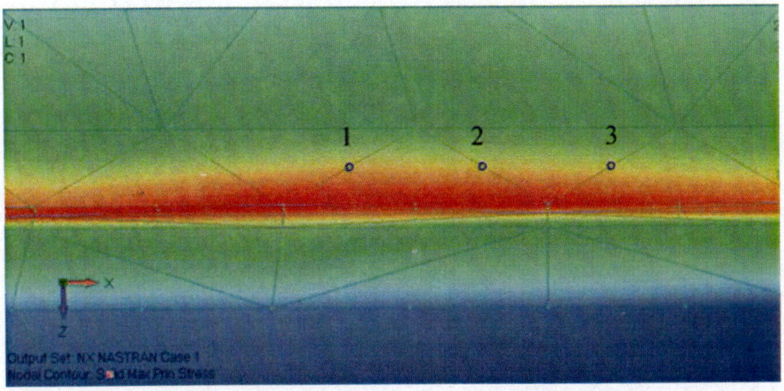

Fig. 11 Points examined in the wear model

exceed the fracture toughness value under the same load, and this means the shaft does not fail but can no longer be used because there has been a reduction in the camshaft diameter.

The analysis of the stress intensity factor around the wear area showed that the value ranges from 1.64 to 2.31 $\text{MPa}\,\text{m}^{1/2}$. This means the stress concentration in the fillet area does not exceed the fracture toughness value of cast iron material which ranges from 14 MPa $\sqrt{\text{m}}$ [11], and this indicates there is no crack potential.

Stress Analysis on Worn Camshafts with Initial Defects. The provision of a parallelogram-shaped defect in the wear area has a very large impact on the stresses concentration which was observed to have led to greater local stresses compared to the applied stresses [15]. The simulation results showed that the maximum normal stress was 252 MPa as shown in Fig. 12, and this exceeds the allowable and tensile limit for the material. This certainly means the components will fail during operation.

Stress Intensity Factor of the Model with Initial Defect. The stress intensity factor *(K_I)* on the worn camshaft with initial defects was calculated to determine the possibility of the crack propagating [14] at three node points as shown in Fig. 13.

The stress intensity value was calculated using Eq. 1, and the results presented in Table 4 showed that the value is greater than the fracture toughness value under the same load, and this means the cracks are propagated and continued until the camshaft failed.

Table 3 Stress intensity factor on wear defects

Point	σ_{xx} (MPa)	r (m)	θ (°)	K_I ($\text{MPa}\,\text{m}^{1/2}$)
1	17.4	0.00075	35	1.64
2	17.2	0.00187	13.2	1.95
3	16.6	0.003	8	2.31
Average				1.97

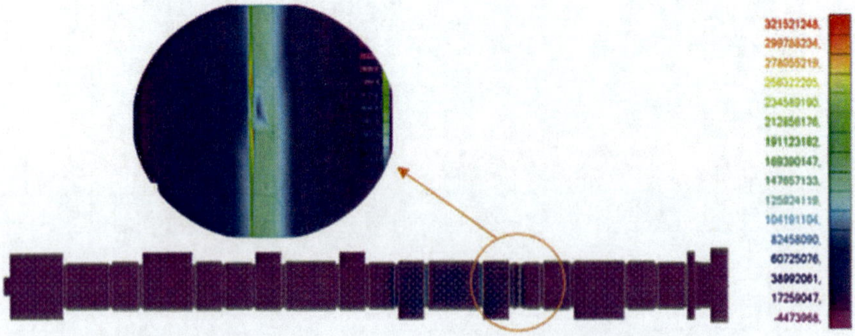

Fig. 12 Area of stress concentration during wear condition with initial defect

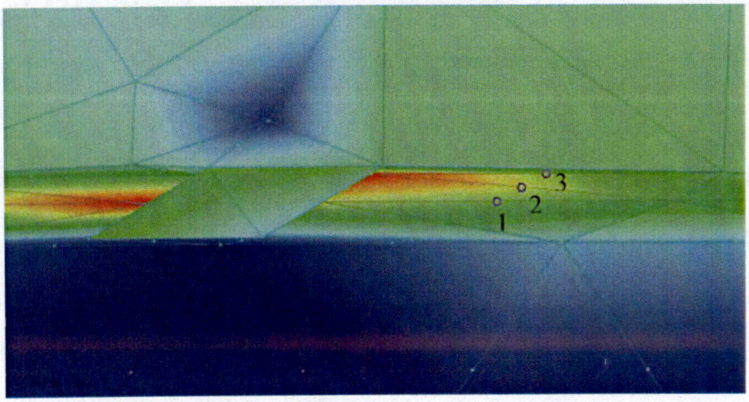

Fig. 13 Points under consideration

Table 4 Calculation of the value of the stress intensity factor (K_I)

Point	σ_{xx} (MPa)	r (m)	θ (°)	K_I ($\text{MPa m}^{1/2}$)
1	220	0.0007	8	14.8
2	252	0.00075	11	17.9
3	230	0.0008	13	17.1
Average				16.6

The table shows that the stress intensity factor value of 17.9 MPa $m^{1/2}$ at the crack tip is greater than the fracture toughness value which was recorded in previous studies to be in the range of 14 MPa $\sqrt{m}$ for cast iron material [11]. This means the crack propagates from the initial defect, and this can lead to the failure of the camshaft.

5 Conclusions

The maximum stress in the fillet area for the model without defects was found to be 11 MPa, and this is within the allowable limits of the material. The stress intensity factor also ranges from 1.3 to 1.5 MPa $m^{1/2}$ which does not exceed the fracture toughness, and this means there is no crack propagation because the stress concentration is within safe limits.

The maximum stress when the camshaft experienced a reduction in diameter due to wear was recorded to be 22 MPa, and the stress intensity ranges from 1.64 to 2.31 MPa $m^{1/2}$ which does not exceed the fracture toughness value. This means it does not have the potential to cause cracks, but the material cannot be used anymore due to the reduction in the camshaft diameter.

The addition of defects to the wear area has a very large impact on the stress as indicated by the maximum stress of 321 MPa recorded which exceeds the allowable limit of the material. Moreover, the stress intensity factor at the crack tip was 17.9 MPa $m^{1/2}$ which exceeds the fracture toughness value which is in the range of 14 MPa $m^{1/2}$ $(K_I \geq K_{IC})$ for cast iron material, thereby causing cracks to propagate and ensuring the camshaft is no longer safe to use.

Acknowledgements The authors are grateful to the Universitas Syiah Kuala for their financial support through Professor Research Grant No. 001/UN11.2.1/PT.01.03/ PNBP/2022. The authors are also grateful to the Department of Mechanical Engineering, Universitas Syiah Kuala, Banda Aceh, for making the facilities accessible.

References

1. Xu XL, Yu ZW, Mao SM (2020) Effect of extra-large compound calcium-aluminosilicate inclusions on cracking of camshafts. Eng Fail Anal 110:104408
2. Bhosale VN, Gaikwad SP (2013) Failure analysis of grey cast iron camshaft by mode of fracture. Int J Eng Res Develop 9:36–42
3. Pulkrabek WW (2004) Engineering fundamentals of the internal combustion engine
4. Bonatesta F, Altamore G, Kalsi J et al (2016) Fuel economy analysis of part-load variable camshaft timing strategies in two modern small-capacity spark ignition engines. Appl Energy 164:475–491
5. Dhavale AS, Muttagi VR (2012) Study of modeling and fracture analysis of camshaft a review. Int J Eng Res Appl (IJERA) 2:835–842
6. Bayrakceken H, Ucun I, Tasgetiren S (2006) Fracture analysis of a camshaft made from nodular cast iron. Eng Fail Anal 13:1240–1245
7. Li P, Li F, Cai A et al (2009) Fracture analysis of chilled cast iron camshaft. China Foundry 6:104–108
8. Husaini, Putra TE, Novriandika S (2020) Study of failure analysis of a fracture crankshaft pulley used on a truck engine. IOP Conf Ser Mater Sci Eng 739(1)
9. Wang G, Taylor D, Bouquin B et al (2000) Prediction of fatigue failure in a camshaft using the crack modelling method. Eng Fail Anal 7:189–197
10. Husaini, Iskandar ST (2001) Analysis of crack toughness on polymethyl methachrilite (PMMA) and nylon. Mech Eg J 4(1)
11. Lee S-C, Chang Y-B (1991) Fracture toughness and crack growth rate of ferritic and pearlitic compacted graphite cast irons at 25 °C and 150 °C
12. ASTM (2000) Standard specification for automotive gray iron castings 03:1–5
13. Rundman KB (2001) Cast irons. Encyclop Mater Sci Technol 1003–1010
14. Munandar D (2021) Study the causes of camshaft failure on car type SUV (Sport Utility Vehicle). Universitas Syiah Kuala
15. Jasdanwalla PH, Mulla S, Salman AM et al (2020) Design and analysis of camshaft using FEA. Int Res J Eng Technol 7:4412–4416
16. Husaini, Kishimoto K (2001) Mixed-mode fracture behavior of PC/ABS blends. In: Proceeding of SPIE—the international society for optical engineering 4317:111–116
17. Husaini (2015) Fundamentals of cracking mechanics. Syiah Kuala University Press, Darusalam, Banda Aceh

Failure Analysis on Leaf Spring of Dump Truck Using Experimental and Finite Element Methods

Husaini, Noor Zaki Alif, Ali Nurdin, and Murtadhi Faris

Abstract A failure in leaf spring usually causes a critical accident because of its importance in transport vehicles such as Dump Trucks. Therefore, this research aims to determine the cause of failure in the leaf spring of the Dump Truck suspension system using experimental and finite element methods. The experimental tests included visual observation of the fracture surface to determine the possibility of fracture or material failure based on the condition of the spring and an SEM test to observe the fracture surface condition of fractography. A hardness test was also carried out to adjust the properties of the spring material according to its standards, chemical composition test to determine the properties of steel, as well as microstructure test, which shows the structure contained in leaf springs. The stress and strain analyses were carried out by modeling components using Autodesk Inventor software. The finite element method analysis used FEMAP 2021.2 software on the maximum stress and strain received by the leaf spring. This was followed by calculating the stress intensity factor (K_I) at several points at each end of the fracture. The average K_I value obtained from the finite element method analysis was 101.3220059 MPa.m$^{1/2}$, while the K_{IC} value was 22 MPam$^{1/2}$. This value met the requirements for fracture mechanics, namely $K_I \geq K_{IC}$.

Keywords Finite element method · Fatigue · Failure analysis · Leaf spring · Stress intensity factor

Husaini (✉) · N. Z. Alif · A. Nurdin · M. Faris
Laboratory of Computational Mechanics, Department of Mechanical Engineering, Universitas Syiah Kuala, Darussalam-Banda, Aceh 23111, Indonesia
e-mail: husainiftm@usk.ac.id

A. Nurdin
e-mail: nurdin.ali@usk.ac.id

M. Faris
e-mail: faris_m@mhs.usk.ac.id

Irwansyah et al. (eds.), *Proceedings of the 4th International Conference on Experimental and Computational Mechanics in Engineering*, Lecture Notes in Mechanical Engineering, https://doi.org/10.1007/978-981-99-7495-5_47

1 Introduction

The automotive industry consists of many parts, each of which has several important characteristics such as leaf springs in suspension systems. This component is mainly used for the introduction of a suspension system to prevent the transmission of vibrations and support the vehicle frame [1, 2]. Goods transport vehicles, namely trucks, have a point of load that tends from the center to the rear of the body, which makes the rear suspension experience a greater deflection. This occurs under repeated compressive loads, causing failure because it has exceeded the fatigue limit of the material [3].

The failure in leaf spring material occurs as a result of fatigue failure due to the weakening of the material caused by the load being applied repeatedly [4]. Material failure is a brittle fracture mechanism, consisting of three stages, namely the occurrence of cracks, propagation, and fracture [5]. Fatigue failure is the most common type in automotive components such as crankshafts and truck springs. Furthermore, leaf springs often experience excessive loads and vibrations due to uneven road surfaces. Since the load occurs suddenly, the wheel rises and falls with irregular frequency. Several reports have shown that the failure of the leaf spring is affected by the loading that varies with time (dynamic load) [6].

Previous research conducted by Clarke and Borowski (2005) indicated that the fracture occurs in the spring eye area. It was carried out using the micrographic method with SEM and stress analysis by the finite element method through ALGOR software. The results showed that leaf springs break due to various factors such as sulfur separation in the midplane, which can weaken the springs, accidents, damaged roads, and the aging process [7].

According to I.B. Eryurek M. Ereke and A. Goksenli (2007), the failure is affected by the lack of safety factors of leaf spring fatigue. Therefore, it causes a sudden brittle fracture when passing through a potholed road surface [8]. Nataraj and Thillikkani also discovered fractographic images of the fracture surface showing a fatigue pattern. This indicated that repeated loads were the cause of the failure of the leaf spring [9].

The survey from the workshop where the specimen was obtained showed that the leaf spring broke around the end of May 2019. Although this component was not built-in, it had been replaced several times and the reason was not known. The incident was discovered by the owner when driving and the truck was immediately taken to the workshop for replacement. The components of the broken leaf spring are presented in Fig. 1.

This research examined the cause of the rear leaf spring fracture number 3 of the Dump Truck vehicle to predict and reduce the occurrence of failures in similar cases.

Fig. 1 Leaf springs on trucks

2 Research Procedure

This research used an experimental test and finite element method. It started with preparing specimens at the fracture section to observe the topography macroscopically and determine the beginning of the crack, followed by the testing of hardness, chemical composition, and microstructure. After experimental testing, leaf spring modeling was carried out with Autodesk Inventor and continued simulation by using FEMAP software [10] to determine the cause of leaf spring failure.

Research Material. The research material was a failed leaf spring of the colt diesel 125 PS number 3 truck on the back. Steel with a carbon content of 0.3–0.55% is known as medium carbon steel, which was included as a material for making shafts, axles, gears, springs, etc. To ensure the type of steel used in the leaf spring specimens, chemical composition tests were carried out. The shape of the leaf spring is shown in Fig. 2.

The fracture shape of leaf spring number 3 is presented in Fig. 3.

Visual Observation. This was conducted on the surface of the damaged material to observe the possibility of failure fracture or the material based on the condition of the spring. It shows the pattern and special signs such as the mechanism due to

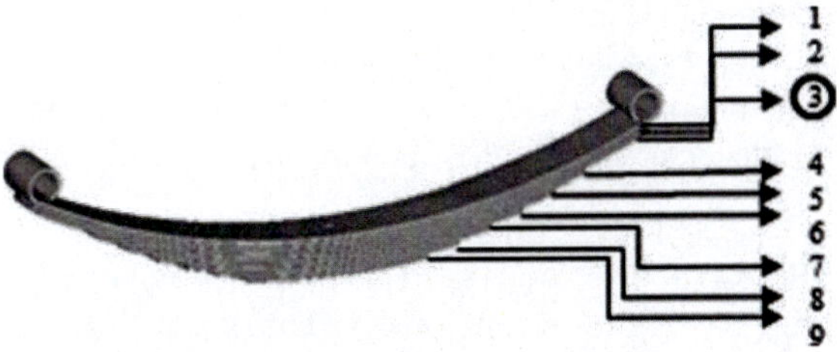

Fig. 2 Shape of the leaf spring based on the serial number

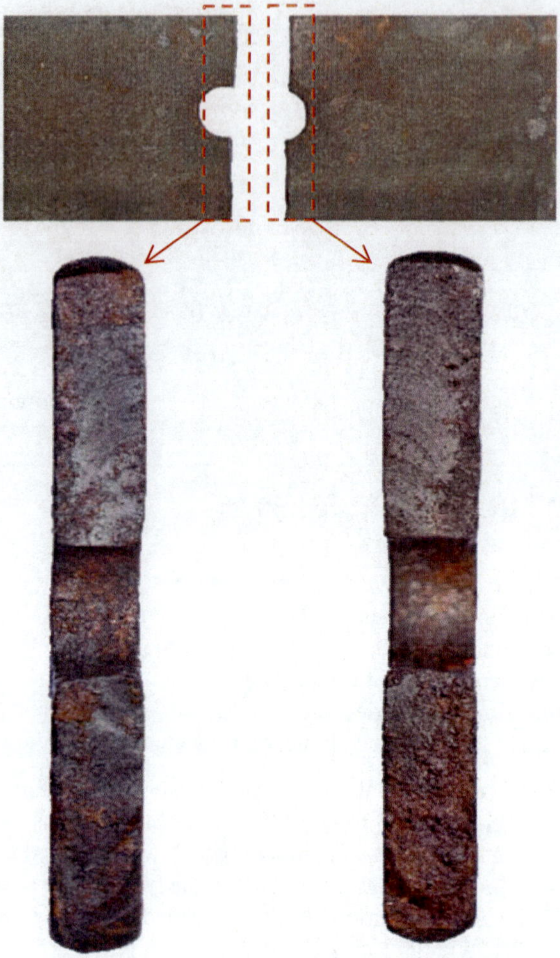

Fig. 3 Surface of the fractured spring

fatigue fracture, which consists of three stages, namely the crack initiation occurrence, propagation, and fracture [11], caused by a certain load received by the leaf spring which leads to a failure. From this observation, the type of failure experienced by the leaf spring specimen to be tested can be estimated, which leads to its initial prediction [12].

This observation was carried out by macroscopic and fractographic methods. The macroscopic approach is used to observe the topography of the surface to identify the characteristics of the fracture surface, while in fractography, observations were made with SEM [13].

Microstructure Analysis. This process was carried out by etching the ground material with HCL + alcohol and visualizing with an optical microscope to determine

Fig. 4 SEM observation on a sample section

the material microstructure. This test observed the characteristics of the grains and phases formed from the leaf spring.

Hardness Test. This test was carried out using the Rockwell Type Zwick/Roell ZHR method, with a Rockwell hardness testing machine. It was conducted by pressing the penetrator to the surface of the specimen with a certain load and depth and giving the value of the difference in depth hardness obtained from major and minor loads [14]. The configuration of direction used was vertical and horizontal in the leaf spring section.

Chemical Composition Test. This test was carried out to determine the chemical composition and mechanical properties of the leaf spring material. The specimen was cut into small parts and tested using a spectrometer.

***Scanning Electron Microscope* (SEM)**. SEM test determined the topography of the specimen surface, including the morphology, shape, and size of the particles making up the object, and the chemical composition. The sample was cut along 10 mm as shown in Fig. 4.

Load Distribution Analysis. Mechanical analysis is a method to determine the cause of damage to leaf springs due to excess load, both static and dynamic. To determine the maximum and minimum loads received by the springs on each wheel, a calculation was performed using Eq. (1) as stated below [16].

$$F = \frac{m \times g}{4}, \tag{1}$$

where F is the force on the spring, m is the mass (Kg) of the vehicle, g is gravity 9.8 m/s^2), and is divided by the four springs in the vehicle.

Maximum allowable stress. The maximum force on the spring needs to be below the structural or yield strength to avoid structural failure due to maximum stress. Considering the safety factor, namely 1.65, the maximum allowable force can be obtained using Eq. (2) as expressed below [17].

$$\sigma_{\max} = \frac{S_y}{S_f}, \tag{2}$$

where S_y is the structural strength and S_f is the safety factor.

Modeling and Analysis Using the Finite Element Method. The finite element method (FEM) is a numerical approach that is suitable to solve engineering problems. Stress analysis is based on the mechanical properties of a material and its ability to withstand applied loads.

In this research, CAD modeling was performed using Autodesk Inventor 2017 software, while Siemens™ FEMAP 2021.2 was used to analyze stress and strain. Before the finite element method analysis, the leaf spring modeling was carried out using the Autodesk Inventor, which was designed according to its original size. Subsequently, it was simulated with FEMAP which describes the behavior of the operating environment with a given load and boundary conditions [10, 16]. Figure 6 shows a detailed research flowchart.

3 Results and Discussion

Visual Observation. This showed that the leaf spring was fractured at the center hole. The fracture shape was ductile, where the surface looked stringy, rough, and gray, as shown in Fig. 5.

Fractographic Fracture Surface Analysis. The SEM observations showed that this fracture was close to the ductile characteristics because the surface was rough, stringy, and gray. Fractography with 5000 × magnification using a Scanning Electron Microscope (SEM) test instrument, initiation originating from the edge of the leaf spring specimen, and crack propagation from the initial area, lead to the spring failure. Furthermore, the cyclic load applied to the spring was perpendicular to the benchmark, which made the direction of crack propagation up and down. Figure 6 shows the results of SEM observations.

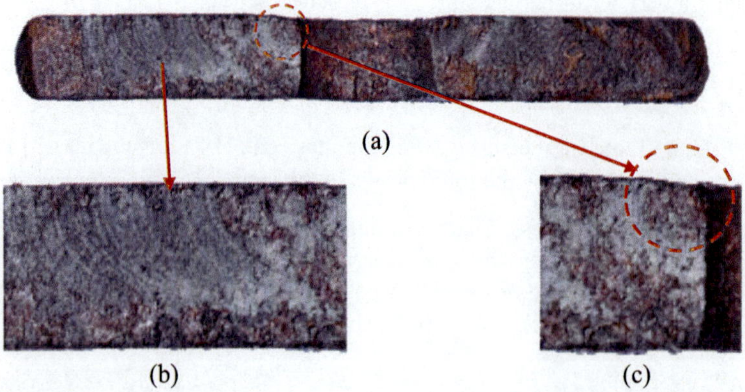

Fig. 5 **a** Leaf spring fracture surface, **b** *beach* mark, **c** *crack initiation*

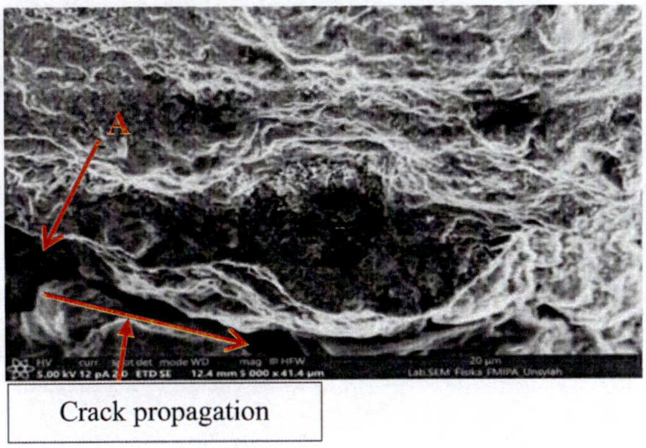

Fig. 6 SEM test results, arrow A indicates the beginning of the crack

Microstructure Test. The microstructure test was carried out using an optical microscope with 100 × magnification, showing the phases contained in the leaf spring specimen in form of dark pearlite and white ferrite phases. The ferrite is weak, soft, and ductile, while pearlite has relatively strong properties, good hardness, and ductility, but low tensile strength and plasticity; hence, its brittleness is high. In the microstructure image, the composition of ferrite is more than pearlite, which makes the material more ductile [18]. The microstructure observation is presented in Fig. 7.

Hardness Value. The hardness value of the leaf spring material was obtained after using a hardness testing machine (Rockwell). The test was carried out in horizontal and vertical directions on a surface that has been finely ground as shown in Fig. 8.

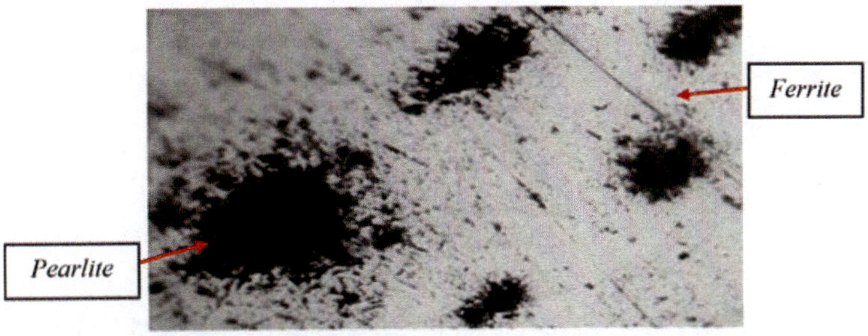

Fig. 7 Microstructure test

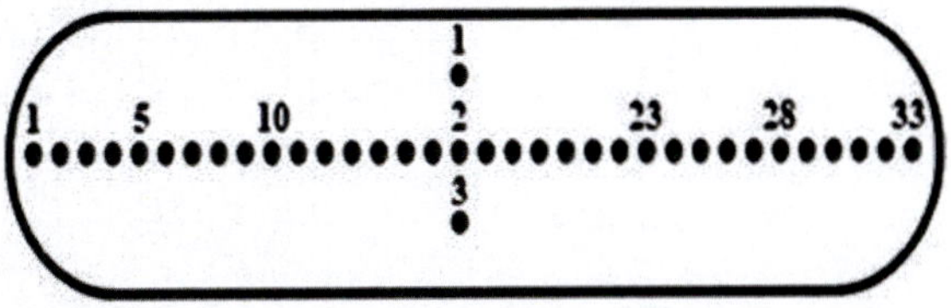

Fig. 8 Hardness test point on leaf spring

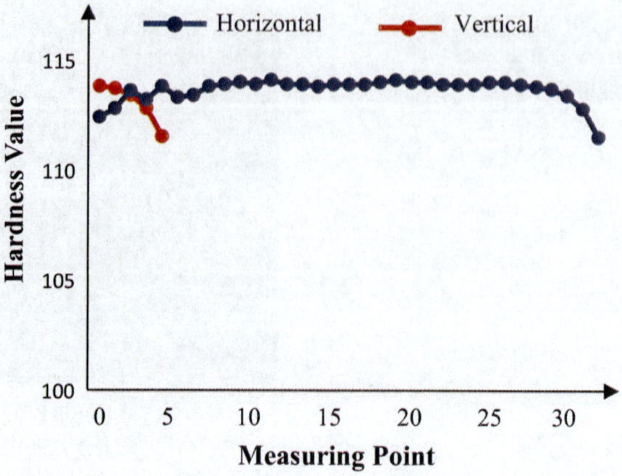

Fig. 9 Graph of spring hardness value in vertical and horizontal directions

The hardness value of the leaf spring can be considered uniform and homogeneous at all points. This shows that the spring material is good in terms of production and follows the standard specifications of spring steel. The graph of hardness values is presented in Fig. 9.

Chemical Composition Test. This test shows that the chemical composition value of the material is in line with the AISI 5160 standard steel. However, the detected Mn value is slightly greater than the standard value. In the process of making alloy steel, these elements are added to slow down the critical cooling rate during the hardening process and increase the hardenability. The properties of Mn in alloys have similarities with carbon, which increases the hardness. Table 1 shows the material chemical composition test on the leaf spring.

Based on the chemical composition, the mechanical properties can also be known as shown in Table 2.

Load Distribution on the Spring. Based on the specifications of the dump truck, the Total Vehicle Weight and Payload (Gross Vehicle Weight) is 8250 kg with nine active springs. Before the fracture occurred, the truck was carrying 10,500 kg of sand.

Table 1 Leaf spring material chemical composition

Element	Test results	AISI 5160
Cr	0.793	0.7–0.9
Mn	1.06	0.75–1
C	0.593	0.56–0.64
Si	0.230	0.15–0.3
S	0.0088	0.04 max
P	< 0.0050	0.035 max

Table 2 Mechanical properties of AISI 5160 steel [19]

Mechanical properties	Value
Tensile strength	724 MPa
Modulus of elasticity	190–210 GPa
Bulk modulus	140 GPa
Shear modulus	80 GPa
Poisson's ratio	0.27–0.30
Hardness, Rockwell B	92 HRB

By using Eq. (1), when all the loads were evenly distributed on the four wheels of the truck, the standard maximum load that can be carried by the springs and the load before the leaf springs fail was obtained.

The maximum load received by one leaf spring of a truck vehicle is 2062.5 kg (20,212.5 N), while the total load received before fracture was 2625 kg (25,725 N). This calculation indicated that the load on the truck exceeded the standard maximum load limit.

Maximum Allowable Stress. AISI 5160 Alloy Steel had a yield strength of 275 MPa and a correction factor of 1.65; therefore, the maximum allowable stress calculated by Eq. (2) was 166.6 MPa. For AISI 5160 material, the maximum allowable stress is 166.6 MPa, which was recalculated with the finite element method software, and both results were compared.

Stress Analysis Using the Finite Element Method. The stress analysis from the finite element method was visualizations in form of color contours that indicate the load value received by the specimen. The analysis was carried out in a static state by providing a load of 25,725 N (2,625 kg). During the analysis, the condition after the load was applied and the results were obtained in form of stress–strain distribution.

The stress simulation shows the distribution of Von Mises Stress on the leaf spring as indicated by color according to the value. The maximum Von Mises Stress result was shown in red at 485.66 MPa and the minimum was shown in purple at 123.88 MPa, as presented in Fig. 10.

The finite element method modeling has been accompanied by a crack initiation of 0.2 mm as marked with a red arrow. The size of the crack was obtained based on the visual observation of the broken leaf spring surface.

Stress Intensity Factor Analysis. According to Danilo et al. [20], the fracture toughness value of AISI 5160 leaf spring steel was 22 $MPa.m^{1/2}$. The stress intensity factor is shown at the crack tip. Failure/crack propagation in the material will occur when the stress intensity factor value due to the applied load is equal to or more than the fracture toughness value ($K_I \geq K_{IC}$). Meanwhile, the stress intensity factor was estimated around the crack tip to determine the maximum stress. The four nodes selected as a reference for calculations are shown in Fig. 11.

K_I at the crack tip for Mode I can be calculated using Eq. (3) [21].

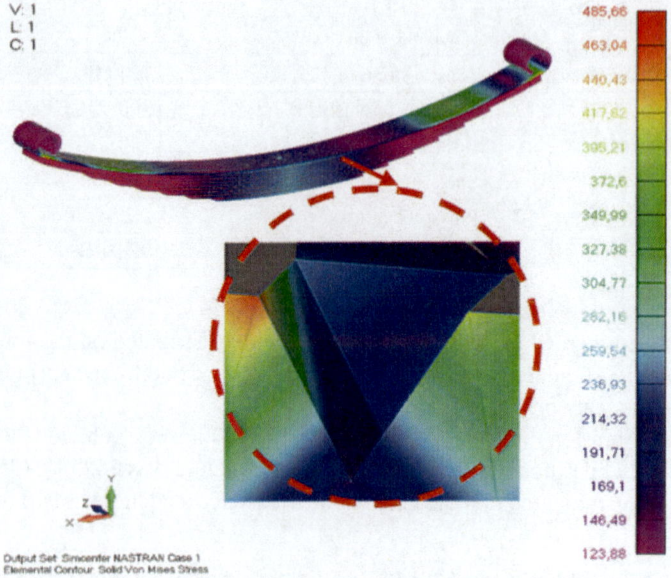

Fig. 10 Maximum stress simulation results around the initial crack

Fig. 11 Four points/nodes observed in the area around the crack

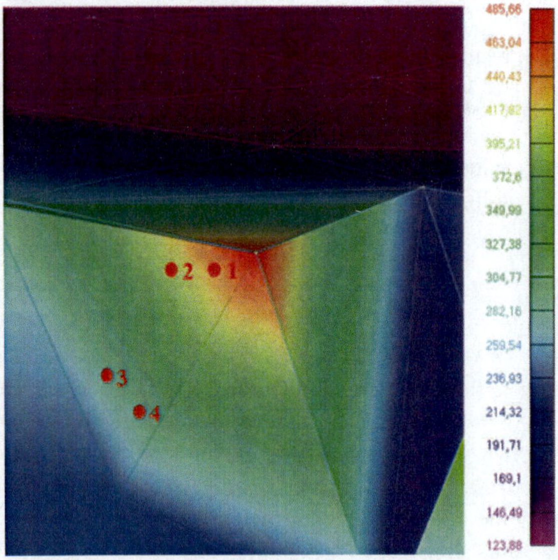

Table 3 K_I calculation results

No.	Nodal point	σ (MPa)	r (m)	θ	K_I (MPa.m$^{1/2}$)
1	44,856	244.41	0.008	30	69.43
2	14,809	171.50	0.012	20	52.36
3	21,171	161.66	0.035	45	126.93
4	44,523	153.59	0.031	60	156.54
	Average				101.32

$$K_I \frac{\sigma_{xx} \cdot \sqrt{2\pi r}}{\left(\cos\frac{\theta}{2}\right)\left[1 - \sin\left(\frac{\theta}{2}\right)\sin\left(\frac{3\theta}{2}\right)\right]}, \tag{3}$$

where

σ = Stress

R = Distance from the crack tip to the nodes observed.

θ = Angle formed.

The calculation of the K_I stress intensity factor at each node observed around the crack is shown in Table 3.

The results in Table 4 show that the average K_I value obtained from the finite element method analysis is 101.32 MPa.m$^{1/2}$, while the KIC value is 98.8 MPa.m$^{1/2}$. Based on the theory of fracture mechanics, when the stress intensity factor (K_I) exceeds the fracture toughness value (K_{IC}), crack propagation and failure occur simultaneously.

Figure 12 shows the relationship between K_I (stress intensity factor) and r (distance from crack tip to node) in the area around the crack. Point 1 is the closest point to the crack tip.

Therefore, crack propagation occurs due to the stress intensity factor that exceeds the fracture toughness of the AISI 5160 material. The large value received by the leaf spring leads to crack propagation that occurs from the initiation to the final fracture.

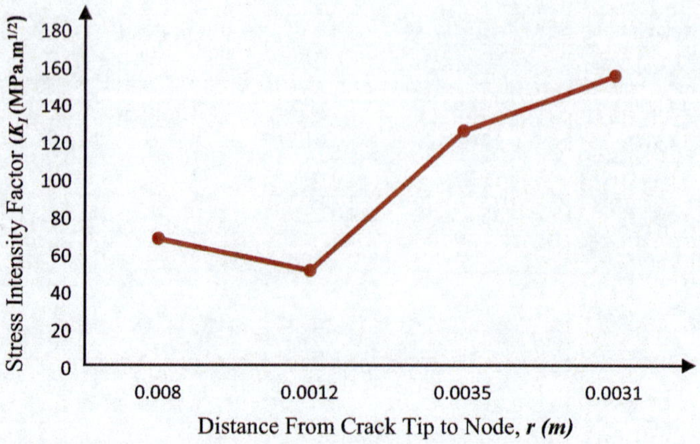

Fig. 12 Relationship between K_I and r

4 Conclusion

From this research on leaf spring failure, several conclusions are obtained, namely:

1. The load received by the leaf spring before failure is greater than the acceptance ability.
2. Fatigue failure is caused by a cyclic load, which involves three stages as shown by the results of visual observations on the fracture surface, namely, crack initiation, propagation, and final fracture.
3. The maximum stress analysis using the finite element method is 485.66 MPa. The calculation of the stress intensity factor (K_I) around the crack tip is 101.3220059 MPa.m$^{1/2}$, while the fracture toughness (KI_C) of AISI 5160 is 22 MPa.m$^{1/2}$. Therefore, these results have met the equation ($K_I \geq K_{IC}$).

Acknowledgements The authors are grateful to the Universitas Syiah Kuala for their financial support through Professor Research Grant No. 001/UN11.2.1/PT.01.03/ PNBP/2022. The authors are also grateful to the Department of Mechanical Engineering, Universitas Syiah Kuala, Banda Aceh, for making the facilities accessible.

References

1. Solanki P, Kaviti AK (2018) Design and computational analysis of semi-elliptical and parabolic leaf spring 19441–19455
2. Basori A, Pasaribu T (2019) Analisis Kegagalan Pegas Daun Kendaraan Bus Dengan Kapasitas 7 ton. Jurnal Konversi Energi Dan Manufaktur UNJ 1:13–21
3. Sepfitrah, & Y. Rizal, 2013. *Analisa peristiwa kegagalan pada "leaft spring" truck colt diesel pengangkut pasir*. Universitas Pasir Pengaraian.

4. Loganathan TG, Vinoth Kumar K, Madhu S (2019) Flexural and fatigue of a composite leaf spring using finite element 1:2214–7853
5. Todinov M (1999) Maximum principal tensile stress and fatigue crack for compression springs. Int J Mech Sci 41:357–370
6. Husaini, Ali N, Riantoni R, Putra TE, Husin H (2019) Study of leaf spring fracture behavior used in the suspension systems in the diesel truck vehicles. In: IOP conference series: materials science and engineering, vol 541, no 1, p 012046
7. Clarke CK, Borowski GE (2005) Evaluation of a leaf spring failure. J Fail Anal Prev 54:1547–7029
8. Eryurek IB, Ereke M, Goksenli A (2005) Failure analysis of the suspension spring of a light duty truck 14:170–178
9. Nataraj M, Thillikkani S (2020) Failure analysis of leaf spring suspension system for heavy load truck vehicle. Int J Heavy Veh Syst 27(1/2)
10. Siemens Digital Industries Software Inc (2020) Simcenter FEMAP. https://www.plm.automation.siemens.com/global/en/products/simcenter/femap.html, diakses pada 11 Oktober 2021
11. Husaini, Nurdin A, Agustian B (2016) Analisa Permukaan Patah Pegas Ulir Suspensi Depan Mobil Sedan. In: Proseding Seminar Nasional Teknologi Rekayasa (SNTR) III Tahun 2016, p 111
12. Galih, S, Cahya S, Efendi M (2016) Fenomena temper embrittlement Pada Baja Martensitik Aisi 410 Untuk Aplikasi Steam Gate Valve 20" class 150 grade WCB; 3
13. Pantazopoulos G, Vazdirvanidis A (2008) Fractographic and metallographic study of spalling failure of steel straightener rolls. J Fail Anal Prev 8:509–514
14. C.C. Yao, M.L. Wang, Y.Y. Ruan, X. M. LV, J.L. Wang, J.M. Ge, & H.W. Jiao, 2020, *Several methods for improving the accuracy of Rockwell hardness testing. In Journal of Physics*: Conference Series (Vol. 1507, No. 10, p. 102010). IOP Publishing
15. Matheus S (2011) Analisis modulus elastisitas dan Angka Poisson Bahan Dengan Uji Tarik. Jurnal Barekeng 5(2):9–14
16. Husaini LR, Handayani N, Ali MS (2021) Failure analysis of a fractured leaf spring as the suspension system applied on the dump truck. Key Eng Mater 892:89–98
17. Sari Y, Dwiyati ST (2015) Korosi H_2S dan CO_2 pada Peralatan Statik di Industri Minyak Dan Gas. Jurnal Konversi Energi dan Manufaktur 1:18–22
18. Wibowo DB, Purwanto D (2007) Pengujian impak besi cor kelabu austemper. Rotasi 9(2):37–41
19. Azom (2021) Material AISI 5160 Alloy Steel (UNS G51600). www.azom.com/article.aspx?ArticleID=6743, diakses 5 Agustus 2021
20. Castro DBV, Ventura JM, Ruckert COFT, Filho WWB (2010) Influence of phosphorus content and quenching/tempering temperatures on fracture toughness and fatigue life of SAE 5160 Steel; 454:445–455
21. Husaini (2015) Dasar—dasar Mekanika Retakan. Universitas Syiah Kuala, Darussalam Banda Aceh
22. Castro DBVD, Ventura JM, Ruckert COFT, Spinelli D, Filho WWB (2010) Influence of phosphorus content and quenching/tempering temperatures on fracture toughness and fatigue life of SAE 5160 steel. Mater Res 13:445–455

Experimental Study of Helical Gear Failure in Off-Road 4 × 4 Gearboxes

Husaini, Darmawan Haris, and Ali Nurdin

Abstract The helical gear in a off-road 4 × 4 gearbox was found to have failed on the surface. Therefore, this study aims to determine the factors causing this failure through the application of visual observation, hardness test, chemical composition test, microstructure observation, Scanning Electron Microscope (SEM) observation, and stress analysis. The chemical composition test showed that the gear material is classified as high carbon steel with AISI 1080 standard, the average hardness value was recorded to be 95.49 HRB which is slightly lower than AISI 1080, and the microstructure is pearlite and cementite. Moreover, the observation of the fracture surface showed some initial defects propagated to a brittle fracture, while the shear stress value was found to be 286.13 MPa which exceeds the allowable shear stress limit of 253.47 MPa. It was concluded that the helical gear failed due to the initial defects in the form of fine cracks and the fact that the shear stress value exceeds the allowable limit.

Keywords Failure analysis · Helical gears · Brittle fracture · Fatigue

1 Introduction

Off-road 4 × 4 is a vehicle used to travel long distances and operates on a manual transmission system. It is usually driven above the average speed and often passes through potholes, thereby, triggering the failure of components, especially the gear in the gearbox which is also known as the transmission system.

Husaini (✉) · D. Haris · A. Nurdin
Department of Mechanical Engineering, Computational Mechanics Labaoratory, UniversitasSyiahKuala, Darussalam, Banda Aceh 23111, Indonesia
e-mail: husainiftm@usk.ac.id

D. Haris
e-mail: haris.tm@mhs.usk.ac.id

A. Nurdin
e-mail: nurdin.ali@usk.ac.id

Irwansyah et al. (eds.), *Proceedings of the 4th International Conference on Experimental and Computational Mechanics in Engineering*, Lecture Notes in Mechanical Engineering, https://doi.org/10.1007/978-981-99-7495-5_48

The gearbox is one of the important components in a vehicle which houses the gear normally used to change the shaft speed and torque [1]. It is also widely used for power transmission in factory machines, machine tools, cars, windmills, and other mechanical systems [2] with the main purpose of changing rotational speed and torque. Moreover, transmission systems are commonly classified into manual and automatic [3], and their performance can be influenced by several factors such as the speed, level of fuel consumption, and the resulting traction force. It is also important to note that fuel consumption and rpm are affected by engine rotation speed [4].

Gears are engine elements in the gearbox that transmit power to each other. Moreover, helical gears are mostly used for power transmission due to their smoother performance which is associated with the larger helic angle [5]. This component often fails and this subsequently has a significant effect on the transmission system because its rotation normally produces a contact force between the teeth which triggers small fine holes or pitting to cause severe damage to the system [6]. The helical gears of a off-road 4 × 4 were observed to have failed severely due to heavy loads and lack of lubrication. Therefore, this study was conducted to examine the main causes of this failure.

A previous study by Husaini et al. [7] showed that the failure of gears can be associated with initial defects which were exacerbated by the impact of dynamic loads when operating on the hard and solid ground [7]. Boonmag et al. [8] also reported that it can be related to the cracks on the outer surface which led to a combination of brittle and ductile fractures as well as the mixture of ferrite and pearlite in the microstructure [8].

Several studies have analyzed causes of failure in machine components with different results which include substandard hardness of the material, fatigue, and the stress intensity factor (K) exceeding the fracture toughness value (Kc) [9–11]. Some also mentioned improper installation procedures [12], high shock loads, and fatigued components [13].

2 Research Methods

The material used in this study is the helical gear observed to have failed on the surface in a off-road gearbox as shown in Fig. 1.

Hardness Test. The hardness value of the gears was determined using a Rockwell hardness tester with type Zwick/Roell ZHR scale B which is normally applied to metal materials that are not too hard. The indenter used was a steel ball with a diameter of 1/16 in. The process involved applying two types of loads which include a minor load of 10 kg and a major load of 100 kgf after which the hardness values were expressed in HRB to be read directly. The test has required only a few seconds and was regulated according to ISO 6568 and ASTM 18 standards. The test points are indicated in the following Fig. 2.

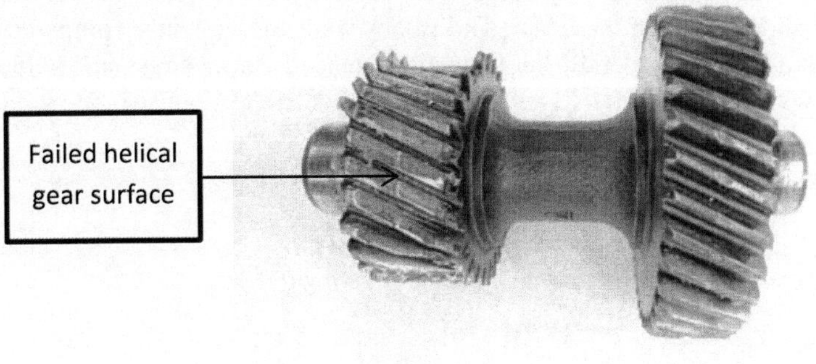

Fig. 1 Failed helical gear

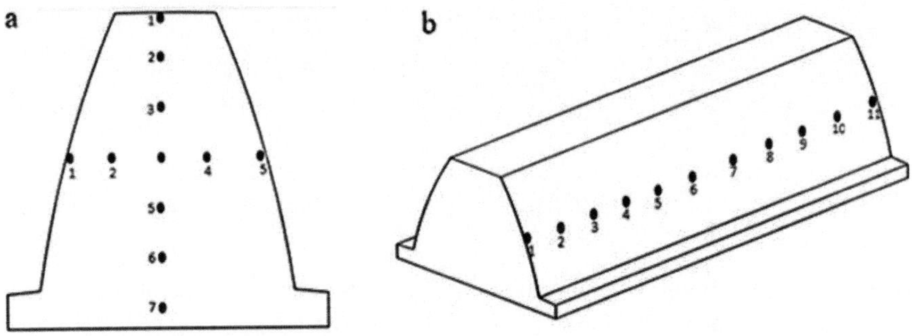

Fig. 2 Hardness test points **a** the middle and **b** the outside

Chemical Composition Test. This test was conducted using a spectrometer to determine the chemical elements contained in the material and their total values. The results were subsequently used as a reference to determine the appropriate standard group of materials which include AISI, ASME, ASTM, JIS, or SAE. The values were also compared with the standards to ensure that the materials used are appropriate.

Microstructure Observation. This was conducted using metallographic techniques to determine the microstructure of the material, and the findings were used to determine the shape of the phase and the amount contained in the material.

Broken Surface Observation. Scanning Electron Microscope (SEM) is an electron microscope that produces an image of a sample by scanning its surface using a focused electron beam with magnification up to a certain scale. The electrons interact with the atoms in the sample to generate different signals containing information about the surface topography and sample composition. The SEM is widely used for study and industrial purposes due to its ability to combine high magnification, large depth of field, good resolution, and capacity to determine the composition and crystallographic information.

Shear Stress Analysis. The stress analysis was conducted to determine the magnitude of the shear stress in the gears. The values were subsequently compared with the allowable limit to ascertain their appropriateness. Shear stress allowable can calculated with Eq. 1 [14].

$$\tau_a = S_o \times \left[\frac{5.6}{5.6 + \sqrt{V}}\right] \tag{1}$$

Shear stress work on helical gear can calculated with Eq. 2.

$$\tau_g = \frac{F_t}{bh} \tag{2}$$

3 Results and Discussion

Gear Material Hardness. The results showed that the hardness value near the outer surface is different from the value recorded at the center as indicated in Fig. 3.

The average value was 95.49 HRB, and this is slightly lower than the 99 HRB recommended for AISI 1080 material.

Gear Material Chemical Composition. The material of the helical gears was found to be medium carbon steel with 0.3–0.8% carbon content, ferrite, and pearlite microstructure, stronger and harder but brittle as presented in the following Table 1.

The results also showed that the inclined gear has a high carbon content and this is different from most helical gears which are classified as low-carbon steel. This inappropriate chemical composition is believed to be one of the factors causing the failure. Moreover, the knowledge of the material's standard as AISI 1080 made it possible to determine its mechanical properties as indicated in the following Table 2.

Gear Material Microstructure. The microstructure test was conducted using an Olympus BX41M optical microscope while etching in the form of HCL and alcohol at a ratio of 2:1 which was applied to determine the grain boundaries. The microstructure was observed at 200 × and 500 × magnification scales, and the results are shown in Fig. 4.

Observation of Gear Fault Surface. Visual observation showed that the damage occurred in more than half of the total tooth profile as presented in Fig. 5. Moreover, the SEM observation showed that the broken surface is a brittle fracture which is characterized by a granular surface, ability to reflect light, and occurrence in materials with a high carbon composition as indicated in Fig. 6.

Shear Stress Analysis. Shear stress allowable can be calculated using Eq. 3.

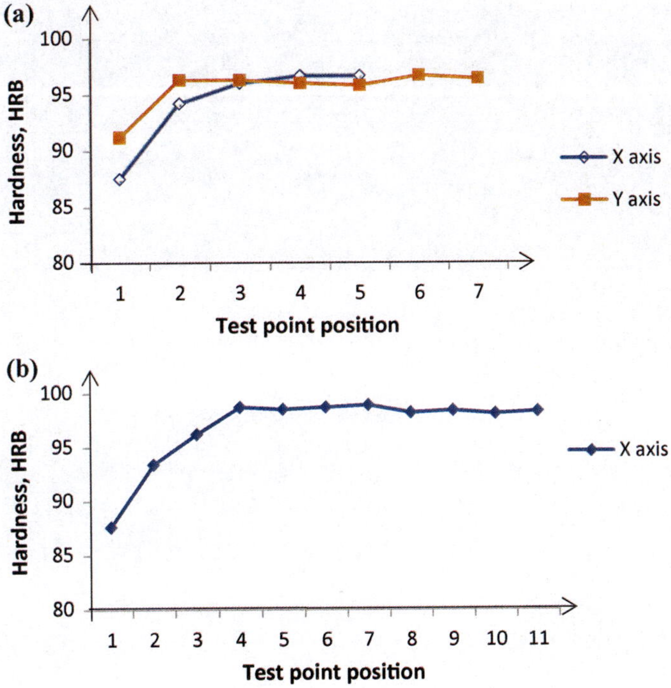

Fig. 3 Graphic of hardness test results **a** the outside and **b** the side

Table 1 Chemical composition test results [15]

Element name	Test result	AISI 1080 standard
Iron, Fe	85.8	98.13–98.65
Carbon, C	0.860	0.75–0.88
Manganese, Mn	0.696	0.60–0.90
Sulfur, S	0.0289	≤ 0.050
Phosphorous, P	0.0312	≤ 0.040

Table 2 Mechanical properties of AISI 1080 materials [15]

Property	Value	Unit
Density (ρ)	7.85	g/cm^3
Ultimate strength (σ_B)	965	MPa
Yield strength (σ_{ys})	585	MPa
Bulk modulus (B)	160	GPa
Shear modulus (G)	80	GPa
Elastic modulus (E)	205	GPa
Poisson ratio (v)	0.29	–
Hardness, Rockwell B Shear stress allowable	99 253.47	HRB MPa

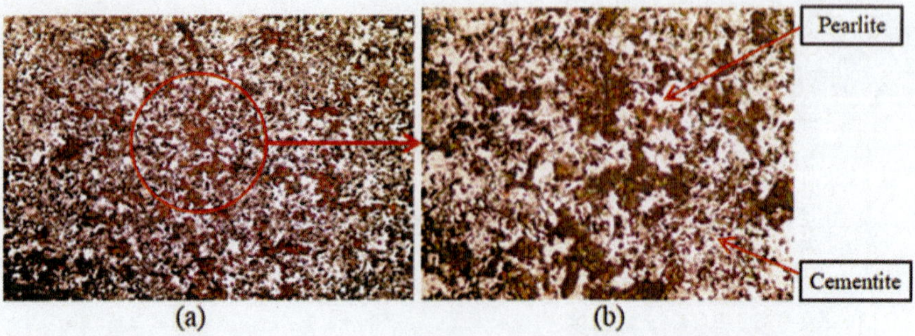

Fig. 4 Microstructure observation results **a** 200 × magnification and **b** 500 × magnification

Fig. 5 Failed surface

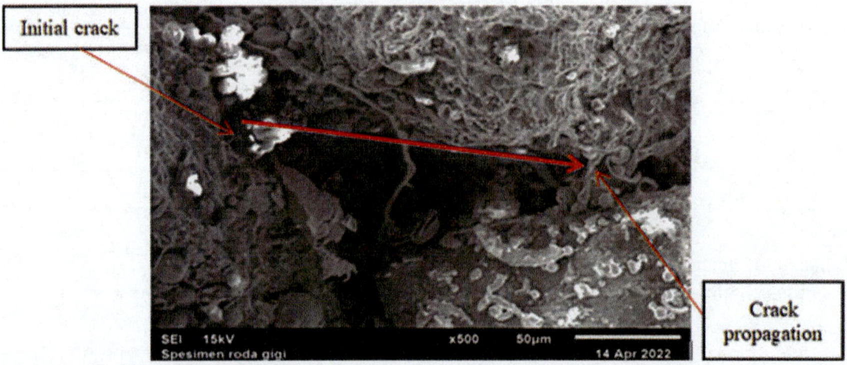

Fig. 6 Observation results using SEM

$$\tau_a = 321,67 \times \left[\frac{5.6}{5.6 + \sqrt{2,24}}\right] = 253,47\,\text{MPa} \tag{3}$$

Shear stress work on helical gear can calculated with Eq. 4.

$$\tau_g = \frac{4292}{0.3 \times 0.05} = \frac{429,2}{0.0015} = 286,13\,\text{MPa} \quad (4)$$

This means that the shear stress was 286.13 MPa and this exceeds the allowable limit of 253.47 MPa, thereby serving as one of the factors causing failure in helical gears.

4 Conclusion

This study shows conducted on the causes of the helical gear failure showed that:

1. It was observed that the hardness on the edge of the gear is lower than on the inside and the average hardness value was recorded to be 95.49 HRB.
2. The carbon content of the gear material was recorded to be 0.860 C, and this is classified as high carbon steel based on AISI 1080 standards.
3. The identified microstructures were pearlite and cementite, while the fracture observed using SEM was brittle.
4. The shear stress (τ_g) calculated was also discovered to be higher than the allowable shear stress (τ_a) with 286.13 MPa > 253.47 MPa, respectively.

Acknowledgements The authors are grateful to the Universitas Syiah Kuala for their financial support through Professor Research Grant No. 001/UN11.2.1/PT.01.03/PNBP/2022. The authors are also grateful to the Department of Mechanical Engineering, Universitas Syiah Kuala, Banda Aceh, for making the facilities accessible.

References

1. Siva K, Balamurali D, Katiresh PVM, Damodaran S, Aravindan PS (2018) Design and optimization of spur gear box parameters. Int Res J Automot Technol 1:34–40
2. Pandey DK, Lim HC (2020) Pinion failure analysis of a helical reduction gearbox in a kraft process. Appl Sci 10:2935
3. Nurdin, Himawanto DA (2019) Studi numerik kekuatan material transmisi roda gigi pico hydro. Jurnal Teknik Mesin Indonesia 14:24–29
4. Nazaruddin, Syafri, Anuar K, Awaluddin A (2018) Analisis Kinerja Transmisi Pada Mobil Hemat Energi Tipe Hybrid, Seminar Nasional Itenas
5. Susanto N, Purwaningsih R (2017) Analisis Pengaruh Transmisi Mobil Manual dan Otomatis terhadap Tingkat Kesulitan yang Dihadapi Pengemudi Pemula. JTI Undip: Jurnal Teknik Industri 12(3):197–204
6. Widodo RA, Satrijo D (2014) Deteksi Kerusakan Roda Gigi dengan Analisis Sinyal Getaran Berbasis Domain Waktu. Jurnal Teknik Mesin Universitas Diponegoro 2(3):182–189
7. Husaini DM, Dawud TE, Putra NA (2020) Failure analysis of spur gear used in transmission system applied on a hand tractor. Key Eng Mater 841:144–149
8. Boonmag V, Wisesook O, Phukaoluan A, Pluphrach G (2018) Microcrack analysis of chromium steel JIS-SCr 420 for helical gear transmission. Key Eng Mater 777:294–299

9. Husaini, Riantoni R, Ali N, Putra TE (2019) Failure analysis of the leaf spring of truck colt diesel using finite element method. IOP Conf Ser Mater Sci Eng 547:012017
10. Husaini, Ali N, Bakhtiar (2016) Analysis on the surface of broken screw spring front car suspension. In: Proceedings of Seminar Nasional teknologi Rekayasa (SNTR) 3:111–117
11. Vukelic G, Breie M (2016) Failure analysis of a motor vehicle coil spring. Procedia Structural Integrity 2:2944–2950
12. Febriyanti E, Suhadi A (2016) Analisis Kegagalan Fire Ring Penyebab Kerusakan Piston Mesin Unit Kendaraan Bermotor, MPI 10(2):99–106
13. Samlawi K, Zaini M (2020) Analisis Kegagalan Roda Gigi Matahari Swing Drive pada Excavator Caterpillar 349 D. Info Teknik 20(2):203
14. Myo Zaw KL (2019) Design of Syncromesh mechanism and stress analysis of gear for Hijet. Iconic Res Eng J 2(8):140–146
15. MatWeb. AISI 1080 Steel, as rolled. http://www.matweb.com/search/datasheet.aspx?matguid=37ca3458dfbe482fbe1efdf59d52e424&n=1&ckck=1. Accessed 27 July 2022

Experimental Investigation of the Impact of Eco-Riding Driving on Urban Motorcycle Fuel Consumption

Muhammad Tadjuddin, Syahriza, Said Amir Azan, Udink Aulia, Fitrika Mita Suryani, and Ragam Iqbal Tawakal

Abstract The efficiency of using fossil fuels for transportation is continuously improving. Increasing vehicle efficiency can be done by better internal combustion engine technology or a method of driving the vehicle itself. Eco-riding is a method of operating a vehicle with better fuel efficiency. This paper discusses how to go with eco-riding and compares it to regular driving. The vehicle being tested is a motorcycle with an eco-riding indicator onboard, and the test is carried out with a travel route within the city of Banda Aceh, Indonesia. The trial took ten days of testing with two different traffic options. The first trip is during rush hour, and the second trip is under regular traffic conditions. From the driving test on rush hour trips, the driving system following eco-riding got a 10.62% fuel reduction, while fuel use was 10.93% more efficient for trips in regular traffic. The driving pattern in this study has been demonstrated to reduce fuel consumption significantly.

Keywords Fuel efficiency · Eco-driving · Urban · Driving method

M. Tadjuddin (✉) · Syahriza · U. Aulia · R. I. Tawakal
Department of Mechanical and Industrial Engineering, Faculty of Engineering, Universitas Syiah Kuala, Banda Aceh, Indonesia
e-mail: m.tadjuddin@usk.ac.id

U. Aulia
e-mail: uaulia@usk.ac.id

S. A. Azan · F. M. Suryani
Department of Civil Engineering, Faculty of Engineering, Universitas Syiah Kuala, Banda Aceh, Indonesia
e-mail: saidamirazan@usk.ac.id

F. M. Suryani
e-mail: Fitrika_mitasuryani@usk.ac.id

Irwansyah et al. (eds.), *Proceedings of the 4th International Conference on Experimental and Computational Mechanics in Engineering*, Lecture Notes in Mechanical Engineering, https://doi.org/10.1007/978-981-99-7495-5_49

1 Introduction

The energy sector is essential in supporting sustainable development because all human activities require energy supplies, either directly or indirectly. Until now, the national energy supply relies on fossil energy sources, namely oil, gas, and coal. In future developments, the sustainability of the energy sector in supporting national development will experience various obstacles related to the imbalance between the energy supply rate and the energy demand rate. Dependence on fossil energy sources is still high, while reserves of alternative energy sources (including renewable energy) have not been widely developed and utilized.

Fuel consumption harms health and the environment, so it should be reduced. One factor that affects the level of energy consumption (fuel oil) of motorized vehicles is driving behavior. Driving behavior is associated with driver behavior that causes acceleration, braking or deceleration, idling, gearing in certain positions, speeding, and turning on and off the vehicle engine. Certain behaviors allow vehicle fuel consumption to be more efficient than other behaviors.

The increasing number of motorized vehicles certainly has an impact on environmental conservation and the use of fossil fuels. The use of public transport is an alternative to reducing emissions. However, the use of public transport is not very attractive both in Indonesia and in several other countries [1]. Traffic signs will also affect the driver's driving, affecting the driver's attitude on the highway [2].

According to government statistics, there are many motorcycles in Indonesia, and the number reaches 115 million units, that is the importance of applying driving techniques oriented toward fuel efficiency, known as eco-riding. In this paper, two conditions are taken, namely normal road conditions and road conditions during peak hours. Previous studies have investigated the relationship between rush hour roads and traffic jams on driving habits [3]. According to dynamic operating limits, driving behavior is classified as aggressive, moderate, and passive. The parameters representing traffic conditions are the percentage of stop time and the average speed of vehicles in urban areas [4].

2 Energy Sector

The energy sector is one of the essential needs for all people in Indonesia. It runs the economy of all people in Indonesia, such as household activities, industry, transportation. When viewed from the aspect of supply, Indonesia is considered as a rich country in energy resources, both unrenewable energy and renewable resources. However, the exploration of energy resources is more focused on fossil energy, which is an unrenewable resource, while renewable energy has not been widely utilized.

Currently, one of the primary human needs besides food, shelter, and clothing is the need for energy. Energy can help to run the economy of all people in Indonesia, such as household activities, industry, transportation. The energy that plays a significant role in Indonesia is oil. In 2015, the largest share of final oil energy consumption was the household sector (35%), followed by the transportation sector (31%), industrial (29%), commercial (4.0%), and others (2.0%). During 2010–2015, the transportation sector experienced the most significant growth reaching 5.2% per year, followed by the household sector (3.8%) and the commercial sector (2.9%). Meanwhile, the growth of the industrial and other sectors decreased by 4.6% and 10%, respectively.

3 Eco-riding

Eco-riding is a way of operating a vehicle that aims to optimize fuel consumption efficiently. Besides being efficient, eco-riding techniques also contribute to environmental conservation programs by reducing CO_2 gas, which has been the number one source of carbon emissions. Eco-driving can be seen as a relatively inexpensive and direct emission reduction effort because it involves modifying driver behavior that supports developing and adopting newer and more efficient technologies [5].

Eco-riding combines three aspects: driving skills that must have the ability to drive smart. The other two are Defensive Riding, which aims to save lives and time, and Safety Riding which prioritizes safe-driving techniques and always thinks of all possibilities. Interestingly, eco-riding combines Safety-Riding techniques with Defensive-Riding psychology. The other two are Defensive Riding, which aims to save lives, time, and Safety Riding, which prioritizes safe-driving techniques and always thinks ahead of all possibilities. Interestingly, eco-riding combines Safety-Riding driving techniques with Defensive-Riding psychology. That is where riding techniques are created that save time and money and are safe and environmentally friendly [6]. Driving speed, acceleration, and pedal use are related to fuel consumption and emissions, so it is necessary to educate the driver. Drivers should also avoid strong acceleration and deceleration when dealing with traffic lights [7]. Not only vehicles with fuel, but electric vehicles also prioritize eco-driving and driving behavior to save the use of electrical energy [4]. The electric vehicle's efficiency depends on its powertrain technology, which has a characteristic of high power density [8].

Eco-riding is a fuel-efficient driving technique. The goal is not only to save fuel consumption ratio but also to maintain optimal engine component life. Eco-riding on motorcyclists can save fuel consumption and routine maintenance of engine components for a long lifespan.

There are several ways to save vehicle fuel that can be done [5], including:

1. Pull the gas slowly while accelerating.
2. Do not operate the engine at high speed if there is no load.
3. Stop the engine if it stops for too long at traffic jams, traffic lights, and other stops.

Fig. 1 Motorcycle

Many variables affect driving efficiency, corresponding to driver behavior. The influence of the driver itself is an aggressive or standard way of driving, while there is also an influence from the nature of other drivers around him [9].

4 Motorcycle

The motorcycle selected in this study, the 2018 HONDA Beat Esp (Fig. 1), was selected because it has an eco-riding indicator on the front panel. The following photos and specifications of the vehicle are used in this study:

This motorcycle was manufactured in 2018. The motorcycle is only used for a short distance within Banda Aceh. The traveling distance from the odometer was 13,000 km. This vehicle service maintenance is carried out regularly so that it can be ascertained that there are no problems with the engine. The specification of the motor cycle can be seen at Table 1 below.

5 Android Application

This research uses an Android-based commercial application. This application is a Trip Recorder, which can record travel tracks connected to Google Map®. It can also record speed, distance, time, and travel routes (Fig. 2).

Table 1 Specifications of motorcycle

Vehicle dimensions	
Long	1856 mm
Wide	666 mm
Tall	1068 mm
Wheelbase	1256 mm
Vehicle weight	92 kg
Latest vehicle condition	
Year of purchase	2018
Vehicle mileage before research	13,000 km
Machine condition and maintenance	Routine service, use in the city
Machine	
Machine type	4 Step, SOHC with air cooling,
Maximum power	6.38 KW (8.68 PS)/7,500 rpm
Maximum torque	9.01 Nm (0.92 kgf m)/6,500 rpm
Step volume	108.2 cc
Fuel supply system	Injection (PGM-FI)
Transmission system	Automatic, V-Matic
Capacity	
Front suspension fuel tank capacity	4.0 L
Lubricating oil capacity	0.7 L at periodic replacement
Fuel	RON92

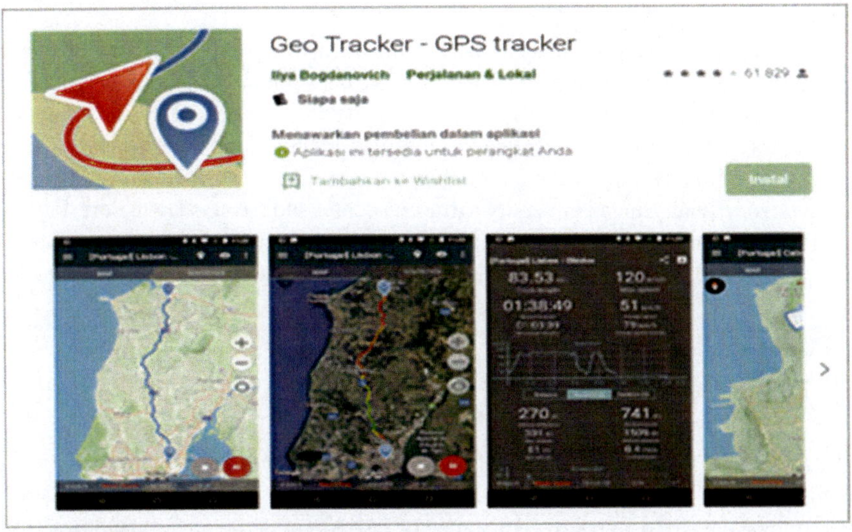

Fig. 2 GPS based on Android Geo Tracker GPS

Fig. 3 Mapping Trip 1

6 Research Methodology

The route was planned and data was recorded along the journey. The data includes speed, distance, and fuel consumption. Data collection is ten working days, during which the driver drives his vehicle with normal behavior for the first five days. Then in the next five days, the driver is educated to keep the eco-riding indicator lit up. Data is taken during peak hours from the starting point to the endpoint on the first trip and back to the start point on the second trip during normal traffic hours. The driver in this study was 21 years old and weighed 55 kg.

Fuel consumption was measured by using measuring cup. Measurements are taken after each trip is completed. The difference between them is the amount of fuel consumed. While the two types of trips in this paper are selected during peak and regular traffic hours, this is intended to represent urban driving conditions.

The first trip starts at 7:00 am, which is the rush hour for Banda Aceh city. The route is from Syiah Kuala University to the city center and so on to Ulele Port (Fig. 3).

The second trip starts at 8 am, which is normal traffic hours for the city of Banda Aceh. The route back to Syiah Kuala University is also through the city center (Fig. 4).

7 Results and Discussion

The test of the Eco-Riding Method took ten working days, five days for normal driving behavior, and five days for driving with the Eco-Riding Method. This study runs two trips in one day; Trip 1 covered a distance of 13.4 km at 07:00 am (considering rush hour) and Trip 2 was 14.4 km at 08:00 am. Fuel consumption data (ml) for every

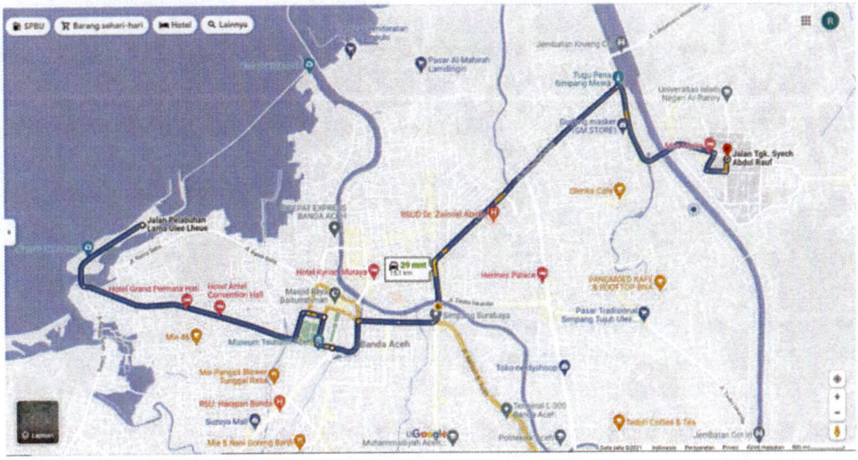

Fig. 4 Mapping Trip 2

trip was then calculated to get an average fuel consumption (km/l). The results show below:

Figure 5 shows the results during rush hour. The Eco-Riding Method can save fuel up to 10.62%. The average value of consumption per liter of normal riding is 54.24 (km/l) and 60 (km/l) for eco-riding. It was found that the wasted km was 52.14 (km/l) on normal riding, and the most economical value for eco-riding was 61.19 (km/l).

Trip 2 data collection was taken from Sunday to Friday weekdays at 08:00 WIB.

Figure 6 shows that the Eco-Riding Method can save fuel up to 10.93%, where the average value of consumption per liter of normal riding is 52.51 (km/l) with

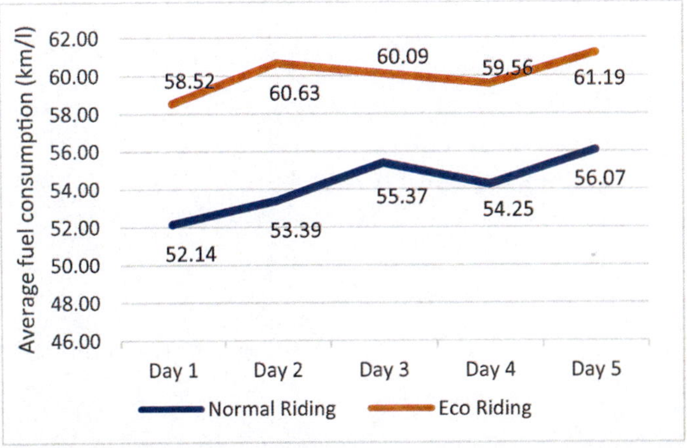

Fig. 5 Average fuel consumption at rush hour

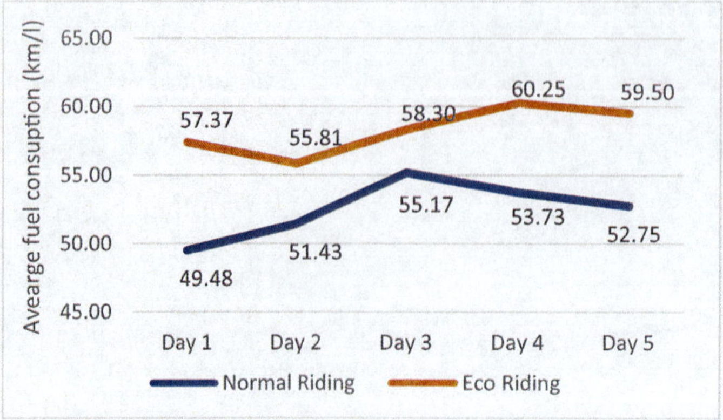

Fig. 6 Average fuel consumption in normal traffic on weekdays

eco-riding at 58.25 (km/l). 1). From 10 days of data collection, the wasted km on normal riding was 49.48 (km/l), and the most economical value was on eco-riding, 60.25 (km/l).

The correlation between average speeds and fuel consumption can be written in mathematical equations.

Figure 7 shows that the average speed with fuel consumption is directly proportional, where the higher the average speed, the higher the fuel consumption. The equations obtained in the Normal Riding Method are $y = -1.85x^2 + 172.68x - 3772.9$ with $R^2 = 0.93$ and the Eco-Riding Method is $y = 1.5x^2 - 118.16x + 2552.7$ with $R^2 = 0.98$. In this equation, it can be found how much fuel consumption is needed at a specified speed.

Figure 8 shows that the average speed with fuel consumption is directly proportional, where the higher the average speed, the higher the fuel consumption. The equation obtained in the Normal Riding Method is $y = 4.18x^2 - 369.49x + 8431.4$ with $R^2 = 0.98$ and the Eco-Riding Method is $y = -6.13x^2 + 528.91x - 11.158$ with $R^2 = 0.99$. From this equation, it also can be found how much fuel consumption is needed at a specified speed.

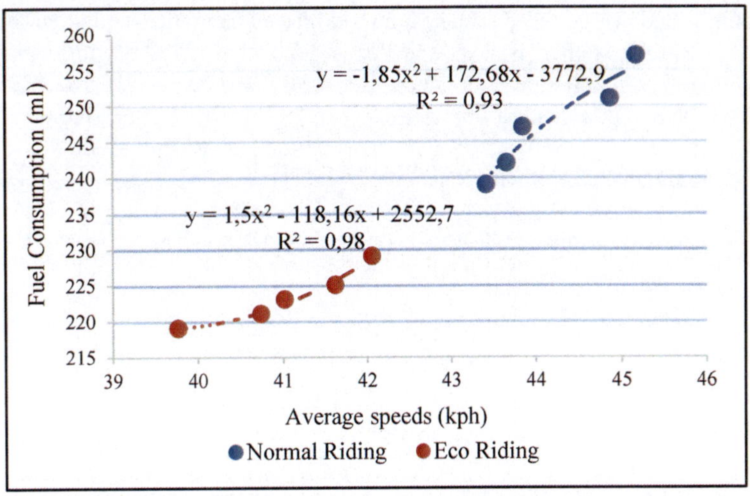

Fig. 7 Relationship between average speed and fuel consumption in Trip 1

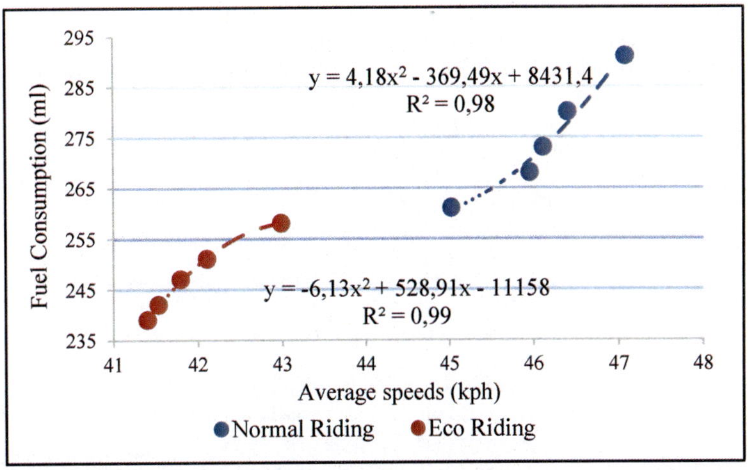

Fig. 8 Relationship between average speed and fuel consumption in Trip 2

8 Conclusion

The results of this study show that driving behavior greatly affects fuel efficiency in the urban environment. The Eco-Riding Method can significantly reduce fuel consumption compared to normal driving. On Trip 1, which is in the rush hour, the Eco-Riding Method can save 10.62% of fuel consumption compared to the Normal Riding Method. In Trip 2, the Eco-Riding Method can save 10.93% of fuel used in normal traffic compared to the Normal Riding Method. The relation between fuel consumption and average speed can be written with a mathematical equation. On Trip 1: Normal Riding Method is $y = -1.85x^2 + 172.68x - 3772.9$ with coefficient relation $R^2 = 0.93$, while the Eco-Riding Method is $y = 1.5x^2 - 118.16x + 2552.7$ with $R^2 = 0.98$. The equation obtained on Trip 2: Normal Riding Method is $y = 4.18x^2 - 369.49x + 8431$ with $R^2 = 0.98$, while the Eco-Riding Method is $y = -6.13x^2 + 528.91x - 11.158$ with $R^2 = 0.99$.

Acknowledgements The authors would like to express the deepest gratitude to the members of the Malemdiwa team at the Design and Manufacture Laboratory of Universitas Syiah Kuala, for their great support of this research.

References

1. Mutiawati C, Suryani FM, Anggraini R (2022) Importance-performance analysis in public transport level of service: a case study of the trans Koetaradja bus in Banda Aceh. Aceh Int J Sci Technol 11(1):70–84
2. Guo Z, Wei Z, Wang H (2016) The expressway traffic sign information volume threshold and AGS position based on driving behavior. Transp Res Procedia 14:3801–3810
3. van Driel CJG, Hoedemaeker M, van Arem B (2007) Impacts of a congestion assistant on driving behavior and acceptance using a driving simulator. Transp Res Part F Traffic Psychol Behav 10(2):139–152
4. Al-Wreikat Y, Serrano C, Sodré JR (2021) Driving behavior and trip condition effects on the energy consumption of an electric vehicle under real-world driving. Appl Energy 297:117096
5. Allison CK, Stanton NA (2019) Eco-driving: the role of feedback in reducing emissions from everyday driving behaviours. Theor Issues Ergon Sci 20(2):85–104
6. Angelo DC, Valentina A, Nuranissa S (2021) Creative strategy for reducing air pollution from motorcycle exhaust by urging urban motorcyclists to implement eco-riding techniques. In: Proceedings of the international conference on economics, business, social, and humanities (ICEBSH 2021), vol 570
7. Rittger L, Schmidt G, Maag C, Kiesel A (2015) Driving behavior at traffic light intersections. Cogn Technol Work 17(4):593–605
8. Tadjuddin M, Hasanuddin I, Fuadi Z, Mukminin A, Arief MR (2020) Power density comparison of multi-air gap axial flux permanent magnet motor for electric vehicle. IOP Conf Ser Mater Sci Eng 931(1)
9. Nguyen S, Cojocaru M, Thommes E (2014) Personal efficiency in highway driving: an agent-based model of driving behavior from a system design viewpoint. In: Canadian conference on electrical and computer engineering, pp 1–6

Experimental Analysis of Acoustics Characteristics of Honeycomb-Backed MPP Panel with Different Material Elasticities

Meifal Rusli, Afdhilla, Dendi Adi Saputra, Gusriwandi, and Mulyadi Bur

Abstract This paper discusses the experimental analysis of the sound absorption coefficient and transmission loss of hybrid panels made of a multi-perforated panel (MPP) backed by a honeycomb structure with different material elasticities. The panels are created using 3D printing and made of thermoplastic polyurethane (TPU) filament, which is more flexible than the second material, polylactic acid (PLA) filament. This study aims to compare the effect of material flexibility on the panel's acoustic properties. The acoustics are measured using a four-channel impedance tube following ASTM E2611-19. The panels have various panel thicknesses, honeycomb wall and MPP thicknesses, and MPP hole densities. The study indicates that as panel thickness increases, the peak of the sound absorption coefficient is shifted to a lower frequency, resulting in enhanced sound insulation. When MPP hole density decreases, the number of MPP holes decreases, and the absorption coefficient and sound insulation increase. Furthermore, TPU panels have a lower sound absorption frequency than PLA and slightly higher insulation than PLA. As a result, material flexibility does not significantly affect the maximum sound absorption coefficient and sound transmission loss; more flexible panel material shifts the sound absorption coefficient peak to a lower frequency.

Keywords Sound absorption · Sound transmission loss · Micro-perforated panel · Honeycomb · Impedance tube

M. Rusli (✉) · Afdhilla · D. A. Saputra · Gusriwandi · M. Bur
Mechanical Engineering Department, Faculty of Engineering, Andalas University, Kampus Limau Manis, Padang 25163, Indonesia
e-mail: meifal@eng.unand.ac.id

D. A. Saputra
e-mail: Dendias@eng.unand.ac.id

Gusriwandi
e-mail: gusriwandi@eng.unand.ac.id

M. Bur
e-mail: mulyadibur@eng.unand.ac.id

Irwansyah et al. (eds.), *Proceedings of the 4th International Conference on Experimental and Computational Mechanics in Engineering*, Lecture Notes in Mechanical Engineering, https://doi.org/10.1007/978-981-99-7495-5_50

1 Introduction

In the modern world, the noise attenuation is becoming ever more crucial. In terms of annoyance, as well as the quality of life and health, it is recognized as one of the pollutants. By taking into account all potential noise generations during the product design process, noise can initially be reduced at its source. Unfortunately, this design procedure has not thoroughly addressed the noise source. In this situation, reducing noise on the transmission path and the receiver side is one choice. Therefore, it is crucial for noise mitigation to develop efficient acoustic or sound absorption materials. Therefore, it is necessary for noise mitigation to develop efficient acoustic or sound absorption materials. Although sound insulation and sound absorption are frequently used interchangeably in common speech, they are fundamentally different. By decreasing reflections and converting absorbed energy into heat, sound-absorbing materials are intended to enhance sound quality, avoid echoes, and minimize unpleasant reverberation inside a space. On the other hand, sound insulation or soundproofing materials serve as an acoustic barrier to the sound source by preventing sound waves from entering or leaving a place [1, 2].

There are now many different sound-absorbing materials that have been developed. Porous material and micro-perforated panel (MPP) are recent technologies frequently used and developed as sound absorber panels [3, 4]. Porous materials have interconnected pores that allow sound waves to move through. Frictional heating between vibrating air molecules and the pore surface primarily dissipates sound energy. The three main types of microscopic structures seen in porous absorptive material are fibrous, cellular, and granular. Fibrous materials that are frequently used in this application include rock wool, glass wool, polyester, and polyurethane. However, their manufacturing and application processes negatively affect human and environmental health [5]. Therefore, some researchers focused on developing the use of natural fibrous materials as sound absorbents, which are biodegradable, naturally renewable, and environmentally friendly [6, 7].

Foams and other porous materials are very simple to process, inexpensive, and capable of absorbing a wide range of sounds [8]. However, they often have poor sound absorption characteristics due to a significant impedance mismatch, particularly at low frequencies (100 to 800 Hz) [9, 10]. In order to improve absorption capabilities at low frequency, acoustics metamaterials such as resonator-type materials like micro-perforated panels (MPPs) can be used as standalone materials or as facings for porous materials [11].

Additionally, a great deal of research is being done on MPPs, which are made up of many small Helmholtz resonators and are effective at absorbing sound energy. The sound energy is converted into heat as it enters an MPP because of friction between the perforated surfaces and the moving air molecules. MPP was first suggested as an alternative to porous sound absorbers, which continue to present dangers to human health and the environment [12]. Despite the resonant frequency range, MPPs continue to function, albeit with minor absorption [13].

Table 1 Mechanical properties of thermoplastic polyurethane and polylactic acid [28]

Mechanical properties	Thermoplastic polyurethane (TPU)	Polylactic acid (PLA)
Hardness	60 (Shore D)	83 (Shore D)
Tensile modulus	53.7 MPa	2346 MPa
Elongation at break	318%	5.2%

To increase the frequency band and sound absorption coefficient of MPP, many researchers have developed various techniques, including inserting a mechanical impedance plate behind the panel [14], dividing different depth cavities [15], combining a membrane cell and mass blocks composite structure [16], and multi-layering elastic MPP plates [17] and a wideband triple-layer micro-perforated panel [18]. Additionally, other researchers suggested that the MPP may be improved by using composite materials, such as wooden panels [19], kenaf and polylactic acid [20], and sandwich panel MPP and natural fiber [21].

Moreover, honeycomb backing and wall partition in the backing air cavity are two commonly used methods to improve the acoustic properties of MPP absorbers [22, 23]. Each air cavity subdivision can achieve the most efficient Helmholtz-type resonance absorption in both scenarios, providing a local one-dimensional sound field and normal incidence into the apertures [24, 25], for example, subdividing the air-back cavity to improve the sound absorption and transmission loss [26] and designing a honeycomb-backed MPP with tailored sound absorption frequencies by adding adjustable perforation baffles [27].

Now, the sound absorption characteristic and sound transmission loss of honeycomb-baked MPP made from two different flexibility materials are investigated experimentally in this paper. The panels are made of thermoplastic polyurethane (TPU), a much more flexible material than polylactic acid (PLA) as the second material (see Table 1). Three-dimensional printing is used to create the panels. This research aims to compare the effect of material flexibility on the acoustic properties of a panel. The acoustics are measured using an ASTM E2611-19-compliant four-channel impedance tube. Panel thicknesses, honeycomb wall thicknesses, and MPP hole densities have varied across the panels.

2 Methodology

2.1 *Material and Specimens*

In this study, the panels are measured and analyzed in circular form with a diameter of 51.2 mm based on the inner diameter of the impedance tube. The specimens' panel thicknesses, honeycomb wall thicknesses, and MPP hole densities are varied. The panels are produced by 3D printing and made of two kinds of material filament: thermoplastic polyurethane (TPU) and polylactic acid (PLA). Table 1 shows the

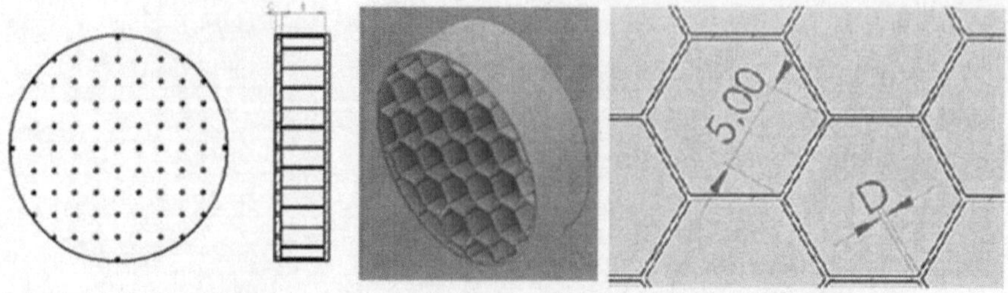

Fig. 1 Specimens' model

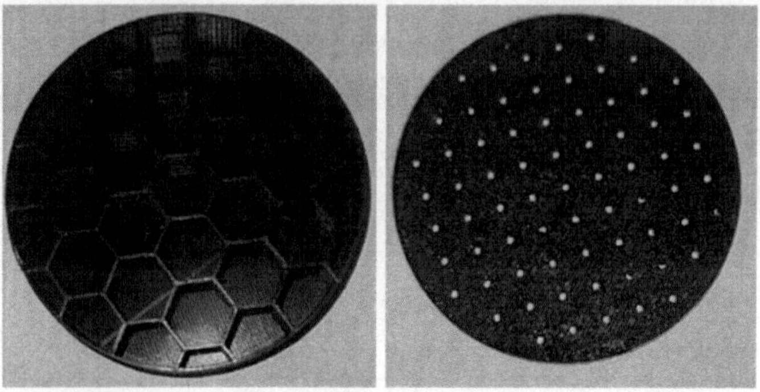

Fig. 2 Sample specimen

mechanical properties of both materials. The TPU has much more flexibility than the PLA, showing a very low tensile modulus and higher elongation at break. The hole diameters of the MPPs are 0.8 mm, and the distance between the holes ranges from 5 to 10 mm. The honeycomb-backed panels' thickness are varied by 10 mm, 15 mm, and 20 mm. Moreover, the honeycomb walls vary by 0.3 mm, 0.7 mm, and 1.0 mm. Figure 1 depicts the panel design.

The Anet A8 3D printing machine, purchased from Shenzen Anet Technology in China, is used for printing. The temperature and printing speed are kept constant at 200 °C and 1440 m/min, respectively, with a 0.1 mm resolution. The printer bed and room temperatures are kept at 60 and 30 degrees Celsius, respectively. Figure 2 depicts the specimens.

2.2 Experimental Setup

A four-channel impedance tube, made of aluminum pipe with a 51.2 mm diameter, is used to measure the acoustics characteristics of the panels within the frequency range from 400 to 3800 Hz. The impedance tube is built in accordance with ASTM E2611-19 for normal incidence of material acoustics characteristics. A random noise with

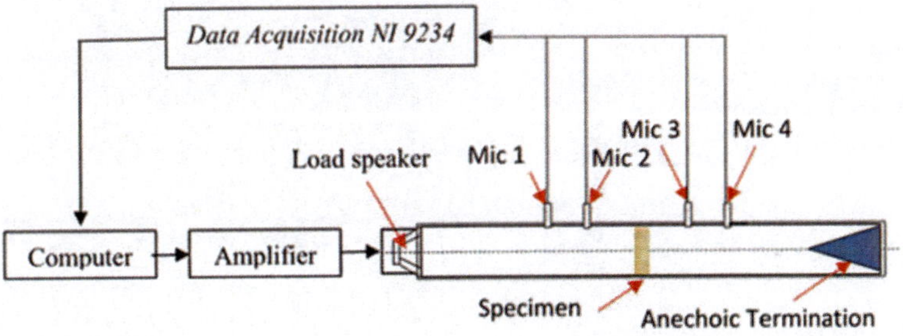

Fig. 3 Four channels' impedance tube for acoustics characteristics' measurement

a frequency range of 0–10 kHz is generated as the sound source in the experiment. Four BSWA MPA 215 microphones are used to measure the sound pressure in four channels. The measured signal from microphones is then connected to a multifunction National Instrument data acquisition card NI-9234. Labview then processes the signals to obtain the transfer function from four microphones. The transfer function generated by four microphones is used to calculate the sound absorption coefficient. Figure 3 illustrates the experimental design. Normal incidence transmission loss and sound absorption coefficient are calculated based on the standard.

3 Result and Discussion

Firstly, the effect of honeycomb panel thickness and different materials is observed. The distance between hole rows of the MPP for the specimens is uniform at 10 mm and the honeycomb wall thickness at 0.7 mm. The honeycomb-backed panels are varied in 10 mm, 15 mm, and 20 mm. The sound absorption coefficient and sound transmission loss (STL) of the specimens are depicted in Fig. 4. As observed in Fig. 4, specimens act as good absorbers with peak absorption coefficient values of almost 1 in a specific frequency, implying that the MPP holes and honeycomb cell act very well as dissipators of sound energy at a different frequency. More honeycomb-backed panel thickness means that deeper cavity cells tend to have lower frequency peaks. The cavity depth has the most significant change rate in affecting the sound absorption capability of the honeycomb-backed panel than other parameters like the plate thickness, aperture, and perforation rate [29].

Moreover, specimens made of TPU filament have lower frequency absorption peaks than stiff PLA panels. TPU has lower modulus elasticity than PLA, which affects the panel stiffness. Panels with lower modulus elasticity have lower stiffness and lower frequency resonance. Material flexibility also significantly affects sound transmission loss (STL). TPU panels tend to have better STL at a frequency lower than 1500 Hz or within 2000 Hz to 3800 Hz. The lower stiffness of TPU panels increases the STL at a lower frequency, and the higher structural damping affects the sound transmission loss at a lower frequency.

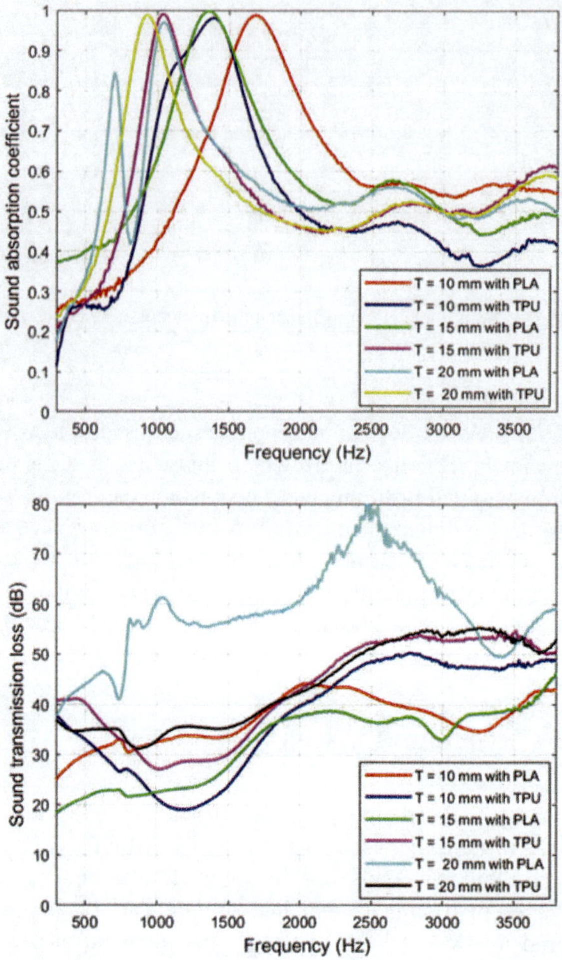

Fig. 4 Sound absorption coefficient (above) and sound transmission loss (below) with various honeycomb panel thicknesses

Next, the effect of honeycomb wall thickness and different materials on the acoustics characteristics of the panel is observed in Fig. 5. The distance between hole rows of the MPP for the specimens is uniform at 10 mm and the wall honeycomb-backed panel thickness of 15 mm. The honeycomb wall thickness is varied from 0.3 mm, 0.7 mm, and 1.0 mm. The sound absorption coefficient and sound transmission loss (STL) of the specimens are depicted in Fig. 5. As observed in Fig. 5, specimens act as good absorbers with peak absorption coefficient values of almost 1 in a specific frequency, implying that the MPP holes and honeycomb cell act very well as dissipators of sound energy at a different frequency. More honeycomb wall thickness means that more flexible cells tend to have no significant change in frequency peaks. The wall thickness has no significant change rate in affecting the sound absorption capability of the honeycomb-backed panel.

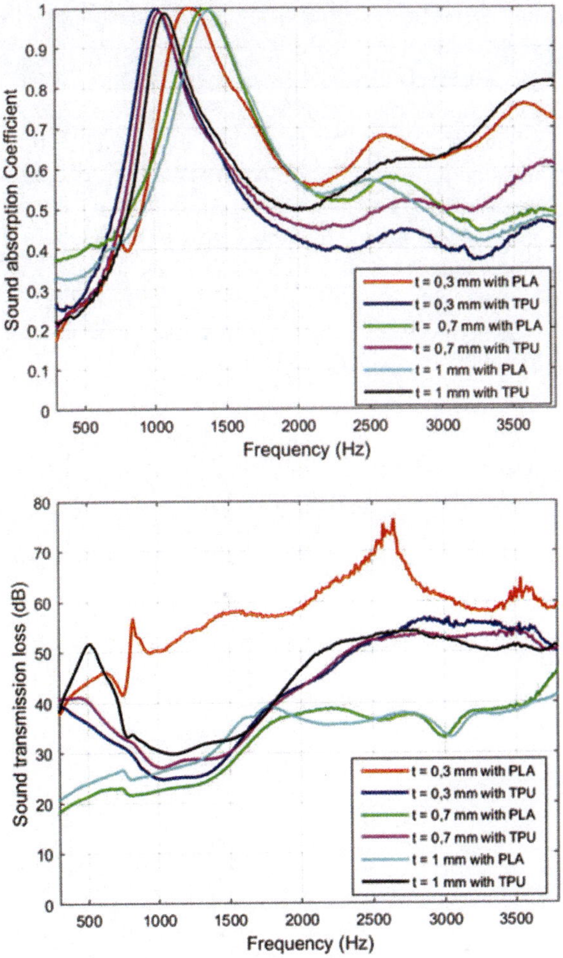

Fig. 5 Sound absorption coefficient (above) and sound transmission loss (below) with various honeycomb wall thicknesses

Furthermore, TPU filament specimens exhibit lower frequency absorption peaks than rigid PLA panels. TPU's lower modulus elasticity than PLA's influences the stiffness of the panel. Lower frequency resonance and stiffness are characteristics of panels with lower modulus elasticity. Sound transmission loss is also greatly influenced by material flexibility (STL). TPU panels often have better STL at frequencies between 1700 and 3800 Hz or less than 1000 Hz. The higher structural damping influences the sound transmission loss at a lower frequency, while the reduced stiffness of TPU panels enhances the STL at a lower frequency.

Next, the influence of MPP hole density and different materials on the acoustics characteristics of the panel is shown in Fig. 6. The distance between hole rows of the MPP for the specimens is varied at 5 mm and 10 mm. The honeycomb-backed panel thickness is 15 mm, and the honeycomb wall and MPP thickness is 0.7 mm. The sound absorption coefficient and sound transmission loss (STL) of the specimens are depicted in Fig. 6.

As observed in Fig. 6, specimens act as good absorbers with peak absorption coefficient values of almost 1 in a specific frequency in hole spacing of 10 mm, implying that the MPP holes and honeycomb cell act very well as dissipators of sound energy at a different frequency. However, MPP with a hole spacing of 5 mm shows a lower sound absorption peak but at a wider frequency band. More honeycomb wall thickness means that more flexible cells tend to have no significant change in frequency peaks. The wall thickness has no significant change rate in affecting the sound absorption capability of the honeycomb-backed panel.

Furthermore, TPU filament specimens exhibit lower frequency absorption peaks than rigid PLA panels. TPU's lower modulus elasticity than PLA's influences the

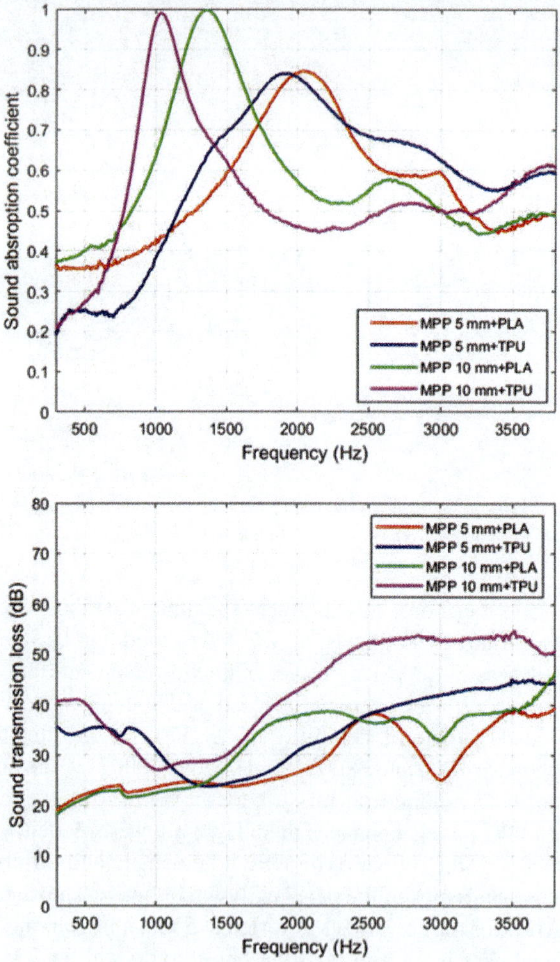

Fig. 6 Sound absorption coefficient (above) and sound transmission loss (below) with various MPP holes' density

stiffness of the panel. Lower frequency resonance and stiffness are characteristics of panels with lower modulus elasticity. Sound transmission loss (STL) is also greatly influenced by material flexibility. TPU panels often have better STL at frequencies between 1500 and 3800 Hz or less than 1200 Hz. The higher structural damping influences the sound transmission loss at a lower frequency, while the reduced stiffness of TPU panels enhances the STL at a lower frequency.

4 Conclusion

The study about the effect of different material flexibilities on acoustics characteristics of honeycomb-backed MPP indicates that as honeycomb-backed panel thickness increases, the peak of the sound absorption coefficient shifts to a lower frequency, resulting in enhanced sound insulation. When MPP hole density decreases, which means that the number of MPP holes decreases, the absorption coefficient and sound insulation increase. Furthermore, TPU panels have a lower frequency of sound absorption than PLA and slightly higher insulation than PLA. Moreover, the TPU panels often have better STL at frequencies between about 1500 Hz and 3800 Hz or less than 1200 Hz. The higher structural damping influences the sound transmission loss at a lower frequency, while the reduced stiffness of TPU panels enhances the STL at a lower frequency. As a result, the more flexible material appears to reduce sound absorption and increase sound insulation than rigid materials.

Acknowledgements The authors gratefully acknowledge the financial support from Andalas University by research grand for Percepatan Guru Besar by contract number T/9/UN.16.17/PP.PRK-PDU-KRP2GB-Unand /2022.

References

1. Benkreira H, Khan A, Horoshenkov KV (2011) Sustainable acoustic and thermal insulation materials from elastomeric waste residues. Chem Eng Sci 66:4157–4171
2. Marques DV, Barcelos RL, Silva HRT, Egert P, Parma GOC, Girotto E, Consoni D, Benavides R, Silva L, Magnago RF (2018) Recycled polyethylene terephthalate-based boards for thermal-acoustic insulation. J Clean Prod 189:251–262
3. Seddeq HS (2009) Factors influencing acoustics performance of sound absorptive materials. Aust J Basic Appl Sci 3(4):4610–4617
4. Arenas JP, Crocker MJ (2010) Recent trends in porous sound-absorbing materials, sound & vibration. 12–17
5. Asdrubali F, Schiavoni S, Horoshenkov KV (2012) A review of sustainable materials for acoustic applications. Building Acoustics 19(4):283–312
6. Berardi U, Iannace G (2015) Acoustic characterization of natural fibers for sound absorption. Build Environ 94–2:840–852
7. Rusli M, Irsyad M, Dahlan H, Gusriwandi, Bur M (2019) Sound absorption characteristics of the natural fibrous material from coconut coir, oil palm fruit bunches, and pineapple leaf. In: IOP conference series: materials science and engineering, vol 602, p 012067

8. Shen L, Zhang H, Lei Y, Chen Y, Liang M, Zou H (2021) Hierarchical pore structure based on cellulose nanofiber/melamine composite foam with enhanced sound absorption performance. Carbohydr Polym 255:117405
9. Chiang YK, Choy YS (2018) Acoustic behaviors of the microperforated panel absorber array in nonlinear regime under moderate acoustic pressure excitation. J Acoust Soc Am 143:538–549
10. Oh J-H, Lee HR, Umrao S, Kang YJ, Oh I-K (2019) Self-aligned and hierarchically porous graphene-polyurethane foams for acoustic wave absorption. Carbon 147:510–518
11. Zhang X, Qu Z, Wang H (2020) Engineering acoustic meta materials for sound absorption: from uniform to gradient structures. Science 23–5 101110
12. Maa DY (2007) Practical single MPP absorber. Int J Acoust Vib 12:3–6
13. Sakagami K, Kobatake S, Kano K, Morimoto M, Yairi M (2007) Sound absorption characteristics of a single microperforated panel absorber backed by a porous absorbent layer. Acoust Aust 39:95–100
14. Xiaodan Z, Xiangqian F (2015) Enhancing low frequency sound absorption of micro-perforated panel absorbers by using mechanical impedance plates. Appl Acoust 88:123–128
15. Guo W, Min H (2015) A compound micro-perforated panel sound absorber with partitioned cavities of different depths. Energy Procedia 78:1617–1622
16. Gai XL, Xing T, Li XH, Zhang B, Wang F, Cai ZN (2018) Sound absorption properties of microperforated panel with membrane cell and mass blocks composite structure. Appl Acoust 137:98–107
17. Kim HS, Ma PS, Kim SR, Lee SH, Seo YH (2018) A model for the sound absorption coefficient of multi-layered elastic micro-perforated plates. J Sound Vib 430:75–92
18. Cobo P, de la Colina C, Roibás-Millán E, Chimeno M, Simón F (2019) A wideband triple-layer microperforated panel sound absorber. Compos Struct 226:111226
19. Song B, Peng L, Fu F, Liu M, Zhang H (2016) Experimental and theoretical analysis of sound absorption properties of finely perforated wooden panels. Materials 9:942
20. Chin DDVS, Yahyaa MN, Dinb MN, Onga P (2018) Acoustic properties of biodegradable composite micro-perforated panel (BC-MPP) made from kenaf fibre and polylactic acid (PLA). Appl Acoust 138:179–187
21. Rusli M, Nanda RS, Dahlan H, Bur M (2020) Sound absorption characteristics of sandwich panel made from double leaf micro-perforated panel and natural fiber. IOP Conf Ser Mater Sci Eng 815:012011
22. Toyoda M, Sakagami K, Takahashi D, Morimoto M (2011) Effect of a honeycomb on the sound absorption characteristics of panel-type absorbers. Appl Acoust 72:943–948
23. Peng X, Ji J, Jing Y (2018) Composite honeycomb metasurface panel for broadband sound absorption. J Acoust Soc Am 144:EL255–EL261
24. Tan WH, Ripin ZM (2016) Optimization of double-layered micro-perforated panels with vibroacoustic effect. J Brazilian Soc Mech Sci Eng 38:745–760
25. Park SH (2013) Acoustic properties of micro-perforated panel absorbers backed by Helmholtz resonators for the improvement of low-frequency sound absorption. J Sound Vib 332:4 895–4 911
26. Carbajo J, Ramis J, Godinho L, Amado-Mendes P (2019) Perforated panel absorbers with micro-perforated partitions. Appl Acoust 149:108–113
27. Zhang X, Wu J, Mao Q, Zhou W, Xiong Y (2020) Design of a honeycomb-microperforated panel with an adjustable sound absorption frequency. Appl Acoust 164:107246
28. Brancewicz-Steinmetz E, Sawicki J, Byczkowska P (2021) The influence of 3D printing parameters on adhesion between polylactic acid (PLA) and thermoplastic polyurethane (TPU). Materials 14:6464
29. Yan S, Wu J, Chen J, Xiong Y, Mao Q, Zhang X (2022) Optimization design and analysis of honeycomb micro-perforated plate broadband sound absorber. Appl Acoust 186:108487

Production of Cellulose Nanocrystal (CNC) Combine with Silane Treatment from Pennisetum Purpureum via Acid Hydrolysis

Ritzky Fachri, Samsul Rizal, Syifaul Huzni, Ikramullah Ikramullah, and Sri Aprilia

Abstract Cellulose nanocrystals, a class of fascinating bio-based nanoscale materials, have received a tremendous amount of interest both in industry and academia owing to their unique structural features and impressive physicochemical. This nanomaterial is a promising candidate for applications in fields such as biomedical, pharmaceuticals, electronics, barrier films, nanocomposites, membranes, supercapacitors. Nanocellulose has the disadvantage of being hydrophilic. The purpose of this study was to characterize nanocellulose from Pennisetum Purpureum fiber that had been given silane treatment. The isolation and characterization of nanocellulose from *Pennisetum Purpureum* were carried out. Chemical pretreatment has included alkali treatment which used 10% NaOH for 35 min and 60 °C temperature. Bleaching process with 12% NaOCL for 24 h and acid hydrolysis with H_2SO_4, 45 °C, 60 min will disperse amorphous part of cellulose. The silane treatment was carried out by soaking and heating the fibers at a temperature of 50 °C for 2 h using APTES (3-aminopropyl-triethoxy) and methanol with a ratio of 9:1. The mixture was centrifuged until neutral pH was obtained. The obtained suspension was sonicated (Branson Sonifier 450) for 5 min with an 800 W and freeze-dried to solid NCC. Based on the fiber characterization, alkali and bleaching pretreatment conditions showed that the surface impurities on the fiber were removed, and peak at 1058 cm^{-1} for C–O–C pyranose ring vibrations indicates the extensive cellulose exposure. The most significant finding from moisture content on dried samples after dialysis showed that nanocellulose only contains ± 1% moisture. These results indicate that *Pennisetum Purpureum* could become a viable source of commercially valuable nanocellulose.

R. Fachri · S. Rizal (✉) · S. Huzni · I. Ikramullah
Department of Mechanical and Industrial Engineering, Universitas Syiah Kuala, Banda Aceh 23111, Indonesia
e-mail: samsul.rizal@unsyiah.ac.id

S. Aprilia
Department of Chemical Engineering, Universitas Syiah Kuala, Banda Aceh 23111, Indonesia

Irwansyah et al. (eds.), *Proceedings of the 4th International Conference on Experimental and Computational Mechanics in Engineering*, Lecture Notes in Mechanical Engineering, https://doi.org/10.1007/978-981-99-7495-5_51

Keywords Characterization · Pennisetum Purpureum · Cellulose Nano Crystal · Silane treatment

1 Introduction

The current growth in environmental awareness has generated increasing interest in the use of natural fibers as alternative reinforcement materials for polymer composites. This is largely owing to their low environmental impact, low cost, and relatively good specific properties. Scientists have been striving to develop biodegradable composites using renewable agro-based materials [1]. Natural fibers derived from plants demonstrate great potentials for use in plastic, automotive, and packaging industries because of their excellent characteristics such as low density, high specific stiffness, good mechanical properties, biodegradability, eco-friendliness, toxicologically harmless, good thermal and acoustic insulation [2, 3]. In addition, these cellulosic fibers can reduce the overall material costs than the starting polymer [4]. Comprehensive reviews conducted by numbers of publications [5–10] have outlined the differences of natural fibers with regard to their mechanical properties and its applications. Several authors documented the use of natural fibers such as bamboo [10], flax [11], coir [12], arundo donax (giant reed) [13], okra [14], jute [15–17], wheat straw [18, 19], and alfa [20] as reinforcement in composite materials.

However, there are concerns regarding the attributes of natural fibers such as their hydrophilic nature, high moisture absorption, poor reactivity, and poor compatibility with polymeric matrices, all of which influence their mechanical properties [21–23]. The hydrophilic nature of natural fiber is known to produce weak interfacial adhesion in polymer–matrix composites [24]. The type of natural fiber can also affect the biological performance of the composites, for example, a composite manufactured from abaca fiber has a much greater moisture content compared to flax-reinforced composites [25]. These problems can be rectified through modification such as alkali treatment to enhance the interfacial adhesion between natural fibers and composite matrices, in addition to enhancing the mechanical, physical, and thermal properties of the fibers [26]. Other modifications during acetylation can modify the surface of the fibers and enhance their hydrophobicity [25].

Pennisetum Purpureum fiber, also locally known as Napier grass, is composed of 46% cellulose, 34% hemicellulose, and 20% lignin. These values will depend on the variety and environmental factors where the plant is grown [27]. The purpose of the alkali treatment is to remove the hemicelluloses, split the fibers in the fibrils, and produce a closely packed cellulose chain owing to the release of the internal strain, which consequently improves the mechanical properties of the fiber [28]. Following the alkali treatment, the fibrillation of the fibers also increases the effective surface area available for wetting by the resin and enhances the bonding between the fiber–matrix interfaces within the polymer composites. The alkali treatment also breaks the hydrogen bonds and increases the number of free hydroxyl groups of the fiber, thus increasing the fiber reactivity [29]. Girisha et al. also studied experimentally sisal fiber which was treated with alkali, silane treatment, and combination of alkali + silane treatment. The results showed that the composite with sisal fiber that was given a combination of alkali + silane treatment had the highest tensile value and better water resistance than only given alkaline treatment or silane treatment alone [24]. To improve the mechanical properties of Pennisetum Purpureum nanocellulose fiber, it is necessary to conduct research to determine the impact of silane treatment on Pennisetum Purpureum cellulose nanocrystal. The aim of this study is to examine experimentally the effect silane treatment to cellulose nanocrystal water content. nano nano cellulose on Pennisetum Purpureum fiber. Liu et al., Rao et al., and Thakur et al. all demonstrated that the natural fibers exhibit great potential for use as an alternative to artificial glass and carbon fibers during the production of thermosetting or thermoplastic composites [30–34].

2 Materials and Methods

The Pennisetum Purpureum grass used was supplied by a farm located in Pango, Banda Aceh, Indonesia. Three to six months' old with a length of 40–60 cm Pennisetum Purpureum fiber was used for this research. Pennisetum Purpureum was used in this study because in Aceh only the leaves of Pennisetum Purpureum are used as animal feed for cows and goats, but the stems are discarded and burned because they are considered waste. Figure 1 shows the Pennisetum Purpureum plant.

Fig. 1 Pennisetum Purpureum

3 Preparation of Cellulose Fibers from Pennisetum Purpureum

Fiber stems of Pennisetum Purpureum with a length of 40–60 cm are separated from the leaves with knife, the fiber chopped into horizontal and dried under the sunray for 40 h until the fiber dried reach a constant weight.

The cellulose was isolated based on previously published techniques with certain modifications [34]. They were treated with 10% aqueous NaOH solution for 45 min at 60 °C in boiling three neck to remove the hemicellulose, lignin, and surface impurities and other greasy materials [35]. The suspension of alkali fiber was washed until pH 7 and dried in furnace with temperature of 60 °C for 24 h.

Bleaching Process. The fiber was broken up in a chopping machine for 10 min soaked into NaOCl 12% for 24 h. The suspension of bleached fiber was washed until it reached pH 7 and dried in furnace with temperature of 60 °C for 24 h.

Acid Hydrolysis. The suspension was hydrolyzed with 64% H_2SO_4 for 60 min at 45 °C, and then, the suspension is given dialysis treatment to transfer dissolved molecules from a solution mixture that occurs due to diffusion on a semi-permeable membrane.

The silane treatment was carried out by soaking and heating the fibers at a temperature of 50 °C for 2 h using 3-aminopropyl-triethoxy (APTES) and methanol with a ratio of 9:1. The mixture was centrifuged until neutral pH was obtained. The obtained suspension was sonicated (Branson Sonifier 450) for 5 min with an 800 W and freeze-dried to solid NCC. Figure 2 shows schematic representation of cellulose nanocrystal production.

4 Characterization of Pennisetum Purpureum Fiber Fibers

4.1 Fourier Transform Infrared

FTIR characterization was performed by using a PerkinElmer FTIR spectrometer Frontier instrument (PerkinElmer, Inc., Waltham, MA, USA) with wave numbers of 0–4000 cm^{-1}. Before characterization, all samples were dried in an oven, drying at 60 ◦C until a constant weight was achieved.

4.2 Water Content

Measurement of water content of cellulose nanocrystal using A.O.A.C method. The first step, dry the empty dish and lid in the oven at 105°C for 3 h and transfer to desiccator to cool. Weigh the empty the dish and lid. Weigh about 3 g of sample to

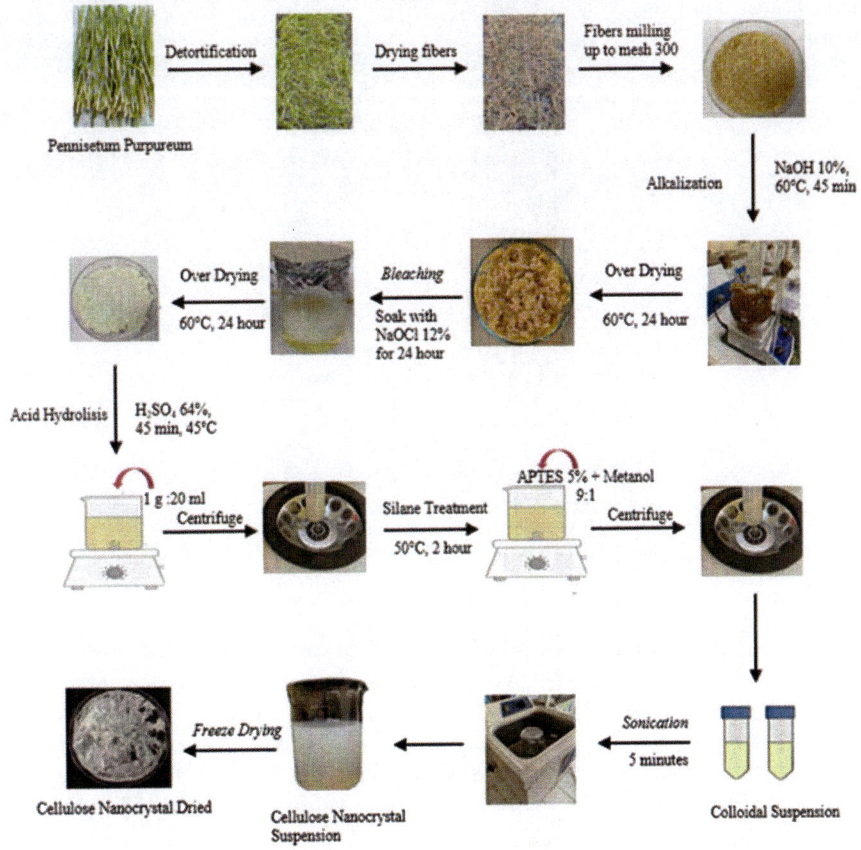

Fig. 2 Schematic representation of cellulose nanocrystal production

the dish. Spread the sample with spatula. Then, place the dish with sample in the oven. Dry for 3 h at 105°C. After drying, transfer the dish with partially covered lid to the desiccator to cool. Reweigh the dish and its dried sample.

5 Results

5.1 Fourier Transform Infrared

The FTIR spectrum of a Cellulose Nanocrystal (CNC) sample is illustrated in Fig. 3. The peak of 3334 cm^{-1} showed that crystalline cellulose was not disturbed with H2SO4 concentration of 64%. The peak of 2900 cm^{-1} indicates C-H stretching vibrations in the cellulose component associated with lignin particles. The band at

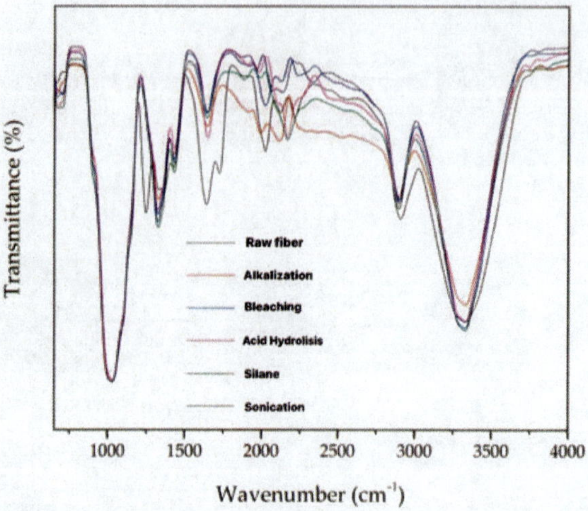

Fig. 3 Fourier Transform Infrared spectra cellulose nanocrystal (CNC)

1640 cm^{-1} is ascribed to O–H bending, because of the absorbed water found in Napier fibers, the strong interaction of cellulose and water makes water removal difficult.

The range of peaks at 1500–1600 cm^{-1} is defined as protein. The absorbance band at 1428 cm^{-1} is associated with CH2 bending, which is a characteristic of polysaccharides [25, 36, 37]. Moreover, the peak at 1428 cm^{-1} represents the C–C stretch [40]. The band located at the peak of 1370 cm^{-1} represents the angular symmetrical deformation of the C–H bond, which can be found in all three samples and is due to the rearrangement of the crystal lattice. A prominent peak at 1110 cm^{-1} was identified as for C–OH stretching and one at 1058 cm^{-1} for C–O–C pyranose ring vibrations which became narrower and sharper due to extensive cellulose exposure. The band at 1030 cm^{-1} correlates with C–O stretching. Finally, for the peak at 896 cm^{-1}, the spectrum has correlated with the B-glycosidic relationship between cellulose glucose units. The characteristic peak of lignin's aromatic ring vibration presents at wavenumber 1200–1300 cm^{-1}. This peak disappeared after chemical and mechanical treatments. This is due to the removal of lignin during the pulping and bleaching [7, 12, 28]. The OH group water absorption peak at around wavenumber 1640 cm^{-1} appears in all samples as expected from other studies [7, 28].

5.2 *Water Content*

The water content of cellulose in dry samples after centrifugation is ± 7%, Crystalline Nanocellulose (CNC) has a water content percentage of ± 13 to 14%. The silane treatment given to nanocellulose was able to reduce the water content of the cellulose to ± 1%. The silane treatment given was able to reduce the water content of the

Table 1 Water content on chemical modification

Sample	Water content after centrifugation (%)	Water content on dried sample (%)
Raw fiber	–	10
Cellulose	97.9	7.3
Nano cellulose	98.7	13.8
Nano cellulose with silane treatment	97.4	1.1

cellulose nanocrystals. Water content after chemical modification can be seen on Table 1.

6 Conclusions

Pennisetum Purpureum nanocellulose has been successfully fabricated combined with silane treatment. Chemical modification is able to remove lignin and hemicellulose in the fiber. This is evidenced by the reduction of the absorbance of the wavelength in the amorphous portion of the fiber. Chemical modification of nanocrystalline cellulose using silane treatment can reduce the water content. The water content of nanocrystal cellulose after silane treatment is ± 1%. This reduction is very significant compared to untreated fiber, which is ± 10%. These properties indicate that nanocrystalline cellulose from Pennisetum Purpureum is an effective reinforcing material for biocomposites.

References

1. Kommula V, Kanchireddy O, Shukla M, Marwala T (2014) Effect of acid treatment on the chemical, structural, thermal and tensile properties of napier grass fibre strands, psrcentre.org., pp 45–51
2. Mohanty AK, Wibowo A, Misra M, Drzal LT (2004) Effect of process engineering on the performance of natural fiber reinforced cellulose acetate biocomposites. Composites: Part A. 35:363–370
3. Thakur VK, Thakur MK, Gupta RK (2013) Synthesis of lignocellulosic polymer with improved chemical resistance through free radical polymerization. Int J Biol Macromol 61:121–126
4. Thakur VK, Thakur MK, Gupta RK (2013) Graft copolymers from cellulose: synthesis, characterization and evaluation. Carbohydr Polym 97:18–25
5. Thakur VK, Thakur MK (2014) Processing and characterization of natural cellulose fibers/thermoset polymer composites. Carbohyd Polym 109:102–117
6. Bongarde US, Shinde VD (2014) Review on natural fiber reinforcement polymer composites. IJESIT 3:431–436
7. Thakur VK, Thakur MK, Gupta RK (2014) Review: raw natural fiber-based polymer composites. Int J Polym Anal Charact 19:256–271

8. Thakur VK, Thakur MK (2015) Recent advances in green hydrogels from lignin: a review. Int J Biol Macromol 72:834–847
9. He L, Li W, Chen D, Zhou D, Lu G, Yuan J (2015) Effects of amino silicone oil modification on properties of ramie fiber and ramie fiber/polypropylene composites. Mater Des 77:142–148
10. Wang F, Shao J, Keer LM, Li L, Zhang J (2015) The effect of elementary fibre variability on bamboo fibre strength. Mater Des 75:136–142 (Accepted Manuscript)
11. Moothoo J, Allaoui S, Ouagne P, Soulat D (2014) A study of the tensile behaviour of flax tows and their potential for composite processing. Mater Des 55:764–772
12. Varma IK, Varma DS, Varma M (1986) Thermal behaviour of coir fibres. Thermochim Acta 108:199–210
13. Fiore V, Scalici T, Valenza S (2014) Characterization of a new natural fiber from Arundo donax L. as potential reinforcement of polymer composites. Carbohydr Polym 106:77–83
14. De Rosa IM, Kenny JM, Puglia D, Santulli C, Sarasini F (2010) Morphological, thermal and mechanical characterization of okra (Abelmoschus esculentus) fibres as potential reinforcement in polymer composites. Compos Sci Technol 70:116–122
15. Prashant K (1986) Mechanical behavior of jute fibers and their composites, Indian. J Technol 24:29–32
16. Pal PK (1984) Jute reinforced plastics: a low cost composite material. Plast Rubber Process Appl 4:215–219
17. Mwaikambo LY, Ansell MP (2002) Chemical modification of hemp, sisal, jute, and kapok fibers by alkalization. J Appl Polym Sci 84:2222–2234
18. Lawther JM, Sun R, Banks WB (1996) Fractional characterization of alkali-labile lignin and alkali-insoluble lignin from wheat straw. Ind Crops Prod 5:291–300
19. Sun R, Lawther MJ, Banks WB (1996) Fractional and structural characterization of wheatstraw hemicelluloses. Carbohydr Polym 29:325–331
20. Paiva MC, Ammar I, Campos AR, Cheikh RB, Cunha AM (2007) Alfa fibres: Mechanical, morphological and interfacial characterization. Compos Sci Technol 67:1132–1138
21. Indran S, Raj RE, Sreenivasan VS (2014) Characterization of new natural cellulosic fiber from Cissus quadrangularis root. Carbohydr Polym 110:423–429
22. Obi Reddy K, Uma Maheswari C, Shukla M, Song JI, Varada Rajulu A (2013) Tensile and structural characterization of alkali treated Borassus fruit fine fibers. Compos Part B Eng 44:433–438
23. Thakur MK, Gupta RK, Thakur VK (2014) Surface modification of cellulose using silane coupling agent. Carbohydr Polym 111:849–855
24. Girisha C, Srinivas GR (2012) Sisal/coconut coir natural fibers—epoxy composites: water absorption and mechanical properties. Int J Eng Innovative Technol (IJEIT) 2:166–170
25. Faruk O, Bledzki AK, Fink HP, Sain M (2012) Biocomposites reinforced with natural
26. Li X, Tabil L, Panigrahi S (2007) Chemical treatments of natural fiber for use in natural fiber-reinforced composites: a review. J Polym Environ 15:25–33
27. Reddy KO, Maheswari CU, Shukla M, Rajulu AV (2012) Chemical composition and structural characterization of Napier grass fibers. Mater Lett 67:35–38
28. Bledzki AK, Gassan J (1999) Composites reinforced with cellulose based fibres. Prog Polym Sci 24:221–274
29. Ray RD, Sarkar BK, Basak RK, Rana AK (2002) Study of the thermal behavior of alkali-treated jute fibers. J Appl Polym Sci 85:2594–2599
30. Reddy KO, Maheswari CU, Reddy DJP, Rajulu AV (2009) Thermal properties of Napier grass fibers. Mater Lett 63:2390–2392
31. Murali Mohan Rao K, Mohana Rao K, Ratna Prasad AV (2010) Fabrication and testing of natural fibre composites: Vakka, sisal, bamboo and banana. Mater Des 31:508–513
32. Liu Z, Erhan SZ, Akin DE, Barton FE (2006) "Green" composites from renewable resources: preparation of epoxidized soybean oil and flax fiber composites. J Agric Food Chem 54:2134–2137
33. Thakur VK, Thakur MK, Gupta RK (2013) Rapid synthesis of graft copolymers from natural cellulose fibers. Carbohydr Polym 98:820–828

34. Candra CSJ, George N, Narayanankutty SK (2016) Isolation and characterization of cellulose nanofibrils from arecanut husk fibre. Carbohydr Polym 142:158–166
35. Ray D, Sarkar BK, Rana AK, Bose NR (2001) Mechanical properties of vinylester resin matrix composites reinforced with alkali-treated jute fibres. Compos Part A Appl Sci Manuf 32:119–127
36. Zuluaga R, Putaux JL, Cruz J, Vélez J, Mondragon I, Gañán P (2009) Cellulose microfibrils from banana rachis: effect of alkaline treatments on structural and morphological features. Carbohydr Polym 76:51–59
37. Gu H (2009) Tensile behaviours of the coir fibre and related composites after NaOH treatment. Mater Des 30:3931–3934

Experimental Study of the Effect of Straight-Section Length on Serpentine Tube Heat Exchanger with Sharp Turns

Dedi Afandi, Ahmad Syuhada, Ratna Sary, and Muhammad Hafiz

Abstract Compact-type heat exchangers are widely used in industry because of their small dimensions and lightweight with high heat transfer rates. This study aims to examine the effect of straight-section length on a serpentine tube heat exchanger with sharp turns. The heat exchanger used is made of galvanized pipe with an inner diameter of 20 mm and an outer diameter of 22 mm. The total length of the heat exchanger is 6 m with variations in the length of the straight section, namely 60, 45, 30, and 15 cm. Water with a temperature of 80 °C is flowed into the heat exchanger using a pump that is integrated with the heater. The flow of water entering the heat exchanger is 0.4 L/s. Water at room temperature is used as a cooling medium for the heat exchanger. The inlet and outlet temperatures of the heat exchanger were measured using a thermocouple, and the cooling water temperature was also measured using a thermocouple. The results showed that a heat exchanger with a straight-section length of 30 cm had the highest heat transfer rate and heat transfer coefficient compared to other heat exchangers. The rate of heat transfer in a heat exchanger with a straight-section length of 30 cm is 15.9% higher than that of a heat exchanger with a straight-section length of 15 cm, 56% higher than the heat exchanger with a straight-section length of 45 cm, and 78.5% higher than the heat exchanger with a straight-section length of 60 cm. The highest convection heat transfer coefficient is obtained in a heat exchanger with a straight-section length of 30 cm, and the average heat transfer coefficient is 1168.5 kW/m^2·K.

Keywords Straight-section length · Serpentine tube · Sharp turns · Heat transfer rate

D. Afandi (✉) · A. Syuhada · R. Sary · M. Hafiz
Department of Mechanical Engineering, Universitas Syiah Kuala, Banda Aceh 23111, Indonesia
e-mail: dediafandi@unsyiah.ac.id

A. Syuhada
e-mail: ahmadsyuhada@unsyiah.ac.id

R. Sary
e-mail: ratnasary_ft@unsyiah.ac.id

Irwansyah et al. (eds.), *Proceedings of the 4th International Conference on Experimental and Computational Mechanics in Engineering*, Lecture Notes in Mechanical Engineering, https://doi.org/10.1007/978-981-99-7495-5_52

1 Introduction

A heat exchanger is a device used to transfer heat from one fluid to another at a different temperature. Heat exchangers are widely used in industries that produce heat in the process, where the heat needs to be removed because it can damage the equipment used [1]. One type of heat exchanger that is widely used in industry is a compact heat exchanger because it has smaller dimensions with a large heat transfer area, so it does not require a large space [2]. The compact heat exchanger has a characteristic ratio of heat transfer surface area to the volume of the heat exchanger greater than 700 m^2/m^3 [3].

Many studies have been carried out to optimize the performance of heat exchangers, including by increasing the heat transfer area [4, 5]. The performance of the heat exchanger can also be improved by increasing the turbulence of the fluid flow [6, 7]. The addition of fins on the outside of the heat exchanger is also able to improve the performance of the heat exchanger [8–12]. Vortex generators can optimize the performance of compact heat exchangers by making the fluid flow more turbulent [13, 14].

M. Awais et al. conducted a study on the effect of straight-section length on a serpentine tube-type heat exchanger on the performance of the heat exchanger. In this study, a heat exchanger with four variations of straight-section length serpentine tube was used, namely uniform serpentine tube, low–high serpentine tube, low–high–low serpentine tube, and high–low–high serpentine tube. The results show that the low–high serpentine tube heat exchanger has the most optimal performance compared to other variations [15].

Based on previous research, this research was conducted to examine the effect of straight-section length on a serpentine tube heat exchanger with sharp turns. Sharp turns are used to optimize the flow turbulence that occurs along the heat exchanger, although it also causes a pressure drop. In this study, four variations of the length of the straight-section were used, namely 60, 45, 30, and 15 cm with a total length of the heat exchanger is 6 m. The aims of this study are to obtain the most optimal straight-section length of the serpentine tube heat exchanger, which has high flow turbulence with low-pressure drop, so that the heat exchanger has optimal performance in removing unnecessary heat energy.

2 Experimental Setup

This research uses a compact serpentine tube heat exchanger with sharp turns made of galvanized pipe. The inside diameter of the heat exchanger tube is 0.02 m, and the outside diameter is 0.022 m. The total length of the heat exchanger is 6 m with four variations of straight-section length, namely 0.6, 0.45, 0.30, and 0.15 m. The configuration of the heat exchanger is shown in Fig. 1. The temperature of the water entering the heat exchanger tube is maintained at 80°C. The specifications of the heat exchanger and other parameters used in this study are shown in Table 1.

The water is heated using a heater to a temperature of 80 °C, then flowed into the heat exchanger tube with a flow rate of 0.4 L/s using a pump integrated with the heater. The heat exchanger is cooled using water in a cooling water tub, and the initial water temperature is measured. The water coming out of the heat exchanger is accommodated in a hot water tub and reheated before flowing back into the heat exchanger. Thermocouples are placed at the inlet and outlet of the heat exchanger to measure the inlet and outlet water temperatures. The thermocouple is also placed in the cooling bath to measure the temperature of the cooling water. The test schematic is shown in Fig. 2.

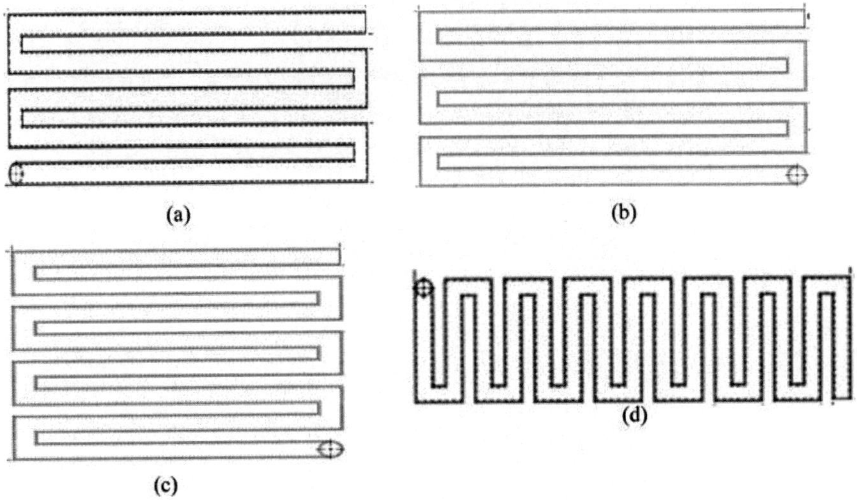

Fig. 1 Heat exchangers' configuration

Table 1 Research parameters

Parameter	Information
Straight-section length configurations	60, 45, 30, and 15 cm
Inner diameter of heat exchanger	20 mm
Outer diameter of heat exchanger	22 mm
Heat exchanger total length	6 m
Cold fluid	Water
Hot fluid	Water
Hot fluid temperature	80^0C
Cold fluid temperature	Ambient temperature (27^0C)
Water flow rate	0.4 L/s

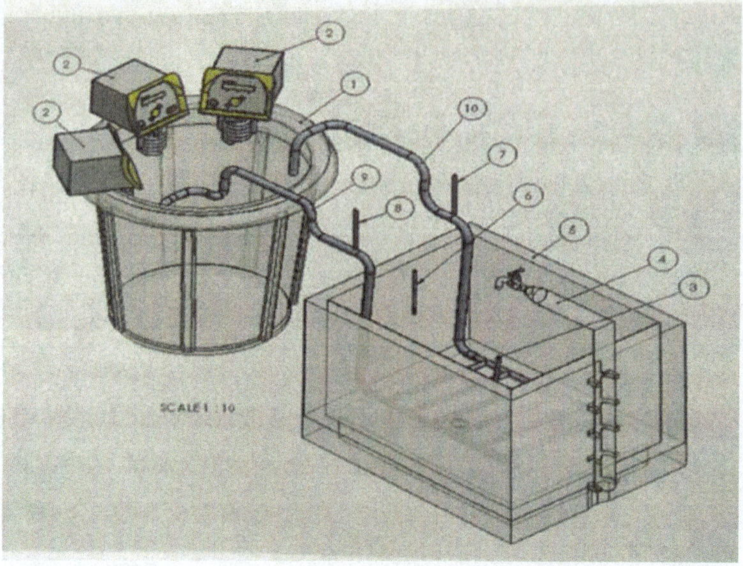

Fig. 2 Test schematic. *Description* 1. Hot water tub. 2. Heater. 3. Thermocouple for cold water. 4. Cold water supply line. 5. Cold water tub. 6. Thermocouple for cold water. 7. Thermocouple for hot water outlet. 8. Thermocouple for hot water inlet. 9. Heat exchanger inlet. 10. Heat exchanger outlet.

3 Data Reduction

The heat transfer in the heat exchanger can be calculated using the following equation:

$$\dot{Q} = \dot{m}C_{\mathrm{p}}(T_{\mathrm{i}} - T_{\mathrm{o}}), \quad (1)$$

where $\dot{Q}$ is the heat transfer rate (W), $\dot{m}$ is the mass flow rate (kg/s) of water flowing in the heat exchanger. C_{p} is the specific heat of water (J/kg · K) at film temperature ($T_f = \frac{T_i + T_o}{2}$). $T_i and T_o$ are the inlet and outlet water temperatures, respectively.

The convection heat transfer coefficient can be calculated by the following equation:

$$h = \frac{\dot{Q}}{A \cdot \Delta T_{\mathrm{m}}}, \quad (2)$$

where h is the convection heat transfer coefficient (W/m^2·K), A is heat transfer area (m^2). ΔT_{m} is the temperature difference between the surfaces of the heat exchanger (T_s) and the ambient temperature (T_∞), which can be calculated using the following equation:

$$T_s = \frac{T_{\text{in, hot}} + T_{\text{out, hot}}}{2} \text{ and } T_\infty = \frac{T_{\text{in, cold}} + T_{\text{out, cold}}}{2} \tag{3}$$

The mass flow rate of the fluid can be calculated by the following equation:

$$\dot{m} = \rho \cdot A \cdot \text{v}, \tag{4}$$

where $\dot{m}$ is the mass flow rate (kg/s), ρ is the fluid density (kg/m^3), and v is the velocity of the fluid flowing through the heat exchanger (m/s)

4 Result and Discussion

4.1 *Heat Transfer Rate*

Figure 3 shows the heat transfer rate and temperature distribution in a serpentine tube heat exchanger with straight-section lengths of 60 cm. In a heat exchanger with a straight-section length of 60 cm, the average heat transfer rate is 12.32 kW. The average difference in inlet temperature and outlet temperature in a heat exchanger with a 60 cm straight-section length is 7.52 °C. The graph trend shows the rate of heat transfer at the beginning of the test showing a significant increase and then slowly decreasing as the duration of the test increases. The longer the test, the heat transfer rate will be more constant until it reaches an equilibrium where the temperature difference no longer increases.

Figure 4 shows the rate of heat transfer in a serpentine tube heat exchanger with a straight-section length of 45 cm. The average heat transfer rate is 14.07 kW, with an

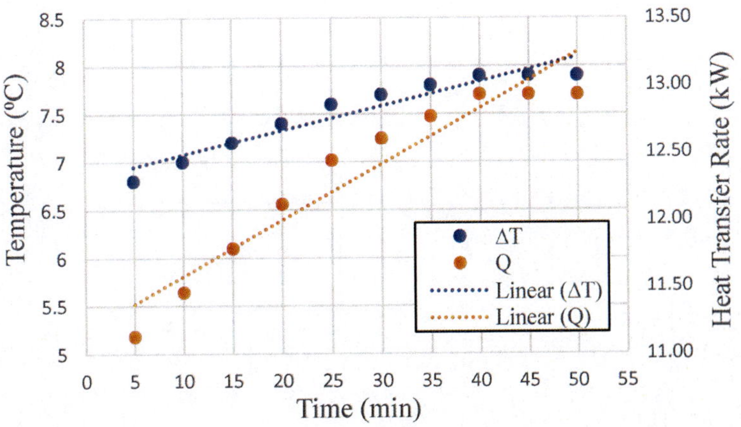

Fig. 3 Heat transfer rate and temperature distribution in a heat exchanger with a straight-section length of 60 cm

average temperature difference of 8.6 °C. The heat transfer rate in a heat exchanger with a straight-section length of 45 cm is 14% higher than that of a heat exchanger with a straight-section length of 60 cm.

Figure 5 shows the rate of heat transfer in a serpentine tube heat exchanger with a straight-section length of 30 cm. The average heat transfer rate is 21.96 kW, with an average temperature difference of 13.4°C. The heat transfer rate of a heat exchanger with a straight-section length of 30 cm is 56% higher than that of a heat exchanger with a straight-section length of 45 cm. This is because in a heat exchanger with a straight-section length of 30 cm, the turbulence of fluid flow in the tube is greater than in a heat exchanger with a straight-section length of 45 cm.

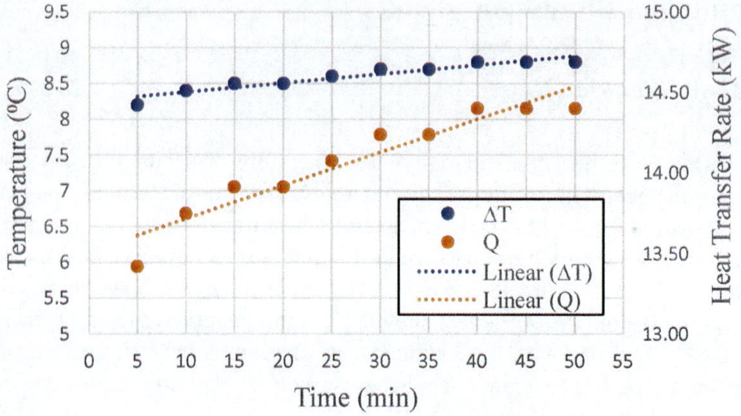

Fig. 4 Heat transfer rate and temperature distribution in a heat exchanger with a straight-section length of 45 cm

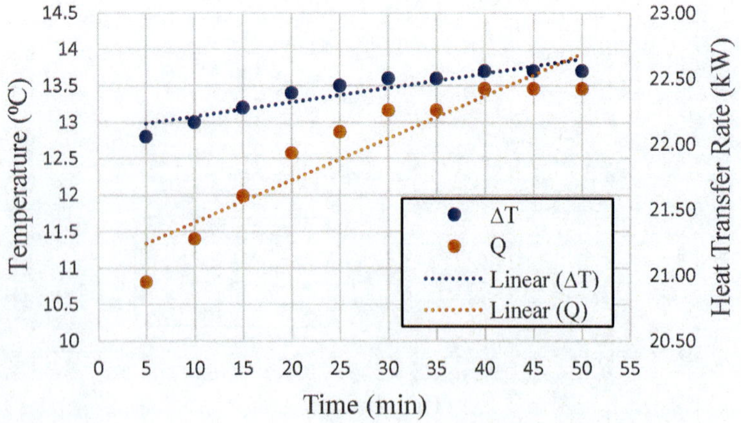

Fig. 5 Heat transfer rate and temperature distribution in a heat exchanger with a straight-section length of 30 cm

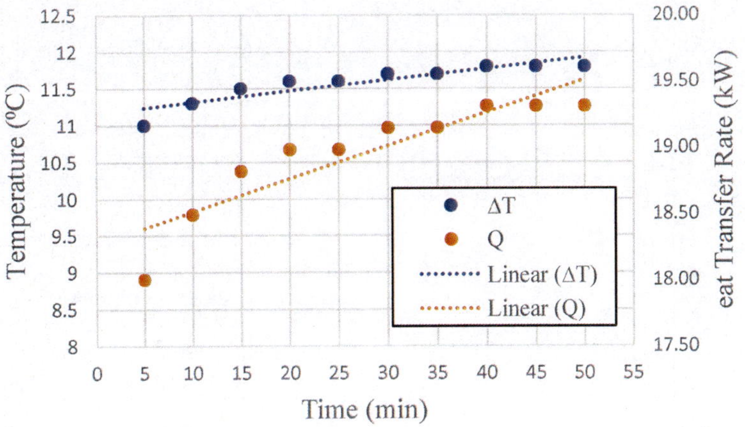

Fig. 6 Heat transfer rate and temperature distribution in a heat exchanger with a straight-section length of 15 cm

Figure 6 shows the heat transfer rate of a serpentine tube heat exchanger with a straight-section length of 15 cm. The average heat transfer rate is 18.95 kW, with the average temperature difference in a heat exchanger with a straight-section length of 15 cm which is 11.6°C. The heat transfer rate in a heat exchanger with a straight-section length of 15 cm is 34.7% higher than that of a heat exchanger with a straight-section length of 45 cm. A heat exchanger with a straight-section length of 15 cm has a heat transfer rate of 15.9% lower than a heat exchanger with a straight-section length of 30 cm. This is because the heat exchanger with a straight-section length of 15 cm has large turbulence of fluid flow and a large pressure drop.

4.2 *Heat Transfer Coefficient*

Figure 7 shows the convection heat transfer coefficient in a serpentine tube heat exchanger with straight-section lengths of 60, 45, 30, and 15 cm. The heat exchanger with a straight-section length of 30 cm has the highest average convective heat transfer coefficient of 1168.5 kW/m^2·K. The average convection heat transfer coefficient in a heat exchanger with a straight-section length of 30 cm is 19.3% higher than in a heat exchanger with a straight-section length of 15 cm. This shows that a heat exchanger with a straight-section length of 30 cm is the most optimal compared to a heat exchanger with a straight-section length of 60, 45, and 15 cm.

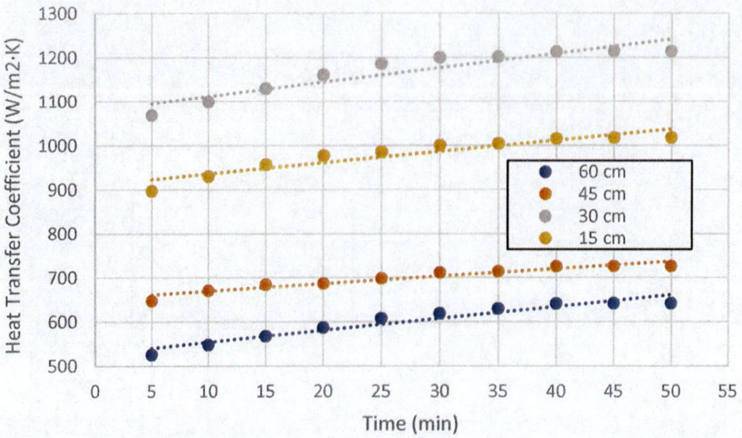

Fig. 7 Heat transfer coefficients in heat exchangers with straight-section length of 60, 45, 30, and 15 cm

5 Conclusion

The results of the research that has been carried out show that heat exchangers with a straight-section length of 30 cm are the most optimal compared to heat exchangers with a straight-section length of 60, 45, and 15 cm. The heat transfer rate in a heat exchanger with a straight-section length of 30 cm is 15.9% higher than that of a heat exchanger with a straight-section length of 15 cm. Compared to a heat exchanger with a straight-section length of 45 cm, a heat exchanger with a straight-section length of 30 cm has a heat transfer rate of 56% higher and 78.5% higher than a heat exchanger with a straight-section length of 60 cm. The highest convection heat transfer coefficient is obtained in a heat exchanger with a straight-section length of 30 cm, and the average heat transfer coefficient is 1168.5 kW/m^2·K.

References

1. Alteneiji M, Ali MIH, Khan KA, dan RK, Al-Rub A (2022) Heat transfer effectiveness characteristics maps for additively manufactured tpms compact heat exchangers. Energy Stor Sav. https://doi.org/10.1016/j.enss.2022.04.005
2. Zilio G et al (2022) Structural analysis of compact heat exchanger samples fabricated by additive manufacturing. Int J Press Vessel Pip 199. https://doi.org/10.1016/j.ijpvp.2022.104714
3. Shah dan RK, Sekuli DP (2007) Selection of heat exchangers and their components
4. Menéndez-Pérez A, Fernández-Aballí Altamirano C, Sacasas Suárez D, Cuevas Barraza C, dan Borrajo-Pérez R (2022) Metaheuristics applied to the optimization of a compact heat exchanger with enhanced heat transfer surface. Appl Therm Eng 214:118887. https://doi.org/10.1016/j.applthermaleng.2022.118887

5. Ji J, Gao R, Shi B, Zhang J, Li F, dan Deng X (2022) Improved tube structure and segmental baffle to enhance heat transfer performance of elastic tube bundle heat exchanger. Appl Therm Eng 200:117703. https://doi.org/10.1016/j.applthermaleng.2021.117703
6. Aboul Khail A, dan Erişen A (2022). Heat transfer and performance enhancement investigation of novel plate heat exchanger. Therm Sci Eng Prog 34:101368. https://doi.org/10.1016/j.tsep.2022.101368
7. Ribeiro F, de Conde KE, Garcia EC, dan Nascimento IP (2020) Heat transfer performance enhancement in compact heat exchangers by the use of turbulators in the inner side. Appl Therm Eng 173:115188. https://doi.org/10.1016/j.applthermaleng.2020.115188
8. Carpio J, dan Valencia A (2021) Heat transfer enhancement through longitudinal vortex generators in compact heat exchangers with flat tubes. Int Commun Heat Mass Transf 120:2021. https://doi.org/10.1016/j.icheatmasstransfer.2020.105035
9. Shi S, Niu J, Wu Z, Luo S, dan Gao X (2022) International journal of heat and mass transfer experimental and numerical investigation on heat transfer enhancement of vertical triplex tube heat exchanger with fractal fins for latent thermal energy storage. Int J Heat Mass Transf 198:123386. https://doi.org/10.1016/j.ijheatmasstransfer.2022.123386
10. Sanchouli M, Payan S, Payan A, dan Nada SA (2022) Investigation of the enhancing thermal performance of phase change material in a double-tube heat exchanger using grid annular fins. Case Stud Therm Eng 34:101986. https://doi.org/10.1016/j.csite.2022.101986
11. Wang M, Sun H, dan Cheng L (2021) Enhanced heat transfer characteristics of nano heat exchanger with periodic fins: a molecular dynamics study. J Mol.Liq 341:116908. https://doi.org/10.1016/j.molliq.2021.116908.
12. Liu X, Wang M, Liu H, Chen W, dan Qian S (2021) Numerical analysis on heat transfer enhancement of wavy fin-tube heat exchangers for air-conditioning applications. Appl Therm Eng 199:117597. https://doi.org/10.1016/j.applthermaleng.2021.117597
13. Modi AJ, dan Rathod MK (2022) Experimental investigation of heat transfer enhancement and pressure drop of fin-and-circular tube heat exchangers with modified rectangular winglet vortex generator. Int J Heat Mass Transf 189:122742. https://doi.org/10.1016/j.ijheatmasstransfer.2022.122742
14. Zheng N, Zhang K, Chen Q, dan Sun Z (2022) Novel self-join winglet vortex generators for enhanced heat transfer of turbulent airflow in round tubes. Int Commun Heat Mass Transf 130:105806. https://doi.org/10.1016/j.icheatmasstransfer.2021.105806
15. Awais M, Saad M, Ayaz H, Ehsan MM, dan Bhuiyan A (2020) Computational assessment of Nano-particulate (Al2O3/Water) utilization for enhancement of heat transfer with varying straight section lengths in a serpentine tube heat exchanger. Therm Sci Eng Prog 20:100521. https://doi.org/10.1016/j.tsep.2020.100521

Experimental Study of Magnetic Force Direction Variation on the Efficiency Effects of the Piezoelectric-Based Energy Harvesting Mechanism

Amir Zaki Mubarak, Muhammad Rizal, and Diyoga Alfarisy

Abstract The development of wireless and microelectronics technology has led to wearable devices where power can be supplied by batteries as well as energy harvesting devices. Despite the advancements made by low-power microdevice technology, the power requirements to apply them are still difficult to find. For this reason, developing new energy harvesting techniques is necessary to maintain self-powered systems. Energy harvesters are also needed as an economically practical alternative to battery charging and reducing greenhouse gas emissions and preserving the environment. One of them is by using piezoelectric-based energy harvester techniques by utilizing rotation. The purpose of this study is to examine the influence of magnetic force direction on the voltage and efficiency produced by piezoelectric-based energy harvester mechanisms using magnetic lighters. In this study, there are two variations of magnetic circuits, the first is that the magnetic circuit is regulated to repel each other (Type A), and the second is the magnetic circuit is regulated by mutual attraction and repulsion (Type B). Data collection is carried out on an energy harvesting mechanism whose piezoelectricity amounts to 4 pieces with a variation of 1000 rpm. The efficiency produced by the Type A magnetic circuit is 0.278% with a resistance of 9780 ohms. The resulting efficiency of the Type B magnetic circuit is 0.460% with a resistance of 9760 rpm. The application of piezoelectric-based energy harvesting mechanisms is more effective when using Type B magnetic circuits.

Keywords Energy harvester · Piezoelectric · Magnetic force · Efficiency

A. Z. Mubarak (✉) · M. Rizal · D. Alfarisy
Department of Mechanical Engineering, Faculty of Engineering, Universitas Syiah Kuala (USK), 23111 Darussalam, Banda Aceh, Indonesia
e-mail: amir_zm@usk.ac.id

M. Rizal
e-mail: muh.rizal@usk.ac.id

Irwansyah et al. (eds.), *Proceedings of the 4th International Conference on Experimental and Computational Mechanics in Engineering*, Lecture Notes in Mechanical Engineering, https://doi.org/10.1007/978-981-99-7495-5_53

1 Introduction

The development of wireless and microelectronics technology has led to wearable devices, where power can be supplied by batteries as well as energy harvesting devices [1]. Despite the advancements made by low-power microdevice technology, the energy density of chemical batteries needs to be increased since the power requirements to apply them are difficult to find [2, 3]. For this reason, developing new energy harvesting techniques is necessary to maintain these self-powered systems. Energy harvesters are also needed as an economically practical alternative to battery charging to reduce greenhouse gas emissions and preserve the environment [4]. One of the devices that can be used to harvest energy is a piezoelectric.

Energy can also be harvested using energy harvester techniques to power small autonomous sensors such as those developed using Micro-electromechanical Systems (MEMO) technology. These systems are often very small and require little power, but the reaction is limited by dependence on battery power. Energy harvesters can be obtained from ambient vibrations, wind, and heat energy. Piezoelectric is a material that can generate electricity when given compressive force [5].

Various kinds of piezoelectric-based energy harvester research that have been carried out include Beeby et al. [6] in their research, namely harvesting vibrational energy using a small electromagnetic generator. Xie et al. [7] examined a marine wave energy harvester from the transverse motion of water particles developed by piezoelectric effects. Mowaviq et al. [8] in their research, piezoelectrics were used as floor-mounted energy harvesters, where energy from human footsteps could be harnessed. Rizal et al. [9] have researched piezoelectric energy harvesters embedded in rotating spindles that are excited by magnetic forces and can charge capacitors.

Adha [10] examined a magnetically pressed piezoelectric-based energy harvester mechanism in which the direction of the magnetic force is regulated only to repel each other. A meticulous energy harvester can produce voltages of up to 8.12 V with motor rotation at 1123 rpm using six piezoelectrics. In his research, magnets are regulated to repel each other, and a drag force makes the energy harvesting mechanism less efficient. Therefore, this study will investigate the effect resulting from variations in the direction of the magnetic force on efficiency in the energy harvester mechanism by utilizing rotation.

2 Method and Experimental Setup

In this study, there are two variations in the direction of magnetic force applied to produce power and efficiency. The first is that the configuration of magnets is arranged to have the same magnetic poles that cause the repulsive force to repel between the magnets. On the magnetic stand, there is only one rotating magnet. This type of magnetic configuration is called Type A and can be seen in Fig. 2a. In the second variation in the direction of the magnetic force, the magnetic configuration is set to have different magnetic poles or intermittent intervals that make the magnet experience a repulsive force and attraction between magnets. On the magnetic stand, there are two rotating magnets. This type of magnetic configuration is called Type B configuration and can be seen in Fig. 1b.

In this study, the energy harvester mechanism was magnetically driven or pressed. A DC motor with a power supply that has a rotary speed controller is used to rotate the magnetic stand, and there are two fixed magnets on the stand. The magnet used is a coin-round neodymium magnet that has a diameter of 10 mm and a thickness of 3 mm. This study used four piezoelectrics attached to the piezoelectric stand, and there were also four magnets arranged with different polar variations along with four suppressors contained in the sleeve. The piezoelectric material used is a type of ceramic that has a diameter of 35 mm.

When the motor rotates, the two fixed magnets on the magnetic stand also rotate and drive the four magnets that remain on the sleeve, so that the repulsive force from the magnet will be passed on to the suppressant, which will press the piezoelectric

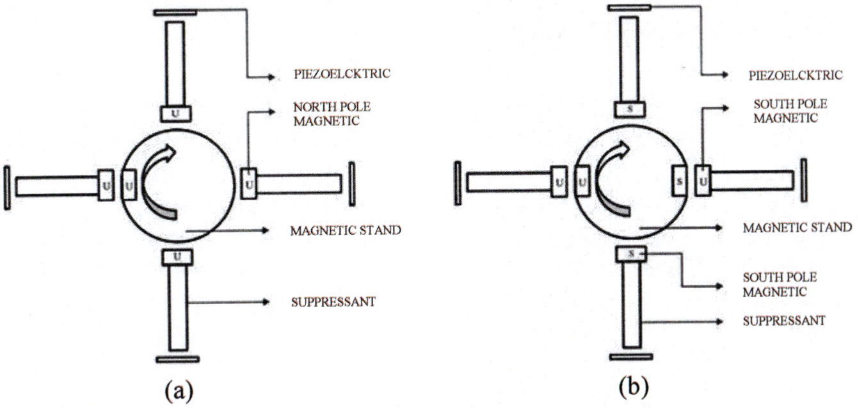

Fig. 1 Magnetic circuit on the energy harvester mechanism

Fig. 2 Experimental setup

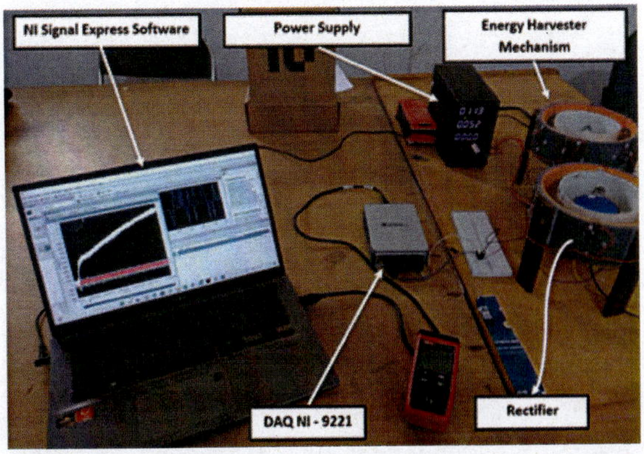

so that the piezoelectric produces voltage. The output voltage is connected with data acquisition NI 9221, which serves to detect objects that are measured and converted into voltage form values and are displayed in the NI Signal Express application. The setup of energy harvester mechanism research can be seen in Fig. 2.

To check the electrical power input used, use a DC power supply (Wanptek-DPS605U) and measure with a handheld tachometer (UNI-T UT373). To measure the output voltage generated by the energy harvester mechanism, it is necessary to use NI 9221 acquisition data to detect objects that are measured and converted into voltage form values and displayed in NI Signal Express applications. The electrical system used in piezoelectric-based energy harvesting mechanisms is in the form of a rectifier circuit. This rectifier circuit is useful for converting alternating electric current (AC) sources into direct current (DC) sources.

3 Results and Discussion

3.1 DC Motor Input Power

Measurements are made by adjusting the input power on the DC power supply so that 1000 rpm is obtained using a tachometer. The results of the DC motor input power measurement show that the input power used in the Type A circuit is greater than the input power used in the Type B circuit and can be seen in Fig. 3.

The difference in the value of the input power in these two magnetic circuits is because in the Type A circuit, there is a drag force that occurs when magnets with the same poles converge, so the DC motor requires more input power to rotate the magnet.

Fig. 3 Result of measuring the input power

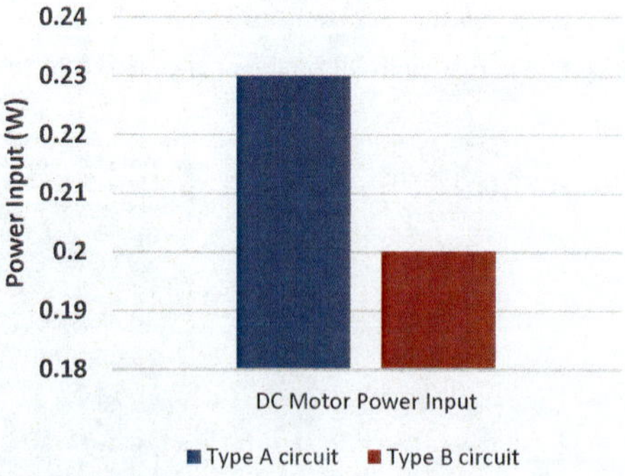

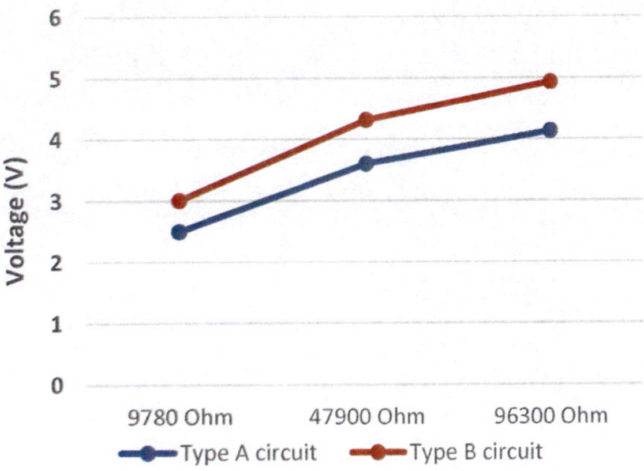

Fig. 4 Output voltage measurement results

3.2 *Output Voltage*

The measurement of the output voltage of the energy harvester mechanism is carried out at a speed of 1000 rpm and can be seen in Fig. 4. Voltage measurements are carried out by adding resistor variations of 9780 Ω, 47,900 Ω, and 96,300 Ω, and the measurement results are displayed on the NI Signal Express software.

The measurement results show that the output voltage produced by the energy harvester mechanism in the Type B circuit produces a voltage value that is greater than the Type A circuit. The highest output voltage value of the two magnetic circuits is at a resistance of 96,300 Ω.

3.3 *Output Power*

The power measurement of the output of the energy harvester mechanism is carried out at a speed of 1000 rpm and can be seen in Fig. 5. Voltage measurements are carried out by adding resistor variations of 9780, 47,900, and 96,300 Ω. To calculate the output power, use the formula $P = \frac{V^2}{R}$. This formula is used because the value of the electric current is unknown [11].

The measurement results show that the power output produced by the energy harvester mechanism in the Type B circuit produces a greater power output value than in the Type A circuit. The highest output voltage value of the two magnetic circuits is at a resistance of 9780 Ω.

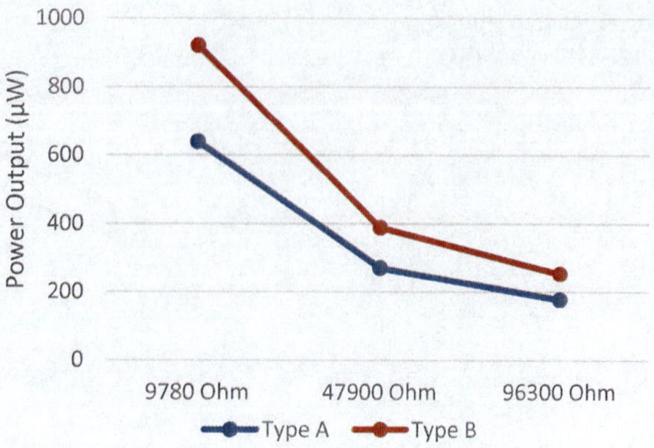

Fig. 5 Output power measurement results

3.4 *Efficiency*

The measurement of the efficiency of the energy harvester mechanism is carried out at a speed of 1000 rpm and can be seen in Fig. 6. Efficiency measurements were carried out by adding resistor variations of 9780, 47,900, and 96,300 Ω. To calculate efficiency, using the formula

$$\eta = \frac{P_{\text{output}}}{P_{\text{input}}} \times 100\%.$$

Based on the figure, the efficiency produced by the energy harvester mechanism in the Type B circuit produces a greater power output value than the Type A circuit. The highest output voltage value of the two magnetic circuits is at a resistance of 9780 Ω.

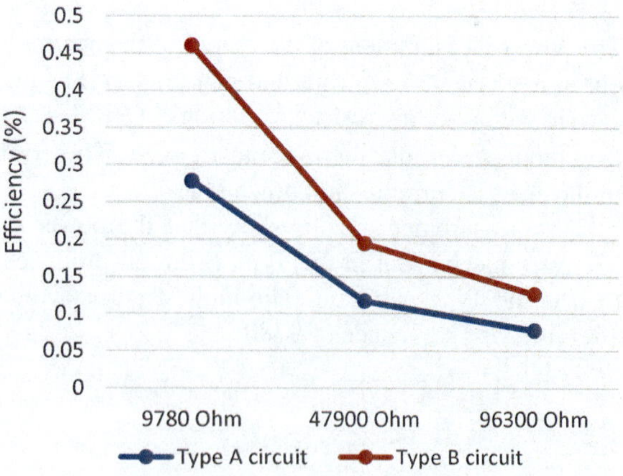

Fig. 6 Efficiency measurement results

Table 1 The measurement result of the energy harvester mechanism

Resistance (ohm)	Type A circuit			Type B circuit		
	Voltage output (V)	Power output (μW)	Efficiency (%)	Voltage output (V)	Power output (μW)	Efficiency (%)
9780	2.5	639.06	0.278	3	920.25	0.460
47,900	3.6	270.56	0.117	4.31	387.81	0.194
96,300	4.13	177.12	0.077	4.93	252.39	0.126

The results of measurements and analysis on the energy harvester mechanism can be seen in Table 1.

The data shows that the voltage output, power output, and efficiency values produced in Type B circuits are greater than those of Type A circuits. The use of resistors greatly affects both types of magnetic circuits. The greater value of voltage output, power output, and efficiency is obtained at a resistance of 9780 Ω, and the lowest value is found at a resistance of 96,300 Ω.

4 Conclusion

Magnetic circuits that are regulated against each other and attraction (Type B) are more efficiently used in energy harvesting mechanisms than Type A magnetic circuits where there is only a repulsive force. This result was then validated with test data showing that the Type A magnetic circuit has an efficiency value of 0.278% and a Type B magnetic circuit with an efficiency value of 0.460%. This is because in the Type A circuit, there is a drag force between magnets when rotating, so the output produced is not optimal. In the Type B circuit, the variation of the magnet is arranged variably or alternately to reduce the drag force so that the output produced is more efficient.

References

1. Rjafallah A, Hajjaji A, Guyomar D, Kandoussi K, Belhora F, Boughaleb Y (2018) Modeling of polyurethane/lead zirconate titanate composites for vibration energy harvesting. J Mater Komposit 53:613–623
2. Zhang Y, Lee C, Zhang X (2019) A novel piezoelectric power generator integrated with a compliant energy storage mechanism. J Phys D: Appl Phys 52(45)
3. Zhou M, Al-Furjan M, Wang B (2018) Modeling and efficiency analysis of a piezoelectric energy harvester based on the flow induced vibration of a piezoelectric composite pipe. Sensors. https://doi.org/10.3390/s18124277
4. Maghsoudi Nia E, Wan Abdullah Zawawi NA, Mahinder Singh BS (2019) Design of a pavement using piezoelectric materials. Mater Werkstoffe. https://doi.org/10.1002/mawe.201900002

5. Kaur B, Agnihotri A, Thapar D, Arora N (2019) Piezoelectric energy harvester design and power conditioning with solar integration. In: 2019 3rd international conference on electronics, materials engineering and nano-technology (IEMENTech), Institute of Engineering & Management, Kolkata
6. Beeby SP, Torah RN, Tudor MJ, Glynne-Jones P, O'Donnell T, Saha CR, Roy S (2007) A micro electromagnetic generator for vibration energy harvesting. J Micromech Microeng 17(7):1257–1265. https://doi.org/10.1088/0960-1317/17/7/007
7. Xie XD, Wang Q, Wu N (2014) Energy harvesting from transverse ocean waves by a piezoelectric plate. Int J Eng Sci 81:41–48. https://doi.org/10.1016/j.ijengsci.2014.04.003
8. Mowaviq MI, Junaidi A, Purwanto S (2018) Lantai pemanen energi listrik menggunakan piezoelektrik. J Energi & Kelistrikan, STT PLN 10(2)
9. Rizal M, Husni A Mubarak Z, Dirhamsyah M, Arhami (2022) Design and experimental study of a piezoelectric energy harvester embedded in a rotating spindle excited by magnetic force. Sens Actuators A Phys 340. https://doi.org/10.1016/j.sna.2022.113521
10. Adha MF (2020) Kajian mekanisme pemanen energi berbasis piezoelektrik dengan pemanfaatan putaran. Program Studi Teknik Mesin, Fakultas Teknik, Universitas Syiah Kuala, Banda Aceh, Tugas Akhir
11. Shanefield DJ (2001) Industrial electronics for engineers, chemists, and technicians. William Andrew Publishing, Norwich, NY

Analysis of Temiki (*Melastoma Candidum*) Extract Compounds as Corrosion Inhibitors Using GC–MS

Joli Supardi, Samsul Rizal, Nurdin Ali, Syarizal Fonna, Ikramullah Ikramullah, and A. K. Arifin

Abstract Temiki (*Melastoma Candidum*) is one of the unutilized wild plants observed to be turning pests on plantations. It has only been used for medicine and not for engineering purposes. Therefore, this study aims to identify and analyze Temiki's chemical compounds, which are useful as corrosion inhibitors. The plant was washed, dried, then blended until it became a fine powder, and was further extracted by employing the maceration method with ethanol as solvent. The extracted compounds were also analyzed using Gas Chromatography-Mass Spectrophotometry (GC–MS) and the results obtained include, 2-Furancarboxaldehyde (CAS) Furfural, 1,2,3-Propanetriol (CAS) Glycerol, Acetic acid,1-(2-methyltetrazole-5-yl) etheny, 4H-Pyran-4one,2,3-dihydro-3,5-dihydroxy, 5-Hydroxymethylfurfural, cis-dimethylmorpholine, 3-penten-1-ol,2,2,4-trimethyl-(CAS)2,2,4, 2-AMINO-9-(3,4-DIHYDROXY-5-HYDORO, D-Allose, beta-D-Glucopyranoside, methyl, Hexadecanoic acid, methyl ester, 9,12,15-Octadecatrienoic acid, methyl ester. In conclusion, these chemical compounds were useful as corrosion inhibitors.

Keywords Temiki (*Melastoma Candidum*) · Inhibitors · Corrosion · GC–MS · Chemical compounds

J. Supardi
Doctoral Program, School of Engineering, Post Graduate Program, Syiah Kuala University, Banda Aceh 23111, Indonesia

Department of Mechanical Engineering, Teuku Umar University, Meulaboh 23615, Indonesia

S. Rizal (✉) · N. Ali · S. Fonna · I. Ikramullah
Department of Mechanical and Industrial Engineering, Syiah Kuala University, Banda Aceh 23111, Indonesia
e-mail: samsul.rizal@unsyiah.ac.id

A. K. Arifin
Department of Mechanical and Materials Engineering, Universiti Kebangsaan Malaysia, 43600 Bangi, Selangor DE, Malaysia

Irwansyah et al. (eds.), *Proceedings of the 4th International Conference on Experimental and Computational Mechanics in Engineering*, Lecture Notes in Mechanical Engineering, https://doi.org/10.1007/978-981-99-7495-5_54

1 Introduction

Temiki (*Melastoma candidum*) is among the 22 species found in Southeast Asia and is considered native to the tropics, subtropics, and the Pacific Islands of Asia. Furthermore, the plant is generally found in bushes, rice fields, as well as mountain slopes and is considered as a herbal medicine by the Chinese, Indian, and Indonesian. It has been scientifically proven that *Melastoma candidum* parts, such as leaves, shoots, bark, seeds, and roots, are useful as a medicine for wounds, diarrhea, hemorrhoids, dysentery, and abdominal pain. Other findings also showed that Temiki was pharmacologically useful as an antiseptic, anti-inflammatory, antitoxic, and antioxidant [1, 2].

Temiki belongs to *Melastomataceae* family, which is in the form of bushes, shrubs, or trees. The leaves are opposite or ragged, single, usually with 3–9 curved bones, and rarely pinnate without support. *Melastomataceae,* also known as *M. malabathricum* L, is a herbaceous plant with a height of 0.5–4 m and young branches with scales. The leaves are stemmed, opposite, elongated oval with a pointed tip, 3–20 by 1–8 cm bony, and hairy on both sides. According to [3], the Temiki plant reproduces by seed and has clustered flowers at the ends of the branches, light purple, and bloom all year-round, while the fruit is buni with a light brown rind and a round shape like a flower vase. The flesh is purple and tastes sweet, while the ripe ones have cracked rinds with many seeds.

Temiki extraction is the process of separating a substance from its mixture using a solvent that is able to extract the desired substance without dissolving the other material. The extraction process generally consists of three basic steps as follows [4]:

1. Addition of solvent mass that comes in contact with the sample, usually through a diffusion process.
2. The separation of solute from the sample and dissolved by the solvent to form an extract phase.
3. Separation of the sample at the extract phase.

It is important to note that a substance's separation process is based on the differences in certain properties, especially its solubility in two different insoluble liquids. Generally, extraction is performed with a solvent based on a component's solubility in other mixture, such as water and the other organic solvents. The material to be extracted is usually dried and crushed into powder or simplicia [5].

The purpose is to extract the chemical components found in natural materials, which is used for corrosion inhibitors. Active materials such as antimicrobial and antioxidant compounds in plants are generally extracted with a solvent. In the extraction process, the compound's number and types entering the liquid are largely determined by the utilized solvent types, and it includes rinsing and extraction phases. In the rinsing phase, the cell content's components were rinsed with the solvent that have been broken in the previous crushing process. Meanwhile, there are swelling of the cell wall and loosening of the cellulose skeleton in the extraction phase; hence, its pores become widened, allowing solvents to easily penetrate. The cell contents are then dissolved in the solvent according to its solubility and diffuse out due to the force caused by the difference in the solutes' concentration inside and outside the cell [6].

Based on the description above, the anti-corrosion technology is further developed to minimize the large losses from corrosion and the lack of organic matter using natural inhibitors, such as plants with no harmful and toxic chemicals [7].

The International Measures of Prevention Application and Economics of Corrosion Technology (IMPACT) estimated global corrosion losses at $2.5 trillion (USD), which is about 3.4% of total Gross Domestic Product (GDP). Based on a two-year global study published at the 2016 CORROSION conference in Vancouver, BC, economic losses were calculated and handled with respect to the management role in the industry. By applying corrosion prevention measures, about 15–35% of damage costs or $375–875 billion were saved [8].

It has been discovered that the corrosion losses have reached 3–4% of GDP in industrialized countries [9]. Furthermore, the US Federal Highway Administration (FHWA) estimated the total annual cost of corrosion in the USA to be $276 billion, which is approximately 3.1% of the country's GDP [10]. In Indonesia, about 20 trillion rupiah was estimated to be lost to the corrosion process yearly [11].

Corrosion is preventable using inhibitors derived from inorganic or organic compounds. This remains the best solution to protect against internal corrosion of metals and is used as the main defense for the oil process and extraction industries. It is important to note that Indonesia is very rich in plant species useful as corrosion inhibitors, but only few other studies explored the benefits of those native plants. For example, efforts have been applied by [12] and [13] to obtain anti-corrosion with inhibitors, such as the use of natural materials from plants.

The method of preventing corrosion includes the addition of inhibitors using inorganic materials that contain hazardous and environmentally-unfriendly compounds. Therefore, further studies are needed on organic inhibitors using Temiki (*Melastoma Candidum*), which is non-toxic and safe for the environment, and also widely useful for medical purposes as well as food coloring [14–16].

In this study, researchers tried to identify the characteristics of the chemical compounds contained in the Temiki extract by using chromatography-mass spectroscopy (GC/MS). GC/MS is a strong analytical technique that is used to measure small molecules in samples, so that we can find out the elements of chemical compounds contained in the Temiki extract in order to be used as a corrosion inhibitor made from natural ingredients that are free of harmful chemicals and friendly to the environment [17].

(a) (b)

Fig. 1 Temiki (*Melastoma Candidum*) plant **a** flowers, **b** fruits

2 Materials and Methods

2.1 Temiki Extract Preparation

Temiki is a wild plant that grows in bushes and it is considered as part of a plant pest that grows on oil palm plantations in Aceh. In the extraction process, dirt and dust were removed from Temiki by washing and drying in the sun for 2 days. Furthermore, it was mashed with a grinder in order to reduce its size for sample extraction purpose [18].

About 30 g of fine powder from the dried sample was weighed, wrapped using filter paper, and inserted into the Soxhlet column. In addition, technical ethanol was added to a 500 ml single-neck flask using a feed-to-solvent ratio of 1:10. The flask containing the sample was assembled with a Soxhlet extraction device at a temperature of 70 °C under reflux for 5 times, while the filtrate obtained was then separated using an evaporator at the same temperature with a pressure of 1 atm to evaporate the ethanol solvent [19, 20] (Fig. 1).

2.2 Temiki Extract Test Using GC–MS

The GC–MS instrument utilized was the Agilent 6980N Network GC system with auto-sampler and Agilent 5973 inert MSD detector. Furthermore, the column used was J&W Scientific, HP-5MS with a size of 0.25 mm × 30 m × 0.25 μm, while the carrier gas was Helium at a flow rate of 1 ml/min. A total of 2 μl was injected into the instrument and the inlet temperature was maintained at 250 °C. It was observed that the initial oven temperature obtained was 50 °C for 5 min, but with an increase of 10 °C/min, it rises to 280 °C. Therefore, the MS quad temperature was 150 °C and the source temperature was 230 °C [5]

2.3 Identification of Temiki Extract Compounds

The GC–MS mass spectrum data have been interpreted by comparing its components with the Wiley 7.0 database. Furthermore, chemical compounds were identified and confirmed based on the peak area as well as retention time. The chemical compounds were measured using peak area normalization [21, 22].

3 Results and Discussions

The analyze and identify the chemical compounds present in Temiki extract used as a corrosion inhibitor, the researcher used Gas Chromatography-Mass Spectrometry (GC/MS) Series QP 2010 Plus—Shimadzu, equipped with a DB-5 ms column (for polar compounds) series 998,847 to identify its components. In GC/MS, the sample is first separated into individual components using Gas Chromatography, and then, these components are ionized and broken down into finer fragments using a mass spectrometer. The fragments are then analyzed to determine the molecular weight and structure of each component. Density and viscosity were analyzed according to ASTM D 4052 and ASTM D 445 [23, 24]. The results of the analysis with GC–MS can be seen in Table 1.

The biological and pharmacological functions of each compound were further analyzed in order to determine its effect on the corrosion reaction [12] (Figs. 2 and 3).

The examination results showed the largest chemical compounds found in Temiki (*Melastoma candidum*) extract, which include 40.73% 5-Hydroxymethylfurfural, beta-D-Glucopyranoside, 13.00% methyl, and 12.05% 2-AMINO-9-(3,4-DIHYDROXY-5-HYDORO. These compounds belong to the glucose and fat derivatives. Table 2 shows the molecular structure and biological activity of the chemical compounds[1, 29].

According to [1], Temiki is a plant rich in secondary metabolism; therefore, it is widely used as medicine for wounds. The GC–MS analytical results showed that most chemical compounds contained in the plant include 5-Hydroxymethylfurfural, beta-D-Glucopyranoside, methyl, and 2-Amino-9-(3,4-Dihydroxy-5-Hydoro as the glucose and fat derivatives. Since several plants contain glucose and fat, they are utilized as medicine to protect body organs and play a role in cell metabolism [1].

As a biodiversity country rich in various plant species, it is very important to identify chemical compounds of plants to be studied and used as corrosion inhibitors. For example, they need to be very environmentally-friendly without any harmful chemical substances [18, 30].

Table 1 Results of chemical compounds in the Temiki extract identified by GC–MS

Peak report TIC

Peak	R. Time	Area	Area%	Height	Height%	A/H	Name
1	1.681	190,309	3.28	42,217	3.01	4.51	2-Furancarboxaldehyde (CAS) Furfural
2	2.055	123,933	2.14	38,844	2.77	3.19	1,2,3-Propanetriol (CAS) Glycerol
3	2.732	206,017	3.55	45,542	3.24	4.52	Acetic acid,1-(2-methyltetrazol-5-yl)etheny
4	3.283	313,353	5.40	82,054	5.84	3.82	4H-Pyran-4one,2,3-dihydro-3,5-dihydroxy;
5	3.808	2,196,895	37.85	571,965	40.72	3.84	5-Hydroxymethylfurfural;
6	4.652	109,615	1.89	18,764	1.34	5.84	cis-dimethylmorpholine;
7	5.041	286,214	4.93	35,828	2.55	7.99	3-penten-1-ol,2,2,4-trimethyl-(CAS)2,2,4;
8	5.767	944,204	16.27	169,230	12.05	5.58	2-amino-9-(3,4-dihydroxy-5-hydoro;
9	6.349	146,107	2.52	42,224	3.01	3.46	D-Allose;
10	7.321	934,665	16.11	182,510	13.00	5.12	Beta-D-Glucopyranoside, methyl;
11	11.150	170,187	2.93	89,327	6.36	1.91	Hexadecanoic acid, methyl ester;
12	12,644	181,909	3.13	85,640	6.10	2.12	9,12,15-Octadecatrienoic acid, methyl ester
		5,803,708	100.00	1,404,145	100.00		

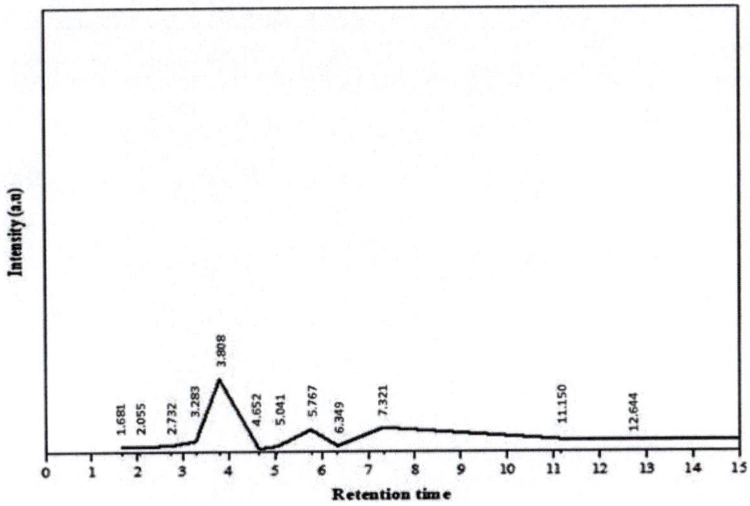

Fig. 2 Temiki extract chromatography

Fig. 3 Temiki (*Melastoma Candidum*) **a** fine powder and **b** extract

4 Conclusion

In this study, 12 chemical elements were discovered in the Temiki (*Melastoma candidum*) extract using GC–MS analysis. Most of the compounds found in the extract were employed as environmentally-friendly corrosion inhibitors.

Table 2 Molecular structure and biological activity of chemical compounds in Temiki extract

No.	Compound name	Molecular structure
1	2-Furancarboxaldehyde (CAS) Furfural [25]	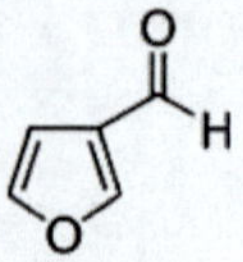
2	1,2,3-Propanetriol (CAS) Glycerol [26]	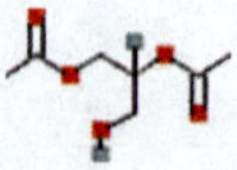
3	Acetic acid,1-(2-methyltetrazol-5-yl)etheny	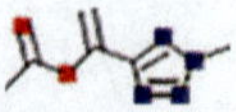
4	4H-Pyran-4one,2,3-dihydro-3,5-dihydroxy [27]	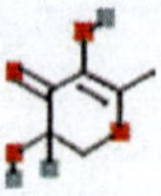
5	5-Hydroxymethylfurfural [28]	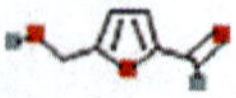
6	Cis-dimethyl morpholine	
7	3-penten-1-ol,2,2,4-trimethyl-(CAS)2,2,4	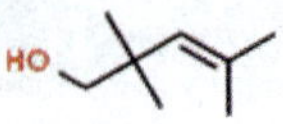
8	2-amino-9-(3,4-dihydroxy-5-hydoro	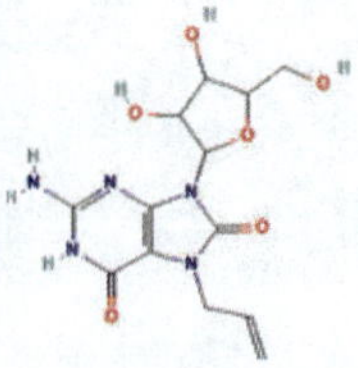
9	D-Allose	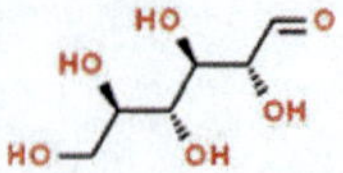
10	Beta-D-Glucopyranoside, methyl	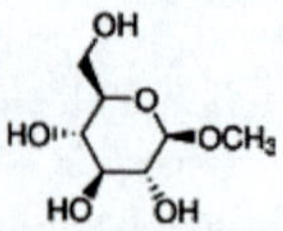

(continued)

Table 2 (continued)

No.	Compound name	Molecular structure
11	Hexadecanoic acid, methyl ester	
12	9,12,15-Octadecatrienoic acid, methyl ester	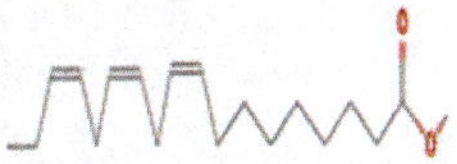

Acknowledgements This study was supported by the Ministry of Education, Culture, Research and Technology, Indonesia, and doctoral dissertation research grant scheme No. 056/SP2H/LT/DPRM/2021.

References

1. Joffry SM et al (2012) *Melastoma malabathricum* (L.) smith ethnomedicinal uses, chemical constituents, and pharmacological properties: a review. Evidence-based Complement. Altern. Med 2012,.https://doi.org/10.1155/2012/258434
2. Handayani M, Lambui dan Nengah Suwastika Lab Bioteknologi OI, Biologi FMIPA Untad Kampus Bumi Tadulako Jl Soekarno-Hatta Km J, Tengah S (2017) Potensi Tumbuhan *Melastoma malabathricum* L. Sebagai Bahan Antibakteri Salmonellosis Ethanol Extracts of *Melastoma malabathricum* L. Leaves Potential as anti-bacterial agent on Salmonella. Nat Sci J Sci Technol 6(2):165–174. [Online]. Available: https://bestjournal.untad.ac.id/index.php/ejurnalfmipa/article/view/8665
3. Tjitrosoepomo G (2007) Taksonomi Tumbuhan (Spermatophyta). Gadjah Mada University Press Yogyakarta, Yogyakarta
4. Wilson ID et al (2000) Encyclopedia of separation science. Academic-Press York, New. Academic-Press York New
5. Sembiring B (2007) Teknologi penyiapan simplisia terstandar tanaman obat. Balitro. Bogor 13(2)
6. Brazilin P, Kayu D, Caesalpinia S, Dengan SL, Untuk A, Kain P (2015) Pemungutan brazilin dari kayu secang (*Caesalpinia sappan* L) dengan metode maserasi dan aplikasinya untuk pewarnaan kain. J Bahan Alam Terbarukan 4(1):6–13. https://doi.org/10.15294/jbat.v4i1.3768
7. Alam PN, Rosnelly CM, Karo karo JA, Ilhamdi A, Soekarno R, Aslam IN (2021) Utilization of natural extracts as corrosion inhibitors in the seawater environment. J Rekayasa Kim Lingkung 16(2):68–74. https://doi.org/10.23955/rkl.v16i2.18156
8. Anonymous (2016) NACE study estimates global cost of corrosion. https://inspectioneering.com/journal
9. Revie RW, Uhlig HH (2008) Corrosion and corrosion control: an introduction to corrosion science and engineering, 4th edn. https://doi.org/10.1002/9780470277270
10. Shreir, Corrosion control. Butterworth-Heinemann College 1999-05-28 1999, 1976
11. Trethewey J, KR, Chamberlain (1991) Korosi Untuk Mahasiswa sain dan Rekayasa. PT. Gramedia Pustaka Utama, Jakarta
12. Oguzie EE (2008) Evaluation of the inhibitive effect of some plant extracts on the acid corrosion of mild steel. Corros Sci 50(11):2993–2998. https://doi.org/10.1016/j.corsci.2008.08.004
13. Sidiq MF (2013) Analisa Korosi Dan Pengendaliannya M. Fajar Sidiq 3(1)
14. Rani BEA, Basu BBJ (2012) Green inhibitors for corrosion protection of metals and alloys: an overview, 2012(i). https://doi.org/10.1155/2012/380217

15. Haryono G, Sugiarto B, Farid H (2010) Ekstrak Bahan Alam sebagai Inhibitor Korosi. Pros. Semin. Nas. Tek. Kim. "Kejuangan" Pengemb. Teknol. Kim. untuk Pengolah. Sumber Daya Alam Indones, pp 1–6
16. Raja PB, Sethuraman MG (2008) Natural products as corrosion inhibitor for metals in corrosive media—a review. Mater Lett 62(1):113–116. https://doi.org/10.1016/j.matlet.2007.04.079
17. Dhivya R, Manimegalai K (2013) Preliminary phytochemical screening and GC–MS profiling of ethanolic flower extract of calotropis. J Pharmacogn Phytochem 2(3):28–32
18. Backer J, Bakhuizen CA, Brink VD (1968) Flora of java. Netherlands Organ Adv Res 3
19. Apriliani N, Ardiansyah A, Siswanti, Sudarmi S (2016) Ekstraksi Daun Kapuk Randu (Ceiba pentandra Gaertn) dengan Pelarut Etanol pp 1–7
20. Fajriati I, Rizkiyah M, Muzakky (2011) Studi Ekstraksi Padat Cair Menggunakan Pelarut HF dan HNO 3 pada Penentuan Logam Cr dan Cu dalam Sampel Sedimen Sungai di Sekitar Calon PLTN Muria A Study of Solid-Liquid Extraction with HF And HNO 3 as Solvent for Determination of Cr and Cu in the River. J ILMU DASAR 12(1):13–22
21. Rompas DEB, Runtuwene MRJ, Koleangan HSJ (2015) Analisis Kandungan Fitokimia dan Uji Aktivitas Antioksidan dari Tanaman Lire (*Hemigraphis repanda* (L) Hall F.). J MIPA 4(2):36 https://doi.org/10.35799/jm.5.1.2016.11410
22. Lakkanasri P, Lothongkum G (2019) Effect of monoethanolamine on corrosion of A283 carbon steel in Propionic acid solution. Eng J 23(4):183–191. https://doi.org/10.4186/ej.2019.23.4.183
23. ASTM D445-06 (2008) Standard test method for kinematic viscosity of transparent and opaque liquids (and calculation of dynamic viscosity). Man Hydrocarb Anal 6th ed, pp 1–10
24. The United States of America (1995) ASTM D4052 : standard test method for density adn relative density of liquids by digital density meter. An am 688–691
25. Left DB, Zertoubi M, Khoudali S, Azzi M (2018) New application of *Chamaerops humilis* L. Extract as a green corrosion inhibitor for reinforcement steel in a simulated carbonated concrete pore solution 36(4):249–257. https://doi.org/10.4152/pea.201804249
26. Identity C, Health EW, Controls S, Information HH, Characterisation R (2020) Zinc chloride human health tier II assessment CAS number : 7646-85-7 no. September 2013, pp 1–15
27. Yuab X, Zhaob M, Liub F, Zengb S, Hub J (2013) Identification of 2,3-dihydro-3,5-dihydroxy-6-methyl-4H-pyran-4-one as a strong antioxidant in glucose–histidine Maillard reaction products
28. Menegazzo F, Ghedini E, Signoretto M (2018) 5-Hydroxymethylfurfural (HMF) production from real biomasses. Molecules 23(9):1–18. https://doi.org/10.3390/molecules23092201
29. Ouici AZH, Tourabi M, Benali O, Selles C, Jama C, Bentiss F (2017) Adsorption and corrosion inhibition properties of 5-amino 1,3,4-thiadiazole-2-thiol on the mild steel in hydrochloric acid medium: thermodynamic, surface and electrochemical studies
30. Starr F, Starr K, Loope L (2003) Melastoma candidum Asian Melastome Melastomataceae. Maui, Hawai'i

The Effects of Welding Speed on Mechanical Properties and Microstructure of Tungsten Inert Gas-Welded Aluminum Alloy 5083 H116

Rela Adi Himarosa, Sudarisman, Mudjijana, Muh. Budi Nur Rahman, and Rahmad Kuncoro Adi

Abstract Welding is one of the oldest joining processes to combine two materials. One of the commonly used welding processes is the tungsten inert gas (TIG) welding. In this study, Aluminum Alloy 5083 H116 is joined using TIG welding. Aluminum Alloy 5083 H116 is chosen because it is a popular material widely used in the industry, thanks to its properties such as good corrosion resistance, resistance in extreme environments, and high impact strength. The welding process of Aluminum Alloy 5083 H116 was performed under several speed variations, which are 1.8, 2 mm/s, and 2.2 m/s. We studied the effect of the welding speed on the mechanical strength and microstructure of the welded joint. The results are that the maximum tensile and bending strength were observed at a speed variation of 1.8 mm/s with a value of 286.75 and 485.19 MPa, respectively, and that the smallest tensile and bending strength values were obtained at a speed variation of 2.2 mm/s with a value of 204.65 MPa and 366.70 MPa, respectively. A good microstructure was attained in the weld joint for a speed variation of 1.8 mm/s.

Keywords Welding · TIG · Tensile · Bending · Microstructure

R. A. Himarosa (✉) · Sudarisman · Mudjijana · Muh. B. N. Rahman
Department of Mechanical Engineering, Universitas Muhammadiyah Yogyakarta, Yogyakarta, Indonesia
e-mail: rela.himarosa@umy.ac.id

Sudarisman
e-mail: sudarisman@umy.ac.id

Mudjijana
e-mail: mudjijana@umy.ac.id

Muh. B. N. Rahman
e-mail: budinurrahman@umy.ac.id

R. K. Adi
Department of Mechanical and Industrial Engineering, Faculty of Engineering, Universitas Gadjah Mada, Yogyakarta, Indonesia
e-mail: rahmad.kuncoro.adi@mail.ugm.ac.id

Irwansyah et al. (eds.), *Proceedings of the 4th International Conference on Experimental and Computational Mechanics in Engineering*, Lecture Notes in Mechanical Engineering, https://doi.org/10.1007/978-981-99-7495-5_55

1 Introduction

Humans have been using welding since ancient times to join two materials, making it one of the oldest joining processes [1]. Welding can be used to join two materials of the same type or different one [2]. The joining of two different materials tends to be more challenging in the welding process than with similar materials because there are changes in the material's chemical composition and heat coefficient during the welding process [3, 4]. Another problem encountered when joining two dissimilar materials is the presence of large residual stresses and the formation of intermetallic compounds in the joint [5]. However, joining two different materials is very important in the industry, making many companies conduct research in combining different types of materials. One main reason is that a joining two different materials cuts production costs and produces lightweight components [6].

One of the welding processes that are widely used today is the tungsten inert gas (TIG) welding. TIG welding is also known as gas tungsten arc welding (GTAW). In TIG welding, the process uses tungsten electrodes to create an electric arc [6]. The advantages of this type of welding are good welding quality, little or no flux required so that there is no potential for flux trapping, and lastly, the finishing results on the welding joint are less than other welding methods [7]. TIG welding can be used effectively for thin welding sections and for thicker sections by adopting the multi-pass method [2].

The welding process always has challenges with welding parameters and the type of material. One material that is challenging for a welding process is that this material is known to have good performance and high strength, so this material is often used in various industries such as the railroad and shipping industries [8–10]. However, this material has several weaknesses after the welding process, such as low joint strength, high porosity in the welding area, and high deformation [11, 12].

In a previous study, Baskoro et al. investigated the effect of welding parameters such as welding current and welding speed on the mechanical and microstructure properties of AA1100 using TIG Welding [13]. In this work, we study the influence of welding speed on the strength of the material. The more heat received by the material during the welding process, the more decrease was observed in the strength of the connection. Therefore, this study aims to determine the appropriate parameters, such as the welding speed, during the TIG welding process of the aluminum AA 5083 H116. In addition, we aim to investigate the characteristics of the AA 5083 H116 material after the welding process by conducting several performance tests, such as tensile testing, hardness, and observations of microstructures.

Table 1 Materials' composition

Chemical		Filler	
Element	Weight (%)	Element	Weight (%)
Si	0.18	Si	0.25
Fe	0.31	Fe	0.4
Cr	0.086	Cr	0.05–0.2
Mn	0.6	Mn	0.05–0.2
Ni	0.0074	Zn	0.1
Zn	0.15	Ti	0.06–0.2
Ti	0.013	Cu	0.1
Cu	0.042	Mg	4.5–5.5
Mg	4.76	Al min	Rem

Table 2 Variations of the welding speed

Specimen code	Welding speed (mm/s)
AA1	1.8
AA2	2.0
AA3	2.2

2 Materials and Method

The material used in this research is Aluminum Alloy 5083 H116, with the dimensions of 300 mm × 150 mm × 3 mm. The chemical composition and filler metal composition are presented in the following Table 1, respectively.

In this work, we perform the welding process at three different speeds. These variations are used to study the effect of welding speed on the mechanical properties and microstructure of the Aluminum Alloy 5083 H116. The variations in speed during the welding process are shown in Table 2. Figure 1 illustrates the schematic of the welding process.

3 Result and Discussion

3.1 Mechanical Properties

The effect of the welding speed on the tensile and bending strength of AA 5083 H116 material is shown in Figs. 2 and 3. Those two figures show that the welding speed influences the tensile and bending strength of the welded joint [14]. The maximum tensile strength value (286.75 MPa) is accomplished in welding with a speed of 1.8 mm/s, which is about 82.73% of weld efficiency [15], while the minimum tensile strength value (204.65 MPa) is obtained in the 2.2 mm/s welding speed. The maximum bending strength value, as shown in Fig. 3, is attained in welding with a speed of 1.8 mm/s, with the value of 479.69 MPa, which is about 74.60% of weld efficiency, while the minimum bending strength value (450.55 MPa) is found in welding speed of 2.2 mm/s. We can observe that as the welding speed increases, the tensile and flexural strengths decrease. The decrease in the tensile and flexural strengths was caused by the two materials not being perfectly connected. This is due to the insufficient penetration of the welding bead into the root gap of the welding [16–18]. The ductility of the materials also increases when the welding speed is decreased. The higher heat generation takes place at lower weld speed, which leads to increase the grain size of the material and restoring the ductility by recrystallization mechanism [19].

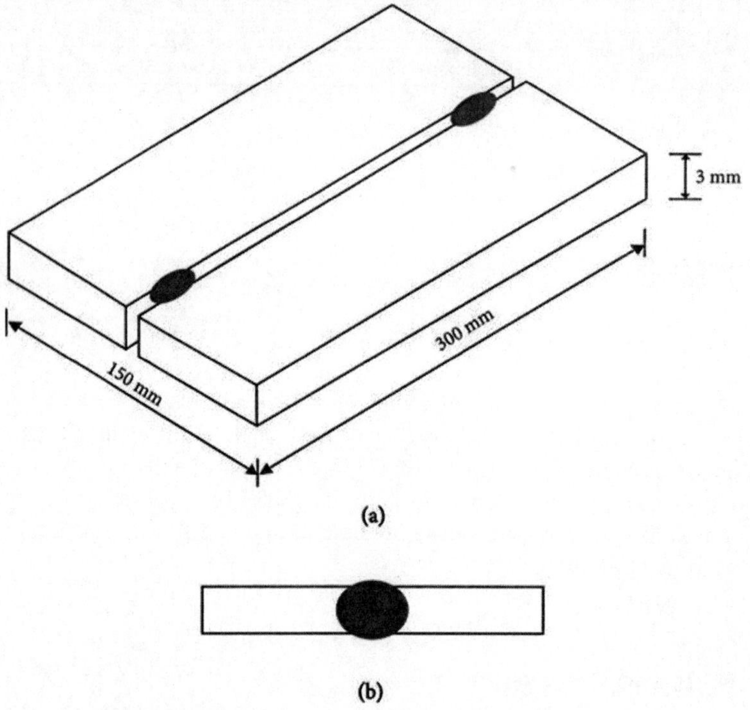

Fig. 1 Schematic illustration of the welding process: **a** the geometry of the welded specimen and **b** one side of the welding

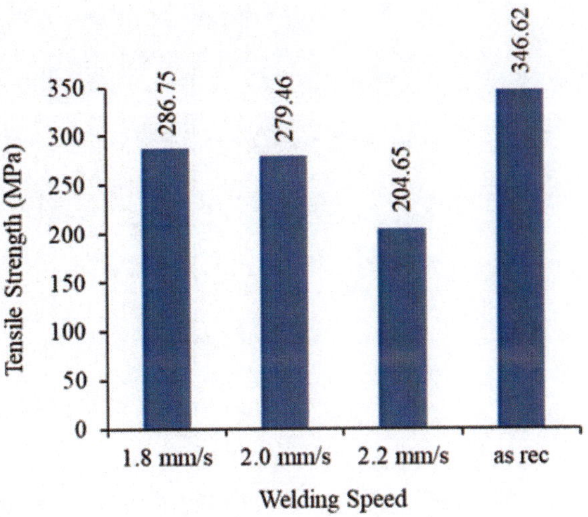

Fig. 2 Effect of welding speed to the tensile strength

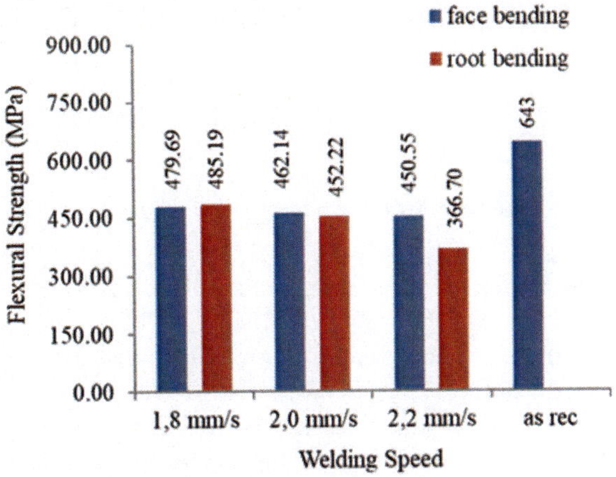

Fig. 3 Effect of welding speed to the flexural strength

3.2 *Micro-hardness*

We performed the micro-hardness test of the welded joint using the Vickers test. The test was carried out by indenting several parts of the joint. The indentation process was carried out in the horizontal direction in the BM, HAZ, and WM sections. The test results are visualized in the form of a hardness profile, as shown in Fig. 4. Table 3 shows the average hardness value of each welding speed variation. The table shows that the welding speed affects the compaction speed of the area undergoing the welding process [20]. We observed that the lowest hardness value was obtained in the weld metal (WM) section in all variations of the welding speed.

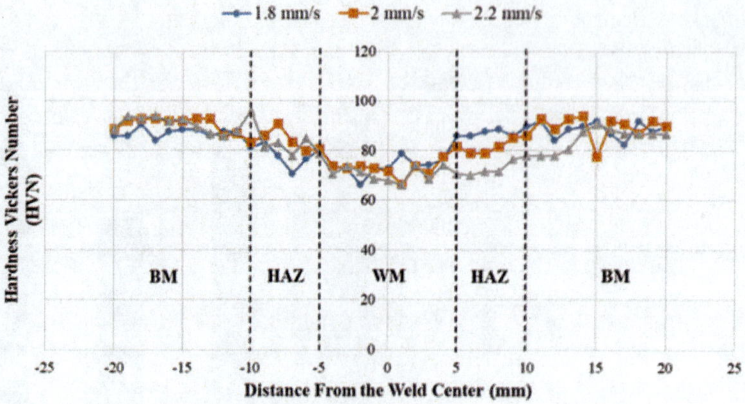

Fig. 4 Vickers hardness profile of TIG-welded joints of AA 5083 H116

Table 3 Average hardness value (VHN)

Welding speed	1.8 (mm/s)	2.0 (mm/s)	2.2 (mm/s)
WM	75.21	73.87	71.28
HAZ	78.31	81.81	83.63
BM	86.64	89.91	90.75

3.3 Microstructure

Figure 5 shows the microstructure of AA 5083 H116 that underwent a welding process using TIG welding. The morphology of the welded specimen was divided into three different visible zones, which are base metal (BM), heat-affected zone (HAZ), and weld metal (WM). The weld metal (WM) is the metal in a welded joint which has been molten during the welding process and then then solidified [21]. The heat-affected zone (HAZ) is the area where the grains grow and coarse grains are formed [22]. Microstructural observation of the HAZ area is important because this section is the weakest part and it is usually the source for initial failure of the joint [23, 24]. According to the observation of the microstructure presented in Fig. 5, the width of the HAZ area at low welding speed variations is better than that of two other welding speed variations. The welding speed of 1.8 mm/s yields the highest value of the tensile and bending strength, compared to other variations. Welding at a low speed also causes heat to be well absorbed in the weld metal area, which causes the material melt completely and produces a good joint [20].

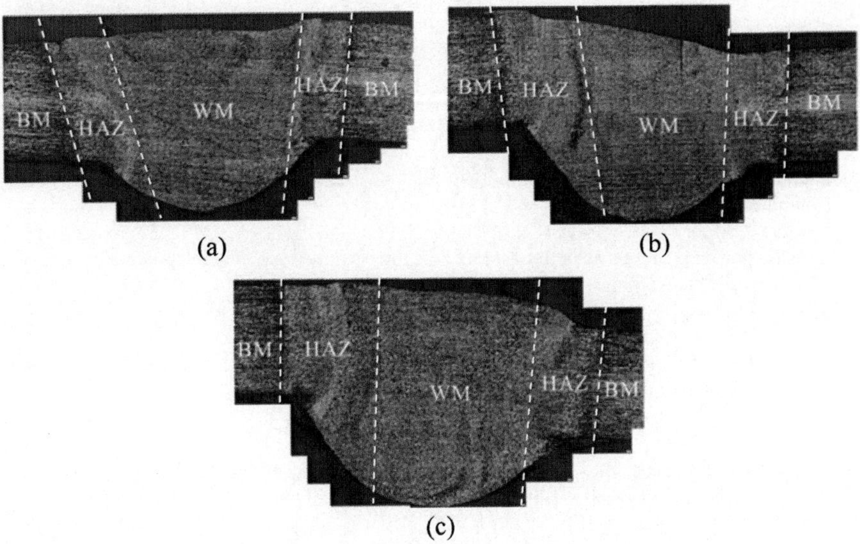

Fig. 5 Microstructure region of the WM, HAZ, and BM in welding speed of **a** 1.8 mm/s, **b** 2.0 mm/s, and **c** 2.2 mm/s

4 Conclusion

Variation in welding speed affects the mechanical strength, hardness, and microstructure of the welded AA 5083 H116. The highest tensile and bending strength values are attained at a speed of 1.8 mm/s, with a value of 286.75 MPa and 479.69 MPa, respectively. We obtain the lowest values of tensile and bending strengths at the welding speed of 2.2 mm/s, with the values of 204.65 MPa and 450.55 MPa, respectively. The hardness values in the HAZ and BM areas increase as the welding speed increases. However, in the WM area, the hardness values decrease as the welding speed increases. According to the microstructure observation, the width of the HAZ area at the welding speed variation of 1.8 mm/s is better than the other speed variations. Based on those experiments, we learned that the welding speed of 1.8 mm/s is the most optimal choice, compared to the other two options, for the welding process of AA 5083 H116 material.

Acknowledgements This research is funded by Universitas Muhammadiyah, Yogyakarta, under the *Hibah Internal Penelitian UMY 2021* program.

References

1. Lancaster JF (1984) The physics of welding. Phys Technol 15(2):73–79
2. Patel D, Jani S (2021) Techniques to weld similar and dissimilar materials by ATIG welding—an overview. Mater Manuf Process 36(1):1–16

3. Wang S, Ma Q, Li Y (2011) Characterization of microstructure, mechanical properties and corrosion resistance of dissimilar welded joint between 2205 duplex stainless steel and 16MnR. Mater Des 32(2):831–837
4. Khan MMA, Romoli L, Fiaschi M, Dini G, Sarri F (2012) Laser beam welding of dissimilar stainless steels in a fillet joint configuration. J Mater Process Technol 212(4):856–867
5. Pourali M, Abdollah-Zadeh A, Saeid T, Kargar F (2017) Influence of welding parameters on intermetallic compounds formation in dissimilar steel/aluminum friction stir welds. J Alloys Compd 715:1–8
6. Assefa AT, Ahmed GMS, Alamri S, Edacherian A, Jiru MG, Pandey V et al (2022) Experimental investigation and parametric optimization of the tungsten inert gas welding process parameters of dissimilar metals. Materials (Basel). 15(13):4426
7. Sharma G, Tyagi R, Priyanshu, Sharma P (2022) Variants of TIG welding process for improvement of weld penetration depth—a review. Mater Today Proc
8. Wu D, Hua X, Li F, Huang L (2017) Understanding of spatter formation in fiber laser welding of 5083 aluminum alloy. Int J Heat Mass Transf 113:730–740
9. Bodukuri AK, Eswaraiah K, Rajendar K (2017) Comparison of Aluminum alloy 5083 properties on TIGW and FSW processes. Mater Today Proc. 4(9):10197–10201
10. Himarosa RA, Adi RK, Budi M, Rahman N, Wisnujati A, Yusuf A et al (2022) Simultaneous double-sided metal inert gas welding of AA5052 aluminum alloy T-Joints. 1067:53–59
11. Atabaki MM, Yazdian N, Kovacevic R (2016) Partial penetration laser-based welding of aluminum alloy (AA 5083–H32). Optik (Stuttg) 127(16):6782–6804
12. Chang B, Allen C, Blackburn J, Hilton P (2013) Thermal and fluid flow characteristics and their relationships with porosity in laser welding of AA5083. Phys Proc 41:478–487
13. Baskoro AS, Milyardi I, Amat MA (2020) The effect of welding parameter on mechanical properties and macrostructure of AA1100 using autogenous TIG welding. Int J Automot Mech Eng 17(1):7562–7569
14. Himarosa RA, Rahman MBN, Adi RK, Putra LA, Nugroho SW (2022) Materials today : proceedings effect of MIG welding speed butt-joint on physical and mechanical properties of materials AA 5083. Mater Today Proc 2022;(xxxx):8–13
15. Venugopal T, Rao KS, Rao KP (2004) Studies on friction stir welded AA 7075 aluminum alloy. Trans Indian Inst Met 57(6):659–663
16. Karunakar P, Kumar P (2020) Influences of TIG welding parameters on tensile behavior of Aluminium AA 6061 alloy V-Groove Joint. J Mater Sci Mech Eng (JMSME) 2(4):317–320
17. Singh L, Singh R, Singh NK, Singh D, Singh P (2013) An evaluation of TIG welding parametric influence on tensile strength of 5083 aluminium alloy. Int J Mech Aerospace, Ind Mechatronics Eng 7(11):1262–1265
18. Javed M, Khalid Hussain A, Lateef A (2010) Influence of welding speed on tensile strength of welded joint in TIG welding process. Int J Appl Eng Res 1(3):518–527
19. Sakthivel T, Sengar GS, Mukhopadhyay J (2009) Effect of welding speed on microstructure and mechanical properties of friction-stir-welded aluminum. Int J Adv Manuf Technol 43(5–6):468–473
20. Zhan X, Zhao Y, Liu Z, Gao Q, Bu H (2018) Microstructure and porosity characteristics of 5A06 aluminum alloy joints using laser-MIG hybrid welding. J Manuf Process 35(July):437–445
21. Darby A, Severn R (2001) Welded design theory and practice. In: Welded design theory and practice. Woodhead Publishing Limited, pp 11–21
22. Lau TW, Bowker JT, Lazor RB (1986) First report of Haz study. Welding for challenging environments. Pergamon Press, pp 167–180
23. Mallick PK (2010) Materials, design and manufacturing for lightweight vehicles related titles, 1st edn. Woodhead Publishing Limited, United Kingdom, p 376
24. Bailey N (1994) Weldability of ferritic steels, 1st edn. Woodhead Publishing Limited, England

Effect of Residual Stress on Corrosion Polarization Mechanism of the ASME SA-179 Material in Corrosive Environment

Reza Putra, Teuku Hafli, Muhammad, and Nurul Islami

Abstract The aim of this research is to investigate the polarization behavior of corrosion on the SA-179 material, which is commonly used in heat exchanger equipment. The existence of tube components as a conduit for cooling fluid such as water or other solutions is the most significant feature of the temperature control of the heat exchanger. The cooling media is frequently a corrosive solution, which may degrade the heat exchanger material's performance. Furthermore, the corrosion process could be accelerated by residual stress generated by the shape of the heat exchanger tube formed by the cold working procedure. The SA-179 Carbon Steel Material, which has been manufactured to represent a heat exchanger tube, metallographic observations, polarization measurements, and macro-observations on the surface of the corroded area have been carried out in this work. To produce the strain and residual stress conditions seen in the elongation and compression segments, the SA-179 material was loaded using the U-Bend method. To generate a corrosion polarization curve that describes corrosion behavior under the conditions of plastic strain and residual stress, the potentiostatic measurement has been applied. Plastic deformation has an effect on grain boundary conditions in the SA-179 material, according to the average grain size discovered in this work. The polarization behavior is also influenced by this circumstance. The corrosion potential value is − 0.766 V at a current of 1.9×10^{-7}. A in the absence of strain and residual stress. Furthermore, the corrosion potential is demonstrated to be − 0.823 V at a current of 6.8×10^{-7}. A under conditions with U-Bend deformation. According to this phenomenon, plastic deformation has a

R. Putra · Muhammad · N. Islami (✉)
Department of Material Engineering, Malikussaleh University, Muara Satu, Indonesia
e-mail: nurulislami@unimal.ac.id

R. Putra
e-mail: reza.putra@unimal.ac.id

Muhammad
e-mail: muhammad.tm@unimal.ac.id

T. Hafli
Department of Mechanical Engineering, Malikussaleh University, Muara Satu, Indonesia
e-mail: hafli@unimal.ac.id

Irwansyah et al. (eds.), *Proceedings of the 4th International Conference on Experimental and Computational Mechanics in Engineering*, Lecture Notes in Mechanical Engineering, https://doi.org/10.1007/978-981-99-7495-5_56

bigger Nobel effect on SA-179 material in a corrosive environment, but has a higher corrosion rate if the corrosion process has already occurred.

Keywords SA-179 Carbon steel · Corrosion potential · Corrosion behavior · Plastic deformation · Residual stress · Corrosive environment · Image histogram

1 Introduction

SA-179 Carbon Steel, commonly known as ASTM A179, is an extensively used material in the construction industry. Many pipe bending procedures are used in the manufacturing business, for example, in the fabrication of heat exchangers, particularly tubular heat exchangers, condensers, and related heat transfer equipment [1]. The method of manufacturing this equipment, as well as the procedure of installing it, has an impact on the internal tension (stress) in the material. Internal stress can take the form of residual stress (residual stress) or strain (strain) on the material that is created and goes through plastic stress [2].

The potential value of the material is impacted by the construction of this equipment, which leaves residual stresses and changes in the shape of the material. If this material is in a corrosive environment, changes in its potential value have an impact on how it behaves. Equipment made of SA-179 material can be used in corrosive environments, such as a plant on the shore or even deep in the ocean.

As a result, it is crucial to investigate the effects of corrosion polarization on heat exchanger materials, particularly in corrosive environments, to understand how much this corrosion polarization behavior affects corrosion rate and equipment performance. Furthermore, the results of this research can be used to make recommendations for the development of materials and the processes used to create them. This research is also expected to contribute to the development of innovative methods and treatments for reducing the impact of strain and residual stress on a material's corrosion behavior.

The experimental material, SA-179 Carbon Steel, will be subjected to mechanical loading according to the ASTM G30 standard using the U-Bend forming method in this study [3]. The corrosion polarization mechanism will then be seen using an electrochemical approach, namely the measurement of anodic potentiostatic polarization in accordance with the ASTM G5 standard. The data were further evaluated using the Tafel extrapolation method to determine the value of the corrosion reaction that happened.

The cathode and anode response curves are combined to form the entire polarization curve. It clearly presents an overview of the polarization behavior related with the occurrence of passivity in the study by providing two polarization curves. Many researchers were conducting studies to analyze polarization on metals with varying pH levels, which could only be noticed using a complete polarization curve [4]. However, because passivity always happens in the anode region, which means when the material undergoes a corrosion reaction, many studies still look at polarization behavior solely at the anode [5].

According to another researcher who focuses solely on the polarization reaction in the anode region, this is true. Pitting corrosion that happens in materials that have undergone a welding procedure can also be seen through the polarization curve in the anode region, according to other researchers [6]. The welding technique in this example has caused a stress that causes pitting corrosion on the austenitic stainless steel [7]. The polarization reaction in the anode region was also reported by Kwok [8], which demonstrates a noticeable variation due to different welding processes. The diagram depicts critical information about polarization in the anode region.

Previous research has shown that when a structure is subjected to mechanical loads, corrosion can occur more quickly. The fissures became more significant due to corrosion occurring in the relevant structure, in addition to stress corrosion cracking [9, 10]. Stress corrosion cracking is also influenced by the corrosive environment. The phenomena of an open pipe receiving high stress from the surrounding environment (material workload), indicating the production of pitting corrosion on its surface, are one example given by A. El-Amoush [11]. According to previous research, if a pipe is exposed to mechanical loads from its environment for more than 10 years, the production of holes and cracks is more severe [11].

Polarization measurements can be performed using the electrochemical cell test method, according to ASTM G-5 and ASTM G-102 standards [12]. The standard required equipment as well as the concept of conducting a polarization test. In this scenario, the implanted resistance serves as a measuring test material. The test findings are in the form of current and potential data, which are plotted to produce a polarization curve, which is a graph illustrating the connection between current density and corrosion potential [13].

The potential measurement begins with the cathode state so that the oxygen in the solution does not form oxides on the material's surface during the measurement. The measurement is conducted in the positive direction to test the influence of the polarization that happens [3]. Sagues et. al. claim that the resistance potential must be regulated in order to produce the correct polarization reaction [14].

When using an electrochemical cell to detect potential, the test material used as a working electrode has its own resistance when no current is provided. Even if no current is provided, the reference electrode could detect the potential resistance of a material when a measurement equipment is attached. The restpot state is the name for this state. A value of 1000 mV is added to the negative direction of the restpot potential value to define the measurement's starting point [15–17]. Furthermore, the value is increased by up to 1000 mV in the positive direction for the measurement's end point. The goal of polarization reaction measurement is to obtain a graph of the polarization curve. The reactions that take place in the anode and cathode areas of the electrochemical cell are used as inputs to create a graph of the material's polarization curve.

Table 1 Composition of material

Composition	% (by weight)
C	0.25
Mn	0.63
P	0.035
S	0.035
Cr	0.07
Fe	Balance

2 Materials and Methods

Specimen, Solution, and Experiment Preparation. The samples were prepared according to the ASTM G-38 Standard. ASTM SA-179 has been chosen due to its extensive usage in manufacturing industries. Table 1 shows the chemical composition and mechanical properties of ASTM SA-179. U-Shape specimens as the working electrode (WE) for the electrochemical test were machined accurately in order to avoid the effect of surface roughness during electrochemical measurements. The specimens were coated using silicone metal after the electric connector was installed.

Electrochemical Measurement. The electrochemical experiments were performed on samples with a working area of 1.0 cm^2 that had been exposed to 3.5 wt% NaCl. A portable model of the potentiostat-type ZIVE PP1 (Korea) was linked to an electrochemical cell for the corrosion polarization measurements. A U-shaped specimen has served as the working electrode (WE). A silver chloride electrode (Ag/AgCl) and the Cu/CuSo4 electrode have served as the reference electrode, and a graphite bar served as the counter electrode (CE). Potentiostatic polarization measurement starts from − 1000 mV from the specimen's natural potential (restpot), to 1000mV above restpot with a silver chloride electrode. The potentiodynamic approach uses a 0.6 V/h (5%) potential sweep rate to continually record the current with a change in potential from − 1000 mV of the restpot. This indicates that the measurement began at the noble level and progressed to the active level [14, 18].

3 Results and Discussion

Granular Condition Post-plastic Deformation. Plastic deformation influences grain boundary conditions in the SA-179 material, according to the average grain size discovered in this work. Metallographic measurements illustrate how strain affects the grain size of a material that has been subjected to plastic deformation. The change in grain size can influence the corrosion behavior of the SA-179 material.

Figure 1a shows the grain boundary conditions in the specimen without receiving load and without deforming, as well as elastic and plastic deformation. Meanwhile, Fig. 1b, c demonstrate the grain boundary conditions of the specimen with plastic

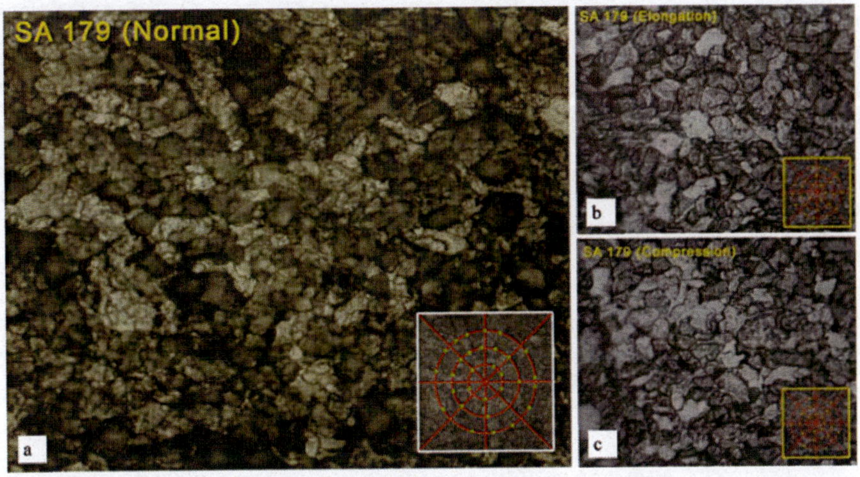

Fig. 1 Intercept point method and grain boundary conditions **a** normal SA-179 material, **b** elongation segment of U-Bend, **c** compression segment of U-Bend

deformation in the elongation and compression segments, respectively. An intercept counting guideline that meets the ASTM E112 standard is used to compute the grain boundary density. Plastic deformation has an effect on grain boundary conditions in the SA-179 material, as shown by the three descriptions of average grain size obtained in this work.

Polarization Curve of SA-179 Material. The current stability (Open-Circuit Potential—OCP) must be measured first during the experimental measurement of the corrosion mechanism. The OCP method entails injecting a 0.0 A current into the surface of the SA-179 material and then analyzing the electrochemical reactions that occurred. When potential stability occurs between the anode and cathode, there is no electric current injected into the material. As a result, the potentiostat measuring apparatus reads the inherent potential of the SA-179 material.

Figure 2 depicts the OCP value, or natural potential value, in the specimen without any material deformation. Using the Cu-CuSo4 reference electrode, OCP stability values ranged from -0.69214 V to 0.70313 V. When translated to the Ag–AgCl reference electrode, this value is 0.87714 V–0.88213 V. Without any plastic deformation on the surface of the material, the OCP stability of the SA-179 material was found to be in a very good OCP range of 11 mV in this investigation.

The natural potential of the material under U-Bend deformation is then measured, as shown in Fig. 3, with an OCP range of 16 mV for a 5-min experiment. Using the Cu-CuSO4 reference electrode, this OCP range is the difference between the minimum potential value of – 0.5954 V and the maximum potential value of – 0.5789 V.

The natural potential results on this SA-179 material reveal a potential difference of 107 mV at the starting value of OCP and 113 mV at the final value of OCP, indicating that the material with U-Bend deformation has a higher potential value than the material without plastic deformation. This condition shows that materials

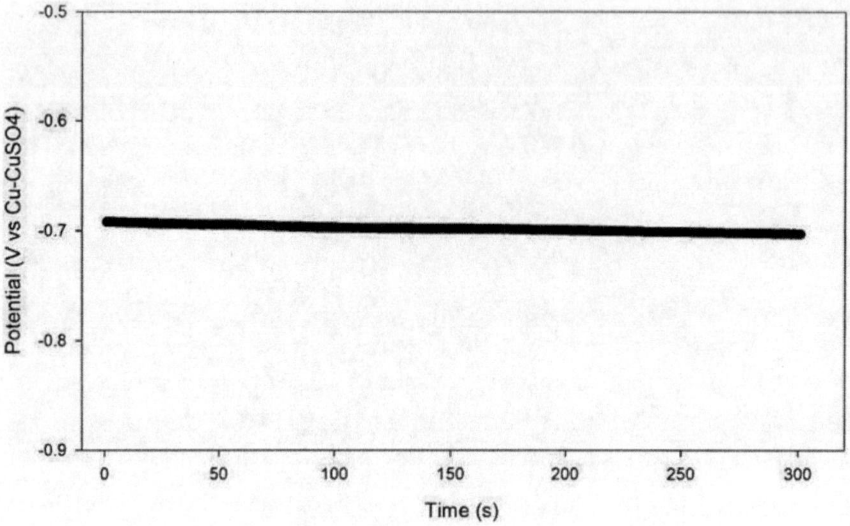

Fig. 2 Natural potential (OCP) of SA-179 material at normal condition

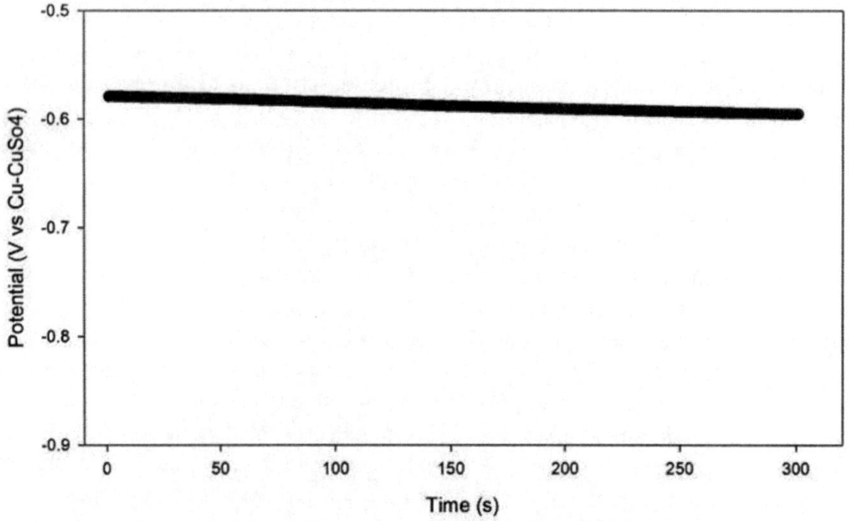

Fig. 3 Natural potential (OCP) of SA-179 material with U-Bend deformation

with plastic deformation have a higher corrosion potential than materials without plastic deformation. Furthermore, the number of corrosion potential that happens in these two sets of test objects may be seen by measuring the polarization curve.

The polarization curve was measured with the ZiveLab PP1 potentiostat using the potentiostatic method, with the measured starting potential 1000 mV lower than

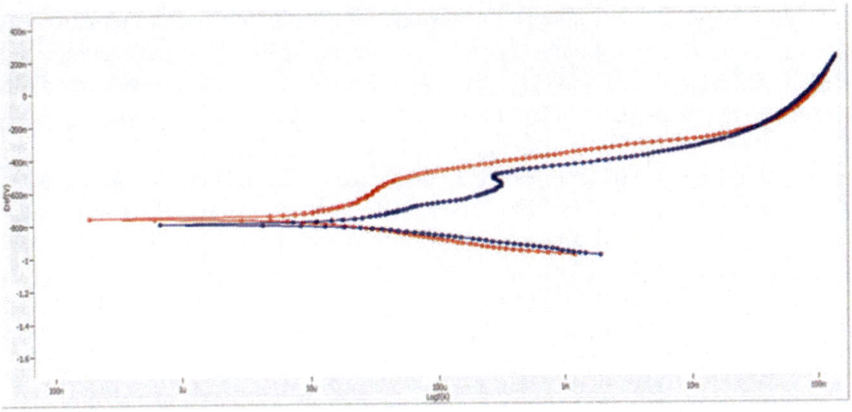

Fig. 4 Polarization curve of SA-179 material in normal condition (red) and under plastic deformation (blue)

the OCP value. The potential readings were also taken using Cu−CuSO41 reference electrode. The current injected during the measurement process is automatically adjusted to the SA-179 material's inherent potential. Figure 4 shows the results of the polarization test.

In a corrosive environment, the polarization curve of the SA-179 material without any plastic deformation, the current density at 1.9×10^{-7} A, and the corrosion potential zone is at − 0.766 V. There were no electrical potential spikes, indicating that there was no major corrosion reaction on the surface of the studied material.

Figure 4 also shows the polarization measurement on the SA-179 material in conjunction with the U-Bend deformation detected on the specimen's outer surface. At the anode area of the polarization curve, the polarization mechanism causes a spike in corrosion potential. This phenomenon is caused by deformation occurring on the surface of the specimen, also the residual stress and grain boundary deformation due to surface plastic deformation. Polarization curve shows the current density was 6.8×10^{-7} A, with the corrosion potential is − 0.823 V.

The pattern of the current density curve for the plastically deformed material is more positive than the normal material, indicating a higher corrosion rate, as can be seen in this image. Normal materials, on the other hand, have a lower corrosion potential than materials with plastic deformation, implying that materials with plastic deformation are slightly nobler than materials without deformation. With these phenomena, it can be deduced that the SA-179 material's plastic deformation has a more noble effect on the corrosive environment, but also has a larger corrosion rate if the corrosion process has begun.

Image Histogram Analysis. To observe the changes in brightness levels in an image, histogram analysis is employed. Histogram analysis is beneficial for evaluating color intensity spreads and can be used to make decisions, such as raising brightness or stretching contrast and color distribution. For image identification, the histogram could offer the percentage of color composition and intensity of texture.

Histogram analysis was employed in the study for corrosion surface observations to observe the distribution of the rust color spectrum caused by corrosion and the basic color of the observed material. It is feasible to get a more precise assessment of the distribution of corrosive rust on the material's surface using this technique [19].

The histogram of the corroded area is extracted using an image processing tool in order to produce a graphic comparison of the image histogram between the corroded surface and the uncorroded surface. In comparison to other metal surface variables, a 100px^2 image of the corroded surface is used as a control histogram. Figure 5 shows the histogram graph of the control picture for the corroded area.

The brightness levels at the four color frequencies, namely Red (R), Green (G), Blue (B), and Greyscale/Luma (L), are on the left side of the image histogram graph, as shown in Fig. 5. The corroded area has a low level of brightness as a result of these phenomena. According to the notion of an image histogram, which describes the brightness level on a scale of 1–256, with one being at the far-left end and 256 being at the far right end of the histogram graph.

Figure 6 shows the histogram graph of the picture on the surface of the specimen without plastic deformation.

Figure 6 shows the RGBL color frequencies' uniform distribution across the image's brightness levels. A non-massive distribution of corrosion on the surface of

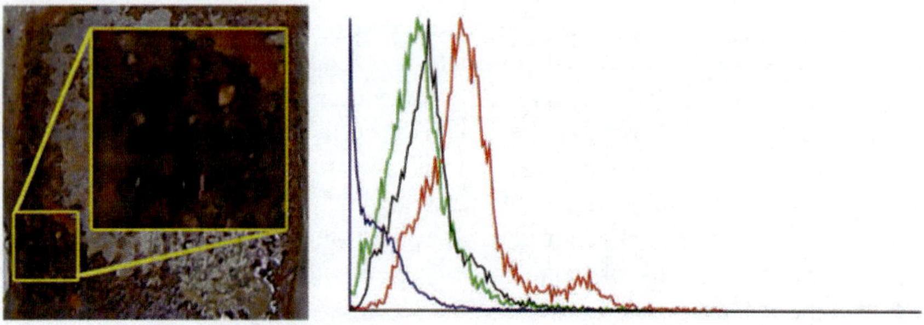

Fig. 5 Image histogram graph for the corroded area as a reference histogram

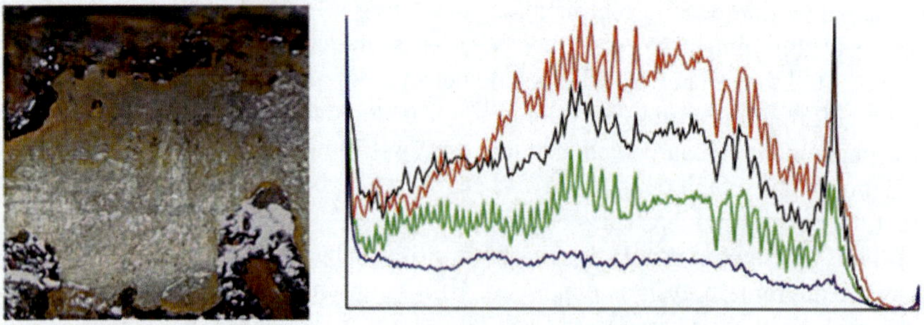

Fig. 6 Image histogram graph for the corroded area without plastic deformation

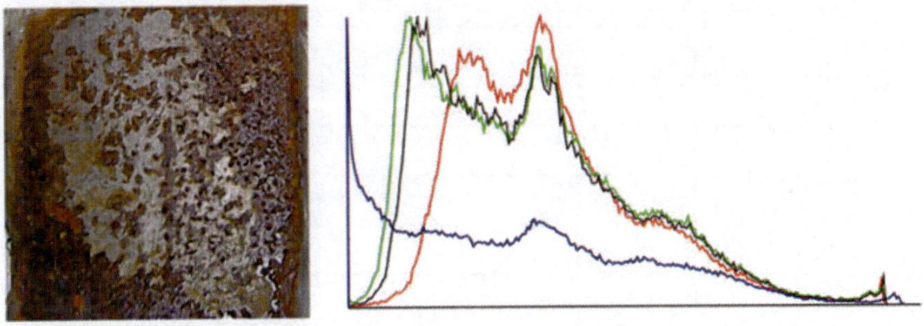

Fig. 7 Image histogram graph for the corroded area with plastic deformation

the SA-179 material that is not plastically distorted is indicated by the histogram graph of the image that is not identical to the histogram of the corroded surface image.

On the other hand, as shown in Fig. 7, the histogram graph of the picture on the plastically deformed surface reveals different phenomena.

Extrapolation Tafel Analysis. The corrosion analysis has used the Tafel extrapolation method to electrochemically determine the corrosion rate on the surface of the SA-179 material [17]. The Tafel extrapolation graph was produced from the results of the polarization measurements using the IVMAN application to make it easier to build an intercept line between the anode and cathode areas from the resulting polarization curve. A Tafel extrapolation analysis on the SA-179 material, which is not plastically damaged, is shown in Fig. 8.

The E_{corr} value is − 682,238 mV, and the I_{corr} value is 9.422 A, according to the Tafel extrapolation analysis in Fig. 8. The intercept line that connects the anode and cathode regions of the polarization curve has the E_{corr} value of − 682,238 mV and the I_{corr} value of 9.422 A. According to the results of the analysis, the corrosion rate in the non-deformed section was 19.72 MPY (Mills per Year). Meanwhile, multiple situations can be visible on the surface of the plastically deformed material. The corrosion rate is 24.75 MPY in this area. Figure 9 shows a Tafel extrapolation analysis for a surface with plastic deformation.

The interception between the cathode region and the anode region on the polarization curve resulted in an E_{corr} value at − 783,639 mV and an I_{corr} value at 11,828 μA, according to the Tafel extrapolation analysis. As a result, the increased corrosion rate in the plastically deformed material differs.

4 Conclusions

Plastic deformation has an effect on grain boundary conditions in SA-179 material, according to the average grain size determined in this study. This disease has an impact on polarization behavior as well. Plastic deformation has a more noble

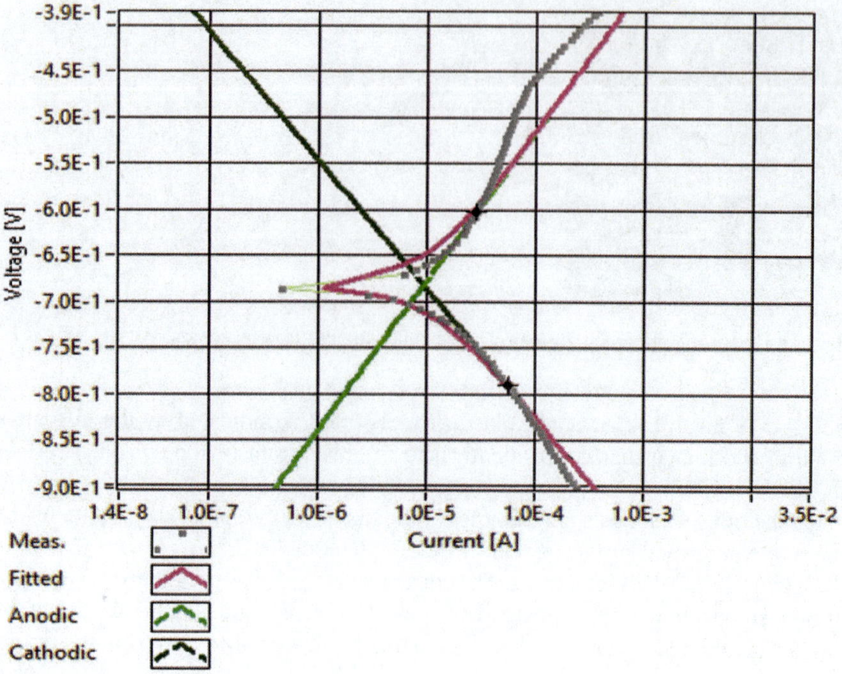

Fig. 8 Tafel extrapolation analysis on SA-179 material without plastic deformation

effect on SA-179 material in a corrosive environment, but has a larger corrosion rate if the corrosion process has occurred, as shown by the corrosion polarization phenomena. The extent of damage caused by corrosion on the specimen's surface can be determined using image histogram analysis.

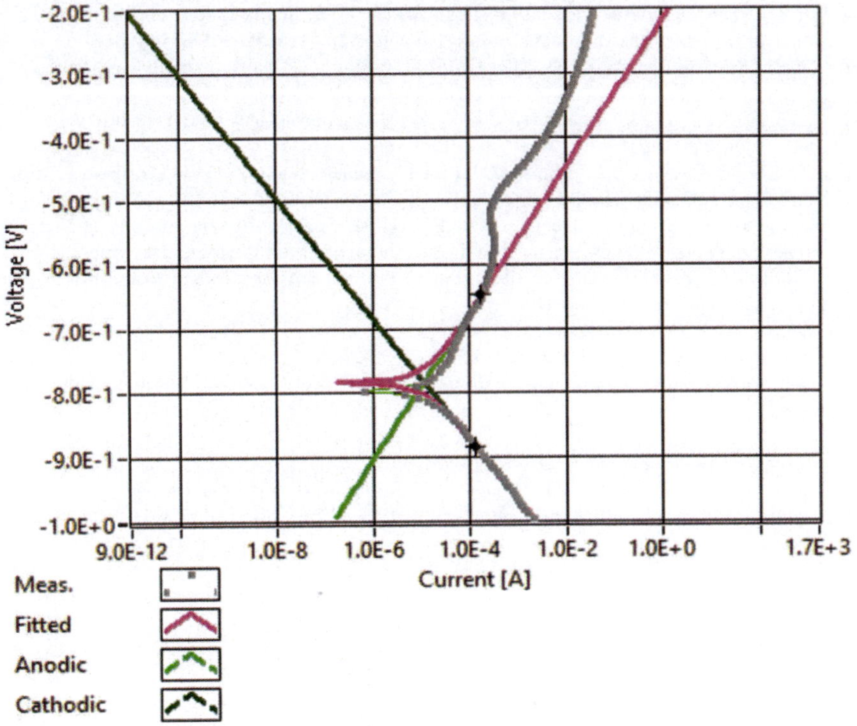

Fig. 9 Tafel extrapolation analysis on surface of SA-179 material with plastic deformation

Acknowledgements The authors would like to thank the Research Center and Community Service (LPPM) of Malikussaleh University, who have funded the research under PNBP Scheme with number 6/PPK-2/SPK-JL/2022.

References

1. Sharma P, Roy H (2015) Mill scale corrosion and prevention in carbon steel heat exchanger. High Temp Mater Proc 34(6):571–576
2. Okuma SO, Eyito JM, Obhuo M (2021) Effect of corrosion on mechanical properties of medium carbon steel in different selected media. Int Res J Appl Sci Eng Tech 7(11):16–24
3. ASTM-G30 (1995) Standard practice for making and using U-bend stress-corrosion test specimens
4. Elsariti SM, Haftirman (2013) Behaviour of stress corrosion cracking of austenitic stainless steels in sodium chloride solutions. Proced Eng 53(0):650–654
5. Meng Y, Gao H, Hu J, Gao L (2019) Effect of pH value on the corrosion and corrosion fatigue behavior of AM60 magnesium alloy. J Mater Res 34(6):1054–1063
6. Yi Y, Cho P, Al Zaabi A, Addad Y, Jang C (2013) Potentiodynamic polarization behaviour of AISI type 316 stainless steel in NaCl solution. Corros Sci 74:92–97

7. Lu BT, Chen ZK, Luo JL, Patchett BM, Xu ZH (2005) Pitting and stress corrosion cracking behavior in welded austenitic stainless steel. Electrochim Acta 50(6):1391–1403
8. Kwok CT, Fong SL, Cheng FT, Man HC (2006) Pitting and galvanic corrosion behavior of laser-welded stainless steels. J Mater Process Technol 176(1–3):168–178
9. Huang Y (2002) Stress corrosion cracking of AISI 321 stainless steel in acidic chloride solution. Bull Mater Sci 25:47–51
10. Zhu LK, Yan Y, Qiao LJ, Volinsky AA (2013) Stainless steel pitting and early-stage stress corrosion cracking under ultra-low elastic load. Corros Sci 77:360–368
11. El-Amoush AS, Zamil A, Jaber D, Ismail N (2014) Stress corrosion cracking of the pre-immersed tin brass heat exchanger tube in an ammoniacal solution. Mater Des 56:842–847
12. ASTM-G102 (1999) Standard practice for calculation of corrosion rates and related information from electrochemical measurements
13. Alvarez MG, Lapitz P, Ruzzante J (2008) AE response of type 304 stainless steel during stress corrosion crack propagation. Corros Sci 50(12):3382–3388
14. Sagues AA, Krane SC, Presuel Moreno FJ, Silva Araya WI (1997) Advanced computational model for sacrificial cathodic protection of submerged reinforced concrete marine footers. Repair and rehabilitation of reinforced concrete structures. The State of the Art, American Society of Civil Engineering, Reston, pp 1–13
15. Islami N, Rashid S, Ariffin AK, Nuawi MZ (2017) Stress corrosion damage on austenitic stainless steel in sodium chloride. Int J Automot Mech Eng 14(1). https://doi.org/10.15282/ijame.14.1.2017.2.0312
16. Rashid S, Islami N, Ariffin AK, Ridha M, Fonna S (2016) The effect of immersion time on the corrosion behavior of SUS304 in brine using half-cell potential measurement. J Teknologi 78(6–9)
17. Badea GE, Caraban A, Sebesan M, Dzitac S, Cret P, Setel A (2010) Polarisation measurements used for corrosion rates determination. J Sustenable Energy 1(1):1–4
18. Chen J, Asmussen RM, Zagidulin D, Noël JJ, Shoesmith DW (2013) Electrochemical and corrosion behavior of a 304 stainless-steel-based metal alloy wasteform in dilute aqueous environments. Corros Sci 66:142–152
19. Xiong J, Yu D, Wang Q, Shu L, Cen J, Liang Q, Chen H, Sun B (2021) Application of histogram equalization for image enhancement in corrosion areas. Shock and Vibration

Analysis of Distinct Substrate Pre-treatment Effects on Flame-Sprayed Hydroxyapatite Coating

Rilo Berdin Taqriban, Prima Syahri Ramadhan, Jamari, Rifky Ismail, and Athanasius Priharyoto Bayuseno

Abstract This paper experiments with stainless-steel AISI 316L substrate pre-treatment techniques in preparation for hydroxyapatite (HAp) flame spray coating. The mechanical, chemical, and combination pre-treatments were compared for the coating thickness and layer adhesion of the HAp to understand the influence of roughness on the layer adhesion of the flame-sprayed HAp layer on the flat substrate. The heat treatment at 800 °C for 1 h with 160 °C/h heating and cooling rate was conducted after spraying. Phase identification by XRD analysis shows that the substrates were partially coated with amorphous HAp. All pre-treated specimens have coating thickness that complies with medical requirements. Sandblasting creates the best hydroxyapatite layer coverage due to the highest surface roughness. On the other hand, the acidic immersion with sulfuric acid lowered the roughness value and caused the hydroxyapatite layer to have poor adhesion.

Keywords Hydroxyapatite · Stainless-steel 316L · Flame spray coating · Pre-treatment · Roughness

1 Introduction

Stainless-steel AISI 316L is the most used metallic implant material with good mechanical strength, wear, and corrosion resistance [1–3]. Unfortunately, this material has a low ability to form a mechanical and chemical bond with the natural bone [4]. Scientists have also found that this metal's nickel and chromium content may become dangerous and carcinogenic after implantation due to ion release [5]. With these drawbacks, the scientist and implant industries are starting to avoid AISI

R. B. Taqriban · P. S. Ramadhan · Jamari · R. Ismail · A. P. Bayuseno (✉)
Department of Mechanical Engineering, Diponegoro University, Semarang, Indonesia
e-mail: apbayuseno@gmail.com

R. B. Taqriban · R. Ismail
Center for Biomechanics, Biomaterial, Biomechatronics, and Biosignal Processing (CBIOM3S), Diponegoro University, Semarang, Indonesia

Irwansyah et al. (eds.), *Proceedings of the 4th International Conference on Experimental and Computational Mechanics in Engineering*, Lecture Notes in Mechanical Engineering, https://doi.org/10.1007/978-981-99-7495-5_57

316L materials and replace them with titanium-based and CoCr materials [1, 6–8]. Meanwhile, in developing countries like Indonesia, the AISI 316L is still the mainly chosen implant material because of its cheap and currently available implant material covered by national health insurance. This problem can be overcome by coating the AISI 316L implant material with hydroxyapatite (HAp).

HAp is an inorganic mineral in the bones with a high biocompatibility property. Coating implants with this material can accelerate osseointegration, bone healing after surgery, reduce the risk of wear and corrosion, and prolong implant life [9, 10]. The flame spray coating method is one of the effective thermal coating methods widely developed for biomedical applications [11–14]. However, this method still has various drawbacks; one of them is the low adhesion of the coating due to low temperature, microcracks, and inadequate surface treatments [13, 15–17]. The specimen's roughness is one factor in successful flame spray coating [18]. To obtain adequate surface roughness, substrate pre-treatment is needed before the coating is done [19].

In this study, two treatments were used to get essential parameters to produce a suitable surface roughness for coating. These treatments were sandblasting and acidic immersion in H_2SO_4 (sulfuric acid) [20, 21]. The experiments were conducted by pre-treating AISI 316L substrate with these two distinct methods and the combination. Before flame spray coating, the surface roughness of the specimens was measured to understand the pre-treatment effect on surface roughness and layer adhesion of the hydroxyapatite coating. This study has also analyzed the XRD, image processing, and coating thickness to obtain beneficial information from the distinct pre-treatment method.

2 Materials and Method

Before pre-treatment, the 50 × 50 × 3 mm AISI 316L substrates from PT. Sutindo Raya Mulia (Surabaya, Indonesia) were sanded with 100, 240, 400, and 1000 grid sandpaper, degreased, and rinsed in deionized water to remove the protective layer and unwanted debris. The different pre-treatment was conducted with sandblasting using #100 mesh of aluminum oxide, sulfuric acid immersion for 5 min, and a combination of the two. After the treatment, the surface roughness test was carried out with the Time TR200 surface roughness tester. Measurements were made at five points on the surface of the specimens to get the average value representing the entire surface. The five points taken are described in Fig. 1.

The oxy-acetylene QHT-7/h flame spray gun from Shanghai Welding and Cutting Tool Works Manufacture (Shanghai, China) was used in this study. The 3 mm powder hole nozzle was used with 7 and 4 bar oxygen and acetylene pressure, respectively. The spray gun was operated manually with a standoff distance of 20 cm. The coating material used was hydroxyapatite powder from Bioworld with product code 40,800,054 (Dublin, Ireland). After coating, the specimens were post-heated using a Tabletop Rapidfire Pro-LP electric furnace (Tacoma, USA) with 800 °C for an

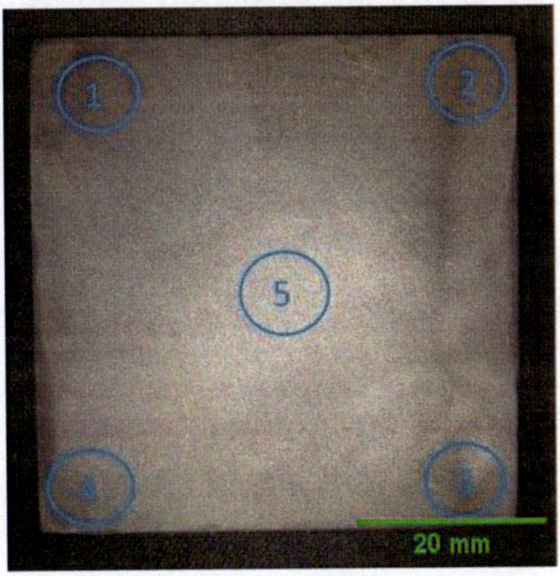

Fig. 1 Surface roughness measuring points

hour holding time with a heating and cooling rate of 160 °C/hour. The X-ray diffraction (XRD) with Shimadzu PV 7000 (Japan) was used to collect crystallographic data. The coating thickness was measured with a UNI-T UT343D coating thickness gauge (Dongguan City, China). The final step is the layer coverage evaluation of the coated specimens using the ImageJ processing software [22].

3 Results

The images of substrates were taken after pre-treatment. The images are used to compare the visible differences in the texture of AISI 316L. The following Fig. 2 shows the visual data obtained.

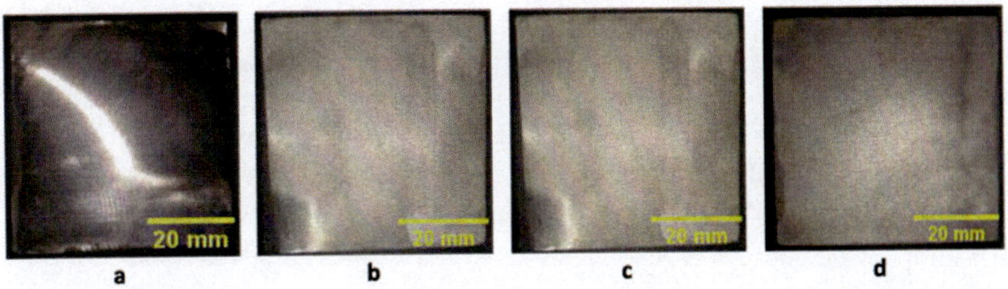

Fig. 2 Surface texture of the untreated (**a**), sandblasted (**b**), acidic immersed (**c**), and sandblast + acid-immersed (**d**) substrates

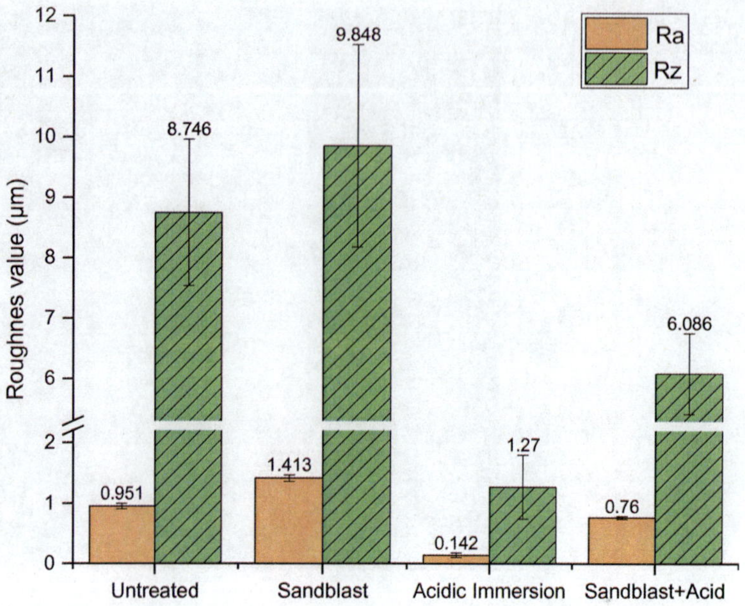

Fig. 3 Average value of Ra and Rz of the specimens

The roughness value is expressed in Ra and Rz. Ra is the most widely used international roughness parameter. Ra is the arithmetic mean and absolute deviation of the roughness profile from the mean, while Rz is the maximum peak-to-valley height or irregularity within sampling [23]. Figure 3 shows the average and standard deviation of surface roughness Ra and Rz values obtained from the surface of the pre-treated specimens at five points described in Fig. 1.

The XRD evaluation was conducted with 30 kV/30 mA Cu-K radiation from 2θ 15° to 90°. The spectra of the coated specimen were stacked with XRD spectra of AISI 316L base material and hydroxyapatite JCPDS 9-0432 to identify the crystal peak of the coated specimens. The stacked XRD data can be seen in Fig. 4.

The coating thickness was measured at five different points of the coated specimens. The average and standard deviations of the measurement are presented in Fig. 5.

The percentage of hydroxyapatite coverage after coating and post-heat on the AISI 316L surface is obtained using ImageJ software. The images captured in this session were under controlled ambient light and not using any flashlight to avoid the shining effect that could be mistaken in the analysis. Table 1 shows the coated specimens' image and the layer adhesion percentage analysis results.

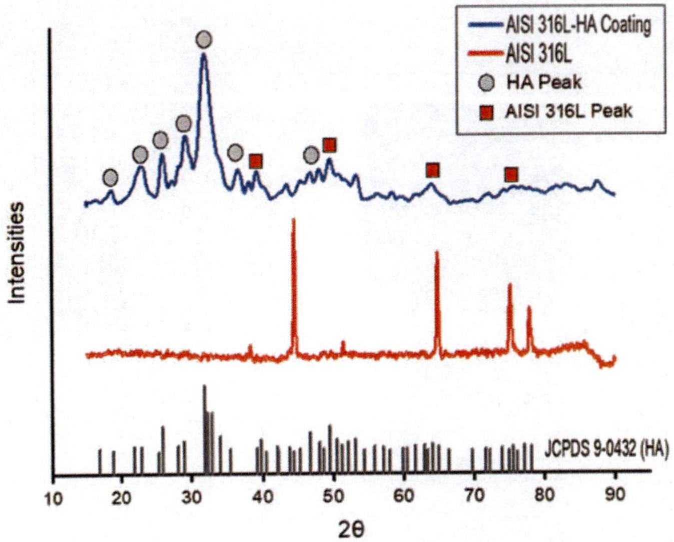

Fig. 4 XRD data of the coated specimens with AISI 316L and JCPDS 9-0432 references

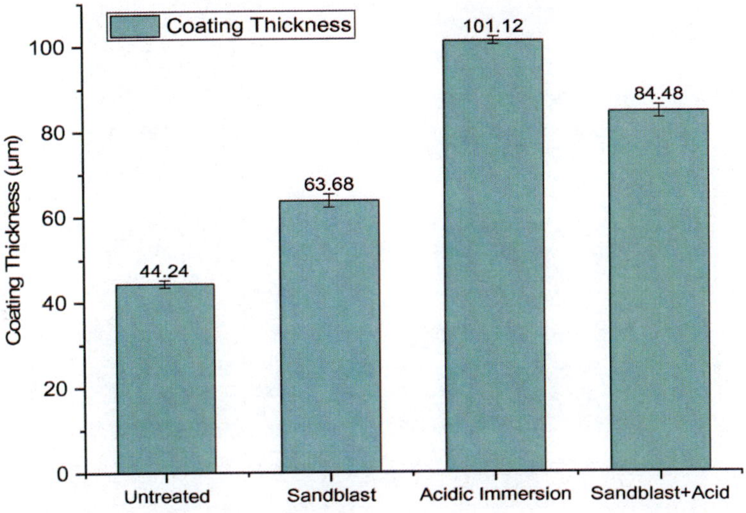

Fig. 5 Coating thickness of coated specimens

4 Discussions

From Fig. 2, the surface textures of the pre-treated substrates were visually less shiny than the untreated substrate. AISI 316L will be less shiny when given more pre-treatment, as is when the sandblasted and acidic immersed AISI 316L in Fig. 2b,

Table 1 As sprayed and image processing results of coated specimens

Specimen	As sprayed	Processed	%
Untreated	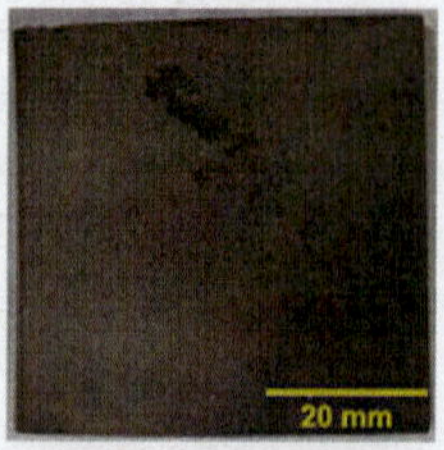		76
Sandblast	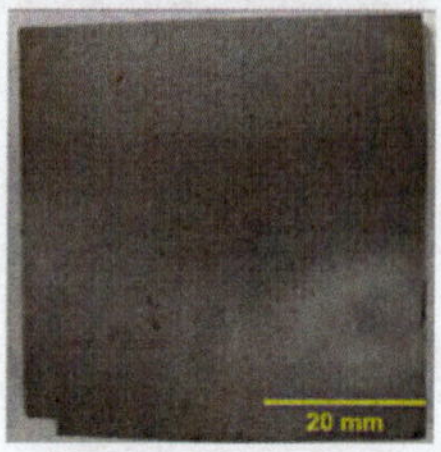	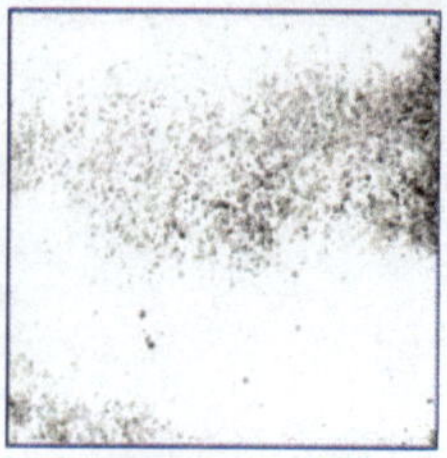	94
Acidic immersion		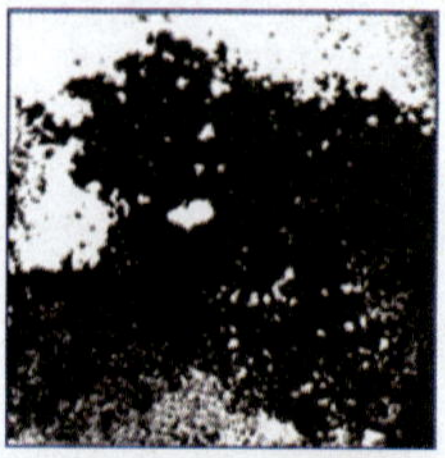	30
Sandblast + acid	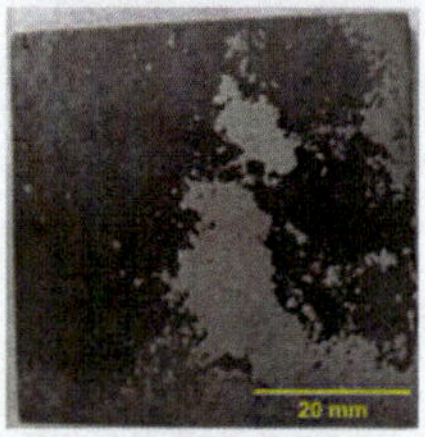	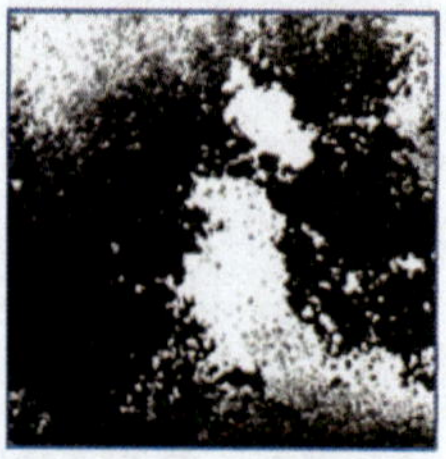	34

c looks less shiny than the untreated in 2a. Meanwhile, the difference between the sandblasted and acidic immersed substrates was almost unnoticeable.

From the roughness data in Fig. 3, the highest roughness was obtained by the sandblasted specimen with a Ra value of 1.413 μm, while the lowest was when AISI 316L was immersed in sulfuric acid with a value of 0.142 μm. The irregularities' value or Rz shows the stability of the surface roughness contour because the smaller the Rz value, the less peak and valley range. AISI 316L after immersion in H_2SO_4 offers the smallest Rz value of 1.270 μm. On the other hand, the highest Rz value occurred in the sandblasted specimen with a value of 9.848 μm. In comparison, the specimens with acidic immersion produce lower Ra and Rz values than the sandblasted and untreated specimens. It can be concluded that sulfuric acid immersion reduces the Ra and Rz values significantly.

Phase identification by XRD in Fig. 4 shows that the substrates are successfully coated with the HAp layer. The peaks of the HAp appear at 2θ of 34.0°, 32.9°, 31.7°, 25.8°, 46.7°, and 49.4° which match with JCPDS data of HAp [24]. The peaks of the AISI 316L substrate also appear at several points, indicating that the coating is porous or there are uncoated areas on the substrate. The non-pointy or valley-like spectra suggest that the layer consists of many amorphous phases, which is not preferable [25, 26]. The crystallinity can be improved with extended post-heat treatment for 2 h or more [27].

The recommended HAp layer thickness for biomedical applications is 50–200 μm [28]. Recently, the recommended thickness could be approximately 50 μm to reduce cost and prevent residual stresses [29]. According to these references, the pre-treated specimens' thickness in Fig. 5 is in accordance with the requirement. The sandblasting and acidic immersion give a higher coating thickness than the untreated specimen. However, the adhesion of the hydroxyapatite layer of chemically pre-treated specimens is poor, where only some parts of the surface were coated.

The coating results show that the sandblasted AISI 316L has the best results with the perfect adhesion of hydroxyapatite powder. Meanwhile, the acid-immersed AISI 316L resulted in unsatisfactory coating layer coverage. The untreated AISI 316L produced a layer with a less clear hydroxyapatite surface, but the coating was still attached. All AISI 316L can be coated, but the acid-immersed substrates mainly experienced layer detachment after coating and post-heat when the temperature started to decrease. This phenomenon can be caused by low mechanical interlock because of low roughness value and thermal expansion [30–33]. However, from the surface roughness data obtained, surface roughness is the main factor of the coating layer detachments. This phenomenon could happen due to the low surface roughness after being given sulfuric acid immersion pre-treatment. The low surface roughness makes the hydroxyapatite not have a solid place to form a mechanical interlocking bond to the substrate.

ImageJ analysis shows that the percentage of hydroxyapatite layer on AISI 316L without pre-treatment was 76% and 94% on AISI 316L after sandblasting treatment. Meanwhile, after immersion in sulfuric acid, the AISI 316L coating only contained 30% of the adherent. The results of the AISI 316L coating after sandblasting and sulfuric acid immersion were only 34% adherent. This result shows that the sandblasting pre-treatment has the best layer adhesion with 94% layer coverage.

5 Conclusions

The research on the effect of post-treatment surface roughness on the hydroxyapatite coating on AISI 316L was successfully conducted. Pre-treatment will determine the surface roughness of AISI 316L. The highest surface roughness can be obtained by sandblasting. The H_2SO_4 immersion pre-treatment reduces the surface roughness value and causes the hydroxyapatite layer to peel off because of the low mechanical interlock between the substrate and hydroxyapatite layer. Despite that, the acidic immersion produces a more stable surface contour indicated by a smaller Rz value. The coating thickness meets the biomedical standard for all four specimens. The use of acidic immersion adds more thickness but less adhesion. For future applications, sandblasting is considered the most recommended pre-treatment for thermal spray coating applications because it can produce the highest percentage of hydroxyapatite attachment to the AISI 316L surface compared to other pre-treatments.

Acknowledgements The authors thank Diponegoro University, Indonesia, for funding this research through Pendidikan Master menuju Doktor untuk Sarjana Unggul (PMDSU) scholarship with Grant Number: 14-03/UN7.P4.3/PP/2020.

References

1. Merola M, Affatato S (2019) Materials for hip prostheses: a review of wear and loading considerations. Materials (Basel) 12. https://doi.org/10.3390/ma12030495
2. Anwar IB, Santoso A, Saputra E, Ismail R, Jamari J, van der Human HE (2017) Bone marrow-derived mesenchymal cell reactions to 316L stainless steel: an in vitro study on cell viability and interleukin-6 expression. Adv Pharm Bull 7:335–338 (2017). https://doi.org/10.15171/apb.2017.040
3. Yang K, Ren Y (2010) Nickel-free austenitic stainless steels for medical applications. Sci Technol Adv Mater 11:014105. https://doi.org/10.1088/1468-6996/11/1/014105
4. Liu X, Chu PK, Ding C (2004) Surface modification of titanium, titanium alloys, and related materials for biomedical applications. Mater Sci Eng R Rep 47:49–121. https://doi.org/10.1016/J.MSER.2004.11.001
5. Brown SA, Simpson JP (1981) Crevice and fretting corrosion of stainless-steel plates and screws. J Biomed Mater Res 15:867–878. https://doi.org/10.1002/jbm.820150611
6. Hua Z, Dou P, Jia H, Tang F, Wang X, Xiong X et al (2019) Wear test apparatus for friction and wear evaluation hip prostheses. Front Mech Eng 5:1–5. https://doi.org/10.3389/fmech.2019.00012
7. Ayu HM, Izman S, Daud R, Krishnamurithy G, Shah A, Tomadi SH et al (2017) Surface modification on CoCrMo alloy to improve the adhesion strength of hydroxyapatite coating. Procedia Eng 184:399–408. https://doi.org/10.1016/j.proeng.2017.04.110
8. Sarker A, Tran N, Rifai A, Brandt M, Tran PA, Leary M et al (2019) Rational design of additively manufactured Ti6Al4V implants to control Staphylococcus aureus biofilm formation. Materialia 5:100250. https://doi.org/10.1016/j.mtla.2019.100250
9. Juliadmi D, Harlendri, Hon Tjong D, Manjas M, Gunawarman (2019) The effect of sintering temperature on bilayers hydroxyapatite coating of titanium (Ti-6Al-4V) ELI by electrophoretic deposition for improving osseointegration. IOP Conf Ser Mater Sci Eng 547. https://doi.org/10.1088/1757-899X/547/1/012005

10. Rani S, Jatolia SN, Scholar R (2018) Electrodeposited hap coatings on titanium alloys: a Review. Int J Res Anal Rev www.ijrar.org 847:847–853
11. Yao HL, Hu XZ, Wang HT, Chen QY, Bai XB, Zhang MX et al (2019) Microstructure and corrosion behavior of thermal-sprayed hydroxyapatite/magnesium composite coating on the surface of AZ91D magnesium alloy. J Therm Spray Technol 28:495–503. https://doi.org/10.1007/s11666-018-0815-y
12. Gadow R, Killinger A, Stiegler N (2010) Hydroxyapatite coatings for biomedical applications deposited by different thermal spray techniques. Surf Coatings Technol 205:1157–1164. https://doi.org/10.1016/j.surfcoat.2010.03.059
13. Cheang P, Khor KA (1995) Thermal spraying of hydroxyapatite (HA) coatings: effects of powder feedstock. J Mater Process Tech 48:429–436. https://doi.org/10.1016/0924-0136(94)01679-U
14. Liu YC, Lin GS, Wang JY, Cheng CS, Yang YC, Lee BS et al (2018) Synthesis and characterization of porous hydroxyapatite coatings deposited on titanium by flame spraying. Surf Coatings Technol 349:357–363. https://doi.org/10.1016/j.surfcoat.2018.06.010
15. Mohseni E, Zalnezhad E, Bushroa AR (2014) Comparative investigation on the adhesion of hydroxyapatite coating on Ti–6Al–4V implant: a review paper. Int J Adhes Adhes 48:238–257. https://doi.org/10.1016/j.ijadhadh.2013.09.030
16. Kahar S, Singh A, Dabhekar S, Vala U, Navadiya B, Patoliya K (2020) Thermal Sprayed Aluminium Coatings : a review
17. Gapsari F, Hidayati NA, Setyarini PH, Alan MPN, Subagyo R, Andoko A (2021) Hydroxyapatite coating on stainless steel 316l using flame spray technique. Int J Eng Trans B Appl 34:493–499. https://doi.org/10.5829/IJE.2021.34.02B.22
18. Alsabti MA, Ciuca I, Ştefan Vasile B, Trusca R, Abou Harb A (2018) Effect of surface roughness Ti6Al4V modified by hydroxyapatite coating. Ann "Dunarea Jos" Univ Galati Fascicle IX. Metall Mater Sci 41:41–48. https://doi.org/10.35219/mms.2018.4.06
19. Wahab HA, Noordin MY, Izman S, Kurniawan D (2013) Quantitative analysis of electroplated nickel coating on hard metal. Sci World J 2013. https://doi.org/10.1155/2013/631936
20. Kadhim MJ, Abdulateef NE, Abdulkareem MH (2018) Evaluation of surface roughness of 316L stainless steel substrate on Nanohydroxyapatite by electrophoretic deposition. Al-Nahrain J Eng Sci 21:28. https://doi.org/10.29194/njes21010028
21. Murakami T, Noda I, Ikeda J, Nakahira A (2015) Investigation of chemical state of vanadium at the interface between thermal sprayed hydroxyapatite coating and Ti-6Al-4V alloy substrate. Funtai Oyobi Fummatsu Yakin/J Japan Soc Powder Powder Metall 62:408–412. https://doi.org/10.2497/jjspm.62.408
22. Chambard M, Remache D, Balcaen Y, Dalverny O, Alexis J, Siadous R et al (2020) Effect of silver and strontium incorporation route on hydroxyapatite coatings elaborated by rf-SPS. Materialia 12. https://doi.org/10.1016/j.mtla.2020.100809
23. Panda A, Sahoo AK, Rout AK (2016) Investigations on surface quality characteristics with multi-response parametric optimization and correlations. Alexandria Eng J 55:1625–1633. https://doi.org/10.1016/j.aej.2016.02.008
24. Saber-Samandari S, Baradaran S, Nasiri-Tabrizi B, Alamara K, Basirun WJ (2018) Microstructural evolution and micromechanical properties of thermally sprayed hydroxyapatite coating. Adv Appl Ceram 117:452–460. https://doi.org/10.1080/17436753.2018.1495895
25. Li TT, Ling L, Lin MC, Peng HK, Ren HT, Lou CW et al (2020) Recent advances in multifunctional hydroxyapatite coating by electrochemical deposition. J Mater Sci 55:6352–6374. https://doi.org/10.1007/s10853-020-04467-z
26. Bose S, Tarafder S, Bandyopadhyay A (2015) Hydroxyapatite coatings for metallic implants 7. Elsevier Ltd.; 2015. https://doi.org/10.1016/b978-1-78242-033-0.00007-9
27. Khor KA, Cheang P, Wang Y (1997) The thermal spray processing of HA powders and coatings. Jom 49:51–57. https://doi.org/10.1007/BF02915482
28. Furlong RJ, Osborn JF (1991) Fixation of hip prostheses by hydroxyapatite ceramic coatings
29. Gross KA, Saber-Samandari S, Heemann KS (2010) Evaluation of commercial implants with nanoindentation defines future development needs for hydroxyapatite coatings. J Biomed Mater Res Part B Appl Biomater 93B:1–8. https://doi.org/10.1002/JBM.B.31537

30. Weiss H (1995) Adhesion of advanced overlay coatings: mechanisms and quantitative assessment. Surf Coatings Technol 71:201–207. https://doi.org/10.1016/0257-8972(94)01022-B
31. Kim W-S, Yun I-H, Lee J-J, Jung H-T (2010) Evaluation of mechanical interlock effect on adhesion strength of polymer–metal interfaces using micro-patterned surface topography. Int J Adhes Adhes 30:408–417. https://doi.org/10.1016/j.ijadhadh.2010.05.004
32. Park K, Kim K, Kim D, Moon B, Park S, Seok CS (2021) Failure mechanism of plasma-sprayed thermal barrier coatings under high-temperature isothermal aging conditions. Ceram Int 47:15883–15900. https://doi.org/10.1016/j.ceramint.2021.02.163
33. Baytak T, Bulut O (2022) Thermal stress in functionally graded plates with a gradation of the coefficient of thermal expansion only. Exp Mech 62:655–666. https://doi.org/10.1007/s11340-021-00818-2

PDE-Constrained Inverse Analysis Using Bayesian Optimization for Finding Hidden Corrosion Given Partial Surface Information

Israr B. M. Ibrahim, Syarizal Fonna, T. Arriessa Sukhairi, Rudi Kurniawan, and Syifaul Huzni

Abstract To ensure infrastructure resilience, regular assessment of infrastructure damage must be carried out. To assess hidden corrosion in reinforced concrete structure, inverse analysis can be employed. Inverse analysis combines partial field data and computational analysis, including machine learning. In this paper, Bayesian optimization was used for inverse analysis to find hidden corrosion in a reinforced concrete corrosion. A simple case study was presented where the hidden corrosion location was known. The partial data on the surface of the concrete was known and used for the inverse analysis. The Bayesian optimization was successfully used to find the location of the hidden corrosion. The Max Expected Improvement acquisition function performs better than Max Probability Improvement one.

Keywords PDE-constrained optimization · Inverse analysis · Reinforced concrete corrosion · Bayesian optimization · Machine learning · Finite element analysis

I. B. M. Ibrahim (✉) · S. Fonna · S. Huzni
CCRG Laboratory, Department of Mechanical Engineering, Universitas Syiah Kuala, Aceh 23111, Indonesia
e-mail: israr@usk.ac.id

S. Fonna
e-mail: syarizal.fonna@usk.ac.id

S. Huzni
e-mail: syifaul@usk.ac.id

T. Arriessa Sukhairi · R. Kurniawan
Department of Mechanical Engineering, Universitas Syiah Kuala, Aceh 23111, Indonesia
e-mail: teuku.arriessa@usk.ac.id

R. Kurniawan
e-mail: kurniawan@usk.ac.id

Irwansyah et al. (eds.), *Proceedings of the 4th International Conference on Experimental and Computational Mechanics in Engineering*, Lecture Notes in Mechanical Engineering, https://doi.org/10.1007/978-981-99-7495-5_58

1 Introduction

The Partial Differential Equation (PDE) and optimization are a staple in engineering analysis, as well as optimization. Optimization typically requires constraints in the form of a function or a constant. The PDE-constrained optimization is a class of optimization where the PDE is the constraint. The PDE-constrained optimization found many applications in engineering, including shape optimization [1–3], fluid mechanics [4, 5], materials engineering [6], biomechanics [7], inverse analysis [8], and others [9]. In this study, the application of PDE-constrained inverse analysis using Bayesian optimization is demonstrated for finding hidden corrosion in reinforced concrete. Reinforced concrete is the main materials of infrastructures worldwide [10]. Degradation of reinforced concrete structures incurs high cost to society [10]. Hence, the development of corrosion monitoring and assessment methods are crucial.

Infrastructure safety and resilience are also one of the main objectives in Sustainable Development Goals (SDG), specifically SDG 9.1, 9.a, and 11.b [11], where both the development of infrastructure and its related policies to ensure reliability and resilience are encouraged. Infrastructure monitoring and assessment involve complex interaction between data, computation, and field measurement. Hence, the application of machine learning techniques is a natural solution. The Government of Indonesia has specifically formulated a national strategic plan for data-driven infrastructure resilience as part of smart city program [12].

Corrosion of rebar is one of the main mechanisms of reinforced concrete degradation. Corrosion is typically modeled as Laplace's equation with polarization functions as its boundary conditions. In this study, the case of corrosion in rebar is formulated as a combinatorial optimization and employed Bayesian optimization for inverse analysis of corrosion profiles. As far as the literature reviews go, this is the first study to employ Bayesian optimization on PDE-constrained formulation for corrosion inverse analysis.

2 Methods

The Hidden Corrosion Problem in Reinforced Concrete Structure. The corrosion in rebar occurs electrochemically. The concrete is a porous structure enabling penetration of species that can trigger electrochemical reaction of corrosion. Once the species reach the rebar, corrosion reaction may occur. The corrosion reaction causes the generation of potential fields across the concrete bulk. This potential field can be measured at the surface of the concrete (Fig. 1). Figure 1 illustrates its mechanism.

This potential field can be modeled using Laplace's equation,

$$\kappa \nabla^2 \phi(x, y) = 0 \tag{1}$$

$$q = \partial \phi / \partial \mathrm{n} \tag{2}$$

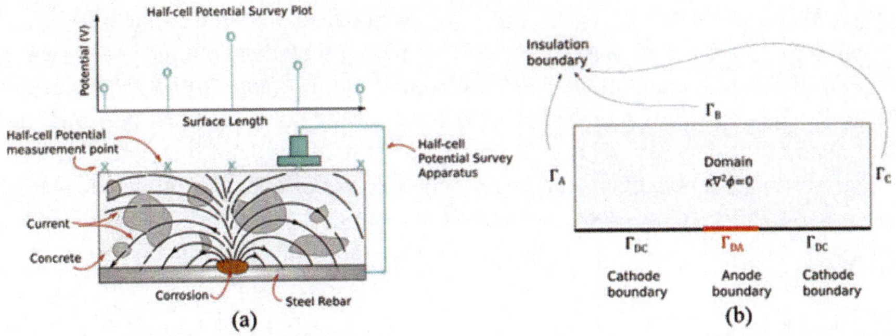

Fig. 1 **a** Half-cell potential survey is an NDT method for hidden corrosion detection in reinforced concrete structures. **b** PDE modeling of hidden corrosion in a reinforced concrete structure

$$q = 0 \text{ at } \Gamma_A, \Gamma_B, \Gamma_C \tag{3}$$

$$\phi = f_a(q) \text{ at } \Gamma_{DA} \tag{4}$$

$$\phi = f_c(q) \text{ at } \Gamma_{DC} \tag{5}$$

where ∇ is Laplacian in terms of Cartesian coordinate, q is the current density, ϕ is electrochemical potential, n is normal vector, κ is conductivity, and Eqs. (3), (4), and (5) are the boundary conditions. Equation (3) is non-electrode boundaries of the system that is modeled as insulation, where the current density is zero due to discontinuity or distant from source. Equation (4) is a boundary condition representing the anode or the corroded part of the rebar, while Eq. (5) represents the rest of the rebar. To capture the polarization phenomenology in corrosion, Eqs. (4) and (5) are functional boundary conditions, where the function f_a (for anode polarization) and f_c (for cathode polarization) are obtained from curve-fitting of experimental polarization data (which depends on the electrodes and electrolyte).

The information obtained from the surface of the concrete structure is enough to infer the condition of corrosion on the rebar inside the concrete. This is termed inverse analysis. The concept is to search through different numerical simulation scenarios to find the one that best matches partial information from real measurement. It can be mathematically stated as an optimization problem,

$$\min\sqrt{\left|\phi_m^2 - \phi_s^2\right|} \tag{6}$$

subject to Eqs. (1)–(5), ϕ_m is experimentally/field-measured potential on the concrete surface, ϕ_s is simulated potential on the concrete surface.

Bayesian Optimization. Bayesian optimization is a model-based, black-box optimization method. It searches for the optima by sampling the objective function. To sample the objective function, it is to build a model of the entire function that it is optimizing, that is a surrogate model. This model includes both the current estimate of that function and the uncertainty around that estimate. It then chooses the next point by computing a posterior distribution of the objective function using the likelihood of the data already acquired and a prior on the type of function.

The surrogate model that is directly compatible with Bayes theorem is the Gaussian process (GP). To model the objective function, $f(x)$, using GP, a number of test points, x_i, were selected over the domain of $f(x)$. The objective function, $f(x)$, as a GP is then,

$$f(x) \sim \mathrm{GP}\big(m(x), k\big(x_i, x_j\big)\big), \tag{7}$$

where $m(x)$ is mean function and $k(x_i, x_j)$ is covariance function. Once a surrogate model has been built, the next task is to select the best candidate solutions through an acquisition function. The best solution is selected through an acquisition function or infill criterion. Common acquisition functions are Expected Improvement (EI, Eq. 9) and Probability of Improvement and (PI, Eq. 8).

$$\mathrm{PI}(x) = 1 - \mathrm{CDF}\left(\frac{f(x_\mathrm{B}) - \mu(x)}{\sigma(x)}\right) \tag{8}$$

$$\mathrm{EI}(x) = (\mu(x) - f(x_\mathrm{B}))\mathrm{CDF}\left(\frac{\mu(x) - f(x_\mathrm{B})}{\sigma(x)}\right) + \mathrm{PDF}\left(\frac{\mu(x) - f(x_\mathrm{B})}{\sigma(x)}\right). \tag{9}$$

For the case of topological optimization in this study, the objective function is Eq. (6) where the variable is ϕ_B which is obtained from the constraints of the objective function (Eqs. 7–10). The optimization was carried out in the Wolfram Mathematica 12.0 environment.

Implementation and Case Studies. The algorithm and the numerical models were developed in Wolfram Mathematica 12.0 environment. In the two-dimensional model, the surface of the concrete and the steel rebar are defined by the model's boundary Γ_B, Γ_DC, Γ_DA, consecutively. Hence, the optimization problem reduces to finding the best combination of Γ_DC and Γ_DA. It is as if the surrogate model developed by the Bayesian optimization relates the ϕ_s and arrangement of Γ_DC and Γ_DA, or,

$$\phi_s = f(\Gamma_\mathrm{DC} \cup \Gamma_\mathrm{DA}). \tag{10}$$

A study case was carried out with a simple rectangular model of a concrete block with dimensions of 2 × 1 m. The hidden corrosion is located 0.6 m from the left side of the block (Fig. 2a). The length of the hidden corrosion is 0.4 m. The model is discretized with 302 triangular elements (Fig. 2b). As in Fig. 1b, the boundary Γ_B is the simulated location of half-potential measurement. Thus, twenty-one potential

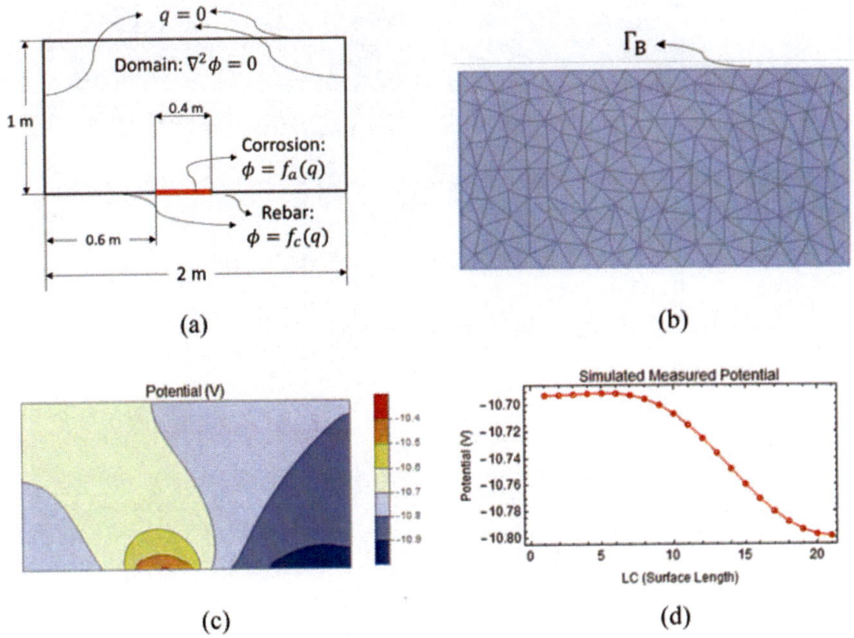

Fig. 2 **a** Boundary conditions of the PDE modeling of hidden corrosion in reinforced concrete structure, **b** the 302 triangular elements as a mesh used in this study, **c** the potential distribution in the domain when a corrosion is located on the left-middle of the steel rebar, **d** twenty-one potential values on the Γ_B boundary, simulating half-cell potential measurement

values were obtained from the FEA results to simulate the half-cell measurement. These 21 potential values are the ϕ_m as in Eq. (6).

3 Results and Discussions

The Bayesian optimization was carried out with termination condition of 300 iterations. The progressions of Bayesian optimization are shown in Fig. 3 along with samples of potential distribution at some iterations. The optimized solution was found at iteration 196th, with objective function value of − 0.011. It can be seen that as the objective function value reduces, the potential distribution becomes more similar to the target potential distribution. Figure 4 shows the comparison between potential values on boundary Γ_B found by the Bayesian optimization and the 21 values of ϕ_m. It can be seen that the optimized solution agrees well with the ϕ_m.

Max Expected Improvement (MEI) Versus Max Probability Improvement (MPI). The results of Bayesian optimization from using different acquisition functions are compared. Figure 5 shows the progression of Bayesian optimization using

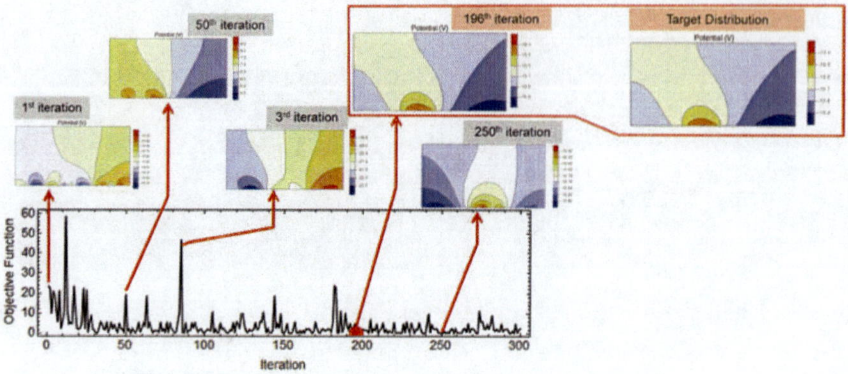

Fig. 3 Iteration of Bayesian optimization, along with potential distributions generated by the optimal solution at some iterations. The optimal solution at iteration 196th is indicated by red circle in the chart

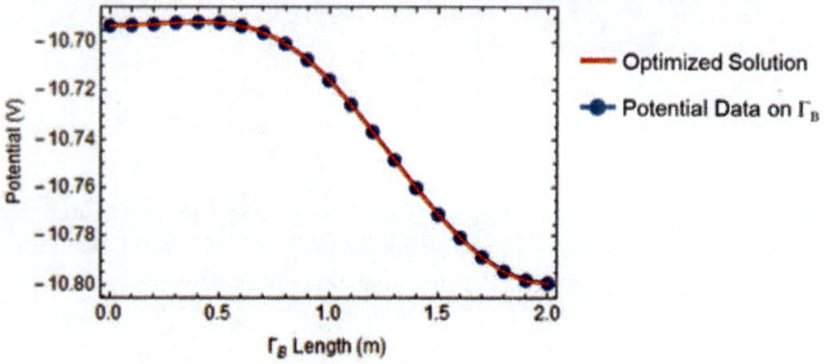

Fig. 4 Comparison of potential values on Γ_B from the optimized solution and the simulated potential data

MEI and MPI. It can be seen that the MEI found the optimal solution earlier (at iteration 157th) than the MPI (at iteration 196th). Thus, for this case of PDE-constrained optimization, the MEI outperforms the MPI.

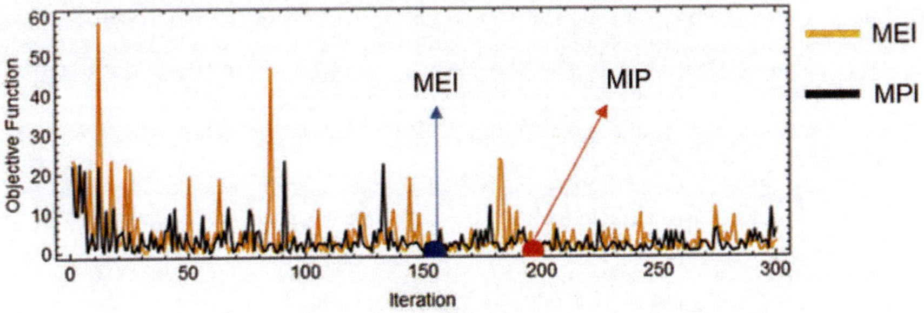

Fig. 5 Performance of Bayesian optimization with MEI and MPI acquisition function

4 Conclusion

A PDE-constrained optimization was formulated for the problem of finding hidden corrosion in reinforced concrete. The objective of the optimization is to find the numerical scenario that matches the given partial information about the corrosion. More specifically, the task is to find the arrangement of boundary conditions to Laplace's equation that yields target potential distribution. Bayesian optimization was used successfully to find the optimal solution. The MEI acquisition function outperforms the MPI in accomplishing the optimization task. However, further investigation is needed to determine if this performance trend applies only to the problem studied in this paper or applies to a broader set of problems.

Acknowledgements The authors thank the Ministry of Education, Culture, Research, and Technology, Republic of Indonesia, Penelitian Dasar (PD) No. 20/UN11.2.1/PT.01.03/DPRM/2022.

References

1. López J, Valizadeh N, Rabczuk T (2022) An isogeometric phase–field based shape and topology optimization for flexoelectric structures. Comput Methods Appl Mech Eng 391:114564. https://doi.org/10.1016/J.CMA.2021.114564
2. Chen W, Gao C, Zhang W, Gong Y (2022) Adjoint-based unsteady shape optimization to suppress transonic buffet. Aerosp Sci Technol 127:107668. https://doi.org/10.1016/J.AST.2022.107668
3. Kambampati S, Chung H, Kim HA (2021) A discrete adjoint based level set topology optimization method for stress constraints. Comput Methods Appl Mech Eng 377:113563. https://doi.org/10.1016/J.CMA.2020.113563
4. Okubo CM, Kiyono CY, Sá LFN, Silva ECN (2021) Topology optimization applied to 3D rotor flow path design based on the continuous adjoint approach. Comput Math Appl 96:16–30. https://doi.org/10.1016/J.CAMWA.2021.05.006
5. Okubo CM, Sá LFN, Kiyono CY, Silva ECN (2022) A discrete adjoint approach based on finite differences applied to topology optimization of flow problems. Comput Methods Appl Mech Eng 389:114406. https://doi.org/10.1016/J.CMA.2021.114406

6. Kumar H, Dash A, Paul A, Bhattacharyya S (2022) A physics-informed neural network-based numerical inverse method for optimization of diffusion coefficients in NiCoFeCr multi principal element alloy. Scr Mater 214:114639. https://doi.org/10.1016/J.SCRIPTAMAT.2022.114639
7. Knopoff DA, Fernández DR, Torres GA, Turner CV (2013) Adjoint method for a tumor growth PDE-constrained optimization problem. Comput Math Appl 66(6):1104–1119. https://doi.org/10.1016/J.CAMWA.2013.05.028
8. Anevlavi D, Belibassakis KA (2022) Analysis of partially cavitating hydrofoils under the free surface using BEM-based adjoint optimization. Appl Math Model. https://doi.org/10.1016/J.APM.2022.07.033
9. Halilovic S, Böttcher F, Kramer SC, Piggott MD, Zosseder K, Hamacher T (2022) Well layout optimization for groundwater heat pump systems using the adjoint approach. Energy Convers Manag 268:116033. https://doi.org/10.1016/J.ENCONMAN.2022.116033
10. Angst UM (2018) Challenges and opportunities in corrosion of steel in concrete. Mater Struct 51(1):1–20
11. United Nations, Do you know all 17 SDGs? https://sdgs.un.org/goals. Accessed 8 Aug 2022
12. BPPT (2020) Strategi Nasional Kecerdasan Artifisial Indonesia [Online]. Available: https://ai-innovation.id/server/static/ebook/stranas-ka.pdf

Experimental Investigation of Non-destructive Vibrational Evaluation Techniques for Determining Elastic Properties of Jute and Glass Fiber Reinforced Composites

Muhammad Rizal, Taufiq Ramadhana Alhaidar, and Amir Zaki Mubarak

Abstract Recently, natural and synthetic fiber reinforced composites may find application in nearly every sector of engineering. In this study, an experimental analysis of the dynamic features of jute fiber reinforced polymer composites (JFRPC) and glass fiber reinforced polymer composites (GFRPC) for the purpose of estimating Young's modulus is presented. The natural frequency value of composite plates made by JFRPC and GFRPC was measured using impulse excitation technique (IET) experiments. The tensile test, however, was carried out to verify the IET test's results. The findings indicate that the strength of a composite made of glass fibers is noticeably better than that of a composite made of jute fibers. However, the mechanical properties of woven jute composite are equivalent to those of glass fiber composite. This is especially true in regard to its Young's modulus, which is almost identical to that of the glass fiber composite. It is anticipated that this method will be able to predict the mechanical characteristics of other materials and that it will have potential as a non-destructive examination method.

Keywords Impulse excitation technique · Young's modulus · Jute and glass fiber composites · Tensile test · Non-destructive test

M. Rizal (✉) · T. R. Alhaidar · A. Z. Mubarak
Department of Mechanical Engineering, Faculty of Engineering, Syiah Kuala University (USK), 23111 Darussalam, Banda Aceh, Indonesia
e-mail: muh.rizal@usk.ac.id

A. Z. Mubarak
e-mail: amir_zm@usk.ac.id

Irwansyah et al. (eds.), *Proceedings of the 4th International Conference on Experimental and Computational Mechanics in Engineering*, Lecture Notes in Mechanical Engineering, https://doi.org/10.1007/978-981-99-7495-5_59

1 Introduction

Implementation of composite materials or structures demonstrates growth in terms of product efficiency, economic effectiveness, and the development of material properties such as strength and modulus. Its application needs are also growing to suit the requirements of sectors such as aircraft building, wind turbines, transportation and automotive, medical equipment, and other industries [1]. The production of composite materials is a complicated process that necessitates the completion of a large number of steps. During this process, the finished composite product may have a wide variety of flaws, which can lead to issues with the user's safety. The identification and evaluation of structural properties can be difficult when dealing with composite materials since these materials are typically inhomogeneous and anisotropic.

The modulus of elasticity is an important feature of a material, and its value is one of the most important input parameters in construction design and quality control of materials. Tests may be performed to determine the modulus of elasticity using a variety of direct and indirect methods. A direct method that takes into account longitudinal and transverse deformations, as well as the effects that are directly connected to those deformations, is utilized in the calculation of the modulus of elasticity. Tests of tensile, flexural, and torsional strength are only a few examples. While the indirect technique relies on determining elastic constants like the shear modulus and modulus of elasticity [2]. The inverse approach, commonly referred to as the indirect technique is based on dynamic experimental data [3].

Previous researchers have used sound waves to use an indirect approach for determining the modulus of elasticity of glass fiber reinforced epoxy composite materials. This method has been successful in determining the value of the modulus of elasticity [4]. When they compared the findings of the destructive test (tensile test) to the results of the modulus of elasticity, they found that there was a 2% discrepancy. Ultrasonic waves have also been utilized for the purpose of determining the elasticity of composite materials [5]. On the other hand, it is not evaluated in comparison to the results of the tensile tests. In a similar manner, Niutta et al. [6] conducted research with the intention of determining the local elasticity characteristics of composite boards through the use of the impulse excitation technique (IET). They report that there is a variation of around 4% between the estimated and nominal elastic properties of the producer data. This difference may be found in the estimated elastic properties. Rizal et al. [7] also conducted research on evaluating the composite material characteristics of jute fiber reinforced polymer using numerical and experimental methods. They come to the conclusion that the variable results of the IET in comparison with the numerical results have relative errors for the calculated material parameters that range from 0.34% all the way up to 8.82%.

However, non-destructive investigations using modal analysis that are compared directly with tensile testing for a variety of composite materials are rare across the research literature. It is very important for the validation of the technique, as well as for ensuring accuracy in the estimate of material properties, to give attention to the amount of uncertainty produced by the use of the approach. The variability in the results of the method's application is important for the validation and accuracy of the estimation of material properties. This work used a non-destructive test method through the impulse excitation technique (IET) to assess the mechanical and dynamic characteristics of the composite structure made of jute and glass fibers, which were validated by a destructive test.

2 Material and Method

2.1 Sample Preparation

In this particular study, the reinforcement was comprised of both woven jute and glass fibers. A mixture of polyester resin and hardener with a ratio of 10:1 was used to form the matrix composition for this experiment. The resin and hardener were mixed into the matrix by hand, and the mixture was then agitated. Before reinforcing the fiber mats in the matrix body, the mixture was completely mixed. Additionally, the ratio of matrix to fiber was set at 70 and 30 wt%, respectively. By using the hand layup technique, the specimen laminates were fabricated with dimensions of 300 × 300 mm, which comes to an average thickness of approximately 5 mm. The compression molding machine was used to press the specimen laminates while the temperature was set to ambient temperature. After curing for one day under pressure, the composite was released from its mold. After being removed from the mold, this cast underwent a further day of air curing. The laminates of woven jute fiber reinforced polymer composite (JFRPC) and glass fiber reinforced polymer composite (GFRPC) were cut into the required ASTM D638 tensile tests after the compression stage was completed. In order to prepare, we made three different samples so that we could meet the requirement for repetition that each test had.

2.2 Tensile Test

The tensile testing was carried out using the Universal Testing Machine (UTM), which has a capacity of 5000 kgf. The axial direction of the test was carried out at a constant speed of 5 mm/min. The machine was doing tests on a total of six specimens, three of which were JFRPC and the remaining three were GFRPC. The ASTM standard D638 was used as a guide for the tensile test, and the dimensions of the specimen are depicted in Fig. 1.

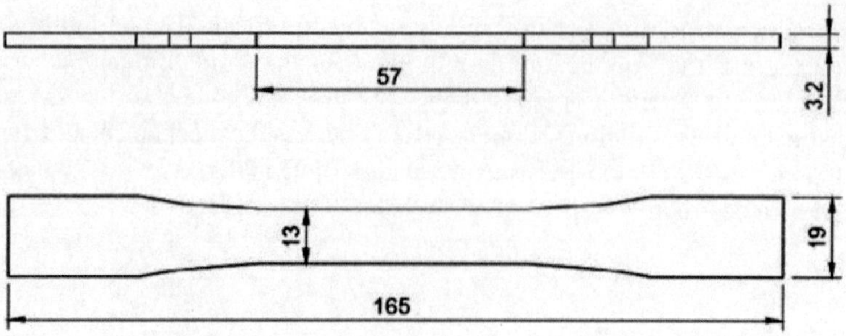

Fig. 1 Dimension of the specimen of the tensile test

2.3 Impulse Excitation Technique

For the characterization of Young's modulus (E) of materials, the impulse excitation technique (IET) is proposed as an indirect identification and non-destructive testing technique. The composite laminates with dimensions of 140 × 170 × 4 mm that were utilized as samples for the IET test were based on ASTM standard 1876-01 and were made of jute and glass fiber reinforced polymer composites (JFRPC and GFRPC, respectively). The measuring setup for this experimental research is shown in Fig. 2. The test's objective is to determine the laminated composite plate's natural frequency by excitation the sample with a PCB modally tuned impact hammer. A PCB 1-axis accelerometer was used to measure the vibration generated on by the mechanical impact on the plate's surface. In order to do the fast Fourier transform (FFT) analysis, the signals are first collected using NI-9250 data acquisition and then transmitted to LabView software. The sample was positioned on two supports that were 0.224 of the sample's entire length from each end, in order to determine the natural frequency in the flexural mode. Young's modulus can be determined using the IET test approach through the following equation [7]:

$$E = 0.9465\left(\frac{mf_f^2}{b}\right)\left(\frac{L^3}{t^3}\right)T, \quad (1)$$

where E is Young's modulus (Pa), m is mass (g), f_f is the flexural resonant frequency (Hz), b is width (mm), L is length (mm), t is thickness (mm), and T is the correction factor for the flexural mode that takes into account the thickness of the plates. When $L/t \geq 20$, T can be determined by the following formula:

$$T = 1.000 + 6.585(t/L)^2 \quad (2)$$

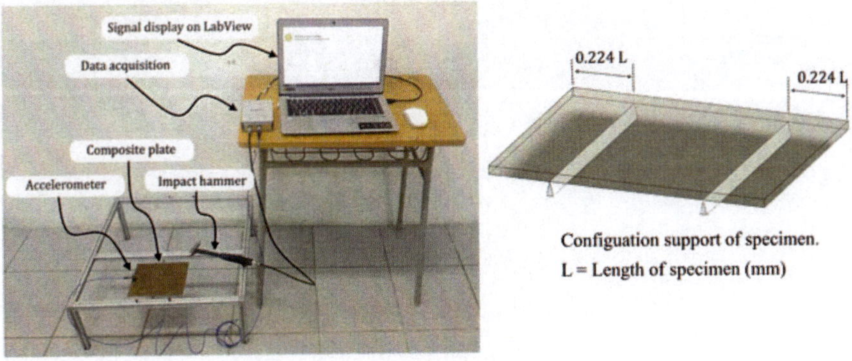

Fig. 2 Experimental setup for IET test in flexural mode

3 Results and Discussion

3.1 Tensile Test

The results of a tensile test that was carried out on an investigated jute fiber reinforced polymer composite (JFRPC), as well as a glass fiber reinforced polymer composite (GFRPC), are shown in Fig. 3. Analyses were performed on three separate specimens of each type of composite material. All of the test curves for the two distinct types of composites demonstrate a linear range at low strain (< 0.5%), and then they change into a hardening region up to the breaking point when they reach a strain of between 1.5 and 2.8%. This occurs when the strain reaches a point that is greater than the linear range at low strain. It is possible to see that the elongation at break of the two different kinds of composites is practically comparable, but the average tensile strength of GFRPC (32.9 MPa) is greater than that of JFRPC (26.7 MPa). This may be confirmed by comparing the experimental results (25.02 MPa). The elongation and strength results from this study are consistent with those from other studies, which found that the tensile strength of glass fiber composite was around 28–78 MPa and that the tensile strength of jute fiber composite material was somewhere between 23 and 40 MPa [8][8].

3.2 Impulse Excitation Technique

The FRF plots from the tested samples composites fabricated from jute (JFRP) and glass (GFRP) fiber were averaged throughout all three trials, and the results can be seen in Fig. 4. It is clear that the vibrational response exhibits itself as a natural frequency in the first mode, which has a frequency range of 256–307.2 Hz for JFRPC,

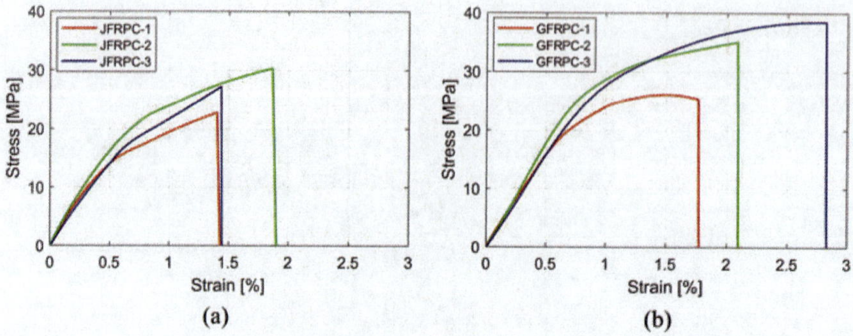

Fig. 3 Stress–strain curve: **a** JFRPC; **b** GFRPC

and 317.4–337.9 Hz for GFRPC. Although the values from each of the three specimens that were repeated implied consistency, the third sample seemed to be slightly different due to the fact that the sample's thickness varied and was not consistent along the edges, which caused a difference in sample mass when compared to the other two samples.

The results of the analysis and measurement of Young's modulus utilizing the natural frequency values of the two different types of composites using Eq. 1 are displayed in Table 1. The results of the tensile test and the IET test are summarized. The simple reality that there was only a little degree of variability in the mechanical properties of the material, as demonstrated by these data, adds validity to the processes that were utilized in preparing and cutting the samples. The results of Young's modulus indicated that the findings of JFRPC and GFRPC are not substantially different from one another.

The data demonstrate that the average value of Young's modulus found from the IET test is between 4.63 and 5.09% greater than the value of Young's modulus obtained from the tensile test. The main reason for this seemingly little gap is due to the main distinctions that exist between the two methods with regard to the theoretical

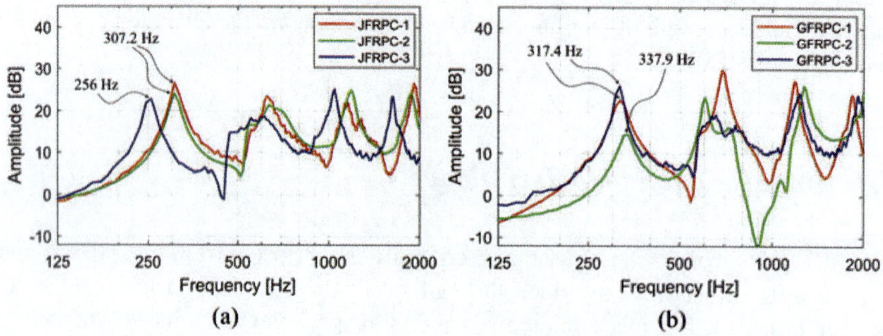

Fig. 4 Frequency response composites from IET test: **a** JFRPC; **b** GFRPC

Table 1 Results of tensile test and impulse excitation technique

Specimens	Maximum tensile strength (MPa)	Natural frequency (Hz)	*E*, Tensile test (MPa)	*E*, IET test (MPa)	Deviation (%)
JFRPC-1	22.68	307.2	3427.52	3660.65	6.80
JFRPC-2	30.28	307.2	3587.48	3476.73	3.09
JFRPC-3	27.15	256	2963.78	3123.56	5.39
Average			**3326.26**	**3420.31**	**5.09**
GFRPC-1	25.24	317.4	3316.81	3561.19	7.37
GFRPC-2	35.04	337.9	3430.09	3629.91	5.83
GFRPC-3	38.45	317.4	3109.66	3131.63	0.71
Average			**3285.52**	**3440.91**	**4.63**

basis as well as the practical situations. The IET is a method that is both dynamic and non-destructive for determining the elastic characteristics of a material through the analysis of a transient free vibration of samples. Conversely, the tensile test is a quasi-static destructive method that is based on the examination of the mechanical behavior of samples that are subjected to high loads with the aim of identifying the material's elastic properties. This method involves exposing the specimens to loads that are greater than their yield strength. The very little disparity in results that can be noticed when comparing the data obtained from the two distinct approaches, however, is probably due to the strain rate factor and the inherent anisotropy of the composite materials. Additionally, although the elastic modulus obtained via the tensile test is reliant on the local strain effect and, as a consequence, represents the local elastic modulus, the elastic modulus collected through the IET test is reflective of the global elastic modulus [10].

4 Conclusion

The non-destructive impulse excitation technique (IET) was successfully used to determine Young's modulus of both jute fiber reinforced polymer composites (JFRPC) and glass fiber reinforced polymer composites (GFRPC). These results were then validated by tensile testing, which showed only a slight difference in the deviation value. As a result of using the impulse excitation technique on jute composite, a value of 3420.31 MPa for the average value of Young's modulus was found. In the meanwhile, the results of the tensile test came in at 3326.26 MPa, which indicates a deviance of around 5.09%. Also, for glass fiber composites, the average Young's modulus IET test is 3440.91 MPa, which is higher than the tensile test of 3285.52 MPa by about 4.63%. The data that was obtained showed that the technique of impulse excitation is suitable for determining the modulus of elasticity or Young's modulus due to the fact that it is non-destructive, rapid, and reliable.

Acknowledgements The authors are grateful to Syiah Kuala University (USK), Ministry of Education, Culture, Research, and Technology of the Republic of Indonesia with Grant No. 145/UN11/SPK/PNBP/2022, for supporting this study.

References

1. Wang B, Zhong S, Lee T, Fancey KS, Mi J (2020) Non-destructive testing and evaluation of composite materials/structures : a state-of-the-art review. Adv Mech Eng 12(4):1–28. https://doi.org/10.1177/1687814020913761
2. Kostic S, Miljojkovic J, Simunovic G, Vukelic D, Tadic B (2021) Uncertainty in the determination of elastic modulus by tensile testing. Eng Sci Technol Int J 5:1–11. https://doi.org/10.1016/j.jestch.2021.05.002
3. Viala R, Placet V, Cogan S (2018) Identification of the anisotropic elastic and damping properties of complex shape composite parts using an inverse method based on finite element model updating and 3D velocity fields measurements (FEMU-3DVF): application to bio-based composite violin soundboards. Compos Part A 106:91–103. https://doi.org/10.1016/j.compositesa.2017.12.018
4. Bucciarelli F, Fierro GPM, Zarrelli M, Meo M (2019) A non-destructive method for evaluation of the out of plane elastic modulus of porous and composite materials. Appl Compos Mater 26:871–896. https://doi.org/10.1007/s10443-018-9754-5
5. Cui R, di Scalea FL (2019) On the identification of the elastic properties of composites by ultrasonic guided waves and optimization algorithm. Compos Struct 223:110969. https://doi.org/10.1016/j.compstruct.2019.110969
6. Niutta CB, Tridello A, Belingardi G, Paolino DS (2021) Nondestructive determination of local material properties of laminated composites with the impulse excitation technique. Compos Struct 262:113607. https://doi.org/10.1016/j.compstruct.2021.113607
7. Rizal M, Mubarak AZ, Razali A, Asyraf M (2019) Free vibration characteristics of jute fibre reinforced composite for the determination of material properties: numerical and experimental studies. In: AIP conference proceedings, vol 2187, p 050020. https://doi.org/10.1063/1.5138350
8. Dobah Y, Bourchak M, Bezazi A, Belaadi A, Scarpa F (2016) Multi-axial mechanical characterization of jute fiber/polyester composite materials. Compos Part B 10:450–456. https://doi.org/10.1016/j.compositesb.2015.10.030
9. El-wazery MS, El-elamy MI, Zoalfakar SH (2017) Mechanical properties of glass fiber reinforced polyester composites. Int J Appl Sci Eng 14(3):121–131. https://doi.org/10.6703/IJASE.2017.14(3).121
10. Pittala RK, Dhanaraju G, Ben Beera S, Ben Beera A (2022) Damping behaviour and self-healing performance evaluation of microcapsules reinforced epoxy composites by impulse excitation technique. J Reinf Plast Compos 1–12. https://doi.org/10.1177/07316844221077952

Failure Analysis of Leaf Spring Used as the Rear Suspension System in Diesel Truck by Numerical and Fracture Mechanics Approach

Rully Anshari, Husaini, and Nurdin Ali

Abstract Leaf springs are a type of spring designed to dampen vibrations caused by loading such as lateral loads, shock loads, and torsion. Due to the main role of this component makes its failure can directly cause a serious accident. This study aims to determine the cause of leaf spring failure in the 110 PS dump truck vehicle suspension system carried out by numerical and fracture mechanic approach. Numerically, leaf springs with initial defects were modeled using the finite element method software called FEMAP to find the value and the stress distribution that occurs in the leaf spring around the location of the initial defect. By fracture mechanics approaching, the stress results obtained are used to calculate the value of the stress intensity factor (K_I) which is then compared with the value of crack toughness (K_{IC}). Based on the results of stress simulation and calculation of stress intensity factor (K_I), it was found that the stress that occurs is below the yield stress of the material and (K_I) $\approx$ (K_{IC}). From the results obtained, the study found that crack propagation was the cause of the leaf spring failure that occurred from the location of the initial defect. This failure occurs by the fatigue mechanism caused by repeated loading.

Keywords Fracture mechanic · Failure · Stress intensity factor · Finite element method · Leaf spring

R. Anshari (✉) · Husaini · N. Ali
Laboratory of Computational Mechanics, Department of Mechanical Engineering, Universitas Syiah Kuala, Darussalam, Banda Aceh 23111, Indonesia
e-mail: rullyanshari@gmail.com

Husaini
e-mail: husainiftm@unsyiah.ac.id

N. Ali
e-mail: nurdin.ali@unsyiah.ac.id

Irwansyah et al. (eds.), *Proceedings of the 4th International Conference on Experimental and Computational Mechanics in Engineering*, Lecture Notes in Mechanical Engineering, https://doi.org/10.1007/978-981-99-7495-5_60

1 Introduction

The spring is a component that is widely used in vehicles, one of which is leaf springs. This type of spring is widely used because it has a simple construction but has great strength [1]. This spring has layers of beam form that is put together and has two springs at the end as a connector to the vehicle frame [2]. As one of the vibration dampers that occurs in trucks, these springs help control the tires when accelerating, turning, and passing when passing through bumpy contoured roads [3]. Crack mechanics is one of the methods that can be used to analyze the causes of failure in a component or construction and to find the character of crack propagation [4]. The finite element method is one of the good methods to be used in failure analysis by studying the stress reactions that arise in complex structures due to the applied load [5].

Research conducted by Nataraj and Thillikkani found that the discovery of a benchmark pattern on the fracture surface of the leaf springs indicated that the failure of the leaf springs occurred with a fatigue mechanism [6]. Husaini et al. in their research in the analysis of leaf spring failure using the finite element method suggested that leaf spring failure can occur if the stress intensity factor KI is greater than the fracture toughness KIC value [7].

In this case, there was a failure of a leaf spring used in a diesel truck. This fracture is known to occur in the spring layer no 4 on the rear truck suspension. This study was conducted to find out what factors are the causes of why this failure occurred so that the same failure does not happen in the future.

2 Research Method

Research Object. In this study, leaf springs were analyzed which broke after 1.5 years of use. The broken leaf spring is shown in Fig. 1.

This spring fracture occurred in the spring layer no 4 of the 5 existing spring layers as shown in Fig. 2.

Visual Inspection. Visual inspection was carried out to determine the type of failure that occurred in the leaf spring through an analysis of the topography of the leaf spring. This process is carried out by taking pictures of the leaf springs with the help of an optical microscope and a scanning electron microscope (SEM).

In previous studies, surface observations have been made using SEM and found the location of the beginning of the crack and the direction of propagation. The results of this observation also indicate that at this location there are several corrosion products that cause porosity. This crack propagation starts from the initial defects that arise which are estimated to have a length of about 0.8 mm. Figure 3 shows the results of SEM observations on the fracture surface of leaf springs.

Fig. 1 Broken leaf spring material

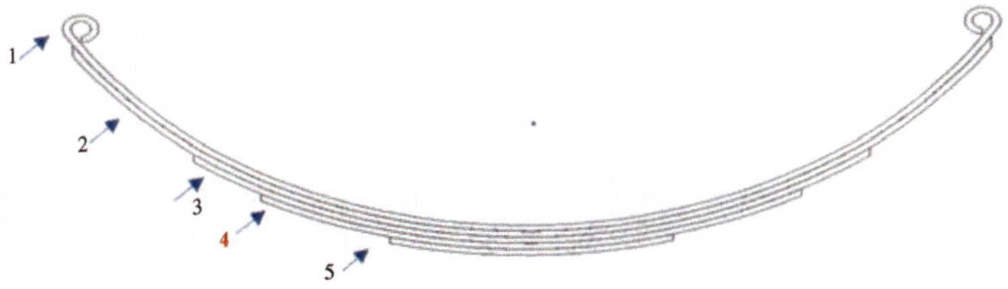

Fig. 2 Failure on spring number four

Microstructure Analysis. In this process, an analysis of the microstructure created on the leaf spring material will be carried out as a result of heat treatment. Leaf spring material will be observed under an optical microscope with a magnification of 100 times. The results of this analysis have been obtained in the previous studies. The microstructure of this spring is dominated by pearlite and ferrite. However, there are inclusions in the leaf springs which can shorten the fatigue life of the spring.

Chemical Composition. Chemical composition testing was carried out to determine the standard of the material used for leaf springs. Then the value of mechanical properties obtained based on material standards will be used in calculating the stress distribution using the finite element method (FEM). The mechanical properties of AISI 4340 steel are listed in Table 1.

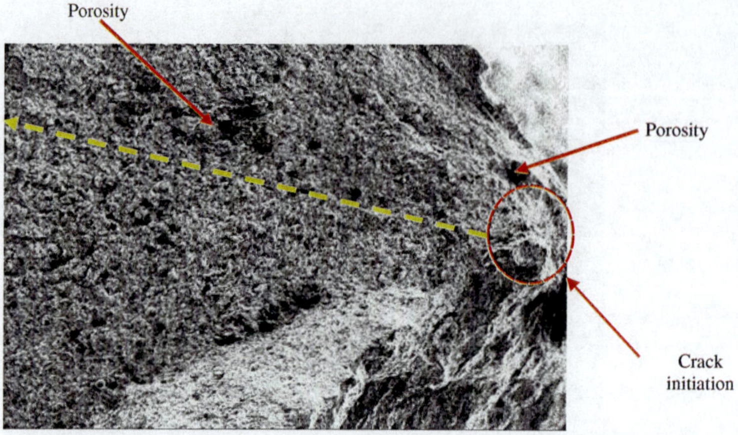

Fig. 3 Observation of fracture surface using SEM

Table 1 Mechanical properties of AISI 4340 [8]

Mechanical properties	Value
Tensile strength	745 MPa
Elastic modulus	190–210 GPa
Bulk modulus	140 GPa
Shear modulus	80 GPa
Poisson's ratio	0.27–0.30
Hardness, Rockwell B	95 HRB

AISI 4340 steel tempered at 410 °C for leaf springs [9] has a yield stress value of 1320 MPa and fracture toughness (KIC) 91.73 ± 3.34 MPa$\sqrt{}$m [10].

Hardness. The hardness test on the spring material aims to determine the magnitude of the hardness value and how it is distributed on the leaf spring. Figure 4 is a graph of the leaf spring hardness test results.

Numerical Analysis. The stress calculation is carried out using the finite element method. This calculation aims to determine the magnitude of the stress that occurs in the leaf spring and how its distribution. This value will be used to calculate the stress intensity factor at the initial defect location modeled on the leaf spring. FEMAP software will be used in this process.

Stress Intensity Factor Calculation. This was used to calculate the distribution of the minimum and maximum bending loads received by the fine cracks in the material which caused the stress intensity factors around the crack tip. It is, however, important to note that the crack propagates when K_I (stress intensity factor) > K_{IC} (fracture toughness) but does not spread when $K_I < K_{IC}$. Moreover, the fracture configuration

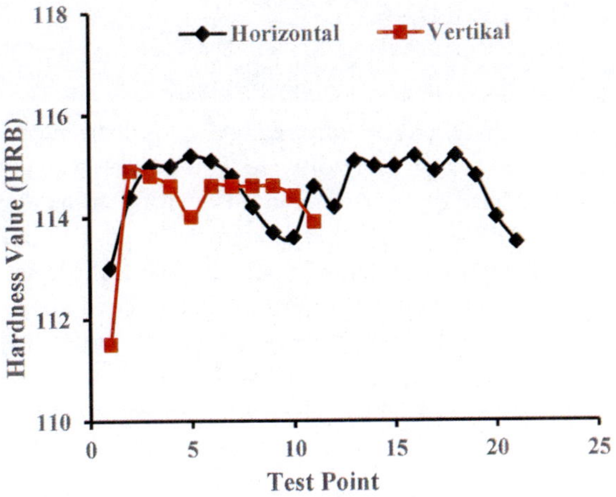

Fig. 4 Hardness test results

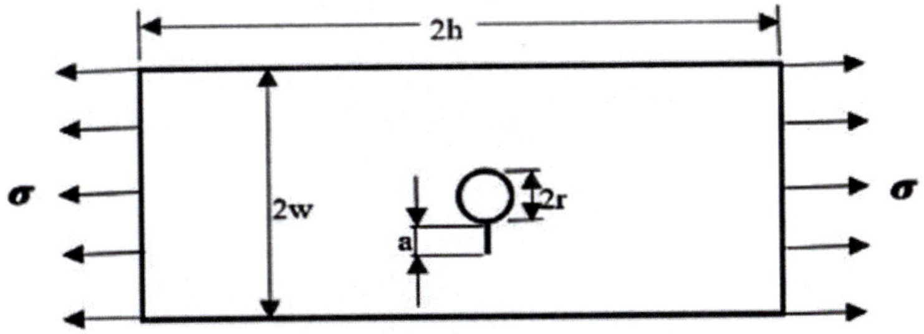

Fig. 5 Single-edge-crack-at-hole specimen schematic model [11]

of the leaf spring was modeled in a single-edge-crack-at-hole specimen schematic and loading model as indicated in Fig. 5.

The value of the stress intensity factor is calculated using Eq. 1:

$$K_I = \sigma\sqrt{\pi a(Fh_s)}, \tag{1}$$

where a is the length of the crack obtained through observations with SEM and is the magnitude of the stress that occurs at the crack tip. Fh_s is the limiting correction factor calculated by equation two referring to the research conducted by Newman Jr and Daniewicz. F_n (finite-width plate correction factor) and F_w (hole influence correction factor).

$$Fh_s = F_n F_w \tag{2}$$

3 Result and Discussion

Analysis of Stress. In this analysis process, a leaf spring model is made using the finite element method to find the stress value and how it is distributed. This process is carried out with the help of the FEMAP software. This process begins with a meshing process with an element size of 7.6 mm (3 In). The shape of the element used is a tetrahedral type because of its good ability to analyze shapes that have a radius and are hollow [12, 13]. Figure 6 shows a model of a meshed leaf spring.

Furthermore, the leaf spring model is given boundary conditions at both ends of the spring eye and the load is located in the middle of the spring [14]. The location of the boundary conditions and the given load is shown in Fig. 7.

After determining the boundary and load conditions, then inputting the mechanical properties of the spring material as a parameter in the calculation process is carried

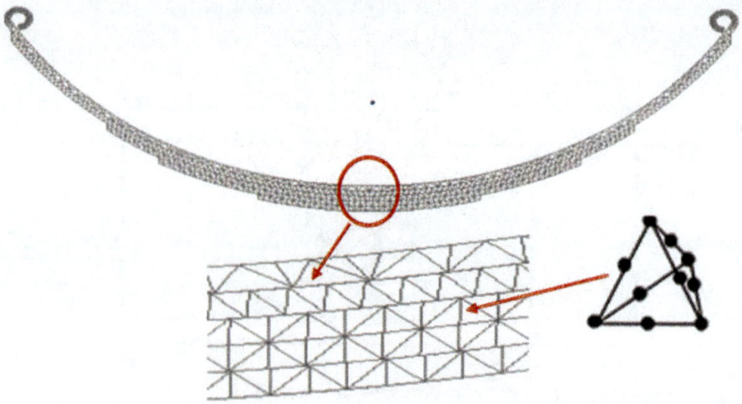

Fig. 6 Mesh model of the leaf spring

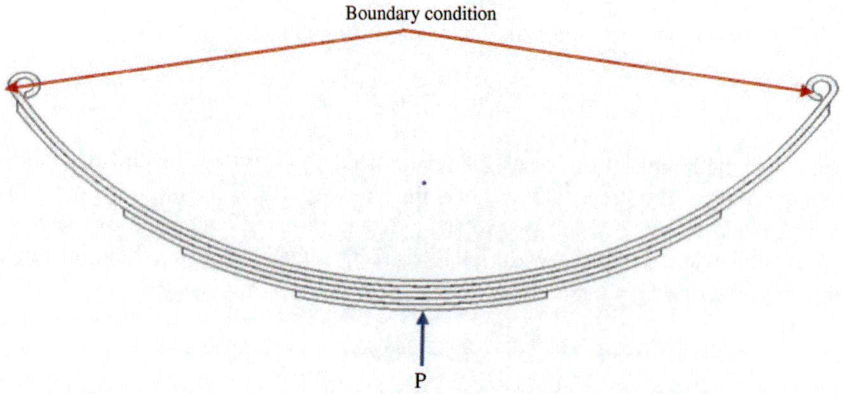

Fig. 7 Location of boundary conditions and loads

out. If the setup process has been completed, then the next process is a simulation process to find out the magnitude of the voltage and how it is distributed.

The results of the simulation obtained the magnitude and distribution of the stress as shown in Fig. 8. The load given to the leaf spring is 29,185 N where this is the load experienced by the left rear spring system.

The highest stress value was recorded at 1003.5 MPa. This stress appears at the location of the leaf spring's initial defect. Figure 9 shows the stresses acting at the location of the initial defect. This stress arises at the location of the initial leaf spring defect whose length is estimated to be 0.8 mm by visual observation.

Fig. 8 Stress distribution in leaf spring

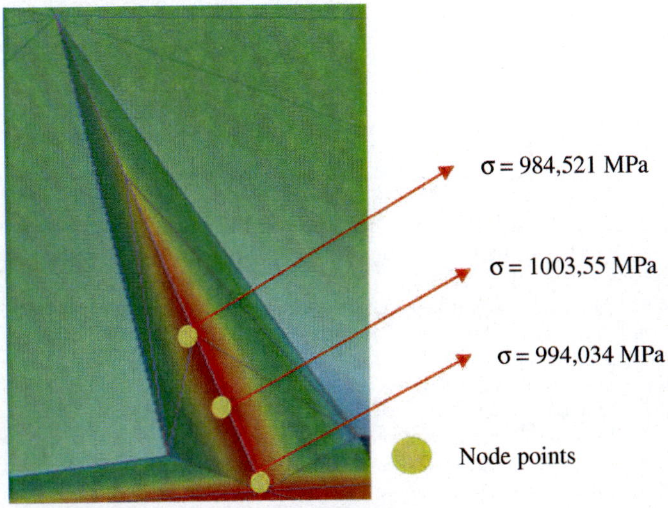

Fig. 9 Stress acting at the location of the initial defect

Table 2 Stress and stress intensity factor value

No.	Nodes	σ (MPa)	a (m)	K_I (MPa $\sqrt{m}$)
1	2,172,488	984.521	0.0008	90.27
2	2,174,083	1003.55		92.01
3	2,184,877	994.034		91.14
	Average			91.14

Stress Intensity Factor Value. Based on the stress value obtained, it is possible to calculate the stress intensity factor value that occurs at the crack tip using Eqs. 1 and 2. The magnitude of the stress intensity factor value is shown in Table 2.

The value of the stress intensity factor is close to the value of the fracture toughness of AISI 4340 steel. As a result, crack propagation occurs which causes the leaf spring to fracture [15–17].

4 Conclusion

The analysis yielded the following conclusions. The failure of the leaf springs was due to the initial defects observed in the inside of the spring bore. As a result of the initial defect that was formed, there was a concentration of stress in the area, where the working stress reached 1003 MPa. By calculating the value of the stress intensity factor (KI) compared to the fracture toughness (KIC), it was found that the value of KI $\approx$ KIC causes crack propagation in the leaf springs which caused the leaf spring fractured.

References

1. Teli M, Chavan U, Phakatkar H (2019) Design, analysis and experimental testing of composite leaf spring for application in electric vehicle. Int J Innov Technol Explor Eng 8:2882–2891
2. Guan H, Kim K, Wang B (2017) Comprehensive path and attitude control of articulated vehicles for varying vehicle conditions. Int J Heavy Veh Syst 24:65. https://doi.org/10.1504/IJHVS.2017.080961
3. Mahanthi D, Murali C (2017) Design and analysis of composite leaf spring for light weight vehicle. Int J Adv Eng Res Sci 4:147–152. https://doi.org/10.22161/ijaers.4.3.23
4. Frankl S, Pletz M, Schuecker C (2022) Improved concept for iterative crack propagation using configurational forces for targeted angle correction. Eng Fract Mech 266:108403. https://doi.org/10.1016/j.engfracmech.2022.108403
5. Thomas D (2016) Using finite element analysis to assess and prevent the failure of safety critical structures. J Fail Anal Prev 17:1–3. https://doi.org/10.1007/s11668-016-0217-8
6. Nataraj M, Thillikkani S (2020) Failure analysis of leaf spring suspension system for heavy load truck vehicle. Int J Heavy Veh Syst 27. https://doi.org/10.1504/IJHVS.2019.10021567

7. Husaini R, Riantoni N, Ali TE (2019) Putra, failure analysis of the leaf spring of truck colt diesel using finite element method. IOP Conf Ser Mater Sci Eng 547:012017. https://doi.org/10.1088/1757-899X/547/1/012017
8. Azom (2012) AISI 4340 Alloy Steel. https://www.azom.com/article.aspx?ArticleID=67
9. Google Patent (2007) Method for Manufacturing a Leaf Spring. https://patents.google.com/patent/US7284308B2/en
10. Li H, Duan Q, Zhang P, Zhou X, Wang B, Zhang Z (2019) The quantitative relationship between fracture toughness and impact toughness in high-strength steels. Eng Fract Mech 211:362–370. https://doi.org/10.1016/j.engfracmech.2019.03.003
11. Newman J, Daniewicz S (2014) Predicting crack growth in specimens with overloads and cold-worked holes with residual stresses. Eng Fract Mech 127:252–266. https://doi.org/10.1016/j.engfracmech.2014.06.012
12. Husaini RH, Liza N, Ali S (2021) Muammar, failure analysis of a fractured leaf spring as the suspension system applied on the dump truck. Key Eng Mater 892:89–98. https://doi.org/10.4028/www.scientific.net/KEM.892.89
13. Wu J, Lee R (1997) The advantages of triangular and tetrahedral edge elements for electromagnetic modeling with the finite-element method. IEEE Trans Antennas Propag 45:1431–1437. https://doi.org/10.1109/8.623133
14. Triveni Z, Babu A (2016) Finite element analysis on leaf spring made of composite material. Int J Adv Res Sci Eng 5:542–552
15. Husaini Z (2016) Mixed mode fracture behavior of an aluminum Alloy A6061 investigated by using compact tension shear specimens. Int J Technol 7:456. https://doi.org/10.14716/ijtech.v7i3.1625
16. Husaini, Kishimoto K, Notomi M (2004) Finite element analysis of rubber particles size distribution on fracture toughness of rubber-modified polymer. Key Eng Mat 261–263. https://doi.org/10.4028/www.scientific.net/KEM.261-263.711
17. Husaini, Putra T, Ali N (2018) The Morlet wavelet transform for reducing fatigue testing time of an automotive suspension signal. AIP Conf Proc 1983:030003. https://doi.org/10.1063/1.5046238

Effect of Perforated Aluminum on the Sound Absorption Ability of *Calotropis gigantea* Fiber Material

Suhaeri, Husaini, Muhammad Dirhamsyah, Iskandar Hasanuddin, and Marthunis

Abstract An excellent example of plant-based fiber used for noise reduction is *Calotropis gigantean* (CG). This research aimed to determine the sound absorption coefficient (SAC) value and sound reduction ability of the CG fiber. The fiber was coated with 0.3 mm thick aluminum and perforated with 1, 1.5, and 2.5 mm diameters. Furthermore, Bruel & Kjaer's Type 4206 impedance tube with a diameter of 100 mm and a thickness of 10 mm was used as the model for the test sample. The sample, which weighed 50 g, was hot pressed in a mold for 10 min at 200 °C, and the test was conducted following the ISO 10534–2 transfer function method using an impedance tube at a frequency of 1/1 octave. The results showed that the test sample without aluminum coating can reduce the SAC by 0.05–0.1 (5–10%) compared to those coated with perforated aluminum. This is because the aluminum surface area is larger than the total hole area that transmits sound to the test sample, so more sound is reflected. Lastly, the uncoated sample's noise reduction coefficient (NRC) value was larger than that of the coated sample and reduced the overall noise by 28%. This proved that the aluminum hole's diameter affects the noise reduction coefficient. The coated sample can be categorized into two classes, class D for uncoated samples and class E for aluminum coated.

Keywords *Calotropis gigantea* fiber · Impedance tube · Sound absorption coefficient · Noise reduction coefficient

Suhaeri
School of Engineering, Universitas Syiah Kuala, Banda Aceh 23111, Indonesia
e-mail: suhaeriabd@usk.ac.id

Suhaeri · Husaini (✉) · M. Dirhamsyah · I. Hasanuddin · Marthunis
Department of Mechanical Engineering, Universitas Syiah Kuala, Darussalam, Banda Aceh 23111, Indonesia
e-mail: husainiftm@usk.ac.id

M. Dirhamsyah
e-mail: Mdirhamsyah@usk.ac.id

I. Hasanuddin
e-mail: iskandarhasanuddin@usk.ac.id

Irwansyah et al. (eds.), *Proceedings of the 4th International Conference on Experimental and Computational Mechanics in Engineering*, Lecture Notes in Mechanical Engineering, https://doi.org/10.1007/978-981-99-7495-5_61

1 Introduction

Noise pollution rises as a result of increased vehicle traffic and the usage of industrial machinery, and this negatively impacts the environment and public health. According to the World Health Organization, noise pollution increases public health problems. Some of these adverse health problems include hearing loss, sleep disturbances, fatigue, cardiovascular and psychophysiological problems, decreased performance, and adverse social behavior [1, 2]. Therefore, it is necessary to reduce the noise level by installing a silencer. In this research, the noise reduction material used was were readily available natural fibers [2, 3], which were obtained from Banda Aceh and Aceh Besar. In the field of acoustics, natural fibers have great potential to replace synthetic fibers and they are derived from animals, minerals, and plants (typically made up of cellulose) [4]. Furthermore, the use of natural fibers in acoustic applications has increased dramatically because of the harmful environmental effects resulting from the use of synthetic fibers. One of them is the fact that most synthetic fibers are non-biodegradable [1]. According to Asdrubali et al. and Da Silva et al., materials for acoustic insulation include kenaf fiber, coconut fiber, reeds, *sisal*, leca (consisting of small, light, and bulging particles of burnt clay), bamboo, glass wool, cotton, jute, recycled rubber and foam, torn carpet, recycled plastic with glass wool, kenaf with polyester, polymer materials, fleece fibers, known roofing systems, and green walls [5, 6].

Following this, the test specimen used in this research is the natural fiber *Calotropis gigantea* (CG), also known as the *Widuri/Biduri* plant in Indonesia or simply, crown flower. The specimen is a Calotropis species that is native to Cambodia, Vietnam, Bangladesh, Indonesia, Malaysia, Thailand, Sri Lanka, India, China, Pakistan, and Nepal and can grow up to 4 m in height [7]. The *Widuri* plant is often found on the coast of Aceh, specifically in Banda Aceh and Aceh Besar where it is also known as *rubek* in the local language (Aceh). In addition, the CG fiber is quite similar to cotton and kapok fibers and it is obtained from the CG plant's fruit Fig. 1. The plant has a gummy white-like substance released by its stems, leaves, and fruit. It is known that the sap is beneficial to health, as it can be used as a herbal remedy for illnesses including dizziness, asthma, bronchitis, dyspepsia, leprosy, tumors, and various digestive disorders [8].

Sukardan et al. also reported that CG fiber has a low density (weight per unit volume). The fiber has the ability to act as a thermal insulator, sound absorber, water, and oil absorber due to the hollow area along its length that can serve as a medium for trapping air or water vapor [8]. Also, it was reported that the diameter of the fiber, its wall thickness, and hollow volume were between 15–26, 0.8–2 μm, and 92.3–94.7%, respectively. Each seed has about 600 strands of fiber, as shown in Fig. 2.

Table 1 shows the similarity between the dimensions of CG, kapok, and cotton fibers [9]. Their physics is shown in Fig. 3.

According to Dilli et al., the CG fruit's tensile strength increases as its fiber content in the polyester matrix also increases. This is because the polyester resin transmits and distributes the stress applied to the CG fibers, resulting in higher strength.

Fig. 1 **a** Cross and longitudinal sections of CG fruit, **b**. CG fruit that has been broken

Fig. 2 *Calotropis gigantea* seed fibers

Table 1 Characteristics of fiber's physical properties [8]

Physical properties	CG fibers	Kapas fibers	Kapok fibers
Length of fiber (mm)	31.75–167.64	24.6–60	16–104.14
Fiber diameter (μm)	15–26	12–45*	10–18
Cellulose (%)	66.5–71.62	85–90	72.86
Fibers Fineness (μ/in)	2.02	2.7	2.02
Fiber Uniformity (%)	84	78	74
Absorption coefficient (α)	–	–	> 0.7**

* Fiber diameter [9], ** At a frequency of 6300 Hz [1]

Furthermore, composites can withstand higher loads before failure, unlike unreinforced polyesters [10]. Another research by Pell explained that the tensile strength of composite materials containing *Calotropis gigantea* fibers was influenced by the volume fraction of the resin material [11]. Sukardan et al. noted the fiber's potential as a sound absorber. Among the many research conducted regarding CG fiber as a composite reinforcement material, none utilized it as a sound-dampening material. Therefore, this research aims to test the sound absorption coefficient (SAC) (α) of CG

Fig. 3 Type of fibers from CG, Kapas, and Kapok

fiber samples with a thickness of 10 mm, as well as the effect of perforated aluminum plate layers 1, 1.5, and 2.5 mm on the test sample's sound absorption ability.

2 Experimental Method

Material. In this research, the plant-based fiber utilized was extracted from the CG fruit. To obtain enough fiber, old fruits were collected from various locations in Banda Aceh and Aceh Besar. It is well known that CG fruits have an overall fiber weight of only 8.9% (0.41 g) of the 4.59 g dry fruit weight, 6.9 mm fruit length, and 0.95 g seed weight [8].

Furthermore, to make the test sample, the fiber from the fruit was extracted, dried, and weighed manually. The weight of each sample was 50 g, and the diameter of the mold was 100 mm which was following the diameter of the impedance tube used.

The test sample was hot pressed in a 10 mm thick mold using a 10 Mpa press (Fig. 4) at 200 °C, enabling the fibers to merge without the use of an adhesive. Also, it was tested to determine the SAC value.

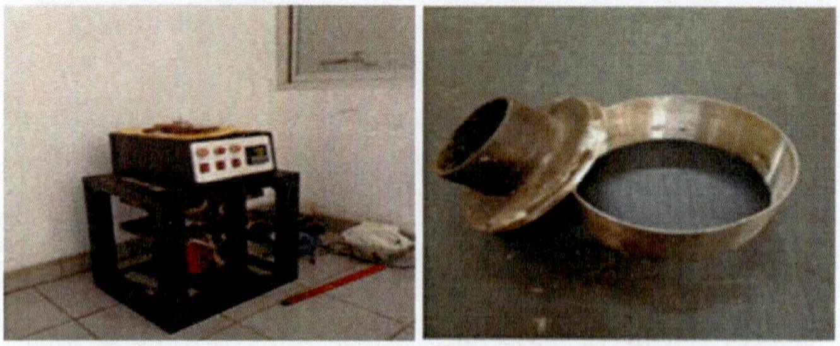

Fig. 4 Hot press tools and molds

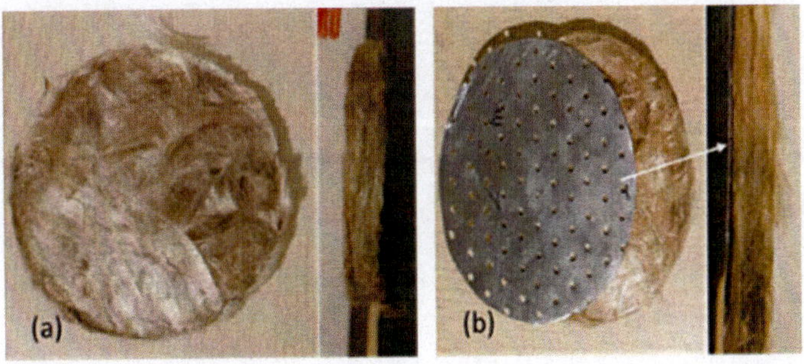

Fig. 5 **a** Test object. **b** Test object with perforated aluminum

Perforated Upholstery. The effects of holes on the specimen coated with a perforated aluminum plate with a 0.3 mm thickness and diameters of 1, 1.5, and 2.5 mm.

Test Specimen. The geometry of the test specimen from CG fibers with a thickness of 10 mm and a diameter of 100 mm coated with a perforated aluminum plate with a thickness of 0.3 mm can be seen in Fig. 5.

Research Parameters. Table 2 shows the parameters used in this research.

Equipment Test. The absorption coefficient (α) test was conducted on the specimen using a Bruel & Kjaer Type 4206 impedance tube having 100 mm in diameter, and a transfer function method (ISO 10534-2) [12]. Furthermore, the sample under test was mounted on the impedance tube in such a way that there was no gap between the sample and the tube's cell wall. The white noise signal sensor was inserted into the tube, and the loudspeaker was mounted close to the end of the impedance tube. Also, broadband, stationery, and random sound waves were produced by the loudspeaker and traveled through the tube as plane waves before striking the sample and reflecting back to the listener. The MATLAB application was then used to calculate the amount of sound pressure on the microphone sensors 1 and 2 in volts at each frequency.

The sound pressure ratio generated by the two microphones 1 and 2 was used to calculate SAC, after which the transfer function between the impedance tube's two microphones was identified. x_1 is the furthest distance between the test sample and the sensor, x_2 is the closest distance between the sensor and the specimen, and S or $(x_1 - x_2)$ is the distance between the two sensors (Fig. 6).

Table 2 Research parameters

Material fibers	Weight (g)	Thick (mm)	Perforated aluminum (mm)	Temp. (°C)
CG*	50	10	Without hole	200
			1	
			1.5	
			2.5	

*CG *Calotropis gigantea* fiber

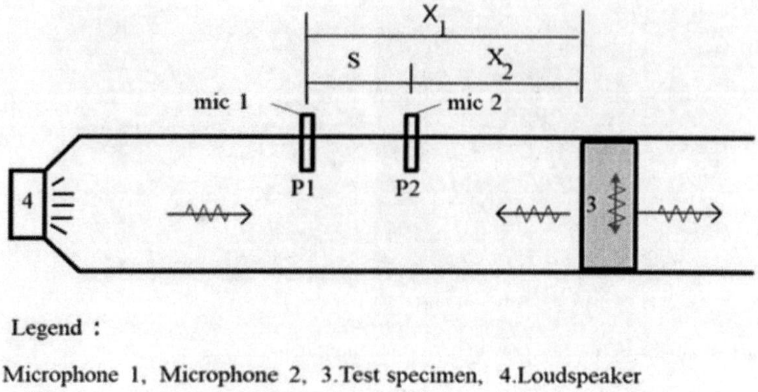

Fig. 6 Microphone positions and distances

The transfer function between the two microphones can be obtained using equation H12 [12]

$$H_{12} = \frac{P_2}{P_1} = \frac{e^{jkx_2} + Re^{-jkx_2}}{e^{jkx_1} + e^{-jkx_1}} \tag{1}$$

The sound reflection factor R can be determined using the following equation.

$$R = \frac{H_{12} - H_1}{H_R - H_{12}} e^{2jkx_1} \tag{2}$$

$$R = \frac{H_{12} - e^{-jks}}{e^{jks} - H_{12}} e^{2jkx_1}. \tag{3}$$

Therefore, the absorption coefficient can be determined using Eq. 4:

$$\alpha = 1 - |R|^2. \tag{4}$$

H_{12} = Transfer function between two microphones
$H_R = e^{jks}$
$H_1 = e^{-jks}$
R = Reflection coefficient
α = Absorption coefficient
P_1 = Sound pressures of the incident wave (volt)
P_2 = Sound pressures of the reflected wave (volt).

Test Equipment. The parameters for the impedance tube research apparatus are shown in Fig. 7 above. To make easy visual comparisons between the different specimens, the noise reduction coefficient (NRC), which is defined as the arithmetic mean value of SAC at 250, 500, 1000, and 2000 Hz, [2, 13], was calculated using

Eq. 5:

$$NRC = \frac{\alpha 250 + \alpha 500 + \alpha 1000 + \alpha 2000}{4}. \tag{5}$$

The tested CG material can be classified into several classes, as seen in Table 3 [14], depending on its SAC value. The best sound absorption is class A, B, C, D, E, and the worst is not classified. Accordingly, the sound absorption classes are determined by calculating the average sound coefficient values, as shown in Table 3.

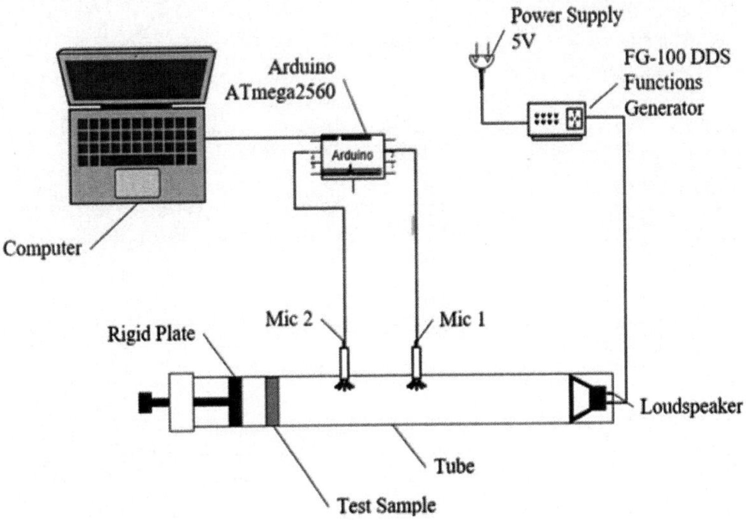

Fig. 7 Experimental setup

Table 3 Sound absorption classes [14]

Sound absorption class	Absorption coefficient (α)
A	0.90; 0.95; 1.00
B	0.80; 0.85
C	0.60; 0.65; 0.70; 0.75
D	0.30; 0.35; 0.40; 0.45; 0.50; 0.55
E	0.25; 0.20; 0.15
Not classified	0.10; 0.05; 0.00

3 Results and Discussions

The SAC of the specimens is shown in Table 4, and the test results from the sound pressure data P_1 and P_2 were calculated using the MATLAB program.

Furthermore, the graph in Fig. 8, which was plotted using the data from Table 4, shows that the diameter of the hole affects the SAC value (α). For sound absorption at a frequency of 7000 Hz, the diameter of the 2.5 mm hole had a lower SAC value (0.59) compared to 1 mm (0.66) and 1.5 mm (0.65) as well as the test sample without aluminum coating (0.69).

Table 4 SAC of CG fibers for difference frequency

	Frequency (Hz)		Perforated aluminum sheet	
	Without hole	1 mm	1.5 mm	2.5 mm
100	0.22	0.13	0.13	0.18
125	0.43	0.37	0.32	0.39
250	0.24	0.15	0.13	0.13
500	0.26	0.19	0.17	0.19
1000	0.27	0.22	0.19	0.17
2000	0.35	0.28	0.27	0.26
4000	0.47	0.39	0.35	0.3
7000	0.69	0.66	0.65	0.59
8000	0.37	0.35	0.36	0.34

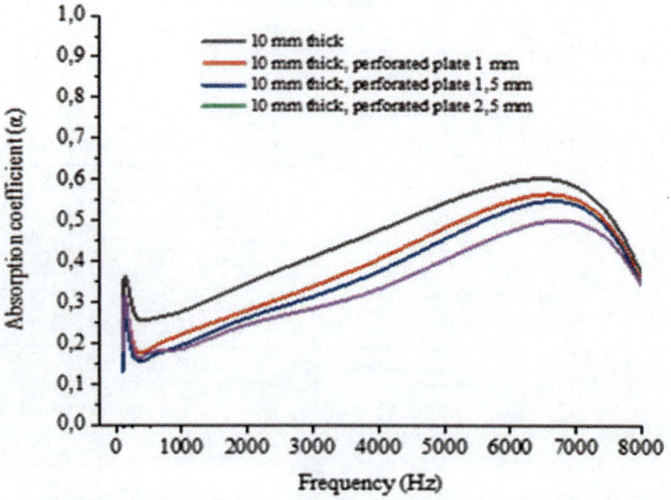

Fig. 8 SAC of different perforated plate

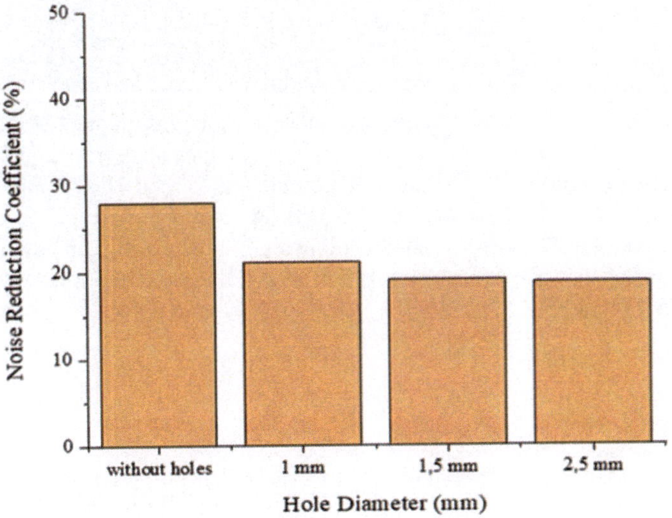

Fig. 9 NRC for different hole diameters

It turned out that the test sample coated with perforated aluminum reduced the SAC ability by an average of 0.05–0.1 (5–10%) compared to those without aluminum coating. This is because the incoming sound is reflected more than it is absorbed.

Accordingly, the noise reduction coefficient (NRC) obtained was the arithmetic mean of the SAC at frequencies of 250, 500, 1000, and 2000, as shown in Table 4. Figure 9 shows a comparison of the computational results. At a 10 mm NRC thickness, the absorption coefficient (α) obtained was 0.28. This indicated a 28% reduction in noise. In addition, the NRC for the test samples coated with aluminum having hole diameters of 1, 1.5, and 2.5 mm were 21, 19, and 18.75%, respectively.

According to the findings, the uncoated test sample was able to reduce noise by 28%, while the coated sample's noise reduction ability was affected by the hole's diameter. The larger the hole diameter, the smaller the noise reduction ability. This is because more sound comes to the surface area of the aluminum sheet and is reflected rather than absorbed by the test object through the total area of the hole in the aluminum.

Furthermore, the sound absorption class was determined by calculating the average sound coefficient value (Table 3). The test sample without an aluminum sheet showed a sound absorption (α) class value of 0.3666. Contrastingly, the coated sample's sound absorption class was equal to 0.2877, 0.2855, and 0.2833 for diameters of 1, 1.5, and 2.5 mm, respectively. Table 4 shows that the CG fiber was categorized in absorption class D. However, they can also be categorized into class E.

4 Conclusion

The test sample coated with perforated aluminum reduced the SAC ability by an average of 0.05–0.1 (5–10%) compared to those without aluminum coating. This is due to the fact that the test sample reflected more sound than it absorbed. Furthermore, at a frequency of 7000 Hz, the uncoated sample's SAC and NRC value obtained was 0.69 and 0.28 (28%), respectively. This proves that CG fiber is a good sound absorption material. The sample tested without aluminum coating can be categorized in classes D and E for perforated sheet diameters of 1, 1.5, and 2.5 mm. Admittedly, it is preferable to use uncoated fibers.

References

1. Bhattacharya SS, Bihola DV (2019) Acoustic properties of kapok fibre. Int J Eng Adv Technol 9(1):2164–2168. https://doi.org/10.35940/ijeat.A9688.109119
2. Shahid NSM, Ahmad MA, Md Tahir FL (2020) Sound absorption coefficient of different green materials polymer on noise reduction. Int J Innov Technol Explor Eng 9(3):2773–2777. https://doi.org/10.35940/ijitee.c9208.019320
3. Jayamani E, Hamdan S, Suid NB (2013) Experimental determination of sound absorption coefficients of four types of Malaysian wood. Appl Mech Mater 315:577–581. https://doi.org/10.4028/www.scientific.net/AMM.315.577
4. Taban E et al (2019) Study on the acoustic characteristics of natural date palm fibres: experimental and theoretical approaches. Build Environ 161:106274. https://doi.org/10.1016/j.buildenv.2019.106274
5. Asdrubali F, Schiavoni S, Horoshenkov KV (2012) A review of sustainable materials for acoustic applications. Build Acoust 19(4):283–312. https://doi.org/10.1260/1351-010X.19.4.283
6. da Silva CCB, Terashima FJH, Barbieri N, de Lima KF (2019) Sound absorption coefficient assessment of sisal, coconut husk and sugar cane fibers for low frequencies based on three different methods. Appl Acoust 156:92–100. https://doi.org/10.1016/j.apacoust.2019.07.001
7. Aiton WT (2022) |*Calotropis gigantea* (L.) Plants of the World Online | Kew Science. Plants of the world online. Retrieved 27 May 2022
8. Sukardan MD, Natawijaya D, Prettyanti P, Cahyadi C, Novarini E (2017) Karakterisasi Serat Dari Tanaman Biduri (Calotropis Gigantea) Dan Identifikasi Kemungkinan Pemanfaatannya Sebagai Serat Tekstil. Arena Tekst 31(2):51–62. https://doi.org/10.31266/at.v31i2.1986
9. Sfiligoj M, Hribernik S, Stana K, Kree T (2013) Plant fibres for textile and technical applications. Adv Agrophys Res. https://doi.org/10.5772/52372
10. Dilli Babu G, Sivaji Babu K, Nanda Kishore P (2014) Tensile and wear behavior of *Calotropis gigentea* fruit fiber reinforced polyester composites. Procedia Eng 97:531–535. https://doi.org/10.1016/j.proeng.2014.12.279
11. Pell YM (2012) Pengaruh Fraksi Volume Terhadap Karakterisasi Mekanik. Semin Nas Sains dan Tek 2012:114–120
12. ISO-10534 (1998) Determination of sound absorption coefficient and impedance in impedance tubes. Part 2 Transf Method ISO 10534, pp 1–27
13. Yang WD, Li Y (2012) Sound absorption performance of natural fibers and their composites. Sci China Technol Sci 55(8):2278–2283. https://doi.org/10.1007/s11431-012-4943-1
14. ISO 11654:1997 (1997) Acoustics: sound absorbers for use in buildings: rating of sound absorption, p. 7

Experimental Study and Simulation of Light Truck Diesel Rear Suspension Failure

Husaini, Faris Murtadhi, and Nurdin Ali

Abstract The rear leaf spring of the light truck diesel vehicle tends to experience failure when passing through a potholed road carrying a load of 11,000 kg. Therefore, it is necessary to determine the factors responsible for these failures using experimental and numerical studies to prevent the reoccurrence of similar cases. The experimental study includes visual inspection to determine the factors responsible for the breakdown based on the surface condition of the broken spring. This is in addition to using SEM observation to observe the fracture surface condition of fractography and the hardness test to evaluate the hardness value of leaf spring. To visualize the structure, a chemical composition test is employed to analyze the steel standard and microstructure observation. Meanwhile, the numerical study includes leaf spring modeling using Autodesk Inventor software, FEMAP 2022.1, and stress analysis. The chemical composition test showed that the leaf spring material complies with the AISI 5160 standard. The visual inspection consists of an initial crack, which is then observed on the SEM. The average hardness of the horizontal and vertical directions is 114.05 and 113.40 HRB, while for the outer surface, it is 114.36 HRB. Based on the observation, the structure of the pearlite phase is less than the ferrite, indicating that the leaf spring has more ductile properties. The stress intensity and toughness factors are 77.99 and 22 MPa $m^{1/2}$, respectively. Therefore, the calculation results have fulfilled the equation ($K_I \geq K_{IC}$).

Keywords Numerical method · Fatigue · Failure analysis · Leaf spring · Stress intensity factor

Husaini (✉) · F. Murtadhi · N. Ali
Laboratory of Computational Mechanics, Department of Mechanical Engineering, Universitas Syiah Kuala, Darussalam-Banda Aceh 23111, Indonesia
e-mail: husainiftm@usk.ac.id

F. Murtadhi
e-mail: faris_m@mhs.usk.ac.id

N. Ali
e-mail: nurdin.ali@usk.ac.id

Irwansyah et al. (eds.), *Proceedings of the 4th International Conference on Experimental and Computational Mechanics in Engineering*, Lecture Notes in Mechanical Engineering, https://doi.org/10.1007/978-981-99-7495-5_62

1 Introduction

The main component of a car's suspension system is the spring, comprising of coil, torsion bar, air, and leaf springs. Light truck diesel transport vehicles consist of leaf spring, which tends to receive lateral, torsional, and shock loads [1]. This spring type is significantly used in heavy vehicles due to its numerous advantages, such as low price, easy maintenance, and large load-carrying capacity. Generally, the leaf spring is designed with an eye that serves to connect it to the vehicle frame. The eye consists of two types, namely Berlin Eye and the Upturned Eye, mainly used by the colt diesel truck [2]. The study by Husaini et al. on leaf spring failure using a chemical composition test, SEM observation, and stress analysis with the finite element method showed that the leaf spring failure occurs due to material fatigue as indicated by the presence of a benchmark pattern on the fracture surface [3]. According to Eryuek et al., the cause of failure is due to the lack of safety factors in leaf spring fatigue; hence when the truck passes through a potholed road, it experiences a sudden fracture [4]. In October 2021, a light truck diesel vehicle experienced a fracture in the middle of the leaf spring number 2, namely the heart bolt section, as shown in Fig. 1.

Therefore, this study aims to determine the cause of the failure in the rear leaf spring number 2 of the light truck diesel vehicle. This will enable drivers to anticipate the occurrence and avoid accidents.

Fig. 1 Leaf spring on truck

2 Study Procedure

This study was carried out using experimental and numerical methods. The experimental method includes visual inspection to observe the topography of the fracture surface carried out macroscopically and fractographically, microstructure observation to determine the phases contained in the specimen, and SEM observation on crack initiation found in visual inspection. Others include hardness test to determine hardness value and chemical composition test to evaluate the type of material used in leaf spring. The numerical method starts by making leaf spring modeling using Autodesk Inventor software and continues with stress simulation using FEMAP software.

Materials. The components used in this study are a leaf spring with a carbon standard of 0.50–1.0%. A chemical composition test was also used to determine the type of material and the fracture of leaf spring number 2 as shown in Fig. 2.

Visual Inspection. This is the process of observing a material to determine the possibility of failure based on the condition of the spring fracture surface. Information obtained through this process is in the form of patterns and special signs, such as the mechanism due to fatigue fracture, which consists of early cracks, propagation, and fracture [5]. Specific loadings experienced by a leaf spring can cause failure, which is determined based on the initial prediction of visual inspection [6].

Microstructure Observation. This is initiated by grinding the specimen until it is smooth using sandpaper and smooth using HCL and alcohol. The surface is also etched before being observed using an optical microscope to determine the material's microstructure. This test is carried out to observe the grain and phase characteristics contained in the leaf spring.

Hardness Test. It starts with cutting the sample close to the fracture surface on the leaf spring and then grounded until the surface is flat. The test is carried out on the outer and inner surfaces of the sample to determine whether the leaf spring has a uniformity of hardness. The hardness Rockwell test is used in the Zwick/Roell ZHV 10 test.

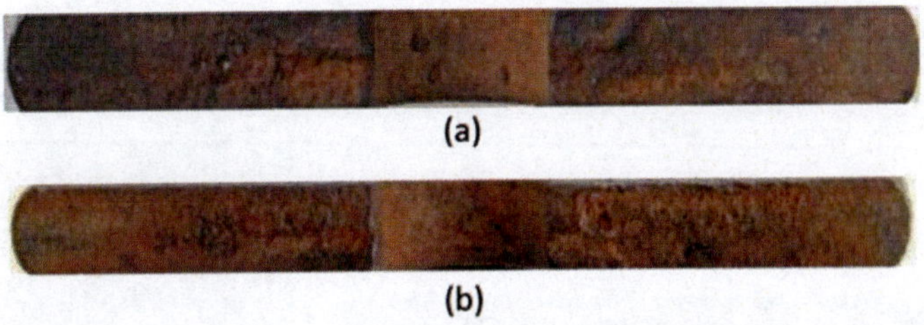

Fig. 2 **a** Left surface of the spring, **b** right surface of the spring

Chemical Composition Test. A chemical composition test is carried out to determine the chemical elements in the leaf spring to be classified according to AISI or ASM standards [7]. This test starts with cutting the sample on the failed leaf spring and ground using sandpaper until the surface is clean and free from corrosion. The specimen is then placed in a spectrometer test machine and vacuumed for $\pm$ 4 h. The chemical composition test aims to determine the chemical elements contained in the sample; hence, the type of material used in the leaf spring can be known.

Scanning Electron Microscope (SEM). This process aims to observe the fracture surface with the help of SEM tools to obtain information in the form of specimen surface topography. The observation starts with cutting the specimen on the leaf spring, which is suspected to be an initial crack, and then the sample is put into the SEM container. The sample is shot with electrons, and the reflected ones are captured by the detector in the SEM, with the results displayed on a computer monitor.

Modeling and Analysis with FEM. The finite element method is a numerical process used to solve engineering problems. Stress analysis is based on the mechanical properties of the material FEMAP 2022.2, while Autodesk Inventor 2021 is used for leaf spring modeling.

3 Results and Discussion

Visual Inspection. The fracture occurs in the number two leaf spring on the center bolt with the visual inspection used to determine its rough, gray, and stringy characteristics, as shown in Fig. 3.

Hardness Value. A hardness test is carried out on the sample and inside the spring surface to determine the homogeneous hardness in each part. The test is carried out horizontally and vertically on the inner surface. While on the surface, it is carried out on the top and bottom of the sample. Horizontal and vertical as well as the top and bottom surface hardness graphs are shown in Figs. 4 and 5.

The average from the spring hardness test in the horizontal and vertical directions is 114.05 and 113.04 HRB, respectively. These hardness values are close to the AISI

Fig. 3 Leaf spring macrostructure, **a** leaf spring fracture surface, **b** initial crack

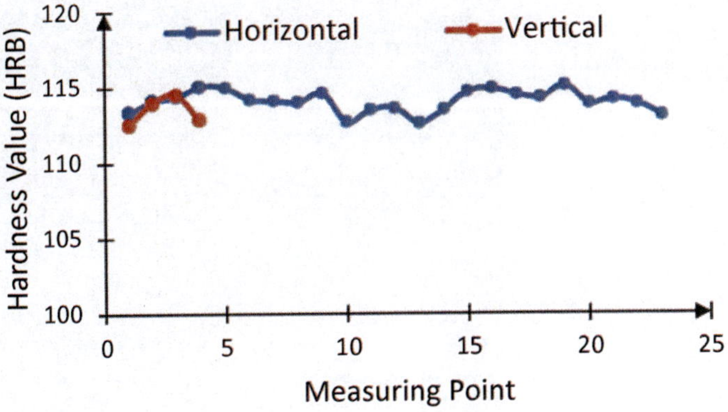

Fig. 4 Horizontal and vertical hardness graph

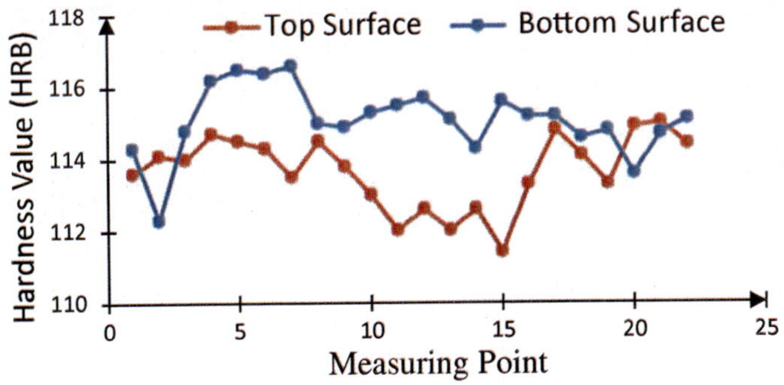

Fig. 5 Top and bottom surface hardness graph

5160 standard hardness of 92 HRB. The difference in the value of 22 HRB is due to heat and spring surface treatments, such as shot peening during the production process. The leaf spring hardness value is evenly and uniformly distributed to all test points and produced in accordance with the steel specification standard. Finally, microstructure observation is made to ensure conformance.

Fracture Surface Observation with SEM. Fractography of the fracture surface from SEM observation uses 1000× magnification. The result showed that the fracture is ductile due to its rough, gray, and stringy characteristics. The magnification results using the SEM test instrument showed that the initiation crack originates from the edge of the leaf spring specimen, and crack propagation occurs from the initial area. The SEM observation results are shown in Fig. 6.

Microstructure Observation. Microstructure observation is carried out with 200× magnification using an optical microscope to obtain the pearlite and ferrite phases, which are dark and light in color, respectively. Pearlite has relatively strong properties, good hardness, and ductility but low tensile strength and plasticity; hence,

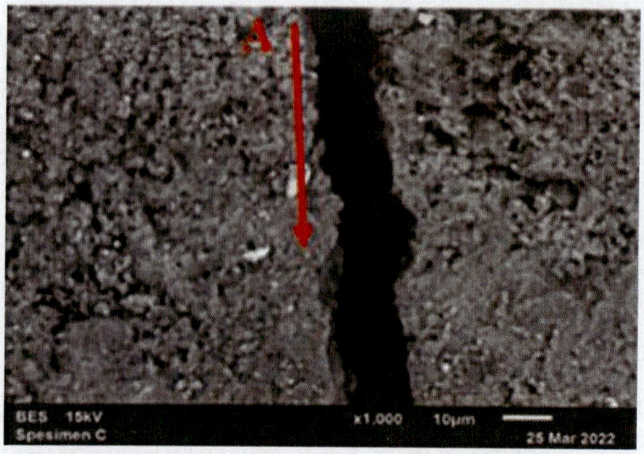

Fig. 6 SEM with 1000× magnification, Arrow A shows the direction of crack propagation

the brittleness value is high [9]. Meanwhile, ferrite has relatively weaker, softer, and more ductile properties. The microstructure observation results showed that the pearlite phase is less than the ferrite, thereby indicating that the leaf spring has more ductile properties. The microstructure of the leaf spring is shown in Fig. 7.

Chemical Composition Test. In the chemical composition test, the carbon value in the leaf spring sample is 0.608, indicating that the steel is medium carbon. Based on the test results, the element's chemical composition complies with the AISI 5160 standard. Medium carbon steel has high ductility and easy machining with low wear resistance and hardness. Therefore, a heat treatment process is needed to change the properties of the material; hence, it can slow down the critical cooling rate during the hardening process and increase the hardenability of the alloy steel. Table 1 shows the test results for the chemical composition of the material on the leaf spring.

Mechanical properties of AISI 5160 steel were obtained based on material standards known from the chemical composition test, as shown in Table 2.

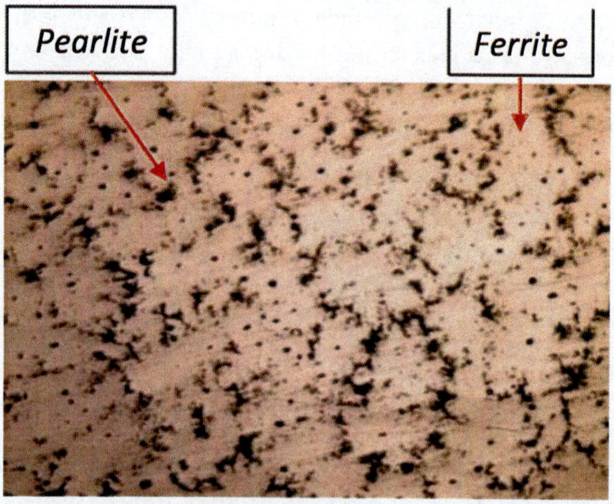

Fig. 7 Leaf spring microstructure

Table 1 Chemical composition of leaf spring material

Elements	Test results	AISI 5160
Iron, Fe	96.0	97.085–97.84
Manganese, Mn	0.939	0.750–1
Chromium, Cr	0.751	0.7–0.9
Carbon, C	0.608	0.560–0.640
Phosphorous, P	0.0318	$\leq$ 0.0350
Silicon, Si	0.174	0.150–0.3
Sulfur, S	0.0318	$\leq$ 0.04

Table 2 Mechanical properties of AISI 5160 steel

Mechanical properties	Value
Tensile strength	724 MPa
Elasticity modulus	190–210 GPa
Bulk modulus	140 GPa
Shear modulus	80 GPa
Poisson's ratio	0.27–0.30
Hardness, Rockwell B	92 HRB

Load Distribution on Spring. The standard maximum load assuming the load is evenly distributed on all four truck wheels and is determined using Eq. (1).

$$F = \frac{m \times g}{4} \tag{1}$$

Before the leaf spring fractures, the truck carries a load of $\pm$ 11,000 kg with ten leaf springs. The chassis weight of the vehicle is 2530 kg, while the gross vehicle weight is 8250 kg.

The maximum load receivable by a spring before and after it fractures are 20,212 and 26,950 N. The calculation results showed that the load on the truck exceeded the maximum standard limit of 8250 kg.

Stress Analysis with FEM. The analysis is carried out in a static state using a spring load of 26.950 N or 2.748 kg. Stress analyses from the finite element method are in the form of color contours with red and purple indicating the highest and lowest values. The analysis results obtained are in the form of stress–strain distribution.

The stress simulation shows the distribution of von Mises stress on the leaf spring as indicated by color, whereby the maximum and minimum values at 272,637 and 0.01092 MPa, are indicated using red and purple. The maximum stress is shown in Fig. 8.

Finite element modeling was accompanied by an initial crack of 0.2 mm marked with a red arrow, obtained based on the visual inspection of the fracture leaf spring surface. The four nodes observed in the area around the crack are shown in Fig. 9.

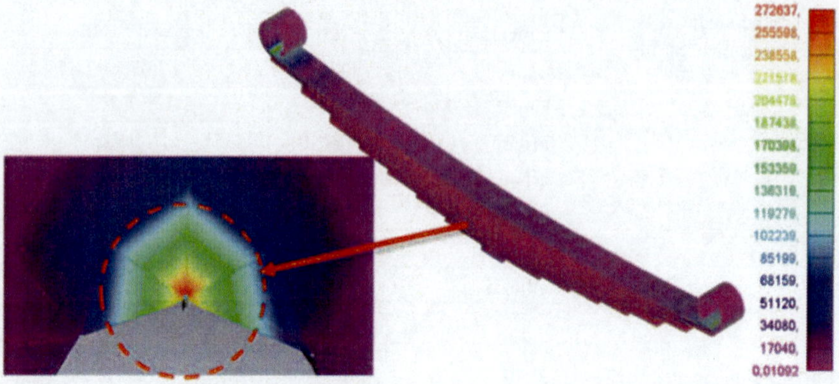

Fig. 8 Maximum stress around the crack

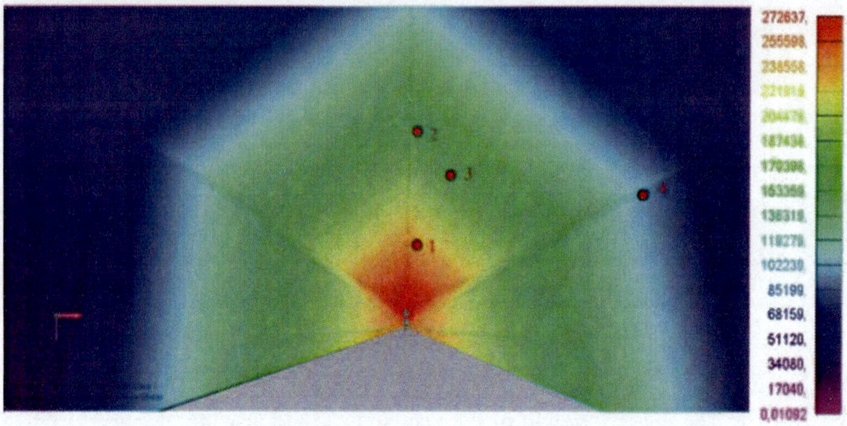

Fig. 9 Nodes observed around the crack

Stress Intensity Factor Analysis. This is a value that indicates the stress intensity at the crack tip. Failure/crack propagation in the material occurs when the stress intensity factor value due to the applied load is equal to or greater than the fracture toughness value ($K_I \geq K_{IC}$). The fracture toughness value in leaf spring steel SAE 5160 is 22 MPa $m^{1/2}$ [10].

The stress intensity factor is carried out around the crack tip to determine the maximum stress. K_I on the crack tip for Mode I can be calculated using Eq. (2) as follows:

$$K_1 = \frac{\sigma_{xx} \cdot \sqrt{2\pi r}}{\left(\cos\frac{\theta}{2}\right)\left[1 - \sin\left(\frac{\theta}{2}\right)\sin\left(\frac{3\theta}{2}\right)\right]} \quad (2)$$

where

Table 3 Stress intensity factor value

No.	Node	σ (MPa)	r (m)	θ	K_I (MPa $m^{1/2}$)
1	7376	152.07	0.0009	98	29.58
2	65,394	647.32	0.0021	93	204.18
3	14,970	162.26	0.0017	107	38.52
4	35,658	115.53	0.0031	153	39.69
Average					77.99

σ = Stress (MPa)
K_I = Stress intensity factor (MPa $m^{1/2}$)
r = Distance from crack tip to node
θ = Angle from crack tip to node.

The calculation results of the stress intensity factor K_I at each point/node observed around the crack are shown in Table 3.

The calculation results in Table 3. show that the average K_I obtained from the finite element method analysis is 77.99 MPa $m^{1/2}$, while the K_{IC} is 22 MPa $m^{1/2}$. This value fulfills the failure requirement where K_I (77.49 MPa $m^{1/2}$) > K_{IC} (22 MPa $m^{1/2}$).

4 Conclusion

Based on the analysis and calculation results of the leaf spring failure, the following conclusions were drawn:

1. The load the leaf spring receives before it fails is 26,950 N, greater than the spring's ability at 20,212 N.
2. The type of fatigue failure based on visual inspection of the fracture surface is crack initiation.
3. The stress intensity factor (K_I) around the crack tip is greater than the fracture toughness value of the material at 77.99 MPa $m^{1/2}$ > (K_{IC}) 22 MPa $m^{1/2}$; hence, it spreads, and the calculation results fulfill the equation ($K_I \geq K_{IC}$).

Acknowledgements The authors are grateful to the Universitas Syiah Kuala for their financial support through Professor Research Grant No. 001/UN11.2.1/PT.01.03/ PNBP/2022. The authors are also grateful to the Department of Mechanical Engineering, Universitas Syiah Kuala, Banda Aceh, for making the facilities accessible.

References

1. Bhatt CD, Nadarajan M, Balaji R, Rohith I, Selokar A (2020) Leaf spring model for heavy load vehicle using solid works and ANSYS analysis. In: Materials today: proceedings
2. Kong YS, Abdullah S, Omar MZ, Haris SM (2016) Failure assessment of a leaf spring eye design under various load cases. Eng Fail Anal 63:146–159
3. Husaini, Riantoni R, Ali N, Putra TE (2019) Failure analysis of the leaf spring of truck colt diesel using finite element method. Mater Eng 547
4. Eryuek IB, Ereke M, dan Goksenli A (2005) Failureanalysis of the suspension spring of a light duty truck. Eng Fail Anal 14:170–178
5. Husaini, Ali N, Bakhtiar A (2016) Fracture surface analysis of saloon car front suspension threaded springs 111
6. Senopati G, Sutowo C, Mabruri E (2016) Temper embrittlement phenomenon in Aisi 410 martensitic steel for steam gate valve 20 applications class 150 Grade WCB, vol 3
7. Husaini DM, Dawud TE, Putra NA (2020) Failure analysis of spur gears used in transmission system applied on a hand tractor. Key Eng Mat 841:144–149
8. Matheus S (2011) Analysis of the modulus of elasticity and Poisson's number of materials with tensile test. J Barekeng 5(2):9–14
9. Wibowo DB,Purwanto D, Austemper gray cast iron impact testing. Rotation 9(2):37–41
10. Sandor LT, Ferreira I (2005) Model for fracture toughness evaluation of the carburized layer for SAE 5115 steel, pp 111–124

Processing and Characterization of Bioepoxy Reinforced Hybrid Composites

K. Balasubramanian and N. Rajeswari

Abstract Material science has shaped the development according to the civilization since the mankind has survived. Composite materials are being used nowadays where less weight, high mechanical performances, resistant to fire, water and chemical resistant, excellent damping, fatigue and other properties are essential. The role of composite materials is significant in more promising areas like aerospace, automotive, constructions even if they have some disadvantages like high cost, high density and huge weight. In this work, a hybrid composite is prepared using bio-based fibres such as ramie, areca and sisal as these bio-reinforcements are eco-friendly, renewable and excellent strength. Mechanical parameters like tensile, compressive, flexural and impact strength are measured to check its suitability to replace the conventional materials.

Keywords Composite material · Fibres · Fabrication · Mechanical property · Analysis

1 Introduction

The introduction of composite materials has changed the world, and those materials are insignificantly employed in many fields like civil, automobile and aerospace as a substitute for conventional materials. Composite materials, having the superior mechanical properties, are divided into three types such as fibre reinforced polymer (FRP's) matrix composite and ceramic matrix composites (CMCs) and metal matrix composites (MMCs).

K. Balasubramanian (✉)
Dr. MGR Educational and Research Institute, Chennai, Tamil Nadu, India
e-mail: balaraji1974@gmail.com

Present Address:
N. Rajeswari
St. Peter's Institute of Higher Education and Research, Chennai, Tamil Nadu, India

Irwansyah et al. (eds.), *Proceedings of the 4th International Conference on Experimental and Computational Mechanics in Engineering*, Lecture Notes in Mechanical Engineering, https://doi.org/10.1007/978-981-99-7495-5_63

Table 1 Properties of natural fibre

Fibre	Density (g/cm^3)	Tensile strength (MPA)	Young's modulus (GPA)	Elongation break (%)	Moisture content (%)
Cotton	1.5	345–110	27.6	2.7–3.2	10.0
Areca	1.5	400	12.0	3.0–10.0	15
Ramie	1.5	400–938	61.4–128	1.2–3.8	12.6
Sisal	1.4	468–640	9.4–220	3.0–7.0	11.0
Coir	131–175	4.0–6.0	4.0–6.0	15–40	8.0
Jute	1.3–1.4	690	30–601.6	11.6	10.8

Metal Matrix Composite, a polymer matrix with at least two constituents, is used in tungsten carbide cutting tools, automobiles, etc. The MMC's high Young's modulus, good processability, economic efficiency and better thermal and electrical conductivity make it suitable to use in space shuttles amidst its high cost and immature technology.

Ceramic Matrix Composites, a subgroup of composite materials and a subgroup of ceramics, have the ceramic fibres embedded in a ceramic matrix which has excellent wear and high hardness. The ceramic matrix, aluminium nitride, silica nitride, etc., possess general properties such as anisotropic, thermal conductivity, lower cost and easy processing.

Fibre Reinforced Composites, in which polymer matrix is embedded with fibres. The fibres are usually glass, carbon, paper, pulp, wood and asbestos. FRPs are mainly used in aerospace, military, marine, sporting goods applications, infrastructure, etc. The main properties of FRPs are non-corroding behaviour, high internal damping, etc.

However, the materials are made by different compositions of different matrices, and composite materials do not withstand at high mechanical loads sometimes due to poor strength and heterogeneous density at outer planes. This may overcome by introducing hybrid composites which is a reinforcement of two or more materials, and the most common hybrid composite material is epoxy fibre–glass and carbon–aramid with epoxy reinforcement which offer high resistance and impact to the materials.

The properties of natural plant fibres are shown in Table 1.

2 Literature Review

Veerarajumaddela et al. [1] had discussed different types of hybrid composites and their mechanical properties, applications, merits and demerits including morphological analysis. In the study, hybrid composites like Al/Sic/FA, Al/Ceramics/BLA and Al/Ceramics particulates/RHA were fabricated and revealed the increase in the content of ceramic increases the hardness of the fibres linearly.

Balasubramanian et al. [2] fabricated composite materials using both synthetic and natural fibres and chemicals used along with natural fibre to yield better strength. From thermal test and water absorption test, it was concluded that synthetic fibre in hybrid composites could be replaced by the natural fibres. Prabhuram et al. [3] had fabricated the materials out of wood sample, cashew powder and combination of husk cashew powder to form matrix. From the experimental analysis of the prepared samples, it is found that the combination of husk and wood sample give better mechanical properties and elongation properties.

Priyanka et al. [4] discussed the various properties of hybrid composites and reported the impact and other mechanical properties of high resistance fibres like carbon glass fibre and Kevlar. In the study, the major parameters, namely geometry of projectile, impact velocity reinforcement of matrix materials, have been considered. Kumar et al. [5] fabricated the material with graphite and epoxy resin and reported that the porosity decreased when the epoxy was increased and also revealed the amalgamation of two materials enhanced the strength of composites.

Shireesha et al. [6] had analysed the properties of natural fibres like cotton, jute and measured the compressive and tensile strength including morphological analysis for different stacking sequences. The results demonstrated the tensile and compressive strength increased with an increase in epoxy content but the flexural strength increased up to some point after which it came down to medium and low ranges with respect to the fibre sequence.

Vieira et al. [7] discussed the effects of fibre treatment, and various manufacturing methods had tested the mechanical properties like tensile, compressive on both treated and untreated fibres. Srinivasa et al. [9] had tested the mechanical properties of areca-based reinforced composites treated with alkali and concluded the increase of treated areca volume fraction increases the strength of composites.

Damodhiran et al. [10] had prepared a hybrid composite using the glass–fibre–coconut coir–human hair, tested the tensile, compressive, flexural properties and reported the hybrid had higher flexibility and strength. Padmaraj et al. [11] tested the mechanical properties of alkali-treated fibre reinforced biodegradable composite material and revealed the alkali treatments improve the adhesion properties between matrixes which led to the superior mechanical properties.

Amico et al. [12] discussed the characterization and fabrication of ramie-glass fibre reinforced polymer of matrix–hybrid composites. The test results showed increase in volume fraction of ramie fibre content increased the thermal stability and other mechanical properties. Chandra Sekhar et al. [13] evaluated the mechanical properties of hemp–ramie–fibres reinforced with epoxy composites and studied the orientation degree of fibre loading at different angles like 30, 90 and 60 and also the damping factor.

Gokulkumar et al. [14] discussed ramie/GFRP fabrication and the test on mechanical properties like tensile and compressive showed the hybrid composite has good tensile and flexural strength as high as 1.03 KN. It is found that the properties of the natural fibre reinforced composites improved by hybridizing it with high-strength synthetic fibres like glass. Mechanical properties of the composites were found to be increased with an increase in the amount of glass fibre in the hybrid. Mohan Kumar

[15] studied the properties and fabricated areca reinforced fibre composites. Various mechanical properties measurement on the prepared specimen showed areca-based phenol formaldehyde composites had good tensile strength.

The literature study reveals that the development of composite materials aims to produce a material that has high specific strength, wear resistance, good strength-to-weight ratio suitable for all applications like where fuel economy with improved engine performance is decisive criteria. This paper concerns with the development of a new hybrid composite using Ramie fibre, Sisal fibre and Areca fibre, and based on the qualities of these selected fibres, the required resin and hardener are chosen. To identify its strength as substitute to other composites which can be made out of other constituents, the mechanical properties of the developed eco-friendly and biodegradable composites with different arrangements are measured and compared.

3 Composites Fabrication and Testing

The composite material is prepared using basic hand layup method which involves laying prepreg piles onto tool by hand to create laminate structures. In this work, tri-directional areca fibre, sisal, ramie fibre is used as reinforcements and epoxy are taken as matrix material. Six specimens are prepared (shown in Fig. 1) placing the selected fibre woven mat in different layers using low-temperature curing epoxy resin and hardener, and the test is done to study its properties like tensile strength, flexural strength, load carrying capacity and deflection. The arrangement of woven mat in the six specimens is tabulated in Table 2.

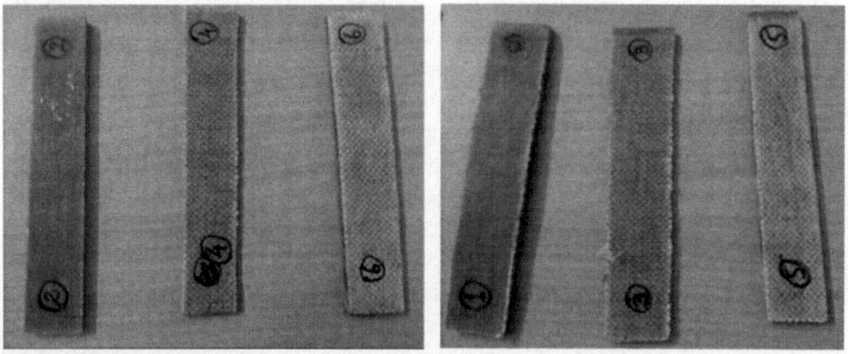

Fig. 1 Fabricated specimen for test

Table 2 Woven mat layer arrangement

Specimen 1	Specimen 2	Specimen 3	Specimen 4	Specimen 5	Specimen 6
Ramie	Ramie	Areca	Areca	Ramie	Areca
Areca	Areca	Ramie	Ramie	Ramie	Areca
Sisal	Sisal	Sisal	Sisal	Sisal	Sisal
Ramie	Areca	Areca	Ramie	Ramie	Areca
Areca	Ramie	Ramie	Areca	Ramie	Areca

4 Result and Discussion

This chapter deals with the behaviour of hybrid composites, and the detailed analysis of fibre content embedded in material reveals its effect on the capability of the specimen's withstanding capabilities under different types of loads.

From Table 3, it is evident that Specimen 5 (ramei-ramei-sisal-ramei-ramei) possess increased tensile strength, flexural strength and impact resistance property and least percentage elongation and hardness when compared to other combination of hybrid composites taken for investigation. This may be because with reference to Table 1, it can be understood that the fibres ramie and sisal, the constituents of Specimen 5, are individually better in tensile, flexural and impact strength property and poor in elongation and hardness property in comparison with areca.

Specimen 6 (areca-areca-sisal-areca-areca) composite has proved to be a better performer in terms of percentage elongation and hardness property and least performer in terms of tensile, flexural and impact resistance property. This is because with reference to Table 1, the areca and sisal, the constituents of Specimen 6 are individually better in elongation and hardness property and lesser in strength in terms of tensile, flexural and impact property when compared with ramei.

Hence, it is inferred from Table 2 that the areca, ramei, sisal hybrid composites (Specimen 1, 2, 3 and 4) have shown improvement in overall mechanical properties in terms of hardness, percentage elongation, tensile, flexural and impact strength. Hence, the further investigation is to derive the best performer among the areca, ramei, sisal hybrid composites (Specimen 1, 2, 3 and 4) which have been fabricated

Table 3 Mechanical properties of different specimens

Specimen ID	Tensile strength (MPa)	Elongation %	Flexural strength (MPa)	Hardness (HRB)	Impact strength (J/m)
Specimen 1	36.01	7.4	77.21	62.6	72.38
Specimen 2	37.81	6.2	79.34	65.5	72.18
Specimen 3	36.19	7.6	77.46	63.3	72.71
Specimen 4	35.86	6.5	76.24	67.3	72.91
Specimen 5	38.51	5.4	80.34	52.2	75.23
Specimen 6	34.39	8.1	68.87	70.1	70.13

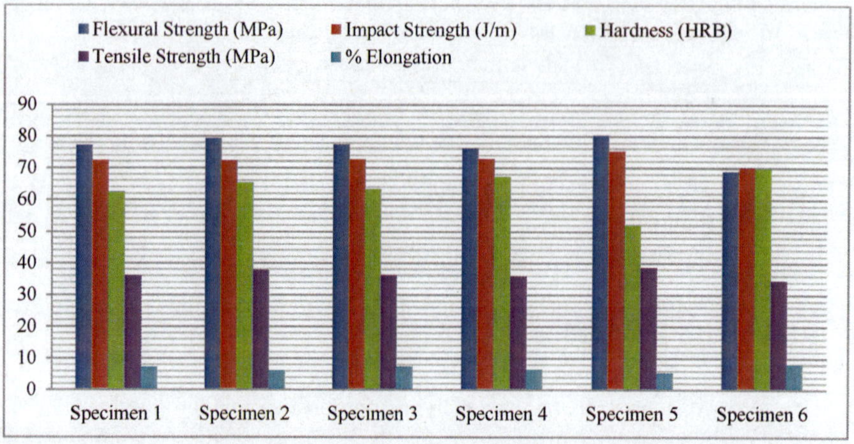

Fig. 2 Comparison of mechanical properties of areca, sisal, ramei altered sandwiched hybrid composites

by altering the layers of sandwich. This inference from Table 2 is clearly depicted in Fig. 2.

Specimen 2 (ramei-areca-sisal-areca-ramei) shows an increased tensile strength of 37.81 MPa and increased flexural strength of 79.34 Mpa, whereas Specimen 4 (areca-ramei-sisal-ramei-areca) is of least tensile and flexural strength when compared among all other altered sandwiched composites of ramei-areca-sisal investigated in this paper.

Specimen 4 (areca-ramei-sisal-ramei-areca) has proved to be better in hardness and impact strength with the values of 67.3 HRB and 72.91 J/m, respectively, whereas Specimen 1 (ramei-areca-sisal-ramei-areca) shows least hardness and Specimen 2 (ramei-areca-sisal-areca-ramei) possess least impact strength, when compared with other altered sandwiched composites of ramei-areca-sisal.

When checked for ductility, the elongation property of Specimen 1 (ramei-areca-sisal-ramei-areca) was found to be more, i.e. 7.4% but the least elongation characteristic is observed in Specimen 2 (ramei-areca-sisal-areca-ramei), when compared with other altered sandwiched composites of ramie-areca-sisal.

It is understood from Fig. 2 which depicts that Specimen 2 (ramei-areca-sisal-areca-ramei) excels in the overall mechanical properties when compared with other five specimens investigated for altered sandwiched hybrid composites in this paper. It is also noted that Specimen 1 (ramei-areca-sisal-ramei-areca) and Specimen 3 (areca-ramei-sisal-areca-ramei) did not show any significant difference when compared with each other for each of their mechanical properties and also their net increase in the overall mechanical properties (Fig. 3).

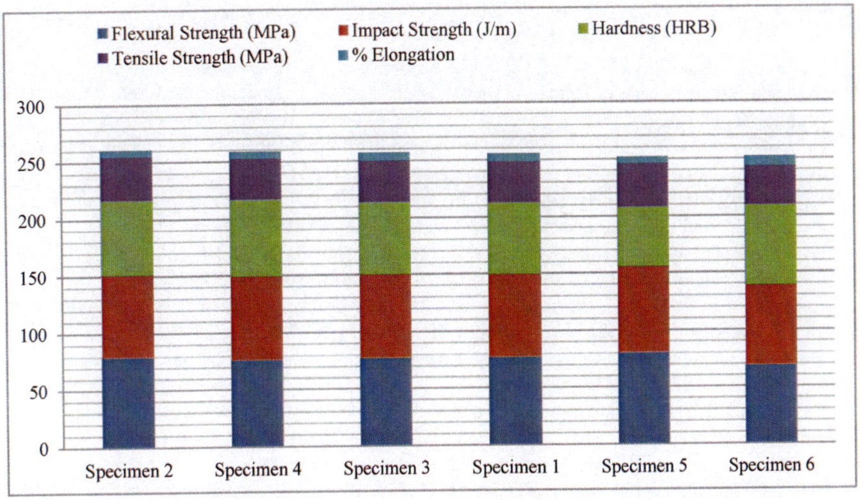

Fig. 3 Excellency hierarchy in overall mechanical properties of areca, sisal, ramei altered sandwiched hybrid composite

5 Conclusion

In the present work, the biofibre composite of areca, ramie and sisal by altered sandwiched fabricated specimens was evaluated for their mechanical properties. Based on the experimental investigation, the following conclusion can be made.

- The ramei-ramei-sisal-ramei-ramei sandwiched composite can be used for application demanding for tensile strength, flexural strength and impact resistance property. But this combination doesn't contribute effectively for ductility and hardness property.
- The areca-areca-sisal-areca-areca combination composite exhibits adequacy in the ductility and hardness property. However, the tensile, flexural and impact strength are lesser when compared to other combination of composites examined in this paper.
- The ramei-areca-sisal-areca-ramei sandwiched combination has exceled in net increase in overall mechanical properties under consideration, when compared to any other combination of composite.
- The ramei-areca-sisal-ramei-areca sandwiched specimen and the areca-ramei-sisal-areca-ramei sandwiched specimen didn't show much difference with each other with regard to each of their mechanical properties tested values and thereby no difference in their net increase in the overall mechanical properties. So both the combination has same effect on their mechanical properties.

References

1. Maddela VR et al (2017) Hybrid composites and their behaviour. Int J Modern Trends Sci Technol 3(11):50–56
2. Balasubramanian K, Rajeswari N, Vaidheeswaran K (2020) Analysis of mechanical properties of natural fibre composites by experimental with FEA. Mater Today Proc 28:1149–1153
3. Prabhuram T et al (2010) Hybrid composite materials. In: 2010 IEEE international conference on frontiers in automobile and mechanical engineering (FAME 2010). pp 27–31
4. Priyanka P et al. High-strength hybrid textile composites with carbon, Kevlar and E-glass fibers for impact strength resistance structures. Mechan Compos Mater 53(5):685–704
5. Kumar K et al (2017) A review on different types of hybrid composite materials with different matrix proportions. Int Res J Eng Technol (IRJET) 4(5):1089–1093
6. Shireesha Y et al (2019) Properties of hybrid composites and its applications. Int J Sci Technol Res (IRJET) 8(8):335–341
7. Vieira CAB et al (2009) Characterization of hybrid composites produced with mats made using different methods. Mater Res 12(4):433–436
8. Satishkumar TP et al (2016) Characterization of sisal/cotton woven mat reinforced polymer hybrid composites. J Ind Text 47:429–452
9. Srinivasa CV et al (2011) Impact and hardness properties of areca and fibre-epoxy reinforced composites. J Mater Environ Sci 2:351–356
10. Damodhiran S et al (2013) Characterization of glass –fibre coconut coir-human hair hybrid composite. Int J Eng Technol 6(1):75–82
11. Padmaraj et al (2013) Development of short areca fibre reinforced biodegradable composite material. Int Conf Eng 64:966–972
12. Amico SC et al (2012) Preparation and characterization of ramie—glass fibre reinforced polymer matrix hybrid composites. Mater Res 15:415–420
13. Chandrasekhar et al (2016) Evaluation of mechanical properties of Hemp-Ramie-fibres reinforced with epoxy reinforced composites. Int J Res Eng Technol 5(13):21–26
14. Gokulkumar S et al (2019) Evaluation of mechanical characterization of Areca/Ramie/GFRP hybrid composites. Int J Res Eng 7(6):ETEDM-2019
15. Cchannabasaapa MKG (2008) A study of short areca fiber reinforced PF composites. In: Proceeding of world congress of engineering, vol 2

The Effect of GTAW Process Parameters for Dissimilar Metals Steel Armour and ASS 304L Using ER316L Filler on Mechanical Properties

Riswanda, Albert Daniel Saragih, Harlian Kadir, and Windi Aprilia Irawan

Abstract The effect of welding parameters on the mechanical properties of weld joints of dissimilar metals (dissimilar metals) between Armour steel and ASS 304L stainless steel greatly affects the quality of the welds. This research was carried out by the GTAW process with a variation of the welding current at 100, 110, and 120 A. Mechanical properties of welded joints were analysed including tensile testing and hardness testing, to determine the strength of the welded joint and the distribution of hardness in the heat-affected area (HAZ). The highest hardness test value was found in the 110 A specimen, which is 527.6 HV in the HAZ Armour area, and there is a correlation to the tensile test results that the fracture occurs in the area of low hardness. The results of the highest tensile strength test were found in the 110 A current specimen at 626,154 MPa with a yield strength of 406.779 MPa. The fractures for all specimens occurred in the fusion line region towards ASS 304L. The welding parameters appropriate for welding dissimilar steel between Armour steel and ASS 304L can be used as a reference for welding military transformation tools and reducing production costs.

Keywords Armour · ASS 304L · Dissimilar-metal · GTAW · Current

Riswanda (✉) · A. D. Saragih · H. Kadir · W. A. Irawan
Mechanical Department, Bandung State Polytechnic, Jl. Gegerkalong Hilir, Ds. Ciwaruga, Kotak Pos 1234, Bandung 40559, Indonesia
e-mail: riswanda@polban.ac.id

A. D. Saragih
e-mail: albertdanielsrgh@polban.ac.id

H. Kadir
e-mail: harliankadir@polban.ac.id

W. A. Irawan
e-mail: windi.aprilia.pmf18@polban.ac.id

Irwansyah et al. (eds.), *Proceedings of the 4th International Conference on Experimental and Computational Mechanics in Engineering*, Lecture Notes in Mechanical Engineering, https://doi.org/10.1007/978-981-99-7495-5_64

1 Introduction

Welding is a technology that plays an important role in the industrial world, especially in the manufacturing sector. Welded joints with different materials (dissimilar metals) have been widely used. Welding of dissimilar metals has different properties due to the reaction between different elements of the two metals. Dissimilar welding is a metal welding process with different physical, mechanical, thermal, and metallurgical properties, so research is needed for the connection [1]. Armour steel is a material that has bullet resistance properties (ballistics) which are usually applied specifically for military vehicle transportation. The characteristics of this metal bullet-resistant material have elements of Nickel (Ni), Chrome (Cr), and Molybdenum (Mo). Armour steel belongs to the medium carbon group with a carbon percentage of 0.34%. Typical Armour steels contain carbide forming elements, usually, molybdenum, chromium, and austenite forming elements, such as manganese and nickel as well as some carbon, and high-toughness steels can withstand ballistic impacts [2].

Austenitic stainless steels are stainless steels containing 18% Cr and 8% Ni. The advantages of materials that have these elements are that they are not easily oxidized by water and air, have anti-rust properties and are resistant to exposure to extreme temperatures. One of the stainless steels that is often used is type 304L, "L" which means low carbon. One of the characteristics of this material is its low thermal conductivity and its susceptibility to thermal expansion during the welding process, these are factors that can cause deformation and crack. Welding unlike Armour steel with ASS 304L stainless steel aims to streamline production costs. Based on these factors, the Gas Tungsten Arc Welding (GTAW) process was chosen to weld austenitic stainless steel because it produces a fairly narrow heat affected zone (HAZ) [3]. The heat-affected zone (HAZ) is the metal in contact with other metals in the welding process resulting in rapid thermal cycling and cooling of the weld joint in the welding process. The grains in the HAZ region change shape from flat and long in the weld metal region, to an elliptical and almost round shape [4].

The GTAW process is welding using a flame arc generated by a fixed tungsten electrode. The filler is made from the same or similar material as the material being welded and separated from the welding torch [5]. Process parameters have an important role in produce quality welds. One of these parameters is the current used during the welding process. The variation of the GTAW process flow results in differences in the physical and mechanical properties of the HAZ material. The current variations used are 100, 110, and 120 A using ER316L filler. This research was conducted to obtain optimal parameters for GTAW joints of non-steel Armour and ASS 304L using ER316L filler.

2 Materials and Methodology

GTAW is a process of joining two metals by melting the added material on the part to be joined using tungsten electrodes [6]. The GTAW process can be carried out using a protective gas, without the application of pressure, and using with the addition of materials in the form of filler metal. The shielding gasses commonly used in this welding process are helium, argon, or a combination of helium and argon. The function of the shielding gas is to avoid oxidation due to the influence of air around the weld which causes porosity [7]. The material used is ASS 304L plate and steel Armour with a size of 300 × 125 × 5 mm. The size formed for the V groove butt weld refers to the ISO 9692-1 standard (Fig. 1).

The welding process requires a shielding gas to protect against impurities [8]. Welded materials that use shielding gas differ from welded materials that do not use shielding gas. Visually, the material that is welded using shielding gas will look cleaner, and there will be no weld defects such as porosity. As for the material being welded without using a shielding gas, porosity defects will appear with an indication of a black smooth round shape. Incomplete penetration will also occur, this is a type of welding defect which indicates that the weld root does not meet the two sides of the weld metal [9]. A filler rod is a filler metal or so-called filler metal used in the GTAW process. Welding with stainless steel material, commonly used fillers are ER308L, ER309L, and ER316L. Alloy ER316L is particularly suitable for welding low-carbon molybdenum austenitic alloys (Table 1).

Armour steel is a material that has bullet resistance properties that are usually applied specifically for military transportation. Table 2 is a table of chemical composition and mechanical properties of Armour steel [11] and the spectro test of ASS 304L and Armour steel. The mechanical properties of Armour steel are shown in Table 3.

Austenitic stainless steels are a group of alloy steels with special properties. The main properties of this material are high corrosion resistance, ease of formability

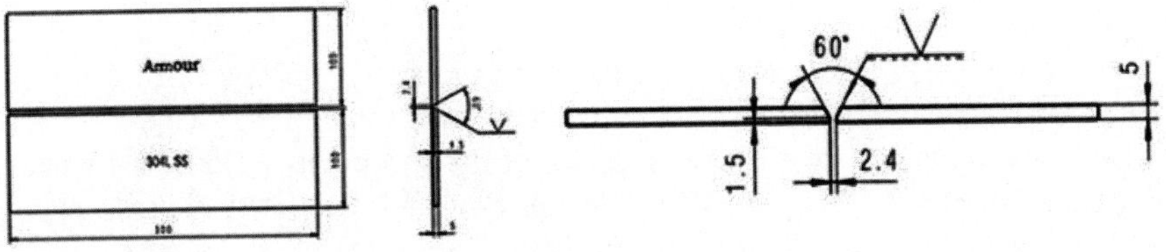

Fig. 1 The welding specimen and the welding angle

Table 1 The chemical compositions of ER316L [10]

Weight%	C	Mn	Si	Cr	Ni	S	P	Mo	Cu
ER316L	0.04–0.08	1.0–2.5	0.30–0.65	18–20	11–14	0.03 max	0.03 max	2–3	0.75 max

Table 2 The chemical compositions of armour steel [11] and the spectro test of ASS 304L and armour steel

Type	Composition, WT%								
	C	Mn	Ni	Cr	Mo	S	Cu	Fe	others
Armour steel	0.29	1.41	0.27	0.55	0.19	0.008	0.083	96.72	0.4
ASS 304L (spectro test)	0.27	1.48	7.82	18.2	0	0.001	0.025	71.8	0.56
Armour steel (spectro test)	0.349	0.308	1.55	0.28	0.58	0.002	0.01	96.6	0.26

Table 3 The mechanical properties of armour steel [12]

Materials	Ultimate tensile strength (MPa)	Elongation (%)	Hardness (HV)
Armour steel	1775	14	495

and high malleability. One of the types of stainless steel that was often used is type 304 L. The chemical composition of stainless steel type 304 L were 17–25% chromium (Cr) and 8–20% nickel (Ni). The material has an Face Centred Cubic (FCC) crystal structure, which are non-magnetic, non-hardenable, formable, and has good weldability [13].

The mechanical properties of a material can be known through a test, one of which is a tensile test. The principle of this test is to provide a sufficiently large tensile force and a strong grip to prevent the material from detaching when a tensile force is applied [14]. Hardness is the ability of a material to resist penetration (stress). Micro Vickers is a method in which the specimen (test object) is emphasized by using a press object (diamond indenter) in the shape of a pyramid with a square base and an angle of 136 degrees facing each other [15].

Welding will be carried out after the tack weld process on bot specimens has been completed. Then, the specimen was put on the welding table in the 1 G position and tightened with clamps for the fixed position of the specimen. Welding parameters are shown in Table 4. The ASS 304 welded material and Armour steel were cut into sections according to standards for the further testing. Figure 2 shows the layout for cutting specimen for the tested.

The hard testing process was carried out using a digital hardness testing machine Mitutoyo HM-122 with a Hardness Vickers (HV) unit loading 200 gf and the distance between points of 250 m. The areas that were tested were weld metal (weld metal), HAZ, and base metal. As shown in Fig. 3.

Tensile testing is carried out in the transverse direction. Tests were carried out to determine the tensile strength of welded specimens of 100, 110, and 120 A. The

Table 4 Welding parameters

Weld parameters			Material	
Welding speed (mm/s)	Current (A)	Filler metal diameter (mm)	Tungsten electrode diameter (mm)	Argon (kgf/cm^2)
1.02	100	2.4	2.4	12
1.02	110			
1.02	120			

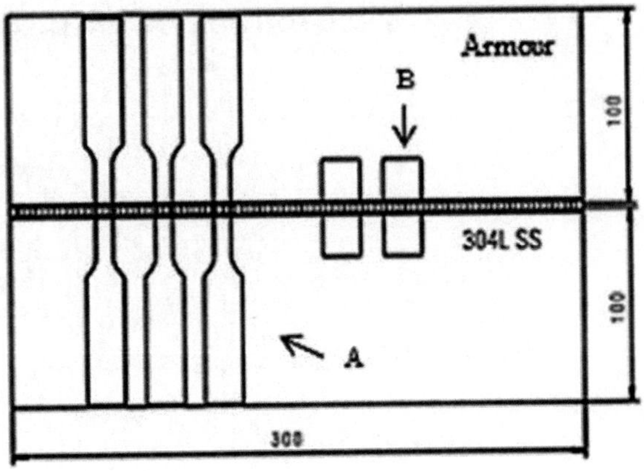

Fig. 2 Specimen test layout (**a** tensile test specimen; **b** hardness and metallographic test specimen)

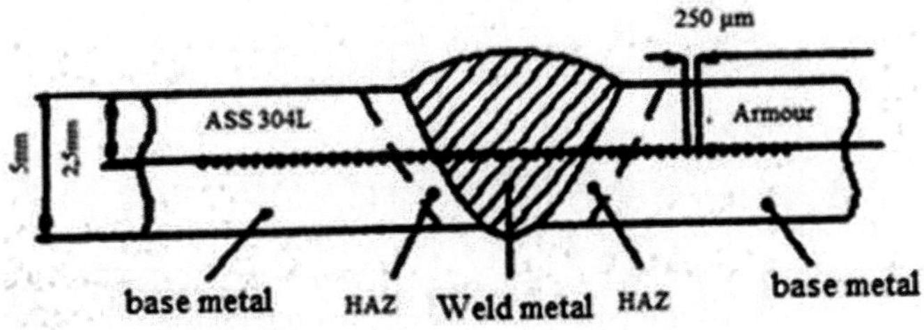

Fig. 3 Hardness test distance and area

tensile test specimen complies with the 1981 JIS Z 2201 standard with dimensions as shown in Fig. 4.

3 Results and Discussion

Hardness testing was carried out with the Vickers Hardness test equipment, and the results of the hardness distribution on the weld specimen were obtained as shown in Fig. 5.

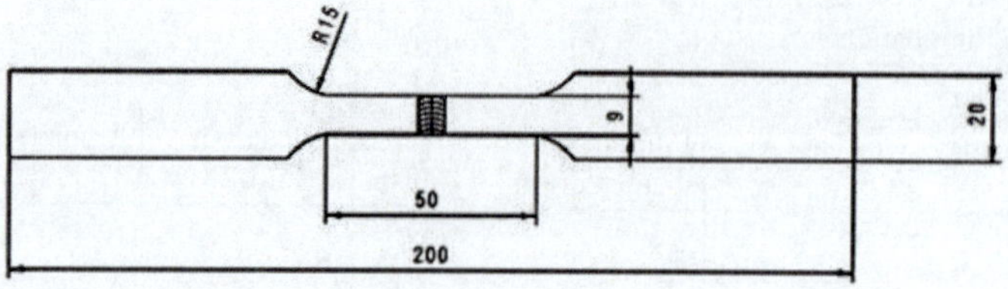

Fig. 4 Tensile test specimen

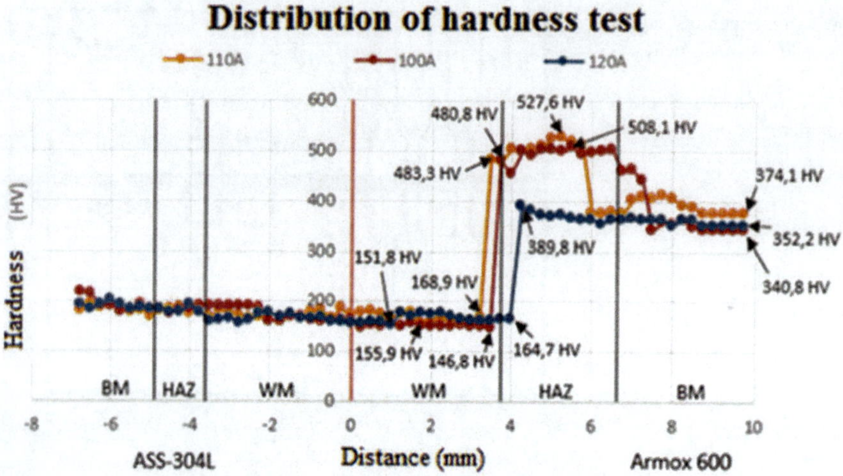

Fig. 5 Hardness test graph

Based on Fig. 5, the distribution of hardness values generally forms the same pattern or graph, namely the hardness in the heat affected zone (HAZ) Armour area is relatively higher than the weld metal (WM) and base metal (BM) areas. This happens because of the rapid freezing process of the liquid phase during the welding process. The highest hardness value occurs at a current of 110 A in the HAZ Armour area of 527.6 HV and the lowest hardness value occurs at a current of 100 A in the WM area which is close to HAZ or can be called a fusion line, which is 146.8 HV.

Tensile testing is carried out to determine the value of the tensile strength of welded specimens with different currents of 100, 110, and 120 A. Figure 6 shows that the graph of the tensile test result with the lowest tensile strength value was 614 MPa and the yield strength of 355 MPa for the current at 100 A. The highest tensile strength value was 626 MPa with the current at 110 A, and the yield strength of 406 MPa. Based on the data, it can be concluded that the optimal tensile strength value was at the 110 A.

Figure 7 shows the tensile test results for specimens 100, 110, and 120 A. The fracture locations of all specimens are in the fusion line area of the ASS 304L material. Overall, the fault position is indicated to be in an area of low hardness.

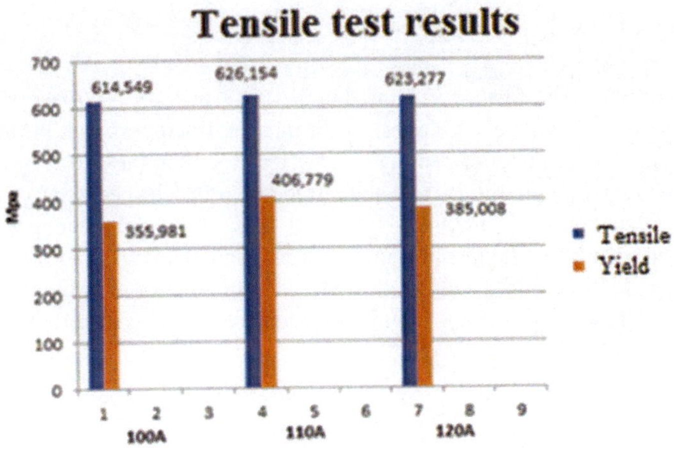

Fig. 6 Tensile test graph

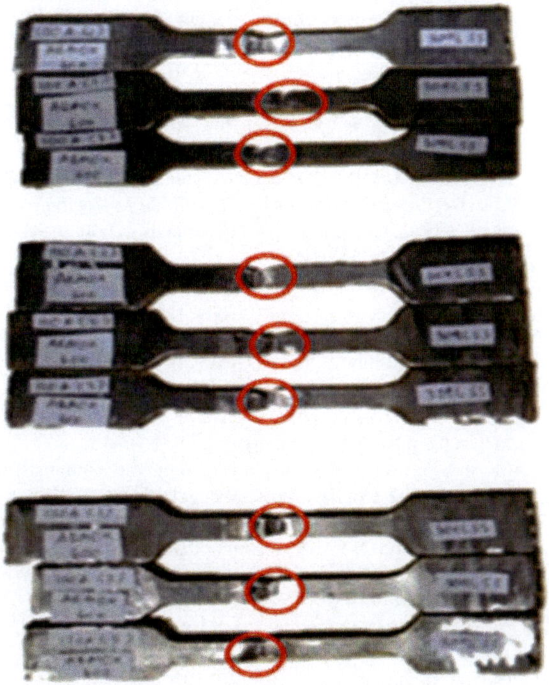

Fig. 7 Fault position of tensile test specimen

4 Conclusions

The highest hardness test value was found in the 110 A in the HAZ Armour area and there is a correlation to the tensile test result that the fracture occurs in the area of low hardness value. The tensile strength value was found at the 110 A of 626 MPa with the yield strength of 406 MPa. The fault location for all specimens occurred in the fusion line area in the direction of ASS 304L, so that from the tensile strength test result and hardness the most optimal current used at 110 A. The results of the hardness test and the tensile test can be concluded that the most optimal current in the GMAW process for metal Armour steel and ASS 304L using ER316L filler at 110 A.

References

1. Wijoyo W, Albab U, Ardika WT, Darojat MW (2019) Karakteristik Kekuatan Tarik Sambungan Las Tak Sejenis Baja Karbon-Stainless Steel. FLYWHEEL J Tek Mesin Untirta V(1):60. https://doi.org/10.36055/fwl.v0i0.5252
2. Purwanto H, Soenoko R, Purnowidodo A, Suprapto A (2016) Pengembangan Material Tahan Balistik Sebagai Bahan Kendaraan Tempur di Indonesia: review. Semin Nas Inov Dan Apl Teknol Di Ind 127–132. ISSN: 208
3. Kadir H, Riswanda S et al (2017) Pengaruh Variasi Arus Proses GTAW Terhadap Sifat Fisik Dan Mekanik Baja Tahan Karat Austenitik AISI 316L, pp 29–35
4. Li L, Deceuster A, Griffiths B (2013) Table 2—Average impact energy values for various simulated HAZs, and the base and filler metals in both the As-welded and heat-treated conditions. Weld Res 92:80–87
5. . Rudy TO, Tumpal Ojahan R (2017) Analisis pengaruh parameter pengelasan gtaw pada. Poros 15:53–62
6. Perkapalan JT, Teknik F, Diponegoro U (2016) Pengaruh Kuat Arus Listrik Dan Sudut Kampuh V Terhadap Kekuatan Tarik Dan Tekuk Aluminium 5083 Pengelasan Gtaw. J Tek Perkapalan 4(1):256–264
7. Sanjaya A, Sutowo C (2007) Pengaruh Hasil Pengelasan Gtaw Dan Smaw Pada Pelat Baja Sa 516 Dengan Kampuh V Tunggal. J Tek 1:10–16
8. Mesin S-T, Teknik F, Surabaya UN, Pengaruh Komposisi Backing Gas Argon (99 %) Dan Gas H2N2 (20%–80%) Pada Proses Pengelasan Gtaw Dengan Material Stainless Steel SA 312 TP 304 Terhadap Struktur Mikro Dan Nilai Kekerasan Nugroho Waluyo Utomo, vol 2
9. Kelautan FT (2018) Analisa Pengaruh Root Gap Pada Pengelasan Material Baja Ah 36 Dengan BackingCeramic Analisa Pengaruh Root Gap Pada Pengelasan Material Baja Ah 36 Dengan Backing Ceramic
10. Aircraft (2013). https://www.aircraftmaterials.com/data/weld/er316l.html. Accessed 13 Feb 2022
11. Padang Yanuar Y, Nugroho S (2014) Karakterisasi Baja Armour Has. Proses Quenching dan Tempering, no. Baja Armour, pp 13–14
12. Fei Z et al (2019) Effect of heat input on weld formation and tensile properties in keyhole mode TIG welding process. Metals (Basel) 9(12). https://doi.org/10.3390/met9121327
13. Rahmatika A, Ibrahim S, Hersaputri M, Aprilia E (2019) Studi Pengaruh Variasi Kuat Arus Terhadap Sifat Mekanik Hasil Pengelasan GTAW Alumunium 1050 Dengan Filler ER 4043. J POLIMESIN 17(1):47–54

14. Salindeho RD, Soukota J, Poeng R (2018) Pemodelan pengujian tarik untuk menganalisis sifat mekanik material. J J-Ensitec 3(1):1–11
15. Subagiyo (2017) Analisis Hasil Kekerasan Metode Vikers Dengan Variasi Gaya Pembebanan Pada Baja. Majapahit Techno 6(2):9–14

Author Index

Irwansyah et al. (eds.), *Proceedings of the 4th International Conference on Experimental and Computational Mechanics in Engineering*, Lecture Notes in Mechanical Engineering, https://doi.org/10.1007/978-981-99-7495-5